Boundary-Value Problems
with
Free Boundaries
for
Elliptic Systems of Equations

TRANSLATIONS OF MATHEMATICAL MONOGRAPHS

VOLUME **57**

Boundary-Value Problems with Free Boundaries for Elliptic Systems of Equations

by V. N. MONAKHOV

American Mathematical Society · Providence · Rhode Island

КРАЕВЫЕ ЗАДАЧИ СО СВОБОДНЫМИ ГРАНИЦАМИ ДЛЯ ЭЛЛИПТИЧЕСКИХ СИСТЕМ УРАВНЕНИЙ

В. Н. МОНАХОВ

«НАУКА», СИБИРСКОЕ ОТДЕЛЕНИЕ
НОВОСИБИРСК 1977

Translated from the Russian by H. H. McFaden
Translation edited by Lev J. Leifman

1980 *Mathematics Subject Classification.* Primary 30C60, 35J65, 35Q15, 35R35, 76-02, 76B10, 76N15, 76S05; Secondary 30C20, 30E20, 31A25, 35A05, 35M05, 46E35, 47G05, 73B99, 73J06, 73F99, 76C05, 76D15, 76D25, 76F99, 76G05.

ABSTRACT. This book is concerned with certain classes of nonlinear problems for elliptic systems of partial differential equations: boundary-value problems with free boundaries. The first part has to do with the general theory of boundary-value problems for analytic functions and applications of it to hydrodynamics. The second presents the theory of quasi-conformal mappings, along with the theory of boundary-value problems for elliptic systems of equations and applications of it to problems in the mechanics of continuous media with free boundaries: problems in subsonic gas dynamics, filtration theory, and problems in elastico-plasticity.

Library of Congress Cataloging in Publication Data

Monakhov, V. N. (Valentin Nikolaevich)
    Boundary value problems with free boundaries for elliptic systems of equations.
    (Translations of mathematical monographs; v. 57)
    Translation of: Kraevye zadachi so svobodnymi graniṫsami ėllipticheskikh sistem uravneniĭ.
    Bibliography: p.
    1. Boundary value problems. 2. Differential equations, Elliptic. 3. Fluid dynamics. I. Title. II. Series.
QA377.M5813   1983   515.3'53   83-2754
ISBN 0-8218-4510-1
ISSN 0065-9282

TABLE OF CONTENTS

PREFACE

The applied sciences (hydrodynamics, geophysics, etc) proposed almost at the same time two types of problems for mathematics: by convention they may be called "direct" and "inverse" problems.

Inverse problems, the theory of which began to take shape only in recent decades, differ from direct problems in that not only the solutions of the differential equations but also the equations themselves or the boundaries (or parts of the boundaries) of their domains are unknown.

The latter type of inverse problems, which (following the terminology adopted in hydrodynamics) we shall call problems with free boundaries, include jet problems in hydrodynamics, elastico-plastic problems, problems of heat propagation in media with a changing phase state (problems of Stefan type), and so on.

This monograph deals with those problems in the mechanics of continuous media with free boundaries which, in the final analysis, reduce to boundary-value problems for quasilinear elliptic systems of first-order equations with operator coefficients.

The book arose from a course of lectures given by the author repeatedly and in several variants first to students at Kazen State University and then (beginning in 1966) to students at Novosibirsk State University.

Therefore, certain sections of the book bear the nature of a textbook and can serve as such in the theory of partial differential equations: boundary-value problems and boundary properties of analytic functions (Chapters I and II), quasiconformal mappings (Chapter V), and boundary-value problems for linear and quasilinear elliptic systems of equations (Chapter VI).

At the end of almost every chapter we state unsolved problems of varying difficulty, some of which may serve as subjects of scientific research or to be assigned as course or degree projects to students specializing in the theory of partial differential equations and its applications to mechanics.

The content of this book is reflected well enough in its Table of Contents. It should also be pointed out that some results of the author and his students are published here for the first time.

The author was greatly aided in the preparation of the book for printing by his students, those on the staff at the Institute of Hydrodynamics, and post-graduate students at the Department of Theoretical Mechanics at Novosibirsk State University: G. V. Alekseev, A. V. Kazhikhov, N. A. Kucher, G. V. Lavrent'ev, A. M. Meĭrmanov, A. A. Oleĭnik, P. I. Plotnikov, and B. G. Putievskiĭ, who read through the manuscript and made many useful remarks. In addition, A. V. Kazhikhov and N. A. Kucher also supplied material used in the writing of §5 in Chapter IV and §4 in Chapter VI.

It would be impossible to overestimate the constant help given the author by the Editor, S. N. Antontsev. He suggested a number of improvements in the presentation, wrote §4 in Chapter VII at the author's request, and wrote §3 in Chapter VI jointly with the author.

The author expresses his sincere gratitude to all these persons.

V. N. Monakhov

Part I

**Boundary-value problems
in the theory of analytic functions
and their application to hydrodynamics**

CHAPTER I

THE BOUNDARY-VALUE PROBLEMS
OF RIEMANN AND HILBERT

§1. Piecewise smooth curves.* Classes of functions on curves

An open smooth arc (contour) is defined to be a connected curve L on the plane $z = x + iy$ with equation

$$z = t(s), \quad s \in [0, s_L],$$

in which $x = \operatorname{Re} t(s)$ and $y = \operatorname{Im} t(s)$ have continuous first derivatives for $s \in [0, s_L]$, and distinct values of the arclength s yield distinct points z, i.e., L does not have self-intersections. The derivative $dt/ds = x_s' + iy_s'$ ($x_s'^2 + y_s'^2 = 1$) can obviously be represented in the form

$$dt/ds = e^{i\theta(s)}, \quad s \in [0, s_L], \tag{1}$$

where $\theta = \theta(s)$ is the angle between the tangent to L at the point $t(s)$ and the OX-axis and is a continuous function. If the endpoints $a = t(0)$ and $b = t(s_L)$ coincide and $\theta(0) = \theta(s_L)$, then the arc is called a closed smooth contour, and a collection** $L = \Sigma_i L_i$ of finitely many smooth contours L_i (closed or not) without common points is called a smooth curve. If the angle $\theta = \theta(s)$ between the line tangent to a continuous arc L and the OX-axis is a piecewise continuous function with finitely many points of discontinuity, then the arc L, or a curve $L = \Sigma_i L_i$ consisting of such arcs, is said to be piecewise smooth.

Let $t_0 = t(s_0)$ and $t = t(s)$ be points on a smooth arc L and α_0 an arbitrary acute angle, $0 < \alpha_0 < \pi/2$. The continuity of the tangent angle $\theta(s)$ to L implies the existence of an $\varepsilon > 0$ such that the nonobtuse angle $\alpha = \alpha(s)$

---

*Editor's note. The author does not consistently distinguish between the terms линия (above) and контур (contour). This translation follows the original.

**Editor's note. The author uses the notation for Σ for the set-theoretic union of arcs.

3

between the tangents to L at the points t_0 and t does not exceed α_0 when $|s_0 - s| < \varepsilon$, i.e.,

$$\alpha(S) = |\theta(s_0) - \theta(s)| \leqslant \alpha_0 < \pi/2. \tag{2}$$

Let $\beta(s)$ be the nonobtuse angle between the chord $t_0 t$ joining the points $t_0 = t(s_0)$ and $t = t(S)$ and the tangent to L at an arbitrary point $t_1 = t(s_1)$, $s_1 \in [s_0, s]$. Then, since L contains a point $t_* = t(s_*)$, $s_* \in [s_0, s]$, at which the tangent to L is parallel to the chord $t_0 t$, the inequality (2) implies that

$$\beta(s_1) = |\theta(s_*) - \theta(s_1)| \leqslant \alpha_0 < \pi/2. \tag{2'}$$

Setting $r = |t_0 - t|$, we find from the obvious equality $|dr/ds| = \cos \beta(s)$ and (2) that

$$k_0 |s - s_0| \leqslant |t(s_0) - t(s)| \leqslant |s_0 - s|, \tag{3}$$

where $0 < k_0 = \cos \alpha_0 < 1$. It is easy to see that (3) is valid for any pair of points $t_1 = t(s_1)$ and $t_2 = t(s_2)$ on L, not necessarily close points, and also when L has corner points which are not cusps ([97], Appendix I, §4°). Let us fix $0 < \alpha_0 < \pi/2$ and an $\varepsilon > 0$ corresponding to it, and consider an arbitrary smooth arc $l \subset L$ with endpoints

$$a = t(s_a), \qquad b = t(s_b) \qquad (0 \leqslant s_0 < s_b = s_a + \varepsilon).$$

Let Π_t be an arbitrary straight line passing through a point $t = t(s) \in L$, $s \in [s_a, s_b]$, and forming a nonobtuse angle β, $0 < \alpha_0 < \beta \leqslant \pi/2$, with the tangent to l at the fixed point $t_0 = t(s_0)$. Obviously, each such line Π_t has only one point $t = t(s)$ in common with l. In particular, the normal n_t, for example, can be taken for the line Π_t at any point of l. At the endpoint a we draw the normal n_a to l and perform a parallel translation of l by a distance ε_0 to both sides of l, keeping the left-hand endpoints of the curves l' and l'' on the normal n_a (see Figure 1). The segment $a'a''$ of n_a and the segment $b'b''$ of the line parallel to it, along with the translated curves l' and l'', bound a domain $\sigma = \sigma(l)$, which will be called a standard neighborhood of the arc l. By assumption, there are no points of self-intersection on the smooth curve L, so there is an $\varepsilon_0 > 0$ such that the corresponding standard neighborhoods σ_1 and σ_2 of any arcs l_1 and l_2 with endpoints a_1, b_1 and a_2, b_2 are disjoint if

$$\min\{|a_1 - a_2|, |a_1 - b_2|, |b_1 - a_2|, |b_1 - b_2|\} \geqslant \varepsilon_0 > 0.$$

Note that the dimensions of a standard neighborhood $\sigma(l)$ depend only on the properties of L.

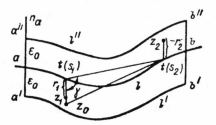

FIGURE 1

The families of translates of l along the normal n_a and straight lines parallel to n_a form a certain coordinate net in the neighborhood $\sigma(l)$. Let $z \in \sigma(l)$ be arbitrary. Draw a line parallel to n_a through it and denote by $t = t(s)$ its point of intersection with l. Let $r = \mp |z - t|$, where the plus sign is taken if z remains to the right of the arc as we go from a to b, and the minus sign is taken otherwise. The quantities s and r play the role of generalized coordinates of a point $z \in \sigma(l)$, and we reflect this in the notation $z = z(s, r)$.

Let $z_1 = z(s_1, r_1)$ and $z_2 = z(s_2, r_2)$ be arbitrary points in the neighborhood $\sigma(l)$, and $\rho(z_1, z_2)$ the shortest distance between them along coordinate curves, i.e.,

$$\rho(z_1, z_2) = |s_1 - s_2| + |r_1 - r_2|.$$

We consider the triangle with vertices at the points $t_1 = t(s_1)$, $t_2 = (s_2)$ and $z_0 = z(s_1, |r_1 - r_2|)$; by construction

$$|z_1 - z_2| = |z_0 - t_2|, \qquad |z_0 - t_1| = |r_1 - r_2|,$$

and, by (2), the angle γ at the vertex t_1 satisfies the inequality $0 < \pi/2 - \alpha_0 < \gamma \leqslant \pi/2$. Then

$$|z - t_2|^2 = |z_0 - t_1|^2 + |t_2 - t_1|^2 - 2\cos\gamma\, |t_1 - t_2| |z_0 - t_1|$$
$$\geqslant \sin^2\frac{\gamma}{2}(|z_0 - t_1| + |t_1 - t_2|)^2,$$

whence, by (3),

$$|z_1 - z_2| = |z_0 - t_2| \geqslant \sin\frac{\gamma}{2}(|r_1 - r_2| + k_0|s_1 - s_2|) \geqslant k(\alpha_0)\rho(z_1, z_2),$$

where $k(\alpha_0) = k_0\sin(\gamma/2) \geqslant \cos\alpha_0 \sin(\pi/4 - \alpha_0/2) > 0$. Thus

$$k\rho(z_1, z_2) \leqslant |z_1 - z_2| \leqslant \rho(z_1, z_2), \qquad k = k(\alpha_0) > 0. \qquad (4)$$

Taking $\rho(z_1, z_2) \geqslant |s_1 - s_2|$, $|r_1 - r_2|$, into account, we get from (4), in particular, that

$$|s_1 - s_2| \leqslant \frac{1}{k}|z_1 - z_2|, \qquad |r_1 - r_2| \leqslant \frac{1}{k}|z_1 - z_2|. \tag{4'}$$

Let us now study the properties of certain classes of functions on smooth contours. A function $\varphi(z)$ is called a Hölder function ($\varphi \in H^\alpha(\mathfrak{M})$) if

$$|\varphi(z') - \varphi(z'')| \leqslant H(\varphi)|z' - z''|^\alpha, \qquad 0 < \alpha \leqslant 1,$$

for any two points z', z'' in its domain $\mathfrak{M}$, where α and $H(\varphi)$ are the Hölder exponent and Hölder constant, respectively, of the function $\varphi(z)$. If $\alpha = 1$, then we also say that $\varphi(z)$ satisfies a Lipschitz condition.

The definition of a Hölder function $\varphi(z_1, \dots, z_n)$ of several complex variables $z_1, \dots, z_n$ is analogous:

$$|\varphi(z'_1, \dots, z'_n) - \varphi(z''_1, \dots, z''_n)| \leqslant \sum_{i=1}^n H_i(\varphi)|z'_i - z''_i|^{\alpha_i}, \qquad 0 < \alpha_i \leqslant 1.$$

Obviously, if $\varphi \in H^\alpha(\mathfrak{M})$ and $f(\varphi) \in H^\beta[\varphi(\mathfrak{M})]$, then for the composite function we have $f[\varphi(z)] \in H^{\alpha\beta}(\mathfrak{M})$, $0 < \alpha\beta \leqslant 1$. And if $\varphi(z) \in H^\alpha(\mathfrak{M})$. And if $\psi(z) \in H^\beta(\mathfrak{M})$, then $\varphi + \psi$, $\varphi\psi \in H^\lambda(\mathfrak{M})$, $0 < \lambda = \min(\alpha, \beta) \leqslant 1$, and, moreover, $[\varphi(z)]^{-1} \in H^\alpha(\mathfrak{M})$.

We shall study the properties of Hölder functions on smooth curves ([97], §§5–6). The inequalities

$$|f|^\alpha + |g|^\alpha \leqslant 2^{1-\alpha}(|f| + |g|)^\alpha, \tag{5}$$

$$0 < |g|^\alpha - |f|^\alpha \leqslant (|g| - |f|)^\alpha \tag{6}$$

are needed, and can be established by determining the maxima of the functions

$$F_1(\xi) = \frac{1 + \xi^\alpha}{(1 + \xi)^\alpha} \leqslant 2^{1-\alpha}, \qquad F_2(\xi) = \frac{1 - \xi^\alpha}{(1 - \xi)^\alpha} \leqslant 1 \qquad \left(\xi = \frac{|f|}{|g|} < 1\right).$$

A direct application of (3) and (5) leads to the following assertion.

$1°$. If $\varphi(t) \in H^\alpha(L_i)$ ($i = 1, 2$), where the L_i have a common endpoint t_0 and $L = L_1 + L_2$ is a smooth arc, and $\varphi(t)$ is continuous at t_0, then $\varphi(t) \in H^\alpha(L)$.

Indeed, the assertion is obvious if the points $t_1 = t(s_1)$ and $t_2 = t(s_2)$ are located on the same side of $t_0 = t(s_0)$; otherwise, i.e. for $s_1 \leqslant s_0 \leqslant s_2$, we have

$$|\varphi(t_1) - \varphi(t_2)| \leqslant |\varphi(t_2) - \varphi(t_0)| + |\varphi(t_1) - \varphi(t_0)|$$

$$\leqslant H(\varphi)(|s_2 - s_0|^\alpha + |s_1 - s_0|^\alpha)$$

$$\leqslant 2^{1-\alpha}H(\varphi)|s_2 - s_1|^\alpha \leqslant 2^{1-\alpha}H(\varphi)k_0^{-\alpha}|t_1 - t_2|^\alpha.$$

2°. Applying inequalities (3) and (6) to the function

$$r^\alpha(t, t_0) = |t - t_0|^\alpha, \qquad 0 < \alpha \leqslant 1,$$

we get that

$$|r^\alpha(t_1, t_0) - r^\alpha(t_2, t_0)| \leqslant [r(t_1, t_0) - r(t_2, t_0)]^\alpha \leqslant k_0^{-\alpha} |t_1 - t_2|^\alpha,$$

i.e., $r^\alpha \in H^\alpha(L)$ with respect to t, and $r^\alpha \in H^\alpha(L)$ also with respect to t_0 because of the symmetry of $r(t, t_0)$ with respect to t and t_0.

3°. Let $\varphi(t, \tau) \in H^\alpha(L \times T)$, where T is some set in which the parameter τ varies. Then

$$\psi(t, t_0, \tau) = \frac{\varphi(t, \tau) - \varphi(t_0, \tau)}{|t - t_0|^\beta} \in H^{\alpha-\beta}(L \times L \times T), \qquad 0 < \beta \leqslant 1,$$

i.e., ψ satisfies a Hölder condition with respect to all three variables. In view of the symmetry of t and t_0 in $\psi(t, t_0, \tau)$, it suffices to prove that ψ is Hölder with respect to one of them; the Hölder property with respect to τ can obviously be established in an elementary way. We set $\omega(s - s_0) = \varphi(t, \tau) - \varphi(t_0, \tau)$, and $\psi(s_i) = \psi[t(s_i)]$, $i = 1, 2$, assuming without loss of generality that $s_0 \leqslant s_1 \leqslant s_2$. Then

$$|\psi(s_1) - \psi(s_2)| \leqslant \frac{|\omega(s_1 - s_0) - \omega(s_2 - s_0)|}{|t_2 - t_0|^\beta}$$

$$+ |\omega(s_1 - s_0)| \left| \frac{|t_2 - t_0|^\beta - |t_1 - t_0|^\beta}{|t_2 - t_0|^\beta |t_1 - t_0|^\beta} \right|.$$

For the first term Δ_1 on the right-hand side we have

$$\Delta_1 \leqslant H(\varphi) k_0^{-\beta} \left(\frac{s_2 - s_1}{s_2 - s_0} \right)^\beta (s_2 - s_1)^{\alpha-\beta} \leqslant H(\varphi) k_0^{-\beta-\alpha} |t_2 - t_1|^{\alpha-\beta}.$$

Applying (6) to the second term Δ_2 under the assumption that $|t_1 - t_0| \leqslant |t_2 - t_1|$, we get

$$\Delta_2 \leqslant H(\varphi) k_0^{-\alpha} \frac{|t_1 - t_0|^\alpha |t_2 - t_1|^\beta}{|t_1 - t_0|^\beta |t_2 - t_0|^\beta} \leqslant H(\varphi) k_0^{-\alpha-\beta} |t_1 - t_2|^{\alpha-\beta}.$$

For $|t_1 - t_0| > |t_2 - t_1|$ we have

$$\Delta_2 \leqslant H(\varphi) k_0^{-\alpha} |t_1 - t_0|^\alpha |t_2 - t_0|^\beta \left| \left| 1 + \frac{t_2 - t_1}{t_1 - t_0} \right|^\beta - 1 \right|.$$

Since

$$|1 - \sigma| \leqslant \left| 1 + \frac{t_2 - t_1}{t_1 - t_0} \right| \leqslant 1 + \left| \frac{t_2 - t_1}{t_1 - t_0} \right| \equiv 1 + \sigma,$$

the obvious inequalities

$$(1 + \sigma)^\beta - 1 \leq \sigma, \qquad 1 - |1 - \sigma|^\beta \leq \sigma,$$

where $0 \leq \beta \leq 1$ and $\sigma \geq 0$, give us that

$$\left| \left| 1 - \frac{t_2 - t_1}{t_1 - t_0} \right|^\beta - 1 \right| \leq \left| \frac{t_2 - t_1}{t_1 - t_0} \right|.$$

Then

$$\Delta_2 \leq H(\varphi) k_0^{-\alpha} |t_1 - t_0|^\alpha |t_2 - t_0|^{-\beta} \left| \frac{t_2 - t_1}{t_1 - t_0} \right|$$

$$= H(\varphi) k_0^{-\alpha} \left| \frac{t_2 - t_1}{t_2 - t_0} \right|^\beta \left| \frac{t_2 - t_1}{t_1 - t_0} \right|^{1-\alpha} |t_2 - t_1|^{\alpha - \beta} \leq H(\varphi) k_0^{-\alpha} |t_2 - t_1|^{\alpha - \beta},$$

which concludes the proof of assertion 3°.

4°. Let $\omega(t)$ be a bounded function on L such that $|d\omega/dt| < c/|t - t_0|$, $t \neq t_0$; by (3), this implies that $|d\omega/dt| < ck_0^{-1}/|s - s_0|$. Consequently, if $\varphi(t) \in H^\alpha(L)$, then also

$$\psi(t) = [\varphi(t) - \varphi(t_0)]\omega(t) \in H^\alpha(L) \qquad (t_1, t_0 \in L).$$

The role of $\omega(t)$ may be played, for example, by the function $\omega(t) = \exp(i\gamma \arg(t - t_0))$ for any $\gamma = \alpha_0 + i\beta_0$. For the first term on the right-hand side of the identity

$$\psi(s_1) - \psi(s_2) = [\varphi(s_1) - \varphi(s_2)]\omega(s_1) + [\varphi(s_2) - \varphi(s_0)][\omega(s_2) - \omega(s_1)]$$
$$= \Delta_1 + \Delta_2$$

we have the estimate

$$|\Delta_1| \leq |\omega| H(\varphi) |s_1 - s_2|^\alpha \leq \text{const} |t_1 - t_2|^\alpha.$$

The second term has a similar estimate for $s_2 - s_0 \leq s_2 - s_1$. Suppose that $s_2 - s_0 > s_2 - s_1$ and $0 < \lambda < 1$. Then the mean value theorem gives us that

$$|\Delta_2| \leq H(\varphi)(s_1 - s_0)^\alpha \frac{ck_0^{-1}(s_2 - s_1)}{|s_1 - s_0 + \lambda(s_2 - s_1)|}$$

$$\leq ck_0^{-1} H(\varphi) \left(\frac{s_2 - s_1}{s_0 - s_1} \right)^{1-\alpha} (s_2 - s_1)^\alpha,$$

which obviously proves assertion 4°.

5°. A direct application of 4° shows that the functions

$$f(t) = (t_0 - t)^{\alpha + i\beta}, \quad \psi(t) = |t_0 - t|^\alpha \frac{t - t_0}{s - s_0}, \quad \varphi(t) = |t - t_0|^\alpha \left| \frac{t - t_0}{s - s_0} \right|^\lambda,$$

where β and λ are arbitrary numbers, satisfy a Hölder condition with exponent $0 < \alpha \leqslant 1$ with respect to the variables $t, t_0 \in L$.

6°. Suppose that $\varphi(t) \in H^\alpha(l)$, $0 < \alpha \leqslant 1$, on an arc l of an arbitrary smooth curve L, and that $s_b - s_a = \varepsilon$, where $a = t(s_a)$ and $b = t(s_b)$ are the endpoints of l; ε is chosen in such a way that (2) holds. We set $\varphi(z) = \varphi(t)$, where $t = t(s)$ is the point where l intersects the straight line passing through z parallel to the normal n_a at a (see Figure 1).

Obviously, $\varphi(z) \in H^\alpha[\sigma(l)]$ for the function extended in this way from the boundary $l \subset L$ to a standard neighborhood $\sigma(l)$ of it.

Indeed, inequalities (3) and (4) give us that

$$|\varphi(z_1) - \varphi(z_2)| = |\varphi(t_1) - \varphi(t_2)| \leqslant H(\varphi)|t_1 - t_2|^\alpha \leqslant H(\varphi)|s_1 - s_2|$$
$$\leqslant H(\varphi)k_0^{-\alpha}|z_1 - z_2|^\alpha,$$

i.e.,

$$|\varphi(z_1) - \varphi(z_2)| \leqslant \tilde{H}(\varphi)|z_1 - z_2|^\alpha, \quad \text{where } \tilde{H}(\varphi) = H(\varphi)k_0^{-\alpha}. \quad (7)$$

Along with the functions in the class $H^\alpha(L)$ we shall consider the piecewise Hölder functions $\varphi(t) \in H_*^\alpha(L)$ that have finitely many points of discontinuity on L and satisfy a Hölder condition on the arcs with these points as endpoints. Denoting by α, $0 < \alpha \leqslant 1$, the smallest Hölder exponent of a piecewise Hölder function $\varphi(t)$ on the intervals of its continuity and by $H_*(\varphi)$ the largest of the Hölder constants, respectively, we arrive at the inequality

$$|\varphi(t_1) - \varphi(t_2)| \leqslant H_*(\varphi)|t_1 - t_2|^\alpha,$$

valid for any points t_1 and t_2 in the intervals of continuity of $\varphi(t)$.

§2. Cauchy type integrals

1°. *Definition and elementary properties of Cauchy type integrals.* A function $\Phi(z)$ defined on the whose z-plane is said to be piecewise analytic with jump curve L it is analytic for all finite z except on some piecewise smooth curve $L = \Sigma_i L_i$ and has finite order at infinity. An elementary example of such a function is the Cauchy integral along a smooth closed contour L

$$\frac{1}{2\pi i}\int_L \frac{\Phi(t)}{t - z}dt = \begin{cases} \Phi(z), & z \in D^+, \\ 0, & z \in D^-, \end{cases}$$

where $\Phi(z)$ is an analytic function in the domain D^+ bounded by L, and D^- is the domain exterior to L. Another example of a piecewise analytic function is the integral

$$\int F(z, t)\, dt \equiv \Phi(z),$$

where $F(z, t)$ has finite order as $z \to \infty$ and is analytic for all $z \notin L$ and fixed $t \in L$, while $L = \Sigma_i L_i$ is a piecewise smooth curve. Indeed, for $z, z + \Delta z \notin L$ we have

$$\left| \frac{\Delta \Phi}{\Delta z} - \int F_z'(z, t) \, dt \right| = \left| \int \left(\frac{F(z, t) - F(z + \Delta z, t)}{\Delta z} - F_z'(z, t) \right) dt \right|$$

$$\leqslant \max \left| \frac{\Delta_z F}{\Delta z} - F_z' \right| \cdot (\text{length of } L),$$

from which it follows that $\Phi(z)$ is differentiable off the contour L. In particular, for $F(z, t) = \varphi(t)/2\pi i(t - z)$ we get that the integral

$$\Phi(z) = \frac{1}{2\pi i} \int_L \frac{\varphi(t) \, dt}{t - z},$$

called a *Cauchy type integral*, is piecewise analytic with jump contour L. If $\varphi(t) \in H^\alpha(L)$, then the Cauchy type integral over L

$$\Phi(t_0) = \frac{1}{2\pi i} \int_L \frac{\varphi(t) \, dt}{t - t_0}$$

exists in the sense of the Cauchy principal value, i.e.,

$$\Phi(t_0) = \frac{1}{2\pi i} \lim_{\varepsilon \to 0} \int_{L_\varepsilon} \frac{\varphi(t) \, dt}{t - t_0},$$

where $L_\varepsilon = L - l$, and l is the arc whose endpoints t' and t'' satisfy the condition $|t_0 - t'| = |t_0 - t''| = \varepsilon$, $t_0 \in l$. To prove this assertion we represent $\Phi(t_0)$ in the form

$$\Phi(t_0) = \frac{1}{2\pi i} \int_L \frac{\varphi(t) - \varphi(t_0)}{t - t_0} \, dt + \frac{\varphi(t_0)}{2\pi i} \int_L \frac{dt}{t - t_0},$$

where the first integral exists as an improper integral in view of the condition $\varphi(t) \in H^\alpha(L)$. Taking account of the facts that $|t' - t_0| = |t'' - t_0|$ and that the angle $\alpha(s) = \arg[(t' - t_0)(t'' - t_0)^{-1}]$ between the chords joining t' to t_0 and t'' to t_0 tends to π as $\varepsilon \to 0$, we get

$$\int_L \frac{dt}{t - t_0} = \ln \frac{b - t_0}{a - t_0} + \lim_{\varepsilon \to 0} \ln \frac{t' - t_0}{t'' - t_0} = \ln \frac{b - t_0}{a - t_0} + \pi i$$

(here a and b are the endpoints of the nonclosed arc L). Thus,

$$\Phi(t_0) = \frac{1}{2\pi i} \int_L \frac{\varphi(t) - \varphi(t_0)}{t - t_0} \, dt + \frac{\varphi(t_0)}{2\pi i} \ln \frac{b - t_0}{a - t_0} + \frac{1}{2} \varphi(t_0), \qquad (1)$$

where the second term vanishes for a closed contour L, i.e., for $a = b$.

$2°$. *Behavior in the proximity of the curve of integration.* To study the behavior of a Cauchy integral close to the curve of integration we represent it in the form

$$\Phi(z) = \frac{1}{2\pi i} \int_L \frac{\varphi(t) - \varphi(z)}{t - z} dt + \frac{\varphi(z)}{2\pi i} \int_L \frac{dt}{t - z}, \qquad (2)$$

where $\varphi(t) \in H^\alpha(L_0)$, $L_0 \subset L$, and $\varphi(t)$ is integrable on $L - L_0$, and $\varphi(z)$ is the extension of it to a standard neighborhood $\sigma(l)$ of an arbitrary arc $l \subset L_0$ (see §1.6°) with endpoints $a_l = t(s_a)$ and $b_l = t(s_b)$, different from the endpoints $a_L = t(0)$ and $b_L = t(s_L)$ of L. We set

$$F(z) = \frac{1}{2\pi i} \int_L \frac{\varphi(t) - \varphi(z)}{t - z} dt$$

and prove the following important theorem.

THEOREM 1. $F(z) \in H^\lambda[\sigma(l)]$, $l \subset L_0$, where $\lambda = \alpha$ for $0 < \alpha < 1$, and $\lambda \in (0, 1)$ is arbitrary for $\alpha = 1$. The Hölder constant $H(F)$ of the function $F(z)$ depends on $H(\varphi)$ on L_0, on the value of the integral of $\varphi(t)$ over $L - L_0$, on the distance to the endpoints of L, and on the geometry of L.

Let us consider the expression

$$2\pi |F(z_1) - F(z_2)| = \left| \int_L \left(\frac{\varphi(t) - \varphi(z_1)}{t - z_1} - \frac{\varphi(t) - \varphi(z_2)}{t - z_2} \right) dt \right| = \left| \int_l + \int_{L-l} \right|,$$

where

$$z_i = z(s_i, r_i) \in \sigma(l) \quad (i = 1, 2), s_1 \leqslant s_2, \qquad |z_1 - z_2| \equiv \delta$$

and the endpoints of the arc $l \subset L$ are located at the points $a_l = t(s_1 - \delta)$ and $b_l = t(s_2 + \delta)$, which do not coincide with the endpoints $a_L = t(0)$ and $b_L = t(s_L)$ of L. Recall that the s_i denote the abcissas of the points $t_i \in l$ of intersection with the straight lines parallel to the normal at a and passing through the z_i, while $r_i = |t_i - z_i| < \varepsilon_0/2$ (see Figure 1 of §1). Applying (1.4) and (1.7), we get

$$\left| \int_l \right| \leqslant \tilde{H}(\varphi) \int_l \left(|t - z_1|^{\alpha - 1} + |t - z_2|^{\alpha - 1} \right) ds$$

$$\leqslant \tilde{H}(\varphi) k^{\alpha - 1} \int_l \left\{ (|s - s_1| + |r_1|)^{\alpha - 1} + (|s - s_2| + |r_2|)^{\alpha - 1} \right\} ds$$

$$\leqslant \tilde{H}(\varphi) k^{\alpha - 1} \int_l \left(|s - s_1|^{\alpha - 1} + |s - s_2|^{\alpha - 1} \right) ds$$

$$= \frac{2k^{\alpha - 1}}{\alpha} \tilde{H}(\varphi) \left[\delta^\alpha + (s_2 - s_1 + \delta)^\alpha \right],$$

from which, using the fact that $|s_2 - s_1| \leqslant k^{-1}\delta$ by (1.4'), we find that

$$\left| \int_l \right| \leqslant c_1 \delta^\alpha = c_1 |z_1 - z_2|^\alpha,$$

where

$$c_1 = \frac{2k^{\alpha-1}}{\alpha} \tilde{H}(\varphi)\left[1 + (1 + k^{-1})^\alpha\right].$$

Let us now estimate the integral along $(L - l)$. To do this we write it in the form

$$\int_{L-l} = \int_{L-l} \frac{\varphi(z_2) - \varphi(z_1)}{t - z_1} dt$$

$$+ \int_{L-l} [\varphi(t) - \varphi(z_2)]\left(\frac{1}{t - z_1} - \frac{1}{t - z_2}\right) dt \equiv I_1 + I_2.$$

We have

$$|I_1| = |\varphi(z_2) - \varphi(z_1)|\left| \ln\frac{b_L - z_1}{a_L - z_1} - \ln\frac{b_l - z_1}{a_l - z_1} \right| \leqslant c_2(\Delta)\hat{H}(\varphi)|z_1 - z_2|^\alpha,$$

where

$$c_2(\Delta) = \max\left| \ln\frac{b_L - z_1}{a_L - z_1} \right| + \max\left| \ln\frac{b_l - z_1}{a_l - z_1} \right| = c_2' + c_2'',$$

and

$$\Delta = \min\{|a_L - a_l|, |a_L - b_l|, |b_l - b_L|, |b_L - a_l|\}.$$

To determine an estimate for $c_2(\Delta)$ we observe that, according to (1.4),

$$k(\delta + |r_1|) \leqslant |a_l - z_1| \leqslant \delta + |r_1|, \qquad k = k(\alpha_0) > 0,$$

$$k(s_2 - s_1 + \delta + |r_1|) \leqslant |b_l - z_1| \leqslant s_2 - s_1 + \delta + |r_1|,$$

whence

$$0 < k \leqslant \left| \frac{b_l - z_1}{a_l - z_1} \right| \leqslant k^{-1}(1 + k^{-1})$$

and consequently $c_2'' < \infty$. Since, on the other hand,

$$\Delta \leqslant |a_L - z_1| \leqslant s_L + \varepsilon_0, \qquad \Delta \leqslant |b_L - z_1| \leqslant s_L + \varepsilon_0,$$

where $2\varepsilon_0$ is the width of the standard neighborhood σ, it follows that $c_2' < \infty$ for $\Delta \neq 0$, and thereby $c_2(\Delta) < \infty$.

It remains to estimate the integral I_2:

$$|I_2| \leqslant \delta\left(\int_0^{s_1-\delta} + \int_{s_2+\delta}^{s_L} \right) \frac{|\varphi(t) - \varphi(z_2)|}{|t - z_1||t - z_2|} ds \equiv \delta(I_2' + I_2'').$$

Suppose first that $s_2 \leqslant \varepsilon_0(\alpha_0)$, i.e., the endpoint $a_L = t(0)$ of L lies along with the points z_1 and z_2 in a standard neighborhood σ_0 of the curve l_a with endpoints a_L and $t(s_2)$. In this case we can use the Hölder property of $\varphi(z)$ in σ_a and the inequality (1.4') in estimating I'_2. Then for $0 < \alpha < 1$

$$I'_2 \leqslant \tilde{H}(\varphi)k^{\alpha-2}\int_0^{s_1-\delta} \frac{ds}{(s_1-s)(s_2-s)^{1-\alpha}} \leqslant \tilde{H}(\varphi)k^{\alpha-2}\int_0^{s_1-\delta} \frac{ds}{(s_1-s)^{2-\alpha}}$$

$$= \frac{\tilde{H}(\varphi)k^{\alpha-2}}{1-\alpha}\frac{s_1^{1-\alpha}-\delta^{1-\alpha}}{s_1^{1-\alpha}\delta^{1-\alpha}} \leqslant \frac{\tilde{H}(\varphi)k^{\alpha-2}}{1-\alpha}\frac{(s_1-\delta)^{1-\alpha}}{s_1^{1-\alpha}\delta^{1-\alpha}},$$

while for $\alpha = 1$

$$I'_2 \leqslant \tilde{H}(\varphi)k^{\alpha-2}\ln(s_1/\delta).$$

Similarly, if $s_L - s_1 \leqslant \varepsilon(\alpha_0)$, i.e., the endpoint $b_L = t(s_L)$ and the points z_1 and z_2 lie in a single standard neighborhood σ_b, then for $0 < \alpha < 1$

$$I''_2 \leqslant \tilde{H}(\varphi)k^{\alpha-2}\int_{s_2+\delta}^{s_L} \frac{ds}{(s-s_2)^{2-\alpha}} \leqslant \tilde{H}(\varphi)\frac{k^{\alpha-2}}{1-\alpha}\frac{(s_L-s_2-\delta)^{1-\alpha}}{(s_L-s_2)^{1-\alpha}\delta^{1-\alpha}},$$

while for $\alpha = 1$

$$I''_2 \leqslant \tilde{H}(\varphi)k^{\alpha-2}\ln\frac{s_L-s_2}{\delta}.$$

Suppose now that $s_2 > \varepsilon(\alpha_0)$. By (1.4') we have

$$s_2 - s_1 \leqslant \frac{1}{k}|z_2-z_1| \leqslant \frac{\varepsilon}{4},$$

when $\delta = |z_1 - z_2| \leqslant k\varepsilon/4$ and z_1 and z_2 are sufficiently close; hence $s_1 - \varepsilon/2 \geqslant \varepsilon/4 > 0$. Let us represent I'_2 in the form

$$I'_2 = \int_0^{s_1-\varepsilon/2} - \int_{s_1-\varepsilon/2}^{s_1-\delta} \equiv \Delta_1 + \Delta_2,$$

where the estimate obtained above is applicable to Δ_2, since the point $t_\varepsilon = t(s_1 - \varepsilon/2)$, which plays the role of the endpoint $a_L = t(0)$, lies by construction in a single standard neighborhood with z_1 and z_2. And the term Δ_1 is bounded by a constant that does not depend on $\delta = |z_1 - z_2|$. Indeed,

$$\Delta_1 \leqslant \max\left\{\left(|t-z_1||t-z_2|\right)^{-1}\left[\left(s_1-\frac{\varepsilon}{2}\right)|\varphi(z_2)|+R(\varphi)\right]\right\}$$

$$\leqslant \frac{4}{\varepsilon^2}\left[\left(s_1-\frac{\varepsilon}{2}\right)\max_{L_0}|\varphi(t)|+R(\varphi)\right] = c_3 < \infty,$$

where $R(\varphi) = \int_{L-L_0}|\varphi(t)||dt|$.

The integral I_2'' can be estimated completely analogously when $s_L - s_1 > \varepsilon(\alpha_0)$. Thus,

$$|I_2| \leqslant \delta(I_2' + I_2'') \leqslant c_4 |z_1 - z_2|^\alpha \quad \text{for } 0 < \alpha < 1$$

and

$$|I_2| \leqslant c_5 |z_1 - z_2| \left| \ln \frac{1}{|z_1 - z_2|} \right| \quad \text{for } \alpha = 1.$$

The theorem is proved.

REMARKS. 1. The estimates obtained above for $|F(z_1) - F(z_2)|$ show that if $\varphi(t) \in H^\alpha(L)$ on a smooth closed contour L, then $F(z) \in H^\alpha$ in a neighborhood of each point of the contour.

2. The theorem remains valid if $L = \Sigma_i L_i$ is a smooth curve, because, when estimating in the proximity to any one of the curves L_i, the derivatives with respect to z of the integrals along the remaining curves are bounded by a constant that depends only on the minimum of the distance between the curves $L_j \subset L$ and on the quantity

$$R(\varphi) = \sum_i \int_{L_i} |\varphi(t)| |dt|$$

(cf. the estimate of Δ_1).

3. If $\varphi(a) = 0$ and $\varphi(t) \in H^\alpha(ab_1)$, $b_1 \in L$, then, by completing L to a closed contour $\tilde{L}$ and setting $\varphi(t) \equiv 0$ on $\tilde{L} - L$, we get from Theorem 1 that $F(z) \in H^\alpha[\sigma(l)]$ (l is an arc containing the point a), i.e., $F(z)$ is a Hölder function in a neighborhood of a.

3°. *Boundary properties of Cauchy type integrals.* Proceeding to an application of Theorem 1 for investigation of the boundary properties of Cauchy type integrals, we introduce some notation. At an arbitrary point t_0 of a smooth curve we locate the center of a disk of radius small enough so that it is divided by L into two parts lying to the right and left, respectively, of L as we move from the endpoint $a = t(0)$ to the endpoint $b = t(s_L)$. Let $\Phi^+(t_0)$ and $\Phi^-(t_0)$ denote the limit values of a function $\Phi(z)$ as we approach $t_0 \in L$ from left-hand and right-hand neighborhoods, respectively, of the curve L. For a closed smooth contour L the quantities $\Phi^\pm(t_0)$ defined in this way are obviously the limit values on L of $\Phi(z)$ from the interior domain D^+ and the exterior domain D^- relative to L. Suppose first that L is a smooth closed contour, $\varphi(t) \in H^\alpha(L_0)$, $L_0 \subset L$, and the point $z \in D^+$ is located in a standard neighborhood $\sigma(l)$ of an arc $l \subset L_0$ containing a fixed point $t_0 \in l$. Taking account of the fact that

$$\frac{1}{2\pi i} \int \frac{dt}{t - z} = 1, \quad z \in D^+,$$

by a property of the Cauchy type integral, we then find from (2) that

$$\Phi(z) = \frac{1}{2\pi i} \int_L \frac{\varphi(t)\,dt}{t-z} = \frac{1}{2\pi i} \int_L \frac{\varphi(t) - \varphi(z)}{t-z}\,dt + \varphi(z), \qquad (3)$$

i.e., $\Phi(z) = F(z) + \varphi(z)$, where $\varphi(z)$ is the extension of $\varphi(t)$ into the neighborhood $\sigma(l)$ (see §1.6°). Passing to the limit in (3) as $z \to t_0$ from D^+ (the existence of this limit follows from the fact that $F(z)$ (Theroem 1) and $\varphi(z)$ are Hölder functions), we get

$$\Phi^+(t) = \frac{1}{2\pi i} \int_L \frac{\varphi(t) - \varphi(t_0)}{t - t_0}\,dt + \varphi(t_0),$$

and so, by the equality

$$\frac{1}{2\pi i} \int_L \frac{dt}{t - t_0} = \frac{1}{2},$$

which follows from (1) for $a = b$ and $\varphi(t_0) = \varphi(t) = 1$, we finally find that

$$\Phi^+(t_0) = \frac{1}{2}\varphi(t_0) + \frac{1}{2\pi i} \int_L \frac{\varphi(t)\,dt}{t - t_0} \equiv \frac{1}{2}\varphi(t_0) + \Phi(t_0). \qquad (4)$$

Suppose now that $z \in D^-$; then

$$\frac{1}{2\pi i} \int_L \frac{dt}{t - z} = 0, \qquad z \in D^-,$$

and, consequently, (2) can be written in the form

$$\Phi(z) = \frac{1}{2\pi i} \int_L \frac{\varphi(t)\,dt}{t - z} = \frac{1}{2\pi i} \int_L \frac{\varphi(t) - \varphi(z)}{t - z}\,dt, \qquad (5)$$

which leads as $z \to t_0$ $(z \in D^-)$ to the equality

$$\Phi^-(t_0) = -\frac{1}{2}\varphi(t_0) + \frac{1}{2\pi i} \int_L \frac{\varphi(t)\,dt}{t - t_0} \equiv -\frac{1}{2}\varphi(t_0) + \Phi(t_0). \qquad (6)$$

The formulas (4) and (6), which are commonly called the Sokhotskiĭ-Plemelj formulas, can be written in a more convenient form:

$$\Phi^+(t_0) - \Phi^-(t_0) = \varphi(t_0), \qquad \Phi^+(t_0) + \Phi^-(t_0) = \frac{1}{\pi i} \int_L \frac{\varphi(t)\,dt}{t - t_0}. \qquad (7)$$

Since in deriving (7) we used the Hölder property of $\varphi(t)$ only in a neighborhood of t_0, these formulas remain true also for a smooth nonclosed arc L. Indeed, the case of a nonclosed arc L reduces to the case just studied by completing L to a smooth closed contour $\tilde{L}$ and setting $\varphi(t) \equiv 0$ on the adjoined arc $(\tilde{L} - L)$. By the local nature of (7), these formulas obviously remain true also for an arbitrary smooth curve $L = \Sigma_i L_i$. Theorem 1 and (2)

enable one to carry out a complete investigation of the properties of the limit
values $\Phi^\pm(t_0)$, i.e., of the Cauchy type integral and its behavior when
approaching the boundary. Let $\sigma^\pm(l)$ be left-hand and right-hand neighborhoods of an arc $L_0 \subset L$, and $\Phi^\pm(z)$ the values of the Cauchy type integral in
these neighborhoods. Let L be a smooth closed contour. Then, by (3) and (5),

$$\Phi^\pm(z) = \frac{1}{2\pi i} \int_L \frac{\varphi(t) - \varphi(z)}{t - z}\, dt + \varphi(z)\delta^\pm \equiv F(z) + \varphi(z)\delta^\pm\,,$$

where $\delta^+ = 1$ for $z \in D^+$ and $\delta^- = 0$ for $z \in D^-$, while $\varphi(z)$ is the extension
to a standard neighborhood $\sigma(l)$ of the arc $l \subset L_0$ and $z \in \sigma(l)$. Theorem 1
implies that $\Phi^\pm(z) \in H^\alpha[\sigma^\pm(l)]$ for any arc $l \subset L_0$, and the quantity $\Phi^\pm(z)$
obviously does not depend on the method of extending $\varphi(t)$ to a standard
neighborhood $\sigma(l)$, $l \subset L_0$. Consequently, $\Phi^\pm(z) \in H^\alpha[\sigma_\varepsilon^\pm(l)]$, where $(\sigma^\pm$
$-\sigma_\varepsilon^\pm)$ is an arbitrarily small neighborhood of the endpoints of L_0. These
properties are preserved also for a nonclosed arc L, since this case reduces to
that just analyzed by completing L to a closed contour $\tilde{L}$ and setting $\varphi(t) \equiv 0$
on $\tilde{L} - L$. Thus, we come to the following theorem of Privalov [74], which we
present in a somewhat more general formulation than his.

THEOREM 2. (PRIVALOV). *Suppose that L is a smooth nonclosed arc and the
density of the Cauchy type integral is a Hölder function on $L_0 \subset L$, $\varphi(t) \in
H^\alpha(L_0)$. Then*

$$\Phi^\pm(t_0) \in H^\alpha(L_0 - l_\varepsilon), \qquad \Phi^\pm(z) \in H^\lambda(\bar{\sigma}_\varepsilon^\pm),$$

*where $\lambda = \alpha$ for $0 < \alpha < 1$ and $0 < \lambda < 1$ is arbitrary for $\alpha = 1$, while l_ε and
$(\sigma^\pm - \sigma_\varepsilon^\pm)$ are arbitrarily small neighborhoods of the endpoints of L_0. If the
contour is also closed and $\varphi(t) \in H^\alpha(L)$, then $\Phi^+(z) \in H^\lambda(\bar{D}^+)$ and $\Phi^-(z) \in
H^\alpha(\overline{K_R - D^+})$ ($K_R \supset D^+$ is a disk of arbitrarily large radius R).*

REMARK. If $\varphi(t) \equiv 0$, $t \in L_0$, on some arc $L_0 \subset L$ with endpoints a_0 and b_0,
and $\varphi(t) \in H^\alpha(L)$, then in a neighborhood $\sigma(l)$ of an arbitrary arc $l \subset L_0$ the
Cauchy type integral can be represented in the form

$$\Phi(z) = \frac{1}{2\pi i} \int_L \frac{\varphi(t) - \varphi(z)}{t - z}\, dt,$$

and therefore, by Theorem 1, $\Phi(z) \in H^\alpha$ in a whole neighborhood of the arc
L_0, with the possible exception of its endpoints. In particular, if $\varphi(a) = 0$ and
$\varphi(t) \in H^\alpha(ab_1)$, then by completing L to a smooth closed contour $\tilde{L}$ and
setting $\varphi(t) \equiv 0$ on $(\tilde{L} - L)$ we get that $\Phi(z) \in H^\alpha(\sigma_1)$ and $\Phi(t_0) \in H^\alpha(l_a)$,
$t_0 \in l_a \subset L$, where σ_a is the disk $|z - a| < \rho_0$ with a cut along the curve L, and
l_a is the arc with endpoints $a_l = a = t(0)$ and $b_l = t(\rho_0)$, and $\rho_0 > 0$ is
sufficiently small (cf. Remark 3 after Theorem 1).

4°. *Behavior of the derivative of a Cauchy type integral when approaching the boundary.* Let us now study the behavior, when approaching the boundary, of the derivative

$$\frac{d\Phi}{dz}(z) = \frac{1}{2\pi i} \int_L \frac{\varphi(t)\,dt}{(t-z)^2},$$

of the Cauchy type integral, where $\varphi(t) \in H^\alpha(L_0)$, $L_0 \subset L$, and $\varphi(t)$ is integrable on $L - L_0$; let z_0 approach a point $t_0 \in L$ that is different from the endpoints $a_0 = t(s_a)$ and $b_0 = t(s_b)$ of L_0. By Remark 1 after Theorem 1, we may assume without loss of generality that the curve L_0 is a smooth closed contour and then represent $d\Phi/dz$ in the form

$$\frac{d\Phi}{dz}(z_0) = \frac{1}{2\pi i} \int_L \frac{\varphi(t) - \varphi(z_0)}{(t-z_0)^2}\,dt,$$

where $\varphi(z_0)$, as always, denotes the extension of $\varphi(t)$ to a standard neighborhood $\sigma(l)$ of an arc $l \subset L_0$. Then

$$2\pi \left|\frac{d\Phi}{dz}\right| \leqslant \int_L \frac{|\varphi(t) - \varphi(z_0)|}{(t-z_0)}\,ds = \int_l + \int_{L-l},$$

where $z_0 = z(s_0, r)$, $0 < s_a + 2\delta < s_0 < s_b - 2\delta$, $|r| < k\varepsilon_0/2$, $0 < \delta < \varepsilon/2$, and $l \subset L_0$ is the arc with endpoints $a_l = t(s_0 - \delta)$ and $b_l = t(s_0 + \delta)$. Using the Hölder condition for $\varphi(z)$ and (1.4), we find that

$$\int_l \leqslant \tilde{H}(\varphi) \int_l \frac{ds}{|t - z_0|^{2-\alpha}} \leqslant \tilde{H}(\varphi)k^{\alpha-2} \int_l \frac{ds}{(|s - s_0| + r)^{2-\alpha}}$$

$$= 2\tilde{H}(\varphi)k^{\alpha-2} \int_{s_0-\delta}^{s_0} \leqslant \frac{c_0(\delta)}{|r|^{1-\alpha}},$$

if $0 < \alpha < 1$, while for $\alpha = 1$

$$\int_l \leqslant c_1(\delta) |\ln|r||\,.$$

Completely similar to the estimate of the integral Δ_1 in the proof of Theorem 1, we get that $|\int_{L-l}| \leqslant c_2(\delta)$, and, consequently,

$$\left|\frac{d\Phi}{dz}(z_0)\right| \leqslant \frac{c_3(\delta)}{|r|^\lambda}, \qquad 0 < \lambda < 1, \tag{8}$$

where $\lambda = 1 - \alpha$ for $0 < \alpha < 1$, $\lambda \in (0,1)$ is arbitrary for $\alpha = 1$, $|r|$ is the distance from z_0 to the boundary, and $\delta > 0$ is connected with the distance from z_0 to the endpoints of the curve L_0 on which $\varphi(t)$ is a Hölder function.

5°. Behavior at the endpoints. In investigating the properties of a Cauchy type integral at the endpoints of the curve of integration and at points of discontinuity of its density we have to deal with the elementary multi-valued functions of a complex argument

$$\ln(z - z_0) = \ln|z - z_0| + i\arg(z - z_0),$$

$$(z - z_0)^\gamma = |z - z_0|^\gamma e^{\gamma i \arg(z - z_0)}, \qquad \alpha + i\beta = \gamma.$$

To isolate the single-valued branches of these functions we shall join their branch points z_0 and $z = \infty$ by a cut and consider them on the slit plane. If nothing special is mentioned, then the cut is assumed to be rectilinear and parallel to the OX-axis, and the respective values of the argument on the upper and lower sides are equal to

$$\arg(z - z_0) = 2\pi k \quad \text{and} \quad \arg(z - z_0) = 2\pi(k + 1).$$

Thus, to fixed values $k = 0, \pm 1, \pm 2, \ldots$ there correspond single-valued branches of the function

$$\arg(z - z_0) = \text{Arg}(z - z_0) + 2\pi k$$

on the slit plane, and thereby also single-valued branches of the functions $\ln(z - z_0)$ and $(z - z_0)^\gamma$. We remark that the function

$$\varphi(t) = (t - z_0)^\gamma, \qquad \gamma = \alpha + i\beta, \qquad 0 < \alpha < 1, z_0 \in D(L),$$

defined on a smooth contour L has a discontinuity at a point $t = t_0$ of L on the cut joining the branch points $z - z_0$ and $z = \infty$ of the function

$$\varphi(z) = (z - z_0)^\gamma.$$

To distinguish a point t_0 of discontinuity of the function $\varphi(t) = (t - z_0)^\gamma$, we shall sometimes use the notation

$$\varphi(t) = \{t - z_0\}_{t_0}^\gamma, \qquad t_0 \in L.$$

Let us consider the Cauchy type integral

$$\Phi(z) = \frac{1}{2\pi i} \int_L \frac{\varphi(t)\, dt}{t - z}, \qquad \varphi(t) \in H^\alpha(L),$$

where L is a nonclosed smooth arc with endpoints a and b. Then for $c = a$ or b

$$\Phi(z) = \frac{1}{2\pi i} \int_L \frac{\varphi(t) - \varphi(c)}{t - z}\, dt + \frac{\varphi(c)}{2\pi i} \int \frac{dt}{t - z} \equiv F(z) + \frac{\varphi(c)}{2\pi i} \ln \frac{z - b}{z - a}.$$

According to the remark after Theorem 2, the Cauchy type integral $F(z)$ satisfies a Hölder condition in a neighborhood of such a c, because its density vanishes at this point. Consequently, for $c = a$ and $c = b$ we have, respectively,

$$\Phi(z) = -\frac{\varphi(a)}{2\pi i} \ln(z - a) + \Phi_1(z), \quad \Phi(z) = \frac{\varphi(b)}{2\pi i} \ln(z - b) + \Phi_2(z), \quad (9)$$

where $\Phi_1(z)$ and $\Phi_2(z)$ are functions satisfying a Hölder condition in a disk $\sigma_c: |z - c| < \rho_0$ ($\rho_0 > 0$, $c = a, b$) with a cut along the curve L. And if $\varphi(z) \in H^\alpha(ac)$ and $\varphi(z) \in H^\alpha(cb)$, where c is an interior point of an arc L with endpoints a, b and $\varphi(c - 0) \neq \varphi(c + 0)$, then by applying (9) to the integrals along ac and cb we get that

$$\Phi(z) = \frac{\varphi(c - 0) - \varphi(c + 0)}{2\pi i}\ln(z - c) + \Phi_0(z), \qquad (10)$$

where $\Phi_0^\pm(z) \in H^\alpha[\overline{\sigma^\pm(l)}]$; $\sigma^\pm(l)$ are right-hand and left-hand neighborhoods, respectively, of an arc l containing c.

Let us now consider a Cauchy type integral whose density has an integrable singularity and depends on a parameter $\tau \in T$:

$$\Phi(z, \tau) = \frac{1}{2\pi i}\int_{ab}\frac{\varphi(t, \tau)\,dt}{(t - a)^\gamma(t - z)}, \gamma = \alpha_0 + i\beta_0, \qquad (11)$$

where $0 \leqslant \alpha_0 < 1$ and $\varphi(t, \tau) \in H^\alpha(ab_1 \times T)$, $0 < \alpha \leqslant 1$, $a = t(0)$, $b_1 = t(s_1)$ $\in ab$ ($s_1 > 0$). For $\delta > 0$ we have

$$(z - a)^{\alpha_0+\delta}\Phi(z) = \frac{1}{2\pi i}\int_{ab}\frac{(z - a)^{\alpha_0+\delta} - (t - a)^{\alpha_0+\delta}}{(t - a)^\gamma(t - z)}\varphi(t, \tau)\,dt$$

$$+ \frac{1}{2\pi i}\int_{ab}\frac{(t - a)^{\delta-i\beta_0}\varphi(t, \tau)\,dt}{t - z},$$

where the second of the Cauchy type integrals is bounded in a neighborhood of a, since its density vanishes at $t = a$ (see the remark after Theorem 2). Applying Hölder's inequality to the density of the first of the integrals, we find that

$$\left|\varphi(t, \tau)\frac{\Delta(z - a)^{\alpha_0+\delta}}{t - z}\right| < c|\varphi|\cdot|t - z|^{\alpha_0+\delta-1},$$

which proves that this integral is bounded at $z = a$. Thus,

$$|z - a|^{\alpha_0+\delta}|\Phi(z, \tau)| < \text{const}, \qquad \delta > 0. \qquad (12)$$

For the particular case

$$\Omega(z) = \frac{1}{2\pi i}\int_{ab}\frac{dt}{(t - a)^\gamma(t - z)}, \qquad (13)$$

of the integral (11) when $\varphi \equiv 1$ the Sokhotskiĭ-Plemelj formulas (7) give us that at any point $t_0 \neq a$ on ab

$$\Omega^+(t_0) - \Omega^-(t_0) = (t_0 - a)^{-\gamma}.$$

Let us make an infinite cut from the point a along the curve ab and assume that on the left-hand side of ab

$$\left[(t-a)^{-\gamma}\right]^+ = (t_0 - a)^{-\gamma}.$$

Then

$$\left[(t_0 - a)^{-\gamma}\right]^- = (t_0 - a)^{-\gamma} e^{-2i\pi\gamma}$$

and, consequently,

$$\omega^+(t_0) - \omega^-(t_0) = (t_0 - a)^{-\gamma},$$

where

$$\omega(z) = \frac{(z-a)^{-\gamma}}{1 - e^{-2i\gamma\pi}} = \frac{e^{i\gamma\pi}}{2i\sin\gamma\pi}(z-a)^{-\gamma}.$$

From the inequality (12) for $\Omega(z)$ and the properties of $\omega(z)$ we have

$$|\Omega(z) - \omega(z)| \leqslant \frac{\text{const}}{|z-a|^{\alpha_0+\delta}}, \qquad \delta > 0,$$

which, together with the continuity (by construction) of $[\Omega(z) - \omega(z)]$ on ab, means that this function is holomorphic in a neighborhood of a. The case of a singularity at b can obviously be reduced to that considered by changing the direction of traverse on ab. Thus,

$$\Omega(z) = -\frac{e^{i\gamma\pi}}{2i\sin\gamma\pi}(z-a)^{-\gamma} + \Omega_0(z),$$

$$\Omega(z) = -\frac{e^{-i\gamma\pi}}{2i\sin\gamma\pi}(z-b)^{-\gamma} - \Omega_1(z), \qquad (14)$$

$$\Omega(t_0) = \pm\frac{1}{2i}\cot\gamma\pi \cdot (t_0 - c)^{-\gamma} + \Omega_2(t_0)(c = a, b),$$

where the functions Ω_0, Ω_1 and Ω_2 are holomorphic in a neighborhood of the point $c = a$ or b.

6°. The theorem of Muskhelishvili. Passing to the general case, we write (11) in the form

$$\Phi(z, \tau) = \frac{1}{2\pi i}\int_{ab} \frac{\varphi(t, \tau) - \varphi(a, \tau)}{(t-a)^\gamma(t-z)}dt + \varphi(a, \tau)\Omega(z)$$

$$\equiv F_0(z, \tau) + \varphi(a, \tau)\Omega(z), \qquad (15)$$

where the behavior of Ω near a is determined by (14). Since $\varphi(t, \tau)$ is a Hölder function with respect to t in a neighborhood of a, the density of the Cauchy type integral $F_0(z, \tau)$ has a singularity of order less than $\alpha_0 = \mathrm{Re}\,\gamma$; therefore, in a neighborhood of a we have, by (12),

$$|F_0(z, \tau)| |z - a|^{\alpha_0 - \delta} < \mathrm{const}, \qquad \alpha_0 = \mathrm{Re}\,\gamma > 0,$$

$$|F_0(z, \tau)| < \mathrm{const}, \qquad \alpha_0 = 0. \tag{16}$$

And if the Hölder exponent $\alpha > 0$ of $\varphi(t, \tau)$ satisfies the inequality $\alpha - \alpha_0 > 0$, $0 < \alpha_0 = \mathrm{Re}\,\gamma < 1$, i.e.,

$$\psi(t) \equiv \frac{\varphi(t, \tau) - \varphi(a, \tau)}{(t - a)^\gamma} \in H^\lambda, \qquad 0 < \lambda < \alpha - \alpha_0, \psi(a) = 0,$$

then, by the remark after Theorem 2,

$$F_0(z, \tau) \in H^{\alpha - \alpha_0 - \delta}(\sigma_a), \qquad 0 < \alpha_0 < \alpha \leq 1,$$

in a neighborhood σ_a of the point $z = a$, where $\delta > 0$ is an arbitrarily small number. Let

$$\Psi(z, \tau) = (z - a)^\gamma F_0(z, \tau) \equiv \frac{(z - a)^\gamma}{2\pi i} \int_{ab} \frac{f(t, \tau)\, dt}{(t - a)^\gamma (t - z)}, \tag{17}$$

where

$$f(t, \tau) = [\varphi(t, \tau) - \varphi(a, \tau)] \in H^\alpha(ab_1 \times T), \qquad b_1 \in ab,$$

and, by (16), $\Psi(a, \tau) = 0$, $\tau \in T$. We prove that $\Psi(z, \tau)$ satisfies a Hölder condition with respect to both arguments in a neighborhood of $z = a$, but in a somewhat different form and by a different method than in Muskhelishvili's proof of it ([97], §§22 − 25).

THEOREM 3 (MUSKHELISHVILI). *Suppose that* $f(t, \tau) \in H^\alpha(ab_1 \times T)$, $0 < \alpha \leq 1$, *with respect to the arguments* t *and* τ, *and* $0 \leq \alpha_0 = \mathrm{Re}\,\gamma < 1$. *Then there is a* $\lambda = \lambda(\alpha, \alpha_0) > 0$ *such that* $\Psi(z, \tau) \in H^\lambda(\sigma_\alpha \times T)$ *and* $\Psi(t_0) \in H^\lambda(l_\alpha \times T)$, *where* σ_α *is the disk* $|z - a| < \rho$ *with a cut along* ab, $l \subset ab$ *is an arc with endpoints* $a_l = a = t(0)$ *and* $b_\rho = t(\rho_0)$, *and* $\rho_0 > 0$ *is a fixed small number. Moreover, the following estimate holds for the Hölder constant* $H(\Psi)$:

$$H(\Psi) \leq \mathrm{const}[\max |f| + H(f)].$$

Let us first dwell on the proof that $\Psi(z, \tau)$ and $\Psi(t_0, \tau)$ are Hölder functions with respect to the first argument, with the second argument τ omitted for convenience. It was proved above that Ψ is a Hölder function for $0 < \alpha_0 < \alpha = 1$; therefore, we assume that $0 < \alpha_0 \leq \alpha < 1$. Without loss of generality it will also be assumed that $\beta = \mathrm{Im}\,\gamma = 0$; otherwise, $f^*(t) = f(t)(t - a)^{-i\beta} \in H^{\alpha - \delta}$ for arbitrary δ, $0 < \delta < \alpha$. We break up the integral appearing in $\Psi(z)$ into two. One, Ψ_1, is taken along the curve l with endpoints $a_l = a = t(0)$ and $b_l = t(\varepsilon)$,

$$2\pi i(z-a)^{-\gamma}\Psi(z) = \left(\int_l - \int_{ab-l}\right)\frac{f(t)\,dt}{(t-a)^{\gamma}(t-z)} \equiv \Psi_1 + \Psi_2,$$

and the other, Ψ_2, is a differentiable function in a neighborhood of $z = a$ (cf. the estimate of Δ_1 in Theorem 1). Therefore, in the proof of the theorem we can assume that the arc ab in (17) is entirely contained in a standard neighborhood $\sigma(a, b)$, i.e., $b = t(s_L)$, $0 < s_L \leqslant \varepsilon$. We then extend $f(t)$ to a standard neighborhood $\sigma(ab)$ and set $f(z) \equiv 0$ for $z \notin \sigma(ab)$. Since $f(a) = 0$, the function $f(z)$ so constructed is in H^α in a neighborhood of $z = a$. We represent $\Psi(z)$ in the form

$$\Psi(z) = \frac{1}{2\pi i}\int \frac{[f(t) - f(z)][(z-a)^{\gamma} - (t-a)^{\gamma}]\,dt}{(t-a)^{\gamma}(t-z)} + \check{\Psi}(z) \equiv \Psi_0 + \check{\Psi},$$

$$\check{\Psi}(z) = \frac{f(z)(z-a)^{\gamma}}{2\pi i}\int_{ab}\frac{dt}{(t-a)^{\gamma}(t-z)} - \frac{1}{2\pi i}\int_{ab}\frac{f(t) - f(z)}{t-z}\,dt$$

$$\equiv \check{\Psi}_1 + \check{\Psi}_2. \tag{18}$$

But since $f(z)$ is a Hölder function and $(z-a)^{\gamma}\Omega(z)$ is analytic, where $\Omega(z)$ is defined by (13), we have that $[f(z)(z-a)^{\gamma}\Omega(z)] \in H^\alpha$ in a neighborhood of $z = a$. Since $f(a) = 0$, what was proved above implies that $\check{\Psi}_2 \in H^{\alpha-\delta}$ in a neighborhood of $z = a$, and so $\check{\Psi}(z) \in H^{\alpha-\delta}$ ($\delta > 0$). It remains to consider the integral $\Psi_0(z) = \Psi(z) - \check{\Psi}(z)$. Let

$$e^{i\arg(t-z)}h(t, z) = \left\{\frac{f(t) - f(z)}{|t-z|^{\delta}}\right\}\left\{\frac{(z-a)^{\gamma} - (t-a)^{\gamma}}{|t-z|^{\gamma-\alpha+\delta}}\right\}$$

(the quantity δ, $0 < \delta < \alpha$, will be specified below). According to the properties of Hölder functions (3° and 5° in §1), the expressions in the curly brackets belong to the class $H^{\alpha-\delta}$ with respect to t and z, and, consequently, $h(t, z) \in H^{\alpha-\delta}$ (§1, 4°). Then

$$2\pi\,|\Psi_0(z_1) - \Psi_0(z_2)|$$

$$= \left|\int_{ab}\frac{\Delta h(t, z)\,dt}{(t-a)^{\gamma}|t-z_1|^{\gamma_0}} + \int_{ab}\frac{h(t, z_0)\Delta|t-z|^{\gamma_0}\,dt}{(t-a)^{\gamma}|t-z_1|^{\gamma_0}|t-z_2|^{\gamma_0}}\right|$$

$$\equiv |I_1 + I_2|,$$

where

$$\Delta h(t, z) = h(t, z_2) - h(t, z_1).$$

$$\Delta |t - z|^{\gamma_0} = |t - z_2|^{\gamma_0} - |t - z_1|^{\gamma_0}, \qquad \gamma_0 = 1 - (\gamma - \alpha + 2\delta),$$

$$0 < \alpha \leqslant \gamma \equiv \alpha_0 < 1.$$

Let δ be chosen so that $\gamma + \gamma_0 < 1$, and hence $0 < \alpha/2 < \delta < \alpha$. For this choice of δ the sum of the zeros of the denominator in I_1 is less than 1. Consequently, using the Hölder condition for $h(t, z)$, we get

$$|I_1| \leqslant c_1 |z_1 - z_2|^{\alpha - \delta},$$

where $c_1 = \text{const}[\max |f(t)| + H(f)]$ ($H(f)$ is the Hölder constant for the function f). Taking account of the fact that $h(t, a) = 0$ and $h(t, z) \in H^{\alpha - \delta}$, we have

$$h_1(t) = \frac{|h(t, z_2)|}{|t - a|^{\alpha - \delta}} < c_2.$$

On the other hand, for an arbitrary integer n the identity

$$\frac{1}{a^n} - \frac{1}{b^n} = (b - a) \sum_{k_1 + k_2 = n + 1} \frac{1}{a^{k_1} b^{k_2}},$$

gives us that

$$\left| \frac{\Delta |t - z_0|^{\gamma_0}}{|t - z_1|^{\gamma_0} |t - z_2|^{\gamma_0}} \right|$$

$$\leqslant |\Delta |t - z|^{\gamma_0/n}| \sum_{k_1 + k_2 = n + 1} |t - z_1|^{-k_1 \gamma_0/n} |t - z_2|^{-k_2 \gamma_0/n}$$

$$\leqslant c_3 |z_1 - z_2|^{\gamma_0/n} \sum_k \frac{1}{\mu_k(z_1, z_2, t)},$$

where each of the μ_k vanishes to an order not greater than $(n + 1)\gamma_0/n$. Then

$$|I_2| \leqslant c_2 c_3 |z_1 - z_2|^{\gamma_0/n} \sum_k \int_{ab} \frac{|dt|}{|t - a|^{\gamma - \alpha + \delta} \mu_k}.$$

Let us choose n so that the integrals on the right-hand side are bounded:

$$\gamma + \alpha + \delta + (n + 1)\gamma_0/n < 1, \qquad \gamma_0 = 1 - (\gamma - \alpha + 2\delta),$$

and so $n + 2 > (1 + \alpha - \gamma)/\delta$. Thus, finally,

$$|\Psi_0(z_1) - \Psi_0(z_2)| \leqslant c_4 |z_1 - z_2|^\lambda, \qquad \lambda = \lambda(\alpha, \gamma) > 0,$$

where $c_4 = \text{const}[\max |f(t)| + H(f)]$.

Passing to the proof that $\Psi(z, \tau)$ is a Hölder function with respect to τ, we use for $\Psi(z, \tau)$ the representation (18), in which it is obviously necessary to show only that $\Psi_0(z, \tau)$ is a Hölder function:

$$\Psi_0(z, \tau) = \frac{1}{2\pi i} \int_{ab} \frac{[f(t, \tau) - f(z, \tau)][(z - a)^\gamma - (t - a)^\gamma]}{(t - a)^\gamma (t - z)} dt$$

$$\equiv \frac{1}{2\pi i} \int_{ab} \frac{h_0(t, z, \tau) dt}{(t - a)^\gamma |t - z|^{1 - \gamma - \delta}}, \qquad 0 < \delta < \alpha,$$

whence, by construction, the function

$$h_0(t, z, \tau) = e^{-i \arg(z - t)} \left\{ \frac{f(t, \tau) - f(z, \tau)}{|t - z|^\delta} \right\} \left\{ \frac{(z - a)^\gamma - (t - a)^\gamma}{|t - z|^\gamma} \right\}$$

belongs to the class $H^{\alpha - \delta}$ with respect to τ for fixed t and z. Since the sum of the zeros of the denominator in the representation of $\Psi_0(z, \tau)$ is equal to $1 - \delta$, which is less than 1, it follows that

$$|\Psi_0(z, \tau_1) - \Psi_0(z, \tau_2)| \leqslant c_5 |\tau_1 - \tau_2|^{\alpha - \delta}.$$

Note that, by what has been proved, $\Psi_0(z, \tau) \in H^\lambda$ with respect to z and τ, and $\lambda > 0$ in a whole neighborhood $|z - a| < \rho_0$, $\rho_0 > 0$, of the point a. But $\check{\Psi}(z, \tau)$ in (18) is a Hölder function in the disk $|z - a| < \rho_0$ slit along ab, and for $z = t_0 \in \Gamma$

$$\check{\Psi}(t_0, \tau) \in H^\lambda(ab_1 \times T), \qquad b_1 = t(\rho_0).$$

Moreover, according to the estimates we have obtained, the Hölder constant $H(\check{\Psi})$ of $\check{\Psi}(z, \tau)$ can be represented in the form

$$H(\check{\Psi}) = \text{const}[\max |f| + H(f)].$$

Theorem 3 is proved.

COROLLARY. *Obviously, the statement of Theorem 3 remains true even for the more general Cauchy type integrals*

$$\Psi(z, \tau) = \frac{\Pi(z)}{2\pi i} \int_{ab} \frac{f(t, \tau) dt}{\Pi(t)(t - z)},$$

where $\Pi(z) = \Pi_1^n (z - t_k)^{\gamma_k}$, $t_k \in [a, b]$, $0 < \text{Re } \gamma_k < 1$, *and* $f(t, \tau) \in H^\alpha\{[t_k, t_{k+1}] \times T\}$.

Indeed, to prove this it suffices to break up the last integral into a sum of integrals over the intervals $[t_k, t_{k+1}]$ and to apply Theorem 3 to each term.

The formulas (14) and (15) for the Cauchy type integral $\Phi(z, \tau)$ in (11) gives us for $c = a$ or b that

$$\Phi(z, \tau) = (z - c)^{-\gamma}\left[\pm e^{\pm i\gamma\pi}\frac{\varphi(c, \tau)}{2i\sin\gamma\pi} + \Psi(z, \tau)\right] + \Omega_0(z),$$

$$\Phi(t_0, \tau) = (t_0 - c)^{-\gamma}\left[\pm\frac{\varphi(c, \tau)}{2i\tan\gamma\pi} + \Psi(t_0, \tau)\right] + \Omega_1(t_0), \qquad (19)$$

where the upper sign is taken for $c = a$, and $\Psi(c, \tau) = 0$. According to Theorem 3, $\Psi(z, \tau)$, $\Psi(t_0, \tau) \in H^\lambda$, $\lambda = \lambda(\alpha, \alpha_0)$ for $\varphi(t, \tau) \in H^\alpha(ab_1 \times T)$, $b_1 \in ab$, while $\Omega_0(z)$ and $\Omega_1(z)$ are holomorphic in a neighborhood of c, and

$$H\{\Phi(z - c)^\gamma\} \leq \text{const}[\max|\varphi| + H(\varphi)].$$

We shall give some examples of the direct application of the Sokhotskiĭ-Plemelj formulas (7), taking into account the properties of Cauchy type integrals at the endpoints of the curve of integration and at points where the density is discontinuous.

7°. *The Poincaré-Bertrand formulas for interchanging the integrations in iterated singular integrals.* Let

$$\varphi(t, t_1) = \frac{\varphi_*(t, t_1)}{\Pi(t, t_1)},$$

where $\varphi(t, t_1)$ is a piecewise Hölder function on a smooth curve L, and

$$\Pi(t, t_1) = \prod_{k=1}^{n}|t - c_k|^{\alpha_k} \cdot |t_1 - c_k|^{\beta_k}, \qquad \alpha_k + \beta_k < 1.$$

Under these conditions the usual formula is valid for interchanging the integrations in an iterated integral, with one of them singular:

$$\int_L \omega(t, z)\,dt\int_L\frac{\varphi(t, t_1)}{t_1 - t}\,dt_1 = \int_L dt_1\int_L\frac{\omega(t, z)\varphi(t, t_1)}{t_1 - t}\,dt, \qquad (20)$$

where $\omega(t, z)$ is integrable on L with respect to t for all z in a given set D. The proof of (20) can be carried out very simply, and we leave it to the reader. As a hint, we note that to prove (20) it is useful to encircle the points t_1 and t by sufficiently small neighborhoods on the curve L and break up the integrals on both sides of (20) into the integrals over these neighborhoods and those over their complements in L. If both iterated integrals are singular, then the usual formulas for interchanging the integrations turn out to be invalid in this case, as shown by the Poincaré-Bertrand interchange formula

$$\frac{1}{\pi i}\int_L \frac{dt}{t-t_0}\frac{1}{\pi i}\int_L \frac{\varphi(t,t_1)}{t_1-t}dt_1 = \frac{1}{\pi i}\int_L dt_1\int_L \frac{\varphi(t,t)dt}{\pi i(t-t_0)(t_1-t)} + \varphi(t_0,t_0),$$

(21)

where t_0 is different from the endpoints of L and from the points of discontinuity of φ. Let us consider the functions

$$\Phi(z) = \frac{1}{\pi i}\int_L \frac{dt}{t-z}\frac{1}{\pi i}\int_L \frac{\varphi(t,t_1)\,dt_1}{t_1-t},$$

$$\Psi(z) = \frac{1}{\pi i}\int_L dt_1\frac{1}{\pi i}\int_L \frac{\varphi(t,t_1)\,dt}{(t-z)(t_1-t)},$$

for which, by (20), we have $\Phi(z) = \Psi(z)$ for $z \notin L$, and, consequently,

$$\Phi^{\pm}(t_0) = \Psi^{\pm}(t_0), \qquad \Phi(t_0) = \tfrac{1}{2}(\Phi^+ + \Phi^-) = \tfrac{1}{2}(\Psi^+ + \Psi^-) = \Psi(t_0).$$

By the Sokhotskiĭ-Plemelj formulas,

$$\Phi(t_0) = \frac{1}{2}(\Phi^+ + \Phi^-) = \frac{1}{\pi i}\int_L \frac{dt}{t-t_0}\int_L \frac{1}{\pi i}\frac{\varphi(t,t_1)}{t_1-t}dt_1.$$

To compute $\Psi(t_0)$ we represent $\Psi(z)$ in the form

$$\Psi(z) = \frac{1}{\pi i}\int_L \frac{dt_1}{t_1-z}\frac{1}{\pi i}\int_L \left(\frac{1}{t-z}-\frac{1}{t-t_1}\right)\varphi(t,t_1)\,dt \equiv \frac{1}{\pi i}\int_L \frac{f(t_1,z)}{t_1-z}dt_1$$

and make use of the Sokhotskiĭ-Plemelj formulas (the dependence of the density $f(t_1, z)$ on the variable z obviously does not hinder this):

$$\Psi^{\pm}(t_0) = \pm f^{\pm}(t_0,t_0) + \frac{1}{\pi i}\int_L \frac{f^{\pm}(t_1,t_0)\,dt_1}{t_1-t_0},$$

whence

$$\Psi(t_0) = \frac{1}{2}(\Psi^+ - \Psi^-) = \frac{1}{2}(f^+ - f^-) + \frac{1}{2\pi i}\int \frac{f^{\pm}+f}{t_1-t_0}dt_1.$$

(22)

But

$$\tfrac{1}{2}\big[f^+(t_0,t_0) - f^-(t_0,t)\big] = \varphi(t_0,t_0),$$

and

$$\frac{1}{2}(f^+ + f^-) = \frac{1}{\pi i}\int_L \left(\frac{1}{t-t_0}-\frac{1}{t-t_1}\right)\varphi(t,t_1)\,dt.$$

Substituting these expressions into (22) and setting $\Phi(t_0) = \Psi(t_0)$, we arrive at (21).

8°. Inverse formulas for Cauchy type integrals. Let L be a collection of closed contours without common points, and let $\Psi(t) \in H^\alpha(L)$. We consider the integral equation

$$\frac{1}{\pi i} \int_L \frac{\varphi(t)\, dt}{t - t_0} = \Psi(t_0), \qquad t_0 \in L, \tag{23}$$

where $\varphi(t) \in H^\alpha$ is an unknown function. Let us multiply both sides of (23) by $1/\pi i (t_0 - t_1)$ and integrate over L, using (21). Then

$$\frac{1}{\pi i} \int_L \frac{\Psi(t_0)\, dt_0}{t_0 - t_1} = \frac{1}{\pi i} \int_L \frac{dt}{t_0 - t_1} \frac{1}{\pi i} \int_L \frac{\varphi(t)\, dt}{t - t_0}$$

$$= \varphi(t_1) + \frac{1}{\pi i} \int_L \varphi(t)\, dt \frac{1}{\pi i} \int_L \frac{dt_0}{(t_0 - t_1)(t - t_0)},$$

but

$$\int_L \frac{dt}{(t_0 - t_1)(t - t_0)} = \frac{1}{t - t_1} \left\{ \int_L \frac{dt_0}{t_0 - t_1} + \int_L \frac{dt_0}{t - t_0} \right\} = 0,$$

and, consequently,

$$\varphi(t_0) = \frac{1}{\pi i} \int_L \frac{\Psi(t)\, dt}{t - t_0}, \qquad t_0 \in L. \tag{24}$$

We call (23) and (24) the inversion formulas for Cauchy type integrals.

9°. The Hilbert inversion formulas. Suppose that an analytic function $w(z) = u + iv$ in the disk $|z| < 1$ is represented as a Schwarz integral in terms of the limit values of its real part:

$$w(z) = \frac{1}{2\pi} \int_0^{2\pi} u(e^{i\gamma}) \frac{e^{i\gamma} + z}{e^{i\gamma} - z}\, d\gamma + iv_0. \tag{25}$$

We use the identity

$$\frac{e^{i\gamma} + z}{e^{i\gamma} - z}\, d\gamma \equiv -d\gamma + \frac{2}{i}\frac{dt}{t - z}, \qquad t = e^{i\gamma},$$

to express the Schwarz integral in terms of a Cauchy type integral,

$$w(z) = \frac{1}{2\pi i} \int_{|t|=1} \frac{2u(t)\, dt}{t - z} - \frac{1}{2\pi} \int_0^{2\pi} u(e^{i\gamma})\, d\gamma + iv_0$$

and apply the Sokhotskiĭ-Plemelj formulas to the latter. Then

$$w^+(t_0) = u(t_0) + \frac{1}{2\pi i} \int_{|t|=1} \frac{2u(t)\, dt}{t - t_0} - \frac{1}{2\pi} \int_0^{2\pi} u(e^{i\gamma})\, d\gamma + iv_0$$

$$= u(t_0) + \frac{1}{2\pi} \int_0^{2\pi} u(e^{i\gamma}) \frac{e^{i\gamma} + e^{i\gamma_0}}{e^{i\gamma} - e^{-i\gamma_0}}\, d\gamma + iv_0.$$

Separating out the imaginary part of the last equality and taking the equality

$$\frac{e^{i\gamma} + e^{i\gamma_0}}{e^{i\gamma} - e^{i\gamma_0}} = \frac{1}{i}\cot\frac{\gamma - \gamma_0}{2},$$

into account we get

$$v(e^{i\gamma_0}) = -\frac{1}{2\pi}\int_0^{2\pi} u(e^{i\gamma})\cot\frac{\gamma - \gamma_0}{2}d\gamma + v_0. \tag{26}$$

Repeating the arguments for the function $\{-iw(z)\}$, we get

$$u(e^{i\gamma_0}) = \frac{1}{2\pi}\int_0^{2\pi} v(e^{i\gamma})\cot\frac{\gamma - \gamma_0}{2}d\gamma + u_0, \tag{27}$$

where $u_0 = u(0,0)$ and $v_0 = v(0,0)$. We call (26) and (27) the Hilbert inversion formulas; they express the limit values of the real and imaginary parts of an analytic function one in terms of the other.

§3. The Riemann boundary-value problem
[the conjunction problem*]

1°. Auxiliary facts. We formulate two theorems which we shall need from the theory of functions of a complex variable (see [32] and [69]).

THEOREM 1 (on analytic continuation). *Suppose that two domains D_1 and D_2 abut along a smooth curve L, and $f_i(z)$ is a function analytic in D_i and continuous up to L. If $f_1(t) = f_2(t)$ for $t \in L$, then these functions constitute analytic continuations of each other, and the function $f(z) = f_i(z)$, $z \in D_i$, is holomorphic in the domain $(D_1 + D_2 + L)$.*

THEOREM 2 (LIOUVILLE). *Suppose that $f(z)$ is analytic on the whole plane with the exception of the points $a_0 = \infty$, a_k $(k = 1,\ldots,n)$, where it has poles, and its expansions in neighborhoods of a_0 and the a_k have the form*

$$P_0(z) = \sum_{i=1}^{m_0} c_i^0 z^i, \qquad P_k\left(\frac{1}{z - a_k}\right) = \sum_{i=1}^{m_k} c_i^k \frac{1}{(z - a_k)^i} \quad (k = 1,\ldots,n).$$

Then

$$f(z) - \left[\sum_{k=1}^{n} P_k\left(\frac{1}{z - a_k}\right) + P_0(z)\right] = \text{const},$$

and $f(z) - P_0(z) = \text{const}$ for $n = 0$.

---

*Editor's note. This term was sometimes translated into English as 'conjugation problem'.

2°. The problem of determining a piecewise analytic function from a jump.
Find a function $\Phi(z)$ that is piecewise analytic on the z-plane with a smooth
jump curve $L = \Sigma_k L_k$ if on this curve

$$\Phi^+(t_0) - \Phi^-(t_0) = \varphi(t), \qquad \varphi(t) \in H^\alpha(L), \tag{1}$$

and one of the following conditions holds in a neighborhood of the point at
infinity:

(a)
$$\left| \Phi(z) - \sum_{k=0}^{n} c_k z^k \right| \to 0 \quad \text{as } |z| \to \infty,$$

(b)
$$|z^n \Phi(z)| < \infty \quad \text{as } |z| \to \infty.$$

According to the Sokhotskiĭ-Plemelj formulas (2.7), the Cauchy type integral

$$\Omega(z) = \frac{1}{2\pi i} \int_L \frac{\varphi(t)}{t - z} dz$$

satisfies the boundary condition (1).

Let us consider the piecewise analytic function $R(z) = [\Phi(z) - \Omega(z)]$, which
satisfies the condition $R^+(t_0) - R^-(t_0) = 0$ on L and, in the case of condition
(a), has finite order at infinity. Then, by the theorem on analytic continuation,
$R(z)$ is holomorphic on the whole plane, and, by Liouville's theorem, $R(z) =
P_n(z) = \Sigma_0^n c_k z^k$. Thus, the general solution of problem (1a) has the form

$$\Phi(z) = \frac{1}{2\pi i} \int \frac{\varphi(t)\, dt}{t - z} + P_n(z); \tag{2}$$

if the solution is bounded at infinity, then $P_n(z) \equiv \text{const}$. In the case of
problem (1b), $P_n(z) \equiv 0$, and the function

$$\Phi(z) = \frac{1}{2\pi i} \int_L \frac{\varphi(t)\, dt}{t - z} \tag{2'}$$

satisfies condition (b) only if the integral on the right-hand side has a zero of
the necessary order at infinity. If we substitute the expansion

$$\frac{1}{t - z} = -\frac{1}{z}\left(\frac{1}{1 - t/z}\right) = -\frac{1}{z}\left(1 + \frac{t}{z} + \cdots\right) = -\sum_{k=0}^{\infty} \frac{t^k}{z^{k+1}}$$

in a neighborhood of $z = \infty$ into (2'), then

$$\Phi(z) = -\frac{1}{2\pi i} \sum_{k=0}^{\infty} \frac{1}{z^{k+1}} \int_L t^k \varphi(t)\, dt.$$

Thus, if condition (b) holds, then

$$\int_L t^k \varphi(t)\, dt = 0 \qquad (k = 0, \ldots, n - 2), \tag{3}$$

i.e., problem (1b) is solvable only under the additional conditions (3) on $\varphi(t)$.

$3°$. *Analytic extension of functions defined on a collection of closed contours.*
Suppose that the collection of closed contours $L = \Sigma L_k$ is the boundary of a
multiply connected domain D^+ whose complement in the whole plane is
denoted by D^-. We shall determine conditions under which a given complex-
valued function $\varphi(t)$ on L is the limit value of a function $\varphi(z)$ analytic in D^+
(D^-), i.e., conditions for the analytic extendability of $\varphi(t)$ to D^+ (D^-). If the
domain D^+ (D^-) is finite, or if it is infinite and $\varphi(\infty)$ is zero, then, by the
well-known Cauchy theorem [69], this condition can be written in the form

$$\int_L \frac{\varphi(t)\,dt}{t - z} = 0, \qquad z \in D^-\,(D^+). \tag{4}$$

The converse assertion is obvious: If (4) holds, then the expression

$$\varphi(z) = \frac{1}{2\pi i} \int_L \frac{\varphi(t)\,dt}{t - z}, \qquad z \in D^+\,(D^-),$$

is a Cauchy integral whose limit values on L coincide with $\varphi(t)$. If $\varphi(t) \in$
$H^\alpha(L)$, then, passing in (4) to the limit from the domain D^- (D^+) as
$z \to t_0 \in L$, we find from the Sokhotskiĭ-Plemelj formulas that

$$\mp \frac{1}{2}\varphi(t_0) + \frac{1}{2\pi i} \int_L \frac{\varphi(t)\,dt}{t - t_0} = 0, \tag{5}$$

where the minus sign corresponds to the case when $\varphi(z)$ is holomorphic in D^-.
The condition (5) is necessary and sufficient for a function $\varphi(t)$ of class H^α on
L to be the boundary value of a function $\varphi(z)$ that is holomorphic in D^+ (D^-)
and, when D^+ (D^-) is infinite, vanishes at infinity.

$4°$. *The index of a continuous function.* Let us consider a smooth closed
contour L and a continuous complex-valued function $G(t) \neq 0$ on L. The
index κ of this function is defined to be the increment of its argument under a
circuit of L in the positive direction (counterclockwise):

$$\kappa = \mathrm{ind}_L\, G(t) = \frac{1}{2\pi}\big[\arg G(t)\big]_L. \tag{6}$$

Since

$$\ln G(t) = \ln |G(t)| + i \arg G(t) \quad \text{and} \quad \big[\ln |G(t)|\big]_L = 0,$$

it follows that

$$\kappa = \frac{1}{2\pi i}\big[\ln G(t)\big]_L = \frac{1}{2\pi i}\int_L d(\ln G(t)), \tag{7}$$

if the Stieltjes integral on the right-hand side makes sense. Obviously, the index is an integer, and

$$\operatorname{ind}_L(G_1 \cdot G_2) = \operatorname{ind}_L G_1 + \operatorname{ind}_L G_2, \qquad \operatorname{ind}_L(G_1/G_2) = \operatorname{ind}_L G_1 - \operatorname{ind}_L G_2,$$

if the $\kappa_i = \operatorname{ind}_L G_i$ are defined. If $G(t)$ is continuously differentiable, then

$$\kappa = \operatorname{ind}_L G(t) = \frac{1}{2\pi i} \int_L d(\ln G(t)) = \frac{1}{2\pi i} \int_L \frac{G'(t)}{G(t)} dt \qquad (8)$$

and, consequently, if $G(z)$ is analytic in $D(L)$ (the $G(t)$ are the limit values of $G(z)$ on L), then the index of $G(t)$ is the logarithmic residue and is determined by the formula

$$\kappa = N - P, \qquad (9)$$

where N is the number of zeros and P is the number of poles of $G(z)$ in $D(L)$ with multiplicity counted. Let $t(s) = x(s) + iy(s)$ be the equation of L. Then $G[t(s)] = \xi(s) + i\eta(s)$ is the equation of a contour L_G in the G-plane, and, since $G(t)$ is continuous and L is closed, it is also closed. The number of loops of L_G around the origin in the G-plane under a circuit of L in the positive direction is the index of $G(t)$, since it coincides with the increment of the argument of $G(t)$. In particular, if $\operatorname{Re} G(t)$ or $\operatorname{Im} G(t)$ equals zero on L, then $\operatorname{ind} G(t) = 0$, since in this case the contour L_G is a segment of the real or imaginary axis, traversed several times. We remark that the index of a function $G(t)$ changes sing when the direction on L is reversed, i.e.,

$$\kappa^+ = \frac{1}{2\pi}[\arg G(t)]_L^+ = -\frac{1}{2\pi}[\arg G(t)]_L^- = -\kappa^-.$$

If $\operatorname{ind}_L G(t) = 0$, then $\ln G(t)$ is a single-valued function on L.

EXAMPLES. a) $\operatorname{ind}_L (t - a_k)^n = n$ for $a_k \in D(L)$ and n an integer, since the $(t - a_k)^n$ are the limit values of the analytic function $(z - a_k)^n$.

b) $\operatorname{ind}_L (t - a_k)^n = 0$ for $a_k \notin D(L)$.

Suppose now that a continuous function $G(t) \neq 0$ is given on $L = \Sigma_0^n L_k$, where the L_k are disjoint smooth closed contours, and the domain $D(L_0)$ contains $L_1, \ldots, L_m$ in its interior. We take a positive circuit in the clockwise direction on the contours L_k for $k = 1, \ldots, m$, and in the counterclockwise direction on L_0. Taking a positive circuit into account, we have

$$\kappa^k = \frac{1}{2\pi}[\arg G(t)]_{L_k} \qquad (k = 0, \ldots, m).$$

The index of $G(t)$ on $L = \Sigma_0^m L_k$ is defined to be the quantity

$$\kappa = \sum_{k=0}^m \frac{1}{2\pi}[\arg G(t)]_L = \sum_{k=0}^m \kappa^k. \qquad (10)$$

Let us fix some points $a_k \in D(L_k)$ in the domains $D(L_k)$, $k = 1, \ldots, m$, and assume for definiteness that $0 \in D(L_0)$ and $0 \notin D(L_k)$, $k = 1, \ldots, m$. We consider the function

$$\Pi(t) = \prod_{k=1}^{m} (t - a_k)^{\kappa^k}.$$

Since for a circuit of L_k ($k = 1, \ldots, m$)

$$\operatorname{ind}_{L_j}(t - a_k) = \begin{cases} 0, & j \neq k, \\ -1, & j = k, \end{cases}$$

it follows that

$$\operatorname{ind}_{L_k}\Pi(t) = -\kappa^k \quad (k = 1, \ldots, m), \qquad \operatorname{ind}_{L_0}\Pi(t) = \sum_{k=1}^{m} \kappa^k.$$

Similarly,

$$\operatorname{ind}_{L_k} t^{-\kappa} = 0 \quad (k = 1, \ldots, m), \qquad \operatorname{ind}_{L_0} t^{-\kappa} = -\kappa.$$

Consequently, if $G(t)$ is a continuous function on $L = \sum_0^m L_k$ whose index κ is defined by (10), then the function $G_0(t) = t^{-\kappa}\Pi(t)G(t)$ has index zero on L. Indeed,

$$\operatorname{ind}_{L_0} G_0(t) = \operatorname{ind}_{L_0} t^{-\kappa} + \operatorname{ind}_{L_0}\Pi(t) - \operatorname{ind}_{L_0} G(t) = -\kappa + \sum_{k=1}^{m} \kappa^k + \kappa_0 = 0.$$

For $k = 1, \ldots, m$,

$$\operatorname{ind}_{L_k} G_0(t) = 0 - \kappa^k + \kappa^k = 0,$$

i.e., $\operatorname{ind}_L G_0(t) = 0$.

5°. The index of a discontinuous function. Suppose that on a smooth closed contour L we are given a piecewise continuous function $G(t) \neq 0$ having finite discontinuities at a finite number of points $t_k \in L$ ($k = 1, \ldots, n$), and $G(t_k - 0) \neq G(t_k + 0)$.

We fix the principal values of the argument of this function at the points $(t_k + 0)$,

$$\arg G(t_k + 0) \in [-\pi/2, \pi/2] \quad (k = 1, \ldots, n)$$

and we determine $\operatorname{Arg} G(t_{k+1} - 0)$ on each of the arcs $\widehat{t_k t_{k+1}}$ by continuity. Let

$$\theta_k = \frac{1}{2\pi}\{\operatorname{Arg} G(t_k - 0) - \arg G(t_k + 0)\},$$

and choose integers ν_k satisfying one of the conditions

$$-1 < \alpha_k = \theta_k - \nu_k \leqslant 0 \tag{11}$$

or

$$0 \le \alpha_k = \theta_k - \nu_k \le 1, \qquad (12)$$

i.e., $\nu_k = [\theta_k] + 1$ in the case of (11) (where $[\theta_k]$ is the largest integer not exceeding θ_k) and $\nu_k = [\theta_k]$ in the case of (12). We define the index κ_p of class $O_p(t_{i_1}, \ldots, t_{i_p})$ of the piecewise continuous function $G(t)$ to be the quantity

$$\kappa_p = \sum_{i=1}^{n} \nu_k, \qquad (13)$$

where t_{i_k} $(k = 1, \ldots, p)$ are the points of discontinuity of $G(t)$ at which $\alpha_{i_k} = \theta_{i_k} - \nu_{i_k} \ne 0$ and α_{i_k} satisfies condition (12). Obviously, a function $G(t)$ in the class O_0 has the largest index, when all the ν_k are chosen from inequality (11). Moreover, if m $(\le n)$ is the number of points t_k at which $\alpha_k = \theta_k - \nu_k \ne 0$ and $p \le m$, then $\kappa_p = \kappa_0 - p$. Indeed, if $\nu'_k = [\theta_k] + 1$ was chosen at the point t_k from condition (11), then the choice $\nu''_k = [\theta_k]$ from condition (12) decreases the index by 1. We construct an elementary $\Omega(t)$ such that the function $G_1(t) = \Omega(t) \cdot G(t)$ is continuous on L and $\text{ind}_{L_0} G_1(t) = \kappa_p$. Suppose first that there is a single point $t = t_1$ of discontinuity of $G(t)$ and

$$\frac{G(t_1 - 0)}{G(t_1 + 0)} = \rho_1 \exp\{\arg G(t_1 - 0) - \arg G(t_1 + 0)\} = \rho_1 e^{i 2 \pi \theta_1}.$$

Let

$$\gamma_1 = \frac{1}{2 \pi i} \ln \frac{G(t_1 - 0)}{G(t_1 + 0)} - \nu_1 = \theta_1 - \nu_1 - \frac{i}{2\pi} \ln \rho_1 \equiv \alpha_1 + i\beta_1,$$

where ν_1 is chosen from (11) or (12), and consider the function

$$\Omega(t) = \{t - z_0\}_{t_1}^{-\gamma_1} \qquad (z_0 \in D(L)),$$

which has a discontinuity at the point $t = t_1 \in L$ (see §2.5°). Then, obviously, the function

$$G_1(t) = \{t - z_0\}_{t_1}^{-\gamma_1} G(t)$$

is continuous on L by construction, and

$$\text{ind}_L G_1(t) = \frac{1}{2\pi i} \left[\ln G_1(t_1 - 0) - \ln G_1(t_1 + 0) \right]$$

$$= -\theta_1 + \nu_1 + \frac{i}{2\pi} \ln \rho_1 + \left(\theta_1 - \frac{i}{2\pi} \ln \rho_1 \right) = \nu_1.$$

When $G(t)$ has several points of discontinuity, we set

$$\frac{G(t_k - 0)}{G(t_k + 0)} = \rho_k e^{i 2 \pi \theta_k}, \qquad \gamma_k = \theta_k - \nu_k - \frac{1}{2\pi} \ln \rho_k$$

and consider the function

$$G_1(t) = \prod_{k=1}^{n} \{t - z_0\}_{t_k}^{-\gamma_k} G(t) \equiv \Omega(t)G(t),$$

where the discontinuity of the kth factor in the product is at the point t_k. The function $G_1(t)$ so constructed is continuous, and, by construction,

$$\operatorname{ind}_L G_1(t) = \sum_{k=1}^{n} \nu_k = \kappa_p.$$

We remark that the function

$$\Omega(t) = \prod_{k=1}^{n} \{t - z_0\}_{t_k}^{-\gamma_k}$$

can be represented as a ratio of the limit values of functions that are analytic in the respective domains $D^+ = D(L)$ and D^- outside L.
    Indeed,

$$\Omega(t) = \prod_{k=1}^{n} \{t - z_0\}_{t_k}^{-\gamma_k} = \frac{\omega^-(t)}{\omega^+(t)}, \tag{14}$$

where

$$\omega^+(z) = \prod_{k=1}^{n} (z - t_k)^{\gamma_k}, \qquad \omega^-(z) = \prod_{k=1}^{n} \left(\frac{z - t_k}{z - z_0}\right)^{\gamma_k},$$

and the functions

$$\omega_k^+(z) = (z - t_k)^{\gamma_k}, \qquad \omega_k^-(z) = \left(\frac{z - t_k}{z - z_0}\right)^{\gamma_k}$$

are single-valued on the plane with a slit along the curve $\overparen{z_0 t_k \infty}$. If, as in 3°, $L = \sum_1^m L_i$, where the L_i are smooth closed contours, and $t_k^i \in L_i$ $(k = 1, \ldots, n_i)$ are the points of discontinuity of a piecewise continuous function $G(t) \neq 0$, then for a positive circuit on the L_i

$$\kappa_p = \operatorname{ind}_L G(t) = \sum_{i=0}^{m} \operatorname{ind}_{L_i} G(t) = \sum_{i=0}^{m} \kappa_p^i. \tag{15}$$

6°. The Riemann problem for Hölder coefficients and a single closed contour. Let L be a smooth closed contour, and let $g(t)$ and $G(t)$ be functions in $H^\alpha(L)$, with $G(t) \neq 0$ on L. It is required to find a function $\Phi(z)$ that is piecewise holomorphic on the whole plane with jump curve L, satisfies on L the condition

$$\Phi^+(t) = G(t)\Phi^-(t) + g(t), \qquad t \in L, \tag{16}$$

and is bounded at infinity. A complete solution of the problem, which is called the Riemann problem or the conjunction problem, was first constructed by Gahov in 1938, and then generalized to the case of several contours by Kvedelidze in 1941 (see [32]). Let us first consider the Riemann problem when $g(t) \equiv 0$ in (16). Suppose first that $\kappa = \operatorname{ind}_L G(t) = 0$; then $\ln G(t)$ is single-valued, and the homogeneous condition (16) can be written in the form

$$[\ln \Phi(t)]^+ - [\ln \Phi(t)]^- = \ln G(t).$$

One solution of the last jump problem is the Cauchy type integral (see 2°)

$$\ln X(z) = \frac{1}{2\pi i} \int_L \frac{\ln G(t)\, dt}{t - z} \equiv \Gamma(z),$$

and a particular solution $X(z)$ of the homogeneous Riemann problem (16) can be expressed in terms of it:

$$X(z) = \exp\left\{ \frac{1}{2\pi i} \int_L \frac{\ln G(t)\, dt}{t - z} \right\} = \exp \Gamma(z); \tag{17}$$

moreover, the piecewise analytic function $X(z)$ does not vanish on the z-plane by construction. Let us represent $G(t)$ in the form

$$G(t) = X^+(t)/X^-(t) \tag{18}$$

and substitute it into (16). Then

$$\left[\frac{\Phi(t)}{X(t)} \right]^+ - \left[\frac{\Phi(t)}{X(t)} \right]^- = 0,$$

and, consequently, the general solution of the homogeneous problem (16) which has finite order at infinity can be written as

$$\Phi(z) = X(z) \sum_{k=0}^{n} c_k z^k \equiv X(z) P_n(z).$$

Suppose now that $\kappa = \operatorname{ind}_L G(t) \neq 0$. We represent the boundary condition (16) in the form

$$\Phi^+(t) = [G(t)t^{-\kappa}](t^\kappa \Phi(t))^-$$

and consider the piecewise analytic function

$$\Omega(z) = \begin{cases} \Phi(z), & z \in D^+, \\ z^\kappa \Phi(z), & z \in D^-. \end{cases}$$

Setting $G_0(t) = t^{-\kappa} G(t)$, we come to a homogeneous Riemann problem with zero index for the determination of the function $\Omega(z)$:

$$\Omega^+(t) = G_0(t)\Omega^-(t).$$

A particular solution $\Omega_0(z)$ of the last problem can be found by (17), where $G(t)$ must be replaced by $G_0(t) = t^{-\kappa}G(t)$, and this can be used to determine a particular solution of the original homogeneous Riemann problem (16):

$$X(z) = \begin{cases} e^{\Gamma(z)}, & z \in D^+, \\ z^{-\kappa}e^{\Gamma(z)}, & z \in D^-; \end{cases} \tag{19}$$

moreover, $X(z)$ does not vanish anywhere in the finite z-plane. Representing $G(t)$ according to (18) as above, we find the general solution of the homogeneous problem (16) with finite order at infinity:

$$\Phi(z) = X(z)P_n(z). \tag{20}$$

The representation (20) shows that, among all the solutions of the homogeneous Riemann problem (16), the particular solution $X(z)$ a) has the smallest order at infinity, and b) does not vanish in the finite plane.

Such a solution of the homogeneous Riemann problem is said to be *canonical*. For $\kappa > 0$ the canonical solution $X(z)$, according to (19), has a zero of order κ at infinity; therefore (20) with $P_n(z) = P_\kappa(z)$ determines a solution of the homogeneous Riemann problem (16) that is bounded at infinity, i.e., it depends on $\kappa + 1$ arbitrary complex constants. Since the canonical solution has a pole of order κ at infinity for $\kappa < 0$, the homogeneous Riemann problem does not have bounded solutions other than the trivial one.

Let us now consider the nonhomogeneous Riemann problem for an arbitrary $\kappa = \mathrm{ind}_L \, G(t)$. Substitution of the representation (18) of $G(t)$ into the boundary condition (16) yields

$$\Phi^+(t) = \frac{X^+(t)}{X^-(t)}\Phi^-(t) + g(t),$$

and hence

$$\left[\frac{\Phi(t)}{X(t)}\right]^+ - \left[\frac{\Phi(t)}{X(t)}\right]^- = \frac{g(t)}{X^+(t)}.$$

Solving the last jump problem, we find that

$$\Phi(z) = \frac{X(z)}{2\pi i} \int_L \frac{g(t)\,dt}{X^+(t)(t-z)} + P_n(z)X(z).$$

For $\kappa > 0$ the canonical function $X(z)$ of the homogeneous problem (16) has a zero of order κ at infinity (it is bounded for $\kappa = 0$); consequently, $P_n(z) \equiv P_\kappa(z)$, and the general solution of the nonhomogeneous problem (16) which is bounded at infinity contains $\kappa + 1$ arbitrary complex constants:

$$\Phi(z) = \frac{X(z)}{2\pi i} \int \frac{g(t)\,dt}{X^+(t)(t-z)} + P_\kappa(z)X(z). \tag{21}$$

The function $X^+(t)$ can be computed by the Sokhotskiĭ-Plemelj formulas and has the form

$$X^+(t) = \exp\{\tfrac{1}{2}\ln G_0(t) + \Gamma(t)\},$$

where $G_0(t) = t^\kappa G(t)$ and

$$\Gamma(t) = \frac{1}{2\pi i}\int_L \frac{\ln G_0(t_0)\,dt_0}{t_0 - t}.$$

For $\kappa < 0$ the canonical function $X(z)$ of the homogeneous problem (16) has a pole of order $|\kappa|$ at infinity; consequently, $P_n(z) = 0$ and

$$\Phi(z) = \frac{X(z)}{2\pi i}\int_L \frac{g(t)\,dt}{X^+(t)(t - z)}, \tag{22}$$

and the Cauchy type integral in (22) must have a zero of order at least $|\kappa|$ at infinity. The condition (3) for a Cauchy type integral to have a zero of the necessary specific order at infinity were obtained in 2° and have the form

$$\int \frac{t^k g(t)\,dt}{X^+(t)} = 0, \qquad k = 0,\ldots,|\kappa| - 2. \tag{23}$$

Thus, for $\kappa < 0$ the nonhomogeneous Riemann problem has a unique bounded solution, which is represented by (22), only if the $2(|\kappa| - 1)$ real solvability conditions (23) on the free term $g(t)$ of the boundary condition (16) hold.

7°. **The Riemann problem with Hölder coefficients in a multiply connected domain.** Let $L = \sum_0^n L_k$, where the L_k are smooth closed contours, and $L_k \subset D(L_0)$, $k = 1,\ldots,m$; and suppose that $g(t)$, $G(t) \in H^\alpha(L)$ are Hölder functions on them, with $G(t) \neq 0$, $t \in L$. It is required to find a solution $\Phi(z)$ of the Riemann problem (16) that is bounded at infinity. In accordance with 4°, we define the index of $G(t)$ by (10):

$$\kappa = \operatorname{ind}_L G(t) = \sum_{k=0}^m \operatorname{ind}_{L_k} G(t) = \sum_{k=0}^m \kappa_k.$$

Suppose for definiteness that the origin O lies in $D(L_0)$, that $O \notin D(L_k)$, $k = 1,\ldots,m$, and that $a_k \in D(L_k)$ are arbitrary fixed points. Then, as shown in 4°,

$$\operatorname{ind}_{L_k} G_0(t) \equiv \operatorname{ind}_{L_k}[\Pi(t)t^{-\kappa}G(t)] = 0, \qquad k = 0,\ldots,m,$$

where $\Pi(t) = \prod_1^m (t - a_k)^{\kappa_k}$. Writing the boundary conditions (16) in the form

$$[\Pi(t)\Phi(t)]^+ = \{\Pi(t)t^{-\kappa}G(t)\}[t^*\Phi(t)]^-$$

and introducing the function

$$\Omega(z) = \begin{cases} \Pi(z)\Phi(z), & z \in D^+, \\ z^\kappa \Phi(z), & z \in D^-, \end{cases}$$

we come to the homogeneous Riemann problem with zero index

$$\Omega^+(t) = G_0(t)\Omega^-(t),$$

the canonical solution of which, similarly to 6°, can be computed by the formula

$$\Omega_0(z) = \exp\left\{ \frac{1}{2\pi i} \int_L \frac{\ln G_0(t)\, dt}{t - z} \right\} \equiv e^{\Gamma(z)}.$$

The respective canonical and general solutions of the original homogeneous Riemann problem (16) take the form

$$X(z) = \begin{cases} [\Pi(z)]^{-1} e^{\Gamma(z)}, & z \in D^+, \\ z^{-\kappa} e^{\Gamma(z)}, & z \in D^-, \end{cases} \tag{24}$$

$$\Phi(z) = P_n(z)X(z).$$

All the properties of the canonical function $X(z)$ in (24), the conclusions about the number of bounded solutions of the homogeneous Riemann problem, and the formulas for the solution of the nonhomogeneous Riemann problem remain the same as for the case of a simply connected domain in the previous subsection.

8°. The Riemann problem with piecewise Hölder coefficients. Let L be a smooth closed contour. Suppose first that $g(t) \in H^\alpha(L)$, and $G(t) \in H^\alpha_*$ is a finite piecewise Hölder function, with $G(t) \neq 0$ on L. It is required to find a solution $\Phi(z)$ of the Riemann boundary-value problem (16) that is bounded everywhere on the z-plane with the possible exception of the points t_k, $k = 1,\ldots,n$, of discontinuity of $G(t)$, where integrable singularities are allowed. In accordance with subsection 5°, we set

$$\theta_k = \frac{1}{2\pi} \left\{ \operatorname{Arg} G(t_k - 0) - \arg G(t_k + 0) \right\},$$

where $\operatorname{Arg} G(t_k - 0)$ are computed by continuity on the arc $\overset{\frown}{t_{k-1}t_k}$, and we choose integers ν_k that satisfy one of the conditions (11) or (12). Supposing for definiteness that the origin lies in the domain $D^+ = D(L)$, we construct the continuous function (see 5°)

$$G_1 = G(t) \prod_{k=1}^{n} \{t\}_{t_k}^{-\gamma_k} \equiv G(t) \frac{\omega^-(t)}{\omega^+(t)},$$

where

$$\gamma_k = \frac{1}{2\pi i}\ln\frac{G(t_k - 0)}{G(t_k + 0)} - \nu_k = \theta_k - \nu_k + \frac{1}{2\pi i}\ln\left|\frac{G(t_k - 0)}{G(t_k + 0)}\right| \equiv \alpha_k + i\beta_k,$$

$$\omega^+(z) = \prod_{k=1}^n (z - t_k)^{\gamma_k}, \qquad \omega^-(z) = \prod_{k=1}^n \left(\frac{z - t_k}{z}\right)^{\gamma_k},$$

with the analytic functions $\omega_k^+ = (z - t_k)^{\gamma_k}$ and $\omega_k^- = ((z - t_k)/z)^{\gamma_k}$ single-valued on the z-plane with a slit along the curve $\widehat{z_0 t_k \infty}$. Moreover, $\operatorname{ind}_L G(t) = \sum_1^n \nu_k \equiv \kappa_p$ depends on the class $O(t_{i_1},\dots,t_{i_p})$, i.e., on the number p of points t_k at which the strict inequality in 5° is satisfied:

$$0 < \alpha_k = \theta_k - \nu_k < 1. \tag{25}$$

We transform the boundary condition of the homogeneous Riemann problem (16) to the form

$$\Phi^+(t) = \left\{t^{-\kappa_p}\prod_{k=1}^n \{t\}_{t_k}^{-\gamma_k}G(t)\right\}\left[t^{\kappa_p}\frac{\omega^+(t)}{\omega^-(t)}\Phi^-(t)\right]$$

and introduce the function

$$\Omega_p(z) = \begin{cases} \Phi(z)[\omega^+(z)]^{-1}, & z \in D^+, \\ z^{\kappa_p}\Phi(z)[\omega^-(z)]^{-1}, & z \in D^-. \end{cases}$$

Then $\Omega_p^+(t) = G_0(t)\Omega_p^-(t)$, where $G_0(t) = t^{-\kappa_p}\prod_{k=1}^n\{t\}_{t_k}^{-\gamma_k}G(t)$ is a continuous function on L with zero index. The canonical solution of the last problem can be represented in the form (17), and this determines the canonical solution of class $O_p(t_{i_1},\dots,t_{i_p})$ for the original homogeneous problem (16):

$$X_p(z) = \begin{cases} \displaystyle\prod_{k=1}^n (z - t_k)^{\gamma_k}e^{\Gamma(z)}, & z \in D^+, \\[2mm] \displaystyle z^{-\kappa_p}\prod_{k=1}^n \left(\frac{z - t_k}{z}\right)^{\gamma_k}e^{\Gamma(z)}, & z \in D^-. \end{cases} \tag{26}$$

By construction, the canonical function $X_p(z) \in O_p(t_{i_1},\dots,t_{i_p})$ can be zero or infinite only at $z = \infty$ and the points t_k, $k = 1,\dots,n$, of discontinuity of $G(t)$. Moreover, at those points t_k for which (25) holds, i.e., at the points $t_{i_1},\dots,t_{i_p}$ singled out by the choice of the class O_p, the canonical solution $X_p(z)$ has a zero of order $\alpha_k - \varepsilon$, where $\varepsilon > 0$ is an arbitrary small number ($\varepsilon = 0$ if $\beta_k = 0$), while at the remaining points of discontinuity of $G(t)$ it has an infinity of integrable order. If we add one more point t_j to the points t_{i_k} for which (25) holds, then the new v_i'' equals $v_i' - 1$, and, consequently,

$$X_{p+1}(z) = (z - t_j)X_p(z). \tag{27}$$

As we found in 5°, the function $G(t)$ has the largest index κ_0 in the class O_0 when (11) holds at all its points of discontinuity. Repeating the arguments used in obtaining (27), we find that

$$X_p(z) = \prod_{k=1}^{n} (z - t_{i_k})X_0(z), \tag{28}$$

where $X_0(z)$ is the canonical solution of class O_0.

A solution of the nonhomogeneous Riemann problem (16) can be represented, as before, by (21) and (22) with $X_p(z)$ defined by (26). Note that, by (2.19) and by Theorem 3 (Muskhelishvili) in §2 on the behavior of the Cauchy type integral at points of discontinuity of the density, the solution $\Phi(z)$ of the nonhomogeneous Riemann problem belongs to some class H^λ, $0 < \lambda < 1$, near those points of discontinuity of $G(t)$ for which (25) holds, and has an infinity of order $\alpha_k + \varepsilon$ at the remaining points, where $\varepsilon > 0$ is an arbitrarily small number ($\varepsilon = 0$ if $\beta_k = \operatorname{Im}\gamma_k = 0$). Thus, a solution of the nonhomogeneous Riemann problem in the class $O_p(t_{i_1}, \ldots, t_{i_p})$ is bounded at the points $t_{i_1}, \ldots, t_{i_p}$ singled out by the choice of the class O_p. Repeating these constructions on each of the contours L_k forming the boundary $L = \Sigma L_k$ of a multiply connected domain, we obviously reduce the Riemann problem with discontinuous coefficients to the Riemann problem with continuous coefficients. The formulas for solving the Riemann problem retain the same form here as in the case of continuous coefficients (7°).

REMARK. All the constructions remain in force when

$$g(t) = g_*(t) \prod_{i=1}^{n} |t - t_i^*|^{-\alpha_i^*}, \qquad 0 < \alpha_i^* < 1,$$

where $g_*(t)$ is a finite piecewise Hölder function. If one of the points t_i^* coincides with a point t_k of discontinuity of $G(t)$, then no solutions of the nonhomogeneous Riemann problem are bounded at this point. Therefore, it makes sense to choose the canonical function of the corresponding homogeneous Riemann problem to be unbounded at the point $t_i^* = t_k$. In this case a solution of the nonhomogeneous Riemann problem has a singularity of order $\max\{\alpha_i^* + \alpha_k\}$ at this point. If

$$\alpha_i^* + \alpha_k < 1, \tag{29}$$

then the canonical function $X(z)$ can be taken to be bounded at $t_i^* = t_k$, and so the solution of the nonhomogeneous Riemann problem will have a singularity of order α_i^* at this point, as in the case $t_i^* = t_k$.

9°. *The Riemann problem for nonclosed contours.* Let $L = \sum_1^m L_k$ be a collection of disjoint nonclosed smooth arcs L_k, and suppose that $g(t)$, $G(t) \in H^\alpha(L)$ are Hölder functions on them, with $G(t) \neq 0$ on L. It is required to find a solution $\Phi(z)$ of the Riemann problem (16) that is bounded at infinity and integrable on L. Let a_k and b_k be the initial point and terminal point of the nonclosed arc L_k, and complete the curve L to a smooth closed contour

$$\tilde{L} = \sum_{k=1}^m \left(L_k + L_k^0 \right) \equiv L + L^0, \quad \text{where } L_k^0 = \overparen{b_k a_{k+1}}$$

is a smooth arc with initial point at b_k and terminal point at a_{k+1} (see the figure). Setting $\Phi^+(t) = \Phi^-(t)$ on L_0, we come to the Riemann problem (studied in 8°) with discontinuous coefficients on a smooth closed contour $\tilde{L}$:

$$\Phi^+(t) = G_1(t)\Phi^-(t) + g_1(t),$$

$$G_1(t) = \begin{cases} G(t), \\ 1, \end{cases} \quad g_1(t) = \begin{cases} g(t), & t \in L = \sum L_k, \\ 0, & t \in L^0 = \sum L^0k. \end{cases}$$

FIGURE

Here each solution of the resulting Riemann problem can, by construction, be extended analytically across the added arcs L_k^0 and, consequently, satisfies the initial Riemann problem for the collection of nonclosed contours L_k.

10°. *The Riemann problem for the upper half-plane.* The Riemann problem for a curve L passing through the point at infinity can be reduced by a linear fractional transformation to the case of a curve $\tilde{L}$ of finite length; however, the case when L is an infinite straight line is encountered very often in applications, and it is useful to give this case special consideration. Suppose that we must find analytic functions $\Phi^+(z)$ and $\Phi^-(z)$ in the upper D^+ and lower D^- half-planes, respectively, that satisfy on the real axis the boundary condition

$$\Phi^+(t) = G(t)\Phi^-(t) + g(t), \qquad t \in (-\infty, \infty), \tag{30}$$

where $G(t), g(t) \in H^\alpha$ for finite t, $G(t) \neq 0$, and

$$|G(t) - G(\infty)| < N_1 |t|^{-\lambda}, \qquad |g(t) - g(\infty)| < N_2 |t|^{-\lambda}, \qquad \lambda > 0.$$

Let

$$\frac{1}{2\pi i}\left[\ln G(t)\right]_{-\infty}^{+\infty} = \operatorname{ind} G(t) = \kappa.$$

Then, since

$$\frac{1}{2\pi i}\left[\ln \frac{t-i}{t+i}\right]_{-\infty}^{+\infty} = 1,$$

it follows that

$$\operatorname{ind} G_0(t) = \operatorname{ind}\left[G(t)\left(\frac{t-i}{t+i}\right)^{-\kappa}\right] = 0.$$

We construct the canonical function in the form

$$X(z) = e^{\Gamma(z)} \text{ for Im } z \geqslant 0 \quad \text{and} \quad X(z) = \left(\frac{z-i}{z+i}\right)^{-\kappa} e^{\Gamma(z)} \text{ for Im } z < 0,$$

$$\Gamma(z) = \frac{1}{2\pi i}\int_{-\infty}^{\infty} \ln\left[\left(\frac{t-i}{t+i}\right)^{-\kappa} G(t)\right] \frac{dt}{t-z}, \tag{31}$$

regarding $z = -i$ as a singular point of this function (the points $z = 0$ and $z = \infty$ cannot be taken, as previously, to be singular, since they lie on the contour). With the singularity at the point $z = -i$ instead of at $z = \infty$ taken into account, the general solution of the nonhomogeneous problem can be determined in terms of $X(z)$ by the same formulas as before:

$$\Phi(z) = X\left\{\Psi - \frac{P_\kappa}{(z+i)^\kappa}\right\}, \quad \kappa \geqslant 0; \qquad \Phi(z) = X(z)(\Psi + c), \quad \kappa < 0, \tag{32}$$

where

$$P_\kappa(z) = \sum_{k=0}^{\kappa} c_k z^k, \qquad \Psi(z) = \frac{1}{2\pi i}\int_{-\infty}^{\infty}\frac{q(t)\,dt}{X^+(t)(t-z)},$$

and for $\kappa < 0$ the solvability conditions take the form

$$\int_{-\infty}^{\infty}\frac{g(t)\,dt}{X(t)(t+i)^k} = 0, \qquad k = 2,\dots,|\kappa|. \tag{33}$$

§4. The Hilbert boundary-value problem

The Hilbert boundary-value problem is commonly defined to be the problem of finding a function $w^+(z) = u + iv$ that is analytic in a domain D^+ with boundary L and satisfies on L the boundary condition

$$a(t)u(t) - b(t)v(t) = g(t), \qquad t \in L, \tag{1}$$

which is more often written in the form

$$\operatorname{Re}[G(t)w^+(t)] = g(t), \qquad t \in L, \tag{2}$$

with $a + ib = G \neq 0$ on L. In the general case $G(t)$ is a bounded measurable function, $g(t)$ is integrable on L to some power $p > 1$, and the initial analytic function $w^+(z)$ belongs to some fixed class as z approaches L. In the case when $G(t)$ is continuous, the quantity $\kappa = \operatorname{ind}_L G(t)$ is called the index of the Hilbert problem.

1°. **The Hilbert problem with Hölder coefficients for the disk.** Suppose that the domain D^+ is the disk $|z| < 1$ and that $G(t), g(t) \in H^\alpha(L)$. In the case of the homogeneous Hilbert problem ($g = 0$) the boundary conditions (2) are written in the form

$$G(t)w^+(t) + \overline{G(t)}\ \overline{w^+(t)} = 0, \qquad t = e^{i\gamma},$$

or, in other words,

$$t^{-\kappa}w^+(t) = -t^{-2\kappa}\frac{\overline{G(t)}}{G(t)}\ \overline{t^{-\kappa}w^+(t)}, \qquad t = e^{i\gamma}.$$

Let

$$\Phi^+(z) = z^{-\kappa}w^+(z), \qquad G_0(t) = -t^{-2\kappa}\frac{\overline{G(t)}}{G(t)},$$

where $\operatorname{ind} G_0(t) = 0$. Then

$$\Phi^+(t) = G_0(t)\overline{\Phi^+(t)}, \qquad t = e^{i\gamma}.$$

Let us consider the function

$$\Phi(z) = \begin{cases} \Phi^+(z), & |z| \leqslant 1, \\ \overline{\Phi^+(1/z)}, & |z| \geqslant 1. \end{cases} \qquad (4)$$

We verify the Cauchy-Riemann relations for $\Phi(z)$ when $|z| > 1$; in complex form they amount to

$$\frac{\partial w}{\partial \bar z} = \frac{1}{2}\left(\frac{\partial u}{\partial x} - \frac{\partial v}{\partial y}\right) + \frac{i}{2}\left(\frac{\partial u}{\partial y} - \frac{\partial v}{\partial x}\right) = 0, \qquad w = u + iv.$$

Let $\zeta = 1/z$ for $|z| > 1$; then

$$\frac{\partial \Phi}{\partial \bar z} = \overline{\frac{\partial}{\partial z}\Phi^+\left(\frac{1}{z}\right)} = \overline{\frac{\partial}{\partial \bar\zeta}\Phi^+(\zeta)\frac{d\bar\zeta}{dz}} = 0,$$

i.e., the function $\Phi(z)$ defined by (4) is piecewise holomorphic on the z-plane. Since $1/\bar t = t$ on the circle, it follows that $\Phi^-(t) = \overline{\Phi^+(t)}$, and, consequently, we arrive at the following Riemann boundary-value problem with index zero for determining the function $\Phi(z)$:

$$\Phi^+(t) = G_0(t)\Phi^-(t), \qquad (5)$$

which is equivalent to problem (3) only under the condition

$$\Phi^-(z) = \overline{\Phi^+\left(\frac{1}{\bar{z}}\right)}, \qquad |z| > 1, \tag{6}$$

which follows from (4).

The canonical function for the homogeneous Riemann problem (5) has the form (see (3.17))

$$\Omega(z) = \exp\left\{\frac{1}{2\pi i}\int_{|t|=1}\frac{\ln G_0(t)\,dt}{t-z}\right\} \equiv e^{\Gamma(z)}. \tag{7}$$

To construct a solution of (5) that satisfies (6) we consider the function

$$\Omega_*(t) = \overline{\Omega(1/\bar{z})} = \begin{cases} \overline{\Omega^-(1/\bar{z})}, & |z| \leq 1, \\ \overline{\Omega^+(1/\bar{z})}, & |z| \geq 1. \end{cases}$$

Taking $G_0(t)\overline{G_0(t)} = 1$, into account, we find from the boundary condition (5) that

$$\Omega_*^-(z) = \overline{\Omega^+(t)} = \overline{G_0(t)}\ \overline{\Omega^-(t)} = \frac{1}{G_0(t)}\overline{\Omega^-(t)} = \frac{1}{G_0(t)}\Omega_*^+(t),$$

$$\Omega_*^+(t) = G_0(t)\Omega_*^-(t),$$

i.e., the function $\Omega_*(z)$ satisfies together with $\Omega(z)$ the boundary-value problem (5). Then it is not hard to see that the function

$$X_0(z) = \{\Omega(z)\Omega_*(z)\}^{1/2} = \exp\left\{\frac{\Gamma(z) + \Gamma_*(z)}{2}\right\} \equiv e^{\Gamma_0(z)}$$

also satisfies the boundary condition (5). But

$$X_0^\pm(z) = \exp\left\{\frac{1}{2}\left(\Gamma^\pm(z) + \overline{\Gamma^\mp\left(\frac{1}{\bar{z}}\right)}\right)\right\},$$

and this also implies condition (6) for $X_0(z)$. Consequently, $X_0^+(z)$ is a solution of problem (3), and the function

$$X(z) = z^\kappa X_0^+(z) \equiv z^\kappa e^{\Gamma_0(z)}, \qquad |z| \leq 1, \tag{8}$$

is a particular solution of the homogeneous Hilbert problem (2) ($g \equiv 0$). Let us compute $\Gamma_0(z)$. We have

$$\Gamma_*(z) = -\frac{1}{2\pi i}\int_{|t|=1}\frac{\overline{\ln G_0(t)}\,d\bar{t}}{\bar{t} - 1/z} = \frac{z}{2\pi i}\int_{|t|=1}\frac{\ln G_0(t)}{t(t-z)}\,dt.$$

Then

$$\Gamma_0(z) = \frac{1}{2}\left[\Gamma(z) + \Gamma_*(z)\right] = \frac{1}{4\pi i}\int_{|t|=1} \frac{\ln G_0(t)}{t}\frac{t+z}{t-z}\,dt. \qquad (9)$$

We now construct the general solution $\Phi(z)$ of (5) that satisfies condition (6), and thereby the general solution of the original homogeneous Hilbert problem (2), with $\Phi(z)$ allowed to have poles of finite order at infinity and, consequently (by (6)), also at the origin. A solution of (5) will be found in the form

$$\Phi(z) = P_n(z)X_0(z) = P_n(z)e^{\Gamma_0(z)},$$

where $P_n(z) = \sum_{-n}^{n} c_k z^k$. We first satisfy (6), taking into account that it is satisfied for $X_0(z) = e^{\Gamma_0(z)}$:

$$P_n \exp\{\Gamma_0^-(z)\} = \overline{P_n\left(\frac{1}{z}\right)} \exp\left\{\overline{\Gamma_0^+\left(\frac{1}{z}\right)}\right\} = \overline{P_n\left(\frac{1}{z}\right)}\exp\{\Gamma_0^-(z)\}.$$

Consequently,

$$P_n(z) = \sum_{k=-n}^{n} c_k z^k = \overline{\sum_{k=-n}^{n} \bar{c}_k z^{-k}} = \sum_{s=-n}^{n} \bar{c}_{-s} z^s \qquad (s=-k),$$

and so

$$c_k = \bar{c}_{-k} \qquad (k = 0, \pm 1, \pm 2, \ldots, \pm n).$$

Thus, if the function $\Phi(z) = P_n(z)e^{\Gamma_0(z)}$ is to satisfy (6), then $P_n(z)$ must be representable in the form

$$P_n(z) = \sum_{k=0}^{n} \left(c_k z^k + \bar{c}_k z^{-k}\right). \qquad (10)$$

Since

$$\operatorname{Im} P_n(e^{i\gamma}) = \operatorname{Im} \sum_{k=0}^{n} \left(c_k e^{ik\gamma} + \bar{c}_k e^{-ik\gamma}\right) = 0$$

on the circle, the function

$$\Phi(z) = \sum_{k=0}^{n} \left(c_k z^k + \bar{c}_k z^{-k}\right)e^{\Gamma_0(z)} \equiv P_n(z)e^{\Gamma_0(z)} \qquad (11)$$

satisfies together with $X_0(z) = e^{\Gamma_0(z)}$ the boundary condition (5). Thus, the general solution of the homogeneous Hilbert problem (2) with a pole of finite order at the origin has the form

$$w^+(z) = z^\kappa P_n(z)e^{\Gamma_0(z)}, \qquad |z| \leqslant 1, \qquad (12)$$

where $\Gamma_0(z)$ and $P_n(z)$ are represented by (9) and (10), respectively. The particular solution $X(z)$ of the homogeneous Hilbert problem (2) represented by (8) is obtained from the general solution $w^+(z)$ when $P_n(z) \equiv 1$ in (12), i.e., it has the smallest order at the origin. Moreover, $X(z)$ has neither a zero nor an infinity at any point of the closed disk $|z| \leqslant 1$ except, perhaps, at $z = 0$. Therefore, by analogy to the Riemann problem we shall call the particular solution $X(z)$ the canonical solution of the homogeneous Hilbert problem (2). The formula (2) implies that for $\kappa < 0$ no solutions of the homogeneous Hilbert problem are bounded in the disk $|z| < 1$. For $\kappa \geqslant 0$ we set

$$P_n(z) \equiv P_\kappa(z) = \sum_{k=0}^{\kappa} \left(c_k z^k + \bar{c}_k z^{-k} \right),$$

and get a solution that is bounded in the disk and depends on $2\kappa + 1$ arbitrary real constants a_0, a_k, b_k $(k = 1, \ldots, 2\kappa)$, where $a_k + ib_k = c_k$. For example, in a particular case of the Hilbert problem, namely the Schwarz problem of determining an analytic function $w^+(z)$ in the disk $|z| < 1$ from its given real part on the circle,

$$\operatorname{Re} w^+(t) = u(t), \qquad t = e^{i\gamma}, \tag{13}$$

where $G(t) \equiv 1$ and the canonical solution of the homogeneous problem is $X(z) = i$, the general bounded solution $w^+(t) = a_0 i$ depends on a single real constant, since $\operatorname{ind} G(t) = 0$. From this it follows, in particular, that two bounded solutions of the Schwarz problem for a fixed function $g(t) \in H^\alpha$ can differ only by a purely imaginary constant. Consequently, the function (called the *Schwarz integral*)

$$w(z) = \frac{1}{2\pi i} \int_{|t|=1} \frac{u(t)(t+z)\,dt}{t(t-z)} + ia_0, \tag{14}$$

which satisfies the boundary condition (13) (it is easy to see this, for example, with the aid of the Sokhotskiĭ-Plemelj formulas), is the general bounded solution of the Schwarz problem (13).

Let us pass to an investigation of the nonhomogeneous Hilbert problem.

Let $X(z)$ be the canonical solution of the homogeneous Hilbert problem (2):

$$\operatorname{Re}[G(t)X(t)] = 0;$$

consequently, the function $\{iG(t)X(t)\}$ is purely real on the circle. We then represent the boundary condition (2) in the form

$$\operatorname{Re}\left[iG(t)\frac{w^+(t)}{iX(t)}X(t) \right] = g(t) \quad \text{or} \quad \operatorname{Re}\left[\frac{w^+(t)}{iX(t)} \right] = \frac{g(t)}{iG(t)X(t)}.$$

Thus, the nonhomogeneous Hilbert problem (2) reduces to the Schwarz problem of determining an analytic function $F(z) = w^+(z)/iX(z)$ in $|z| < 1$ that has at the origin a pole of order κ if $\kappa \geqslant 0$ and a zero of order $|\kappa|$ if $\kappa < 0$.

From (14) we find a particular solution

$$\frac{w^+(z)}{iX(z)} = \frac{1}{2\pi i} \int_{|t|=1} \frac{g(t)(t+z)\,dt}{iG(t)X(t)t(t-z)},$$

of the Schwarz problem, and thereby also a particular solution

$$w^+(z) = \frac{X(z)}{2\pi i} \int_{|t|=1} \frac{g(t)(t+z)\,dt}{G(t)X(t)t(t-z)}$$

of the nonhomogeneous Hilbert problem (2).

The difference between any two solutions of the nonhomogeneous Hilbert problem (2) obviously satisfies the homogeneous boundary condition (2) and, consequently, is equal to the general solution of the homogeneous Hilbert problem. Thus, in the case $\kappa \geq 0$ the general solution of the nonhomogeneous Hilbert problem (2) can be represented in the form

$$w^+(z) = \frac{z^\kappa e^{\Gamma_0(z)}}{2\pi i} \int_{|t|=1} \frac{g(t)(t+z)\,dt}{G(t)X(t)t(t-z)} + z^\kappa P_\kappa(z) e^{\Gamma_0(z)} \qquad (15)$$

and depends on $2\kappa + 1$ real arbitrary constants, where $X(t) = t^\kappa e^{\Gamma_0(t)}$, and $\Gamma_0(z)$ and $P_\kappa(z)$ are determined from (9) and (10). If a bounded solution exists for $\kappa < 0$, then it is unique and can be represented in the form

$$w^+(z) = \frac{z^\kappa e^{\Gamma_0(z)}}{2\pi i} \int_{|t|=1} \frac{g(t)(t+z)\,dt}{G(t)X(t)t(t-z)}, \qquad (16)$$

and for it to be bounded the integral on the right-hand side must have a zero of order at least $|\kappa|$. Expanding the integral on the right-hand side of (16) with the help of the identity

$$\frac{t+z}{t-z} = 1 + \frac{2z}{t-z} = 1 + 2 \sum_{k=1}^{|\kappa|-1} \frac{z^k}{t^k} + \frac{2z^{|\kappa|} t^{1-|\kappa|}}{t-z} \qquad (17)$$

and equating the coefficients of the powers z^k $(k = 0, 1, \ldots, |\kappa| - 1)$ to zero, we arrive at the following relations:

$$\int_{|t|=1} \frac{g(t)}{G(t)X(t)} t^{-k-1}\,dt = 0, \qquad k = 0, \ldots, |\kappa| - 1, \qquad (18)$$

which imply that the $w^+(z)$ in (16) is bounded. For $k = 0$ the equalities (18) give us that

$$\int_0^{2\pi} g(e^{i\gamma})\left[G(e^{i\gamma})X(e^{i\gamma})\right]^{-1} d\gamma = 0,$$

and, since $\operatorname{Re}[G(t)X(t)] = \operatorname{Im} g = 0$, this equality is a single real condition on the function $g(t)$. Thus, in the case $\kappa < 0$ the nonhomogeneous Hilbert problem (2) is solvable in the class of bounded functions only if $2|\kappa|-1$ real solvability conditions on $g(t)$ are satisfied. In the case of a positive index $\kappa > 0$ it is possible to construct a solution of the nonhomogeneous Hilbert problem that vanishes at previously given fixed points z_i, $i = 1,\ldots,\kappa$, of the disk $|z| < 1$. To do this it suffices in constructing the general solution of the nonhomogeneous problem to take instead of the canonical function $X(z) = z^\kappa e^{\Gamma_0(z)}$ the function

$$\tilde{X}(z) = z^\kappa P_\kappa(z) e^{\Gamma_0(z)},$$

where the coefficients of the polynomial

$$Q_{2\kappa} = z^\kappa P_\kappa(z) = \sum_{k=0}^{\kappa} \left(c_k z^{\kappa+k} + \bar{c}_k z^{\kappa-k} \right)$$

are chosen from the conditions $Q_{2\kappa}(z_i) = 0$, $i = 1,\ldots,\kappa$. Then the general solution of the nonhomogeneous Hilbert problem (2) that satisfies for $\kappa > 0$ the conditions

$$w^+(z_i) = 0, \qquad i = 1,\ldots,\kappa, \tag{19}$$

has the form

$$w^+(z) = \frac{z^\kappa P_\kappa(z) e^{\Gamma_0(z)}}{2\pi i} \left\{ \int_{|t|=1} \frac{g(t)(t+z)\,dt}{\tilde{X}(t)G(t)t(t-z)} + iC \right\}, \tag{20}$$

where $\tilde{X}(t) = t^\kappa P_\kappa(t) e^{\Gamma_0(t)}$, and C is an arbitrary real constant which can be chosen in such a way that the solution $w^+(z)$ has in addition a zero at a fixed point of the circle. For $\kappa < 0$ the identity (17) can be used to write (16) in the form

$$w^+(z) = \frac{z^\kappa e^{\Gamma_0(z)}}{2\pi i} \int_{|t|=1} \frac{g(t)}{X(t)G(t)t} \left\{ 1 + 2 \sum_{k=1}^{|\kappa|-1} \frac{z^k}{t^k} - \frac{2z^{|\kappa|}t^{1-|\kappa|}}{t-z} \right\} dt,$$

from which, under condition (18), we get

$$w^+(z) = \frac{e^{\Gamma_0(z)}}{\pi i} \int_{|t|=1} \frac{t^\kappa g(t)\,dt}{e^{\Gamma_0(t)}G(t)(t-z)}. \tag{21}$$

By construction, the function $w^+(z)$ represented by (21) is a solution of the nonhomogeneous Hilbert problem (2) only under the conditions (18) on $g(t)$.

2°. **The Hilbert problem with piecewise Hölder coefficients.** Suppose that the functions $g(t)$ and $G(t)$ in the boundary condition (2) of the Hilbert problem are in H_*^α, i.e., they satisfy a Hölder condition on arcs $[\gamma_k, \gamma_{k+1}]$ of the circle and have finite discontinuities at the points $t_k = e^{i\gamma_k}$, $k = 1,\ldots,n$, and, moreover, that $|G(e^{i\gamma})| \neq 0$, $\gamma \in [0, 2\pi]$ and $g(t)$ does not have points of discontinuity that are not points of discontinuity of $G(t)$.

It is required to find an analytic function $w^+(z)$ in $|z| < 1$ that satisfies the boundary condition (2) of the Hilbert problem on the circle and belongs to the class $O_m(t_{k_1},\ldots,t_{k_m})$, $0 \leqslant m \leqslant n$, i.e., is bounded at the distinguished points t_{k_i}, $i = 1,\ldots,m$, and has integrable singularities at the remaining points of discontinuity of $G(t)$. Considering first the homogeneous Hilbert problem, we write the boundary condition (2) in the form

$$w^+(t) = -\frac{G(t)}{\overline{G(t)}}\,\overline{w^+(t)}. \tag{22}$$

Let us write $G_1(t) = -\overline{G(t)}/G(t)$ and compute the index of this function in the class $O_m(t_{k_1},\ldots,t_{k_m})$. We mention that at some of the points t_{i_k}, $i = 1,\ldots,m$, of discontinuity of $G(t)$ the function $G_1(t)$ may turn out to be continuous, and then the condition for boundedness of the desired solution of problem (22) and also of the original problem (2) is automatically satisfied. Let

$$\theta_k = \frac{1}{2\pi}\{\operatorname{Arg} G_1(t_k - 0) - \arg G_1(t_k + 0)\},$$

where $\operatorname{Arg} G_1(t_k - 0)$ is computed by continuity on the arc $\overparen{t_{k-1}, t_k}$, and choose integers ν_k from the condition

$$0 \leqslant \alpha_k = \theta_k - \nu_k < 1, \qquad k = k_1,\ldots,k_m, \tag{23}$$

for those points t_i^k, $i = 1,\ldots,m$, of discontinuity of $G(t)$ at which the solution is desired to be bounded, and from the condition

$$-1 < \alpha_k = \theta_k - \nu_k \leqslant 0 \tag{24}$$

for the remaining points t_k. Suppose first that

(a) $\kappa \equiv \kappa_m = \frac{1}{2}\sum_1^n \nu_k$ is an integer, called the index of the Hilbert problem O_m in this case, and let

$$\Phi^+(z) = z^{-\kappa} \prod_{k=1}^{n} (z - t_k)^{-\alpha_k} w^+(z). \tag{25}$$

Then condition (22) can be written in the form

$$t^\kappa \prod_{k=1}^{n} (t - t_k)^{\alpha_k}\Phi^+(t) = -\frac{G(t)}{\overline{G(t)}} \prod_{k=1}^{n} \left(\frac{1}{t} - \bar{t}_k\right)^{\alpha_k} t^{-\kappa}\,\overline{\Phi^+(t)}$$

or

$$\Phi^+(t) = G_0(t)\,\overline{\Phi^+(t)}, \tag{26}$$

$$G_0(t) = -t^{-2\kappa} \prod_{k=1}^{n} \{-tt_k\}_{t_k}^{-\alpha_k}\frac{G(t)}{\overline{G(t)}}, \tag{27}$$

where, by construction, each factor

$$\{-tt_k\}_{t_k}^{-\alpha_k} = \left(\frac{1}{t} - \bar{t}_k\right)^{\alpha_k}(t - t_k)^{-\alpha_k}$$

has a discontinuity at the corresponding point t_k and, just as the function $\{t\}_{t_k}^{-\alpha_k}$, cancels the discontinuity of $G_1(t) = -\overline{G(t)}/G(t)$ at this point (see §3.8°). But, as shown in the definition of the index of a discontinuous function (§3.5°),

$$\mathrm{ind}\left[\prod_{k=1}^{n}\{-tt_k\}_{t_k}^{-\alpha_k}\frac{\overline{G(t)}}{G(t)}\right] = \sum_{k=1}^{n}\nu_k = 2\kappa,$$

and so the index of the continuous function $G_0(t)$ defined by (27) is zero. For a continuous function $G_0(t)$ of index zero we constructed in the preceding subsection a solution $X_0(z)$ of problem (25),

$$X_0(z) = e^{\Gamma_0(z)},$$

where $\Gamma_0(z)$ is defined by (9). Substituting the solution $X_0(z)$ found for problem (26) into the left-hand side of (25), we find a particular solution of the homogeneous Hilbert problem of class O_m,

$$X(z) = z^\kappa \prod_{k=1}^{n}(z - t_k)^{\alpha_k}e^{\Gamma_0(z)}; \tag{28}$$

as in the case of the Hilbert problem with continuous coefficients, we shall call it the canonical solution. By construction, this solution vanishes at the points $t_{k_1}, \ldots, t_{k_m}$, where (23) holds for $\alpha_k \neq 0$, and is unbounded at the remaining points of discontinuity of $G(t)$ such that $\alpha_k \neq 0$. As in the case in 1°, the general bounded (in $|z| < 1$) solution of the homogeneous Hilbert problem can be represented for $\kappa \geq 0$ in the form

$$w^+(z) = P_\kappa(z)X(z), \tag{28*}$$

where $P_\kappa(z) = \sum_0^\kappa(c_k z^k + \bar{c}_k z^{-k})$ and depends on $2\kappa + 1$ arbitrary real constants. For $\kappa < 0$ the homogeneous Hilbert problem (2) does not have solutions in the class O_m that are bounded in the disk. The canonical solution $X(z)$ of the homogeneous Hilbert problem can be used with the same formulas as for continuous coefficients $g(t)$ and $G(t)$ to construct a solution of the nonhomogeneous Hilbert problem (2) also in the case of discontinuous coefficients. For $\kappa \geq 0$ the solution of the nonhomogeneous Hilbert problem has the form

$$w^+(z) = \frac{z^\kappa \Pi(z)e^{\Gamma_0(z)}}{2\pi i}\int_{|t|=1}\frac{g(t)(t + z)\,dt}{G(t)X(t)t(t - z)} + z^\kappa P_\kappa(z)\Pi(z)e^{\Gamma_0(z)} \tag{29}$$

or, similarly to (20),

$$w^+(z) = \frac{P_\kappa(z)X(z)}{2\pi i}\left\{\int_{|t|=1}\frac{g(t)(t+z)\,dt}{G(t)\tilde{X}(t)t(t-z)} + iC\right\}, \qquad (30)$$

where

$$\tilde{X}(z) = z^\kappa P_\kappa \Pi(z)e^{\Gamma_0(z)}, \qquad \Pi(z) = \prod_{k=1}^{n}(z - t_k)^{\alpha_k}.$$

For $\kappa < 0$

$$w^+(z) = \frac{z^\kappa \Pi(z)e^{\Gamma_0(z)}}{2\pi i}\int_{|t|=1}\frac{g(t)(t+z)\,dt}{G(t)X(t)t(t-z)}, \qquad (31)$$

and the $2|\kappa| - 1$ solvability conditions (18) on the function $g(t)$ are necessary for the solution to be bounded at the origin. Similarly to the preceding subsection (see (21)), we can construct for $\kappa < 0$ a function

$$w^+(z) = \frac{\Pi(z)e^{\Gamma_0(z)}}{\pi i}\int_{|t|=1}\frac{t^\kappa g(t)\,dt}{\Pi(t)e^{\Gamma_0(t)}G(t)(t-z)} \qquad (32)$$

satisfying the boundary condition (2) of the nonhomogeneous Hilbert problem only when the conditions (18) hold for $g(t)$. By the results of the investigations of the Cauchy type integral at points of discontinuity of the density (§2.5°) and by Theorem 3 (Muskhelishvili) in §2, the functions $w^+(z)$ constructed according to (29)–(32) are Hölder functions in a neighborhood of the points t_k (singled out in the choice of the class O_m) for which (23) holds, and have integrable singularities at the remaining points of discontinuity of $G(t)$ with $\alpha_k \neq 0$. But if (23) holds for all the points of discontinuity of $G(t)$, then the functions $w^+(z)$ in (29), (30), and (32) belong, according to Theorem 3 in §2, to some class H^λ ($0 < \lambda < 1$) in the closed disk, while $w^+(z)$ in (31) belongs to H^λ everywhere in $|z| \leq 1$ except in a neighborhood of $z = 0$. The formulas (29)–(32) provide a solution also when $g(t)$ can be reprsented in the form

$$g(t) = \frac{g_*(t)}{\Pi|t - t_k|^{\alpha_k^*}}, \qquad 0 < \alpha_k^* < 1, \qquad (33)$$

where $g_*(t) \in H_*^\alpha$, i.e., is a piecewise Hölder function. If some point t_{k_i} ($i = 1,\ldots,s$) of discontinuity of $g(t)$ coincides with a point t_{k_j} ($j = 1,\ldots,n$) of discontinuity of $G(t)$, then the canonical function $X(z)$ of the homogeneous problem is chosen to be unbounded at this point. The function $X(z)$ can also be taken to be bounded at this point if

$$\alpha_{k_i}^* + \alpha_{k_j} < 1; \qquad (34)$$

of course, the index of the problem is thereby decreased (cf. the analogous case of the Riemann problem, §3.8°).

Let us now consider the case when condition (a) does not hold, i.e.,

(b) $\sum_1^n \nu_k = 2\kappa + 1$, κ an integer.

Considering the homogeneous Hilbert problem (2), we let

$$\Phi^+(z) = z^{-\kappa} \prod_{k=1}^n (z - t_k)^{-\alpha_k} w^+(z), \qquad (25)$$

as in the case when (a) holds, and this leads us to the previous boundary condition (26), with the continuous function $G_0(t)$ represented by (27); but now ind $G_0(t) = 1$.

We extend $\Phi^+(z)$ to the whole plane by (4) and get a Riemann problem with index equal to 1,

$$\Phi^+(t) = G_0(t)\Phi^-(t), \qquad (35)$$

which is equivalent to problem (26) only when

$$\Phi^-(z) = \overline{\Phi^+(1/z)}. \qquad (36)$$

The canonical solution of the homogeneous Riemann problem (35) can be taken in the form

$$X_0(z) = \begin{cases} e^{\tilde{\Gamma}_0(z)}, & |z| \leqslant 1, \\ z^{-1} e^{\tilde{\Gamma}_0(z)}, & |z| > 1 \end{cases}$$

($\tilde{\Gamma}_0(z)$ is determined by (9), in which the role of $G_0(t)$ is played by the continuous function of zero index $\tilde{G}_0(t) = t^{-1}G_0(t)$). Indeed, according to the constructions in 1°, the function $\Phi_0(z) = e^{\tilde{\Gamma}_0(z)}$ satisfies conditions (36) and the boundary-value problem

$$\Phi_0^+(t) = \tilde{G}_0(t)\Phi_0^-(t) \equiv [G_0(t)t^{-1}]\Phi_0^-(t),$$

whence

$$X_0^+(t) = \Phi_0^+(t) = [t^{-1}G_0(t)]\Phi_0^-(t) = G_0(t)X^-(t).$$

We write the general solution of the homogeneous Riemann problem (35) with a pole at the origin and at infinity in the form

$$\Omega(z) = (az + b/z + c)X_0(z)$$

and choose the coefficients a, b, and c from condition (36):

$$z^{-1}(az + b/z + c) = \bar{a}/z + \bar{b}z + \bar{c},$$

from which we find that $b = 0$ and $a = c = \rho e^{i\alpha}$. Thus, the general solution of (35) satisfying (36) in the class of functions with poles at the points $z = 0$ and $z = \infty$ is the function

$$\Phi(z) = \rho(e^{i\alpha}z - e^{-i\alpha})e^{\tilde{\Gamma}_0(z)}. \tag{37}$$

Then the function

$$X(z) = z^\kappa(e^{i\alpha}z - e^{-i\alpha})\Pi(z)e^{\tilde{\Gamma}_0(z)}, \tag{38}$$

where $\Pi(z) = \prod_1^n (z - t_k)^{\alpha_k}$, is a solution of the homogeneous Hilbert problem and vanishes at some point $t = e^{-2i\alpha}$ of the circle (α is an arbitrary constant). As shown in the derivation of (37), the homogeneous Riemann problem (35) does not have solutions in the class O_m which satisfy (36) and do not have a zero of the first order on the circle. Consequently, if condition (a) fails to hold, i.e., if $\Sigma \nu_k = 2\kappa + 1$ (κ an integer), then the homogeneous Hilbert problem (2) does not have solutions in the given class O_m that do not have a zero of the first order on the circle. It is natural to call the function $X(z)$ defined by (38) the canonical solution of the homogeneous Hilbert problem in the class O_m. All the subsequent constructions and formulas for solution of the nonhomogeneous Hilbert problem remain formally the same as for case (a). However, the singular integrals in the solutions of the nonhomogeneous problem should now, on the basis of the identity

$$\frac{z - e^{-2i\alpha}}{(t - e^{-2i\alpha})(t - z)} = \frac{1}{t - z} - \frac{1}{t - e^{2i\alpha}},$$

be understood as the difference of two integrals in the Cauchy principal-value sense. A simple example of a homogeneous Hilbert problem in which case (b) is realized is given by the boundary-value problem

$$\text{Re}\left\{e^{i\gamma/2}w^+(t)\right\} = 0, \qquad t = e^{i\gamma},$$

which can also be written in the form

$$e^{i\gamma/2}w^+(t) + e^{-i\gamma/2}\overline{w^+(t)} = 0$$

or $w^+(t) = -t^{-1}\overline{w^+(t)}$. According to the preceding constructions, the canonical solution of this problem has the form $X(z) = z^{-1}(e^{i\alpha}z - e^{-i\alpha})$ and vanishes at the point $t = e^{-2i\alpha}$.

3°. The Hilbert problem for the upper half-plane. The Hilbert problem for an arbitrary domain $D(L)$ with smooth boundary L can be reduced to the Hilbert problem for the disk by means of a conformal mapping of this domain onto the disk. However, the solution for the Hilbert problem in the upper half-plane (which is often encountered in applications) can be constructed directly, similarly to the case of the disk. Let D^+ and D^- be the upper and lower half-planes, respectively, and L the real axis; moreover, the coefficients $G(t)$ and $g(t)$ are in H^α for all finite t, and

$$|G(t) - G(\infty)| < N_1 |t|^{-\lambda}, \qquad |g(t) - g(\infty)| < N_2 |t|^{-\lambda}, \qquad \lambda > 0. \tag{39}$$

In the case of the homogeneous Hilbert problem we write the boundary condition (2) in the form

$$w^+(t) = -\frac{G(t)}{\overline{G(t)}} \overline{w^+(t)}$$

or, setting

$$\Phi^+(z) = \left(\frac{z-i}{z+i}\right)^{-\kappa} w^+(z), \qquad G_0(t) = -\left(\frac{t-i}{t+i}\right)^{-2\kappa} \frac{G(t)}{\overline{G(t)}},$$

in the equivalent form

$$\Phi^+(t) = G_0(t) \overline{\Phi^+(t)}, \tag{40}$$

where $\kappa = \operatorname{ind} \overline{G(t)}$ is the Hilbert index of the problem; since

$$\operatorname{ind}((t-i)/(t+i)) = 1,$$

it follows that $\operatorname{ind} G_0(t) = 0$. Extending $\Phi^+(t)$ to the whole plane by the formulas

$$\Phi(z) = \begin{cases} \Phi^+(z), & \operatorname{Im} z \geq 0, \\ \overline{\Phi^+(\bar{z})}, & \operatorname{Im} z < 0, \end{cases} \tag{41}$$

we arrive at the Riemann problem with zero index for the half-plane (§3.10°):

$$\Phi^+(t) = G_0(t)\Phi^-(t), \tag{42}$$

which is equivalent to problem (40) only if

$$\Phi^-(z) = \overline{\Phi^+(\bar{z})}. \tag{43}$$

The canonical function $X_0(z)$ of the homogeneous Riemann problem (42) has the form

$$X_0(z) = \exp\left\{ \frac{1}{2\pi i} \int_{-\infty}^{\infty} \ln G_0(t) \frac{dt}{t-z} \right\} = e^{\Gamma_0(z)}, \tag{44}$$

and, since

$$\overline{\Gamma_0(z)} = -\frac{1}{2\pi i} \int_{-\infty}^{\infty} \overline{\ln G_0(t)} \frac{dt}{t-z} = \Gamma_0(z),$$

it satisfies condition (43). Consequently, the canonical function of the homogeneous Hilbert problem (2) can be written in final form as follows:

$$X(z) = \left(\frac{z-i}{z+i}\right)^{\kappa} e^{\Gamma_0(z)}, \qquad \operatorname{Im} z \geqslant 0. \tag{45}$$

The subsequent constructions and formulas for solution of the nonhomogeneous Hilbert problem differ from the case of the disk domain D considered in subsections 1° and 2° only in the different form of the Schwarz integral

$$w(z) = u(x, y) + iv(x, y) = \frac{1}{2\pi i} \int_{-\infty}^{\infty} \frac{u\,dt}{t-z} \tag{46}$$

and the fact that for $\kappa \geqslant 0$ the arbitrary polynomial $P_\kappa(z)$ has the form

$$P_\kappa(z) = \sum_{k=0}^{\kappa} \left[c_k \left(\frac{z-i}{z+i}\right)^{k} + \bar{c}_k \left(\frac{z-i}{z+i}\right)^{-k} \right]$$

(the form of this polynomial is, in essence, connected with the mapping $\zeta = (z-i)/(z+i)$ of the upper half-plane onto the disk $|\zeta| < 1$).

4°. The Schwarz-Christoffel formula for polygons. Let us first determine the behavior of a conformal mapping at corner points of domains [69]. Suppose that the function $z = F(\xi)$ maps the upper half-plane $\operatorname{Im} \xi > 0$ conformally onto the domain D, and the point ξ_0 of the real axis corresponds to a corner point z_0 with internal angle $\alpha\pi$ (Figure 1).

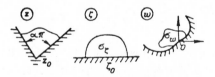

FIGURE 1

For simplicity it is assumed that the boundary of D in a neighborhood of the corner point z_0 consists of rectilinear segments (the general case is studied in §3 of Chapter II).

Let us consider the auxiliary function $w = (F(\xi) - z_0)^{1/\alpha} = w(\xi)$, which maps a neighborhood σ_ξ of the point $\xi = \xi_0$ conformally onto a neighborhood σ_w of the point $w = 0$, with a segment of the axis $\operatorname{Im} \xi = 0$ passing into a rectilinear segment. Extending $w(\xi)$ by the symmetry principle to a full neighborhood of ξ_0 and expanding it in a Taylor series there, we find that

$$w(\xi) = c_1(\xi - \xi_0) + c_2(\xi - \xi_0)^2 + \cdots,$$

where $c_1 \neq 0$ because the mapping is conformal. Then

$$F(\xi) - z_0 = [w(\xi)]^\alpha = (\xi - \xi_0)^\alpha \{c_1 - c_2(\xi - \xi_0) + \cdots\}^\alpha,$$

whence

$$dz/d\zeta = (\zeta - \zeta_0)^{\alpha-1} f(\zeta), \qquad f(\zeta_0) = c_1^\alpha \alpha \neq 0,$$

and, consequently, $(dz/d\zeta)(\zeta_0) = \infty$ for $\alpha < 1$, while $(dz/d\zeta)(\zeta_0) = 0$ for $\alpha > 1$. Let us now consider a polygon P_n without points of self-intersection and with angles $\alpha_k \pi$ $(k = 1,\ldots,n)$ at the vertices (Figure 2). Suppose that the points t_k $(k = 1,\ldots,n)$ of the real axis are images of the vertices z_k of P_n under the conformal mapping $z = F(\xi)$ of the upper half-plane $\operatorname{Im} \xi > 0$ onto it. From the geometric meaning of the derivative of a conformal mapping we have

$$\beta(t) \equiv \arg \frac{dz}{dt}(t) = \beta_k \pi, \qquad t \in [t_k, t_{k+1}],$$

where the $\beta_k \pi$ are the angles formed by the sides of the polygon with the OX-axis. Then, recovering the analytic function $-i \ln(dz/d\zeta)$ by the Schwarz formula (46), we get

$$-i \ln \frac{dz}{d\zeta} = \frac{1}{\pi i} \int_{-\infty}^{+\infty} \frac{\beta(t)\,dt}{t - \zeta} - i \ln c_0$$

$$= \sum_{k=1}^{n} \frac{\beta_{k-1} - \beta_k}{i} \ln(\zeta - t_k) + \beta_0 - i \ln c_0,$$

and so

$$z = c_0 e^{i\beta_0} \int_{t_1}^{\zeta} \prod_{k=1}^{n} (\zeta - t_k)^{\alpha_k - 1} \, dt + z_1 \qquad (\alpha_k = \beta_{k-1} - \beta_k + 1).$$

The last formula is called the Schwarz-Christoffel formula. To determine the unknown constants c_0, β_0 and t_k $(k = 1,\ldots,n)$ it is necessary to use the known lengths l_k $(k = 1,\ldots,n-1)$ of the sides of the polygon

$$l_k = \int_{t_k}^{t_{k+1}} \left| \frac{dz}{dt} \right| dt \qquad (k = 1,\ldots,n-1);$$

three of the constants here can be chosen arbitrarily. We remark that the derivative

$$\frac{dz}{d\zeta} = c_0 e^{i\beta_0} \prod_{k=1}^{n} (\xi - t_k)^{\alpha_k - 1}$$

of the Schwarz-Christoffel integral satisfies the following homogeneous Hilbert problem with discontinuous coefficients in the boundary condition:

$$\frac{dy}{dt} - \tan(\beta_k \pi) \frac{dx}{dt} = 0, \qquad t \in [t_k, t_{k+1}],$$

which fact we shall use later in solving the analogous nonhomogeneous problem (§1 of Chapter III).

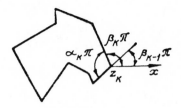

FIGURE 2

5°. *The mixed boundary-value problem for the upper half-plane.* Suppose that a_k and b_k ($k = 1, \ldots, n$) are points of the real axis L with $-\infty < a_1 < b_1 < \alpha_2 < \cdots < b_n < \infty$, and $\widehat{a_k b_k}$, $\widehat{b_k a_{k+1}} \subset L$ are the corresponding segments on L. Let $L_1 = \Sigma_1^n \widehat{a_k b_k}$, $L_2 = \Sigma_1^n \widehat{b_k a_{k+1}}$, where $a_{n+1} = a_1$, and $\widehat{b_n a_1} = (b_n, \infty) + (-\infty, a_1)$ is an infinite segment on L. The mixed boundary-value problem is defined to be the problem of finding an analytic function $w^+(z) = u + iv$ in the upper half-plane that is bounded at infinity and satisfies on the real axis the boundary conditions

$$\operatorname{Re} w^+(t) = f_1(t), \quad t \in L_1, \qquad \operatorname{Im} w^+(t) = f_2(t), \quad t \in L_2, \qquad (47)$$

where $f_i(t) \in H^\alpha(L_i)$, and

$$|f_2(t) - f_2(\infty)| < N_1 |t|^{-\lambda}, \qquad \lambda > 0, t \in \widehat{b_n a_1}. \qquad (48)$$

The mixed boundary-value problem is a particular case of the Hilbert problem (2) with discontinuous coefficients, where

$$G(t) = \begin{cases} 1, & t \in L_1, \\ -i, & t \in L_2, \end{cases} \qquad g(t) = f_i(t), \quad t \in L_i. \qquad (49)$$

The canonical function of class $O_n(a_1, \ldots, a_n)$ for the homogeneous problem (47), i.e. the one bounded at the points a_k and unbounded at the b_k, can be taken in the form

$$X(z) = \left(\prod_{k=1}^n \frac{z - a_k}{z - b_k} \right)^{1/2}. \qquad (50)$$

The function $X(z)$ is easily seen to satisfy the boundary condition (47). Indeed,

$$\frac{t - a_i}{t - b_i} > 0, \quad i \neq k, \qquad \frac{t - a_k}{t - b_k} < 0 \quad \text{for } t \in [a_k, b_k],$$

i.e., $\operatorname{Re} X(t) = 0$ for $t \in [a_k, b_k]$; and, similarly, $\operatorname{Im} X(t) = 0$ for $t \in [b_k, a_{k+1}]$. Then the general solution of the nonhomogeneous mixed problem that is in the class O_n and bounded at infinity has the form

$$w^+(z) = \frac{1}{\pi i} \prod_{k=1}^n \left(\frac{z - a_k}{z - b_k} \right)^{1/2} \left\{ \int_{-\infty}^\infty \prod_{k=1}^n \left(\frac{t - a_k}{t - b_k} \right)^{-1/2} \frac{f(t)\, dt}{t - z} + i\beta \right\}, \quad (51)$$

where, by (49),

$$f(t) = \frac{g(t)}{G(t)} = \begin{cases} f_1(t), & t \in L_1, \\ if_2(t), & t \in L_2. \end{cases} \quad (52)$$

A solution that is bounded at all the points a_k and b_k can be represented in the form

$$w^+(z) = \frac{\Pi(z)}{\pi i} \int_{-\infty}^\infty \frac{f(t)\, dt}{\Pi(t)(t - z)}, \quad (53)$$

where $\Pi(z) = \{\prod_1^n (z - a_k)(z - b_k)\}^{1/2}$ has a pole of order n at infinity. Therefore, the following solvability conditions on $f(t)$ are necessary for $w^+(z)$ to be bounded at infinity:

$$\int_{-\infty}^\infty \frac{f(t)}{\Pi(t)} t^k\, dt = 0 \qquad (k = 0, \dots, n - 2). \quad (54)$$

In particular, for $n = 1$, i.e., when the coefficients in the boundary condition (47) have only two points of discontinuity, the solvability conditions (54) do not arise, and, consequently, there is always a solution of the mixed boundary-value problem that is bounded in the upper half-plane in this case. The formulas (51) and (53) are called the Keldysh-Sedov formulas (they are often called the Signorini formulas in the non-Russian literature), after the authors who first gave a complete solution of the mixed boundary-value problem.

6°. The mixed boundary-value problem with parameters. Let L_1^p and L_2^q be collections of p arbitrary intervals $[a_k, b_k] \subset L_1$ and q arbitrary intervals $[b_k, a_{k+1}] \subset L_2$, respectively, and consider the following boundary-value problem.

PROBLEM P. *Find a bounded analytic function* $w^+(z) = u + iv$ *in the upper half-plane and* $n - 1$ *real constants* u_k *and* v_k *that are related by the boundary conditions*

$$\operatorname{Re} w^+ = f_1(t), \quad t \in (L_1 \setminus L_1^p); \quad \operatorname{Im} w^+ = f_2(t), \quad t \in (L_2 \setminus L_2^q); \quad (55)$$

$$\operatorname{Re} w^+ = f_1 + e^a u_k, \qquad t \in [a_k, b_k] \subset L_1^p;$$

$$\operatorname{Im} w^+ = f_2(t) + e^a v_k, \qquad t \in [b_k a_{k+1}] \subset L_2^q,$$

where $a(t)$, $f_i(t) \in H^\alpha(L_i)$, $i = 1, 2$, *are given functions satisfying* (48) (*in particular*, $a \equiv 0$).

Let us show that the boundary-value problem P is uniquely solvable for any given functions $a(t)$ and $f_i(t)$, $i = 1, 2$. According to 4°, to prove this assertion it suffices to establish the conditions (54), which amount to a system of linear algebraic equations in the unknown parameters u_k and v_k:

$$\sum_p u_k \int_{a_k}^{b_k} g_l(t)\, dt + \sum_q v_k \int_{b_k}^{a_{k+1}} g_l(t)\, dt = -\int_{-\infty}^{\infty} (f_1 + if_2) e^{-\alpha} g_l\, dt, \quad (56)$$

where $g_l(t) = e^a |\Pi(t)|^{-1} t^l$; the integrals over all the intervals in the collections L_1^p and L_2^q appear in Σ_p and Σ_q, respectively.

The determinant of the system (56) is nonzero. Indeed, otherwise we can form a linear combination of its rows with λ_i ($i = 0, \ldots, n - 2$) not all equal to zero and get the equalities

$$\int_{a_k}^{b_k} e^{a(t)} \frac{P_{n-2}(t)}{|\Pi(t)|}\, dt = \int_{b_k}^{a_{k+1}} e^{a(t)} \frac{P_{n-2}(t)}{|\Pi(t)|}\, dt = 0, \qquad P_{n-2} = \sum_{k=0}^{n-2} \lambda_k t^k,$$

where $[a_k, b_k] \subset L_1^p$ and $[b_k, a_{k+1}] \subset L_2^q$. For these relations to hold the polynomial $P_{n-2}(t)$ must have at least one zero in each of the $n - 1$ intervals in the collection $L_1^p + L_2^q$, and this is possible only if $\lambda_i = 0$, $i = 0, \ldots, n - 2$. Consequently, system (56) has a unique solution for arbtrary $a(t)$ and $f_i(t)$.

REMARK. Mapping the upper half-plane conformally onto the disk of unit radius, we get the unique solvability of Problem P also in the case of the disk.

§5. Stability of the Hilbert problem

Without dwelling separately on the stability of the Hilbert problem with Hölder coefficients in the boundary condition, we consider at once the more general case of piecewise Hölder coefficients. Let us introduce some notation. We single out $2n$ points $t_k^1 = e^{i\gamma_k^1}$ and $t_k^2 = e^{i\gamma_k^2}$ ($k = 1, \ldots, n$) on the circle $|z| = 1$ that satisfy the inequalities

$$\inf_{k,m} |t_k^m - t_{k+1}^m| > \delta_0 > 0, \qquad \max_k |t_k^1 - t_k^2| < M_0 \varepsilon, \quad (1)$$

and denote by $\tau = \tau(t)$ a continuously differentiable homeomorphism of the circle onto itself with the properties

$$\tau(t_k^1) = t_k^2 \quad (k = 0, \ldots, n), \qquad \max_k |d\tau/dt - 1| < \varepsilon. \quad (2)$$

where $t_0^1 = t_0^2 = 1$ and $M_0 = M_0(\delta) = \text{const.}$ If $\gamma = \arg t$, $\theta = \arg \tau(t)$, then a homeomorphism $\tau = e^{i\theta(\gamma)}$, $\gamma \in [0, 2\pi]$, satisfying (2) can be constructed, for example, in the form

$$\tau(e^{i\gamma}) = \exp\left\{ i\gamma + i \sum_{k=0}^{n} \left(\gamma_k^2 - \gamma_k^1 \right) \prod_{k \neq 1} \left(\frac{\gamma - \gamma_i^1}{\gamma_k^1 - \gamma_i^1} \right) \right\}. \tag{3}$$

Indeed, $\tau(e^{i\gamma_k^1}) = e^{i\gamma_k^2}$, $\tau(1) = 1$ by construction, and for sufficiently small $\varepsilon > 0$ we have

$$\frac{d\theta}{d\gamma} \geq 1 - M_0 \varepsilon \sum_{k=0}^{n} \left| \frac{d}{d\gamma} \prod_{k \neq i} \left(\frac{\gamma - \gamma_i^1}{\gamma_k^1 - \gamma_i^1} \right) \right| > 0,$$

which implies that the mapping $\tau = \tau(t)$ is a homeomorphism. It is sometimes convenient to represent the homeomorphism $\tau = \tau(t)$ defined by (3) in the form

$$\tau(e^{i\gamma}) = \exp\left\{ i\gamma + i \int_0^\gamma f(\gamma)\, d\gamma \right\}, \tag{4}$$

where

$$f(\gamma) = \sum_{k=0}^{n} \left(\gamma_k^2 - \gamma_k^1 \right) \frac{d}{d\gamma} \prod_{k \neq 1} \left(\frac{\gamma - \gamma_k^1}{\gamma_k^1 - \gamma_i^1} \right),$$

and, by construction,

$$\int_0^{\gamma_k^1} f(\gamma)\, d\gamma = \gamma_k^2 - \gamma_k^1 \qquad (k = 1, \dots, n).$$

Suppose that $F(t) \in H^\beta$, $\beta > 0$, on the circle, and

$$C(F) = \max_t |F(t)|, \qquad H(F, \beta) = \sup_{t', t''} \frac{|F(t') - F(t'')|}{|t' - t''|^\beta}.$$

Let us consider two functions $\tau(t)$ and $\tau^0(t)$ that are close in H^λ, $\lambda > 0$; this means that $C(\tau - \tau^0) < \varepsilon$ and $H(\tau - \tau^0, \lambda) < \varepsilon$ for a sufficiently small $\varepsilon > 0$. Then the functions $F[\tau(t)]$ and $F[\tau^0(t)]$ are close in $H^{\beta\lambda - \delta}$, $\beta\lambda - \delta > 0$, and, moreover,

$$C(\Delta_\tau F) < H(F, \beta)\varepsilon^\beta, \qquad H(\Delta_\tau F, \lambda\beta - \delta) \leq N\varepsilon^\delta, \tag{5}$$

where

$$\Delta_\tau F = F[\tau(t)] - F[\tau^0(t)], \qquad N = 2H(F, \beta)\max\{1, H^\beta(\tau, \lambda)\}.$$

Indeed, the first of the inequalities (5) is obvious, and to establish the second we assume first that $|t' - t''| < \varepsilon$. Then

$$\tilde{H} = \frac{|\Delta_\tau F(t') - \Delta_\tau F(t'')|}{|t' - t''|^{\beta\lambda - \delta}}$$

$$\leq H(F, \beta)\left\{\frac{|\tau(t') - \tau(t'')|^\beta + |\tau^0(t') - \tau^\theta(t'')|^\beta}{|t' - t''|^{\beta\lambda - \delta}}\right\}$$

$$\leq 2H(F, \beta)H^\beta(\tau, \lambda)|t' - t''|^\delta \leq N\varepsilon^\delta.$$

For $|t' - t''| \geq \varepsilon$ we have

$$\tilde{H} \leq \varepsilon^{\delta - \beta\lambda}\left\{\left|F[\tau(t')] - F[\tau^0(t')]\right| + \left|F[\tau(t'')] - F[\tau^0(t'')]\right|\right\}$$

$$\leq 2\varepsilon^{\delta - \beta\lambda}H(F, \beta)\max_t |\tau(t) - \tau^0(t)|^\beta \leq 2H(F, \beta)\varepsilon^{\delta + (1-\lambda)\beta} \leq N\varepsilon^\delta.$$

Inequality (5) is proved.

Suppose now that $F(t) \in H_*^\beta$, i.e., $F(t)$ is piecewise Hölder on the circle, with points of discontinuity t_k $(k = 1, \ldots, n)$; $l_k = \widehat{t_k t_{k+1}}$ are arcs of the circle. Let

$$H_*(F, \beta) = \sup_{t', t'' \in l_k} \frac{|F(t') - F(t'')|}{|t' - t''|^\beta},$$

where $0 < \beta < 1$, and β is the smallest of the Hölder exponents of $F(t)$ on its intervals of continuity. We introduce the concept of closeness also for discontinuous functions.

DEFINITION 1. Piecewise Hölder functions $F^1(t)$, $F^2(t) \in H_*^\beta$ are said to be *close in* $H_*^{\beta_0}$, $0 < \beta_0 < \beta_1$, if their points of discontinuity t_k^1 and t_k^2 $(k = 1, \ldots, n)$ are close and subject to the inequalities (1), and, moreover,

$$C(\Delta_\tau F) \leq \varepsilon_1(\varepsilon), \qquad H_*^{\beta_0}(\Delta_\tau F, \beta) \leq \varepsilon_1(\varepsilon), \qquad (6)$$

where $\Delta_\tau F^1[\tau(t)] - F^2(t)$, $\tau = \tau(t)$, is a homeomorphism of the form (3) that carries the points of discontinuity of $F^1(t)$ into those of $F^2(t)$, and $\varepsilon_1(\varepsilon) \to 0$ as $\varepsilon \to 0$.

Let us proceed to investigate the stability of the Hilbert problem

$$\text{Re}\left[G(t)w(t)\right] = g(t), \qquad t = e^{i\gamma}, \qquad (7)$$

where $G(t) \in H_*^\beta$ and $t_k = e^{i\gamma_k}$ $(k = 1, \ldots, n)$ are the points of discontinuity of the piecewise Hölder function $G(t)$, $|G(t)| \neq 0$, and

$$g(t) = g_*(t)\prod_{k=1}^{l}|t - \rho_k|^{-\alpha_k^*}, \qquad 0 \leq \alpha_k^* < 1,$$

with $g_*(t) \in H_*^\beta$, and ρ_k $(k = 1,\ldots,l)$ are the points of finite and infinite discontinuity of $g(t)$. According to (4.28) and (4.28*), the canonical function of the class $O_m(t_{k_1},\ldots,t_{k_m})$, $0 \le m \le n$, with a pole of order $|\kappa|$ at the origin if $\kappa < 0$ or with κ zeros in the disk $|z| < 1$ if $\kappa \ge 0$, can be taken in the form

$$X(z) = \begin{cases} z^\kappa \Pi(z) e^{\Gamma_0(z)}, & \kappa < 0, \\ z^\kappa \Pi(z) e^{\Gamma_0(z)} P_\kappa(z), & \kappa \ge 0. \end{cases} \tag{8}$$

Here $\Pi(z) = \Pi_1^n (z - t_k)^{\alpha_k}$, $0 \le \alpha_{k_i} < 1$ for $i = 1,\ldots,m$, $-1 < \alpha_k \le 0$ for the remaining $k \ne k_i$,

$$\Gamma_0(z) = \frac{1}{4\pi i} \int_{|t|=1} \frac{\ln[G_0(t)](t + z)\, dt}{t(t - z)}, \tag{9}$$

$$G_0(t) = -t^{-2\kappa} \frac{\overline{G(t)}}{G(t)} \prod_{k=1}^{n} \{-tt_k\}_{t_k}^{-\alpha_k} \in H^\beta, \qquad \beta > 0,$$

and, finally, $P_\kappa(z) = \Sigma_0^n (c_k z^k + \bar{c}_k z^{-k})$, where the constants c_k are chosen in such a way that $P_\kappa(z_i) = 0$ $(i = 1,\ldots,\kappa)$, $z_i \ne z_j$, $i \ne j$, $|z_i| < 1$, and will be regarded as fixed in what follows. Depending on the sign of the index κ, the solution of the nonhomogeneous Hilbert problem (7) can be represented in the form (see (4.30) and (4.32))

$$w(z) = \frac{X(z)(z - 1)}{\pi i} \int_{|t|=1} \frac{g(t)\, dt}{G(t)X(t)(t - 1)(t - z)}, \qquad \kappa \ge 0, \quad (10)$$

$$w(z) = \frac{\Pi(z) e^{\Gamma_0(z)}}{\pi i} \int_{|t|=1} \frac{t^\kappa g(t)\, dt}{\Pi(t) e^{\Gamma_0(t)} G(t)(t - z)}, \qquad \kappa < 0, \quad (11)$$

where the integral in (10), on the basis of the identity

$$\frac{z - 1}{(t - 1)(t - z)} = \frac{t + z}{2t(t - z)} - \frac{t + 1}{2t(t - 1)}, \tag{12}$$

is understood as a difference of singular integrals, while the function $w = w(z)$ in (11) satisfies (7) only under the $2|\kappa| - 1$ solvability conditions (4.18). If a point ρ_i of infinite discontinuity of $g(t)$ coincides with a point t_k of discontinuity of $G(t)$, then the inequality $\alpha_k + \alpha_k^* < 1$ will be required for the solutions constructed to be integrable at this point $t = t_k = \rho_i$. Let

$$u(t) = \ln G_0(t), \qquad v(t) = \frac{g(t)\hat{\Pi}(t)}{X(t)G(t)}, \tag{13}$$

where $\hat{\Pi}(t)$ is an integrable function chosen in such a way that $v(t) \in H^\beta$, $\beta > 0$, on the circle. For example,

$$\hat{\Pi}(t) = \prod_{k=1}^{n} (t - t_k)^{\hat{\alpha}_k} \prod_{i=1}^{l} (t - \rho_i)^{\hat{\alpha}_i^*},$$

where $\hat{\alpha}_i^* = \varepsilon_0$ for the points of finite discontinuity of $g(t)$, $\hat{\alpha}_k = \alpha_k + \varepsilon_0$, $\alpha_k > 0$, for the points of discontinuity of $G(t)$ that do not coincide with points of infinite discontinuity of $g(t)$, and $\hat{\alpha}_k + \hat{\alpha}_i^* = \alpha_k + \alpha_i^* + \varepsilon_0$ otherwise. For the remaining points at which $g(t)$ is unbounded, we have $\hat{\alpha}_i^* = \alpha_k^*$ if $g_*(t)$ is continuous at this point and $\hat{\alpha}_k^* = \alpha_k^* + \varepsilon_0$ otherwise. The number ε_0, $0 < \varepsilon_0 < 1$, is chosen from the conditions $0 < \hat{\alpha}_k < 1$ and $0 < \hat{\alpha}_i^* < 1$, and for $\rho_i = t_k$ it is required in addition that $0 \leqslant \hat{\alpha}_k + \hat{\alpha}_i^* < 1$.

Thus, the singularities of the density of the Cauchy type integral in (10) and (11) are separated out in the function $[\hat{\Pi}(t)]^{-1}$, and these formulas take the form

$$w(z) = \frac{z^\kappa}{\pi i} P_\kappa(z) \Pi(z) e^{\Gamma_0(z)} \int_{|t|=1} \frac{(z-1)v(t)\,dt}{\hat{\Pi}(t)(t-1)(t-z)} \qquad (\kappa \geqslant 0),$$

$$(10^*)$$

$$w(z) = \frac{1}{\pi i} e^{\Gamma_0(z)} \Pi(z) \int_{|t|=1} \frac{t^\kappa v(t)\,dt}{\hat{\Pi}(t)(t-z)} \qquad (\kappa < 0). \qquad (11^*)$$

We index the points of discontinuity of the functions $G(t)$ and $g(t)$ in the order of increase of their arguments, using the common notation t_k ($k = 1,\ldots,n_0$, $n_0 \leqslant n + l$), $\delta_0 = \inf_k |t_k - t_{k+1}| > 0$ for all of them. For $f(t) \in H^\beta$, $\beta > 0$, let

$$\| f(t) \|_\beta = C(f) + H(f, \beta) \equiv \max_t |f(t)| + \sup_{t',t''} \frac{|f(t') - f(t'')|}{|t' - t''|^\beta}.$$

LEMMA 1. *The functions $u(t)$ and $v(t)$ in (13) and the function*

$$\Omega(z) = \hat{\Pi}(z)[\Pi(z)]^{-1} w(z), \qquad (14)$$

where $w(z)$ is defined by (10) or (11), satisfy the inequalities

$$\| u(t) \|_{\beta_0} \leqslant M_u, \qquad \| v(t) \|_{\beta_1} \leqslant M_v, \qquad \| \Omega(z) \|_{\beta_2} \leqslant M_\Omega \qquad (15)$$

with constants $\beta_i > 0$ and $M < \infty$ depending only on the properties of $g(t)$ and $G(t)$. If $\inf |t_k^i - t_{k+1}^i| \geqslant \delta_1 > 0$ ($k = 1,\ldots,n_0$; $i = 1,\ldots,\infty$) and $t_k^i \to t_0^k$, $\| u^i - u^0 \|_{\beta_p} \to 0$, and $\| v^i - v^0 \|_{\beta_1} \to 0$ as $i \to \infty$, then $\| \Omega^i - \Omega^0 \|_{\beta_2} \to 0$ for $|z| < 1$.

The first two inequalities in (15) are satisfied by construction, and to prove the last one we break up the integrals in (10) and (11) into a sum of integrals

$$\int_{|t|=1} = \sum_{k=1}^{n_0} \int_{\widehat{t_k t_{k+1}}}$$

taken over the arcs $\widehat{t_k t_{k+1}}$ of the circle $|t|=1$. Those integrals in which the density is unbounded only at one endpoint of $\widehat{t_k t_{k+1}}$ will be left unchanged. We break up each of the remaining integrals into a sum of two integrals with density unbounded only at one endpoint of $\widehat{t_k t_{k+1}}$, and for this we use the identity

$$\frac{1}{(t-a)^\alpha (t-b)^\beta} = \frac{1}{(b-a)} \left[\frac{(t-a)^{1-\alpha}}{(t-b)^\beta} - \frac{(t-b)^{1-\beta}}{(t-a)^\alpha} \right].$$

Thus, with (14) taken into account, the function $\Omega(z)$ can be represented in the form

$$\Omega(z) = \sum_k \Phi_k(z)(z-t_k)^{\lambda_k} \int_{\widehat{t_k t_{k+1}}} \frac{g_k(t)\, dt}{(t-t_k)^{\lambda_k}(t-z)} \equiv \sum \Omega_k(z)\Phi_k(z),$$

where $0 < \lambda_k < 1$ and the functions Φ_k and g_k satisfy a Hölder condition for $|z| \leqslant 1$ and $|t| = 1$, respectively. And Theorem 3 (Muskhelishvili) in §2 is applicable to the functions $\Omega_k(z)$ and tells us that they satisfy a Hölder condition in the closed disk $|z| \leqslant 1$. We remark that for $\kappa \geqslant 0$ the integrals in the products

$$\Phi_k(z)\Omega_k(z) = \tilde{\Phi}_k(z)(z-1)(z-t_k)^{\lambda_k} \int_{\widehat{a_k a_{k+1}}} \frac{\tilde{g}_k(t)\, dt}{(t-1)(t-t_k)^{\lambda_k}(t-z)}$$

corresponding to the arcs $\widehat{a_k a_{k+1}}$ containing the point $t=1$ must be broken up into sums of singular integrals with the help of (12). To prove the second part of the lemma we write $\Phi_k(z)$ in the form

$$\Phi_k(z) = e^{\Gamma_0(z)} \prod_{j \neq k} (z-t_j)^{\lambda_j} Q_\kappa(z),$$

where $Q_\kappa(z) = z^\kappa P_\kappa(z)(z-t_k)$, $\kappa \geqslant 0$, and $Q_\kappa(z) = \text{const}$ for $\kappa < 0$. Then for $\|u^i(t) - u^0(t)\|_\beta \to 0$ and $t_k^i \to t_k^0$ formula (9) implies the convergence $\|\Gamma_0^i(z) - \Gamma_0^0(z)\|_{\beta_0} \to 0$, and thereby, since the functions $\prod_{j \neq k}(z - t_j^i)^{\lambda_j}$ have the Hölder property with respect to the arguments t_j^i and z, also the convergence $\|\Phi_k^i - \Phi_k^0\|_\beta \to 0$ for some $0 < \beta \leq \beta_0$. We next reduce the integrals appearing in $\Omega_k^i(z)$ to integrals over the upper semicircle

$$\Phi_k^i[z^i(\zeta)]\Omega_k^i[z^i(\zeta)] = \check{\Phi}_k(\xi)(\xi - 1)^{\lambda_k}\int_{|\tau|=1,\mathrm{Im}\,\tau>0} \frac{\hat{g}_k^i(\tau)\,d\tau}{(\tau - 1)^{\lambda_k}(\tau - \xi)}$$

$$\equiv \tilde{\Phi}_k^i(\zeta)\tilde{\Omega}_k^i(\xi)$$

by means of the substitution

$$z = z^i(\zeta) = \frac{\zeta - \zeta_0^i}{1 - \zeta\bar{\zeta}_0^i}; \qquad t = z^i(\tau),$$

where the parameters ζ_0^i ($|\zeta_0^i| < 1$) are chosen from the conditions

$$t_k^i = \frac{1 - \zeta_0^i}{1 - \bar{\zeta}_0^i}, \qquad t_{k+1}^i = \frac{1 + \zeta_0^i}{1 + \bar{\zeta}_0^i}.$$

Because of the convergence $t_k^i \to t_k^0$, the mappings $z = z^i(\zeta)$ constructed obviously converge uniformly together with their derivatives in the disk $|z| \leq 1$ to the function $z = z^0(\zeta) = \zeta$. Then, according to (5),

$$\left\|\Phi_k^i[z^i(\zeta)] - \Phi_k^i(\zeta)\right\|_{\beta-\delta} \to 0, \qquad \left\|g_k^i[z^i(\tau)] - g_k^i(\tau)\right\|_{\beta-\delta} \to 0,$$

from which, by the convergence $\|\Phi_k^i(\zeta) - \Phi_k^0(\zeta)\|_\beta \to 0$ proved above, it follows that

$$\left\|\Phi_k^i[z^i(\zeta)] - \Phi_k^0(\zeta)\right\|_{\beta-\delta} \to 0.$$

In a completely analogous way, we get

$$\|\hat{g}_k^i(\tau) - \hat{g}_k^0(\tau)\|_{\beta_1} \to 0, \quad \text{and so} \quad \|\hat{\Omega}_k^i(\zeta) - \hat{\Omega}_k^0\|_{\beta_2} \to 0$$

with some $\beta_1, \beta_2 > 0$, which ends the proof of Lemma 1.

LEMMA 2. *Suppose that* $t^i = t^i(\tau) \in C^\lambda$, $0 < \lambda < 1$, *is a sequence of homeo-morphisms of the circle* $|\tau| = 1$ *onto itself, and*

$$\lim_{i \to \infty} \|t^i(\tau) - t^0(\tau)\|_\lambda = 0.$$

Then the following sequences converge to zero for some $\beta_i > 0$ ($i = 0, 1, 2$):

$$\left\|u[t^i(\tau)] - u[t^0(\tau)]\right\|_{\beta_0}, \quad \left\|v[t^i(\tau)] - v[t^0(\tau)]\right\|_{\beta_1}, \quad \|\Omega^i(z) - \Omega^0(z)\|_{\beta_2}.$$

The convergence to zero of the sequence $\|u^i - u^0\|_{\beta_0}$ ($\beta_0 = \beta_\lambda - \delta > 0$) follows directly from (5), which implies the convergence $\|\Gamma_0^i(z) - \Gamma_0^0(z)\|_{\beta_0} \to 0$, and thereby also the convergence $\|v^i - v^0\|_{\beta_1} \to 0$ ($\beta_1 > 0$). Consequently, by Lemma 1, $\|\Omega^i - \Omega^0\|_{\beta_2} \to 0$ $(\beta_2 > 0)$ in the disk $|z| \leqslant 1$, which concludes the proof of Lemma 2.

If $g(t)$ is bounded and has no points of discontinuity different from those of $G(t)$, then (15) and the assertions of Lemma 2 hold for $\Omega(z) \equiv w(z)$, where $w(z)$ is a bounded solution of the form (10) or (11) for the Hilbert problem (7).

DEFINITION 2. The solution $w = w(z)$ of class $O_m^-(t_{k_1}, \ldots, t_{k_m})$ for the Hilbert problem (7), represented by (10) or (11) (depending on the sign of the index), is said to be *stable in the class* H^{β_2}, $\beta_2 > 0$, *for* $|z| \leqslant 1$ if coefficients $G^1(t)$, $g_*^1(t)$ and $G^2(t)$, $g_*^2(t)$ of the boundary condition (7) that are close in the sense of Definition 1 lead to values of the functions

$$\Omega^i(z) = \hat{\Pi}^i(z)\left[\Pi^i(z)\right]^{-1}w^i(z), \qquad i = 1, 2,$$

for $|z| \leqslant 1$ that are close in H^{β_2}, i.e., $C(\Delta\Omega) < \varepsilon_1(\varepsilon)$ and $H(\Delta\Omega, \beta_2) < \varepsilon_1(\varepsilon)$, where $\Delta\Omega = \Omega^1(z) - \Omega^2(z)$ and $\varepsilon_1(\varepsilon) \to 0$ as $\varepsilon \to 0$.

REMARK. In the case when $w(z)$ is stable in H^{β_2} for $|z| \leqslant 1$ the formula (14) connecting $w(z)$ and $\Omega(z)$ implies that the inequalities

$$C(\Delta w) < \varepsilon_1(\varepsilon), \qquad H^{\beta_2}(\Delta w, \beta_2) < \varepsilon_1(\varepsilon)$$

hold in the disk $|z| \leqslant 1$ with deleted neighborhoods of the points t_k contained in $\hat{\Pi}(z)$ (in the domain $\overline{D}$), i.e., in D the solutions $w^1(z)$ and $w^2(z)$ of the Hilbert problem (7) are themselves stable in H^{β_2}. In particular, the bounded solutions $w^1(z)$ and $w^2(z)$ are close in the closed disk $|z| \leqslant 1$. Let $\tau = \tau(t) \in H^\lambda$, $0 < \lambda \leqslant 1$, be a homeomorphism of the circle onto itself that is close to the identity transformation $\tau = t$, i.e.,

$$C(\Delta\tau) < \varepsilon, \qquad H(\Delta\tau, \lambda) < \varepsilon \qquad (\Delta\tau = \tau(t) - t).$$

Let us consider the boundary-value problems (7) with coefficients

$$G^1(t) = G[\tau(t)], \quad g^1(t) = g[\tau(t)] \quad \text{and} \quad G^2(t) = G(t), \quad g^2(t) = g(t).$$

According to Lemma 2,

$$\|\Omega(z) - \Omega^1(z)\|_{\beta_2} < \varepsilon_1(\varepsilon), \qquad \beta_2 > 0,$$

i.e., the solution $w(z)$ of the Hilbert problem (7) is stable in H^{β_2}, $\beta_2 > 0$, with respect to transformations of the boundary condition (7) by homeomorphisms that are close to the identity. Before proceeding to an investigation of the stability of solutions of the Hilbert problem (7) in the general case, we remark that if in the boundary condition (7)

$$\|g_*(t)\|_\beta = \|g(t)\Pi|t - \rho_k|^{-\alpha_k^0}\|_\beta < \varepsilon \qquad (\beta > 0),$$

then

$$\|\Omega(z)\|_{\beta_2} = \|w(z)\hat{\Pi}(z)\left[\Pi(z)\right]^{-1}\|_{\beta_2} < \varepsilon_1(\varepsilon) \qquad (\beta_1 > 0), \qquad (16)$$

where $\varepsilon_1(\varepsilon) \to 0$ as $\varepsilon \to 0$, and $w(z)$ is represented in the form (10) or (11), depending on the sign of the index κ. Indeed, as above, let us represent $\Omega(z)$ in the form

$$\Omega(z) = \sum_k \Phi_k(z)(z - t_k)^{\lambda_k} \int_{\widehat{t_k t_{k+1}}} \frac{g_k(t)\, dt}{(t - t_k)^{\lambda_k}(t - z)} = \sum_k \Phi_k(z)\Omega_k(z),$$

where

$$g_k(t) = v(t)\tilde{\Pi}_k(t) = \frac{g_*(t)\hat{\Pi}_k(t)}{X(t)G(t)}$$

and the products $\tilde{\Pi}_k(t)$ and $\hat{\Pi}(t)$ consist of factors of the form $(t - t_i)^{\lambda_i}$ $(i \neq k, k + 1)$, $0 \leqslant \lambda_i < 1$. Then, according to Theorem 3 in §2,

$$\|\Omega_k(z)\|_{\beta_2} \leqslant \mathrm{const} \|g_*(t)\|_\lambda,$$

which proves (16). Suppose now that the pairs $G^1(t)$, $g^1_*(t)$ and $G^2(t)$, $g^2_*(t)$ of functions are close in the sense of Definition 1, and $\tau = \tau(t)$ is a homeomorphism of the form (3) that carries the points t^1_k $(k = 1,\dots,n_0)$ of discontinuity of the functions $G^1(t)$ and $g^1_*(t)$ into the points t^2_k $(k = 1,\dots,n_0)$ of discontinuity of the functions $G^2(t)$ and $g^2_*(t)$. Let $w^i(z)$ $(i = 1, 2)$ be the solution of (7) corresponding to the given $G^i(t)$ and $g^i(t)$, and $w^1_\tau(z)$ the solution of (7) with the coefficients $G^1[\tau(t)]$ and $g^1[\tau(t)]$. Then the proven stability of the solutions of (7) with respect to a transformation of the boundary condition (7) by the homeomorphism $\tau = \tau(t)$ implies that

$$\|\Omega^1_\tau - \Omega^1\|_{\beta_2} < \varepsilon_1(\varepsilon), \qquad \varepsilon_1 \to 0 \text{ as } \varepsilon \to 0, |z| \leqslant 1. \tag{17}$$

Computing termwise the right-hand and left-hand sides of the boundary conditions corresponding to the coefficients $G^2(t)$, $g^2(t)$ and $G^1[\tau(t)]$, $g^1[\tau(t)]$, which have the same points t^2_k $(k = 1,\dots,n_0)$ of discontinuity by construction, we next find that

$$\mathrm{Re}\left[G^2(t)w_\tau(t)\right] = g_\tau(t),$$

where

$$w_\tau(z) = w^2(z) - w^1_\tau(z),$$

$$g_\tau(t) = \Delta_\tau g + \mathrm{Re}\left[\delta_\tau G^1 w_\tau\right] \equiv \frac{g^*_\tau(t)}{\Pi\, |t - t_k|^{\lambda_k}}, \qquad \lambda_k < 1,$$

$$\Delta_\tau g = g^2(t) - g^1[\tau(t)], \qquad \Delta_\tau G = G^2(t) - G^1[\tau(t)].$$

By construction, $g^*_\tau(t) \in H^{\beta_1}$, $\beta_1 > 0$, and since the functions $G^1(t)$ and $g^1_*(t)$ are assumed to be close in H^{β_0} to the functions $G^2(t)$ and $g^2_*(t)$, we have

$\| g_\tau^*(t) \|_{\beta_1} < \varepsilon_2(\varepsilon)$ $(\varepsilon_2 \to 0$ as $\varepsilon \to 0)$. Then (16) gives us that

$$\| \Omega_\tau^1 - \Omega^2 \|_{C^{\beta_2}} = \| \Omega_\tau \|_{\beta_2} \leqslant \varepsilon_3(\varepsilon) \qquad (\varepsilon_3 \to 0 \text{ as } \varepsilon \to 0). \tag{18}$$

The inequalities (17) and (18) enable us to get the estimate

$$\| \Omega^1(z) - \Omega^2(z) \|_{\beta_2} \leqslant \| \Omega^1 - \Omega_\tau^1 \|_{\beta_2} + \| \Omega_\tau^1 - \Omega^2 \|_{\beta_2} \leqslant \varepsilon_1(\varepsilon) + \varepsilon_2(\varepsilon)$$
$$(\varepsilon_1 + \varepsilon_2 \to 0 \text{ as } \varepsilon \to 0).$$

Thus, it is proved that the solutions of the Hilbert problem (7) are stable in H^{β_2}, $\beta_2 > 0$, in the sense of Definition 2 under arbitrary small changes in the coefficients of the boundary condition (7). We formulate the result as a theorem.

THEOREM 1. *The solutions $w(z)$ of the Hilbert boundary-value problem (7) represented in the form (10) or (11), depending on the sign of the index, are stable in H^β, $\beta > 0$, under small changes in the coefficients of the boundary condition (7).*

REMARK. It should be pointed out that for the case of a negative index in the Hilbert problem (7) we have proved the stability of the function $w(z)$ represented by (11) and satisfying the boundary condition (7) only under the additional solvability conditions (4.28). Since it is natural in the proof of stability to assume that solutions exist, the conditions (4.28) were always assumed to be satisfied. However, we shall see later in the investigation of the Hilbert boundary-value problem for elliptic systems of equations (§§3 and 4 of Chapter V) that the proven stability of functions of the form (11) under a change in the boundary coefficients $G(t)$ and $g(t)$ is of significant interest also when the solvability conditions (4.28) are not satisfied.

§6. Singular integral equations. Generalizations of the Riemann and Hilbert boundary-value problems

1°. A linear integral equation

$$\varphi(t_0) + \int_L K(t_0, t)\varphi(t) \, dt = f(t_0), \tag{1}$$

with kernel of the form

$$K(t_0, t) = \frac{M(t_0, t)}{(t - t_0)^\alpha} \qquad (0 \leqslant \alpha < 1), \tag{2}$$

where $M(t_0, t)$ is a continuous function, is called a *Fredholm equation*. Let us consider the singular integral equation with Cauchy kernel

$$a(t_0)\varphi(t_0) + \frac{1}{\pi i} \int_L \frac{M(t_0, t)}{t - t_0} \varphi(t) \, dt = f(t_0),$$

where the integral is understood in the sense of the Cauchy principal value, $L = \Sigma_0^m L_k$ is a smooth curve, and $a(t_0)$, $M(t_0, t) \in H^\alpha$ on L. Denoting

$$M(t_0, t_0) = b(t_0), \qquad \frac{1}{\pi i} \frac{M(t_0, t) - M(t_0, t_0)}{t - t_0} = k(t_0, t),$$

we can write the equation in the form

$$a(t_0)\varphi(t_0) + \frac{b(t_0)}{\pi i} \int_L \frac{\varphi(t)\, dt}{t - t_0} + \int_L k(t_0, t)\varphi(t)\, dt = f(t_0), \qquad (3)$$

where the kernel $k(t_0, t)$ can obviously be represented in the form (2).

Assume that

$$a^2 - b^2 \neq 0, \qquad t \in L. \qquad (4)$$

For $k(t_0, t) \equiv 0$ the corresponding equation (3)

$$a(t_0)\varphi(t_0) + \frac{b(t_0)}{\pi i} \int_L \frac{\varphi(t)\, dt}{t - t_0} = f(t_0) \qquad (5)$$

is called the characteristic equation.

The equation

$$a(t_0)\psi(t_0) - \frac{1}{\pi i} \int_L \frac{b(t)\psi(t)}{t - t_0}\, dt + \int_L k(t, t_0)\psi(t)\, dt = h(t_0) \qquad (6)$$

obtained from the homogeneous equation (3) by interchanging the variables in the kernel is called the transposed equation (3), while the corresponding equation

$$a(t_0)\psi(t_0) - \frac{1}{\pi i} \int_L \frac{b(t)\psi(t)}{t - t_0}\, dt = h(t_0) \qquad (7)$$

is called the transposed characteristic equation (5). The characteristic equation (5) can be solved in closed form. Indeed, let us consider the function

$$\Phi(z) = \frac{1}{2\pi i} \int_L \frac{\varphi(t)\, dt}{t - z}.$$

According to the Sokhotskiĭ-Plemelj formulas,

$$\varphi(t_0) = \Phi^+(t_0) - \Phi^-(t_0), \qquad \frac{1}{\pi i} \int_L \frac{\varphi(t)\, dt}{t - t_0} = \Phi^+(t_0) + \Phi^-(t_0).$$

Substituting these expressions into (5), we come to the Riemann problem

$$\Phi^+(t) = G(t)\Phi^-(t) + g(t),$$

where $G(t) = (a(t) - b(t))/(a(t) + b(t))$ (by condition (4), $0 < |G(t)| < \infty$), and $g(t) = f(t)/(a(t) + b(t))$. The index

$$\kappa = \mathrm{ind}\, \frac{a(t) - b(t)}{a(t) + b(t)}$$

of this Riemann problem is called the index of the characteristic equation (5). Well-known formulas yield a solution $\Phi(z)$ of the Riemann problem, and its limit values on the contour L can be represented in the form

$$\Phi^\pm(t_0) = X^\pm(t_0) \left\{ \pm \frac{1}{2} \frac{g(t)}{X^+(t_0)} + \Psi(t_0) - \frac{1}{2} P_{\kappa-1}(t) \right\}$$

for $\kappa \geqslant 0$, where

$$\Psi(t_0) = \frac{1}{2\pi i} \int_L \frac{g(t)\, dt}{X^+(t)(t - t_0)}$$

and $P_{\kappa-1}(t_0)$ is a polynomial of degree $\kappa - 1$. Forming the expression $\varphi(t_0) = \Phi^+(t_0) - \Phi^-(t_0)$, after elementary transformations we get

$$\varphi(t_0) = a(t_0)f(t_0) - \frac{G(t_0)Z(t_0)}{\pi i} \int_L \frac{f(t)\, dt}{Z(t)(t - t_0)} + b(t_0)Z(t_0)P_{\kappa-1}(t_0)$$

$$= Rf, \tag{8}$$

where

$$Z(t_0) = [a(t_0) + b(t_0)X^+(t_0)] = [a(t_0) - b(t_0)]X^-(t_0).$$

For $\kappa < 0$ it is necessary to set $P_{\kappa-1}(t_0) \equiv 0$ in (8) and to require the solvability conditions of the Riemann problem:

$$\int_L \frac{f(t)t^{k-1}\, dt}{Z(t)} = 0 \quad (k = 1, \ldots, |\kappa|). \tag{9}$$

The homogeneous equation (5) has only the trivial solution for $\kappa \leqslant 0$, and κ linearly independent solutions for $\kappa > 0$. The transposed equation (7) can be solved similarly. Assuming that $b(t_0) \neq 0$, we set $\omega(t) = b(t)\psi(t)$; then (7) takes the form

$$a(t_0)\omega(t_0) - \frac{b(t_0)}{\pi i} \int_L \frac{\omega(t)\, dt}{t - t_0} = b(t_0)h(t_0), \tag{7*}$$

and, moreover, by (7*), $\omega(t_0) = 0$ at the points where $b(t_0) = 0$, which also justifies the substitution when $b(t_0)$ vanishes. But (7) is a characteristic equation of the form (5) with index

$$\kappa' = \operatorname{ind} \frac{a(t) + b(t)}{a(t) - b(t)} = -\operatorname{ind} \frac{a(t) - b(t)}{a(t) + b(t)} = -\kappa. \tag{10}$$

Consequently, if the index κ of the characteristic equation (5) is nonpositive, then $\kappa' = -\kappa \geqslant 0$, and the homogeneous equation (7) has $|\kappa|$ linearly independent solutions, while for $\kappa > 0$ and $\kappa' = -\kappa < 0$ equation (7) does not have nontrivial solutions. Thus, if n is the number of solutions of the homogeneous characteristic equation (5) and n' is the number of solutions of the transposed equation (7) then

$$n - n' = \kappa. \tag{11}$$

This formula holds also for the general homogeneous singular equation (3) and determines the main difference between singular equations and Fredholm equations, for which $n - n' = 0$. The general singular equation (3) can be reduced in various ways to an equivalent Fredholm equation (regularized). We dwell very briefly only on the regularization method of I. N. Vekua (see [97]). Let us rewrite (3) in the form

$$a(t_0) + \frac{b(t_0)}{\pi i} \int_L \frac{\varphi(t)\, dt}{t - t_0} = F(t_0), \tag{12}$$

where

$$F(t_0) = -\int_L k(t_0, t)\varphi(t)\, dt + f(t_0).$$

Substituting the value $F(t_0)$ into the formula (8) for the solution of the characteristic equation (12) and taking into account that the interchange of the singular integral and the improper integral

$$\int_L \frac{dt}{t - t_0} \int_L \frac{N_0(t, \tau)\varphi(\tau)\, d\tau}{(\tau - t)^\alpha} = \int_L N_1(\tau, t_0)\varphi(\tau)\, d\tau$$

leads to an improper integral, i.e., $N_1(\tau, t_0)$ has the form (2), we arrive at the Fredholm equation

$$\varphi(t_0) + \int_L N(t_0, t)\varphi(t)\, dt = g(t_0). \tag{13}$$

The kernel $N(t_0, t)$ of equation (13) is a known function, but $g(t_0)$ depends on κ arbitrary constants for $\kappa > 0$, while $|\kappa|$ solvability conditions must be satisfied in addition for $\kappa \leq 0$, and together with them (13) is equivalent to (3).

The case of piecewise Hölder $a(t_0)$ and $b(t_0)$ in (3) is handled in a completely analogous way. The reader can find more details on the theory of singular equations in the monographs of Muskhelishvili [97] and Gakhov [32].

2°. We formulate some boundary-value problems that generalize the Riemann and Hilbert problems.

I. *The Riemann problem with a shift (the Haseman problem).* Suppose that $\alpha = \alpha(t)$ is a function transforming the curve L onto itself or onto a part of L. It is required to construct a piecewise holomorphic function $\Phi(z)$ on the z-plane with jump curve L on which the following relation holds:

$$\Phi^+[\alpha(t)] = G(t)\Phi^-(t) + g(t). \tag{14}$$

We shall study this problem in Chapter VIII in the class of elliptic systems of equations, which is more natural for it; we will see, in particular, that it is solvable also for analytic functions.

II. *The Carleman problem.* This problem differs from the preceding one only in that the conjugation operation appears in the boundary condition

$$\Phi^+[\alpha(t)] = G(t)\overline{\Phi^{\pm}(t)} + g(t), \tag{15}$$

a circumstance which, however, has a considerable effect on the methods for its investigation (see Chapter VIII).

III. *The Riemann and Hilbert problems with derivatives.*

$$\sum_k \left[a_k(t)\frac{d^k\Phi^+(t)}{dt^k} - b_k(t)\frac{d^k\Phi^-(t)}{dt^k} \right] = g(t, \Phi^{\pm}), \tag{16}$$

$$\mathrm{Re}\left\{ \sum_k [a_k(t) + ib_k(t)]\frac{d^k\Phi^+(t)}{dt^k} \right\} = g(t, \Phi^+), \tag{17}$$

where $g(t, \Phi^{\pm})$ and $g(t, \Phi^+)$ are integral operators (see [70]).

IV. *The Riemann and Hilbert problems for vector-valued functions.* Suppose that $\Phi^{\pm}(z) = \{\Phi_1^{\pm}, \ldots, \Phi_n^{\pm}\}$ and $g(t) = \{g_1, \ldots, g_n\}$ are vector-valued functions, the components $\Phi_k^{\pm}(z)$ are piecewise holomorphic functions with jump curve L, and $G(t) = \|G_{ij}\|$ is an $n \times n$ square matrix. The Riemann and Hilbert boundary-value problems for the vector-valued function $\Phi(z) = \{\Phi_1, \ldots, \Phi_n\}$ are written in the same form as for a single function. These problems will be studied in Chapter VIII for elliptic systems of equations.

CHAPTER II

SINGULAR OPERATORS IN SPACES OF SUMMABLE FUNCTIONS. APPLICATION OF BOUNDARY-VALUE PROBLEMS AND TO THE STUDY OF BOUNDARY PROPERTIES OF ANALYTIC FUNCTIONS

§1. Banach spaces, fixed-point principles

We shall restrict ourselves to complete normed complex linear spaces (B-spaces) in what follows.

A complex linear space is defined to be a set E of elements satisfying the following axioms.

1. E is an abelian group with respect to a group operation of addition, i.e., for any elements f, $g \in E$ an element $f + g \in E$ is defined, and the addition operation satisfies the conditions: a) $f + g = g + f$; b) $f + (g + p) = (f + g) + p$; c) there is an element 0 (the zero) such that $f + 0 = f$ for any $f \in E$; and d) for each element $f \in E$ there exists an element $-f$ such that $f + (-f) = 0$.

2. For each $f \in E$ and any complex number λ an element $\lambda f \in E$ is defined, and: a) $\lambda(\mu f) = (\lambda \mu)f$; b) $\lambda(f + g) = \lambda f + \lambda g$, $(\lambda + \mu)f = \lambda f + \mu f$; and c) $1 \cdot f = f$.

A complex linear space E is said to be normed if each element $f \in E$ is assigned a real number $\| f \|$ (the norm of f) satisfying the conditions a) $\| f \| \geqslant 0$, and $\| f \| = 0$ only for $f = 0$; b) $\| f \| + \| g \| \geqslant \| f + g \|$; and c) $\| \lambda f \| = | \lambda | \, \| f \|$.

In a normed linear space we can introduce the concept of norm convergence by saying that a sequence converges in the norm to an element $f_0 \in E$ if $\| f_n - f_0 \| \to 0$ as $n \to \infty$. A sequence $f_n \in E$ is called a Cauchy sequence if for any $\varepsilon > 0$ there is an $N(\varepsilon)$ such that $\| f_n - f_m \| < \varepsilon$ when $n, m \geqslant N(\varepsilon)$. A normed complex linear space E is said to be complete if each Cauchy sequence converges in the norm to an element of this space.

Complete normed complex linear spaces are commonly called Banach spaces (B-spaces). Suppose that $D \subset E^2 \equiv E$ is a domain in the plane of $z = x + iy$, and $f(z) \equiv f(x, y)$ is a complex-valued function defined in the closed domain $\overline{D}$. We consider examples of B-spaces of functions on the plane.

The space $C(\overline{D})$ of continuous functions in $\overline{D}$, with the norm

$$\| f \|_C = \max_{z \in \overline{D}} |f(z)|.$$

The space $C^m(\overline{D})$ of functions that are continuous in the closed domain $\overline{D}$ together with their partial derivatives of order up through m, with the norm

$$\| f \|_{C^m} = \sum_{k=0}^{m} \sum_{l=0}^{k} \left\| \frac{\partial^k f}{\partial x^{k-l} \partial y^l} \right\|_C, \qquad m \geqslant 0, \ C^0 \equiv C.$$

The space $H^\alpha(\overline{D})$ of functions that satisfy in $\overline{D}$ a Hölder condition with exponent $0 < \alpha \leqslant 1$, with the norm

$$\| f \|_{H^\alpha} = \sup_{z_1, z_2 \in \overline{D}} \frac{|f(z_1) - f(z_2)|}{|z_1 - z_2|^\alpha}.$$

The space $C^\alpha(\overline{D})$ of continuous functions satisfying in D a Hölder condition with exponent $0 < \alpha \leqslant 1$, with norm $\| f \|_{C^\alpha} = \| f \|_{C} + \| f \|_{H^\alpha}$. In the case of a bounded domain D it is obvious that $C^\alpha(\overline{D}) \equiv H^\alpha(\overline{D})$, but when D is unbounded we have only $C^\alpha(\overline{D}) \subset H^\alpha(\overline{D})$; for example, the function $f(z) = |z|^\alpha$ belongs to H^α, but not to C^α.

The space $C^{m+\alpha}(\overline{D})$ of functions having continuous partial derivatives in $\overline{D}$ of order up through m, with all the mth-order partial derivatives satisfying a Hölder condition with exponent $\alpha > 0$, with the norm

$$\| f \|_{C^{m+\alpha}} = \| f \|_{C^m} + \sum_{k=0}^{m} \left\| \frac{\partial^m f}{\partial x^{m-k} \partial y^k} \right\|_{H^\alpha}.$$

The space $L_p(\overline{D})$, $p \geqslant 1$, of p-integrable functions in the domain D, with the norm

$$\| f \|_{L_p} = \left\{ \int_D \int |f|^p \, dx \, dy \right\}^{1/p}.$$

The space $L_p^\alpha(\overline{D})$ with the norm

$$\| f \|_{L_p^\alpha} = \| f \|_{L_p} + \sup \frac{\| f(z + \Delta z) - f(z) \|_{L_p}}{|\Delta z|^\alpha}, \qquad 0 < \alpha < 1.$$

We mention some properties of the space $L_o(\overline{D})$, $p \geqslant 1$. The functions of the class $L_p(\overline{D})$, $p \geqslant 1$, are continuous in the $L_p(\overline{D})$-norm, i.e., for every $\varepsilon > 0$

there is a $\delta(\varepsilon) > 0$ such that

$$\| f(z + \Delta z) - f(z) \|_{L_p(\bar{D})} < \varepsilon \quad \text{if } |\Delta z| < \delta(\varepsilon).$$

HÖLDER'S INEQUALITY. If $f_k(z) \in L_{p_k}(\bar{D})$ $(k = 1, \ldots, n)$ and $1/p = 1/p_1 + \cdots + 1/p_n$, then

$$\{ f_1 \cdots f_n \} \in L_p(D) \quad \text{and} \quad \| f_1 \cdots f_n \|_{L_p} \leq \prod_{k=1}^{n} \| f_k \|_{L_{p_k}}. \tag{1}$$

MINKOWSKI'S INEQUALITY. Let $f_k \in L_p(\bar{D})$, $p > 1$. Then

$$\left\| \sum_k f_k \right\|_{L_p(\bar{D})} \leq \sum_k \| f_k \|_{L_p(\bar{D})}. \tag{2}$$

If $D = \Sigma_1^n D_k$ and $f \in L_p(\bar{D})$, $p > 1$, then

$$\| f \|_{L_p(\bar{D})} \leq \sum_{k=1}^{n} \| f \|_{L_p(\bar{D}_k)}. \tag{3}$$

THE CONVERSE HÖLDER AND MINKOWSKI INEQUALITIES.

$$\| f \cdot g \|_{L_1} \geq \left\| \frac{1}{f} \right\|_{L_p}^{-1} \cdot \| g \|_{L_{p/p+1}}, \qquad p > 0, \tag{4}$$

$$\| |f| + |g| \|_{L_p} \geq \| f \|_{L_p} + \| g \|_{L_p}, \qquad 0 < p < 1. \tag{5}$$

We consider the set of vector-valued functions $f = (f_1, \ldots, f_n)$ whose components each belong to some particular B-space, $f_k \in B_k$. This set $B = (B_1 \times \cdots \times B_n)$ becomes a B-space if the norm of an element in it is defined, for example, by $\| f \|_B = \Sigma_1^n \| f_k \|_{B_k}$.

A set $\mathfrak{M}$ in a Banach space B is said to be relatively compact if every infinite sequence of elements of it contains a subsequence convergent in the norm of B.

THEOREM 1 (ARZELÀ-ASCOLI). *A subset $\mathfrak{M} \subset C(\bar{D})$ is relatively compact if and only if it is uniformly bounded and equicontinuous with respect to the norm of C, i.e.:*

1) $\| f \|_{C(\bar{D})} \leq M < \infty$ *for any function $f(z) \in \mathfrak{M}$; and*

2) *for each $\varepsilon > 0$ there is a $\delta(\varepsilon) > 0$ such that for $|\Delta z| < \delta(\varepsilon)$ any $f(z) \in \mathfrak{M}$ satisfies*

$$\| f(z + \Delta z) - f(z) \|_C < \varepsilon.$$

THEOREM 2 (M. RIESZ). *A subset $\mathfrak{M} \subset L_p(\bar{D})$, $p > 1$, is relatively compact if and only if it is uniformly bounded and equicontinuous with respect to the L_p-norm, i.e.:*

1) $\| f \|_{L_p} \leq M < \infty$ *for any $f(z) \in \mathfrak{M}$; and*

2) *for each $\varepsilon > 0$ there is a $\delta(\varepsilon) > 0$ such that for $|\Delta z| < \delta(\varepsilon)$ and $f(z) \in \mathfrak{M}$ satisfies*

$$\| f(z + \Delta z) - f(z) \|_{L_p} < \varepsilon.$$

EXAMPLES of compact sets in B-spaces: Any bounded set $\mathfrak{M} \subset C^\alpha(\overline{D})$, $\alpha > 0$, is relatively compact in $C(\overline{D})$; any bounded set $\mathfrak{M} \subset C^\beta(\overline{D})$, $1 \geqslant \beta > \alpha > 0$, is relatively compact in $C^\alpha(\overline{D})$; any bounded set $\mathfrak{M} \subset L_p^\alpha(\overline{D})$, $0 < \alpha \leqslant 1$, is relatively compact in $L_p(\overline{D})$; and any bounded set $\mathfrak{M} \subset C^{m+\beta}(\overline{D})$, $1 \geqslant \beta > \alpha > 0$, is relatively compact in $C^{m+\alpha}(\overline{D})$, $m > 0$, $0 \leqslant \alpha < 1$.

A subset $\mathfrak{M}$ of a Banach space B is said to be dense in B if for each $f \in B$ and $\varepsilon > 0$ there is an $f_\varepsilon \in \mathfrak{M}$ such that $\| f - f_\varepsilon \|_B < \varepsilon$.

An example of a dense subset of the spaces $L_p(\overline{D})$, $p > 1$, is the set $\mathring{C}^m(\overline{D})$, $m \geqslant 0$, of functions $f(z) \in C^m(\overline{D})$ that vanish for $z \notin D_f$, $\overline{D}_f \subset D$ a compact set (functions with compact support).

Let us consider the set of transformations (operators) F from a Banach space B_1 to a Banach space B_2. We denote it by $(B_1 \to B_2)$. A transformation $F \in (B_1 \to B_2)$ is said to be bounded if it carries each bounded set $\mathfrak{M} \subset B_1$ into a bounded set $F(\mathfrak{M}) \subset B_2$. A transformation $F \in (B_1 \to B_2)$ is continuous if it carries each norm-convergent sequence in B_1 into a norm-convergent sequence in B_2. A transformation $F \in (B_1 \to B_2)$ is said to be linear (additive and homogeneous) if $F(\lambda f_1 + \mu f_2) = \lambda F f_1 + \mu F f_2$ for any $f_1, f_2 \in B_1$ and any complex numbers λ and μ. In the case of linear transformations $F \in (B_1 \to B_2)$ boundedness implies continuity, and conversely. From this assertion it follows, in particular, that for a continuous linear transformation $F \in (B_1 \to B_2)$ there is a number $k \geqslant 0$ such that $\| F f \|_{B_2} \leqslant k \| f \|_{B_1}$ for all $f \in B_1$ for which F is defined. The smallest number k for which the last inequality holds is called the norm of the transformation F and denoted by $\| F \|_{(B_1 \to B_2)}$. A transformation $F \in (B_1 \to B_2)$ is said to be compact if it maps each bounded set $\mathfrak{M} \subset B_1$ into a relatively compact set $\mathfrak{N} \subset B_2$.

A transformation $F \in (B_1 \to B_2)$ is said to be completely continuous if it is continuous and compact.

If for a transformation $F \in (B_1 \to B_2)$ there exists a transformation $F^{-1} \in (B_2 \to B_1)$ such that $F^{-1}(Ff) = f$ for any $f \in B_1$ and $F(F^{-1}g) = g$ for any $g \in B_2$, then F and F^{-1} are said to be mutually inverse transformations.

We state a theorem of F. Riesz on extension of bounded linear operators (see [31] or [136]).

THEOREM 3 (F. RIESZ). *Let $F \in (B_1 \to B_2)$ be a bounded linear operator defined on a dense subset $\mathfrak{M}$ of B_1. Then F admits an extension to the whole space with preservation of norm, and such an extension is unique.*

We give a theorem of M.Riesz (see [31]) on the convexity of the norm of a bounded linear operator in the spaces $L_p(\overline{D})$, $p > 1$.

THEOREM 4 (M. RIESZ). *Suppose that* $F \in (L_p \to L_p)$ *is a bounded linear operator in the spaces* $L_p(\overline{D})$, $p > 1$. *Then the norm* Λ_p *of this operator is a continuous function of* p, *and* $\ln \Lambda_p = \lambda(p)$ *is a convex function.*

Let us consider a transformation $F \in (B \to B)$ carrying a Banach space B into itself. A point f_* is called a fixed point of F if $f_* = Ff_*$; i.e., the fixed points of a transformation are the solutions of the equation $f = Ff$.

We state the main theorems for existence of fixed points of transformations of a Banach space into itself (fixed-point principles).

THEOREM 5 (BANACH). *Let* $F \in (B \to B)$ *be a contraction, i.e.,* $\| Ff - Fg \| \leqslant k \| f - g \|$, *where* $0 < k < 1$ *and* $f, g \in B$. *Then the transformation* F *has a unique fixed point, which can be constructed as the limit of the sequence* $\{ f_n \}$, $f_{n+1} = Ff_n$ ($n = 0, 1, \ldots, f_0$ *an arbitrary element of* B).

A set $\mathfrak{M} \subset B$ is said to be convex if it contains along with any two of its elements f_1 and f_2 also the elements $f_t = (1 - t)f_1 + tf_2 \in \mathfrak{M}$ for all $t \in [0, 1]$. It is easy to see that the ball $\| f - f_0 \| < r$, $f, f_0 \in B$, is a convex set in a B-space.

THEOREM 6 (SCHAUDER). *If a completely continuous transformation* $F \in (B \to B)$ *carries a closed convex bounded set* $\mathfrak{M}$ *into itself, then this transformation has at least one fixed point* f_* ($f_* \equiv Ff_*$).

The following theorem contains a more general fixed-point principle.

THEOREM 7 (LERAY-SCHAUDER). *Let* F_λ, $\lambda \in [0, 1]$, *be a one-parameter family of transformations of a Banach space into itself and* $\mathfrak{M} \subset B$ *a bounded connected open set. Then the equation* $f = F_\lambda f$ *has at least one solution in* $\mathfrak{M}$ *for all* $\lambda \in [0, 1]$, *provided that the following conditions hold:*

1) *The transformations* F_λ, $\lambda \in [0, 1]$, *are completely continuous on* $\overline{\mathfrak{M}}$.

2) *The* F_λ *are equicontinuous with respect to* λ, *i.e., for any* $\varepsilon > 0$ *there exists a* $\delta(\varepsilon) > 0$ *such that* $\| F_{\lambda_1} f - F_{\lambda_2} f \| < \varepsilon$ *for all* $f \in \overline{\mathfrak{M}}$ *when* $| \lambda_1 - \lambda_2 | < \delta(\varepsilon)$.

3) *The transformations* F_λ *do not have fixed points on the boundary* $\Gamma = \partial \mathfrak{M}$, *i.e.,* $f \neq F_\lambda f$ *for* $f \in \Gamma$ *and* $\lambda \in [0, 1]$.

4) *For* $\lambda = 0$ *the equation* $f = F_0 f$ *has a unique solution* $f_0 \in \mathfrak{M}$, *and in some neighborhood of* f_0 *the transformation* $G_0 f = f - F_0 f$ *is locally one-to-one.*

THEOREM 8 (on small perturbations). *Suppose that* $(F_\lambda + \varepsilon A)$ *is a two-parameter mapping of a Banach space* B *into itself* ($\lambda \in [0, 1]$, $\varepsilon > 0$) *and* F_λ *satisfies the conditions of the Leray-Schauder theorem on some closed bounded set* $\mathfrak{M}$.

Assume one of the following conditions holds:

(i) *A is a completely continuous mapping of B into itself.*

(ii) *A satisfies a Lipschitz condition on each bounded set* $\mathfrak{N} \subset B$, *i.e.,*
$\| Af_1 - Af_2 \| \leqslant q \| f_1 - f_2 \|, f_1, f_2 \subset \mathfrak{N}, q = q(\mathfrak{N})$.

Then there exists an $\varepsilon_0 > 0$ *such that for* $\varepsilon < \varepsilon_0$ *the transformation* $(F_\lambda + \varepsilon A)$ *has at least one fixed point in* $\mathfrak{M}$ *for all* $\lambda \in [0, 1]$.

The first assertion of the theorem is an immediate consequence of the Leray-Schauder fixed-point principle. Indeed, condition 3) of this principle implies the existence of an $\alpha > 0$ such that $\| \varphi - F_\lambda(\varphi) \| \geqslant \alpha$ for all $\lambda \in [0, 1]$ and $\varphi \in \partial \mathfrak{M}$. Then, choosing $\varepsilon_0 < \frac{1}{2} \alpha \max \| A\varphi \|$ for $\varphi \in \partial \mathfrak{M}$, we arrive at the transformation

$$\Phi_\lambda = F_\lambda + \lambda \varepsilon A / \lambda_1, \qquad \lambda \in [0, 1],$$

and the Leray-Schauder theorem is applicable to it for $\varepsilon < \varepsilon_0$ and each fixed $\lambda_1 \in [0, 1]$. A proof of the second assertion can be found, for example, in Krasnosel'skiĭ's book [57] (Chapter III, §2.2).

THEOREM 9 (on weakly coupled systems). *Consider the system of equations*

$$\varphi = \Phi_\lambda(\varphi) + \varepsilon P(\varphi, \psi), \qquad \psi = F_\lambda(\psi) + \varepsilon Q(\varphi, \psi),$$

where $\lambda \in [0, 1]$, *and* Φ_λ *and* F_λ *are completely continuous transformations mapping the respective Banach spaces* B_1 *and* B_2 *into themselves.*

Suppose that Φ_λ *and* F_λ *satisfy the conditions of the Leray-Schauder theorem and that one of the following conditions holds:*

(i) *P and Q are completely continuous transformations of* $(B_1 \times B_2)$ *into* B_1 *and* B_2, *respectively.*

(ii) *P and Q satisfy a Lipschitz condition on each bounded subset of* $B_1 \times B_2$.

Then there exists an $\varepsilon_0 > 0$ *such that for* $0 < \varepsilon \leqslant \varepsilon_0$ *the original system of equations has at least one solution.*

This theorem is a simple consequence of the previous one, since it is easily shown that the transformation $(B_1 \times B_2) \to (B_1 \times B_2)$,

$$(\varphi, \psi) = (\Phi_{\lambda\varphi}, F_{\lambda\psi}) \equiv \Lambda_\lambda(\varphi, \psi),$$

satisfies the conditions of the Leray-Schauder theorem in $B_1 \times B_2$.

For completely continuous linear transformations $F \in (B_1 \to B_2)$ there are analogues of the Fredholm theorems which generalize the well-known theorems on linear integral equations. We state one of them.

THEOREM 10 (F.RIESZ). *The equation* $f - Ff = g$, *where* F *is a completely continuous linear transformation of a Banach space* B *into itself, is solvable for any* $g \in B$ *if and only if the corresponding homogeneous equation* $f - Ff = 0$ *has* *only the trivial solution $f \equiv 0$. In this case the solution of the original equation is uniquely determined, and the transformation $I - F$ has a bounded inverse.*

§2. Singular integrals in spaces of summable functions

1°. Cauchy-Lebesgue type integral and Poisson-Lebesgue integral. A Cauchy-Lebesgue type integral is defined to be an expression

$$\Phi(z) = \frac{1}{2\pi i} \int_L \frac{\varphi(t)\, dt}{t - z}, \qquad z \in D^{\pm}(L),$$

where L is a simple closed contour and $\varphi(t)$ is an integrable function on L. As always, we define the integral in the sense of the Cauchy principal value to be

$$\Phi(t_0) = \frac{1}{2\pi i} \int_L \frac{\varphi(t)\, dt}{t - t_0} = \frac{1}{2\pi i} \lim_{\varepsilon \to 0} \int_{L_\varepsilon} \frac{\varphi(t)\, dt}{t - t_0},$$

where $L_\varepsilon = L - l$, with the endpoints of the arc $l \subset L$ located at the points $a = t(s_0 - \varepsilon)$ and $b = t(s_0 + \varepsilon)$. Let us consider the expression

$$\Psi(\varepsilon, t_0) = \frac{1}{2\pi i} \left\{ \int_L \frac{\varphi(t)\, dt}{t - z} - \int_{L_\varepsilon} \frac{\varphi(t)\, dt}{t - t_0} \right\}, \tag{1}$$

where z is an arbitrary point of $D^+(L)$ or $D^-(L)$ lying on a ray $\widehat{t_0 z}$ forming a nonobtuse angle γ with the tangent to L at the point t_0, and, moreover,

$$|z - t_0| = \varepsilon \quad \text{and} \quad \gamma \geqslant \gamma_0 > 0. \tag{2}$$

We prove that if $\varphi(t)$ is continuous at t_0, then

$$\lim_{\varepsilon \to 0} \Psi(\varepsilon, t_0) = \pm \tfrac{1}{2} \varphi(t_0), \tag{3}$$

where the sign $+$ corresponds to z approaching t from the domain $D^+(L)$. We suppose first that $\varphi(t_0) = 0$ and establish the equality

$$\lim_{\varepsilon \to 0} \Psi(\varepsilon, t_0) = 0.$$

Let us represent $\Psi(\varepsilon, t_0)$ in the form

$$\Psi(\varepsilon, t_0) = \frac{1}{2\pi i} \int_l \frac{\varphi(t)\, dt}{t - z} + \frac{z - t_0}{2\pi i} \int_{L_\varepsilon} \frac{\varphi(t)\, dt}{(t - t_0)(t - z)} \equiv \Psi_1 + \Psi_2$$

($l = (ab)$, $a = t(s_0 - \varepsilon)$, $b = t(s_0 + \varepsilon)$). With the help of the law of cosines for the triangle with vertices at the point t_0, z, and $t(s)$, $s \in [s_0 - \varepsilon, s_0 + \varepsilon]$, and with angle $\alpha > \tfrac{1}{2}\gamma_0 > 0$ at the apex we find that

$$|t - z|^2 = |t - t_0|^2 + |t_0 - z|^2 - 2|t_0 - z||t - t_0| \cos \alpha$$

$$\geqslant \sin^2 \frac{\alpha}{2} \left(|t - t_0| + |t_0 - z| \right)^2 \geqslant k_1^2 \left(|s - s_0| + k_0^{-1}\varepsilon \right)^2,$$

$$k_1 = k_0 \sin \frac{\gamma_0}{4},$$

where we have used the inequality $|t - t_0| \geq k_0 |s - s_0|$, $k_0 > 0$ ((1.3) in Chapter I). Consequently,

$$|\Psi_1(\varepsilon, t_0)| \leq \frac{1}{2\pi k_1} \int_{s_0 - \varepsilon}^{s_0 + \varepsilon} \frac{|\varphi(t)|\, ds}{|s - s_0| + \varepsilon k_0^{-1}} \leq \frac{1}{\pi k_1} \max_{t \in l} |\varphi(t)| \ln\left(\frac{1 + k_0^{-1}}{k_0^{-1}}\right),$$

and, by the condition $\varphi(t_0) = 0$,

$$\lim_{\varepsilon \to 0} \Psi_1(\varepsilon, t_0) = 0.$$

For $\Psi_2(\varepsilon, t_0)$ we have

$$|\Psi_2(\varepsilon, t_0)| \leq \frac{\varepsilon}{2\pi} \int_{L_\varepsilon} \frac{|\varphi(t)|\, ds}{|t - t_0||t - z|}$$

$$= \frac{\varepsilon}{2\pi} \left(\int_{s_0 + \varepsilon}^{s_0 + \delta} + \int_{s_0 - \delta}^{s_0 - \varepsilon} + \int_{L_\delta} \right) = I_1 + I_2 + I_\delta.$$

If $\delta = \delta(\varepsilon) > \varepsilon$ is sufficiently small, then for I_1 (I_2) we get

$$I_1(\varepsilon, t_0) \leq \frac{\varepsilon}{k_0 2\pi k_1} \int_{s_0 + \varepsilon}^{s_0 + \delta} \frac{\max |\varphi(t)|\, ds}{(s - s_0)(s - s_0 + k_0^{-1}\varepsilon)}$$

$$\leq \frac{\varepsilon}{2\pi k_0 k_1} \max |\varphi(t)| \left(\frac{1}{\varepsilon} - \frac{1}{\delta} \right).$$

We suppose that $\varepsilon/\delta(\varepsilon) \to 0$ as $\varepsilon \to 0$, and, taking account of the fact that $\max |\varphi[t(s)]|$ over $s \in [s_0 + \varepsilon, s_0 + \delta]$ goes to zero as $\varepsilon \to 0$, we find that

$$\lim_{\varepsilon \to 0} I_1(\varepsilon, t_0) = 0 \qquad \left(\lim_{\varepsilon \to 0} I_2(\varepsilon, t_0) = 0 \right).$$

On the other hand, for sufficiently small $\delta(\varepsilon) > \varepsilon > 0$,

$$|z - t(s)| \geq |z - t(s_0 + \delta)| \geq k_1(\delta + k_0^{-1}\varepsilon) \geq k_1 \delta$$

and

$$|t(s) - t(s_0)| \geq |t(s_0 + \delta) - t(s_0)| \geq k_0 \delta, \qquad t = t(s) \in L_\delta;$$

consequently,

$$I_\delta \leq \frac{\varepsilon}{2\pi k_0 k_1 \delta^2} \int_{L_\delta} |\varphi(t)|\, ds \to 0$$

as $\varepsilon \to 0$ if it is assumed that $\lim \varepsilon/\delta^2(\varepsilon) = 0$.

Thus (3) is proved. If now $\varphi(t_0) \neq 0$ at the Lebesgue point t_0 of $\varphi(t)$, then

$$\Psi(\varepsilon, t) = \frac{\varphi(t_0)}{2\pi i} \left[\int_L \frac{dt}{t - z} - \int_{L_\varepsilon} \frac{dt}{t - t_0} \right]$$

$$+ \frac{1}{2\pi i} \left[\int_L \frac{f(t)\, dt}{t - z} - \int_{L_\varepsilon} \frac{f(t)\, dt}{t - t_0} \right] = A_\varepsilon + F,$$

where $f(t) = \varphi(t) - \varphi(t_0)$. Since $f(t_0) = 0$ by construction, what was proved above implies that

$$\lim_{\varepsilon \to 0} F(\varepsilon, t_0) = 0,$$

and, by the properties of the singular integrals in the expression for A_ε, we have

$$\lim_{\varepsilon \to 0} A_\varepsilon(t_0) = \varphi(t_0) \lim_{\varepsilon \to 0} \left\{ \frac{1}{2\pi i} \int_L \frac{dt}{t - z} - \frac{1}{2\pi i} \int_{L_\varepsilon} \frac{dt}{t - t_0} \right\} = \pm \frac{1}{2} \varphi(t_0),$$

where the plus sign corresponds to z approaching t_0 from $D^+(L)$; and this finishes the proof of (3). If we consider the difference

$$\Psi_0(\varepsilon, t_0) = \Psi^+(\varepsilon, t_0) - \Psi^-(\varepsilon, t_0),$$

where $\Psi^\pm(\varepsilon, t_0)$ is defined by (1), in which $z = z_1 \in D^+(L)$ and $z = z_2 \in D^-(L)$, respectively, and $|z_1 - t_0| = |z_2 - t_0| = \varepsilon$, then (3) implies that

$$\lim_{\varepsilon \to 0} \Psi_0(\varepsilon, t_0) = \varphi(t_0).$$

Thus, we have proved the following theorem of Privalov [109].

THEOREM 1 (PRIVALOV). *Let $\Phi(z)$ be a Cauchy-Lebesgue type integral with summable density $\varphi(t)$,*

$$\Phi^\pm(z) = \frac{1}{2\pi i} \int_L \frac{\varphi(t)\, dt}{t - z}, \qquad z \in D^\pm(L).$$

Then the following equality holds almost everywhere on the boundary L when points $z_1 \in D^+$ and $z_2 \in D^-$ ten to a point t_0 of L along nontangential paths in such a way that 66G: $|z_1 - t_0| = |z_2 - t_0|$

$$\Phi^+(t_0) - \Phi^-(t_0) = \varphi(t_0). \tag{4}$$

If the singular integral

$$\Phi(t_0) = \frac{1}{2\pi i} \int_L \frac{\varphi(t)\, dt}{t - t_0}$$

exists almost everywhere on L, then the Cauchy-Lebesgue type integral has limit values along tangential paths defined almost everywhere on L and equal to

$$\Phi^\pm(t_0) = \pm \frac{1}{2\pi i} \varphi(t_0) + \frac{1}{2\pi i} \varphi \int_L \frac{\varphi(t)\, dt}{t - t_0}. \tag{5}$$

COROLLARY. *If the function $\varphi(t)$ satisfies a Lipschitz condition*

$$|\varphi(t_1) - \varphi(t_2)| \le H(\varphi)|t_1 - t_2|, \qquad t_1, t_2 \in L,$$

then almost everywhere on L the derivative $\Phi'(z)$ of the Cauchy-Lebesgue type integral satisfies the relation

$$\left[\Phi'(t_0)\right]^+ - \left[\Phi'(t_0)\right]^- = \varphi'(t_0). \tag{6}$$

Indeed, we extend $\varphi(t)$ to a neighborhood of an arc l containing a point t_0 of continuity of $\varphi'(t)$ and represent $\Phi'(z)$ in the form

$$\Phi'(z) = \frac{1}{2\pi i} \int_L \frac{f(t, z) \, dt}{t - z},$$

where $f(t, z) = (\varphi(t) - \varphi(z))/(t - z)$ is bounded and continuous with respect to t and z in a neighborhood of t_0, and $\lim f(t, z) = \varphi'(t_0)$ as $t, z \to t_0$. Then (6) is obtained as a consequence of the first part of Theorem 1, which is obviously true also for the integral with density $f(t, z)$. In the case when L is the unit circle the obvious equality

$$\frac{t}{t - z} - \frac{t}{t - z_*} = \frac{1}{2}\left(\frac{t + z}{t - z} - \frac{t + z_*}{t - z_*}\right) = \operatorname{Re} \frac{t + z}{t - z},$$

where $t = e^{i\gamma}$, $z = re^{i\gamma_0}$, and $z_* = r^{-1}e^{i\gamma_0}$, leads to the relation

$$\frac{1}{2\pi i} \int_{|t|=1} \frac{\varphi(t) \, dt}{t - z} - \frac{1}{2\pi i} \int_{|t|=1} \frac{\varphi(t) \, dt}{t - z_*} = \frac{1}{2} \int_0^{2\pi} \varphi(e^{i\gamma})k_1(r, \gamma - \gamma_0) \, d\gamma,$$

$$k_1(r, \gamma - \gamma_0) = \operatorname{Re} \frac{t + z}{t - z} = \frac{1 - r^2}{1 + r^2 - 2r\cos(\gamma - \gamma_0)}. \tag{7}$$

Since $|z_*| = 1/|z| \to 1$ as $|z| \to 1$, and $|e^{i\gamma_0} - z| \to |e^{i\gamma_0} - z_*|$, the next statement follows from an application of the first part of Theorem 1 to the integrals on the left-hand side of (7).

THEOREM 2 (FATOU). *If the density $\varphi(t)$ of the Poisson-Lebesgue integral on the right-hand side of (7) is summable on the circle, then the integral has radial limits (as $r \to 1$) almost everywhere on the circle, and they are equal to*

$$\lim_{r \to 1} \frac{1}{2\pi} \int_0^{2\pi} \varphi(e^{i\gamma})k_1(r, \gamma - \gamma_0) \, d\gamma = \varphi(e^{i\gamma_0}). \tag{8}$$

COROLLARY. *If $\varphi(t)$ satisfies a Lipschitz condition, then*

$$\lim_{r \to 1} \frac{d}{d\gamma_0}\left\{\frac{1}{2\pi} \int_0^{2\pi} \varphi(e^{i\gamma})k_1(r, \gamma - \gamma_0) \, d\gamma\right\} = \frac{d\varphi}{d\gamma_0} \tag{9}$$

almost everywhere on the circle $|z| = 1$.

To prove (9) we differentiate both sides of (7) with respect to γ_0. Then

$$\frac{d}{d\gamma_0}\left\{\frac{1}{2\pi}\int_0^{2\pi}\varphi(e^{i\gamma})k_1(r,\gamma-\gamma_0)\,d\gamma\right\}$$

$$= iz\Phi'(z) - iz_*\Phi'(z_*) = iz_*\left[\Phi'(z) - \Phi'(z_*)\right] + i(z - z_*)\Phi^*(z).$$

But the limit as $r \to 1$ of the first term on the right-hand side of this equality is, by (6), equal almost everywhere to

$$it_0\frac{d\varphi(t_0)}{dt_0} = \frac{d\varphi}{d\gamma_0}.$$

And, by inequality (2.8) in Chapter I for the derivative $d\Phi/dz$ of the Cauchy type integral, the second term has the estimate

$$|z - z_*|\,|\Phi'(z)| \leqslant 2|z - t_0|\frac{\text{const}}{(z - t_0)^\lambda}, \qquad 0 < \lambda \geqslant 1,$$

which implies that its limit as $r \to 1$ is zero.

THEOREM 3 (FATOU). *A bounded analytic function $\Phi(z)$ on the disk $|z| < 1$ has definite limit values almost everywhere on the circle $|z| = 1$ as $r \to 1$.*

We consider the function $F(z) = \int_0^z \Phi(z)\,dz$, which is analytic in the disk $|z| < 1$ and satisfies a Lipschitz condition in the closed disk, because $|\Phi(z)|$ is bounded. Representing the real and imaginary parts of $F(z) = u(x, y) + iv(x, y)$ by Poisson integrals and taking (9) into account for each of them, we arrive at the statement of the theorem.

2°. Singular operators defined on the circle and on the real axis.

THEOREM 4 (F. RIESZ). *Let $f(x) \in \overset{\circ}{C}{}^1$ for $x \in (-\infty, \infty)$, and define*

$$U(f|\xi) = \frac{1}{\pi}\int_{-\infty}^{\infty}\frac{f(x)\,dx}{x - \xi}, \qquad \xi \in (-\infty, \infty),$$

where the operator is understood in the principal-value sense. Then

$$\|Uf\|_{L_p} \leqslant A_p\|f\|_{L_p}, \qquad p > 1, \tag{10}$$

where $A_p = A(p) < \infty$ and $A_2 = 1$.

Let

$$F(\zeta) = u + iv = \frac{1}{\pi}\int_{-\infty}^{\infty}\frac{f(x)\,dx}{x - \zeta}, \qquad \zeta = \xi + i\eta, \eta > 0.$$

The imaginary part $v(\xi, \eta)$ is a Poisson integral; consequently, $v(\xi, 0) = f(\xi)$. The real part is

$$u(\xi, \eta) = \frac{1}{\pi} \int_{-\infty}^{\infty} \frac{x - \xi}{(x - \xi)^2 + \eta^2} f(x)\, dx$$

$$= \frac{1}{\pi} \int_{0}^{\infty} \frac{f(\xi + x) - f(\xi - x)}{x} \cdot \frac{x^2}{x^2 + \eta^2}\, dx, \qquad \eta > 0,$$

from which, letting $\eta \to 0$, we get $u(\xi, 0) = U(f\,|\,\xi)$. Let

$$\Delta = (\partial^2/\partial\xi^2 + \partial^2/\partial\eta^2).$$

Then, taking the Cauchy-Riemann relations for u and v and the equalities $\Delta u = \Delta v = 0$ into account, we find that

$$\Delta\,|\,u\,|^p = p(p - 1)\,|\,u\,|^{p-2}\left(u_x^2 + u_y^2\right); \quad \Delta\,|\,v\,|^p = p(p - 1)\,|\,v\,|^{p-2}\left(u_x^2 + u_y^2\right);$$

$$\Delta\,|\,F\,|^p = p^2\,|\,F\,|^{p-2}\left(u_x^2 + u_y^2\right).$$

Let us first consider the case $2 \leqslant p < \infty$.

The last equalities then imply that

$$\Delta\!\left(|\,F\,|^p - \frac{p}{p - 1}\,|\,u\,|^p\right) = p^2\left(|\,F\,|^{p-2} - |\,u\,|^{p-2}\right)\left(u_x^2 + u_y^2\right) \geqslant 0,$$

since $|\,u\,| \leqslant |\,F\,|$. Applying Green's formula

$$\iint_{D} (\varphi\Delta\psi - \psi\Delta\varphi)\, d\xi\, d\eta = \int_{\Gamma}\!\left(\varphi\frac{\partial\psi}{\partial n} - \psi\frac{\partial\varphi}{\partial n}\right) ds$$

to the functions $\varphi \equiv 1$ and $\psi = |\,F\,|^p - p\,|\,u\,|^p/(p - 1)$ in the domain $D = \{\eta > \eta_0, \xi^2 + \eta^2 < R^2\}$, we get

$$0 \leqslant \iint_{D} \Delta\!\left(|\,F\,|^p - \frac{p}{p - 1}\,\bigg|\,|\,u\,|^p\right) d\xi\, d\eta$$

$$= \int_{\alpha_0}^{\pi - \alpha_0} \frac{\partial}{\partial R}\!\left(|\,F\,|^p - \frac{p}{p - 1}\,|\,u\,|^p\right)_{\xi = Re^{i\theta}} d\theta$$

$$- \int_{-R_0}^{R_0} \frac{\partial}{\partial\eta}\!\left(|\,F\,|^p - \frac{p}{p - 1}\,|\,u\,|^p\right)_{\xi = \xi + i\eta_0} d\xi,$$

where $\alpha_0 = \arcsin(\eta_0/R)$ and $R_0 = R\cos\alpha_0$.

Obviously, the first integral on the right-hand side of the equality tends to zero as $R \to \infty$; therefore,

$$\frac{\partial}{\partial\eta} \int_{-\infty}^{\infty}\!\left(|\,F\,|^p - \frac{p}{-p - 1}\,|\,u\,|^p\right) d\xi \equiv \frac{d\psi(\eta)}{d\eta} \leqslant 0$$

for $\eta = \eta_0 \geqslant 0$. Since $\psi(\eta) \to 0$ as $\eta \to \infty$, the last inequality gives us that $\psi(\eta) \geqslant 0$ for $\eta > 0$, i.e.,

$$\int_{-\infty}^{\infty} |F(\xi + i\eta)|^p \, d\xi \leqslant \frac{p}{p-1} \int_{-\infty}^{\infty} |u(\xi + i\eta)|^p \, d\xi, \qquad \eta > 0.$$

This inequality and the obvious relation

$$\left(\int_{-\infty}^{\infty} |F|^p \, d\xi \right)^{2/p} = \|u^2 + v^2\|_{L_{p/2}} \leqslant \|u^2\|_{L_{p/2}} + \|v^2\|_{L_{p/2}}$$

give us that

$$\left(\frac{p}{p-1} \right)^{2/p} \|u^2\|_{L_{p/2}} \leqslant \|u^2\|_{L_{p/2}} + \|v^2\|_{L_{p/2}},$$

and hence

$$\|u\|_{L_{p/2}}^2 \leqslant \frac{1}{(p/(p-1))^{2/p} - 1} \|v^2\|_{L_{p/2}},$$

i.e.,

$$\int_{-\infty}^{\infty} |u|^p \, d\xi \leqslant \left[\frac{1}{(p/(p-1))^{2/p} - 1} \right]^{p/2} \int_{-\infty}^{\infty} |v|^p \, d\xi.$$

Letting $\eta \to 0$, we arrive at the required inequality (10). The case $1 < p \leqslant 2$ is treated in a completely analogous way, and the theorem is proved.

COROLLARY. *The operator*

$$U_a f = \int_{-a}^{a} \frac{f(\xi) \, d\xi}{\xi - x}$$

is linear and bounded in $L_p(\overline{D})$, $D = \{-a < x < a\}$, and

$$\|U_a f\|_{L_p(\overline{D})} \leqslant A_p \|f\|_{L_p(\overline{D})}, \qquad A_2 = 1.$$

Indeed, considering the set of functions $f(x) \in \overset{\circ}{C}{}^1(\overline{D})$ and setting $f \equiv 0$ for $|x| > a$, we get $U_a f = Uf$. Then

$$\|U_a f\|_{L_p(\overline{D})} \leqslant \|Uf\|_{L_p(E)} \leqslant A_p \|f\|_{L_p(E)} = A_p \|f\|_{L_p(\overline{D})},$$

from which, taking the fact that $\overset{\circ}{C}{}^1(\overline{D})$ is a dense subset of $L_p(\overline{D})$ into account, we get the required inequality for all $f(x) \in L_p(\overline{D})$ from Theorem 3 in §1.

The next theorem follows from a complete repetition of the arguments in Theorem 4 as applied to the disk, with the density of the set of functions $f(t) \in \overset{\circ}{C}{}^1[0, 2\pi]$ in $L_p[0, 2\pi]$, $p > 1$, taken into account.

THEOREM 5 (F. RIESZ). *The operator $\overset{\circ}{U}(f \mid \gamma_0)$ defined by*

$$\overset{\circ}{U}(f \mid \gamma_0) = \frac{1}{2\pi} \int_0^{2\pi} f(e^{i\gamma}) \cot \frac{\gamma - \gamma_0}{2} d\gamma \qquad (11)$$

is bounded in $L_p[0, 2\pi]$, $p > 1$, and

$$\| \overset{\circ}{U}(f \mid \gamma) \|_{L_p} \leqslant \overset{\circ}{A}_p \| f \|_{L_p}, \qquad \overset{\circ}{A}_p = 2\left(\frac{p}{p-1}\right)^{1/p}.$$

The identity

$$\frac{1}{\pi i} \int_{|t|=1} \frac{f(t)\, dt}{t - t_0} = \frac{1}{2\pi i} \int_0^{2\pi} f(e^{i\gamma}) \cot \frac{\gamma - \gamma_0}{2} d\gamma + \frac{1}{2\pi} \int_0^{2\pi} f(e^{i\gamma})\, d\gamma$$

and the inequality in Theorem 5 give us the inequality

$$\| U(f \mid t_0) \|_{L_p} \leqslant \left(\overset{\circ}{A}_p + 1\right) \| f \|_{L_p}, \qquad p > 1,$$

for the operator U in Theorem 4 defined on the circle by

$$U(f \mid t_0) = \frac{1}{\pi i} \int_{|t|=1} \frac{f(t)\, dt}{t - t_0}, \qquad t_0 = e^{i\gamma_0}. \qquad (11^*)$$

With a view to a sharper computation of the norm of this operator in $L_2[0, 2\pi]$ we expand a function $f(e^{i\gamma}) \in L_2$ in a Fourier series $f(t) = \sum_{-\infty}^{\infty} C_k t^k$, $t = e^{i\gamma}$, and, taking

$$\int_0^{2m} t^k t^{-m}\, d\gamma = \begin{cases} 0, & k \neq m, \\ 2\pi, & k = m, \end{cases}$$

into account, we compute its norm in terms of the coefficients:

$$\| f(t) \|_{L_2}^2 = \int_0^{2\pi} \left(\sum_{-\infty}^{\infty} C_k t^k \sum_{-\infty}^{\infty} \overline{C}_k \bar{t}^k \right) d\gamma = 2\pi \sum_{-\infty}^{\infty} |C_k|^2.$$

The last relation is called Parseval's equality. Applying next the Sokhotskiĭ-Plemelj formulas to the analytic functions

$$\Phi_k(z) = \frac{1}{2\pi i} \int_{|t|=1} \frac{t^k\, dt}{t - z} = \begin{cases} z^k\ (k = 0, 1, \ldots)\ \text{and}\ 0\ (k = -1, \ldots)\ \text{for}\ |z| < 1, \\ 0\ (k = 0, 1, \ldots)\ \text{and}\ -z^k (k = -1, \ldots)\ \text{for}\ |z| > 1, \end{cases}$$

we get that

$$U(t^k \mid t_0) = \Phi_k^+ + \Phi_k^- = \begin{cases} t_0^k, & k = 0, 1, \ldots, \\ -t_0^k, & k = -1, -2, \ldots. \end{cases}$$

Then Parseval's equality gives us that

$$\|U(f)\|_{L_2}^2 = \left\|\sum_{-\infty}^{\infty} C_k^* t_0^k\right\|_{L_2}^2 = 2\pi \sum_{-\infty}^{\infty} |C_k^*|^2 = \|f\|_{L_2}^2,$$

where $C_k^* = C_k$ for $k = 0, 1, \ldots$, and $C_k^* = -C_k$ for $k = -1, -2, \ldots$. We state this result as a theorem.

THEOREM 6. *The operator* $U(f\,|\,t_0)$ *given by* (11*) *is bounded in the spaces* $L_p[0, 2\pi]$, $p > 1$, *and*

$$\|U(f\,|\,t_0)\|_{L_p} \leqslant \overline{A}_p \|f\|_{L_p}, \qquad A_2 = 1.$$

We remark that, as follows from the assertions in the second part of Privalov's theorem (Theorem 1), the existence almost everywhere on l of the singular integral

$$\Phi(z_0) = \frac{1}{2\pi i} \int_l \frac{\varphi(t)\,dt}{t - z_0}, \qquad z_0 = t_0 \in l,$$

implies the existence almost everywhere on l of the limit values $\Phi^{\pm}(t_0)$ of the Cauchy-Lebesgue type integral $\Phi(z)$ along any nontangential paths, as well as the validity of the Sokhotskiĭ-Plemelj formulas everywhere on l. Consequently, by Theorems 4–6, many of those properties of a Cauchy type integral with a piecewise Hölder density that follow from the Sokhotskiĭ-Plemelj formulas can be extended to Cauchy-Lebesgue type integrals defined on the real axis and on the circle with a density $\varphi(t)$ in $L_p(l)$, $p > 1$. In particular, the limit values of the real part $\varphi = \varphi(z)$ of an analytic function $w(z) = \varphi(z) + i\psi(z)$, $\operatorname{Im} w(0) = 0$ in the disk $|z| < 1$, that is represented by a Schwarz integral with real density

$$\varphi(e^{i\gamma}) \in L_p[0, 2\pi], \quad p > 1, \qquad w(z) = \frac{1}{2\pi} \int_0^{2\pi} \varphi(e^{i\gamma}) \frac{e^{i\gamma} + z}{e^{i\gamma} - z} d\gamma, \qquad (12)$$

coincide with $\varphi(e^{i\gamma})$ almost everywhere on the circle.

Moreover, the imaginary part $\psi = \psi(e^{i\gamma})$ of this analytic function also has limit values almost everywhere on the circle, and they are related to $\varphi(e^{i\gamma})$ by the well-known Hilbert formulas (§2.9° in Chapter I):

$$\psi(e^{i\gamma_0}) = -\frac{1}{2\pi} \int_0^{2\pi} \varphi(e^{i\gamma}) \cot\frac{\gamma - \gamma_0}{2} d\gamma,$$

$$\varphi(e^{i\gamma}) = \frac{1}{2\pi} \int_0^{2\pi} \psi(e^{i\gamma}) \cot\frac{\gamma - \gamma_0}{2} d\gamma. \qquad (12^*)$$

THEOREM 7 (ZYGMUND). *If* $\varphi(e^{i\gamma})$, $\gamma \in [0, 2\pi]$, *is a real function and*

$$\sup |\varphi(e^{i\gamma})| = \pi/2m, \qquad m > 0, \tag{13}$$

then the following functions are integrable for $\gamma \in [0, 2\pi]$:

$$\exp\{\pm p\mathring{U}(\varphi \mid \gamma_0)\}, \qquad \exp\{p \mid \mathring{U}(\varphi \mid \gamma_0)\mid\}, \qquad 0 < p < m,$$

where $\mathring{U}$ *is the operator defined by* (11). *If* $\varphi(e^{i\gamma})$ *is continuous and* 2π-*periodic, then these functions are integrable for any* $p > 0$.

Let $w(z) = \varphi(z) + i\psi(z)$ be the analytic function defined by (12) in the disk $|z| < 1$. Applying the Cauchy formula

$$F(0) = \frac{1}{2\pi i} \int_{|z|=r} \frac{F(z)}{z} dz = \frac{1}{2\pi} \int_0^{2\pi} F(re^{i\gamma}) d\gamma, \qquad r > 1,$$

to the function $F(z) = \exp\{\pm ipw(z)\}$, analytic in the disk, we get

$$\frac{1}{2\pi} \int_0^{2\pi} \exp\{\pm p\psi(re^{i\gamma})\} \exp\{\pm ip\varphi(re^{i\gamma})\} d\gamma = \exp\{\pm ipw(0)\}.$$

The real part of the equality has the form

$$\frac{1}{2\pi} \int_0^{2\pi} \cos\{p\varphi(re^{i\gamma})\} \exp\{\pm p\psi(re^{i\gamma})\} d\gamma = \cos\{p\varphi(0)\}.$$

The maximum principle for a function $\varphi = \varphi(re^{i\gamma})$ harmonic in the disk and condition (13) on the circle give us that

$$\sup_{|z|<1} |\varphi(z)| \leqslant \pi/2m, \qquad m > 0;$$

therefore, the last equality implies that for $0 < p < m$

$$\frac{1}{2\pi} \int_0^{2\pi} \exp\{\pm p\psi(re^{i\gamma})\} d\gamma \leqslant \left[\cos \frac{p\pi}{2m}\right]^{-1}.$$

Observing also that $\exp(p \mid \psi \mid) \leqslant \exp(p\psi) + \exp(-p\psi)$, we get

$$\frac{1}{2\pi} \int_0^{2\pi} \exp\{p \mid \psi(re^{i\gamma})\mid\} d\gamma \leqslant 2\left[\cos \frac{p\pi}{2m}\right]^{-1}.$$

Passing to the limit as $r \to 1$ in the above inequalities and taking account of the fact that, since $w = \varphi + i\psi$ can be represented by the Schwarz integral (12), the function $\psi(re^{i\gamma})$ has a limit almost everywhere, and it is expressed by the first of the Hilbert formulas (12*). Then

$$\frac{1}{2\pi} \int_0^{2\pi} \exp\left\{\pm p \frac{1}{2\pi} \int_0^{2\pi} \varphi(e^{i\gamma}) \cot \frac{\gamma - \gamma_0}{2} d\gamma\right\} d\gamma_0 \leqslant \left[\cos \frac{p\pi}{2m}\right]^{-1},$$

$$\frac{1}{2\pi} \int_0^{2\pi} \left\{\left|\frac{p}{2\pi} \int_0^{2\pi} \varphi(e^{i\gamma}) \cot \frac{\gamma - \gamma_0}{2} d\gamma\right|\right\} d\gamma_0 \leqslant 2\left[\cos \frac{p\pi}{2m}\right]^{-1}. \tag{14}$$

Let us now assume that $\varphi(e^{i\gamma})$ is continuous and periodic. We knoiw from classical analysis that for each $\varepsilon > 0$ there is a trigonometric polynomial

$$Q(\gamma) = \frac{a_0}{2} + \sum_{k=1}^{n} (a_k \cos k\gamma + b_k \sin k\gamma)$$

such that

$$\max_{\gamma \in [0,2\pi]} |\varphi(e^{i\gamma}) - Q(\gamma)| < \varepsilon.$$

Then from (14) we find that

$$\frac{1}{2\pi} \int_0^{2\pi} \exp\left\{ \pm \frac{p}{2\pi} \int_0^{2\pi} [\varphi(e^{i\gamma}) - Q(\gamma)] \cot\frac{\gamma - \gamma_0}{2} d\gamma \right\}$$

$$\times \exp\left\{ \pm \frac{p}{2\pi} \int_0^{2\pi} Q(\gamma) \cot\frac{\gamma - \gamma_0}{2} d\gamma \right\} d\gamma_0 \leqslant \left[\cos\frac{p\pi}{2m} \right]^{-1} M,$$

where the second factor under the integral sign is bounded by a number $M(p)$, while (by the foregoing) the first is integrable for $p < \pi/2\varepsilon$, i.e., for all $p > 0$.
The theorem is proved (see [136], Vol. I, Chapter VII, §2).

3°. *Singular operators defined on Lyapunov curves.*

THEOREM 8. *The operator U defined by*

$$U(f|t_0) = \frac{1}{\pi i} \int_l \frac{f(t)\, dt}{t - t_0}, \qquad t_0 \in l, \qquad (15)$$

on a Lyapunov curve is bounded in each space $L_p(l)$, $p > 1$, and

$$\|U(f|t_0)\|_{L_p(l)} \leqslant \tilde{A}_p \|f\|_{L_p(l)}. \qquad (16)$$

Let the domain $D(l)$ be mapped conformally onto the disk $|\zeta| < 1$. Since the resulting correspondnece $t = t(\tau)$ of boundary points of the circle and the arc l has the properties (Theorem 3 in §3)

$$\frac{dt}{d\tau} \neq 0, \quad t(\tau) \in C^{1+\alpha} \quad \text{for } |\tau| = 1,$$

the operator U can be recast in the form

$$U(f|t_0) = \frac{h(\tau_0, \tau_0)}{\pi i} \int_{|\tau|=1} \frac{f(\tau)\, d\tau}{\tau - \tau_0} + \frac{1}{\pi i} \int_{|\tau|=1} \frac{\Delta h f(\tau)\, d\tau}{\tau - \tau_0} \equiv I_1 + I_2,$$

where $f(\tau) = f[t(\tau)] \in L_p$ on the circle, and

$$h(\tau_0, \tau) = \frac{dt}{d\tau} \cdot \frac{\tau - \tau_0}{t(\tau) - t(\tau_0)}$$

satisfies a Hölder condition with respect to both arguments, and $\Delta h = h(\tau, \tau_0) - h(\tau_0, \tau_0)$. For $I_1(f \mid t_0)$ the needed estimate follows from Theorem 6, while for $I_2(f \mid t_0)$ Hölder's inequality gives us that

$$| I_2^\varepsilon(f \mid t_0) |^p \leqslant \pi^{-p} \| h \|_{C^\alpha}^p \left\{ \int_{\Gamma_\varepsilon} \frac{|f(\tau)| \, d\gamma}{|\tau - \tau_0|^{1-\alpha}} \right\}^p$$

$$\leqslant N_1 \int_{\Gamma_\varepsilon} |f(\tau)|^p \, |\tau - \tau^0|^{\alpha-1} \, d\gamma \left\{ \int_{|\tau|=1} |\tau - \tau_0|^{\alpha-1} \, d\gamma \right\}^{p/q}$$

$$\leqslant N_2 \int_{\Gamma_\varepsilon} |f(\tau)|^p \, |\tau - \tau_0|^{\alpha-1} \, d\gamma,$$

where Γ_ε is the unit circle with an ε-neighborhood of the point $\tau = \tau_0$ deleted. Integrating both sides of the last inequality and passing to the limit as $\varepsilon \to 0$, we get

$$\int_{|\tau_0|=1} | I_2(\tau_0) |^p \, d\gamma_0 \leqslant N_2 \int_{|\tau|=1} |f(\tau)|^p \, d\gamma \max_{|\tau|=1} \left\{ \int_{|\tau_0|=1} |\tau - \tau_0|^{\alpha-1} \, d\gamma_0 \right\}$$

$$\leqslant | N_3 \| f(\tau) \|_{L_p}^p,$$

which concludes the proof of the theorem.

REMARK. Since the singular integral (15) exists almost everywhere on l for $f(t) \in L_p(l)$, $p > 1$, Theorem 1 (Privalov) implies, in particular, that the Sokhotskiĭ-Plemelj formulas hold almost everywhere on l.

Let us now prove that the following operators with special singularities are bound in $L_p(l), p > 1$:

$$U_n(\varphi \mid t_0) = \frac{\Pi(t_0 - c_k)^{\alpha_k}}{\pi i} \int_l \frac{\varphi(t) \, dt}{\Pi(t - c_k)^{\alpha_k}(t - t_0)},$$

$$0 \leqslant \alpha_k \leqslant \frac{1}{q} = \frac{p-1}{p} \quad \text{for } k = 1, \ldots, n_0, \tag{17}$$

$$-\frac{1}{p} < \alpha_k \leqslant 0 \quad \text{for } k = n_0, \ldots, n \ (0 \leqslant n_0 \leqslant n),$$

where l is a closed Lyapunov curve.

We remark that the assumption that l is closed is not a restriction, since we can arrive at this case by completing a nonclosed arc l to a smooth closed contour $\tilde{l}$, with $\varphi(t) \equiv 0$ on $\tilde{l} - l$.

Consider first the integral

$$I(\alpha, \beta) = \int_l |t - t_0|^{-\alpha} |t - c|^{-\beta} \, ds, \quad 0 \leqslant \alpha, \quad \beta < 1, \quad t_0, c \in l.$$

It is obviously bounded for $\alpha + \beta < 1$. For $\alpha + \beta \geqslant 1$ we fix an arbitrary $\delta > 0$ and break up the integral I:

$$I(\alpha, \beta) = \int_{\tilde{s}-\delta}^{\tilde{s}} + \int_{\tilde{s}}^{\tilde{s}+\delta} + \int_{l-l_\delta} \equiv I_1 + I_2 + I_3,$$

where $c = t(\tilde{s})$ and l_δ is the arc on l with endpoints $t(\tilde{s} + \delta)$. The integral I_3 is bounded. Taking account of the fact that

$$k_0 |s - s_0| \leqslant |t_0 - t| \leqslant |s - s_0|, \qquad k_0 = k_0(\delta) > 0, s \in [\tilde{s} - \delta, \tilde{s} + \delta],$$

we get for the integral I_1 (I_2) that

$$I_1 \leqslant k_0^{-\alpha-\beta} \int_{\tilde{s}-\delta}^{\tilde{s}} (\tilde{s} - s)^{-\beta} |s - s_0|^{-\alpha} ds$$

$$= k_0^{-\alpha-\beta} |\tilde{s} - s_0|^{1-\alpha-\beta} \int_0^{\sigma_*} \sigma^{-\beta} |\sigma \pm 1|^{-\alpha} d\sigma,$$

where $\sigma_* = \delta |\tilde{s} - s_0|^{-1}$ and $\sigma = (\tilde{s} - s)|\tilde{s} - s_0|^{-1}$. But the last integral is bounded for $\alpha + \beta > 1$, while for $\alpha + \beta = 1$

$$\int_0^{\sigma_*} \leqslant N_1 + N_2 |\ln|\hat{s} - s_0||.$$

Thus,

$$I(\alpha, \beta) = \begin{cases} M_0, & \alpha + \beta < 1, \\ M_1 + M_2 |\ln|t_0 - c||, & \alpha + \beta = 1, \\ M_3 |t - c|^{1-\alpha-\beta}, & \alpha + \beta > 1. \end{cases} \tag{18}$$

We let

$$U_\alpha(\varphi | t_0) = \frac{(t_0 - c)^\alpha}{\pi i} \int_l \frac{\varphi(t) dt}{(t - c)^\alpha (t - t_0)}, \qquad 0 \leqslant \alpha < \frac{1}{q} = \frac{p-1}{p},$$

and represent this operator in the form

$$U_\alpha(\varphi | t_0) = \frac{1}{\pi i} \int_l \frac{(t_0 - c)^\alpha - (t - c)^\alpha}{(t - c)^\alpha (t - t_0)} \varphi(t) dt + U(\varphi | t_0) \equiv \tilde{U}_\alpha + U,$$

where U is defined by (15). Taking account of the fact that $(z - c)^\alpha$ is a Hölder function and applying Hölder's inequality (1.1), we get

$$|\tilde{U}_\alpha|^p \leqslant N(\alpha) \left\{ \int_0^{s_1} \frac{|\varphi(t)| ds}{|t - c|^\alpha |t - t_0|^{1-\alpha}} \right\}^p$$

$$= N \left\{ \int_l |\varphi| |t - t_0|^{(\alpha-1)(1/p+1/q)} |t - c|^{\varepsilon-(\alpha-\varepsilon)} ds \right\}^p$$

$$\leqslant N \left[I\left(1 - \alpha, \frac{p(\alpha + \varepsilon)}{p-1}\right) \right]^{p-1} \int_l |\varphi|^p |t - t_0|^{\alpha-1} |t - c|^{\varepsilon p} ds.$$

Since $\alpha < (p-1)/p$, we have $p(\alpha + \varepsilon)/(p-1) < 1$ for sufficiently small $\varepsilon > 0$; consequently, (18) gives us that

$$I^{p-1} \leqslant M_3^{p-1} |t_0 - c|^{-\alpha - p\varepsilon}.$$

Then

$$\int_0^{s_l} |\tilde{U}_\alpha|^p \, ds_0 \leqslant N_1 \int_0^{s_l} |\varphi|^p |t - c|^{\varepsilon p} \left(\int_0^{s_l} |t_0 - t|^{-\alpha - p\varepsilon} \, ds_0 \right) ds,$$

and, since $\alpha + p\varepsilon < 1$ for sufficiently small $\varepsilon > 0$, (18) implies that

$$\int_0^{s_l} |\tilde{U}_\alpha|^p \, ds_0 \leqslant N_1 M_3 \int_0^{s_l} |\varphi(t)|^p \, ds,$$

which, by (16), leads to the estimate

$$\|U_\alpha(\varphi \mid t_0)\|_{L_p(l)} \leqslant M_\alpha \|\varphi(t)\|_{L_p(l)}. \tag{19}$$

Let us now consider the operator

$$U_{-\alpha}(\varphi \mid t_0) = \frac{1}{\pi i (t_0 - c)^\alpha} \int_l \frac{\varphi(t)(t - c)^\alpha}{t - t_0} \, dt, \qquad 0 < \alpha < 1/\beta,$$

which can be represented in the form

$$U_{-\alpha} = \frac{1}{\pi i} \int_\Gamma \frac{(t - c)^\alpha - (t_0 - c)^\alpha}{(t_0 - c)^\alpha (t - t_0)} \varphi(t) \, dt + U \equiv \tilde{U}_{-\alpha} + U.$$

Estimates analogous to the preceding ones give us

$$|\tilde{U}_{-\alpha}|^p \leqslant N_1 [I(1 - \alpha, \varepsilon q)]^{p-1} \int_0^{s_l} |\varphi|^p \, ds.$$

Let $\varepsilon > \alpha/q$. Then, according to (18),

$$[I(1 - \alpha, \varepsilon q)]^{p-1} \leqslant M_3^{p-1} |t_0 - c|^{\alpha(p-1) - p\varepsilon},$$

and, consequently,

$$\int_0^{s_l} |\tilde{U}_{-\alpha}|^p \, ds_0 \leqslant N_1 M_3^{p-1} \int_0^{s_l} |\varphi|^p |t - c|^{\varepsilon p} \left(\int_0^{s_l} |t - t_0|^{\alpha - 1} |t - c|^{-\alpha - p\varepsilon} \, ds_0 \right) ds.$$

We require that $\alpha/q < \varepsilon < (1 - \alpha)/p$, which is possible because $\alpha < 1/p$. Then $\alpha + p\varepsilon < 1$. By (16) and (18),

$$\|U_{-\alpha}(\varphi \mid t_0)\|_{L_p(l)} \leqslant M_{-\alpha} \|\varphi(t)\|_{L_p(l)}. \tag{20}$$

We represent the operator U_n defined by (17) in the form

$$U_n = \sum \frac{\Pi_k(t_0)(t_0 - c_k)^{\alpha_k}}{\pi i} \int_{l_k} \frac{\varphi(t) \, dt}{\Pi_k(t)(t - c_k)^{\alpha_k}(t - t_0)} \equiv \sum U_n^k,$$

where $\Pi_k(t) = \Pi_{i \neq k}(t - c_i)^{\alpha_i}$, l_k $(l = \Sigma l_k)$ are the arcs on l with endpoints $a_k = t(s_k - \delta_k)$, $b_k = t(s_k - \delta_{k-1})$, $\delta_k = |s_k - s_{k-1}|/2$, and $c_k = t(s_k)$. By Minkowski's inequality (1.3),

$$\|U_n(\varphi \,|\, t_0)\|_{L_p(l)} \leqslant \sum \|U_n^k(\varphi \,|\, t_0)\|_{L_p(l_k)},$$

so it suffices to get an estimate of U_n on one of the arcs l_k. But $0 < |\Pi_k(t_0)| < \infty$ for $t_0 \in l_k$, and so (19) and (20) give us

$$\|U_n^k(\varphi \,|\, t_0)\|_{L_p(l_k)} \leqslant M^k \|\varphi\|_{L_p(l)}.$$

If $\alpha_k \geqslant 0$, then $0 < |\Pi_i(t_0)| < \infty$ for $t_0 \in l_k$ and for all i; consequently, (19) and (20) are applicable to all the operators U_n. Suppose that $-1/q < \alpha_k < 0$ and $i \neq k - 1, k, k + 1$. Then

$$|t_0 - c_i|, |t - t_0| \geqslant k_0 \delta/2 > 0, \qquad t \in l_i,$$

whence

$$\|U_n^i(\varphi \,|\, t_0)\|_{L_p(l_k)} \leqslant \tilde{N} \left\| |t_0 - c_k|^{-|\alpha_k|} \int_{l_i} \frac{|\varphi(t)| \, ds}{|t_i - t|^{\alpha_i}} \right\|_{L_p} \leqslant M^i \|\varphi\|_{L_p(l)}.$$

Estimates for U_n^{k+1} and U_n^{k-1} can be obtained from the representation

$$U_n^{k+1} = \frac{\Pi(t_0)}{\pi i} \left(\int_{l_0} + \int_{l_{k-1} - l_0} \right) \frac{\varphi(t) \, dt}{\Pi(t)(t - t_0)} \equiv U' + U'',$$

where l_0 is the arc with endpoints

$$\tilde{a}_{k+1} = t\left(s_{k+1} - \frac{\delta_{k+1}}{2} \right) \quad \text{and} \quad \tilde{b}_{k+1} = t\left(s_{k+1} - \frac{\delta_{k+2}}{2} \right),$$

since $t_0 \notin l_0$, the estimate for U_n^i with $i \neq k - 1, k, k + 1$ is applicable to U'. Next,

$$\|U''\|_{L_p(l_k)} \leqslant \|U''\|_{L_p(l_k^0)} + \|U''\|_{L_p(l_k - l_k^0)},$$

where l_k^0 is an arc with endpoints

$$a_k^0 = t\left(s_k - \frac{\delta_k}{2} \right) \quad \text{and} \quad b_k^0 = t\left(s_k + \frac{\delta_{k+1}}{2} \right),$$

and there is an obvious estimate for each term in terms of $\|\varphi\|_{L_p(l)}$. Thus, the theorem is proved.

THEOREM 9 (KVEDELIDZE). *The operator U_n defined by* (17) *is bounded and linear in* $L_p(l), p > 1$:

$$\|U_n(\varphi \,|\, t_0)\|_{L_p(l)} \leqslant M \|\varphi(t)\|_{L_p(l)}.$$

4°. *The interpolation theorem of E. M. Stein.* We first present some auxiliary propositions.

THEOREM 10 (Phragmén-Lindelöf principle). *Let $f(z)$ be a bounded analytic function in the strip $B = \{\alpha \leqslant x \leqslant \beta, |y| < \infty\}$ with $|f(z)| \leqslant M$ (Re $z = \alpha\beta$) on the boundary of B. Then $|f(z)| \leqslant M$ at each interior point of B, and if $|f(z)| = M$ at some interior point, then $f(z) = Me^{i\gamma}$, $\gamma = $ const.*

If $f(z)$ tends uniformly (in x) to zero as $y \to \pm\infty$, then the assertion follows from the maximum principle for analytic functions. Indeed, for any point $z_0 \in B$ there is a rectangle $B_\eta = \{\alpha \leqslant x \leqslant \beta, |y| < \eta\}$ such that $z_0 \in B_\eta$. But $f(x + i\eta) \to 0$, so for sufficiently large η we have $|f(z)| \leqslant M$ for $z \in \Gamma_\eta = \partial B_\eta$, and, consequently, for all $z \in B_\eta$. In the general case we consider the sequence

$$f_n(z) = f(z)e^{z^2/n} = f(z)e^{(x^2-y^2)/n}e^{i2xy/n}, \qquad n = 1, 2, \ldots.$$

For each fixed n we have $f_n(z) \to 0$ as $|y| \to \infty$ uniformly in $x \in [\alpha, \beta]$, while on the straight lines $x = \alpha, \beta$

$$|f_n(z)| \leqslant Me^{\gamma^2/n}, \qquad \gamma = \max\{|\alpha|, |\beta|\}.$$

By what was proved, $|f_n(z)| \leqslant Me^{\gamma^2/n}$ for $z \in B$, and this gives the desired inequality in the limit as $n \to \infty$. If $|f(z_0)| = M$ at some interior point $z_0 \in B$ and $f(z) \not\equiv M$, then we would have $|f(z)| \leqslant |f(z_0)|$ in some neighborhood of z_0 (the opposite inequality is impossible by what was proved), which contradicts the maximum modulus principle.

The next theorem is a corollary to this.

THEOREM 11. *Suppose that the conditions of Theorem* 10 *hold and on $\Gamma = \partial B$ the function $f(z)$ satisfies the inequalities*

$$|f(\alpha + iy)| \leqslant M_1, \qquad |f(\beta + iy)| \leqslant M_2, \qquad |y| < \infty.$$

Then at each interior point $z_0 \in B$

$$|f(z_0)| \leqslant M_1^{L(x_0)}M_2^{1-L(x_0)}, \qquad L(t) = \frac{t}{\alpha - \beta} + \frac{\beta}{\beta - \alpha}. \qquad (21)$$

If equality is attained in (21) *at some interior point, then*

$$f(z) \equiv M_1^{L(z)}M_2^{1-L(z)}e^{i\gamma}, \qquad \gamma = \text{const}. \qquad (22)$$

The proof follows from Theorem 10, applied to the function

$$F(z) = f(z)M_1^{-L(z)}M_2^{L(z)-1}.$$

Let us consider the collection of functions $f(x)$ defined on some set E and measurable with respect to some measure $\mu \geq 0$ on E. It is assumed that

$$\| f \|_{p,\mu} \equiv \left\{ \int_F |f(x)|^p \, d\mu \right\}^{1/p} < \infty. \tag{23}$$

This collection forms a normed linear space $L_{p,\mu}(E) \equiv L_p(\mu, E)$.

Let S be the collection of simple, finite, μ-measurable functions on E. By definition, such a function is zero outside some set of finite measure (if $\mu(E) = \infty$), takes finitely many values (different from $\pm\infty$) on E, and is measurable with respect to the measure $\mu(x)$. The last condition is equivalent to requiring that the sets E_i on which a simple function takes one of its values y_i are measurable with respect to μ.

Letting $X_{E_i}(x)$ be the characteristic function of the set E_i, we represent a simple function in the form

$$f(x) = \sum_i y_i X_{E_i}(x). \tag{24}$$

We remark that S is dense in any space $L_p(\mu, E)$, $1 \leq p < \infty$.

Since it is assumed that $\mu(x) \geq 0$, Hölder's inequality is valid:

$$|l(g)| \equiv \left| \int_E f(x)g(x) \, d\mu(x) \right| \leq \| f \|_{p,\mu} \| g \|_{p',\mu} \tag{25}$$

for any $f \in L_{p,\mu}$ and $g \in L_{p',\mu}$, where $p' = p/(p-1)$.

It follows from (25) that the integral $l(g)$ on the left-hand side can be regarded for fixed $f \in L_{p,\mu}$ as a linear functional on $L_{p',\mu}$ with norm

$$\| l \|_{p',\mu} = \sup_{g \in s} \left| \int_E f(x)g(x) \, d\mu(x) \right|, \qquad \| g \|_{p',\mu} = 1. \tag{26}$$

The formulas (25) and (26) imply that $\| l \|_{p',\mu} \leq \| f \|_{p,\mu}$, and, moreover, it is easy to see that equality holds here.

Indeed, consider the funcion

$$g(x) = \frac{|f(x)|^{p-1}\bar{f}(x)}{\| f \|_{p,\mu}^{p-1} |f(x)|}. \tag{27}$$

By construction, $g(x) \in L_{p',\mu}(E)$, $\| g \|_{p',\mu} = 1$, and equality is attained in (25) for $g(x)$, i.e.,

$$\| f \|_{p',\mu} = \sup_{g \in s} \left| \int_E f(x)g(x) \, d\mu(x) \right|, \qquad \| g \|_{p',\mu} = 1. \tag{28}$$

This formula holds also if $\| f \|_{p,\mu} = +\infty$.

To see this we assume that $\mu(E) < \infty$ and construct the following sequence of functions: $f_n(x) = f(x)$ wherever $|f(x)| \leq n$, and $f_n(x) = 0$ elsewhere. By (27), functions $g_n(x)$ correspond to them which from some index n on belong to $L_{p',\mu}(E)$ and satisfy $\| g_n \|_{p',\mu} = 1$. Moreover,

$$\int_E f(x) g_n(x)\, d\mu(x) = \int_E f_n(x) g_n(x)\, d\mu(x) \leq \| f_n \|_{p,\mu} \to \infty,$$

so that the right-hand side of (28) is also equal to $+\infty$. If $\mu(E) = \infty$, then the functions $f_n(x)$ must vanish outside some sets of finite measure.

We shall consider additive homogeneous operators T acting from one function space $L_p(\mu, E)$ to another similar space $L_q(\nu, E)$. If, moreover, T is bounded, i.e.,

$$\| Tf \|_{q,\nu} \leq \| f \|_{p,\mu} M, \tag{29}$$

then we say it is of type (p, q). The infimum of the numbers M in (29) is called the norm of T and denoted by $\| T \|$. Each operator satisfying (29) admits a unique extension to the closure of its domain with preservation of norm.

In particular, if T is bounded on a dense subset of $L_{p,\mu}$, for example, on S, then it admits a unique extension to all of $L_{p,\mu}$ with preservation of norm.

THEOREM 12 (E. M. STEIN). *Let u_1 and u_2 be nonnegative functions on a set E that are measurable with respect to a measure $\mu \geq 0$, and k_1 and k_2 nonnegative functions that are measurable on a set E' with respect to a measure $\nu \geq 0$. Consider an additive and homogeneous operator T defined on the set S of simple functions on the space (μ, E) and transforming the functions in S into measurable functions on the space (ν, E'). Suppose that for any simple function f*

$$\| Tf \cdot k_i \|_{1/\beta_i,\nu} \leq M_i \| f \cdot u_i \|_{1/\alpha_i,\mu}, \qquad M_i = \text{const}, i = 1, 2, \tag{30}$$

where (α_1, β_1) and (α_2, β_2) are points in the square $0 \leq \alpha, \beta \leq 1$.

Consider also the functions $k = k_1^{1-t} k_2^t$, $u = u_1^{1-t} u_2^t$ and the numbers $\alpha = (1-t)\alpha_1 + t\alpha_2$ and $\beta = (1-t)\beta_1 + t\beta_2, 0 \leq t \leq 1$.

Then T can be uniquely extended to a linear (bounded) operator on the whole space of functions f that satisfy the condition

$$\| f \cdot u \|_{1/\alpha,\mu} = \left(\int_E |u(x)|^{1/\alpha} |f(x)|^{1/\alpha}\, d\mu(x) \right)^{\alpha} < \infty, \tag{31}$$

and

$$\| Tf \cdot k \|_{1/\beta,\nu} \leq M_t \| f \cdot u \|_{1/\alpha,\mu}, \qquad M_t = M_1^{1-t} M_2^t. \tag{32}$$

Let us consider the measurable sets

$$E_\varepsilon = \{ x \mid x \in E, \varepsilon \leq u_1, u_2 \leq 1/\varepsilon \}, \qquad E'_\varepsilon \{ x \mid x \in E', \varepsilon \leq k_1, k_2 \leq 1/\varepsilon \},$$

where $\varepsilon > 0$ is an arbitrarily specified number, and the operators V and V' are defined as follows:

$$Vf(x) = \begin{cases} f(x), & x \in E_\varepsilon, \\ 0, & x \notin E_\varepsilon, \end{cases} \qquad V_g'(x) = \begin{cases} g(x), & x \in E_\varepsilon', \\ 0, & x \notin E_\varepsilon'. \end{cases}$$

It follows from (30) that T can be uniquely extended to the space of functions satisfying one of the conditions

$$\| f \cdot u_1 \|_{1/\alpha_1,\mu} < \infty, \qquad \| f \cdot u_2 \|_{1/\alpha_2,\mu} < \infty,$$

and the norm of T in these spaces is bounded by the respective constants M_1 and M_2. In particular, this operator is defined on the functions of the form $V(f \cdot u_1^{z-1} u_2^{-z})$ for any value of the complex parameter z and for any simple function $f(x)$.

Let us consider simple functions $f(x)$ and $g(x)$ satisfying the conditions

$$\| f \|_{1/\alpha,\mu} = 1, \qquad \| g \|_{(1/\beta)',\mu} = 1, \tag{33}$$

where $\alpha(t)$ and $\beta(t)$ are defined in the theorem. Representing these functions in the form $f(x) = |f(x)| e^{i\theta}$, $g(x) = |g(x)| e^{i\tau}$ and extending the values $\alpha(t)$ and $\beta(t)$ to the complex domain $(t = z)$, we construct the functions

$$F_z = |f(x)|^{\alpha(z)/\alpha} e^{i\theta}, \quad \alpha > 0, \qquad \text{and} \qquad F_z = f(x), \quad \alpha = 0;$$
$$G_z = |g(x)|^{(1-\beta(z))/(1-\beta)} e^{i\tau}, \quad \beta < 1, \qquad \text{and} \qquad G_z = g(x), \quad \beta = 1.$$

Since f and g are taken on the unit spheres in $L_{1/\alpha,\mu}$ and $L_{(1/\beta)',\mu}$, it follows that

$$\| F_{iy} \|_{1/\alpha,\mu} = 1, \qquad \| G_{iy} \|_{(1/\beta)',\nu} = 1,$$

and from the representation for simple functions we have

$$F_z(x) = \sum_j e^{i\theta_j} |y_j|^{\alpha(z)/\alpha} X_{E_j}(x)$$

for $\alpha > 0 \ (\theta_j = \arg y_j)$, and

$$F_z(x) = \sum_j e^{i\theta_j} |y_j| X_{E_i}(x)$$

for $\alpha = 0$.

Consider the analytic function

$$\int_{E'} G_z u_z(f) \, d\nu(x) \equiv \Phi(z) = \int_{E'} G_z V' \{ k_1^{1-z} k_2^z T [V (F_z u_1^{z-1} u_2^{-z})] \} \, d\nu(x). \tag{34}$$

Since $g(x)$ is a simple function on E', the integral here extends over a set of finite ν-measure, contained in E'_ε. The inequalities (30) imply that the integral (34) is finite for any $z \in B = \{0 \leqslant x \leqslant 1, |y| < \infty\}$, and the properties of the functions $F_z(x)$ and $G_z(x)$ give us that $\Phi(z)$ is continuous for $z \in \bar{B}$ and is analytic at interior points of B.

Let us now use a formula of type (28) applied to the function $u_z(f)(x)$, $x \in E'$, along with the fact that the operators V and V' are bounded and have norms not exceeding 1. Setting $z = iy$ in (34) and assuming that $\|q\|_{\delta,\nu} = 1$ and $\delta = (1/\beta_1)'$, we get successively that

$$|\Phi(iy)| = \left| \int_{E'} G_{iy} u_{iy}(f) \, d\nu(x) \right| \leqslant \sup_{\|G_{iy}\|_{\delta,\nu} = 1} \left| \int_{E'} G_{iy} u_{iy}(f) \, d\nu(x) \right|$$

$$= \left\| V' \left\{ k_1^{1-iy} k^{iy} k_2^{iy} T \left[V \left(F_{iy} u_1^{iy-1} u_2^{-iy} \right) \right] \right\} \right\|_{1/\beta_1,\nu}.$$

The conditions of the theorem and the properties of the functions then give us that

$$|\Phi(iy)| \leqslant \left\| k_1 T \left[V \left(F_{iy} u_1^{iy-1} u_2^{-iy} \right) \right] \right\|_{1/\beta_1,\nu}$$

$$\leqslant M_1 \left\| V \left(F_{ij} u_1^{iy-1} u_2^{-iy} \right) \right\|_{1/\alpha,\mu} \leqslant M_1 \| F_{iy} \|_{1/\alpha,\mu} = M_1.$$

The arguments are similar for $z = 1 + iy$.

Thus, on the boundary of the strip B the function $\Phi(z)$ satisfies the inequalities

$$|\Phi(iy)| \leqslant M_1, \qquad |\Phi(1 + iy)| \leqslant M_2,$$

which permit us to apply the Phragmén-Lindelöf principle (Theorem 11) to it; and this gives

$$|\Phi(t)| \leqslant M_1^{1-t} M_2^t, \qquad 0 \leqslant t \leqslant 1. \tag{35}$$

By definition, the functions $F_z(x)$ and $G_z(x)$ coincide with $f(x)$ and $g(x)$ for $z = t$. Therefore, (34) and (35) imply that

$$\left| \int_{E'} g(x) u_t(f) \, d\nu(x) \right| \equiv \left| \int_{E'} g(x) V' \left\{ k_1^{1-t} k_2^t T \left[V \left(f(x) u_1^{1-t} u_2^{-t} \right) \right] \right\} \, d\nu(x) \right|$$

$$\leqslant M_1^{1-t} M_2^t.$$

Here $f(x)$ and $g(x)$ are any simple functions satisfying (33). Again using (29) for an arbitrary f, we set $f_0 = f/\| f \|$ and find that

$$\| u_t(f) \|_{1/\beta,\nu} \leqslant M_t \| f \|_{1/\alpha,\mu}. \tag{36}$$

Recalling that $k(x) = k_1^{1-t} k_2^t$ and $u(x) = u_1^{1-t} u_2^t$, and taking the meaning of the operators V and V' into account, we rewrite (36) in the form

$$\left(\int_{E'_\varepsilon} |k(x) T[V(fu^{-1})]|^{1/\beta} \, d\nu(x) \right)^\beta \leqslant M_t \left(\int_E |f(x)|^{1/\alpha} \, d\mu(x) \right)^\alpha.$$

Passage to the limit as $\varepsilon \to 0$ leads to an inequality equivalent to (32), and the theorem is proved.

§3. Boundary properties of conformal mappings, and criteria for univalence

A curve Γ is said to be piecewise Lyapunov ($\Gamma \in C_*^{1+\alpha}$), $\alpha > 0$ if the angle $\theta(x)$ between the tangent to Γ and the OX-axis is a piecewise Hölder function.

The following auxiliary theorem establishes the existence of conformal mappings of domains of a special form (a particular case of the Riemann theorem) and at the same time investigates the smoothness of the mappings.

THEOREM 1. *Let D be a simply connected domain (finite or infinite) with finite boundary $\Gamma \in C_*^{1+\alpha}$, $\alpha > 0$, such that the angle $\theta = \theta(x) \in C_*^\alpha$ between the tangent to Γ and the OX-axis satisfies the condition*

$$|\theta(x)| \leqslant \pi/2 - \delta, \qquad \delta > 0, x \in [x_0, x_1]. \tag{1}$$

Then there exists a conformal mapping $z = F(\zeta)$ of the disk $|\zeta| < 1$ onto D that carries the points $\zeta = \pm 1$ into the extreme points z_0 and z_1 of Γ with respect to the OX-axis ($\operatorname{Re} z_i = x_i$). The derivative of this mapping has the property that

$$\left\| \ln \frac{dz}{d\zeta}(e^{i\gamma}) \right\|_\alpha \leqslant M(\delta, \varepsilon) \tag{2}$$

everywhere on the circle except at the points corresponding to points of discontinuity of $\theta(x)$.

It is assumed that we know the boundary correspondence $x = x(\gamma) \in C^\beta$, $\beta(\delta) > 0$, between the arguments $\gamma \in [0, 2\pi]$ of points on the circle and the abscissae $x \in [x_0, x_1]$ of points on the boundary Γ. Taking account of the geometric meaning of the argument of the derivative of a conformal mapping, we then have

$$\arg \frac{dz}{dw}(e^{i\gamma}) = \theta[x(\gamma)], \qquad \gamma \in [0, 2\pi],$$

where $z = F[\zeta(w)]$, and $\zeta = \zeta(w)$ is a conformal mapping of the plane with a finite cut onto the disk $|\zeta| < 1$; $w = \frac{1}{2}(\zeta + 1/\zeta)$.

If the function $\ln(dz/dw)(\zeta)$ is recovered by Schwarz's formula, then the following representation for the derivative $dz/d\zeta = dz/dw \cdot dw/d\zeta$ of the desired mapping $z = F(\zeta)$ results:

$$\frac{dz}{d\zeta} = \frac{k}{2\zeta}\left(\zeta - \frac{1}{\zeta}\right)\exp\frac{i}{2\pi}\left\{\int_0^{2\pi}\frac{\theta[x(\gamma)](t + \zeta)\,d\gamma}{t - \zeta}\right\}, \qquad (3)$$

where k is a real constant that is unknown for the present. Consideration of relation (3) on the circle allows us to obtain an equation for the unknown boundary correspondence $x = x(\gamma)$:

$$x = k\int_0^\gamma \sin\gamma\cos\theta(x)\exp[U(i\theta\,|\,\gamma)]\,d\gamma + x_0 = \Omega(x/\gamma). \qquad (4)$$

in which the constant k must be chosen from the condition $\Omega(x\,|\,\pi) = x_1$, i.e.,

$$k = (x_1 - x_0)\left\{\int_0^\pi \sin\gamma\cos\theta(x)[\exp U(i\theta\,|\,\gamma)\,d\gamma]\right\}^{-1} \equiv \frac{x_1 - x_0}{k_0}. \qquad (5)$$

Here $U(i\theta\,|\,\gamma)$ is the value of the Schwarz operator occurring in (3) at the point $\zeta = e^{i\gamma}$. Taking condition (1) into account and using the converse Hölder inequality (1.4), we get

$$k_0 \geqslant \cos\left(\frac{\pi}{2} - \delta\right)\int_0^\pi e^{U(i\theta|\gamma)}\sin\gamma\,d\gamma$$

$$\geqslant \cos\left(\frac{\pi}{2} - \delta\right)\left\{\int_0^\pi e^{-pU}\,d\gamma\right\}^{-1/p}\left\{\int_0^\pi \frac{d\gamma}{(\sin\gamma)^{p/p+1}}\right\}^{(p+1)/p}$$

$$\geqslant \cos\left(\frac{\pi}{2} - \delta\right)\left[\frac{2\pi}{\cos(\pi/2 - \delta)}\right]^{-1/p}\left\{\int_0^\pi \frac{d\gamma}{(\sin\gamma)^{p/p+1}}\right\}^{(p+1)/p} \equiv m(\delta),$$

where the constant p is chosen from the condition $1 < p < \pi/(\pi - 2\delta)$ for Zygmund's theorem (Theorem 7 in §2) to be applicable to the function $e^{U(i\theta|\gamma)}$. Then for any two points $t = e^{i\gamma}$ and $t' = e^{i\gamma'}$ we find that

$$|z(e^{i\gamma}) - z(e^{i\gamma'})| = \left|\int_{t'}^t \frac{dz}{dt}\,dt\right| \leqslant \frac{x_1 - x_0}{m(\delta)}\int_{\gamma_1}^\gamma |e^{U(i\theta|\gamma)}|\,d\gamma$$

$$\leqslant \frac{x_1 - x_0}{m(\delta)}\left[\frac{2\pi}{\cos(\pi/2 - \delta)}\right]^{1/p}|\gamma - \gamma'|^{p-1/p} \equiv M_0(\delta)|\gamma - \gamma'|^\beta,$$

$$\beta = \frac{p-1}{p}.$$

Thus, the following a priori estimates hold for solutions of (4):

$$\|\Omega(x\,|\,\gamma)\|_{H^{\beta_0}} \leqslant (2\pi)^{\beta-\beta_0} M_0(\delta), \tag{6}$$

$$x_0 \leqslant \Omega(x\,|\,\gamma) \leqslant x_1, \qquad \gamma \in [0,2\pi] \tag{7}$$

for all β_0, $0 \leqslant \beta_0 \leqslant \beta = (p-1)/p$.

The transformation $\Omega(x\,|\,\gamma)$ obviously maps C^{β_0} continuously into C^{β} and, consequently, is compact in C^{β_0} for $0 < \beta_0 < \beta$; moreover, it carries the closed convex set $\mathfrak{M} \subset C^{\beta_0}$ determined by (6) and (7) into itself. Therefore, by Schauder's theorem (Theorem 6 in §1), $\Omega(x\,|\,\gamma)$ has at least one fixed point in $\mathfrak{M}$ (with our estimate (6) taken into account). The inequality (2) can be established by a direct application of Privalov's theorem (Theorem 4 in §2 of Chapter I) to the Schwarz integral in (3). The theorem is proved.

We remark that, since any other mapping of $|\zeta| < 1$ onto D can be obtained from the conformal mapping $z = F(\zeta)$ constructed in the theorem by a linear fractional transformation, estimate (2) remains true for all these mappings.

An important variational property of conformal mappings is contained in the next theorem.

THEOREM 2 (Lindelöf principle). *Let D^1 and $D^2 \subset D^1$ be two simply connected domains. Suppose that the boundaries Γ^i of the D^i have a common arc $\Gamma \subset \Gamma^i$ and are analytic in a neighborhood of it (Γ may degenerate into a point). If $w = w^i(z)$ are functions mapping the D^i conformally onto the disk $|w| < 1$ and carrying a common point $z_0 \in D^i$ (for definiteness, $z_0 = 0$) into the point $w = 0$, then*

$$\left|\frac{dw^2}{dz}(z)\right| \leqslant \left|\frac{dw^1}{dz}(z)\right|, \quad z \in \Gamma \subset \Gamma^i, \qquad \left|\frac{dw^2}{dz}(0)\right| \geqslant \left|\frac{dw^1}{dz}(0)\right|. \tag{8}$$

Let $D_\zeta^2 = w^1(D^2)$ and $\Gamma_\zeta = w^1(\Gamma)$ be the images of the domain $D^2 \subset D^1$ and the arc $\Gamma \subset \Gamma^i$, respectively, under the conformal mapping $\zeta = w^1(z)$ of D^1 onto $|\zeta| < 1$ with normalization $w^1(0) = 0$. We consider the following bounded harmonic function in D_ζ^2:

$$\varphi(\zeta) = \ln\left|\frac{w^2(\zeta)}{\zeta}\right|, \qquad \zeta \in D_\zeta^2,$$

where $w = w^2(\zeta)$ is the conformal mapping of D_ζ^2 onto $|w| < 1$ with normalization $w^2(0) = 0$. By construction, $\varphi(e^{i\gamma}) = 0$, $\zeta = e^{i\gamma} \in \Gamma_\zeta$, and $\varphi(re^{i\gamma}) > 0$ on $(\Gamma_\zeta^2\,|\,\Gamma)$, so the maximum principle for harmonic functions implies that $\varphi(re^{i\gamma}) > 0$, $re^{i\gamma} \in D_\zeta^2$, and the function $\varphi(re^{i\gamma})$ attains an absolute minimum on an arc of the circle. Consequently, $\varphi(re^{i\gamma})$ is a nonincreasing function of r as $r \to 1$ for each $e^{i\gamma} \in \Gamma_{\zeta}$, and, therefore,

$$\frac{d\varphi}{dr}(e^{i\gamma}) \leq 0, \qquad e^{i\gamma} \in \Gamma_{\zeta}.$$

Then, with the help of one of the Cauchy-Riemann relations on Γ_{ζ}, written in polar coordinates and applied to the function

$$\omega = \ln \frac{w^2(\zeta)}{\zeta} = \varphi + i(\theta - \gamma), \qquad \theta = \arg w^2(\zeta),$$

which is analytic in D_{ζ}^2, we find that

$$\frac{d\varphi}{dr} = \frac{d(\theta - \gamma)}{d\gamma} = \frac{d\theta}{d\gamma} - 1 \leq 0.$$

Next, since $|w^2(\zeta)| = |\zeta| = 1$ for $\zeta \in \Gamma_{\zeta}$ and, consequently,

$$\frac{d\theta}{d\gamma} = \left| \frac{dw^2(\zeta)}{d\zeta} \right| = \left| \frac{dw^2[w^1(z)]}{dz} \right| \left(\frac{dw^1}{dz} \right)^{-1}, \qquad z \in \Gamma,$$

the inequality $d\theta/d\gamma \leq 1$ reduces to the form (8).

The second inequality in (8) follows directly from the fact that the harmonic function

$$\varphi[\zeta(z)] = \ln \left| \frac{w^2(z)}{w^1(z)} \right| = \ln \left| \frac{w^2(z)}{z} \right| - \ln \left| \frac{w^1(z)}{z} \right| \geq 0$$

in D^2 is nonnegative.

REMARK. It follows from the boundary properties of conformal mappings established in Theorem 3 that the Lindelöf principle remains true also when $\Gamma^i \in C^{1+\alpha}$, $\alpha > 0$, in a neighborhood of the common arc Γ. To see this it suffices to pass to the limit in the inequalities (8) corresponding to analytic arcs $\{\Gamma_k^i\}$ approximating the Γ^i.

THEOREM 3 (KELLOGG). *Suppose that the function $z = F(\zeta)$ maps the disk $|\zeta| < 1$ univalently onto a domain D whose boundary includes a finite arc $l \in C^{1+\alpha}$, $\alpha > 0$. Then,*

$$\left| \ln \frac{dF(e^{i\gamma_1})}{d\zeta} - \ln \frac{dF(e^{i\gamma_2})}{d\zeta} \right| \leq k(\delta, \varepsilon) |\gamma_1 - \gamma_2|^{\alpha} \tag{9}$$

everywhere on the circular arc $l_{\zeta} = F^{-1}(l)$ (the preimage of l) except in an ε-neighborhood of its endpoints.

Since the assertion of the theorem has a local character, the arc l can be assumed to be small enough that

$$|\theta(s) - \theta_0| \leqslant \pi/2 - \delta, \qquad \delta > 0, \qquad (1^*)$$

for some fixed value θ_0, where $\theta = \theta(s) \in C^\alpha$, $\alpha > 0$, is the angle between the tangent to l and the real axis, and s is the arclength parameter of l. We use the function $z = \Phi(w)$ to map the disk $|w| < 1$ conformally onto the z-plane with a slit along l, with the normalization indicated in Theorem 1. By condition (1), the inequality (2) in Theorem 1 holds for the function $(dz/dw)e^{-i\theta_0}$. The image $D^2 = \Phi(D)$ of D lies inside $D^1 = D^1\{|w| < 1\}$ and has the analytic arc $l_w = \{w = e^{i\gamma}, \gamma \in [\theta, \pi]\}$ in common with it. Then, by Lindelöf's principle, the derivatives of the conformal mappings $\zeta = \zeta^i(w)$ of the D^i onto $|\zeta| < 1$ satisfy (8), i.e.,

$$\left| \frac{d\zeta(w)}{dw} \right| \leqslant \left| \frac{d\zeta^1(w)}{dw} \right| = 1, \qquad w \in lw.$$

On the other hand, we choose the domain $D^0 \subset D^2$ in such a way that part of its boundary Γ^0 lies on the arc l_w and is at a distance of $\varepsilon > 0$ from its endpoints, while the angle $\theta(\varphi)$, $\varphi = \operatorname{Re} w$, between the tangent to Γ^0 and the real axis satisfies condition (1).

Then, again from the Lindelöf principle, together with inequality (2) for a conformal mapping $\zeta = \zeta(w)$ of D^0 onto $|\zeta| < 1$, we find that

$$M_0(\delta, \varepsilon) \leqslant \left| \frac{d\zeta^0(w)}{dw} \right| \leqslant \left| \frac{d\zeta^2(w)}{dw} \right|, \qquad w \in l_2^* \subset l_w.$$

Passing to the inverse functions in the inequalities we have obtained and taking the properties of the mapping $z = \Phi(w)$ into account, we get

$$M_1(\delta, \varepsilon) \leqslant \left| \frac{dF}{d\zeta}(e^{i\gamma}) \right| = \left| \frac{dz}{dw} \right| \left| \frac{dw}{d\zeta} \right| \leqslant M_2(\delta, \varepsilon).$$

The last estimates imply, in particular, that $s = s(\gamma) \in C^1$ on the image of l. Consequently, (9) can be obtained by applying Privalov's theorem (Theorem 4 in §2 of Chapter I) to the Schwarz integral in the representation (3) for $dz/d\zeta = dF/d\zeta$, and the theorem is proved.

Theorem 3 makes it easy for us to extend the property established in §4.4° of Chapter I for the derivative of a conformal mapping at a corner boundary point formed by intersecting straight line segments to the case of the intersection of arbitrary Lyapunov curves. Indeed, let $\alpha\pi$, $0 < \alpha \leqslant 2$, be the interior angle at a common point z_0 (for definiteness $z_0 = 0$) of two arcs Γ^1, $\Gamma \in C^{1+\beta}$, $\beta > 0$, lying on the boundary Γ of a domain D. We draw the tangents to the Γ^i

at the point $z = 0$ and map the interior of the resulting angle of measure $\alpha\pi$ onto the upper half-plane $\operatorname{Im} w > 0$ by the function $w = z^{1/\alpha} \equiv \Phi(z)$. If we then map the disk $|\zeta| < 1$ onto the image $D_w = \Phi(D)$ of D and take account of the fact that $\Gamma_w = \Phi(\Gamma^1 + \Gamma^2) \in C^{1+\beta}$ in a neighborhood of $w = 0$, we get

$$\frac{dz}{d\zeta} = \alpha\left[w(\zeta)^{\alpha-1}\frac{dw}{d\zeta} \right] \equiv (\zeta - \zeta_0)^{\alpha-1}f(\zeta),$$

where $f(\zeta_0)$ is neither 0 nor ∞, and ζ_0 is the image of the point $w = 0$, $w(\zeta_0) = 0$. Summing up, we arrive at the following theorem.

THEOREM 4. *Let* $z = F(\zeta)$ *be a conformal mapping of the disk* $|\zeta| < 1$ *onto a domain* D, *with the point* $\zeta_0 = e^{\gamma i_0}$ *going into the corner point* $z_0 = F(e^{i\gamma_0})$ *formed by arcs* $\Gamma^1, \Gamma^2 \in C^{1+\beta}$, $\beta > 0$, *lying on the boundary* Γ *of* D. *Then the derivative* $dF/d\zeta$ *has the representation*

$$dz/d\zeta = (\zeta - \zeta_0)^{\alpha-1}f(\zeta) \tag{10}$$

at this point, where $f(\zeta)$ *is a continuous function in a neighborhood of* $\zeta = \zeta_0$, *and* $f(\zeta_0) \neq 0$.

We now prove two simple sufficient criteria for univalence in the case of closed convex domains.

THEOREM 5. *Suppose that* $z = F(\zeta)$ *is an analytic function in a finite or infinite convex domain* D, *and that its derivative is continuous in* $\overline{D}$ *except for a finite number of points* $\zeta = \sigma_k$, *in a neighborhood of which it is integrable. If for some fixed* α

$$\operatorname{Re}\left[e^{i\alpha}\frac{dF}{d\zeta}(t) \right] \geq 0, \qquad t \in \Gamma = \partial D, \tag{11}$$

then either $dF/d\zeta \equiv 0$ *or* $F(\zeta)$ *is univalent in* D.

Since D is a convex domain, we have

$$|F(\zeta_1) - F(\zeta_2)| = \left| \int_{\zeta_1}^{\zeta_2}\frac{dz}{d\zeta}d\zeta \right| \geq |\zeta_1 - \zeta_2|\left| \int_0^1 \operatorname{Re}\left[e^{i\alpha}\frac{dF(\tau)}{d\zeta} \right]dt \right| \equiv C_0|\zeta_1 - \zeta_2|$$

for any $\zeta_1, \zeta_2 \in \overline{D}$, where $C_0 \neq 0$ and $\tau(t) \equiv \zeta + t(\zeta_1 - \zeta_2)$. This inequality proves that $F(\zeta)$ is univalent in D.

THEOREM 6 (KAPLAN). *The function*

$$z = F(\zeta) = \frac{1}{2\pi}\int_0^{2\pi} g(\gamma)\frac{e^{i\gamma} + \zeta}{e^{i\gamma} - \zeta}d\gamma \tag{12}$$

is univalent in the disk $|\zeta| < 1$ *if* $g(\gamma) \in C[0, 2\pi]$ *is nondecreasing in* $[0, \pi]$ *and nonincreasing in* $[\pi, 2\pi]$. *If, moreover,* $g(\gamma) \in C^{1+\alpha}$, *then* $F(\zeta)$ *is univalent in* $\overline{D}$.

Suppose first that $g(\gamma) \in C^{1+\alpha}$. We map $|\zeta| < 1$ conformally onto the strip $D = \{-\pi/2 < \operatorname{Im} w < \pi/2\}$ by the function $w = \ln((\zeta - 1)/(\zeta + 1))$. Let $\zeta = \zeta(w)$ be the inverse mapping. Then for all $\gamma \in [0, 2\pi]$

$$\operatorname{Re}\left\{ \frac{dF}{dw}[\zeta(w)] \right\} = \operatorname{Re}\left\{ \frac{\zeta^2 - 1}{2} \frac{dF(\zeta)}{d\zeta} \right\} = \sin\gamma \frac{dg(\gamma)}{d\gamma} \geqslant 0.$$

Consequently, by the preceding theorem, the function $z = F[\zeta(w)]$ is univalent in the strip $\overline{D}$, and the function $z = F(\zeta)$ is also univalent in the closed disk $|\zeta| \leqslant 1$, because $\zeta = \zeta(w)$ is univalent in $\overline{D}$. In the general case we approximate $g(\gamma)$ by functions $g_n(\gamma) \in C^{1+\alpha}$ and observe that (similarly to the preceding) the inequality

$$|F_n(\zeta_1) - F(\zeta_2)| \geqslant C_0\left(D_\zeta^0\right)|\zeta_1 - \zeta_2|,$$

which ensures that the limit function $z = F(\zeta)$ is univalent in $\overline{D}_\zeta^0 \subset D_{\zeta_n}$, holds uniformly with respect to n in any domain $\overline{D}_\zeta^0 \subset D_\zeta = \{|\zeta| < 1\}$. The theorem is proved.

§4. The classes H_p and E_p of analytic functions

The theorems proved in the preceding section now enables us to construct solutions in the class of analytic functions for boundary-value problems with summable (not with piecewise Hölder) coefficients in the boundary conditions. Moreover, it is only necessary to require that the boundary conditions be satisfied almost everywhere on the boundary. However, for a close connection between the number of solutions of boundary-value problems and their indices it does not suffice to require, for example, only that the limit values of their solutions be integrable to some power on l. Let us consider an elementary boundary-value problem as an example:

$$\Phi^+\left(e^{i\gamma}\right) - \Phi^-\left(e^{i\gamma}\right) = 0, \qquad \gamma \in [0, 2\pi].$$

This problem admits a whole class of nontrivial solutions

$$\Phi(z) = \exp\left(\lambda \frac{1 + z}{1 - z}\right)$$

(λ a real parameter) that are bounded on the whole plane except for the singular point $z = 1$ on the boundary. Here

$$\Phi^{\pm}\left(e^{i\gamma}\right) = \exp\left(\lambda \frac{1 + e^{i\gamma}}{1 - e^{i\gamma}}\right) = \exp\left(i\lambda \cot\frac{\gamma}{2}\right), \qquad \gamma \neq 0, 2\pi,$$

i.e., $|\Phi^{\pm}(e^{i\gamma})| = 1$, and, consequently, $\Phi^{\pm}(e^{i\gamma}) \in L_p[0, 2\pi]$ for any p.

Thus, the homogeneous jump problem admits a whole family of nontrivial solutions that are summable on the circle to any power p. In order to exclude such cases we must restrict the class of functions in which the solutions of the boundry-value problems are sought. Such a restriction must obvously involve the requirement that the convergence of the desired solutions to their limit values have a certain regularity.

1°. *The classes H_p.* The well-known classes H_p, $p > 0$, (so named in honor of Hardy, who first studied them) turned out to be most closely associated with boundary-value problems.

DEFINITION 1. An analytic function $\Phi(z)$ in the unit disk $|z| < 1$ is said to *belong to the class H_p, $p > 0$,* if there exists a positive number $H_p(\Phi)$ such that

$$H_p(\Phi, r) = \frac{1}{2\pi} \int_0^{2\pi} |\Phi(re^{i\gamma})|^p \, d\gamma \leqslant H_p(\Phi), \qquad 0 \leqslant r \leqslant 1. \qquad (1)$$

It is easy to see that in the above example

$$\Phi(z) = \exp\left(\lambda \frac{1+z}{1-z}\right) \notin H_p$$

for every $p > 0$.

It turns out that the quantity $H_p(\Phi, r)$ in (1) is a nondecreasing function of r:

$$H_p(\Phi, r) \leqslant H_p(\Phi, \rho) \qquad \text{for } 0 \leqslant r < \rho < 1, \qquad (2)$$

and, consequently, the class H_p is characterized by the property that

$$H_p(\Phi) = \lim_{r \to 1} \frac{1}{2\pi} \int_0^{2\pi} |\Phi(re^{i\gamma})|^p \, d\gamma < \infty. \qquad (1^*)$$

To prove (2) we construct the function

$$f_n(z) = \frac{1}{n} \sum_{k=1}^n |\Phi(\xi_k z)|^p = \frac{1}{2\pi} \sum_{k=1}^n |\Phi(re^{i(\gamma + \gamma_k)})|^p \Delta\gamma,$$

where $\xi_k = e^{i\gamma_k}$, $k = 1, \ldots, n$ are the nth roots of unity, and $\Delta\gamma = 2\pi/n$. The last sum can obviously be regarded as a Riemann integral sum for the integral

$$H_p(\Phi, r) = \frac{1}{2\pi} \int_0^{2\pi} |\Phi(re^{i\gamma})|^p \, d\gamma,$$

and, therefore, $f_n(re^{i\gamma}) \to H_p(\Phi, r)$ uniformly with respect to γ.

We prove that the function

$$f_n(z) = \sum_{k=1}^n |\Phi_k(z)|^p, \qquad \Phi_k(z) = \frac{1}{n}\Phi(\xi_k z),$$

cannot attain a local maximum at interior points of an arbitrary domain D in the disk $|z| < 1$. This implies, in particular, that

$$f_n(re^{i\gamma}) \leqslant \max_\gamma f_n(\rho e^{i\gamma}) = f_n(\rho e^{i\gamma_0}), \qquad 0 \leqslant r < \rho < 1,$$

and passage to the limit as $n \to \infty$ in this leads us to the desired inequality (2). Suppose the opposite: The function $f_n(z) \neq 0$ takes a maximal value at some interior point $z_0 \in D$. Suppose that

$$\Phi_k(z_0) \neq 0, \qquad k = 1,\ldots,n_0;$$

$$\Phi_k(z_0) = 0, \qquad k = n_0 + 1, n_0 + 2,\ldots,n, \; 1 \leqslant n_0 < n.$$

There obviously exists a disk σ_ε: $|z - z_0| \leqslant \varepsilon$, $\sigma_\varepsilon \in D$, such that $\Phi_k(z) \neq 0$ for all $z \in \sigma_\varepsilon$, and $f_n(z_1) < f_n(z_0)$ at some point z_1 with $|z_1 - z_0| = \varepsilon$. Let

$$F(z) = \sum_{k=1}^{n} \Phi_k^p(z) \exp\{-i \arg \Phi_k^p(z_0)\},$$

where the $\Phi_k^p(z)$ are arbitrarily chosen branches that are single-valued in σ_ε. By construction,

$$F(z_0) = \sum_{k=1}^{n} |\Phi_k^p(z_0)| = f_n(z_0),$$

and interior to σ_ε we hve

$$|F(z)| \leqslant \sum_{k=1}^{n_0} |\Phi_k(z)|^p \leqslant \sum_{k=1}^{n} |\Phi_k(z)|^p = f_n(z) \leqslant f_n(z_0) = |F(z_0)| \; ;$$

since $f_n(z_1) < f_n(z_0)$, it follows that also $|F(z_1)| < |F(z_0)|$.

The last inequalities contradict the maximum modulus principle for analytic functions. The inequality (2) is proved.

2°. Blaschke functions. Let ζ_k be a sequence of points in the unit disk $|z| < 1$, indexed in order of nondecreasing moduli, i.e., $|\zeta_k| \leqslant |\zeta_{k+1}|$, such that

$$\sum_{k=1}^{\infty} (1 - |\zeta_k|) < \infty, \qquad |\zeta_k| < 1, k = 1,2,\ldots. \tag{3}$$

DEFINITION 2. The *Blaschke function* of a sequence $\{\zeta_k\}$ satisfying condition (3) is defined to be the infinite product

$$b(z) = \prod_{k=1}^{\infty} \frac{\zeta_k - z}{1 - \bar{\zeta}_k z} \cdot \frac{|\zeta_k|}{\zeta_k}. \tag{4}$$

Let us consider the sequence

$$B_n(z) = \sum_{k=1}^{n} \ln\left\{\frac{|\zeta_k|}{\zeta_k} \cdot \frac{\zeta_k - z}{1 - \bar{\zeta}_k z}\right\},$$

which for any choice of branch of the logarithm is connected with the sequence

$$b_n(z) = \prod_{k=1}^{n} \frac{\zeta_k - z}{1 - \bar{\zeta}_k z} \cdot \frac{|\zeta_k|}{\zeta_k} \tag{4*}$$

by the relation $b_n(z) = \exp B_n$, $n = 1,2,\ldots.$

We remark that the limit of $b_n(z)$ as $n \to \infty$ exists only if the limit of $B_n(z)$ exists. We write the general term of the sum $B_n(z)$ in the form

$$\ln \frac{|\zeta_n|}{\zeta_n} \cdot \frac{\zeta_n - z}{1 - \bar{\zeta}_n z} = \ln(1 - c_n),$$

where

$$|c_n| = \left| 1 - \frac{|\zeta_n|}{\zeta_n} \cdot \frac{\zeta_n - z}{1 - \bar{\zeta}_n z} \right| \leqslant \frac{1 + r}{1 - r}(1 - |\zeta_n|), \qquad r = |z|. \qquad (5)$$

By (3), the series $\sum_1^\infty c_n$ converges, and, since $c_n^{-1} \ln(1 - c_n) \to -1$ as $n \to \infty$, the series $\sum_1^\infty \ln(1 - c_n)$ also converges. The estimate (5) shows that $\sum \ln(1 - c_n)$ even converges uniformly in any domain of the disk $|z| < 1$ that does not contain the points ζ_k, and, consequently, the Blaschke function $b(z) = \lim_{n \to \infty} b_n(z)$ is analytic at each point $z \neq \zeta_k$. Since the $b_n(z)$ are bounded in the disk $|z| < 1$, with $|b_n| \leqslant 1$, it follows that also $|b(z)| \leqslant 1$, and this implies that $b(z)$ is analytic in the whole disk $|z| < 1$.

The boundedness of $b(z)$ together with Fatou's theorem (Theorem 3 in §2) implies the existence of limit values $b(e^{i\gamma})$ almost everywhere on the circle. We prove that $|b(e^{i\gamma})| = 1$ almost everywhere on the circle, which is obviously equivalent to proving that

$$\lim_{r \to 1} \frac{1}{2\pi} \int_0^{2\pi} |b(re^{i\gamma})| \, d\gamma \equiv \lim_{r \to 1} H_1(b, r) = 1. \qquad (6)$$

Without loss of generality we assume that $\zeta_k \neq 0$ in the expression (4) for $b(z)$; otherwise we can separate out the factor z^m corresponding to the roots $\zeta_k = 0$. Since the function $H_1(b, r)$ is monotonic in r,

$$H_1(b, r) \geqslant H_1(b, 0) \geqslant \prod_{k=1}^\infty |\zeta_k|.$$

For fixed n we have $\lim_{r \to 1} |b_n(z)| = 1$. Therefore, setting $b(z) = b_n R_n$, we get

$$\lim_{r \to 1} H_1(b, r) = \lim_{r \to 1} H_1(b_n R_n, r) = \lim_{r \to 1} H_1(R_n, r)$$

$$\geqslant H_1(R_n, 0) \geqslant \prod_{k=n+1}^\infty |\zeta_k|.$$

The tail $\prod_{n+1}^\infty |\zeta_k|$ of the convergent product $\prod_1^\infty |\zeta_k| = H_1(b, 0)$ tends to 1; hence,

$$\lim_{r \to 1} H_1(b, r) \geqslant \prod_{k=n+1}^\infty |\zeta_k| \geqslant 1;$$

but, on the other hand, the inequality $|b(z)| \leq 1$ gives us

$$\lim_{r \to 1} H_1(b, r) \leq 1,$$

which proves (6).

3°. *Representation of H_p-functions.* We write out the zeros of a function $\Phi(z) \in H_p$, $p > 0$, as a sequence $\{\zeta_k\}$, $|\zeta_k| \leq |\zeta_{k+1}|$, with each zero repeated as many times as its multiplicity, and we form the finite product $b_n(z)$ according to (4). Let us consider the quotient $\Phi(z)/b_n(z)$, which is analytic in $|z| < 1$ for each fixed $n < \infty$. Since $|b_n(e^{i\gamma})| = 1$, it follows that

$$H_p(\Phi) = \lim_{r \to 1} H(\Phi, r) = \lim_{r \to 1} H_p(\Phi/b_n, r),$$

i.e., the function $\Phi(z)/b_n(z)$ belongs to H_p, $p > 0$, and, by the monotonicity of $H_p(\Phi/b_n, r)$ with respect to r,

$$H_p(\Phi/b_n, r) \leq H_p(\Phi), \qquad r \geq 0. \tag{7}$$

Separating out the factor z^m, $m \leq n$, corresponding to the points $\zeta_k = 0$ in $b_n(z)$, and setting $r = 0$ in the last inequality, we get

$$\left| \frac{\Phi(z)}{z^m} \right|_{z=0}^p \left(\prod_{k=1}^n |\zeta_k| \right)^{-p} \leq H_p(\Phi),$$

and so

$$1 \geq \sum_{k=m}^n |\zeta_k| \geq \{H_p(\Phi)\}^{-1/p} \left| \frac{\Phi(z)}{z^m} \right|_{z=0} > 0.$$

Since the product $\tilde{b}_n = \prod_m^n |\zeta_k|$ decreases with increasing n ($|\zeta_k| < 1$), the limit $\tilde{b}_0 = \lim b_n$ exists ($0 < b_n < 1$), and with it also the finite limit

$$\lim_{n \to \infty} \left(\ln \tilde{b}_n \right) = \lim_{n \to \infty} \ln \prod_{k=m}^n |\zeta_k| = \lim_{n \to \infty} \sum_{k=m}^n \ln[1 - (1 - |\zeta_k|)].$$

But as shown above, the convergence of the series with general term

$$\ln[1 - (1 - |\zeta_k|)]$$

implies the convergence of the series with general term $(1 - |\zeta_k|)$, i.e., condition (3) holds.

Thus, the product

$$b(z) = \prod_{k=1}^\infty \frac{\zeta_k - z}{1 - \bar{\zeta}_k z} \cdot \frac{|\zeta_k|}{\zeta_k},$$

where the ζ_k are the zeros of the function $\Phi(z) \in H_p$, $p > 0$, is a Blaschke function, and all the properties proved above are true for it. By construction, the function $\Phi_0(z) = \Phi(z)/b(z)$ does not have zeros in the domain $|z| < 1$.

Passing to the limit as $n \to \infty$ in (7), we find that

$$H_p(\Phi_0, r) \leqslant H_p(\Phi),$$

which means that $\Phi_0(z)$ belongs to H_p, $p > 0$. This result will be formulated as a theorem.

THEOREM 1 (F. RIESZ). *Each function* $\Phi(z) \not\equiv 0$ *in* H_p, $p > 0$, *admits a representation*

$$\Phi(z) = b(z)\Phi_0(z), \qquad |z| < 1, \tag{8}$$

where $b(z)$ *is a Blaschke function* (4), *while* $\Phi_0(z) \in H_p$ *and does not have zeros in* $|z| < 1$; *moreover,* $H_p(\Phi_0, r) \leqslant H_p(\Phi)$, $r \in [0, 1]$.

4°. Boundary properties of H_p-functions. The representation (8) for analytic functions of the class H_p, $p > 0$, enables us to prove the following theorem.

THEOREM 2 (F. RIESZ). *For each function* $\Phi(z)$ *of the class* H_p, $p > 0$, *as* $r \to 1$ *limit values* $\Phi(e^{i\gamma}) \in L_p[0, 2\pi]$ *exist almost everywhere, and*

$$\lim_{r \to 1} \int_0^{2\pi} |\Phi(re^{i\gamma})|^p \, d\gamma = \int_0^{2\pi} |\Phi(e^{i\gamma})|^p \, d\gamma, \tag{9}$$

$$\lim_{r \to 1} \int_0^{2\pi} |\Phi(re^{i\gamma}) - \Phi(e^{i\gamma})|^p \, d\gamma = 0. \tag{10}$$

Let us represent $\Phi(z)$ according to (8) and consider the functions

$$h(z) = \ln|\Phi_0(z)| \equiv \ln|\Phi(z)/b(z)|,$$

$$h^+(z) = \begin{cases} h(z) & \text{for } h \geqslant 0, \\ 0 & \text{for } h < 0, \end{cases}$$

where $h(z)$ is harmonic in $|z| < 1$. We apply to the functions $\varphi(\gamma) = e^{ph^+}$ and $g(\gamma) \equiv 1$ the known ([90], p. 167) inequality

$$\exp\left\{ \frac{1}{L(q)} \int_0^{2\pi} q(\gamma) \ln \varphi(\gamma) \, d\gamma \right\} \leqslant \frac{1}{L(q)} \int_0^{2\pi} q(\gamma)\varphi(\gamma) \, d\gamma,$$

which is valid for nonnegative functions $\varphi(\gamma)$ and $q(\gamma)$ that satisfy the conditions

$$L(q) = \int_0^{2\pi} q(\gamma) \, d\gamma > 0, \qquad \int_0^{2\pi} q(\gamma)\varphi(\gamma) \, d\gamma < \infty.$$

This inequality can be obtained in the limit from the numerical inequality

$$\prod_{k=1}^n \varphi_k^{q_k} \leqslant \left(\frac{\sum_1^n q_k \varphi_k}{\sum_1^n q_k} \right)^{\sum_1^n q_k},$$

which connects the geometric and arithmetic means ([45], Chapter II, §2.5 (2.5.1)). Then, setting

$$E = \left\{ \gamma : h^+ \left(e^{i\gamma} \right) \geqslant 0 \right\} \quad \text{and} \quad CE = \left\{ [0, 2\pi] \backslash E \right\},$$

we get

$$\exp\left\{ \frac{p}{2\pi} \int_0^{2\pi} h^+ \left(re^{i\gamma} \right) d\gamma \right\} \leqslant \frac{1}{2\pi} \left\{ \int_E | \Phi(re^{i\gamma}) |^p \, d\gamma + \int_{CE} d\gamma \right\}$$

$$= \left[H_p(\Phi) + 1 \right] < \infty.$$

Consequently,

$$\frac{1}{2\pi} \int_0^{2\pi} h(re^{i\gamma}) \, d\gamma \leqslant \frac{1}{p} \ln\left[H_p(\Phi) + 1 \right] < \infty.$$

We consider an increasing sequence of numbers r_k, $r_k \to 1$, and in each disk $|z| < r_k$ we construct a harmonic function $u_k(z)$ taking the continuous values $h^+ \left(r_k e^{i\gamma} \right)$ on the circle $|z| = r_k$. By the maximum principle for harmonic functions we have $h(z) \leqslant u_k(z)$, $|z| \leqslant r_k < 1$, and

$$u_k(0) = \frac{1}{2\pi} \int_0^{2\pi} u_k \left(r_k e^{i\gamma} \right) d\gamma \leqslant \frac{1}{p} \ln\left[H_p(\Phi) + 1 \right] < \infty.$$

At those points of the circle $|z| = r_k$ where $h(z) \geqslant 0$ we have

$$u_k \left(r_k e^{i\gamma} \right) = h \left(r_k e^{i\gamma} \right) \leqslant u_{k+1} \left(r_k e^{i\gamma} \right),$$

and, since $u_{k+1}(r_k e^{i\gamma}) \geqslant 0$ for all γ, it follows that

$$u_k \left(r_k e^{i\gamma} \right) \leqslant u_{k+1} \left(r_k e^{i\gamma} \right)$$

on the remaining part of the circle, where $u_k(r_k e^{i\gamma}) = 0$. Consequently, $u_k(z) \leqslant u_{k+1}(z)$ everywhere in the disk $|z| \leqslant r_k < r_{k+1}$.

Thus, in each disk $|z| < \rho < 1$ we have constructed a nondecreasing sequence of harmonic functions $u_k(z)$ (the corresponding r_k are greater than p) that converges by virtue of boundedness at the origin:

$$u_k(0) \leqslant (1/p) \ln\left[H_p(\Phi) + 1 \right].$$

By Harnack's theorem [69], such a sequence converges uniformly on compact subsets of $|z| < 1$ to a harmonic function $u(z)$, and, by construction, $h(z) \leqslant u(z)$ for $u(z) \geqslant 0$. Setting $u(z) - h(z) = u_0(z) \geqslant 0$, we arrive at a representation of the harmonic function $h(z)$ as a difference of nonnegative harmonic functions:

$$\ln | \Phi_0(z) | = h(z) = u(z) - u_0(z), \qquad u, u_0 \geqslant 0.$$

Consequently, $\Phi_0(z)$ can be represented in the form

$$\Phi_0(z) = \Phi_1(z) \left[\Phi_2(z) \right]^{-1},$$

where $\Phi_1(z)$ and $\Phi_2(z)$ are analytic in the disk $|z| < 1$ and are connected with the functions $u_0(z)$ and $u(z)$ by the relations

$$|\Phi_1(z)| = \exp\{-u_0(z)\}, \qquad |\Phi_2(z)| = \exp\{-u(z)\}.$$

Since $\Phi_1(z)$ and $\Phi_2(z)$ are bounded in $|z| < 1$, Fatou's theorem (Theorem 3 in §2) tells us that they have limit values almost everywhere on the circle, and the original analytic function

$$\Phi(z) = b(z)\Phi_0(z) \equiv b(z)\Phi_1(z)[\Phi_2(z)]^{-1}$$

therefore has limit values $\Phi(e^{i\gamma})$ almost everywhere.

To prove (9) we represent $\Phi(z)$ in the form

$$\Phi(z) = b(z)\Phi_0(z) \equiv b(z)\{F(z)\}^{2/p},$$

where $\Phi_0(z)$ does not have zeros in $|z| < 1$. Then

$$\frac{1}{2\pi} \int_0^{2\pi} |F(re^{i\gamma})|^2 \, d\gamma = \frac{1}{2\pi} \int_0^{2\pi} |\Phi_0(re^{i\gamma})|^p \, d\gamma \le H_p(\Phi) < \infty,$$

i.e., the analytic function $F(z)$ belongs to H_2. For fixed $r < 1$ let us expand $F(re^{i\gamma})$ in the series

$$F(re^{i\gamma}) = \Sigma c_k r^k e^{ik\gamma}$$

and use Parseval's equality ([99], Chapter VII, §3, formula (5))

$$H_2(F, r) = \frac{1}{2\pi} \int_0^{2\pi} |F(re^{i\gamma})|^2 \, d\gamma = \sum_{k=0}^{\infty} |c_k|^2 r^{2k}.$$

The series $\Sigma_0^{\infty} |c_k|^2$ converges, because $\lim_{r \to 1} H_2(F, r)$ is bounded. Consequently, by the Riesz-Fischer theorem ([99], Chapter VII, §3, Theorem 3), there exists a function $F(e^{i\gamma})$ whose Fourier series as $\Sigma_0^{\infty} c_k e^{ik\gamma}$, and

$$\frac{1}{2\pi} \int_0^{2\pi} |F(e^{i\gamma})|^2 \, d\gamma = \sum_{k=0}^{\infty} |c_k|^2.$$

We fix arbitrarily a branch of the function $\{F(e^{i\gamma})\}^{2/p} \equiv \Phi(e^{i\gamma})b^{-1}(e^{i\gamma})$. Then

$$\lim_{r \to 1} \frac{1}{2\pi} \int_0^{2\pi} |\Phi(re^{i\gamma})|^p \, d\gamma = \lim_{r \to 1} \frac{1}{2\pi} \int_0^{2\pi} |b(re^{i\gamma})|^p \, |F(re^{i\gamma})|^2 \, d\gamma$$

$$\le \lim_{r \to 1} \frac{1}{2\pi} \int_0^{2\pi} |F(re^{i\gamma})|^2 \, d\gamma = \frac{1}{2\pi} \int_0^{2\pi} |\Phi(e^{i\gamma})^p \, d\gamma.$$

On the other hand, by Fatou's lemma ([99], Chapter VI, §1, Theorem 9) on passing to the limit under the Lebesgue integral sign, we have

$$\frac{1}{2\pi} \int_0^{2\pi} |\Phi(e^{i\gamma})|^p \, d\gamma \le \lim_{r_n \to 1} \frac{1}{2\pi} \int_0^{2\pi} |\Phi(r_n e^{i\gamma})|^p \, d\gamma,$$

which, together with the preceding inequality, proves (9). To prove (10) we choose an arbitrary sequence of numbers $r_k < 1$, $r_k \to 1$, and use Minkowski's inequality (1.3):

$$\left\{ \int_0^{2\pi} |\, \Phi(r_n e^{i\gamma}) - \Phi(e^{i\gamma})\,|^p \, d\gamma \right\}^{1/p} \leqslant \left(\int_E |\, \Phi(r_n e^{i\gamma}) - \Phi(e^{i\gamma})\,|^p \, d\gamma \right)^{1/p}$$

$$+ \left(\int_l |\, \Phi(r_n e^{i\gamma})\,|^p \, d\gamma \right)^{1/p} + \left(\int_l |\, \Phi(e^{i\gamma})\,|^p \, d\gamma \right)^{1/p},$$

where the measurable set $E = [0, 2\pi] - l$ is chosen according to Egorov's theorem ([99], Chapter IV, §3, Theorem 5) in such a way that the sequence $\Phi(r_n e^{i\gamma})$ converges uniformly to $\Phi(e^{i\gamma})$ on it. Then the first term on the right-hand side of the resulting inequality tends to zero as $r_n \to 1$, while the limits

$$\lim_{r_k \to 1} \left\{ \int_l |\, \Phi(r_k e^{i\gamma})\,| \right\}^{1/p} = \left\{ \int_l |\, \Phi(e^{i\gamma})\,|^p \, d\gamma \right\}^{1/p}$$

of the second term and of the third term can be made as small as desired by making the measure of the set l small. The theorem is proved.

THEOREM 3 (F. Riesz and G. Fikhtengol'ts). *The function $\Phi(z)$ is in the class H_p, $p \geqslant 1$, if and only if it is represented by a Cauchy-Lebesgue integral*

$$\Phi(z) = \frac{1}{2\pi i} \int_{|t|=1} \frac{\Phi(t)\, dt}{t - z} \tag{11}$$

whose limit values $\Phi^+(t_0)$ are equal almost everywhere to $\Phi(t)$ and are pth power summable.

Since the Cauchy-Lebesgue integral is identically zero outside the disk $|z| < 1$, it follows that

$$\Phi(z) = \frac{1}{2\pi i} \int_{|t|=1} \frac{\Phi(t)\, dt}{t - z} - \frac{1}{2\pi i} \int_{|t|=1} \frac{\Phi(t)\, dt}{t - z_*}$$

$$= \frac{1}{2\pi} \int_0^{2\pi} \Phi(t) K_1(r, \gamma - \gamma_0)\, d\gamma,$$

where $z = re^{i\gamma_0}$ $(r > 1)$, $z_* = 1/z$, and K_1 is the Poisson kernel (see (2.7)),

$$K_1(r, \gamma - \gamma_0) = \operatorname{Re} \frac{t + z}{t - z}.$$

Then

$$\int_0^{2\pi} |\Phi(re^{i\gamma_0})|^p \, d\gamma_0 \leq \int_0^{2\pi} \left[\frac{1}{2\pi} \int_0^{2\pi} |\Phi(e^{i\gamma})| \, K_1(r, \gamma - \gamma_0) \, d\gamma \right]^p d\gamma_0$$

$$= \left(\frac{1}{2\pi} \right)^p \int_0^{2\pi} \left[\int_0^{2\pi} |\Phi(e^{i\gamma})| \, K_1^{1/p} \, d\gamma \right]^p d\gamma_0$$

$$\leq \left(\frac{1}{2\pi} \right)^p \int_0^{2\pi} \left\{ \int_0^{2\pi} |\Phi(e^{i\gamma})|^p K_1(r, \gamma - \gamma_0) \, d\gamma \left[\int_0^{2\pi} K_1(r, \gamma - \gamma_0) \, d\gamma \right]^{p/q} \right\} d\gamma_0$$

$$= \int_0^{2\pi} |\Phi(e^{i\gamma})|^p \, d\gamma,$$

i.e., $\Phi(z) \in H_p$, $p \geq 1$, where we have used the easily verified equality

$$\frac{1}{2\pi} \int_0^{2\pi} K_1(r, \gamma - \gamma_0) \, d\gamma = 1.$$

Thus, the first part of the theorem is proved.

Suppose now that the function $\Phi(z) = \Sigma_0^\infty c_k z^k$ belongs to the class H_p, $p \geq 1$. Then, by Theorem 2, it has limit values $\Phi(e^{i\gamma}) \in L_p$, $p \geq 1$, almost everywhere. Let us consider the Cauchy-Lebesgue type integral

$$F(z) = \frac{1}{2\pi} \int_{|t|} \frac{\Phi(t) \, dt}{t - z} = \sum_{k=0}^{\infty} z^k \left\{ \frac{1}{2\pi} \int_0^{2\pi} \Phi(e^{i\gamma}) e^{-ik\gamma} \, d\gamma \right\},$$

where the series on the right-hand side converges in $|z| < 1$ because the Fourier coefficients

$$a_k = \frac{1}{2\pi} \int_0^{2\pi} \Phi(e^{i\gamma}) e^{-ik\gamma} \, d\gamma$$

of the summable function $\Phi(e^{i\gamma})$ converge to zero. On the other hand, by the Cauchy formulas for the function $\Phi(z) = \Sigma_0^\infty c_k z^k$ in the disk $|z| < r < 1$,

$$c_k r^k = \frac{1}{2\pi} \int_0^{2\pi} \Phi(re^{i\gamma}) e^{-ik\gamma} \, d\gamma.$$

Then the equalities (10) give us that

$$\lim_{r \to 1} |c_k r^k - a_k| \leq \lim_{r \to 1} \frac{1}{2\pi} \int_0^{2\pi} |\Phi(re^{i\gamma}) - \Phi(e^{i\gamma})| \, d\gamma = 0,$$

i.e., $a_k = c_k$, whence $F(z) \equiv \Phi(z)$, and, consequently, $\Phi(z)$ is a Cauchy-Lebesgue integral. The theorem is proved.

COROLLARY. *The Cauchy theorem holds for each function* $\Phi(z) \in H_p$, $p \geq 1$:

$$\int_{|t|=1} \Phi^+(t) \, dt = 0.$$

Indeed, for each $r < 1$ the Cauchy theorem gives us that

$$\int_0^{2\pi} \Phi(re^{i\gamma}) \, de^{i\gamma} = 0.$$

And since

$$\left| \int_0^{2\pi} r\Phi(re^{i\gamma}) \, de^{i\gamma} - \int_0^{2\pi} \Phi^+(e^{i\gamma}) \, de^{i\gamma} \right|$$

$$\leqslant \int_0^{2\pi} |\Phi(re^{i\gamma}) - \Phi^+(e^{i\gamma})| \, d\gamma + (1-r)\int_0^{2\pi} |\Phi^+(e^{i\gamma})| \, d\gamma,$$

the relation (10) with $r \to 1$ leads to the Cauchy theorem for $r = 1$.

THEOREM 4 (V. I. SMIRNOV). *Suppose that* $\Phi(z) \in H_p, p > 0$. *If* $|\Phi(e^{i\gamma})| \leqslant M$ *for almost all* $\gamma \in [0, 2\pi]$, *then* $|\Phi(z)| \leqslant M$ *for all* $|z| < 1$. *If* $\Phi(e^{i\gamma}) \in L_q[0, 2\pi]$ *for* $q > p$, *then* $\Phi(z) \in H_q$.

We represent $\Phi(z)$ according to (8):

$$\Phi(z) = b(z)\Phi_0(z), \qquad \Phi_0(z) \neq 0 \text{ for } |z| < 1,$$

and let $\Phi_0^p(z) \in H^1$ be a fixed single-valued branch. By the preceding theorem, the analytic function $[\Phi_0(z)]^p$ in $|z| < 1$ is representable by a Cauchy-Lebesgue integral, and, by (12), also by a Poisson-Lebesgue integral:

$$\Phi_0^p(re^{i\gamma_0}) = \frac{1}{2\pi} \int_0^{2\pi} \Phi_0^p(e^{i\gamma}) K_1(r, \gamma - \gamma_0) \, d\gamma.$$

Since $|\Phi(z)| \leqslant |\Phi_0(z)|$ for $|z| < 1$, and

$$|\Phi(e^{i\gamma_0})| = |\Phi_0(e^{i\gamma_0})|$$

almost everywhere in $[0, 2\pi]$, it follows that

$$|\Phi(re^{i\gamma_0})|^p \leqslant \frac{1}{2\pi} \int_0^{2\pi} |\Phi(e^{i\gamma})|^p K_1(r, \gamma - \gamma_0) \, d\gamma \leqslant M^p,$$

which proves the first part of the theorem. Applying Hölder's inequality with the exponents $p_0 = q/p < 1$ and $q_0 = q/(q-p)$, we get

$$|\Phi(re^{i\gamma_0})|^p \leqslant \frac{1}{2\pi} \int_0^{2\pi} |\Phi(e^{i\gamma})|^p K_1^{p/q} K_1^{(q-p)/q} \, d\gamma$$

$$\leqslant \frac{1}{2\pi} \left(\int_0^{2\pi} |\Phi(e^{i\gamma})|^q K_1(r, \gamma - \gamma_0) \, d\gamma \right)^{p/q} \left(\int_0^{2\pi} K_1(r, \gamma - \gamma_0) \, d\gamma \right)^{(q-p)/p}$$

$$= \frac{1}{(2\pi)^{p/q}} \left\{ \int_0^{2\pi} |\Phi(e^{i\gamma})|^q K_1(r, \gamma - \gamma_0) \, d\gamma \right\}^{p/q},$$

and so

$$|\Phi(re^{i\gamma_0})|^q \leqslant \frac{1}{2\pi} \int_0^{2\pi} |\Phi(e^{i\gamma})|^q K_1(r, \gamma - \gamma_0)\, d\gamma,$$

which after integration along the circle $|z| = r < 1$ gives

$$\int_0^{2\pi} |\Phi(re^{i\gamma_0})|^q\, d\gamma \leqslant \int_0^{2\pi} |\Phi(e^{i\gamma})^q|\, d\gamma < \infty,$$

i.e., $\Phi(z) \in H_q$.

THEOREM 5 (SMIRNOV). *Let* $\Phi(z) = \varphi(z) + i\psi(z)$ *be an analytic function in* $|z| < 1$ *whose real part* $\varphi(z)$ *is continuous for* $|z| \leqslant 1$. *Then* $\exp\{i\Phi(z)\} \in H_p$ *for any* $p > 0$.

The condition (2.13) of Theorem 7 in §2 (Zygmund) for some $m > 0$ follows from the continuity of $\varphi(e^{i\gamma})$, and so, by virtue of Theorem 7 in §2, $|\exp[i\Phi(z)]|^p$ is integrable for $0 < p < m$, which means that $\exp[i\Phi(z)]$ is in H_p. By the second assertion of the same theorem, the imaginary part $\psi(re^{i\gamma})$ of $\Phi(z)$ has limit values $\psi(e^{i\gamma})$ that are summable to any power; consequently,

$$\exp[i\Phi(e^{i\gamma})] \in L_p[0, 2\pi]$$

for any $p > 0$. Then the assertion to be proved follows from Theorem 4.

THEOREM 6 (SMIRNOV). *A function* $\Phi(z)$ *that is analytic in the unit disk and has a positive real part belongs to the class* H_p, *where* p *is any positive number less than* 1.

Indeed, if $|\arg \Phi(z)| < \pi/2$, then the relation

$$\Phi^p(r) = |\Phi(r)|^p e^{ip \arg \Phi}$$

implies that

$$|\Phi(z)|^p \leqslant \frac{\mathrm{Re}\, \Phi^p(z)}{\cos(p\pi/2)}, \qquad 0 < p < 1.$$

By the last inequality,

$$\frac{1}{2\pi} \int_0^{2\pi} |\Phi(\rho e^{i\gamma})|^p\, d\gamma \leqslant \frac{\mathrm{Re}\, \Phi^p(0)}{\cos(p\pi/2)}$$

which proves that $\Phi(z)$ is in H_p.

5°. The classes E_p. Suppose that D is a domain bounded by a Jordan rectifiable curve Γ.

DEFINITION 3. An analytic function $\Phi(z)$ in D *belongs to the class* E_p, $p > 0$, if there exists a sequence of rectifiable curves Γ_n that converge to Γ and are such that, uniformly with respect to n,

$$\int_{\Gamma_n} |\Phi(z)|^p |dz| \leqslant M. \tag{13}$$

The functions in E_p have many of the properties of the functions in H_p. We show first of all that condition (13) implies that

$$\int_{\Gamma_r} |\Phi(z)|^p |dz| \leqslant M, \tag{14}$$

where Γ_r is the image of the circle $|w| = r$ under a conformal mapping of the disk $|w| < 1$ onto D. Indeed, let $z = F_n(w)$ $(F_n(0) = z_0, \; F_n'(0) > 0)$ be the function conformally mapping the disk $|w| < 1$ onto D. Then, by (13),

$$\int_{\Gamma_n} |\Phi(z)|^p |dz| = \int_{|w|=1} |\Phi[F_n(w)]|^p |F_n'(w)| \, |dw| < M, \tag{15}$$

which means that $|\Phi(F_n)| |F_n'|^{1/p}|$ is in H_p. On the other hand, taking account of the facts that (by a theorem of Carathéodory ([41], Chapter II, §5, Theorem 1)) the mapping functions $F_n(w)$ converge to $F(w)$ uniformly in each disk $|w| < r < 1$ and that $\{\Phi(F_n) | F_n'|^{1/p}\} \in H_p$, we find that

$$\int_{\Gamma_r} |\Phi(z)|^p |dz| = \int_{|w|=r} |\Phi(F)|^p |F'| \, |dw| = \lim_{n\to\infty} \int_{|w|=r} |\Phi(F_n)|^p |F_n'| \, |dw| \leqslant M,$$

which proves (14). In all of what follows, we restrict ourselves to the use of functions of the class E_p in domains with Lyapunov boundaries. Therefore, $F'(e^{i\gamma}) \neq 0$ everywhere on the circle Γ_1, and, by the properties of H_p-functions, the function

$$\varphi(w) = \Phi[F(w)][F'(w)]^{1/p} \in H_p$$

has limit values $\varphi(e^{i\gamma})$ for almost all w, $|w| = 1$. This implies that $\Phi[F(w)]$ have limit values

$$\Phi[F(e^{i\gamma})] = \varphi(e^{i\gamma})[F'(e^{i\gamma})]^{-1/p}$$

almost everywhere on the circle, and thereby $\Phi(z)$ also has the limit values for almost all $z \in \Gamma = \partial D$.

We remark that Theorem 3 (F. Riesz and G. Fikhtengol'ts) and Theorems 4–6 of Smirnov obviously hold for E_p-functions in domains with Lyapunov boundaries, since the operation described above can be used to place these functions in correspondence with functions in H_p.

THEOREM 7. *An analytic E_{p_0}-function $\Phi(z)$ $(p_0 > 1)$ in a domain D with Lyapunov boundary Γ is summable in D to any power $p < 2p_0$.*

Representing a function $\Phi(z) \in E_{p_0}$, $p_0 > 1$, by a Cauchy-Lebesgue integral and using Hölder's inequality with exponents $p = (2 - \varepsilon)p_0$, $r = p_0/\alpha$, and $\lambda = p_0/p_0 - 1$, $1/p + 1/r + 1/\lambda = 1$, we get

$$(2\pi)^p \iint_{D_0} |\Phi(z)|^p \, dx \, dy \leqslant \iint_{D_0} \left(\int_\Gamma |\Phi(t)| \, |t - z|^{-1} \, ds \right)^p dx \, dy$$

$$\leqslant \iint_{D_0} \left[\int_\Gamma |\Phi(t)|^{1-\alpha} |t - z|^{-\beta} \left(|\Phi(t)|^\alpha |t - z|^{\beta-1} \right) ds \right]^p dx \, dy$$

$$\leqslant \|\Phi(t)\|_{L_{p_0}(\Gamma)}^{p/r} \iint_{D_0} \left(\int_\Gamma |t - z|^{-\beta\gamma} \, ds \right)^{p/\lambda} \left(\int_\Gamma |\Phi(t)|^{p_0} |t - z|^{(\beta-1)p} \, ds \right) dx \, dy,$$

where D_0 is any subdomain of D with $\overline{D}_0 \subset D$, and

$$\alpha = \frac{1 - \varepsilon}{2 - \varepsilon}, \qquad \beta = \frac{p - 2 + \varepsilon}{p + \varepsilon}, \qquad 0 < \varepsilon < 1.$$

But,

$$\beta\lambda = \frac{p}{p + \varepsilon} < 1, \qquad (1 - \beta)p = \frac{2p}{p + \varepsilon} < 2,$$

and, consequently,

$$\sup_{z \in \overline{D}} \int_\Gamma |t - z|^{-\beta\lambda} \, ds = M_1 < \infty, \qquad \sup_{t \in \Gamma} \iint_D |t - z|^{(\beta-1)p} \, dx \, dy = M_2 < \infty.$$

Thus, finally

$$(2\pi)^p \iint_D |\Phi(z)|^p \, dx \, dy \leqslant M_1^{p/\lambda} M_2 \|\Phi(t)\|_{L_{p_0}(\Gamma)}^{p/r+1} \equiv M < \infty, \qquad p = (2 - \varepsilon)p_0.$$

§5. The Hilbert boundary-value problem with piecewise continuous coefficients. Stability of the problem

The first proofs of the Gakhov formulas for solving the Riemann problem in the case of continuous coefficients were given by Mikhlin, Gokhberg, Kvedelidze, V. V. Ivanov, and Simonenko (see [121] for the references).

Somewhat later Kvedelidze considered the case of piecewise continuous coefficients. Further relaxations of the assumptions were obtained almost simultaneously and independently in the investigations of Simonenko, Widom, and Danilyuk (see [121]).

1°. Let us consider the Hilbert boundary-value problem

$$\mathrm{Re}[G(t)w(t)] = g(t), \qquad t = e^{i\gamma}, \tag{1}$$

with piecewise continuous coefficients $G(t)$ and $g(t)$, assuming, as usual, that $G(t) \neq 0$. The solution of the boundary-value problem (1) will be sought in the class of analytic functions $w(z) \in H_p$, $p > 1$, in the disk $|z| < 1$, whose limit values satisfy the boundary condition (1) almost everywhere on the circle. Proceeding analogously to the case of piecewise Hölder coefficients (§4.2° of Chapter I), we set $\theta_k = 1/2\pi\{\operatorname{Arg} G_1(t_k - 0) - \arg G_1(t_k + 0)\}$ and $G_1(t) = -\overline{G(t)}/G(t)$, where $\operatorname{Arg} G(t_k - 0)$ is computed from continuity on the arc $(t_{k-1}t_k)$. Let integers ν_k be chosen from the condition

$$-1 < \alpha_k = \theta_k - \nu_k \leqslant 0 \qquad (k = 1,\ldots,n) \tag{2}$$

and assume that $\kappa = \frac{1}{2}\Sigma_1^n \nu_k$ is an integer; in this case κ is called the index of the Hilbert problem (1), and the function $X(z)$ defined by the equalities

$$X(z) = z^\kappa \prod_{k=1}^n (z - t_k)^{\alpha_k} e^{\Gamma_0(z)}, \tag{4}$$

$$\Gamma_0(z) = \frac{1}{4\pi i} \int_{|t|=1} \ln[G_0(t)] \frac{(t + z)\, dt}{(t - z)t}, \tag{5}$$

is called the canonical function of the homogeneous Hilbert problem (1), where

$$G_0(t) = -t^{-2\kappa} \frac{\overline{G(t)}}{G(t)} \prod_{k=1}^n (-tt_k)^{-\alpha_k},$$

which is a continuous function (cf. §4.2° of Chapter I). Since the density of the Schwarz integral (5) is continuous, Theorem 5 in §4 gives us that $\exp\{\pm\Gamma_0(z)\} \in H_p$ for any $p > 0$. Since the α_k are nonpositive, each of the functions $(z - t_k)^{\alpha_k}$ is in the class H_{p_k}, $1 < p_k < 1/|\alpha_k|$, and, consequently,

$$\prod_{k=1}^n (z - t_k)^{\alpha_k} \equiv \Pi(z) \in H_p, \qquad 1 < p < \min_k \frac{1}{|\alpha_k|},$$

$$[\Pi(z)]^{-1} \in H_p, \qquad p > 0 \text{ arbitrary.}$$

Thus, we arrive at the relations

$$[z^{-\kappa}X(z)] \in H_p, \qquad 1 < p < \min_k \frac{1}{|\alpha_k|}, \tag{6}$$

$$[z^{-\kappa}X(z)]^{-1} \in H_p, \qquad p > 0 \text{ arbitrary.}$$

As shown in §2, the Schwarz integral $\Gamma_0(z)$ has limit values because its density is integrable to the power $p > 1$, and they coincide almost everywhere on the circle with the density $\ln[G_0(t)]$. Therefore, the limit values $X(t)$ (which exist almost everywhere) satisfy the homogeneous boundary condition (1) (with $g \equiv 0$) by construction, as in the case of piecewise Hölder coefficients. To construct the general solution $w = w(z)$ of the homogeneous Hilbert problem we consider the function

$$\Phi(z) = \frac{w(z)}{X(z)}, \qquad w(z) \in H_p, p > 1.$$

Since, according to (6), $[X(z)z^{-\kappa}]^{-1} \in H_p$ for arbitrary $p > 0$, it follows that $[z^\kappa \Phi(z)] \in H_p, p > 1$. The limit values of $\Phi(z)$, which exist almost everywhere on the circle, satisfy the condition

$$\mathrm{Re}[\Phi(t)G(t)X(t)] = [-iX(t)G(t)]\,\mathrm{Re}[i\Phi(t)] = 0,$$

since $X(t)$ satisfies the homogeneous condition (1). Thus,

$$\mathrm{Im}\,\Phi(t) = 0 \tag{7}$$

almost everywhere on the circle. For negative κ we have

$$\Phi(z) = \frac{w_0(z)}{X(z)} \in H_p, \quad p > 1 \quad \text{and} \quad \Phi(0) = 0.$$

Therefore, the function $[i\Phi(z)]$ can be represented by a Cauchy-Lebesgue integral and can thereby be represented also by a Schwarz integral with density $\varphi(t) = \mathrm{Re}[i\Phi(t)] = 0$, so that $\Phi(z) \equiv 0$. For $\kappa \geq 0$ the function $\Phi(z)$, which solves the Schwarz problem (7), admits a pole of order κ at the origin, and $[z^\kappa \Phi(z)] \in H_p, p > 1$. Therefore,

$$\Phi(z) = P_\kappa(z) = \sum_{k=0}^{\kappa} \left(c_k z^k + \bar{c}_k z^{-k} \right).$$

Thus, for $\kappa < 0$ the homogeneous Hilbert problem (1) does not have solutions in the class $H_p, p > 1$, while for $\kappa \geq 0$ the general solution $w_0(z) \in H_p, p > 1$, of the homogeneous problem (1) can be written in the form

$$w_0(z) = z^\kappa P_\kappa(z) = \prod_{k=1}^{n} (z - t_k)^{\alpha_k} e^{\Gamma_0(z)}. \tag{8}$$

If in the case of the nonhomogeneous Hilbert problem (1) we represent a solution $w(z)$ in the form $w(z) = \Phi(z)X(z)$, where $[z^\kappa \Phi(z)] \in H_p, p > 1$, then we arrive at the problem

$$\mathrm{Re}\{i\Phi(t)\} = g(t)[-iG(t)X(t)]^{-1} \equiv \nu(t),$$

where, by (6) and the piecewise continuity of $g(t)$, we have $\nu(t) \in L_p$ for any $p > 1$. Solving the Schwarz problem thus obtained for the function $i\Phi(z)$, we find by the usual formulas the general solution

$$w(z) = \frac{X(z)}{2\pi i} \int_{|t|=1} \frac{g(t)(t+z)\,dt}{X(t)G(t)(t-z)t} + P_\kappa(z)X(z) \tag{9}$$

of the nonhomogeneous Hilbert problem (1), where $P_\kappa(z) \equiv 0$ for $\kappa < 0$, and from the conditions for the function $w(z)$ in (9) to be bounded at the origin we get the usual solvability condition on $g(t)$:

$$\int_{|t|=1} \frac{g(t)}{G(t)X(t)} t^{k-1} \, dt = 0 \qquad (k = 0, \dots, |\kappa| - 1). \tag{10}$$

Let us state this result as a theorem.

THEOREM 1. *The Hilbert boundary-value problem* (1) *with piecewise continuous coefficients in the boundary condition and with* $\kappa < 0$ *has a unique solution in* H_p, $p > 1$, *only under the* $2|\kappa| - 1$ *solvability conditions* (10). *For* $\kappa \geq 0$ *the general solution in* H_p, $p > 1$, *of the Hilbert problem* (1) *depends on* $2\kappa + 1$ *arbitrary real constants.*

2°. We now investigate the stability of the Hilbert problem (1) with piecewise continuous coefficients under small changes in the coefficients of the boundary condition. A sequence of functions $w_n(z) \in H_p$, $p > 0$, is said to converge in H_p if $H_p(w_n)$ (see (1*) in §4) converges to $H_p(w_0)$ for a limit function $w_0(z) \in H_p$. If the terms of the sequence $\{w_n(z)\}$ are represented as Cauchy-Lebesgue type integrals

$$w_n(z) = \frac{1}{2\pi i} \int_{|t|=1} \frac{\varphi_n(t) \, dt}{t - z}, \qquad \varphi_n(t) \in L_p, p > 1, \tag{11}$$

then for it to converge in H_p to a function $w_0(z)$ it suffices that

$$\| \varphi_n(t) - \varphi_0(t) \|_{L_p} \to 0 \quad \text{as } n \to \infty,$$

where $\varphi_0(t)$ is the density in the representation of $w_0(z)$ by a Cauchy-Lebesgue type integral.

We introduce a norm for an element $w(z)$ in H_p, $p > 1$, by setting

$$\| w(z) \|_{H_p} = \lim_{r \to 1} \left\{ \frac{1}{2\pi} \int_0^{2\pi} | w(re^{i\gamma}) |^p \, d\gamma \right\}^{1/p}. \tag{12}$$

It can be shown ([136], Vol. I, Chapter VII, §7, Theorem 7.60) that the set H_p ($p > 1$) of functions forms a Banach space under this norm. If a function $w(z) \in H_p$, $p > 1$, is represented by a Cauchy-Lebesgue type integral (11), then, by taking estimates in L_p ($p > 1$) of such integrals for $z = t_0 = e^{i\gamma_0}$ into account (the theorems of F. Riesz in §2), we get

$$\| w(z) \|_{H_p} \leq M_p \| \varphi(t) \|_{L_p}, \qquad p > 1.$$

Recall that, according to Definition 1 in §5 of Chapter I, piecewise continuous functions $F^1(t)$, $F^2(t) \in C_*$, on the circle $|t| = 1$ are close in C_* if the points of discontinuity t_k^1 and t_k^2 $(k = 1,\ldots,n)$ are subject to the inequalities

$$0 < \delta_0 < \inf | t_{k+1}^i - t_k^i |, \qquad \max_k | t_k^1 - t_k^2 | < M_0 \varepsilon \qquad (13)$$

and there is a homeomorphism $\tau = \tau(t) \in C^\beta, \beta > 0$, that carries the points of discontinuity of $F^1(t)$ into those of $F^2(t)$ and is such that

$$\max_t | F^1[\tau(t)] - F^2(t) | < \varepsilon_1(\varepsilon), \qquad (14)$$

where $\varepsilon_1(\varepsilon) \to 0$ as $\varepsilon \to 0$.

LEMMA 1. *Let $w(z)$ be a solution of the Hilbert problem* (1) *represented by* (9), *and*

$$u(t) = \ln[G_0(t)], \qquad v(t) = g(t)[X(t)G(t)]^{-1}. \qquad (15)$$

Then

$$\|u(t)\|_C \le M_u, \qquad \|v(t)\|_{L_p} \le M_v, \qquad \|w(z)\|_{H_p} \le M_w \qquad (16)$$

for $t = e^{i\gamma}$, $\gamma \in [0, 2\pi]$, and for sufficiently large $p_0 > 1$ and $p > 1$ ($p = p(\alpha_k, p_0)$), where the constants $M < \infty$ depend only on the properties of $G(t)$ and $g(t)$. If

$$\inf_{i,k} | t_k^i - t_{k+1}^i | \ge \delta_0 > 0 \qquad (k = 1,\ldots,n; i = 1,2,\ldots),$$

and $t_k^i \to t_k^0$, $\|u^i - u^0\|_C \to 0$ and $\|v^i - v^0\|_{L_p} \to 0$ as $i \to \infty$, then

$$\|w^i(z) - w^0(z)\|_{H_p} \to 0, \qquad p > 1.$$

The inequalities (16) follow directly from the definitions of $G_0(t)$ and $v(t)$ and from the representation (9) for $w(z)$. If we take account of the convergence in C and L_{p_0} ($p_0 > 1$ is arbitrarily large) of the respective densities $u^i(t) = \ln[G_0^i(t)]$ and $v^i(t) = g(t)[X^i(t)G^i(t)]^{-1}$ of the Cauchy-Lebesgue type integrals in the representations (9) for the $w^i(z)$, then the last assertion of the lemma also becomes obvious.

The following assertion can be established similarly to §5 of Chapter I.

LEMMA 2. *If $t^i = t^i(\tau) \in C^\beta, 0 < \beta < 1$, is a sequence of homeomorphisms of the circle $|\tau| = 1$ onto itself, and*

$$\lim_{i \to \infty} \|t^i(\tau) - t^0(\tau)\|_\beta = 0, \qquad (17)$$

then the following sequences converge to zero:

$$\|u[t^i(\tau)] - u[t^0(\tau)]\|_C,$$

$$\|v[t^i(\tau)] - v[t^0(\tau)]\|_{L_{p_0}} \quad and \quad \|w^i(z) - w^0(z)\|_{H_p},$$

where u, v, and w are defined by the respective equalities (15) *and* (9).

By (17), the sequence $\| u[t^i(\tau)] - u[t^0(\tau)] \|_C$ obviously converges to zero, and this implies immediately that $\| \Gamma_0^i(z) - \Gamma_0^0(z) \|_{H_{p_0}} \to 0$, where $\Gamma_0(z)$ is the Schwarz integral defined by (5), and $p_0 > 1$ is an arbitrarily large number. Then the sequence $\| v[t^i(\tau)] - v[t^0(\tau)] \|_{L_{p_0}}$ also converges to zero, and therefore, by Lemma 1, $\| w^i(z) - w^0(z) \|_{H_p} \to 0.$

DEFINITION 1. The Hilbert boundary-value problem (1) is said to be *stable in* H_p, $p > 1$, if (9) gives solutions $w^1(z)$ and $w^2(z)$ of this problem that are close in H_p whenever the coefficients $G^1(t)$, $g^1(t)$ and $G^2(t)$, $g^2(t)$ in the boundary condition (1) are close in C_* (in the sense of (13) and (14)), i.e., $\| w^1(z) - w^2(z) \|_{H_p} < \varepsilon_1(\varepsilon)$, $\varepsilon_1(\varepsilon) \to 0$ as $\varepsilon \to 0$.

THEOREM 2. *The Hilbert boundary-value problem* (1) *with piecewise continuous coefficients of the boundary condition is stable in* H_p, $p > 1$, *under small changes of the coefficients in* C_*.

As in the corresponding Theorem 1 in §5 of Chapter I for the Hilbert problem with piecewise Hölder coefficients, the stability of the boundary-value problem (1) with piecewise continuous coefficients is an immediate consequence of Lemmas 1 and 2 proved above.

However, Theorem 2 can also be established directly from a consideration of the boundary-value problem for the difference $w(z) = w^1(z) - w^2(z)$ of the solutions of the Hilbert boundary-value problem (1) that correspond to coefficients $G^1(t)$, $g^1(t)$ and $G^2(t)$, $g^2(t)$ that are close in C_*:

$$\operatorname{Re}[G^1(t)w(t)] = \tilde{g}(t), \tag{18}$$

where $\tilde{g}(t) = g^1 - g^2 + \operatorname{Re}[w^2(t)(G^2 - G^1)]$. Indeed, $|g^1(t) - g^2(t)| < \varepsilon$ ($|G^2 - G^1| < \varepsilon$) on common intervals of continuity of the functions g^1 and g^2 (G^1 and G^2), and the lengths of intervals where the difference $|g^1 - g^2|$ is just bounded are small. Taking account of the fact that $\| w^2(z) \|_{H_p} < M_2 < \infty$ for some $p > 1$, we find that $\| \tilde{g}(t) \|_{L_p} < \varepsilon_1(\varepsilon)$, $\varepsilon_1(\varepsilon) \to 0$ as $\varepsilon \to 0$. Then

$$\left\| \frac{\tilde{g}(t)}{G^1(t)X^1(t)} \right\|_{L_q} < \varepsilon_2(\varepsilon)$$

for $1 < q < p$, and $\varepsilon_2 \to 0$ as $\varepsilon \to 0$, since $[X^1(t)]^{-1} \in L_{p_0}$ for any $p_0 > 1$. Consequently, the formula (9) for the solution $w(z) = w^1(z) - w^2(z)$ of the boundary-value problem (18) gives us that

$$\| w(z) \|_{H_q} \leqslant \varepsilon_3(\varepsilon) \to 0 \quad \text{as } \varepsilon \to 0,$$

which proves Theorem 2.

§6. Boundary-value problems with measurable coefficients

We formulate a Riemann boundary-value problem. Let $\Gamma \in C^{1+\alpha}$, $\alpha > 0$, consist of finitely many closed disjoint contours Γ_k that bound a domain D. Denote by D^- the (generally speaking) nonconnected domain complementing $D^+ + \Gamma$ in the whole plane.

It is required to find analytic functions $\Phi^{\pm}(z) \in E_\lambda(D^{\pm})$, in $D^{\pm}$ that satisfy on the contour $\Gamma \in C^{1+\alpha}$ the boundary condition

$$\Phi^+(t) = G(t)\Phi^-(t) + g(t), \tag{1}$$

where $G(t)$ and $g(t)$ are given measurable functions on Γ, with $g(t) \in L_\lambda$, $\lambda > 1$. Following Simonenko [121], we assume that the coefficient $G(t)$ satisfies the following conditions:

(i) $0 < M_1 \leqslant |G(t)| \leqslant M_2 < \infty$.

(ii) There exists a $\delta > 0$ such that for each point $t_0 \in \Gamma$ there is a circle in which the values of $G(t)$ are contained in a sector with vertex at the origin and with angular opening $(\pi/p - \delta), p > 1$.

We remark that $\lambda > 1$ can be chosen arbitrarily in $[2p/(2p - 1), 2p]$. By applying the Heine-Borel covering theorem, it is not hard to see that condition (ii) can be replaced by

(iii) There exists a finite covering (B) of Γ by intervals such that the values of $G(t)$ on each of them are contained in a sector with opening $[\pi/p - \delta_1]$, $0 < \delta_1 < \delta$.

Bearing this in mind, we take a point t_k on each contour Γ_k and cut it at this point, i.e., we take the point t_k to be two points t_k^+ and t_k^-. From the covering (B) of the uncut contours Γ_k we form a covering of the cut contours Γ_k by replacing the intervals containing t_k by half-open intervals with closed ends at t_k. Fixing arbitrarily the value of $\arg G(t)$ at the points t_k to which there correspond fixed points $G(t)$ on the G-plane, and following the direction of circuit of the contour Γ along the chain of intervals, we successively determine $\arg G(t)$ on the intervals encountered in such a way that

$$|\arg G(t) - \arg G(t')| \leqslant \pi,$$

when t and t' belong to a single interval. As a result we get a completely defined branch of the argument $\varphi(t)$ that has two values at the points t_k: $\varphi(t_k^-)$ before traversing and $\varphi(t_k^+)$ after traversing. With this definition the function $\varphi(t)$ obviously depends on the t_k and does not depend on the choice of the covering (B) that satisfies (ii).

However, the difference

$$\frac{1}{2\pi}\{\varphi(t_k^+) - \varphi(t_k^-)\} = \kappa_k \tag{2}$$

does not depend in addition on the choice of the t_k.

The number

$$\sum_{k=0}^{m} \kappa_k = \kappa = \operatorname{ind} G \tag{3}$$

is defined to be the index of the Riemann problem (1). The values $\varphi(t_k^{\pm})$ correspond to one and the same point $G(t_k)$ on the G-plane, so κ is an integer. In order to see more easily the construction of $\arg G(t) = \varphi(t)$, we approximate $G(t)$ on each interval of the covering (B) by continuous functions $G^i(t)$ that converge almost everywhere to $G(t)$. In passing to each next interval B_{j+1} in (B) we use the value of $\arg G^i(t)$ chosen on the preceding interval at the point $t_{j,j+1} \in B_j \cap B_{j+1}$ with which the determination of $\arg G^i$ in B_{j+1} begins. For the continuous functions G^i on the intervals in (B) the above construction coincides with the usual definition of the index and does not depend on the choice of the initial point t_k. In the limit as $i \to \infty$ we arrive at the definition of κ_k given above, since the values of $G(t)$ on a set of measure zero are obviously not affecting the value of the index.

Writing $r(t) = |G(t)|$, we now represent $G(t)$ in the form $r(t)e^{i\varphi(t)}$.

LEMMA 1. *Every measurable function $G(t)$ satisfying conditions* (i) *and* (ii) *can be represented in the form*

$$G(t) = G_1(t)G_2(t). \tag{4}$$

Here $G_1(t)$ satisfies a Hölder condition, $\operatorname{ind} G_1(t) = \kappa$, and $|G_1(t)| = 1$ on Γ, while $G_2(t)$ satisfies (i) *and* (ri), *and*

$$|\arg G_2(t)| < \pi/2p - \delta_1, \qquad 0 < \delta_1 < \delta. \tag{5}$$

For a proof it suffices to construct a function $\psi(t) \in C^\alpha$ satisfying

$$|\varphi(t) - \psi(t)| \leqslant \pi/2p - \delta_1,$$

with $\psi(t_k^+) - \psi(t_k^-) = 2\pi\kappa_k$. Then

$$G_1(t) = e^{i\psi(t)}, \qquad G_2(t) = e^{i[\varphi(t) - \psi(t)]}|G(t)|.$$

Proceeding to the construction of $\psi(t)$, we observe that there exists a finite covering (B) such that the values of $\psi(t)$ on each interval B_i of it are contained in a segment of length less than $\pi/p - \delta_1$.

Let $\delta_2 > 0$ be so small that every arc of length less than δ_2 is entirely contained in some interval B_i of the covering (B) ($0 < \delta_2 < \max_i \operatorname{meas} B_i \cap B_{i+1}$). We assume that the contour Γ is cut at the points t_k, and divide it into a finite number of half-open arcs Γ_k of length less than $\delta_2/3$. The piecewise constant functions $f_1(t)$ and $f_2(t)$ are defined by

$$f_1(t) = \sup_{\tau \in \Gamma_l} \varphi(\tau), \quad t \in \Gamma_l, \qquad f_2(t) = \inf_{\tau \in \Gamma_l} \varphi(\tau), \quad \tau \in \Gamma_l.$$

The function $\varphi(t)$ is between the piecewise constant functions $f_1(t)$ and $f_2(t)$, and $0 \leqslant f_1(t) - f_2(t) \leqslant \pi/p - \delta_1$. Therefore, it suffices to construct a function $\psi(t)$ satisfying the conditions

$$|f_1(t) - \psi(t)| \leqslant \pi/2p - \delta_1, \qquad |f_2(t) - \psi(t)| \leqslant \pi/2p - \delta_1.$$

To do this we set

$$\psi(t_{l,l+1}) = \frac{1}{2}\{\max[f_1(\Gamma_l), f_1(\Gamma_{l+1})] + \min[f_2(\Gamma_l), f_2(\Gamma_{l+1})]\}$$

at the points $t_{l,l+1}$ common to the arcs Γ_l and Γ_{l+1}, while at the cut points t_k^+ and t_k^-

$$\varphi(t_k^{\pm})$$

$$= \frac{1}{2}\{\max[f_1(\tilde{\Gamma}_k^{\pm}), f_1(\tilde{\Gamma}_k^{\pm}) \pm 2\pi\kappa_k] + \min[f_2(\tilde{\Gamma}_l^{\pm}), f_2(\tilde{\Gamma}_{l+1}^{\pm})] \pm 2\pi\kappa_k\},$$

where $\tilde{\Gamma}_k^+$ and $\tilde{\Gamma}_k^-$ are the arcs abutting at the points t_k^+ and t_k^-. At the remaining points $\psi(t)$ is obtained by linear interpolation on the arc. The function $\psi(t)$ thus constructed satisfies the conditions of the lemma.

LEMMA 2. *The function $G_2(t)$ can be represented in the form*

$$G_2 = \frac{X^+(t)}{X^-(t)}, \qquad X(z) = \exp\left\{\frac{1}{2\pi i}\int_{\Gamma}\frac{\ln G_2(t)\,dt}{t-z}\right\}, \tag{6}$$

where $x^{-1}(z) - 1$, $X(z) - 1 \in E_{p+\varepsilon}(D^{\pm})$, $\varepsilon = \varepsilon(\delta_1) > 0$. Moreover, the homogeneous problem (1) *with $G \equiv G_2$ has only the trivial solution in the class $E_2(D^{\pm})$.*

We map the disk $|w| < 1$ conformally onto the domain D_k^+ by the function $z = \alpha(w)$ and represent $\ln X_k(z)$ in the form

$$\ln X_k = \frac{1}{2\pi i}\int_{|t|=1}\frac{\ln G_2|\alpha(t)|\,dt}{t-w} + \frac{1}{2\pi i}\int_{|t|=1}\ln G_2\left\{\frac{\alpha'(t)}{\alpha(t) - \alpha(w)} - \frac{1}{t-w}\right\}dt,$$

where the second integral is bounded because the kernel is integrable. Since the first of the integrals with density $\ln|G_2|$ has bounded real part, it remains to consider only the integral

$$\frac{1}{2\pi}\int_{|t|=1}\frac{\arg G_2\,dt}{t-w} = U(i\arg G_2|w) + \text{const.}$$

Here U denotes the Schwarz operator. Then Zygmund's theorem (Theorem 7 in §2), which is applicable to the function $\exp\{U(i\arg G_2)\}$ by virtue of (5), gives us the first part of the lemma. If we then represent $G_2(t)$ in the form (6), we arrive at the equality

$$\left[\frac{\Phi}{X}(t)\right]^{+} - \left[\frac{\Phi}{X}(t)\right]^{-} = 0, \qquad t \in \Gamma,$$

where the functions in both terms are representable by Cauchy integrals, since $\Phi^{\pm} \in E_p(D^{\pm})$ and $(X^{\pm})^{\pm 1} - 1 \in E_{p/(p-1)}(D^{\pm})$.
Consequently, $\Phi^{\pm}(z) = 0$, and the lemma is proved.

LEMMA 3. *The nonhomogeneous problem* (1) *with* $G(t) \equiv G_2(t)$ *and* $g(t) \in L_2(\Gamma)$ *is uniquely solvable in the class* $E_2(D^{\pm})$, *and its solution* $\Phi^{\pm}(z)$, $z \in D^{\pm}$, *can be represented by Gahov's formula*

$$\Phi^{\pm}(z) = \frac{X^{\pm}(z)}{2\pi i} \int_{\Gamma} \frac{g(\tau)\,d\tau}{X^{+}(\tau)(\tau - z)}, \qquad (7)$$

where $X^{\pm}(z)$ *is defined by* (6).

Note first that it is always possible to select a number $\gamma \neq 0$ such that

$$|1 - \gamma G_2(t)| \leq q < 1. \qquad (8)$$

Indeed, the values of $G(t)$ are contained in an annular sector $0 < M_1 \leq |G_2| \leq M_2 < \infty$ (condition (i)) with opening less than π (condition (5)). Consequently, a dilation with coefficient γ can be used to transform the latter into a new sector whose points are all at a distance less than 1 from 1.

We then consider the following problem, which is equivalent to the original one:

$$\Phi_1^{+} = \gamma G_2 \Phi_1^{-} + g, \qquad (9)$$

where $\Phi_1^{+} = \Phi^{+}$, $\Phi^{-} = \gamma \Phi_1^{-}$, and the number γ satisfies (7).
Taking the formulas

$$\Phi_1^{\pm} = \pm \frac{1}{2}\Phi_1 + \frac{1}{2\pi i} \int_{\Gamma} \frac{\Phi_1(t)\,dt}{t - z}, \qquad \Phi_1 = \Phi_1^{+} = \Phi_1^{-},$$

into account, we rewrite (9) in the form

$$\Phi_1 = (\gamma G_2 - 1)\left[-\frac{1}{2}\Phi_1 + \frac{1}{2\pi i} \int_{\Gamma} \frac{\Phi_1(\tau)\,d\tau}{\tau - t}\right] + g. \qquad (10)$$

Let the curves Γ_k be carried in a one-to-one fashion by conformal mappings into disjoint unit circles, and let $\beta(t) \in C^{1+\alpha}$, $\beta = \beta(t)$, be the corresponding boundary correspondence. With this taken into acocunt, (10) is transformed as follows:

$$\Phi_1^* = (\gamma G_2^* - 1)\left\{ -\frac{1}{2}\Phi_1^* + \frac{1}{2\pi i}\int_{\Gamma_k^*} \frac{\Phi_1^*(\tau)\,d\tau}{\tau - t}\right\}$$

$$+ \frac{\gamma G_2^* - 1}{2\pi i}\left\{\int_{\Gamma^* - \Gamma_k^*} \frac{\Phi_1^*(\tau)\,d\tau}{\tau - t} + \int_{\Gamma^*}\Phi_1^*(\tau)\left[\frac{\beta'(\tau)}{\beta(\tau) - \beta(t)} - \frac{1}{\tau - t}\right]d\tau\right\} + g^*.$$

Here we have introduced the following notation: $f^*(t) = f[\beta(t)]$; $\Gamma_k^* = \beta(\Gamma_k)$ is the unit circle that is the image of the curve Γ_k; and $\Gamma_1^* = \Sigma_k \Gamma_k^*$. By the thoerem of F. Riesz (Theorem 6 in §2),

$$\left\|\frac{1}{2\pi i}\int_{\Gamma_k^*} \frac{\Phi_1^*\,d\tau}{\tau - t}\right\|_{L_2(\Gamma_k^*)} \leq \frac{1}{2}\|\Phi_1^*\|_{L_2(\Gamma_k^*)},$$

so the norm of the the first term, as an operator in $L_2(\Gamma^*)$, is less than 1, and all the remaining terms are completely continuous operators. The proof of the first part of the lemma is concluded by applying the Fredholm alternative (Theorem 9 in §1) to the equation obtained for Φ_1^*, which, according to Lemma 2, has only the zero solution for $g^*(t) = 0$. Proceeding further as in the proof of Lemma 2, we transform the original Riemann problem to the jump problem

$$(\Phi/X)^+ = (\Phi/X)^- + g/X^+,$$

where, by the property of the functions we have $X^\pm(z)$, $(\Phi/X)^+ \in E_1(D^\pm)$, while $g/X^+ \in L_{1+\varepsilon}$ $(\varepsilon > 0)$. Recovering its solution with the help of the Cauchy-Lebesgue type integral, we arrive at (7), and the lemma is proved.

DEFINITION 1. The class of functions $\rho(t)$ that give rise to bounded operators

$$T\varphi = \rho(t)\int_\Gamma \rho^{-1}(\tau)\frac{\varphi(\tau)\,d\tau}{\tau - t}, \qquad \varphi \in L_p(\Gamma),$$

in the space $L_p(\Gamma)$ is denoted by A_p, $p > 0$.

Here are some obvious properties of these classes:
1) If $\rho \in A_p$, $\rho_1^{\pm 1} \in L_\infty$, and $\rho_1\rho \in A_p$.
2) If $\rho \in A_p$, then $\rho^{-1} \in A_{p/(p-1)}$.
3) If $\rho \in A_p$, then $|\rho| \in A^p$.
4) If $\rho \in A_p$, $d\rho_1^{\pm 1}/dt \in L_\infty$, then $\rho\rho_1 \in A_p$.

LEMMA 4 (on weights). *The functions*

$$X^{\pm}(t) = \exp\left\{ \pm \frac{1}{2} G_2 + \frac{1}{2\pi i} \int_{\Gamma} \frac{\ln|G_2(\tau)|\, d\tau}{\tau - t} \right\} \exp\left\{ \frac{1}{2\pi} \int_{\Gamma} \frac{\arg G_2(\tau)\, d\tau}{\tau - t} \right\}$$

(11)

introduced in Lemma 2 belong to A_λ *for* $\lambda \in [2p/(2p-1), 2p]$, $p \geqslant 1$.

The lemma is obvious for $\lambda = 2$ and $p = 1$. Indeed, passing in (7) to the limit values on the contour Γ, we get

$$\frac{X^+(t)}{2\pi i} \int_{\Gamma} \frac{g(\tau)\, d\tau}{X^+(\tau)(\tau - t)} = \left[\Phi^+(t) - \frac{1}{2} g(t) \right] \in L_2,$$

which implies that $X^+(t) \in A_2$. But since the first factor in (11) is bounded above and below, it follows at once that $X^-(t) \in A_2$.

Now let l be a number large enough so that

$$p < |\sigma| = \frac{p(l-2)}{l-2p} < \frac{\pi p}{\pi - p\delta_1},$$

and consider the weight

$$\rho^{\sigma} \equiv \rho_0 \exp\left\{ \frac{1}{2\pi i} \int_{\Gamma} \frac{\sigma \arg G_2\, d\tau}{\tau - t} \right\},$$

where $\rho(t) = X^{\pm}(t)$, and the meaning of $\rho_0(t)$ is clear from (11).

By the assumptions made about σ,

$$|\sigma \arg G_2| \leqslant \left(\frac{\pi}{2p} - \delta_1 \right) \frac{\pi p}{\pi - p\delta_1} = \frac{\pi}{2} \cdot \frac{\pi - 2p\delta_1}{\pi - p\delta_1} \equiv \frac{\pi}{2p_0} - \delta_0,$$

where $p_0 > 0$ and $\delta_0 > 0$; consequently, the assertions of Lemmas 2 and 3 remain valid for $G_2^* = G_2^{\sigma}$. By what was just proved, we then have for the corresponding $(X^*)^{\pm}$ that $(X^*)^{\pm} \in A_2$, and, since $|\rho_0^{\pm 1}| \in L_{\infty}$, it follows that also $\rho^{\sigma} \in A_2$.

To prove that $\rho^{\pm 1} \in A_{p_2}$ we use Stein's theorem (Theorem 12 in §2), with

$$\frac{1}{\alpha} = \frac{1}{\beta} = 2p;$$

$$Tf = \int_{\Gamma} \frac{f(\tau)\, d\tau}{\tau - t}; \qquad k_1 = u_1 = \rho^{\sigma}; \qquad \beta_1 = \alpha_1 = \frac{1}{2};$$

$$k_2 = u_2 = 1; \qquad \beta_2 = \alpha_2 = \frac{1}{l}.$$

Then

$$\|\rho^{\pm 1} T(\rho^{\pm 1}\varphi)\|_{L_{2p}} \leqslant M \|\varphi\|_{L_{2p}}, \quad \text{i.e., } \rho^{\pm 1} \in A_{2p},$$

but, by property 2) of the weights, $\rho \in A_{2p/(2p-1)}$ if $\rho^{-1} \in A_{2p}$. Using Stein's theorem once more, we get

$$\rho(t) = X^{\pm}(t) \in A_{\lambda}, \qquad \lambda \in \left[\frac{2p}{2p-1}, 2p \right].$$

The lemma is proved.

THEOREM 1. *The Riemann boundary-value problem* (1) *is unconditionally solvable in the case* $\kappa \geqslant 0$, *and its solution depends on* κ *arbitrary complex constants. In the case* $\kappa < 0$ *the Riemann problem is solvable only under the* $|\kappa|$ *solvability conditions*

$$\int_{\Gamma} \frac{g(t)}{X^{+}(t)} t^{k} \, dt = 0, \qquad k = 0, 1, \ldots, |\kappa| - 1, \tag{12}$$

and it then has a unique solution.

Using (4) and (6), we represent $G(t)$ in the form

$$G(t) = G_1(t)G_2(t) = \frac{X_1^{+}(t)}{X_1^{-}(t)} \cdot \frac{X_2^{+}(t)}{X_2^{-}(t)} \equiv \frac{X^{+}(t)}{X^{-}(t)}.$$

Here $X_1^{\pm}$ are the limit values of the canonical function $X_1(z)$ of the homogeneous problem (1) for $G(t) \equiv G_1(t) \in C^{\alpha}$; $X_1(z)$ is continuous and does not have zeros in the open plane, but it has a zero of order κ at infinity if $\kappa > 0$, while $X_1(\infty) = 1$ if $\kappa = 0$, and it has a pole of order $|\kappa|$ at infinity if $\kappa < 0$ (§3.7° in Chapter I).

This representation for $G(t)$ enables us to reduce (1) to the following jump problem:

$$(\Phi/X)^{+} - (\Phi/X)^{-} = g/X^{+}. \tag{13}$$

Suppose that $\kappa = 0$. In this case $\Phi^{\pm}/X^{\pm} \in E_1(D^{\pm})$, and the solution of (13) must have the form

$$\Phi^{\pm}(z) = \frac{X^{\pm}(z)}{2\pi i} \int_{\Gamma} \frac{g(\tau) \, d\tau}{X^{+}(\tau)(\tau - z)}. \tag{14}$$

We remark that $g \in L_p$, $[X^{\pm}(t)]^{\pm 1} \in L_{p/(p-1)}$, and, consequently, $\Phi^{\pm} \in E_{p_0}$ for some $p_0 > 0$. On the other hand, passing to the limit values in (14), we get

$$\Phi^{\pm}(t) = \pm \frac{1}{2} g(t) + \frac{X^{\pm}(t)}{2\pi i} \int_{\Gamma} \frac{g(\tau) \, d\tau}{X^{+}(\tau)(\tau - t)},$$

and since $X^{\pm} \in A_p$ by Lemma 4, it follows that $\Phi^{\pm}(t) \in L_{\lambda}$. Then, by a theorem of Smirnov (Theorem 4 in §4), which is valid for the classes E_p according to §4.5°, the properties $\Phi^{\pm}(z) \in E_{p_0}$ and $\Phi^{\pm}(t) \in L_p$, $p > p_0$, imply that $\Phi^{\pm}(z) \in E_p$. Consequently, (14) gives the solution of the original Riemann problem, and it is unique.

Suppose that $\kappa > 0$. In this case $\Phi^{+}/X^{+} \in E_1(D^{+})$, and the function Φ^{-}/X^{-} belongs to $E_1(D^{-})$ if we subtract from it some polynomial of degree $\kappa - 1$. Let us rewrite problem (13) in the form

$$\left(\frac{\Phi}{X}\right)^{+} - \left(\frac{\Phi}{X} - P_{\kappa-1}\right)^{-} = \frac{g}{X^{+}} + P_{\kappa-1},$$

where the expressions in parentheses belong to $E_1(D^{+})$ and $E_1(D^{-})$, respectively. Solution of the last problem and the use of elementary transformations yield

$$\Phi^{\pm}(z) = \frac{X^{\pm}(z)}{2\pi i}\left[\int_{\Gamma} \frac{g(\tau)\,d\tau}{X^{+}(\tau)(\tau - z)} + P_{\kappa-1}(z)\right]. \tag{15}$$

Just as for $\kappa > 0$ we can check that $\Phi^{\pm}(z) \in E_p$ for any polynomial $P_{\kappa-1}(z)$.

Suppose that $\kappa < 0$. In this case (14) does not give a solution of the problem (13) for every free term, since $X^{-}(z)$ has a pole of order $|\kappa|$ at infinity. Therefore, as in the case of Hölder coefficients, the Riemann problem (1) is solvable if and only if the solvability conditions (12) hold. The theorem is proved.

Let us now consider the Hilbert boundary-value problem

$$\operatorname{Re}[G(t)\Phi(t)] = g(t), \qquad t = e^{i\gamma}, \tag{16}$$

where $\Phi(z) \in H_p$ is an unknown function, $g(t) \in L_p$, $G(t) = e^{i\theta(t)}$, $|\theta(t)| \leqslant \pi/4p - \delta$, $p > 1$, and $\delta > 0$.

Under these conditions the Hilbert problem reduces to the Riemann problem with coefficient $G_{*}(t) = \overline{G(t)}/G(t)$ satisfying the conditions for applicability of Theorem 1. Taking this into account, we arrive at the following theorem.

THEOREM 2. *If* $\operatorname{ind}[G(t)] = \kappa \geqslant 0$, *then the homogeneous Hilbert problem* ($g \equiv 0$) *and the nonhomogeneous one are unconditionally solvable, and the solution of the homogeneous problem depends on* $2\kappa + 1$ *real constants.*

If $\operatorname{ind}[G(t)] = \kappa < 0$, *then the homogeneous problem is not solvable, and the nonhomogeneous one is solvable uniquely only if the usual* $2|\kappa| + 1$ *solvability conditions are satisfied* (*the conditions* (4.18) *in Chapter I*).

CHAPTER III

THE MIXED BOUNDARY-VALUE PROBLEM
WITH FREE BOUNDARY

§1. Statement of the problem. Investigation of representations
of the solution in the case of a polygonal line

1°. *Statement of the problem.* Let D_z be the simply connected domain inside a boundary L_z consisting of a polygonal line L_z^1 passing through given points z_i, $i = 1,\dots,n$, and an unknown arc L_z^2.

The conditions

$$\varphi = f_1(\tau), \qquad \psi = f_2(\tau) \tag{1}$$

are specified on L_z^2, where $w(\tau) = f_1(\tau) + if_2(\tau)$ is the boundary value of an analytic function $w(z)$, $0 \neq |\,dw/dz\,| < \infty$, and τ denotes one of the following parameters: $x = \operatorname{Re} z$, $\alpha = \arctan \frac{d\psi}{dx}$, $\theta = \arg z$, $r = |z|$, or s (the arclength parameter of L_z^2); here an increase (or decrease) in τ from τ_n to τ_1 corresponds to a traverse of L_z^2 in the positive direction.

To the arc L_z^2 there corresponds an arc L_w^2 with equation (1) in the plane $w = \varphi + i\psi$ of the complex potential. The equation of the arc L_w^1 complementing L_w^2 to form a closed contour L_w is assumed to be given:

$$\Phi(\varphi, \psi) = 0. \tag{2}$$

This equation connecting the real and imaginary parts of the analytic function $w(z)$ is the boundary condition on L_z^1 for determining $w(z)$ inside D_z.

The derivative of the unknown function $w(z)$ is assumed to satisfy the condition $|\ln|\,dw/dz\,\|| < \infty$ on the whole contour L_z with the possible exception of a finite number of points at which

$$dw/dz = (z - z_k)^{\gamma_k} F(z), \qquad |\ln|F(z)\|| < \infty, |\gamma_k| < 1.$$

Let us illustrate the problem by a sketch in the case of the parameter $x = \operatorname{Re} z$.

133

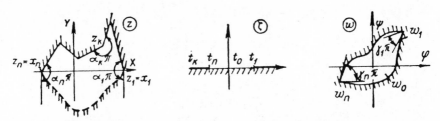

The $f_i(\tau)$ and $\Phi(\varphi, \psi)$ are assumed to be single-valued twice continuously differentiable functions of their arguments. Suppose for definiteness that a positive traverse of the arc L_w^2 corresponds to increasing τ. We map the upper half-plane ζ onto D_w conformally by the function $w = \omega(\zeta)$ in such a way that certain points t_n, t_0 and t_1 ($t_n < t_0 < t_1$) of the real axis are carried into the fixed point $w_0 \in L_w^2$ and the points $w_i = w(\tau_i)$, $i = 1$, respectively. Here it is obvious that the image of the point at infinity lies on L_w^1, since the finite segment $[t_n, t_1]$ corresponds to the arc L_w^2.

Comparing the boundary values of the functions $w(\tau) = f_1(\tau) + if_2(\tau)$ and $w = \omega(t)$ on L_w^2, we get the identity $f_1(\tau) + if_2(\tau) \equiv \omega(t)$, which gives us a function

$$\tau = H(t, t_0, t_1, t_n). \tag{3}$$

Differentiating with respect to t, we find that

$$\frac{d\tau}{dt} = \frac{\omega_t'(t)}{w_\tau'[H(t)]} = h(t, t_0, t_1, t_n), \qquad t \in [t_n, t_1], \tag{4}$$

and $h(t) \geqslant 0$ for $t \in [t_n, t_1]$, since Δt and $\Delta \tau$ have the same sign by virtue of the fact that increasing t and τ correspond to traverse of the arcs L_z^2 and L_w^2 in the positive direction. The function $h(t)$ will obviously be nonpositive if τ is decreasing. Because L_w^2 is sufficiently smooth, the function $d\omega/dt$ does not have singularities in the interval (t_n, t_1), i.e., $0 \neq |d\omega/dt| < \infty$ for $t \in (t_n, t_1)$. Thus, $0 < h(t) < \infty$, $t \in (t_n, t_1)$.

Let $\gamma_i \pi$ ($0 < \gamma_i < 2$) denote the interior angles between L_w^1 and L_w^2 at the points $w_i = w(\tau_i)$, $i = 1, n$. Then, taking account of the behavior of the derivative $d\omega/dt$ of the mapping function at the corner points t_n and t_1 (see §3 of Chapter II), we represent $h(t)$ in the form

$$h(t) = h^*(t)(t_1 - t)^{\gamma_1 - 1}(t - t_n)^{\gamma_n - 1}, \tag{5}$$

where $0 < h^*(t) < \infty$ for $t \in [t_n, t_1]$.

For all the values of τ except $\tau = \arg z$ the choice of the coordinate system on the z-plane does not influence the statement nor the solution of the corresponding problems. Therefore, unless something special is mentioned in what follows, it will be assumed for simplicity that the polygonal line L_z^1 is located above the OX-axis, and $z_i = x_i$ $(i = 1, n)$, $x_n < 0 < x_1$.

The equations of the links of the polygonal line are

$$xk_i - y = b_i \qquad (i = 1,\dots,n-1),$$

$$k_i = \frac{y_{i+1} - y_i}{x_{i+1} - x_i}, \qquad b_i = \frac{y_{i+1}x_i - x_{i+1}y_i}{x_{i+1} - x_i}. \tag{6}$$

It will be shown below that under the above conditions on the functions f_i and Φ there exist a unique finite contour L_z including the given polygonal line L_z^1 and a function $w(z)$ analytic in the interior of D_z that satisfies the boundary conditions (1) and (2); moreover, if L_z^1 is a convex polygonal line, then $w(z)$ is univalent in D_z. The main difficulty in proving this assertion is to find an analytic function $z = F_n(\zeta)$ that maps the upper half-plane onto D_z and satisfies the boundary-value problem (4), (6) on the real axis. Indeed, if such a function is known, then the equation $z = F_n(t)$ of the contour L_z is also known, and the function $w(z)$ is reconstructed by the formula

$$w = \omega\left[F_n^{-1}(z)\right], \tag{7}$$

where $\zeta = F_n^{-1}(z)$ is the function inverse to $z = F_n(\zeta)$. We remark that in a neighborhood of the points $t_i = F_n^{-1}(z_i)$ $(i = 2,\dots,n-1)$ the function $dF_n/d\zeta$ must be representable in the form

$$\frac{dF_n}{d\zeta} = \prod_{k=2}^{n-1} (\zeta - t_k)^{\alpha_k - 1} R(\zeta),$$

where $0 \neq |R(t_k)| < \infty$, and $\alpha_k \pi$ are the interior angles of the polygonal line L_z^1.

2°. Construction of a solution. Let $\tau = x$ (the case of other values will be taken up in §7). We write the boundary-value problem (4), (6) in the form

$$k_i \frac{dx}{dt} - \frac{dy}{dt} = 0; \qquad t \in [t_i, t_{i+1}] \, (i = 1,\dots,n-1), \tag{8}$$

$$\frac{dx}{dt} = h(t, t_0, t_1, t_n), \qquad t \in [t_n, t_1]. \tag{9}$$

The point at infinity falls in one of the intervals (t_i, t_{i+1}), which is thereby divided into two intervals: (t_i, ∞) and $(-\infty, t_{i+1})$. This problem is a particular case of the Hilbert boundary-value problem with discontinuous coefficients for the half-plane:

$$\operatorname{Re}(a + ib)\frac{dz}{dt} = c(t), \qquad t \in (-\infty, \infty),$$

or

$$a(t)u - b(t)v = c(t),$$

with

$$u = dx/dt, \qquad v = dy/dt,$$

$$a(t) = \begin{cases} k_i, & t \in [t_i, t_{i+1}] \ (i = 1,\ldots,n-1), \\ 1, & t \in [t_n, t_1], \end{cases}$$

$$b(t) = \begin{cases} 1, & t \notin [t_n, t_1], \\ 0, & t \in [t_n, t_1], \end{cases} \qquad c(t) = \begin{cases} 0, & t \notin [t_n, t_1], \\ h(t), & t \in [t_n, t_1]. \end{cases}$$

Starting from the commonly known properties of the derivative of a mapping function at corner points of a contour (see §3 of Chapter II), we must seek a solution of problem (8), (9) in the class of functions $dz/d\zeta$ that are bounded at the angles of the polygonal line that are greater than π, and unbounded at the remaining angles.

Let us construct the canonical function of the homogeneous Hilbert problem corresponding to problem (8), (9).

We consider the derivative of a function mapping the half-plane onto a polygon consisting of a polygonal line with links parallel to the links of the given polygonal line, and of the straight lines $x = x_i$ $(i = 1, n)$; let $c_0 \Pi(\zeta) = \Pi_{i=1}^n (\zeta - t_i)^{\alpha_i - 1}$, where $\alpha_i \pi$ are the angles at the vertices of the polygon and $c_0 = i\,|c_0|\exp\{-i\beta_m\}$, with $\beta_m = \arctan k_m$ the angle between the link of the polygonal line corresponding to the infinite segment $\{[t_m, \infty) + (-\infty, t_{m+1}]\}$ and the OX-axis. The real and imaginary parts of $c_0 \Pi(\zeta)$ satisfy condition (8), and $\mathrm{Re}[c_0 \Pi(\zeta)] = 0$ along the links parallel to the OY-axis. Thus, the function $c_0 \Pi(\zeta)$ satisfies the homogeneous problem and the necessary conditions at the angles of the polygonal line, i.e., it can be taken as the canonical function of the homogeneous problem; consequently, using the general solution of the Hilbert boundary-value problem (§4.2° of Chapter I), we get that

$$\frac{dz}{d\zeta} = \frac{\Pi(\zeta)}{\pi i} \int_{t_n}^{t_1} \frac{h(t)dt}{\Pi(t)(t - \zeta)} + p(\zeta)\Pi(\zeta),$$

where $p(\zeta)$ is a polynomial, and so

$$z = \frac{1}{\pi i} \int_{t_1}^{\zeta} \Pi(\zeta) \int_{t_n}^{t_1} \frac{h(t)dtd\zeta}{\Pi(t)(t - \zeta)} + \int_{t_1}^{\zeta} p(\zeta)\Pi(\zeta)d\zeta + z_1.$$

Let us join the endpoints of the polygonal line L_z^1 by a straight line. The sum S_n^1 of the interior angles of the polygon thus obtained is equal to $\pi(n-2)$, and for the polygon constructed above the sum S_n at the finite vertices is $S_n = S_n^1 + \pi = \pi(n - 1)$. Then at infinity we have the estimate

$$|\Pi(\zeta)| \leqslant k_1 |\zeta|^{\sum_{k=1}^{n}\alpha_k - n} = k_1 / |\zeta| .$$

Therefore, for the desired contour to be finite we set $p(\zeta) \equiv 0$. Then

$$z = F_n(\zeta) = \frac{1}{\pi i} \int_{t_1}^{\zeta} \prod_{j=1}^{n} (\zeta - t_j)^{\alpha_j - 1} \int_{t_n}^{t_1} \frac{h(t,t_0,t_1,t_n)dt d\zeta}{\Pi_{j+1}^{n}(t - t_j)^{\alpha_j - 1}(t - \zeta)} + z_1.$$

$$(10)$$

If we map the half-plane onto the disk $|\zeta| \leqslant 1$ by the function

$$\zeta = (\zeta_1 - i)/(\zeta_1 + i),$$

then (10) takes the form

$$z = F_n^1(\zeta) = \frac{1}{\pi} \int_{t_1}^{\zeta} \Pi(\zeta) \int_{L_z^2} \frac{h_1(\tau,\tau_0,\tau_1,\tau_n)d\tau d\zeta}{\Pi(\tau)(\tau - \zeta)} + z_1, \qquad (11)$$

where the $\tau_k = (t_k - i)/(t_k + i)$ $(k = 0,\ldots,n)$ are the points of the circle corresponding to the points t_k, and

$$h_1(\tau,\tau_0,\tau_1,\tau_n) \frac{2}{(\tau - 1)^2} h(t,t_0,t_1,t_n),$$

with $0 \neq |h_1| < \infty$ for $t \in L_z^2$.

3°. *Some properties of the solution.* In (10) let

$$M(\zeta) = \frac{1}{\pi i} \int_{t_n}^{t_1} \frac{h(t)dt}{\Pi(t)(t - \zeta)}.$$

By the Sohockiĭ-Plemelj formula, we find for $\zeta \to t_0 \in (t_n, t_1)$ that

$$M + (t_0) = \frac{h(t_0)}{\Pi(t_0)} + M(t_0),$$

and since $0 \neq |h(t_0)| < \infty$, it follows that $0 \neq |M^+(t_0)| < \infty$. For $t_0 \notin [t_n, t_1]$

$$M + (t_0) = M(t_0) = \frac{\pm c_0}{\pi} \int_{t_n}^{t_1} \frac{h(t)dt}{|\Pi(t)|(t - t_0)},$$

since $\mathrm{Re}[c_0\Pi(t)] = 0$, $t \in [t_n, t_0]$. Consequently, if $t \neq t_1, t_n$, then

$$0 \neq |M^+(t)| < \infty,$$

i.e., the mapping defined by (10) can fail to be conformal only at the points t_1 and t_n.

In particular, as $\zeta \to \infty$,

$$0 \neq \left| \zeta^2 \frac{dz}{d\zeta} \right| < \infty.$$

Let us now investigate the behavior of $dz/d\zeta$ at t_1 and t_n.

Let $h(t) = h^*(t)(t - t_1)^{\gamma_1 - 1}(t - t_n)^{\gamma_n - 1}$, where $0 \neq |h^*(t)| < \infty$. Then, if $\gamma_i > \alpha_i$ $(i = 1, n)$, the function

$$\frac{h(t)}{\Pi(t)} = h^*(t) \prod_{k=2}^{n-1} (t - t_k)^{1 - \alpha_k}(t - t_1)^{\gamma_1 - \alpha_1}(t - t_n)^{\gamma_n - \alpha_n}$$

vanishes at the corresponding point t_i, and, consequently, $M(\zeta)$ is bounded at this point. If $\gamma_i < \alpha_i$ $(i = 1, n)$, then at this point

$$\left| \frac{h(t)}{\Pi(t)} \right| < k_3 |t - t_i|^{\gamma_i - \alpha_i}.$$

Taking the behavior of a Cauchy integral at the endpoints (§2 of Chapter I) into account, we conclude that

$$|M(\zeta)| \leq k_4 |\zeta - t_i|^{\gamma_i - \alpha_i}.$$

For $\gamma_i = \alpha_i$ we obviously have that $|M(\zeta)| \leq k_5 |\ln(\zeta - t_i)|$.

Thus,

$$|M(\zeta)| \leq k_6 \quad \text{for } \gamma_i > \alpha_i \ (i = 1, n),$$

$$|M(\zeta)| \leq k_4 |\zeta - t_i|^{\gamma_i - \alpha_i} \quad \text{for } \gamma_i < \alpha_i \ (i = 1, n), \tag{12}$$

$$|M(\zeta)| \leq k_5 |\ln(\zeta - t_i)| \quad \text{for } \gamma_i = \alpha_i \ (i = 1, n).$$

Since $dz/d\zeta = \Pi(\zeta)M(\zeta)$, it follows that, respectively,

$$|dz/d\zeta| \leq k_7 |\zeta - t_i|^{\alpha_i - 1} \quad \text{for } \gamma_i > \alpha_i,$$

$$|dz/d\zeta| \leq k_8 |\zeta - t_i|^{\gamma_i - 1} \quad \text{for } \gamma_i < \alpha_i, \tag{13}$$

$$|dz/d\zeta| \leq k_9 |\zeta - t_i|^{\alpha_i - 1} |\ln(\zeta - t_i)| \quad \text{for } \gamma_i = \alpha_i.$$

If $\alpha_i < 1$ $(i = 1, n)$, then the representation (10) is unique in the class of bounded functions $z = z(\zeta)$. Indeed, with the necessary requirements on $dz/d\zeta$ at the points $t_2, \ldots, t_{n-1}$ and $\zeta = \infty$ taken into account, other representations of the solutions could be obtained only by changing the behavior of the canonical function at t_1 and t_n, i.e., by considering the following canonical functions of the homogeneous problem (8), (9) along with $\Pi(\zeta)$:

$$\frac{(\zeta - t_n)^{\delta}\Pi(\zeta)}{\zeta - t_1}, \qquad \frac{(\zeta - t_1)^{\delta}\Pi(\zeta)}{\zeta - t_n}, \qquad \frac{\Pi(\zeta)}{(\zeta - t_1)(\zeta - t_n)}, \qquad \delta = 0, 1.$$

However, since $\alpha_i < 1$ $(i = 1, n)$, to such canonical functions would correspond a solution $z = z(\zeta)$ that is unbounded at one of the points t_1 or t_n, or at both points simultaneously.

Suppose now that only one of the α_i $(i = 1, n)$, say α_n, is greater than 1 $(\alpha_n > 1)$. In this case the representation (10) may fail to be unique, since the canonical function can be taken in one of the forms

$$c_0\Pi(\zeta), \quad \Pi_n(\zeta) = c_0\Pi(\zeta)(\zeta - t_n)^{-1}, \quad \text{or } \Pi'_n(\zeta) = c_0\Pi(\zeta)(\zeta - t_1)(\zeta - t_n)^{-1}$$

(for definiteness we set $|c_0| = 1$). To the first of the canonical functions there corresponds the solution (10), and to the other two there correspond

$$dz/d\zeta = \Pi_n(\zeta)[M_n(\zeta) + k], \tag{14}$$

$$dz/d\zeta = \Pi'_n(\zeta)M'_n(\zeta), \tag{15}$$

respectively, where k is a real constant, and

$$M_n^m(\zeta) = \frac{1}{\pi i}\int_{t_n}^{t_1}\frac{h(t)dt}{\Pi_n^m(t)(t - \zeta)} = -\frac{1}{\pi}\int_{t_n}^{t_1}\frac{h(t)|t - t_n|\,dt}{|\Pi(t)|\,|t - t_1|^m(t - \zeta)}$$

($m = 0$ and 1 for M_n and M'_n, respectively). In the representations (14) and (15) $dz/d\zeta$ has a zero of the second order at infinity (for $k \neq 0$), i.e., the corresponding mappings are conformal at infinity. It is obvious that $|M'_n(\zeta)| \neq 0$ for $\zeta \neq t_1, t_n$, and, consequently, $|dz/d\zeta| \neq 0$ in (15) for $\zeta \neq t_i$ $(i = 1, n)$. It is easy to see that $M_n(\zeta) + k$ can vanish only on the intervals $[t_1, \infty)$ and $(-\infty, t_n]$.

If $\alpha_1 > \gamma_1$ and $\alpha_n - 1 > \gamma_n$, then (12) and (13), in which $\alpha_n - 1$ now plays the role of α_n, imply that the function

$$M_n(t) = -\frac{1}{\pi}\int_{t_n}^{t_1}\frac{h(\tau)d\tau}{|\Pi_n(\tau)|(\tau - t)}, \qquad t \notin (t_n, t_1),$$

is unbounded at the points $t = t_i$ $(i = 1, n)$. But $M_n(t) \geqslant 0$, $t \in [t_n, \infty)$, and $M_n(t) \leqslant 0$, $t \in (-\infty, t_n]$, i.e., $M_n(t) \in (-\infty, \infty)$ for $t \in (t_n, t_1)$, and, consequently, for any k there exists a point $t^* \notin [t_n, t_1]$ at which $M_n(t^*) + k = 0$. Since the corresponding mapping fails to be conformal at the point t^*, the representation (14) does not satisfy the requirements of the original problem in this case.

Suppose that at least one of the conditions $\alpha_1 > \gamma_1$ or $\alpha_n > 1 + \gamma_n$ is not satisfied. Then, generally speaking, there can exist k's such that $M_n(\zeta) + k \neq 0$ at every point ζ.

We determine conditions under which this is possible.

Suppose that $\alpha_1 > \gamma_1$ and $\alpha_n - 1 < \gamma_n$. Then $0 \leqslant M_n(t) \leqslant \infty$ for $t \in [t_1, \infty)$ and $M_n(t) \leqslant 0$ for $t \in (-\infty, t_n]$. The function $M_n(t)$ has a bound $-M_n^- \leqslant M_n(t)$

for $t \in (-\infty, t_n]$. If we take $k > M_n^-$, then $M_n(t) + K > 0$, and, consequently, the representation (14) is possible for all $k > M_n^-$. If $\alpha_1 < \gamma_1$ and $\alpha_n - 1 > \gamma_n$, then, similarly, $-\infty \leqslant M_n(t) \leqslant M_m^+$ for $[t_n, t_1]$, and, if $k < -M_n^+$, then $M_n(t) + k < 0$, i.e., the representation (14) is possible for $k < -M_n^+$. But if the inequalities $\alpha_1 > \gamma_1$ and $\alpha_n - 1 > \gamma_n$ hold simultaneously, then $-M_n^- \leqslant M_n(t) \leqslant M_n^+$ for $t \notin (t_n, t_1)$, and, hence, the representation (14) is possible for $k \notin [-M_n^-, M_n^+]$.

In some cases we can control the angles at the points z_1 and z_n by choosing one or the other representation, and in other cases the representations are geometrically indistinguishable, i.e., the angles at z_1 and z_n coincide for the corresponding contours. Apparently, it can be proved that if two representations are geometrically indistinguishable, then the contours L_z corresponding to them coincide, the unknown arcs L_z^2 included. We have not investigated this question.

Let us consider the case when $\alpha_1 > 1$ and $\alpha_n > 1$ at the same time. Here $dz/d\zeta = \Pi(\zeta)M(\zeta)$ can have the representations

$$dz/d\zeta = c_0\Pi_i(\zeta)[M_i(\zeta) + k] \qquad (i = 1, n)$$

and

$$dz/d\zeta = c_0\Pi_{1,n}(\zeta)[M_{1,n}(\zeta) + k_1 + k_2\zeta],$$

where

$$\Pi_{1,n}(\zeta) = \Pi(\zeta)(\zeta - t_1)^{-1}(\zeta - t_n)^{-1}, \qquad M_{1,n}(\zeta) = \int_{t_n}^{t_1} \frac{h(t)dt}{\Pi_{1,n}(t)(t - \zeta)},$$

and k, k_1 and k_2 are arbitrary real constants. The first two representations coincide with the representations (10) and (14) considered above. We shall not study the question of whether there exist constants k_1 and k_2 for which the equality $M_{1,n}(\zeta) + k_1 + k_2\zeta = 0$ does not hold for any ζ, i.e., whether a new representation is possible.

In what follows we carry out all the arguments for the representation (10) for the sake of convenience, although they are valid also for the remaining representations, provided, of course, that these representations are admissible. Therefore, the uniqueness proved below for the solution must be understood as absolute only if the representation (10) is unique, as, for example, when $\alpha_i < 1$, $(i = 1, n)$.

Otherwise, the uniqueness of the solution is understood in the sense that there are not two solutions of our problem having one and the same representation (10).

REMARK. It is sometimes convenient to consider the normalization of the mapping $z = F_n(\zeta)$ in which the image of the point $\zeta = \infty$ lies on the unknown arc L_z^2 instead of the normalization adopted above. We can arrive at this case by transforming the formula (10) by means of the substitution $\zeta = 1/\chi$, $t_k = 1/\tau_k$ ($t_n < 0 < t_1$):

$$\Pi(\zeta) = \prod_{k=1}^{n} (\zeta - t_k)^{\alpha_k - 1} = c_1 \chi \prod_{k=1}^{n} (\chi - \tau_k)^{\alpha_k - 1} \equiv c_1 \chi \Pi(\chi),$$

$$M(\zeta) = \frac{1}{\pi i} \int_{t_1}^{t_n} \frac{h(t)dt}{\Pi(t)(t - \zeta)} = \frac{c_1^{-1}\chi}{\pi i} \int_{t_n \infty t_1} \frac{h\left(\dfrac{1}{\tau}\right)d\tau}{\Pi(\tau)(\tau - \chi)} \equiv c_1^{-1}\chi M(\chi),$$

$$\frac{dz}{d\chi} = \frac{dz}{d\zeta} \cdot \frac{dz}{d\chi} = \Pi(\chi)M(\chi),$$

where $(\tau_n \infty \tau_1)\widehat{} = [\tau, \infty) + (-\infty, \tau_1]$ and $c_1 = \Pi_1^n(-t_k)^{\alpha_k - 1}$. Since $0 \neq |M(\zeta)| < \infty$ it follows that $\lim_{|\chi| \to \infty} |M(\chi)\chi| \neq 0$ nor ∞; consequently,

$$\lim_{|\chi| \to \infty} \left| \frac{dz}{d\chi}\chi^2 \right| \neq 0, \infty.$$

Returning to the original notation, we get

$$z = F_n(\zeta) = \int_{t_1}^{\zeta} \frac{\Pi(\zeta)}{\pi i} \int_{(t_n \infty t_1)\widehat{}} \frac{h(t)dtd\zeta}{\Pi(t)(t - \zeta)} + z_1, \tag{16}$$

where the images of the vertices of the polygonal line are connected by the inequalities $t_1 < t_2 < \cdots < t_n$.

§2. Proofs of existence and uniqueness theorems
by the continuity method

1°. *The system of equations for parameters.* As we saw in the preceding section, the mapping $z = F_n(\zeta)$ of the upper half-plane onto the domain D_z can be represented in the form

$$z = \int_{t_1}^{\zeta} \frac{\Pi(\zeta)}{\pi i} \int_{(t_n t_1)\widehat{}} \frac{h(t, t_0, t_1, \ldots, t_n)dtd\zeta}{\Pi(t)(t - \zeta)} + z_1 \equiv F_n(\zeta, t_k) \tag{1}$$

in the case of the mixed boundary-value problem with free boundary L_z^2 for a polygonal line L_z^1, where $\Pi(\zeta) = \prod_{k=1}^{n}(\zeta - t_k)^{\alpha_k - 1}$, $\alpha_k \pi$ are the interior angles of L_z^1, the points t_k ($k = 1, \ldots, n$) are unknown images of the vertices of the polygonal line, and t_0 is the image of a point $z_0 \in L_z^2$ with fixed abscissa x_0. Here the inside integral is taken over the finite interval $(t_n t_1)\widehat{} \equiv [t_n, t_1]$ in the case of a normalization of the mapping $z = F_n(\zeta)$ for which $t_k < t_{k+1} < \cdots < t_n < t_1 < \cdots < t_{k-1}$, and over the infinite "interval" $(t_n t_1) \equiv [t_n, \infty) + (-\infty, t_1]$ in the case $t_1 < t_2 < \cdots < t_n$. If the constants t_k $(k = 0, \ldots, n)$ in (1) are fixed arbitrarily, with it assumed for definiteness that $t_n < t_1$ and $t_k \notin [t_n, t_1]$ $(k = 2, \ldots, n - 1)$, then the equation $z = F_n(t, t_0, \ldots, t_n)$ for $t \notin (t_n, t_1)$ will give a polygonal line P whose links are parallel to the links of the given polygonal line L_z^1.

Suppose that the constants are selected in such a way that

$$\int_{t_i}^{t_{i+1}} \left| \frac{dF_n}{dt} \right| dt = l_i \qquad (i = 1, \ldots, n - 2), \tag{2}$$

where the l_i are the lengths of the links of L_z^1, and one of the integrals is taken over the infinite interval $[t_k, \infty) + (-\infty, t_{k+1}]$. The function $x(t)$ can be represented in the form $x(t) = \int_{t_1}^{t} h(t)\, dt + x_1$ on the segment $[t_n, t_1]$, in view of the equality $F_n(t_1, t_0, \ldots, t_n) = z_1$ and (1.9). Consequently,

$$x(t_n) = \int_{t_1}^{t_n} h(t)\, dt + x_1 = H(t_n) = x_n$$

according to the definition of $h(t)$ (see (1.3) and (1.4)). The system (2) implies that $n - 1$ vertices of the polygonal lines P and L_z^1 coincide, i.e., $F_n(t_i, t_0, \ldots, t_n) = z_i$ $(i = 1, \ldots, n - 1)$. Then the fact that the links of P and L_z^1 are parallel and the equality $x(t_n) = \operatorname{Re} F_n(t_n) = x_n$ give us that $F_n(t_n) = z_n$, i.e., P coincides completely with the given polygon.

Thus we have proved the following lemma.

LEMMA 1. *If the system of equations* (2) *is solvable for the constants* t_k $(k = 0, \ldots, n)$ *(three of them can be chosen arbitrarily), then the corresponding function* $z = F_n(\zeta, t_0, t_1, \ldots, t_n)$ *defined by* (1) *maps the upper half-plane onto a domain* D_z *whose boundary contains the given polygonal line* L_z^1.

2°. *A priori estimates of the solution.* We assume that the image of the point at infinity lies on the free boundary L_z^2, i.e., the constants t_k are related by the inequalities $-\infty < t_1 < t_2 < \cdots < t_n < \infty$, where t_1 and t_n are fixed.

LEMMA 2. *If some of the paramters* t_k *in* (1) *converge to a common value, then the length* l_k *of at least one of the links of* L_z^1 *tends either to zero or to infinity.*

Suppose first that T_1 and t_n are not among the converging parameters, and that for definiteness t_k, $k = 2, \ldots, p$, $p < n$, are converging. We write the length of the pth side in the form

$$l_p = \int_{t_p}^{t_{p+1}} |\Pi(t)|\, |M(t)|\, dt = \int_{t_p}^{t_{p+1}} \prod_{k=2}^{p+1} |t - t_k|^{\beta_k}\, |M_1(t)|\, dt,$$

where

$$\beta_k = \alpha_k - 1 \quad (k = 2,\dots,p+1), \qquad 0 \neq |M_1(t)| < \infty \quad \text{for } t \in [t_2, t_{p+1}].$$

Let $\nu = -\Sigma_2'^p \beta_k$, and $\mu = \Sigma_2''^p \beta_k$ where Σ' contains all negative β_k and Σ'' contains all positive β_k.

We first consider the case $\mu - \nu + 1 \leqslant 0$ and show that $l_p \to \infty$. Observe that the equation corresponding to the link $l_p = l_{n-1}$ is missing in (2), but that, by Lemma 1, it is also satisfied for $p = n - 1$ when the equations (2) are solvable.

For $t \in [t_p, t_{p+1}]$ and $k = 2,\dots,p$ we have

$$|t - t_p| \leqslant |t - t_k| \leqslant |t - t_2|. \tag{3}$$

On the other hand, since t_2 and t_{p+1} do not converge, for any $\varepsilon \in (0, |t_{p+1} - t_2|)$ we must have $t_p + \varepsilon < t_{p+1}$ as $\Delta t_p = t_p - t_2 \to 0$.

Then, for $t \in [t_p, t_p + \varepsilon]$,

$$|M_1(t)| \, |t - t_{p+1}|^{\beta_{p+1}} \geqslant A \neq 0.$$

This estimate and (3) give us that

$$l_0 \geqslant \int_{t_p}^{t_{p+\varepsilon}} |M_1(t)| \prod_{k=2}^{p+1} |t_k - t|^{\beta_k} \, dt \geqslant A \int_{t_p}^{t_{p+\varepsilon}} |t - t_2|^{-\nu} |t - t_p|^{\mu} \, dt.$$

In the integral on the right-hand side we make a change of variables, setting $t = (t_p - t_2)s + t_p = \Delta t_p s + t_p$. This gives

$$l_p \geqslant A(\Delta t_p)^{\mu - \nu + 1} \int_0^{\varepsilon/\Delta t_p} s^{\mu} (s + 1)^{-\nu} \, ds.$$

If $\mu - \nu + 1 < 0$, then $(\Delta t_p)^{\mu - \nu + 1} \to \infty$ as $\Delta t_p \to 0$, and, hence, $l_p \to \infty$, since the integral obviously is not equal to zero. If $\mu - \nu + 1 = 0$, then

$$l_p \geqslant A \int_0^{\varepsilon/\Delta t_p} s^{\mu - \nu} \left(1 + \frac{1}{s}\right)^{-\nu} ds \geqslant A \int_1^{\varepsilon/\Delta t_p} s^{-1} \left(1 + \frac{1}{s}\right)^{-\nu} ds \geqslant \frac{A}{2^\nu} \int_1^{\varepsilon/\Delta t_p} \frac{ds}{s},$$

i.e., $l_p \geqslant (A/2^\nu) \ln |\varepsilon/\Delta t_p| \to \infty$ as $\Delta t_p \to 0$.

Suppose now that $\mu - \nu + 1 > 0$. In the upper half-plane $\operatorname{Im} \zeta > 0$ we consider the semicircle K_r of radius $r = \Delta t_p = t_p - t_2$ with center at the point $\zeta_0 = (t_p + t_2)/2$. For sufficiently small Δt_p it can be assumed that the points t_k $(k \neq 2,\dots,p)$ lie outside the segment $[\zeta_0 - \Delta t_p, \zeta_0 + \Delta t_p]$ which the semicircle K_r cuts out on the axis $\operatorname{Im} \zeta = 0$. Consequently, $0 \neq a < |\zeta - t_k| < \tilde{a}$ for $\zeta \in K_r$ and $k \neq 2,\dots,p$, and so

$$|M_1(\zeta)| \cdot |\zeta - t_{p+1}|^{\beta_{p+1}} \leqslant A < \infty \qquad (\zeta \in K_r).$$

On the other hand,

$$\frac{r}{2} \leqslant |\zeta - t_k| < 2r \qquad (\zeta \in K_r, k = 2,\ldots,p).$$

The semicircle K_r is mapped by the function $z = F_n(\zeta)$ into some smooth "bridge" Λ_p lying inside D_z and joining certain interior points of the links L_1 and L_p of L_z^1. We show that the length of the "bridge" Λ_p tends to zero as $r = \Delta t_p \to 0$, and this implies, in particular, that the length l_k of any link L_k $(k = 2,\ldots,p - 1)$ tends to zero:

$$\Lambda_p = \left| \int_{k_r} \prod_{k=2}^{p+1} (\zeta - t_k)^{\beta_k} M_1(\zeta)\, d\zeta \right| \leqslant \pi r \max \left\{ \left| M_1(\zeta) \prod_{k=2}^{p+1} (\zeta - t_k)^{\beta_k} \right| \right\}$$

$$\leqslant A\pi r \max \prod_{k=2}^{p}{}' |\zeta - t_k|^{\beta_k} \prod_{k=2}^{p}{}'' |\zeta - t_k|^{\beta_k} \leqslant A\pi r (2r)^{\mu} \left(\frac{r}{2}\right)^{-\nu} \leqslant 2^{\mu+\nu} A\pi r^{\mu-\nu+1};$$

$\Lambda_p \to 0$ also as $r = \Delta t_p \to 0$, because $\mu - \nu + 1 > 0$ by assumption.

Suppose now that the converging parameters t_k include, say, t_1, and that the $t_k, k = 1,\ldots,p$ ($p < n$, since t_1 and t_n are fixed) converge. As above, let

$$\nu = -\sum_{k=1}^{p}{}' \beta_k, \qquad \mu = \sum_{k=1}^{p}{}'' \beta_k, \qquad \beta_k = \alpha_k - 1 \qquad (k = 1,\ldots,p),$$

where Σ' contains the negative β_k, and Σ'' the positive ones.

Since $|M(t)| \geqslant a \neq 0$, $t \in (t_1, t_n)$, the case $\mu - \nu + 1 \leqslant 0$ does not differ at all from that considered at the beginning of the subsection, and, consequently, $l_p \to \infty$ as $t_p - t_1 = \Delta t_p \to 0$.

If $\beta = \mu - \nu + 1 - \gamma_1 \geqslant 1$, then, taking account of the equality

$$\lim_{\Delta t_p \to 0} |\Pi(t)| |t - t_1|^{1-\gamma_1} = |f(t)| |t - t_1|^{\beta}, \qquad |f(t_1)| < \infty,$$

we get for all finite ζ that

$$|M(\zeta)| = \frac{1}{\pi} \left| \int_{(t_n t_1)} \frac{\tilde{h}(t)\,dt}{\Pi(t)|t - t_1|^{1-\gamma_1}(t - \zeta)} \right| \to \infty \qquad \text{as } \Delta t_p \to 0,$$

and hence

$$l_p \geqslant \int_{t_{p+1}-\varepsilon}^{t_{p+1}} |\Pi(t)| |M(t)|\, dt \to \infty, \qquad 0 < \varepsilon < t_{p+1} - t_p,$$

where

$$\tilde{h} = h(t)|t - t_1|^{1-\gamma_1} \qquad \tilde{h}(t_1) \neq 0.$$

It remains to consider the case $-\gamma_1 < \beta = \mu - \nu + 1 - \gamma_1 < 1$. Let K_r be the semicircle of radius $r = \Delta t_p = t_p - t_1$ with center at $\zeta_0 = (t_p + t_1)/2$ and lying in the upper half-plane $\operatorname{Im} \zeta > 0$. Since $|M(\zeta)| \, | \zeta - t_1|^\delta \leqslant A < \infty$ for all finite ζ ($\delta = 0$ for $\beta < 0$ and $\delta = \beta + \varepsilon$ for $\beta \geqslant 0$), and since $\mu - \nu - \delta + 1 = \gamma_1 - \varepsilon > 0$ for $\beta \geqslant 0$ and sufficiently small $\varepsilon > 0$ and $\mu - \nu - \delta + 1 = \mu - \nu + 1 > 0$ for $\beta < 0$ (as in the corresponding case when only the t_k ($k = 2, \ldots, p$) converge), we have

$$|\Lambda_p| = \left| \int_{k_r} \Pi(\zeta) \, |\zeta - t_1|^{-\delta} M(\zeta) \, |\zeta - t_1|^\delta \, d\zeta \right| \to 0 \quad \text{for } \Delta t_p \to 0.$$

Further, as $r = \Delta t_p \to 0$

$$\left| F_n\left(t_1 - \frac{r}{2}\right) - z_1 \right| = \left| \int_{t_1 - r/2}^{t_1} \Pi(t) \, |t - t_1|^{-\delta} M(t) \, |t - t_1|^\delta \, dt \right|$$

$$\leqslant A \int_{t_1 - r/2}^{t_1} \Pi(t) \, |t - t_1|^{-\delta} \, dt \to 0,$$

by the inequality $\mu - \nu - \delta = \gamma_1 - \varepsilon - 1 > -1$ ($\beta \geqslant 0$) or $\mu - \nu - \delta = \mu - \nu > -1$ ($\beta < 0$).

Thus, the left-hand endpoint of the "bridge" Λ_p, which lies at the point $z = F_n(t_1 - \frac{1}{2}\Delta t_p)$, converges to z_1 as $\Delta t_p \to 0$, and its length Λ_p tends to zero. As previously, this implies that the lengths l_k ($k = 1, \ldots, p$) of the links converge to zero, and Lemma 2 is proved.

The polygonal line L_z^1 is said to be nondegenerate if its angles $\alpha_k \pi$ at the vertices satisfy the conditions $0 < \alpha_k \pi < 2\pi$ and the lengths l_k of the links are finite and nonzero.

We give another statement of Lemma 2.

LEMMA 2*. *For each solution* $\{t_2, t_3, \ldots, t_{n-1}\}$, $t_k \in [t_1, t_n]$ ($k = 2, \ldots, n - 1$), *of the system of equations* (2) *that corresponds to a given nondegenerate polygonal line* L_z^1 *the following inequalities hold*:

$$t_{k+1} - t_k > \varepsilon > 0, \qquad k = 1, \ldots, n - 1, \tag{4}$$

where $\varepsilon > 0$ *depends only on the geometry of the polygonal line.*

3°. *Local uniqueness of solutions.* To arbitrary values of the parameters t_k ($k = 2, \ldots, n - 1$), connected by the inequalities $t_1 < t_2 < \cdots < t_n$ there corresponds a certain nondegenerate polygonal line L_z^1 according to formula (1). Thus, the system (2) is solvable for certain values of l_k ($k = 1, \ldots, n - 1$). The system (2) turns out to have the remarkable property that there are no other solutions in a small neighborhood of each of its solutions, i.e., the solutions of (2) are locally unique. This is equivalent to the following assertion.

LEMMA 3. *If the system* (2) *corresponding to some nondegenerate polygonal line* L_z^1 *is solvable for the constants* t_k $(k = 2,\ldots,n - 1)$, *then its Jacobian*

$$\frac{D(l_1,l_2,\ldots,l_{n-2})}{D(t_2,t_3,\ldots,t_{n-1})}$$

is nonzero at this solution.

Let us first show that the expressions $\partial l_i/\partial t_k$ $(i = 1,\ldots,n - 2, k = 2,\ldots,n - 1)$ are continuous. For $k \neq i, i + 1$ the continuity of

$$\frac{\partial l_i}{\partial t_k} = \int_{t_i}^{t_{i+1}} \frac{\partial}{\partial t_k}\left|\frac{dF_n}{dt}\right| dt$$

follows from that of

$$\frac{\partial}{\partial t_k}\left|\frac{dF_n}{dt}\right| \qquad t \in [t_i, t_{i+1}].$$

Let $k = i$. Then l_i can be written in the form

$$l_i = \int_{t_i}^{t_{i+1}} (t - t_i)^{\alpha_{i+1}-1}(t - t_{i+1})^{\alpha_{i+1}-1} f(t, t_2,\ldots,t_{n-1})\, dt,$$

where $\partial f/\partial t$, $\partial f/\partial t_i$ and $\partial f/\partial t_{i+1}$ are continuous for $t \in [t_i, t_i + 1]$. Let

$$s = \frac{t - t_i}{t_{i+1} - t_i} = \frac{t - t_i}{\Delta t_i}.$$

Then

$$l_i = (\Delta t_i)^{\alpha_i + \alpha_{i+1}-1}\int_0^1 s^{\alpha_i-1}(1 - s)^{\alpha_{i+1}-1} f(\Delta t_i s + t_i, t_2,\ldots,t_{n-1})\, ds.$$

Differentiating the last equality, which is possible by the continuity of $\partial f/\partial s$ and $\partial f/\partial t_i$, we get

$$\frac{\partial l_i}{\partial t_i} = (\Delta t_i)^{\alpha_i + \alpha_{i+1}-2}(\alpha_i + \alpha_{i+1} - 1)$$

$$\cdot \int_0^1 s^{\alpha_i-1}(1 - s)^{\alpha_{i+1}-1} f(\Delta t_i s + t_i, t_2,\ldots,t_{n-1})\, ds$$

$$+ (\Delta t_i)^{\alpha_i + \alpha_{i+1}-1}\int_0^1 s^{\alpha_i-1}(1 - s)^{\alpha_{i+1}-1}\left[\frac{\partial f}{\partial t_i} + \frac{\partial f}{\partial t}(1 - s)\right] ds.$$

From this it is clear that $\partial l_i/\partial t_i$ is continuous in its arguments, since $\Delta t_i = t_{i+1} - t_i > \varepsilon > 0$ by (4). The continuity of $\partial l_i/\partial t_{i+1}$ is proved similarly. Suppose that the parameters t_k $(k = 2,\ldots,n - 1)$ in (2) are varied, and

$t_1 < t_2 < \cdots < t_n$, where t_1 and t_n are fixed. We compute the variations δl_i $(i = 1, n - 2)$, assuming that all the varied polygonal lines pass through the point z_1, i.e., $\delta z_1 = 0$, and

$$\sum_{k=2}^{n-1} \frac{\partial l_i}{\partial t_k} \delta t_k = \delta l_i \qquad (i = 1, \ldots, n - 2). \tag{5}$$

To prove the lemma it suffices to show that the homogeneous system corresponding to (5) has only the zero solution $\delta t_k = 0$ $(k = 2, \ldots, n - 1)$, i.e., the variations δt_k of the constants vanish when the variations δl_i $(i = 1, \ldots, n - 2)$ of the geometric quantities determining the polygonal line vanish.

Suppose the opposite: The system (5) has a nonzero solution δt_k $(k = 2, \ldots, n - 1)$ for $\delta l_i = 0$ $(i = 1, \ldots, n - 2)$. Then we can use (16) to compute the variation of the function $z = F_n(\zeta, t_2, \ldots, t_{n-1})$ for fixed ζ:

$$\delta z = \int_{t_1}^{\zeta} \Pi(\zeta) \sum_{k=2}^{n-1} \delta t_k \left[(1 - \alpha_k)(\zeta - t_k)^{-1} M(\zeta) + \frac{\partial M}{\partial t_k} \right] dt, \tag{6}$$

where

$$M(\zeta) = \frac{1}{\pi i} \int_{t_n t_1} \frac{h(t) dt}{\Pi(t)(t - \zeta)} \quad \text{and} \quad \delta z_1 = 0.$$

The variation of the inverse function $\zeta = F_n^{-1}(z, t_2, \ldots, t_{n-1})$ for fixed z can be determined from the same formula (1):

$$\frac{dz}{d\zeta} \delta \zeta + \int_{t_1}^{\zeta} \Pi(\zeta) \sum_{k=2}^{n-1} \delta t_k \left[(1 - \alpha_k)(\zeta - t_k)^{-1} M(\zeta) + \frac{\partial M}{\partial t_k} \right] d\zeta = 0,$$

where $dz/d\zeta$ and $\delta \zeta$ are computed as functions of ζ. From the last equality and (6) we find that

$$\delta z + \frac{dz}{d\zeta} \delta \zeta = 0. \tag{7}$$

Let us consider this equality on the real axis. If we take dz along the links of the polygonal line, then $d\zeta$ will be directed along the real axis $\operatorname{Im} \zeta = 0$, i.e., $\operatorname{Im} d\zeta = 0$. We decompose the unknown variation δz in the directions of the tangent and normal to the polygonal line L_z and denote the components by $(\delta z)_s$ and $(\delta z)_n$, respectively. If $\delta z_i = \delta l_i = 0$ $(i = 1, \ldots, n - 2)$, then the polygonal parts of the varied contours coincide, and, consequently, $(\delta z)_n = 0$. If $(\delta z)_s \neq 0$, then $\delta \zeta / d\zeta = -(\delta z)_s / dz$ by (7), and, since $(\delta z)_s$ and dz have the same direction, i.e., $\arg(\delta z)_s = \arg dz$, it follows that $\operatorname{Im}((\delta z)_s / dz) = 0$, and so

$$\operatorname{Im} \delta \zeta = -((\delta z)_s / dz) \operatorname{Im} d\zeta = 0 \quad \text{for } t \in [t_1, t_n].$$

The last equality holds also for $(\delta z)_s = 0$, since $\delta z \equiv 0$ in this case, whence $\delta\zeta = -\delta z \frac{d\zeta}{dz}$. Then (7) implies

$$\delta x = \operatorname{Re} \delta z = -\frac{dz}{dt}\delta\zeta; \qquad \delta y = \operatorname{Im} \delta z = -\frac{dy}{dt}\delta\zeta \quad \text{for } t \in [t_1, t_n],$$

from which, taking the boundary condition (1.8) into account, we get

$$k_i\delta x - \partial y = -\delta\zeta\left(k_i\frac{dx}{dt} - \frac{dy}{dt}\right) = 0, \qquad t \in [t_i, t_{i+1}].$$

On the other hand, from the boundary condition (1.9), which can be written in the form

$$x = \int_{t_1 t} h(t, t_0, t_1, t_n)\, dt + x_1, \qquad t \in \{[t_n, \infty) + (-\infty, t_1]\},$$

it follows that

$$\delta x = 0, \qquad t \in \{[t_n, \infty) + (-\infty, t_1]\},$$

since $\delta t_0 = \delta t_1 = \delta t_n = \delta x_1 = 0$.

Thus, the function δz satisfies the boundary problem

$$k_i\delta x - \delta y = 0, \qquad t \in [t_i, t_{i+1}], \qquad \delta x = 0, \qquad t \in \{[t_n, \infty) + (-\infty, t_1]\}.$$
$$(8)$$

According to (6), the function δz is bounded everywhere in the upper half-plane (including the point at infinity) except possibly at $t_1, \ldots, t_n$. Consequently, the solution of the boundary-value problem (8) can be written in the form

$$\delta z = c_0 p_m(\zeta) \prod_{k=1}^{n} (\zeta - t_k)^{\alpha_k + \varepsilon_k}, \qquad (9)$$

where $p_m(\zeta)$ is an mth-degree polynomial with real coefficients, and the ε_k are integers, with $\Sigma_1^n(\alpha_k + \varepsilon_k) + m = -\varepsilon_0$ ($\varepsilon_0 = 0$ for $\delta z\,|_{\zeta=0} = \text{const} \neq 0$; $\varepsilon_0 \geqslant 1$ for $\delta z\,|_{\zeta=\infty} = 0$). But $\Sigma_1^n\alpha_k = n - 1$, and so $\Sigma_1^n\varepsilon_k = -(n + m + \varepsilon_0 - 1)$. Then a comparison of the quantities δz and $d\delta z/d\zeta$, computed from (6) and (9), in a neighborhood of the points t_k ($k = 1, \ldots, n$) shows that $\varepsilon_k \geqslant -1$ if the corresponding δt_k is not 0, and $\varepsilon_k \geqslant 0$ otherwise. From this it follows, in particular, that $\varepsilon_1 = \varepsilon_n = 0$, and, hence,

$$\sum_{k=2}^{n-1} \varepsilon_k = -(n - 1 + \varepsilon_0 + m),$$

which is not possible even in the case $m = \varepsilon_0 = 0$, since $\varepsilon_k \geqslant -1$. Thus, $\delta z = 0$.

It is clear from (6) that $\delta t_k = 0$ if δz is bounded at the points t_k, where $\alpha_k < 1$. For those t_k where $\alpha_k > 1$ the equality $\delta t_k = 0$ follows from the boundedness of the function $d\delta z/d\zeta$.

This concludes the proof of the lemma.

4°. An existence theorem. We fixed the constants t_k arbitrarily, setting

$$t_k = t_k^0 \quad (k = 2,\dots,n-1), \qquad t_1 < t_2^0 < ,\dots, < t_{n-1}^0 < t_n.$$

Then to the function $z = F_n(\zeta, t_0, t_1, t_2^0,\dots,t_{n-1}^0, t_n)$ with $\zeta = t \in [t_1, t_n]$ there corresponds a certain nondegenerate polygonal line passing through the point z_1 with lengths of links $l_k = l_k^0$ $(k = 1,\dots,n-2)$. By construction, the system (2) has for $l_k = l_k^0$ $(k = 1,\dots,n-2)$ at least one solution $t_k = t_k^0$ $(k = 2,\dots,n-1)$. Let $\hat{l}_k$ $(k = 1,\dots,n-2)$ be the values of the right-hand sides of (2) for which the solvability of the system must be proved.

Let us consider the set $\{l_k\}$ of right-hand sides of (2), determined by the inequalities

$$l_k^- = \min(l_k^0, \hat{l}_k) - \delta_k < l_k < \max(l_k^0, \hat{l}_k) + \delta_k = l_k^+, \tag{10}$$

where the $\delta_k > 0$ are chosen from the conditions $\min(l_k^0, \hat{l}_k) - \delta_k > 0$. Obviously, all the polygonal lines L_z^1 with angles $\alpha_k \pi$, $0 < \alpha_k < 2$, at the corresponding vertices and with lengths l_k of links subject to the inequalities (10) are nondegenerate. Consequently, by Lemma 2, there is an $\varepsilon > 0$ such that the inequalities (4) hold for each solution of (2) corresponding to such polygonal lines. Let

$$l = (l_1,\dots,l_{n-2}), \qquad \tau = (\tau_1,\dots,\tau_{n-2}) \equiv (t_2,\dots,t_{n-1})$$

and

$$f = (f_1,\dots,f_{n-2}) \equiv \left\{ \int_{t_1}^{t_2} \left|\frac{dF_n}{dt}\right| dt,\dots, \int_{t_{n-2}}^{t_{n-1}} \left|\frac{dF_n}{dt}\right| dt \right\},$$

where l, τ, and f are vectors in $(n-2)$-dimensional Euclidean space E_{n-2}. We write (2) in the form

$$f_k(\tau_1,\dots,\tau_{n-2}) = l_k \quad (k = 1,\dots,n-2) \tag{11}$$

or in the vector form

$$f(\tau) = l. \tag{11*}$$

Let $\mathfrak{M}_l$ be the set of vectors $l = (l_1,\dots,l_{n-2})$ whose projections l_k $(k = 1,\dots,n-2)$ satisfy (10), and $\mathfrak{M}_\tau$ the set of vectors $\tau = (\tau_1,\dots,\tau_{n-2})$ whose projections $\tau_k \equiv t_{k-1}$ are subject to the inequalities (4). Then, by Lemmas 2 and 3, system (11) has the following properties.

1. At the points τ where (11) is solvable

$$\frac{D(f_1,\dots,f_{n-2})}{D(\tau_1,\dots,\tau_{n-2})} \neq 0.$$

2. The image $\overline{\mathfrak{M}}_\tau \subset E_{n-2}$ of a closed bounded set $\overline{\mathfrak{M}}_l \subset E_{n-2}$ is a closed bounded set.

3. For $l = l^0 = (l_1^0, \ldots, l_{n-2}^0)$ the system (11) has by construction at least one solution $\tau = \tau^0 = (t_1^0, \ldots, t_{n-2}^0)$.

According to Weinstein's method of continuity [134], these properties imply the existence of at least one solution τ of the system (11) for any given vector $l \in \mathfrak{M}_l$. Indeed, it follows from Young's implicit function theorem that the set of $l = (l_1, \ldots, l_{n-2})$ such that the system (11) is solvable is open. On the other hand, the continuity of the functions $f_k = f_k(\tau_1, \tau_2, \ldots, \tau_{n-2})$ with respect to their arguments in any finite domain in E_{n-2} implies that this set is closed. In view of the solvability of (11) for $l = l^0$, this set is not empty; therefore, it coincides with the whole space, and, consequently, (11) is solvable also for the given values $(\tilde{l}_1, \ldots, \tilde{l}_{n-2}) = \tilde{l}$.

We illustrate the proof of this assertion by more detailed elementary arguments, following the idea of Weinstein's method of continuity. Let us join the points l^0 and $\tilde{l}$ of the set $\mathfrak{M}_l$ by a smooth curve Γ that lies entirely in $\mathfrak{M}_l$. Since (11) is solvable for $l = l^0$, property 1 implies that

$$\frac{D(f_1, \ldots, f_{n-2})}{D(\mathring{\tau}_1, \ldots, \mathring{\tau}_{n-2})} \neq 0$$

for the Jacobian at the point $\mathring{\tau} = (\mathring{\tau}_1, \ldots, \mathring{\tau}_{n-2})$; hence, the system is also solvable in a neighborhood of l^0 by Young's implicit function theorem. We take a point $l^1 \neq l^0$ on the curve Γ in this neighborhood. The system is solvable at this point; consequently,

$$\frac{D(f_1, \ldots, f_{n-2})}{D(\tau_1', \ldots, \tau_{n-2}')} \neq 0$$

for the Jacobian at the corresponding point τ', and this implies that the system is also solvable in a neighborhood of l'. In this neighborhood on the curve $\Gamma' \subset \Gamma$ joining l and $\tilde{l}$ we choose a point $l^2 \neq l^1$ and repeat the process of moving along Γ towards the point $\tilde{l}$. Finally, traversing the whole curve Γ, we hit the point $\tilde{l}$ and thereby prove the solvability of (11) for $l = \tilde{l}$. This process of motion cannot terminate before $\tilde{l}$ is reached. Indeed, suppose the opposite: the process breaks off at a point $l^* \in \Gamma$, and τ^* is the point corresponding to it. Then (11) is solvable at l^*; consequently,

$$\frac{D(f_1, \ldots, f_{n-2})}{D(\tau_1^*, \ldots, \tau_{n-2}^*)} \neq 0,$$

which implies solvability in a neighborhood of the point. Thus, on the curve $\Gamma^* \subset \Gamma$ joining l^* and $\tilde{l}$ we can choose a point $l^* + \Delta l \neq l^*$ at which (11) is solvable, which contradicts the initial assumption. Thus, the solvability of (11) at the given point $\tilde{l}$ has been proved. Thereby the solvability of the system (2) for the given polygonal line has been proved.

5°. *Global uniqueness of the solution.* To prove global uniqueness of the solution of system (2) for given $\tilde{l}_k$ $(k = 1, \ldots, n - 2)$ it suffices to find l_k^0 $(k = 1, \ldots, n - 2)$ such that the system is uniquely solvable. Then by moving from the values l_k^0 $(k = 1, \ldots, n - 2)$ to $\tilde{l}_k$ $(k = 1, \ldots, n - 2)$ according to the method of continuity, we arrive at the unique solvability of this system for given right-hand sides. Let us construct such l_k^0 $(k = 1, \ldots, n - 2)$. To do this we add the identities $\alpha_i \pi \equiv \alpha_i \pi$ $(i = 2, \ldots, n - 1)$ to the system (2), where the $\alpha_i \pi$ are the interior angles of the polygonal line L_z^1, and, using the new notation, we write the augmented system in the form

$$f_k(\tau_1, \ldots, \tau_{2(n-2)}) = l_k \qquad [k = 1, \ldots, 2(n - 2)]. \tag{12}$$

Here τ_k, f_k, and l_k $(k = 1, \ldots, n - 2)$ have the same meanings as in (11), while $\tau_k = l_k = f_k = \pi \alpha_{k+3-n}$ for $k = n - 1, \ldots, 2(n - 2)$.

Obviously, Lemmas 2 and 3 are also valid for the extended system (12); consequently, properties 1–3 of the system (11) remain true for it.

Let $\tau = (\tau_1, \ldots, \tau_{2(n-2)})$ and $l = (l_1, \ldots, l_{2(n-2)})$ be vectors in the Euclidean space $E_{2(n-2)}$, and let $\tilde{l} = (\tilde{l}_1, \ldots, \tilde{l}_{n-2}, \alpha_1, \ldots, \alpha_{n-2})$ be the vector for which it is necessary to prove that (12) is solvable. We construct a vector l^0 for which (12) is uniquely solvable. Suppose that all $\alpha_i \pi = \alpha_i^0 \pi = \pi$ $(i = 2, \ldots, n - 1)$, i.e., consider the mixed boundary-value problem with free boundary for the case when the polygonal line L_z^1 degenerates into a segment. This problem is uniquely solvable, since the system (2) corresponding to it is absent. On the segment $z_1 z_2$ we choose arbitrary points z_k^0 $(k = 2, \ldots, n - 1)$, $z_1^0 = z_1, z_n^0 = z_n$, and let $l_k^0 = |z_{k+1}^0 - z_k^0|$ $(k = 1, \ldots, n - 2)$. Let t_k^0 $(k = 2, \ldots, n - 1)$ be the images of the points z_k^0 under the conformal mapping $z = F_n^0(\zeta)$, where, by construction, $F_n^0(\zeta)$ does not depend on t_k.

Consider for arbitrary l_k $(k = 1, \ldots, n - 2)$ the system of equations

$$\int_{t_k}^{t_{k+1}} \left| \frac{dF_n^0(t)}{dt} \right| dt = l_k \qquad (k = 1, \ldots, n - 2),$$

$$\alpha_k \pi = \alpha_k^0 \pi = \pi \qquad (k = 2, \ldots, n - 1). \tag{12*}$$

By the inequalities

$$0 < \frac{\partial l_k}{\partial t_{k+1}} = \left| \frac{dF_n^0(t_{k+1})}{dt} \right| < \infty \qquad (k = 1, \ldots, n - 2)$$

the first of the equations (12*) is uniquely solvable for the single unknown t_2 appearing in it. Substituting the value found in the second equation of (12*), we determine t_3 uniquely, and so on. Thus, all the t_k $(k = 2, \ldots, n - 1)$ are determined uniqely. In particular, for $l_k = l_k^0$ the system (12), which coincides with (12*) in this case, has a unique solution

$$\tau^0 = \left(t_2^0, \ldots, t_{n-1}^0, \alpha_2^0 \pi, \ldots, \alpha_{n-1}^0 \pi \right).$$

We join the constructed point l^0 to $\tilde{l}$ by a curve Γ lying inside $\mathfrak{M}_l \supset l^0, \tilde{l}$, and, moving by the method of continuity along Γ from l^0 towards $\tilde{l}$, we prove the uniqueness of the solution of (12) for $l = \tilde{l}$. According to Lemma 2, the set $\mathfrak{M}_l$ is mapped into the interior of the set $\mathfrak{M}_\tau$; therefore, nonuniqueness could arise only due to branching of the image $\Gamma_\tau \in \mathfrak{M}_\tau$ of the curve $\Gamma \subset \mathfrak{M}_l$ at some point τ^*.

Suppose that this happens, and let l^* be the image of τ^*. The system (12) is uniquely solvable at l^*; hence, by Young's theorem, it is also uniquely solvable in a neighborhood of l^*, but this contradicts the assumption that the curve Γ_τ branches at τ^*. This proves the global uniqueness of solutions of (12).

6°. *Univalence of solutions.* We shall prove that the function is univalent when the polygonal line L_z^1 is convex.

The unit disk $|\zeta| < 1$ will be taken as a canonical domain. Since the function $w(\zeta)$ conformally mapping the disk $|\zeta| < 1$ onto the domain D_w is univalent by construction, to prove that $w = w(z)$ is univalent it suffices to show that $z = F_n(\zeta)$ is univalent. Suppose that the constants $t_j \equiv e^{i\theta_j}$ $(j = 1, \ldots, n)$ are fixed in such a way that the polygonal line L_z^1 is taken to the upper half of the circle $|\zeta| = 1$, while the unknown arc L_z^2 is taken to the lower half.

Let us consider the function $x = \operatorname{Re} F_n(e^{i\theta})$ for $\theta \in [0, \pi]$. It can be represented in the form

$$x = \int_{t(0)}^{t(\theta)} \left| \frac{dF_n}{dt} \right| \cos \arg\left(\frac{dF_n}{dt} \right) dt + x_1 \equiv g(\theta), \qquad \theta \in [0, \pi],$$

where $t = t(\theta)$ $(dt/d\theta > 0)$ is the boundary correspondence under the conformal mapping of $|\zeta| < 1$ onto the upper half-plane.

Since the polygonal line L_z^1 is assumed to be convex, it follows that

$$\cos\left(\arg \frac{dF_n}{dt} \right) = \cos \beta_i \pi \leqslant 0 \quad \text{for } \theta \in [\theta_i, \theta_{i+1}],$$

where $\beta_i \pi$ is the angle between the ith link and the OX-axis. Consequently, the function $x = g(\theta)$ decreases on the inteval $[0, \pi] = \Sigma_1^{n-1}[\theta_i, \theta_{i+1}]$ and, by the conditions on $f_1(x)$ and $f_2(x)$, increases on $[\pi, 2\pi]$. With the help of the function $x = g(\theta)$, $\theta \in [0, 2\pi]$, $z = F_n(\zeta)$ can be represented by the Schwarz integral

$$z = F_n(\zeta) = \frac{1}{2\pi} \int_0^{2\pi} \frac{e^{i\theta} + \zeta}{e^{i\theta} - \zeta} g(\theta) \, d\theta + iy_0. \tag{13}$$

Then the theorem of Kaplan (Theorem 6 in §3 of Chapter II) implies the univalence of the function $z = F_n(\zeta)$, and with it also the univalence of the solution $w = \omega[F_n^{-1}(z)]$ of the mixed boundary-value problem stated at the beginning of §1.

We now formulate the results obtained in $1°$–$6°$ as the following theorem.

THEOREM 1. *The system of equations* (2) *corresponding to a given nondegenerate polygonal line L_z^1 is uniquely solvable, and thereby so is the mixed boundary-value problem with free boundary* (1.1) *and* (1.2) *(§1.1°). Moreover, if L_z^1 is convex, then the function $z = F_n(\zeta)$ mapping the upper half-plane onto the unknown domain D_z is univalent for* Im $\zeta \geqslant 0$.

$7°$. *Remarks.* 1. In the case of the absence of a free boundary, Theorem 1 implies the existence theorem for a conformal mapping of the upper half-plane onto the domain bounded by a given polygon that does not use the Riemann existence theorem. Such a theorem was first proved by Weinstein [134] by the method of continuity presented above.

2. In the statement of the problem we assume that the functions $f_i(\tau)$ are given on the segment $[\tau_n, \tau_1]$ ($\tau_n < \tau_1$). The following generalization is possible:

$$\begin{aligned} \varphi &= f_1^i(\tau) \\ \psi &= f_2^i(\tau) \end{aligned} \qquad \tau \in [\tau_n^i, \tau_1^i] \ (\tau_n^i < \tau_1^i), \tag{14}$$

where $\tau_1^1 = \tau_1^2$ and $\tau_n^1 < \tau_n^2$, i.e., the original interval is traversed twice. Suppose here that the arc L_w^2 with equation (14) has no points of self-intersection. Then all the computations remain as before, except that $h(t)$ changes sign in passing through the point $t^* = t(\tau_1^i)$ corresponding to the value $\tau \equiv \tau_1^1$. The unknown arc L_z^2 will be smooth if $|dz/dt|_{t=t^*} \neq 0$; otherwise $z(t^*)$ will be a corner point or a cusp of L_z^2.

A similar argument is used in the case when there are several intervals in which the parameter τ varies, with the intervals traversed twice or more, but continuously.

Geometrically, these generalizations correspond to lack of uniqueness of the equation of the unknown arc L_z^2 as a function of the parameter τ. For example, for $\tau \equiv x$ it is possible to find the unknown arc L_z^2 in the class of curves similar to that pictured below in the figure in §6 of this chapter, where $L_z^2 \equiv \overline{L}_z^5 + \overline{L}_z^6 + \overline{L}_z^7$.

3. It was possible to solve the boundary-value problem

$$k_i x - y = b_i, \qquad t \in [t_i, t_{i+1}] \quad (i = 1, \ldots, n - 1),$$

$$x = H(t, t_0, t_1, t_n), \qquad t \in [t_n, t_1],$$

for determination of the analytic function $z(\zeta)$ at once, without differentiating the equations of the links of the polygonal line. Then

$$z(\zeta) = \frac{\Pi(\zeta)}{\pi i} \int_{-\infty}^{\infty} \frac{g(t)dt}{\Pi(t)(t - \zeta)},$$

where

$$g(t) = \begin{cases} b_i, & t \in [t_i, t_{i+1}] \\ H(t), & t \in [t_n, t_1] \end{cases} \quad (i = 1, \ldots, n - 1),$$

$\Pi(\zeta) = \Pi_1^n (\zeta - t_k)^{\alpha_k}$, and the constants t_k must be determined from the conditions for the function $z(\zeta)$ to be bounded at infinity.

§3. Proof of the existence theorem solely on the basis of a priori estimates of the solution

In many problems with free boundaries local uniqueness of solutions of systems of the form (2.2) either is not valid in general, or is valid only under stringent restrictions on the boundary data of the problem. Therefore, it is of considerable interest to prove existence theorems without using local uniqueness of a solution.

We now present one such method for proving solvability, based solely on a priori estimates of the solution, for the example of a mixed boundary-value problem with free boundary. Here we use a modification of a well-known fixed-point principle (an elementary analytic proof of which can be found in Nagumo's article [98]) for finite-dimensional transformations. For infinite-dimensional Banach spaces it is called the Leray-Schauder principle (Theorem 7 in §1 of Chapter II), and we keep this name for the case of finite-dimensional spaces.

The system (2.2) for determination of the constants can be written in the form

$$l_k = \int_{t_k}^{t_{k+1}} \left| \frac{dF_n(t, \tau, \alpha)}{dt} \right| dt \equiv f_k(\tau, \alpha), \tag{1}$$

where $\alpha = (\alpha_1, \ldots, \alpha_n)$ and $\tau = (\tau_1, \ldots, \tau_n)$. We look for a solution of (1) that corresponds to a given polygonal line with lengths of links $l_k = \tilde{l}_k$ and angles $\tilde{\alpha}_k \pi$ $(k = 1, \ldots, n)$.

Let us construct a one-parameter family of polygonal lines that depend continuously on the parameter $\lambda \in [0, 1]$ in such a way that $\lambda = 1$ corresponds to the original polygonal line L_z^1, while $\lambda = 0$ corresponds to the segment joining its endpoints. Such a family can be constructed, for example, as follows. On the segment $z_1 z_n$ ($z_1 \equiv x_1$, $z_n \equiv x_n$) we take points $z_k^0 = x_k^0 = x_n + k(x_1 - x_n)/(n - 1)$ and join them to the corresponding vertices z_k ($k = 2,\ldots,n - 1$) of L_z^1 by disjoint finite smooth curves Γ_k that do not intersect the polygonal line except at endpoints (see the figure).

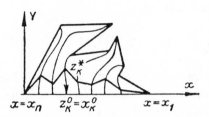

Let s be the arclength parameter of these curves, starting from the points z_k^0, and let σ_k be the length of Γ_k. We write the equation of Γ_k in the form $z = g_k(s)$ and $s \in [0, \sigma_k]$. Then $z_k^0 = g_k(0) = x_k^0$, $z_k = g_k(\sigma_k)$ ($k = 2,\ldots,n - 1$). Let us consider the family of polygonal lines L_z^λ with vertices at the points z_1, z_n and $z_k^n = g_k(\lambda \sigma_k)$ ($k = 2,\ldots,n - 1$), $\lambda \in [0, 1]$.

Denote by $\alpha_k^\lambda \pi = \alpha_k(\lambda)\pi$ ($k = 1,\ldots,n$) the interior angles of the polygon formed by the polygonal line L_z^λ and the rays $z = x_1$ and $x = x_n$ ($y \leqslant 0$), and by $\tilde{l}_k(\lambda)$ the lengths of the links of L_z^λ. Since the functions $g_k(s)$ are smooth, the quantities $\alpha_k(\lambda)$ and $\tilde{l}_k(\lambda)$ are continuously differentiable functions of the parameter $\lambda \in [0, 1]$. And since the Γ_k are disjoint and each of the links $\tilde{l}_k(\lambda)$ travels to the segment $x_k^0 x_{k+1}^0$ in its own "corridor" between the curves Γ_k and Γ_{k+1}, it follows that none of the polygonal lines L_z^λ are degenerate.

The system of equations for determining the constants t_k for each of the L_z^λ takes the form

$$l_k = \int_{t_k}^{t_{k+1}} \left| \frac{dF_n(t,\tau,\alpha^\lambda)}{dt} \right| dt \equiv f_k(\tau,\alpha^\lambda). \tag{2}$$

Since the family of polygonal lines L_z^λ is nondegenerate, Lemma 2* on a priori estimates in §2 gives us the existence of an $\varepsilon > 0$ independent of λ such that for all solutions of (2)

$$t_{k+1} - t_k < \varepsilon < 0 \qquad (k = 1,\ldots,n - 1). \tag{3}$$

Consequently, for fixed constants $t_1, t_n > 0$

$$0 < t_1 < t_2 < \cdots < t_n < \infty. \tag{3*}$$

On the set $\mathfrak{M}_\tau$ of elements $\tau = (t_2, \ldots, t_{n-1})$ satisfying (3) we define a transformation $T(\tau, \lambda) = \{T_1, \ldots, T_{n-1}\}$ by the relations

$$T_{k+1} = t_{k+1} \frac{f_k(\tau, \alpha^\lambda)}{\tilde{l}_k(\lambda)} \qquad (k = 1, \ldots, n-2), \tag{4}$$

where the $\tilde{l}_k(\lambda)$ are the lengths of the links of L_z^λ.

For fixed $\tau \in \mathfrak{M}_\tau$ we have

$$\frac{\partial f_k(\tau, \alpha)}{\partial \alpha_j} = \int_{t_k}^{t_{k+1}} \left\{ \ln|t - t_j| \cdot |\Pi(t)| \cdot |M(t)| + \Pi(t) \frac{\partial M}{\partial \alpha_j} \right\} dt,$$

and the integrals on the right-hand side are bounded for all $\lambda \in [0, 1]$ and $\tau \in \mathfrak{M}_\tau$. Since the functions $\alpha_j^\lambda = \alpha_j(\lambda)$ are continuously differentiable with respect to λ by construction, it follows that

$$\left| \frac{\partial f_k}{\partial \lambda} \right| = \left| \sum_{j=2}^{n-1} \frac{\partial f_k}{\partial \alpha_j} \cdot \frac{d\alpha_j}{d\lambda} \right| < N < \infty,$$

where the constant N does not depend on $\lambda \in [0, 1]$, nor on $\tau \in \overline{\mathfrak{M}}_\tau$.

Thus, the transformation $T(\tau, \lambda)$, along with $f_k(\tau, \alpha^\lambda)$ and $\tilde{l}_k(\lambda)$, is continuously differentiable with respect to λ for fixed $\tau \in \overline{\mathfrak{M}}_\tau$. On the other hand, it was shown in the proof of Lemma 3 in §2 that the functions $l_k = f_k(\tau, \lambda)$ are continuously differentiable with respect to t_k on the set $\overline{\mathfrak{M}}_\tau$. Consequently, the transformation $T(\tau, \lambda)$ satisfies the first condition of the Leray-Schauder theorem. By construction, $T = T(\tau, \lambda)$ can have fixed points $T = \tau$ on $\mathfrak{M}_\tau$ for a particular $\lambda \in [0, 1]$ only at points where (2) is solvable with the given $l_k = \tilde{l}_k(\lambda)$. Indeed, $t_k \neq 0$ on $\mathfrak{M}_\tau$ by virtue of (3), so for $T = \tau$ the relations (4) become the system (2). But according to the a priorir estimates (3), the system (2) does not have solutions on the boundary $\overline{\mathfrak{M}}_\tau \backslash \mathfrak{M}_\tau$ of $\overline{\mathfrak{M}}_\tau$, and this implies the absence of fixed points for $T(\tau, \lambda)$ when $\tau \in (\overline{\mathfrak{M}}_\tau \backslash \mathfrak{M}_\tau)$ and $\lambda \in [0, 1]$. On the other hand, the mixed boundary-value problem with free boundary obtained for the segment $x_1 x_n$ when $\lambda = 0$ (§2.5°) is uniquely solvable together with its system (2) for any left-hand sides l_k ($k = 1, \ldots, n-2$); this means, in particular, the unique solvability of the equation $\tau - T(\tau, 0) = \eta$ for arbitrary η. Consequently, all the conditions of the Leray-Schauder theorem are satisfied for $T(\tau, \lambda)$, and it has at least one fixed point on $\mathfrak{M}_\tau$ for each $\lambda \in [0, 1]$. Thus, we have proved that the system (1) obtained from (2) when $\lambda = 1$ is solvable, and thereby so is the mixed boundary-value problem with free boundary (1.1), (1.2) (§1.1°) for a given polygonal line L_z^1. The proof that (1) is solvable can be carried out especially simply if the length l_j of one of the links of L_z^1 is large enough so that the lengths of the remaining links satisfy the inequality

$$\sum_{k=1}^{n-1} l_k - l_j \leqslant x_1 - x_n. \tag{5}$$

Suppose for definiteness that $l_j = l_{n-1}$. Then the transformation $T(\tau)$ is defined as follows:

$$\int_{T_k}^{T_{k+1}} \left| \frac{dF_n(t,\tau,\alpha)}{dt} \right| dt = \tilde{l}_k \qquad (k = 1,\dots,n-2). \tag{6}$$

This transformation is defined on the set $\mathfrak{M}_\tau$ of elements $\tau = (t_2,\dots,t_{n-1})$, satisfying (3). Indeed, let $\tau^0 \in \mathfrak{M}_\tau$ be an arbitrary point and l_k^0 ($k = 1,\dots,n-1$) the lengths of the links of the corresponding polygonal line L_z^{01}, which obviously satisfy the inequality

$$\sum_{k=1}^{n-1} l_k^0 \geqslant x_1 - x_n. \tag{7}$$

Starting from the point z_1 we lay off the distance $\sum_1^k \tilde{l}_i$ ($k = 1,\dots,n-2$) on L_z^{01} and take the preimage of the resulting point $z_k \in L_z^{01}$ under the mapping $z = F_n(\zeta, \tau^0)$ as the desired $\mathring{T}_k$ ($k = 2,\dots,n-1$). The possibility of this construction is ensured by (5) and (7); and the resulting element $\mathring{T} = (\mathring{T}_2,\dots,\mathring{T}_{n-1})$ of the Euclidean space E_{n-2} belongs to $\mathfrak{M}_\tau$. Since the transformation $T = T(\tau)$ defined by (6) is continuous in τ on the set $\mathfrak{M}_\tau$ and maps $\mathfrak{M}_\tau$ into itself, the familiar Brouwer fixed point principle for finite-dimensional transformations [57] implies that it has at least one fixed point in $\mathfrak{M}_\tau$.

§4. An existence theorem in the case of a curvilinear boundary

1°. We present some needed facts from function theory. Let $z = g(\sigma)$ be a complex-valued function that is continuous on $[a, b]$. Such a function determines a continuous curve L. We draw a polygonal line P_n through arbitrary points $z = g(\sigma_k)$, $\sigma_k \in [a, b]$, of this curve and consider its length

$$\sum_{k=1}^{n} |g(\sigma_{k+1}) - g(\sigma_k)| \equiv \sum_{k=1}^{n} l_k.$$

The curve L is said to be rectifiable if

$$l = \sup \sum_{k=1}^{n} |g(\sigma_{k+}) - g(\sigma_k)| < \infty \tag{1}$$

(for all n and arbitrary polygonal lines P_n), and the quantity l is its length. Functions $z = g(\sigma)$ satisfying condition (1) are called functions of bounded variation, and one of their important properties is that they are differentiable almost everywhere on $[a, b]$. Obviously, the length l of L does not depend on the method of parametrization. Choose the parameter σ to be the arclength s of this curve, measured from the initial point. Then the modulus of the derivative $dz/ds = dg/d\sigma$ (which exists almost everywhere) is equal to 1 almost everywhere on $[0, l]$, and consequently, a rectifiable curve L has a tangent at almost all its points. We shall say that a sequence $\{f_n\}$ of analytic functions in a domain D in the extended z-plane converges uniformly inside D if it converges uniformly in any closed subdomain $\overline{D}_0 \subset D$. A sequence $\{f_n(z)\}$ of analytic functions in D is said to be relatively compact in D if it contains a subsequence that converges uniformly inside D.

A COMPACTNESS PRINCIPLE. *A sequence $\{f_n(z)\}$ of analytic functions in a domain D is relatively compact if and only if it is uniformly bounded inside this domain, i.e.,*

$$|f_n(z)| \le M < \infty$$

in each closed subdomain $\overline{D}_0 \subset D$ for all n.

2°. We now prove that the mixed boundary-value problem with free boundary (1.1), (1.2) (§1.1°) is solvable in the case when the given part L_z^1 of the boundary L_z is a simple (without self-intersections) rectifiable arc. Let us inscribe in L_z^1 a polygonal line P_n with vertices at the points $z_k^n \in L_z^1$ ($k = 1, \ldots, n$), including the endpoints $z_1 = z_1^n$ and $z_0 = z_n^n$ of the arc. Let $\alpha_k^n \pi$ ($k = 1, \ldots, n$) be the interior angles of the polygon formed from P_n and the half-lines $x = x_1^n$ and $x = x_n^n$ emanating from z_1^n and z_n^n and directed so that $\sum_1^n (\alpha_k^n - 1) = -1$.

Then, by what was proved in §§1–3, for each fixed polygonal line P_n there exists a function $z = F_n(\zeta)$ that maps the upper half-plane Im $\zeta > 0$ conformally onto the domain $D_n(P_n + L_n^2)$ (L_n^2 is the unknown free boundary) and is representable in the form

$$z = F_n(\zeta) \equiv \int_{t_1}^{\zeta} \Pi_n(\zeta) M_n(\zeta)\, d\zeta + z_1, \qquad (2)$$

where

$$\Pi_n(\zeta) = c_0 \prod_{k=1}^{n} (\zeta - t_k^n)^{\alpha_k^n - 1}, \qquad M_n(\zeta) = \frac{1}{\pi i} \int_{t_n^n t_1^n} \frac{h(t)\, dt}{\Pi_n(t)(t - z)},$$

and the t_k^n are the images of the vertices z_k^n of P_n. The constants t_k^n are normalized as follows:

$$-1 = t_1^n < t_2^n < \cdots < t_n^n = 1. \qquad (3)$$

We increase the number n of vertices of P_n by adding new vertices in such a way that the lengths of all the links of the polygonal line P_n decrease. Since L_z^1 is a rectifiable arc, the family $\{P_n\}$ of polygonal lines so constructed has the unique limit L_z^1, and, by the definition (1) of the length l of L_z^1, the sum of the lengths of their links is less than l for all n. We find an estimate for $z = F_n(t)$ for any n and any $t \in (-\infty, \infty)$. Since

$$[-1, 1] = [t_1^n, t_n^n] = \sum_{k=1}^{n-1} [t_k^n, t_{k+1}^n],$$

we have for $t \in [-1, 1]$ that

$$|F_n(t)| \leqslant \int_{-1}^{1} |\Pi_n(t)| \cdot |M_n(t)| \, dt + |z_1|$$

$$\leqslant \sum_{k=1}^{n} \int_{t_k^n}^{t_{k+1}^n} |\Pi_n(t)| \cdot |M_n(t)| \, dt + |z_1|.$$

Next, observe that for $|t| \geqslant 1$

$$\operatorname{Re} F_n(t) = x_1 + \int_{t_1^n t} h \, dt + x_1 \equiv H(t), \tag{4}$$

where $H(t)$ does not depend on n and satisfies a Hölder condition for $|t| \geqslant 1$, including a neighborhood of $t = \pm\infty$. Let us separate out the imaginary part $f_n(t)$ of $F_n(t)$ on $[-1, 1]$:

$$\operatorname{Im} F_n(t) = f_n(t), \qquad t \in [-1, 1], \tag{5}$$

where, in view of the proof above that the functions $F_n(t)$ are bounded for $t \in [-1, 1]$,

$$|f_n(t)| \leqslant l + |z_1|, \qquad t \in [-1, 1].$$

Solving the mixed problem (4), (5) in the class of bounded functions, we find the following representation for the functions $F_n(\zeta)$:

$$z = F_n(\zeta) = \frac{\sqrt{1 - \zeta^2}}{\pi i} \left\{ \int_{|t| \geqslant 1} \frac{H(t) \, dt}{\sqrt{1 - t^2}\,(t - \zeta)} + i \int_{-1}^{1} \frac{f_n(t) \, dt}{\sqrt{1 - t^2}\,(t - \zeta)} \right\}, \tag{6}$$

and for $\zeta \to t_0 \notin [-1, 1]$ this gives

$$F_n^+(t_0) = H(t_0) + \frac{\sqrt{1 - t_0^2}}{\pi i} \left\{ \int_{|t| \geqslant 1} \frac{H(t) \, dt}{\sqrt{1 - t^2}\,(t - t^0)} + i \int_{-1}^{1} \frac{f_n(t) \, dt}{\sqrt{1 - t^2}\,(t - t_0)} \right\}$$

$$= A^+(t_0) + B_n(t_0),$$

where

$$B_n(t_0) = \frac{\sqrt{1-t_0^2}}{\pi i} \int_{-1}^{1} \frac{f_n(t)\,dt}{\sqrt{1-t^2}\,(t-t_0)}.$$

Since $H(t)$ is a Hölder function, the properties of the Cauchy type integral (§1.3 in Chapter I) imply that $A^+(t_0) \in C^\alpha$ for all $|t_0| \geq 1$. On the other hand, for $t_0 \leq -1$,

$$|B_n(t_0)| \leq \max_t |f_n(t)| \frac{\sqrt{1-t_0^2}}{\pi} \int_{-1}^{1} \frac{dt}{\sqrt{1-t_0^2}\,(t-t_0)}$$

$$= \max_t |f_n(t)| \leq l + |z_1|.$$

A similar estimate holds also for $t_0 \geq 1$. Thus, for all $t \in (-\infty, \infty)$

$$|F_n(t)| \leq l + |z_1| + C(A), \tag{7}$$

where $C(A) = \max_{t_0} |A^+(t_0)|$ is a constant depending only on the properties of $H(t)$. Then, by the maximum modulus principle for analytic functions, (7) holds in the whole domain $\mathrm{Im}\,\zeta > 0$, and, consequently, the sequence $\{F_n(z)\}$ is relatively compact in the upper half-plane. Let $\{F_{n_k}(z)\}$ be a subsequence of it that converges uniformly inside $\mathrm{Im}\,\zeta > 0$ to a limit function $z = F_0(\zeta)$. In the upper half-plane the analytic function $z = F_0(\zeta)$ satisfies the inequality

$$|F_0(\zeta)| \leq l + |z_1| + C(A), \qquad \mathrm{Im}\,\zeta > 0,$$

which is valid for all the terms of the sequence $\{F_n(\zeta)\}$. According to Fatou's theorem (Theorem 6 in §2 of Chapter I), $F_0(\zeta)$ has limit values almost everywhere on the real axis, and, since the given rectifiable arc L_z^1 is the unique limit of the polygonal lines P_n, it follows that $z = F_0(\zeta) \in L_z^1$ for $t \in [-1, 1]$. Moreover, since the function $z = g(s)$ of bounded variation in the equation of L_z^1 has a derivative almost everywhere, it can be shown that the derivative $dF_0(\zeta)/d\zeta$ also has limit values for almost all $t \in [-1, 1]$. However, we shall not dwell on the proof of this assertion, since it requires a more detailed study of the boundary properties of Cauchy type integrals (see §2 in Chapter II).

Let us consider the rectangle

$$R_\varepsilon: \{|\mathrm{Re}\,\zeta| \leq 1 + \varepsilon, \mathrm{Im}\,\zeta \leq \varepsilon\}, \qquad \varepsilon > 0,$$

and let D be the upper half-plane with the rectangle R_ε deleted for a sufficiently small $\varepsilon > 0$. Then, since the $f_n(t)$ are bounded and $H(t)$ is differentiable, the functions

$$A(\zeta) = \frac{\sqrt{1 - \zeta^2}}{\pi i} \int_{|t| \geq 1} \frac{H(t)\,dt}{\sqrt{1 - t^2}\,(t - \zeta)},$$

$$B_n(\zeta) = \frac{\sqrt{1 - \zeta^2}}{\pi} \int_{-1}^{1} \frac{f_n(t)\,dt}{\sqrt{1 - t^2}\,(t - \zeta)}$$

in (6) are differentiable in $\overline{D}$, and

$$\left| \frac{dA(\zeta)}{d\zeta} \right| \leq N(\varepsilon), \qquad \left| \frac{d^k B_n(\zeta)}{d\zeta^k} \right| \leq N_k(\varepsilon),$$

with constants not dependent on n. Consequently, taking (by the compactness principle) a subsesquence of $\{dF_n(\zeta)/d\zeta\}$ that converges inside the domain and taking account of the fact that $F_{n_k} \to F_0$, we get

$$\left| \frac{dF_0}{d\zeta} \right| \leq N(\varepsilon) + N_1(\varepsilon).$$

By Fatou's theorem (Theorem 6 in §2 of Chapter I), this inequality implies that the function $dF_0(\zeta)/d\zeta$ has limit values for almost all $|t| \geq 1 + \varepsilon$. Similarly, if $H(t)$ is sufficiently smooth, it can be proved that the limit values of $F_0(\zeta)$ have the same smoothness as $H(t)$ for $|t| \geq 1 + \varepsilon$. Thus, the function $z = F_0(\zeta)$ just constructed, together with the function $w = \omega(\zeta)$ mapping the upper half-plane onto the given domain in the plane of $w = \varphi + i\psi$, solves the mixed boundary-value problem with free boundary (1.1), (1.2) (§1.1°) for the given rectifiable curve L_z^1.

We formulate this result as a theorem.

THEOREM 1. *For a given rectifiable arc the mixed boundary-value problem with free boundary* (1.1), (1.2) (§1.1°) *has at least one solution.*

A uniqueness theorem for a larger class of problems that includes, in particular, a mixed boundary-value problem with free boundary, will be proved later (Chapter VIII) by other methods.

REMARK. If the free boundary L_z^2 is absent, then Theorem 1 implies the Riemann theorem on the existence of a conformal mapping of a domain bounded by a simple closed rectifiable curve onto the canonical domain similarly to the way the existence theorem for mappings of polygons followed from Theorem 1 in §2 (see Remark 1 after that theorem).

§5. Examples of hydrodynamics problems that reduce
to a mixed boundary-value problem with free boundary

1°. *The filtration problem for a fluid in an earthen dam.* Since filtration problems are dealt with in a separate chapter (Chapter VIII), we shall not dwell here on a physical analysis of them, but only give a mathematical formulation of one such problem. Suppose that filtration of a fluid occurs in the body of an earthen dam (Figure 1) bounded by an impermeable base $M_1 M_4$, permeable walls $M_1 M_2$ and $M_3 M_4$, and an unknown depression curve $M_2 M_3$ (the boundary of the wet and dry earth).

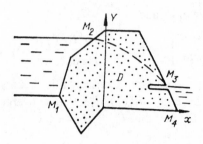

FIGURE 1

It is required to determine the complex filtration potential

$$w = \varphi(x, y) + i\psi(x, y),$$

which is an analytic function in the domain D, from the following conditions on the boundary: $\varphi = \varphi_1$ on $M_1 M_2$, $\varphi = \varphi_2$ on $M_3 M_4$; $\psi = 0$ on $M_1 M_4$; $\psi = \psi_0$ and $\varphi - ky = \varphi_0$ on $M_2 M_3$, where k is the filtration coefficient and ψ_0 is a quantity called the seepage (which is usually unknown, while the remaining parameters are given). Obviously, the problem is a particular case of the mixed boundary-value problem with free boundary studied in §§1–4. The parameters to be determined are the images of the vertices of the polygonal line $M_2 M_1 M_4 M_3$, as well as either the seepage ψ_0 or the filtration coefficient k. And if the location of one of the walls of the dam is not specified, then the parameters k and ψ_0 can be fixed. Polubarinova-Kochina first (see [106]), studied this problem and similar ones for particular shapes of the given boundaries using the methods of the analytic theory of ordinary differential equations.

Much work (see the survey in [106]) has been devoted to the study of these problems for particular shapes of the boundaries of the filtration domain by the hodograph method, in which geometric parameters determining the sizes of the boundaries and often also their form are found only after the problem is solved.

2°. Determination of the underground contour of a hydraulic structure from a given distribution of pressures. In the domain bounded by the permeable parts AB and CD of the bottom of a river bed and the unknown impermeable base BC of a dam (Figure 2) it is required to determine a complex filtration potential $w = \varphi + i\psi$ satisfying the following conditions:

$$\varphi = 0 \quad \text{on } CD, \quad \varphi = -kH \quad \text{on } AB,$$
$$\psi = 0, \quad \varphi = -kH(x), \quad x \in [0, l] \quad \text{on } BC.$$

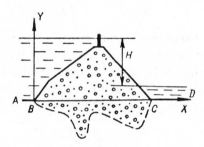

FIGURE 2

A given half-strip in the plane of the complex potential $w = \varphi + i\psi$ corresponds to the filtration domain. This problem is a particular case of the mixed boundary-value problem with free boundary. Since the given polygonal line L_z^1 in this problem is the infinite "segment" $CDAB$ of the line $y = 0$, the formula (1.10) corresponding to this case gives at once a complete solution of the problem due to the absence of unknown parameters in it.

N. B. Il'inskiĭ has devoted a large amount of work (a survey can be found in [102]) to the investigation of this problem and various modifications of it in which the role of the polgonal line L_z^1 is played as before by an infinite "segment" $CDAB$.

We remark that the existence theorems proved in §§1–4 give us the solvability of this problem for an arbitrary curvilinear shape of the river bed $CDAB$.

3°. The inverse problem of the theory of impact. In [112] V. S. Rogozhin stated and solve the following interesting problem of determining the shape of a body upon the vertical impact of it against a fluid (Figure 3).

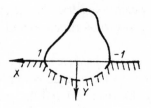

FIGURE 3

Let $\varphi(x, y) = P_t/\rho$ be the velocity potential of the motion arising from the rest state as a result of the impact, where P_t is the impact pressure. Suppose that the impact pressure on the surface of the body is given, which is equivalent to giving the potential

$$\varphi = V_0 \Omega(x), \qquad x \in [-1, 1], \tag{1}$$

where V_0 is the projection of the translational velocity on the y-axis. If we also taken into account that $\partial \varphi / \partial n = V_n$ on the surface of the body, where V_n is the projection of the velocity on the normal, and assume that the angular velocity and the projection of the translational velocity on the x-axis are equal to zero, then for the stream function $\psi(x, y)$ we obtain

$$\psi = -V_0 x, \qquad x \in [-1, 1]. \tag{2}$$

On the free boundary

$$\varphi = 0, \qquad |x| \geqslant 1. \tag{3}$$

Condition (1)–(3) determine a domain in the plane of the complex potential $w = \varphi + i\psi$, and, consequently, this problem is a particular case of the mixed boundary-value problem with free boundary. The role of the polygonal line L_z^1 in this problem is played by an infinite "segment" of the line $y = 0$, and the solution is determined to within an arbitrary constant.

4°. The inverse problem of transonic flow of a gas. Let $\varphi = \varphi(x, y)$ be the flow potential of a gas; $q = \{\varphi_x^2 + \varphi_y^1\}^{1/2}$ and $\theta = \arctan(\varphi_y/\varphi_x)$ are the magnitude and slope angle, respectively, of the flow velocity; $\rho = \rho(q)$ is the density, and $\psi = \psi(x, y)$ is the stream function. It is known (see, for example, [16], formula (3.26)) that in the case of a Chaplygin gas the stream function satisfies the equation

$$K(\eta)\psi_{\xi\xi} + \psi_{\eta\eta} = 0, \tag{4}$$

where $\eta = \int_{-q}^{q} \rho_0^{-1} \rho q^{-1} dq$, $\xi = \theta/\theta_0$, θ_0 is a fixed parameter; q is the critical velocity, $\rho_0 = \rho(q_0)$, and the given function $K(\eta)$ changes sign on passing through the sonic line $\eta = 0$. In the Poritsky model ([16], formula (16.14)), where

$$K(\eta) = \begin{cases} 1, & \eta > 0, \\ -1, & \eta < 0, \end{cases}$$

the equation (4) reduces to the Lavrent'ev equation

$$(\text{sgn}\,\eta)\psi_{\xi\xi} + \psi_{\eta\eta} = 0. \tag{5}$$

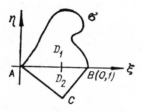

FIGURE 4

Suppose that first the Tricomi problem is solved in the domain $D_1 + D_2$ (Figure 4), i.e., the solution $\psi(\xi, \eta)$ of (5) is sought from the following boundary conditions:

$$\psi = f(s) \text{ on } \sigma, \qquad \psi = g(\xi) \text{ on } AC, \qquad \xi \in [0, \tfrac{1}{2}]. \tag{6}$$

Here continuous $\psi(\xi, \eta)$ is sought in $\overline{D}_1 + \overline{D}_2$, with $\partial\psi/\partial\xi$ and $\partial\psi/\partial\eta$ continuous in $D_1 + D_2$. The Tricomi problem (5), (6) reduces (see [18]) to the determination of a function $\psi(\xi, \eta)$ that is harmonic in D_1, while on $\Gamma_1 = (AB + \sigma)$:

$$\psi = f(s) \quad \text{on } \sigma,$$
$$\left(\frac{\partial\psi}{\partial\xi} - \frac{\partial\psi}{\partial\eta}\right) - \frac{d}{d\xi}g\left(\frac{\xi}{2}\right) \quad \text{on } AB, \xi \in [0, 1]. \tag{7}$$

We let $\partial\psi/\partial\eta = u$ and $\partial\psi/\partial\xi = v$, and consider the Tricomi problem with free boundary in which the boundary conditions

$$u = f_1(\xi), \qquad \xi \in [0, 1], \qquad v = f_2(\xi), \qquad \xi \in [0, 1] \tag{8}$$

are given on the unknown curve σ, and the function $g(\xi)$ is equal to $a\xi + b$ (a and b are constants) on the curve AC. According to the second of the relations (7), this is equivalent to the following boundary condition on AB:

$$v - u = \tfrac{1}{2}a \quad \text{on } AB, \xi \in [0, 1]. \tag{9}$$

The problem (8), (9) is a mixed boundary-value problem with free boundary for the analytic function $w = u + iv$ in the domain D, and the corresponding formula (1.10) gives its complete solution.

§6. The mixed boundary-value problem with free boundary for a system of polygonal lines

1°. Suppose that on each of the known polygonal lines L_z^i $(i = i_1, \dots, i_m)$ (with $n_i - 1$ links) on the unknown contour L_z we are given a boundary condition

$$\Phi_i(\varphi, \psi) = 0, \tag{1}$$

while on the unknown arcs L_z^k of this contour $(k = k_1, k_2)$ we are given

$$\begin{aligned}\varphi &= f_1^k(x) \\ \psi &= f_2^k(x)\end{aligned} \qquad x \in [x_0^k, x_1^k]. \tag{2}$$

It is required to find a closed contour L_z that includes the given polygonal lines, and an analytic function $w(z) = \varphi + i\psi$ in the domain D_z that satisfies conditions (1) and (2).

It is assumed that the functions $f_i^k(x)$ can be multi-valued functions of their arguments on the whole segment $x_0^k x_1^k$ or on some parts of it. Then by breaking up the segment into parts, selecting single-valued branches of the functions $f_i^k(x)$ $(i = 1, 2)$ on each of them, and indexing the endpoints of the new segments in the order of traverse along the unknown arc $\bar{L}_z^k$, we get that each of them corresponds to a part of $\bar{L}_z^k$. It can thereby be assumed that the functions $f_i^k(x)$ $(i = 1, 2)$ in (2) are single-valued on their segments by allowing some of the unknown arcs to be joined with one another, i.e., by assuming that there are more unknown than known arcs $(s \geqslant m)$, their number being fixed in advance and determined by the multivaluedness of the given functions $f_i^k(x)$.

Geometrically, this corresponds to the lack of uniqueness of the equations for some of the unknown arcs $\bar{L}_z^i$ as functions of the parameter x.

Let us index the polygonal lines and the unknown arcs L_z^i and $\bar{L}_z^k$ in the order of their succession on the unknown contour L_z, and specify on each of the polygonal lines L_z^i the direction of traverse. We fixed the location of one of the given polygonal lines, for example with respect to the origin, by assuming that its endpoints are located on the OX-axis at the points $z_i^1 \equiv x_i^1$ $(i = 1, n; x_n^1 < 0 < x_1^1)$.

The equations of the remaining polygonal lines are given to within a translation along the OY-axis, and, moreover, each polygonal line obviously lies in the strip $x_1^i \leqslant x \leqslant x_n^i$ (the index i of n_i will be omitted), where $x_1^i = x_1^{i-1}$ and $x_n^i = x_0^{i+1}$. We set $x_1^k = x_0^{k+1}$ if the unknown arc $\bar{L}_z^k$ is followed again by an unknown arc $\bar{L}_z^{k+1}$. The inequality $x_1^i > x_n^i$ is possible, depending on the direction of the traverse of L_z^i in the corresponding segments (analogously, $x_0^k > x_1^k$ is possible for $[x_0^k, x_1^k]$). Of course, we assume that the problem is geometrically well-posed on the whole, i.e., it is possible to join the endpoints of the given polygonal lines L_z^i by certain curves located in the corresponding strips $x_1^i \leqslant x \leqslant x_n^i$ between neighboring polygonal lines (with regard to the multivaluedness of the equations of the unknown arcs) in such a way that the resulting contour (which preserves the direction of traverse on the polygonal lines L_z^i) does not have points of self-intersection. The contour L_w formed from the arcs L_w^i with the equations (1) and the arcs with the equations (2) is assumed to be closed and without points of self-intersection.

We present a sketch to clarify the statement of the problem.

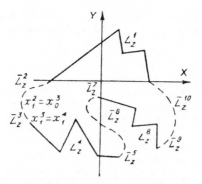

The respective polygonal lines L_z^4 and L_z^8 can be moved along the straight lines $x = x_n^4$, $x = x_1^4$ and $x = x_n^8$, $x = x_1^8$.

The twice continuously differentiable functions $f_1^i(x)$, $f_2^i(x)$, and $\Phi_i(\varphi, \psi)$ are assumed to be compatible in the sense that the directions of traverse coincide on the contours L_z and L_w.

$2°$. As in the case of a single unknown arc, by mapping the upper half-plane Im $\zeta \geqslant 0$ onto the domain D_w and comparing the values of the function $w = \omega(\zeta)$ obtained with the conditions (2), we obtain the boundary-value problem

$$k_j^i \frac{dx}{dt} - \frac{dy}{dt} = 0, \qquad t \in \left[t_j^i; t_{j+1}^i \right] \ (j = 1, \dots, n_i - 1; \ i = i_1, \dots, i_m),$$

$$\frac{dx}{dt} = h_k(t), \qquad t \in \left[t_0^k, t_1^k \right] \ (k = k_1, \dots, k_s), \tag{3}$$

where the k_j^i are the slopes of the links of L_z^i, $t_n^i = t_0^{i+1}$ ($t_1^i = t_1^{i+1}$) if this point corresponds to the juncture of the arcs L_z^i and L_z^{i+1} (respectively, the arcs $\bar{L}_z^i$ and L_z^{i+1}), and $t_1^i = t_0^{i+1}$ if this point corresponds to the juncture of $\bar{L}_z^i$ and $\bar{L}_z^{i+1}$. Obviously, under the conditions (3) the function $h_i(t)$ can be represented in the form

$$h_i(t) = h_i^*(t) \left| t - t_0^i \right|^{\gamma_0^i - 1} \left| t - t_1^i \right|^{\gamma^i - 1},$$

where $0 \neq |h_i^*(t)| < \infty$ for $t \in [t_0^i, t_1^i]$, and $\gamma_0^i \pi$ and $\gamma^i \pi$ are the angles between the arcs of the contour $L_\omega = \Sigma_{i,j}(L_w^i + L_w^j)$ at their juncture points. For example, we fix the points t_n^1, t_1^1, and t_0^1 and assume that the point at infinity passes into some point of L_z^1, i.e., one of the segments $[t_j^1, t_{j+1}^1]$ is infinite. These conditions uniquely determine the function $\omega(t)$ and, consequently, also the points t_0^k and t_1^k ($k = k_1, \dots, k_s$). We can take the canonical function of the homogeneous Hilbert problem corresponding to problem (3) to be the derivative of an analytic function mapping the upper half-plane onto the interior of some polygon $P(z)$ formed from a polygonal line $P'(z)$ with links parallel to the links of the given polygonal lines L_z^i and with line segments $\bar{P}_z^i$ parallel to the OY-axis and joining the endpoints of neighboring polygonal lines. Consequently, this function can be written in the form

$$\Pi(\zeta) = C_0 \prod_{i=i_1}^{i_m} \prod_{k=1}^{n_i} \left(\zeta - t_k^i \right)^{\alpha_k^i - 1}. \tag{4}$$

If the polygon P_z taken is finite, then the function $\Pi(\zeta)$ has a zero of order two at infinity. It is possible to choose the juncture angles of the polygonal lines P_z^i and the segments $\bar{P}_z^j$ in such a way that $\Pi(\zeta)$ has a pole of order $m - 2$ at infinity; however, the polygon P_z will then be infinite, and m vertices of it corresponding to endpoints of the polygonal lines P_z^i will lie at infinity. Obviously, all new representations of the canonical function that satisfy the requirements of our problem can be obtained from the single fixed canonical function $\Pi(\zeta)$ by multiplying it by

$$\prod_1^m \left(\zeta - t_n^k\right)^{\varepsilon_n^k} \left(\zeta - t_1^k\right)^{\varepsilon_1^k}$$

where ε_n^k and ε_1^k are integers satisfying the conditions $|\alpha_i^k - 1 + \varepsilon_i^k| < 1$. We construct one of the canonical functions by starting out from a polygon P_z with only one vertex at infinity. Such a canonical function $\Pi(\zeta)$ has a first-order zero at infinity, and, consequently, $\Sigma_{i=i_1}^{i_m} \Sigma_{k=1}^{n_i} (\alpha_k^i - 1) = -1$ in (4). The number κ of distinct representations of the solution depends on the number of polygonal lines and their mutual locations (to within a translation along the OY-axis and a dilation); therefore, it is difficult to compute in the general case. However, from the construction of the canonical function it is clear that $\kappa > 1$. By choosing this or that representation of the canonical function it is possible, as in the case of a single arc, to control the juncture angles of the polygonal lines L_z^i and the unknown arcs $\bar{L}_z^k$ and to single out classes of problems for which the given representation is unique. All the subsequent arguments will be carried out only for the single representation of the canonical function $\Pi(\zeta)$ that has a first-order zero at infinity. Then the general solution of the boundary-value problem (2) can be written in the form

$$z = F(\zeta) = \int_{t_n}^{\zeta} \frac{\Pi(\zeta)}{\pi i} \sum_{k=k_1}^{k_s} \int_{t_0^k}^{t_1^k} \frac{h_k(t)\,dt}{\Pi(t)(t - \zeta)} + z_n^1, \tag{5}$$

where the limits of integration in the inside integrals are fixed.

3°. Let us write a system for determining the constants that were left arbitrary:

$$l_k = \int_{t_k^i}^{t_{k+1}^i} \left| \frac{dF}{dt} \right| dt. \tag{6}$$

We prove that this system completely determines the problem, i.e., if it can be solved for the constants t_k^i ($k = 2,\dots,n_i - 1$; $i = i_1,\dots,i_m$), then (5) with $\zeta = t$ gives us a contour that includes the given polygonal lines L_z^i. Taking the first equation in (6) into account, we find from (5) that

$$x(t) = \int_{t_n^1}^{t} \left\{ \sum_k h_k(t) + \sum_i \frac{dx_i}{dt} \right\} dt + x_n', \tag{7}$$

with $h_k(t) = 0$ for $t \in (t_0^k, t_1^k)$, and $dx_i/dt = 0$ for $t \notin (t_1^i, t_n^i)$; moreover, by the definition of $h_k(t)$,

$$\int_{t_0^k}^{t_1^k} h_k(t)\,dt = x_1^k - x_0^k. \tag{8}$$

Since $z(t_n^1) = z_n^1$, formulas (7) and (8) and the equality $x(t_1^1) = x_1^1 = x_0^2$ imply that $x(t_1^2) = x_1^2$. If the unknown arc $\bar{L}_z^2$ is followed by the unknown arc $\bar{L}_z^3$, then $x_1^2 = x_0^3$, and, according to (8), $x(t_1^3) = x_1^3$. But if the arc $\bar{L}_z^2$ is followed by the polygonal line L_z^3, then $x_1^2 = x_1^3$. Suppose that the constants are chosen so that the lengths of the links of the polygonal line P_z^3 obtained from (5) coincide with l_k^3 ($k = 2,\dots,n_3 - 1$), i.e., we have the equations in (6) corresponding to the polygonal line L_z^3. Then, since the first link of P_z^3 lies on the given line $x = x_1^3$ and $n_3 - 2$ links of it coincide with the corresponding links of the given polygonal line L_z^3, equation (8) tells us that the endpoint of the last $((n_3 - 1)$th) link of it falls on the given line $x = x_n^3$. Consequently, (7) can be rewritten in the form

$$x(t) = \int_{t_n^3}^{t} \left\{ \sum_k h_k(t) + \sum_i \frac{dx_i}{dt} \right\} dt + x_n^3.$$

Repeating the cycle of arguments from the start, we arrive at the equality $x(t_1^4) = x_1^4$, and so on. It is obvious that if we go through the whole system (6) in this way, we get a solution of the problem that satisfies all the necessary requirements.

From this it follows, in particular, that if we arbitrarily fix the constants t_k^i ($k = 2, \ldots, n_i - 1$; $i = i_1, \ldots, i_m$), then (5) with $\zeta = t$ gives us a contour L_z that includes certain polygonal lines P_z^i with links parallel to the links of the given L_z^i. The main difficulty in proving the existence of the constants consists in proving the local uniqueness of the solution of (6); we proceed to a proof of this.

4°. We make some additional assumptions.

1. All the polygonal lines except, perhaps, the one that remains in place are convex (the total number of polygonal lines is $m \geq 2$).

2. The slope angles of these polygonal lines are $\gamma(\zeta) \neq 0, \pi$.

3. At least $m + 1$ of the $2m$ angles $\alpha_1^i \pi$, $\alpha_{n_i}^i \pi$ formed by the terminal links of the polygonal lines L_z^i ($i = 1, \ldots, m$) with the OY-axis are less than π. Then, by construction, the canonical function $\Pi(\zeta)$ is unbounded at the corresponding points $\zeta = t_1^i, t_{n_i}^i$.

Suppose that the system (6) has at least two infinitesimally close solutions. Then (5) gives us that

$$\delta z = \int_{t_n^1}^{\zeta} \Pi(\zeta) \sum_{k,j} \left[\left(1 - \alpha_k^j\right)\left(\zeta - t_k^j\right)^{-1} M(\zeta) + \frac{\partial M}{\partial t_k^j} \right] \delta t_k^j \, d\zeta, \tag{9}$$

where

$$M(\zeta) = \frac{1}{\pi i} \sum_{k=k_1}^{k_s} \int_{t_0^k}^{t_1^k} \frac{h_k(t)\,dt}{\Pi(t)(t - \zeta)}.$$

Let us transform (9):

$$\frac{\partial M(\zeta)}{\partial t_k^j} = \frac{1}{\pi i} \sum_{i=i_1}^{i_s} \int_{t_0^i}^{t_{i+1}^i} \frac{h_i(t)}{t - \zeta} \cdot \frac{\partial}{\partial t_k^j}\left\{ \frac{1}{\Pi(t)} \right\} dt = \frac{\alpha_k^j - 1}{\pi i} \sum_{i=i_1}^{i_s} \int_{t_0^i}^{t_{i+1}^i} \frac{h_i(t)\,dt}{\Pi(t)\left(t - t_k^j\right)(t - \zeta)}.$$

But

$$\left(\alpha_k^j - 1\right)\left[M(\zeta) - M\left(t_k^j\right)\right]\left(\zeta - t_k^j\right)^{-1}$$

$$= \frac{\alpha_k^j - 1}{\pi i\left(\zeta - t_k^j\right)} \sum_{i=i_1}^{i_s} \int_{t_0^i}^{t_{i+1}^i} \frac{h_i(t)}{\Pi(t)}\left\{ \frac{1}{t - \zeta} - \frac{1}{t - t_k^j} \right\} dt = \frac{\partial M(\zeta)}{\partial t_k^j}.$$

Therefore, the term in the square brackets in (9) is equal to

$$\frac{M\left(t_k^j\right)\left(1 - \alpha_k^j\right)}{\zeta - t_k^j},$$

and so this formula can be written in the form

$$\delta z = \int_{t_n^1}^{\zeta} \Pi(\zeta) \sum_{k,i} \frac{\left(\alpha_k^j - 1\right) M\left(t_k^i\right)}{\zeta - t_k^j} \delta t_k^j \, d\zeta. \tag{10}$$

From this it is clear that δz is bounded at the points corresponding to the angles of the polygonal line that are greater than π, as well as at the points t_0^i and t_1^i, since $\delta t_0^i = \delta t_1^i = 0$; but at the remaining points it is unbounded.

On the other hand, since L_z^1 remains in place under the variation, δz satisfies (as in the case of a single arc; cf. (8) in §2) the relation

$$k_j^1 \delta x - \delta y = 0, \qquad t \in \left[t_j^1, t_{j+1}^1\right] \qquad (j = 1, \ldots, n_1 - 1). \tag{11}$$

Moreover, since $h_k(t)$ is independent of the unknown constants, it follows that $\delta h_k = 0$, and for $\delta z_1 = \delta l_k = 0$ ($k = 2, n_i - 1$; $i = i_1, \ldots, i_m$) this gives us

$$\delta x = 0, \quad t \in \left[t_0^i, t_1^{i+1} \right] \quad (i = i_1, \ldots, i_s). \tag{12}$$

For the remaining polygonal lines L_z^i we have

$$k_j^i \delta x - \delta y = \delta y_1^i, \quad t \in \left[t_j^i, t_{j+1}^i \right] \quad (j = 1, \ldots, n_i - 1). \tag{13}$$

Indeed, considering the functions $F^*(\zeta) = F(\zeta) - \delta y_1^i$, we get that for them the polygonal line L_z^i remains in place under the variation, and, consequently, these functions will satisfy a relation analogous to that established for L_z^1. The canonical function of the homogeneous problem corresponding to the problem (11)–(13) obviously has the form

$$\tilde{\Pi}(\zeta) = \Pi(\zeta) \prod_{i,j} \left(\zeta - t_{n_i}^i \right) \left(\zeta - t_1^j \right), \tag{14}$$

where $\Pi_{i,j}$ depends on the points t_1^j, $t_{n_i}^i$ at which $\Pi(\zeta)$ is unbounded, and there are at least $m + 1$ such points, by condition 3. Denote the number of points $t_{n_i}^i$ and t_1^j appearing in $\Pi_{i,j}$ by $p \geqslant m + 1$. Then the function $\tilde{\Pi}(\zeta)$ has a pole of order $p - 1 \geqslant m \geqslant 2$ at infinity. The solution of the boundary-value problem (11)–(13) can be written in the form

$$\delta z = \frac{\tilde{\Pi}(\zeta)}{\pi i} \sum_{k=2}^{m} \delta y_1^k \int_{t_1^k}^{t_0^k} \frac{dt}{\tilde{\Pi}(t)(t - \zeta)}, \tag{15}$$

where the integrals are taken over the segments

$$\left[t_1^k, t_0^k \right] = \sum_{k=1}^{n_j - 1} \left[t_j^k, t_j^{k+1} \right].$$

By construction, the behavior of the function δz, expressed in the form (15), at the points t_j^i coincides with the behavior of the δz computed by (10); therefore, the δz in (15) is bounded at infinity. It is necessary that $p - 2$ additional conditions hold, of which the first $m - 1$ have the form

$$\sum_{k=2}^{m} \delta y_1^k \int_{t_1^k}^{t_0^k} \frac{t^j dt}{\tilde{\Pi}(t)} = 0 \quad (j = 0, 1, 2, \ldots, m - 2). \tag{16}$$

The conditions (16) can be regarded as a system of equations for finding the variations δy_1^k ($k = 2, \ldots, m$). We show that (16) has only the trivial solution $\delta y_1^k = 0$ ($k = 2, \ldots, m$). We rewrite (16) in the form

$$\sum_{k=2}^{m} \delta y_1^k \int_{t_1^k}^{t_0^k} \frac{t^j e^{-i\gamma(t)}}{|\Pi(t)|} dt = 0 \quad (j = 0, \ldots, m - 2), \tag{17}$$

where $\gamma(t) = \arg \tilde{\Pi}(t)$. Multiplying the real and imaginary parts of (17) by $\sin \varphi_0$ and $\cos \varphi_0$, respectively, and adding, we get

$$\sum_{k=2}^{m} \delta y_1^k \int_{t_1^k}^{t_0^k} \frac{t^j \sin[\gamma(t) + \varphi_0]}{|\tilde{\Pi}(t)|} dt = 0. \tag{18}$$

Since $\tilde{\Pi}(\zeta)$ satisfies the homogeneous problem corresponding to the boundary-value problem (11)–(13), it follows that $\gamma(t) = \arg \tilde{\Pi}(t) = \text{const}$ for $t \in [t_i^k, t_{i+1}^k]$, and, according to conditions 1 and 2 stated at the beginning of this subsection, on each of the segments

$$\left[t_1^k, t_0^k \right] = \sum_{i=1}^{n_i - 1} \left[t_i^k, t_{i+1}^k \right]$$

one of the following relations holds:

$$\arg \tilde{\Pi}(t) = \gamma(t) \in [0, \pi] \quad \text{or} \quad \gamma(t) \in (\pi, 2\pi).$$

For φ_0 small enough so that $[\gamma(t) + \varphi_0] \in (0, \pi)$ or $[\gamma(t) + \varphi_0] \in (\pi, 2\pi)$, $\sin[\gamma(t) + \varphi_0]$ does not change sign in the intervals $[t_1^k, t_0^k]$.

Let

$$\left| \frac{\sin[\gamma(t) + \varphi_0]}{\tilde{\Pi}(t)} \right| = |Q(t)|, \qquad \delta y_1^k \operatorname{sgn}[\sin\{\gamma(t) + \varphi_0\}] = a_k$$

and write (18) in the form

$$\sum_{k=2}^m a_k \int_{t_1^k}^{t_0^k} t^j |Q(t)| \, dt = 0 \qquad (j = 0, \ldots, m-2). \tag{19}$$

We prove that the determinant Δ of (19) is nonzero, and, consequently, $a_k = \delta y_1^k = 0$ ($k = 2, \ldots, m$). But then $\delta z = 0$, as is clear from (15). Suppose not: Suppose that the determinant Δ of (19) is 0. Let

$$\mu_k = \int_{t_1^k}^{t_0^k} \sum_{j=1}^{m-2} c_j t^j |Q(t)| \, dt \qquad (k = 2, \ldots, m)$$

be the elements of a row in Δ obtained as a linear combination of the rows of Δ. Since $\Delta = 0$, there exist c_j, not all equal to zero, such that all $\mu_k = 0$. But if $\mu_k = 0$ ($k = 2, \ldots, m$), then the polynomial $R_{m-2}(t) = \sum_0^{m-2} c_j t^j$ changes sign in each of the $m-1$ intervals (t_1^k, t_0^k), i.e., it must have at least $m-1$ real roots, which is impossible since the degree of this polynomial is equal to $m-2$. Thus, $\Delta \neq 0$, and so

$$a_k = \pm \delta y_1^k = 0 \qquad (k = 2, \ldots, m) \quad \text{and} \quad \delta z = 0.$$

According to (10), the last identity is possible only if $\delta t_k^j = 0$ ($k = 1, \ldots, n_j$, $j = 1, \ldots, m$). The local uniqueness of solutions of (6) has been proved. The proof of the lemma that the parameters cannot converge is obviously completely analogous to the case of a single polygonal line.

5°. If all the polygonal lines L_z^i degenerate into segments, then the solution of the corresponding problem is unique. Consequently, adjoining the identities

$$\alpha_k^j \pi \equiv \alpha_k^j \pi \qquad \left(k \neq 1, n_j \right)$$

to (6), we find (similarly to the case of a single polygonal line) the values of the right-hand sides in the augmented system (6) to which corresponds the unique solution of this system. Then, proceeding by the continuity method from the values found for the right-hand sides, we prove the unique solvability of the augmented system (6) also for the values of the right-hand sides corresponding to the given polygonal lines L_z^i ($i = 1, \ldots, m$). As in the case of a single polygonal line, it is possible to see that the function $x(t) = \operatorname{Re} F^m(t)$, $t \in (-\infty, \infty)$, increases on some segment $[t', t'']$ and decreases for $f \notin [t', t'']$, which implies that the solution is univalent. The case of arbitrary rectifiable arcs L_z^i ($i = 1, \ldots, m$) is treated in a way completely analogous to that in §4.

§7. Assignment of the boundary conditions in terms of functions
of the other parameters

We recall the statement of the mixed boundary-value problem with free boundary. Let D_z^+ be a simply connected domain interior to a boundary L_z made up of the polygonal line L_z^1 passing through given points z_i ($i = 1, \ldots, n$), and an unknown arc L_z^2.

On L_z^2 it is given that

$$
\begin{aligned}
u &= f_1(\tau), \\
v &= f_2(\tau),
\end{aligned}
\qquad \tau \in [\tau_n, \tau_1],
\tag{1}
$$

where $w(\tau) = f_1(\tau) + if_2(\tau)$ is the boundary value of an analytic function $w(z)$, $0 < |\,dw/dz\,| < \infty$, and τ stands for one of the parameters $x = \operatorname{Re} z$, $\alpha = \arctan y'_x$, $\theta = \arg z$, or the arclength s of L_z^2, for which an increase of τ from τ_n to τ_1 corresponds to a traverse of L_z^2 in the positive direction.

To the arc L_z^2 corresponds the arc L_w^2 with equation (1) in the plane of the complex potential w. The equation

$$
\Phi(u, v) = 0
\tag{2}
$$

of an arc L_w^1 completing L_w^2 to form a closed contour L_w is assumed to be given.

Suppose that the function $w = \omega(\zeta)$ maps the half-plane onto d_w^+ in such a way that the points t_0, t_1 and t_n ($t_n < t_1$) on the real axis pass, respectively, into the fixed point w_0 and the points $w_i = w(\tau_i)$ ($i = 1, n$); obviously, the image of the point at infinity lies then on L_w^1.

The identity

$$
f_1(\tau) + if_2(\tau) \equiv w \equiv \omega(t)
$$

gives us a functional dependence $\tau = H(t, t_0, t_n, t_1)$, $t \in [t_n, t_1]$.

Then

$$
\frac{d\tau}{dt} = \frac{\omega_t'(t)}{\omega_\tau'[H(t)]} = h(t, t_0, t_1, t_n), \qquad t \in [t_n, t_1],
\tag{3}
$$

where $0 < |h(t)| < \infty$ for $t \in (t_n, t_1)$, by the correspondence between an increase in τ and a traverse of the arcs L_z^2 and L_w^2 in the positive direction.

If the interior angles between L_w^1 and L_w^2 at the points $w(\tau_i)$ ($i = 1, n$) are denoted by $\gamma_i \pi$ ($0 < \gamma_i < 2$), then $h(t)$ can be represented in the form

$$
h(t) = \frac{h^*(t)}{(t_1 - t)^{1-\gamma_1}(t - t_n)^{1-\gamma_n}},
$$

where $0 < |h^*(t)| < \infty$ for $t \in [t_n, t_1]$.

For certain choices of the parameter τ, for example, $\tau = \arg z$, the choice of the coordinate system in the z-plane has a significant influence on the formulation of the corresponding problem. In order not to complicate the subsequent arguments, we fix once and for all the position of the polygonal line L_z^1 with respect to the coordinate axes, assuming for simplicity that it is located above the OX-axis and that $z_i \equiv x_i$ ($i = 1, n$), $x_n < 0 < x_1$.

Let us write the equation of the links of the polygonal line:

$$
\begin{aligned}
& xk_i - y = b_i \qquad (i = 1, \dots, n-1), \\
& k_i = \frac{y_{i+1} - y_i}{x_{i+1} - x_i} \quad \text{and} \quad b_i = \frac{y_{i+1}x_i - x_{i+1}y_i}{x_{i+1} - x_i}.
\end{aligned}
\tag{4}
$$

THEOREM 1. *Under the above conditions on the functions f_i and Φ there exist a unique finite contour L_z including the given polygonal line L_z^1 and an analytic function $w(z)$ in D_z^+ that satisfies the boundary conditions* (1) *and* (2).

The theorem was proved in §§1–3 for $\tau \equiv x$.

1°. Suppose that $\tau \equiv \alpha \in [\alpha_n, \tilde{\alpha}_1]$, and $\alpha_n > 0$. Then the boundary-value problem (3), (4) can be rewritten in the form

$$
\begin{aligned}
& \arg \frac{dz}{dt} = \alpha_i \quad \text{for } t \in [t_i, t_{i+1}] \ (i = 1, \dots, n-1), \\
& \arg \frac{dz}{dt} = H(t) \quad \text{for } t \in [t_n, t_1],
\end{aligned}
\tag{5}
$$

where the $\alpha_i = \arctan k_i \geqslant 0$ are computed for a traverse of L_z^1 in the positive direction. Since $\arg(dz/dt) = \operatorname{Im}\ln(dz/dt)$, we recover the analytic function $\ln(dz/d\zeta)$ by the Schwarz formula for the half-plane and find that

$$z = F_n(\zeta) \equiv c_0 e^{i\beta} \int_{t_1}^{\zeta} M_n(\zeta) \prod_{k=1}^{n} (\zeta - t_k)^{\beta_k - 1}\, d\zeta + z_1. \tag{6}$$

Here the following notation is introduced:

$$M_n(\zeta) \equiv M_n(\zeta, t_0, t_1, t_n) = \exp\left\{ \frac{1}{\pi}\int_{t_n}^{t_1} \frac{H(t)\,dt}{t - \zeta} + \frac{\alpha_n}{\pi}\ln(\zeta - t_n) \frac{\tilde{\alpha}_1}{\pi}\ln(\zeta - t_1)\right\},$$

$$\beta_1 = \frac{\tilde{\alpha}_1 - \alpha_1}{\pi} - 1; \qquad \beta_n = \frac{\alpha_{n-1} - \alpha_n}{\pi} + 1.$$

It is not hard to compute that the estimate $|dF_n/dt| < k_1/t^2$ holds at infinity, and this immediately implies that the contour with equation $z = F_n(t)$ is finite.

The representation (6) of the solution of problem (5) is obviously unique in the class of bounded functions $z(t)$.

To prove Theorem 1 it remains to show that the contour $z = F_n(t)$ includes the given polygonal line L_z^1, i.e., that for an arbitrary choice of three of the constants t_i ($i = 0, \ldots, n$) the remaining ones can be uniquely determined from the conditions $z_i = F_n(t_i)$, $i = 1, \ldots, n$.

Let us fix the constants t_0, t_1 and t_n, and construct a system for determining the constants that remain arbitrary:

$$l_i = \int_{t_i}^{t_{i+1}} \left| \frac{dF_n(t)}{dt} \right| dt \qquad (i = 1, \ldots, n-1). \tag{7}$$

This system of equations contains one more constant than the corresponding system in the case $\tau \equiv x$, since the specification $\alpha_n = H(t_n)$ is unrelated to the location of the end point of L_z^2.

The variations of l_i ($i = 1, \ldots, n-1$) when the c_0 and t_i ($i = 2, \ldots, n-1$) change in (7) are equal to

$$\delta l_i = \sum_{k=2}^{n-1} \frac{\partial l_i}{\partial t_k}\delta t_k + \frac{\partial l_i}{\partial c_0}\partial c_0 \qquad (t = 1, \ldots, n-1). \tag{8}$$

We prove that if (7) has a solution for certain values of $\bar{l}_i$ ($i = 1, \ldots, n-1$), then in a sufficiently small neighborhood of the arbitrary constants the solution is unique, i.e., we prove that for $\delta l_i = 0$ ($i = 1, \ldots, n-1$) the system (8) has only the zero solution $\delta t_i = 0$ ($i = 2, \ldots, n-1$).

Suppose that there are two solutions of (7), and that $\zeta = f(z)$ and $\zeta = f^*(z)$ are the corresponding functions mapping D_z^+ onto the upper half-plane.

Comparing these functions at the same points of the half-plane, we have $f(z) = f^*(z + \delta z)$. If we restrict ourselves to the first approximation, we get the equality $\delta f + (df/dz)\delta z = 0$, which in a sufficiently small neighborhood of the constants c_0 and t_i ($i = 2, \ldots, n-1$) allows us to assume that $\operatorname{Im}\delta\zeta = 0$ and $\arg \delta z = \alpha_t$ for $t \notin [t_n, t_1]$. But, on the other hand,

$$\delta\left(\arg\frac{dz}{d\zeta}\right) = \delta H(t, t_0, t_1, t_n) = 0$$

for $t \in [t_n, t_1]$. Then, considering the function $\delta\zeta$, which is regular in the upper half-plane except at the points t_i ($i = 2, \ldots, n-1$), where it can have integrable singularities, while its imaginary part is equal to zero to $t \in [t_n, t_1]$, we get that $\delta\zeta = \text{const}$; since $\delta\zeta|_{\zeta = t_i} = 0$ ($i = 1, n$), it follows that $\delta\zeta \equiv 0$.

It can be shown that constants t_i that give a solution to (7) cannot converge (cf. Lemma 2 in §2).

Then, fixing the t_i ($i = 0, \ldots, n$) arbitrarily, we find from the system (7) values $\bar{l}_i$ ($i = 1, \ldots, n-1$) which determine a certain polygonal line $\bar{L}_z^1$; proceeding from them by the method of continuity (§2.4), we arrive at the solution of (7) for the given l_i.

2°. We set $\tau = \arg z = \theta \in [\pi, 2\pi]$ and require that the interior angles between the arcs L_z^1 and L_z^2 are at least π. The system (3), (4) can be written in the form

$$
\begin{aligned}
x \tan H(t) - y &= 0, && t \in [t_n, \infty) + (-\infty, t_1], \\
x k_i - y &= b_i, && t \in [t_i, t_{i+1}] \quad (i = 1, \ldots, n-1).
\end{aligned}
\tag{9}
$$

The canonical function $X(\zeta)$ of the homogeneous problem corresponding to the Hilbert problem (9) is sought from the conditions

$$
\arg X = \arctan k_i \quad \text{for } t \in [t_i, t_{i+1}] \qquad \arg X = H(t) \quad \text{for } t \in [t_n, \infty) + (-\infty, t_1]
$$

with the requirement of boundedness at the t_i $(i = 1, \ldots, n)$. Obviously,

$$
\arg X = \arctan K_i \quad \text{for } t \in [t_i, t_{i+1}],
$$

and $\arg X = H(t)$ for $t \in [t_n, \infty) + (-\infty, t_1]$ with the requirement of boundedness at the t_i $(i = 1, \ldots, n$. Obviously,

$$
X(\zeta) = M_n(\zeta, t_0, t_1, t_n) \prod_{k=1}^{n} (\zeta - t_k)^{\beta_k},
$$

where the notation is the same as in (6), except that $\beta_1 = (2\pi - \alpha_1)/\pi$ and $\beta_n = \alpha_{n-1}$.

Then, using (4.29) of Chapter I, we get

$$
z = F_n(\zeta) \equiv \frac{X(\zeta)}{\pi i} \sum_{j=1}^{n-1} \frac{b_j}{k_j + i} \int_{t_j}^{t_{j+1}} \frac{dt}{X^+(t)(t - \zeta)}.
\tag{10}
$$

It is easy to verify that as we approach the points t_i $(i = 1, \ldots, n)$ from the upper half-plane we have $\lim F_n(\zeta) = z_i$, the point of intersection of the corresponding links of the polygonal line, and consequently $F_n(t)$ can be defined at these points by $F_n(t_i) = z_i$.

The derivative $dF_n(\zeta)/d\zeta$ is representable in the form

$$
\frac{dF_n(\zeta)}{d\zeta} = \prod_{k=1}^{n} (\zeta - t_k)^{\beta_k - 1} R(\zeta), \qquad 0 < |R(\zeta)| < \infty.
$$

The function $X(\zeta)$ has a pole of order $n - 1$ at infinity; therefore, the $n - 2$ conditions $(n \geq 3)$ which determine the t_i must hold for $F_n(\zeta)$ to be bounded at infinity. The system for their determination can be written in the form

$$
\sum_{j=1}^{n-1} \frac{b_j}{k_j + i} \int_{t_j}^{t_{j+1}} \frac{t^{m-1} dt}{X^+(t)} = 0 \qquad (m = 1, \ldots, n-2).
$$

Obviously, not all the b_j are equal to zero for $n \geq 3$. Suppose, for example, that $b_{n-1} \neq 0$. We prove that this system can be solved for the b_i $(i = 1, \ldots, n-2)$ with arbitrary t_i. Since $\arg X^+(t) = (\alpha_j + n - j)\pi$ on $[t_j, t_{j+1}]$, and $1/(k_j + i) = -i \cos \alpha_j \pi e^{i\alpha_j \pi}$, it can be rewritten as

$$
\sum_{j=1}^{n-2} (-1)^{n-j} b_j \cos \alpha_j \pi \int_{t_j}^{t_{j+1}} \frac{t^{m-1} dt}{|X^+(t)|} = b_{n-1} \cos \alpha_{n-1} \pi \int_{t_{n-1}}^{t_n} \frac{t^{m-1} dt}{|X^+(t)|}.
\tag{11}
$$

Suppose that $\alpha_j \neq 1/2$, i.e., none of the links of the polygonal line L_z^1 is parallel to the OY-axis. Then some of the terms on the right-hand sides of (11) must be nonzero. Let

$$
a_{mj} = (-1)^{n-j} \cos \alpha_j \pi \int_{t_j}^{t_{j+1}} \frac{t^m dt}{|X^+(t)|}
$$

be the elements of the determinant Δ of (11), and let

$$
\bar{a}_{lj} = \cos \alpha_j \pi \int_{t_j}^{t_{j+1}} \frac{P_{n-3(t)} dt}{|X^+(t)|} \qquad (j = 1, \ldots, n-2)
$$

be the elements of the lth row, obtained as a linear combination of other rows.

If $\bar{a}_{lj} = 0$, then the interval $[t_j, t_{j+1}]$ $(j = 1,\dots,n-2)$ must contain at least one root of the polynomial $P_{n-3}(t)$, and, since there are $n-2$ intervals while the degree of the polynomial is at most $n-3$, it follows that all the $\bar{a}_{lj}$ cannot be zero at the same time.

Thus, the rows of the determinant Δ are linearly independent, so it is not equal to zero.

Let us fix the constants t_0, t_1, and t_n. The system for determining the remaining constants takes the form

$$b_i = b_i(b_{n-1}, t_2, t_3, \dots, t_{n-1}) \qquad (i = 1, \dots, n-2). \tag{12}$$

If we vary the constants $t_2, \dots, t_{n-1}$ in this system, then the contours $z = F_n(t)$ corresponding to the b_i found from it will include the polygonal lines with links parallel to the links of the given one. Arguments analogous to those in 1° lead to the boundary-value problem

$$\arg \delta z = \alpha_i, \qquad t \in [t_i, t_{i+1}],$$
$$\arg \delta z = H(t, t_0, t_1, t_n), \qquad t \in [t_1, t_n].$$

Taking account of the fact that the function δz must be bounded at infinity and at the points t_1 and t_n, and assuming that $\beta_k < 1$ $(k = 2, \dots, n-1)$, we get

$$\delta z = P_m(\zeta) X(\zeta) \prod_{k=1}^{n} (\zeta - t_k)^{-1}(\zeta - t_i) \qquad (i = 1, \dots, n).$$

Then the function

$$\delta\zeta = \delta z \frac{d\zeta}{dz} = (\zeta - t_i) R(\zeta) \qquad (R(\zeta) \neq 0)$$

which is bounded in the upper half-plane, has a second-order pole at infinity, and its imaginary part is equal to zero for $t \in [t_n, t_1]$. It follows from elementary arguments that $\delta\zeta \equiv 0$. The case when some of the β_k are greater than 1 is treated similarly.

It is now completely obvious that the system (12) can be solved for fixed b_i $(i = 1, \dots, n)$ (cf. 1°).

3°. Let $\tau \equiv s/l^k \in [0, l^{1-k}]$ $(k = 0, 1)$. In this case the problem as initially stated is, generally speaking, not solvable. Suppose that $k = 0$ and l is fixed. We require that the polygonal line pass through the fixed points z_i and specify the angles α_i $(i = 1, \dots, n-1)$ between its links and the OX-axis, along with the lengths of its links to within a similarity coefficient λ.

If $k = 1$, then l is assumed to be unknown, and the polygonal line is fixed.

We write the boundary-value problem (3), (4) in the form

$$\arg \frac{dz}{dt} = \alpha_i, \qquad t \in [t_i, t_{i+1}] \qquad (i = 1, \dots, n-1), \tag{13}$$

$$\ln\left|\frac{dz}{dt}\right| = \ln h_k(t, t_0, t_1, t_n), \qquad t \in [t_n, t_1],$$

where $h_0(t) = h(t)$ and $h_1(t) = lh(t)$.

Solving the mixed boundary-value problem (13) for the half-plane, we find that

$$\ln\frac{dz}{d\zeta} = \frac{1}{\pi i} \sqrt{\frac{\zeta - t_n}{\zeta - t_1}} \left\{ i \sum_{k=1}^{n-1} \alpha_k \int_{t_k}^{t_{k+1}} \sqrt{\frac{t - t_1}{t - t_n}} \frac{dt}{t - \zeta} + \int_{t_n}^{t_1} \sqrt{\frac{t - t_1}{t - t_n}} \frac{\ln h_k(t)\,dt}{t - \zeta} + C_1 \right\}. \tag{14}$$

One of the integrals in the sum is taken over the infinite interval $[t_k, \infty) + (-\infty, t_{k+1}]$. It is easy to see that in a neighborhood of the points t_k $(k = 2, \dots, n-1)$

$$\ln\frac{dz}{d\zeta} = (\beta_k - 1)\ln(\zeta - t_k) + \Phi_k(\zeta),$$

where $\beta_k \pi$ is the interior angle of the polygonal line, and $\Phi_k(\zeta)$ is a function that is bounded at $\zeta = t_k$. We assume that $h_k(t)$ satisfies the inequality

$$h_k(t) \leqslant \frac{k_1}{(t_1 - t)^{1-\beta_1}(t - t_n)^{1-\beta_n}}.$$

Close to t_n we have

$$\ln \frac{dz}{d\zeta} = \frac{\beta_n - 1}{\pi i} \sqrt{\frac{\zeta - t_n}{\zeta - t_1}} \int_{t_n}^{t_1} \sqrt{\frac{t - t_1}{t - t_n}} \frac{\ln(t - t_n)dt}{t - \zeta} + \Phi_n(\zeta).$$

Thus, it is clear that

$$\ln \frac{dz}{d\zeta} = (\beta_n - 1)\ln(\zeta - t_n) + \tilde{\Phi}_n(\zeta).$$

The constant C_1 can be chosen in such a way that close to t_1

$$\ln \frac{dz}{d\zeta} = (\beta_1 - 1)\ln(\zeta - t_1) + \Phi_n(\zeta).$$

Accordingly,

$$\ln \frac{dz}{d\zeta} = \ln \left[\prod_{i=1}^{n} (\zeta - t_i)^{\beta_i - 1} M_h^k(\zeta) \right],$$

where $M_h^k(\zeta)$ does not have singularities at the points $\zeta = t_j$ $(j = i, \ldots, n)$ and can be computed either directly from (14) or by solving the boundary-value problem

$$\arg M_h^k(t) = \alpha_i - \arg \prod_{j=1}^{n} (t - t_j)^{\beta_j - 1}, \qquad t \in [t_i, t_{i+1}] \, (i = 1, \ldots, n-1),$$

$$\ln |M_h^k(t)| = \ln \left[h_k(t) \prod_{j=1}^{n} |t - t_j|^{1-\beta_j} \right], \qquad t \in [t_n, t_1].$$

Then

$$z = F_n^k(\zeta) = \int_{t_1}^{\zeta} \prod_{i=1}^{n} (\zeta - t_i)^{\beta_i - 1} M_h^k(\zeta) \, d\zeta + z_1. \tag{15}$$

Suppose that the constants t_0, t_1, and t_n are fixed. We write the system for determining the remaining constants:

$$l_i \lambda^{1-k} = \int_{t_i}^{t_{i+1}} \left| \frac{dF_n^k(t)}{dt} \right| dt \qquad (i = 1, \ldots, n-1), \tag{16}$$

where the l_i are given.

The proof that the constants exist is carried out just as in 1°. The corresponding boundary-value problem for $dz/d\zeta$ has the form

$$\delta \arg \frac{dz}{d\zeta} = 0, \quad t \in (t_n, t_1), \qquad \delta \ln \left| \frac{dz}{d\zeta} \right| = 0, \quad t \notin [t_n, t_1].$$

The analytic function $\delta \ln(dz/d\zeta)$ in the upper half-plane is obviously bounded at the points t_1 and t_n and has a zero of order not less than 1 at infinity. Then $\delta \ln(dz/d\zeta) = 0$, and, since $\delta z |_{\zeta = t_i} = 0$ $(i = 1, n)$, we have $\delta z = 0$. The subsequent arguments are completely analogous to those in 1°.

THEOREM 2. *The assertions of Theorem* 1 *remain true if the polygonal line is replaced by an arbitrary rectifiable curve* L_z^1.

This is proved similarly to Theorem 1 in §4.

§8. On boundary-value problems with free boundaries
for the case of an unknown domain in the plane of the complex potential

In this section we study a number of problems that differ in principle from those investigated in §§1–7; some of them are direct generalizations of problems in hydrodynamics.

1°. Let us consider the following boundary-value problem with free boundary:

$$\arg\frac{dw}{dz} = f(\alpha) \quad \text{on } L_z^1 \tag{1}$$

$$\varphi = f_1(x),$$
$$\psi = f_2(x), \quad x \in [x_0, x_1] \text{ on } L_z^2, \tag{2}$$

where $\alpha = \arctan(dy/dx)$ is the angle between the tangent to the known smooth convex arc L_z^1 and the OX-axis. Let L_x^2 be an unknown arc that completes L_z^1 to form a closed contour $L_z = L_z^1 + L_z^2$. We seek the unknown arc L_z^2 and an analytic function $w(z) = \varphi + i\psi$ in the domain D_z that satisfies conditions (1) and (2). Let us inscribe in L_z^1 a polygonal line passing through fixed points z_i $(i = 1, \ldots, n)$ (z_1 and z_n are the endpoints of L_z^1), and let us assume that condition (1) is preserved along its links. Then (1) can be written in the form

$$\arg\frac{dw}{dz} = f(\alpha_j) \quad (i = 1, \ldots, n - 1), \tag{3}$$

where the α_i are the angles between the links of the polygonal line and the OX-axis. In the arc L_w^2 with equation (2) we inscribe a polygonal line P_w^2 passing through fixed points w_i $(i = n, \ldots, m + n - 1)$ of L_w^2, where w_n and w_{n+m-1} are the endpoints of L_w^2. We denote the slopes of the links of P_w^2 by k_i $(i = n, \ldots, n + m - 2)$ and rewrite (2) in the form

$$\frac{d\psi}{d\varphi} = k_i \quad (i = n, \ldots, m + n - 2).$$

Inscribing the polygonal line is equivalent to a piecewise linearization of equation (2):

$$\varphi = a_i x + a_i', \quad \psi = b_i x + b_i', \quad (i = n, \ldots, m + n - 2), \tag{4}$$

with $k_i = b_i/a_i$.

We introduce a function $w = w(\zeta)$ mapping the domain D_w onto the upper ζ-half-plane, where $L_w = L_w^1 + L_w^2$, and L_w^1 is determined by solving our original problem.

Conditions (3) and (4) give us that

$$\arg\frac{dw}{dt} = \arg\frac{dw}{dz} + \arg\frac{dz}{dt} = f(\alpha_i) + \alpha_i, \quad t \in [t_i, t_{i+1}] \ (i = 1, \ldots, n - 1)$$

and

$$\frac{d\psi}{dt} = k_i\frac{d\varphi}{dt}, \quad t \in [t_i, t_{i+1}] \ (i = n, \ldots, n + m - 2),$$

where the t_i are the images of the z_i $(i = 1, \ldots, n)$ and the w_i $(i = n, \ldots, n + m - 2)$ under the mappings $z = z(\zeta)$ and $w = w(\zeta)$ of the upper half-plane Im $\zeta > 0$ onto the domains D_z and D_w. Finally, with the notation $k_i = f(\alpha_i) + \alpha_i (i = 1, \ldots, n - 1)$,

$$\frac{d\psi}{dt} = k_i\frac{d\varphi}{dt}, \quad t \in [t_i, t_{i+1}] \ (i = 1, \ldots, m + n - 2). \tag{5}$$

Here the points t_1 and t_n are fixed ($t_n < t_1$) and the polygonal line P_z^1 passes into $[t_1, \infty) +$ $(-\infty, t_n]$, while P_w^2 passes into $[t_n, t_1]$, i.e.,

$$t_k \in [t_n, t_1] \qquad (k = n, \ldots, m + n - 1)$$

and

$$t_k \in \{[t_1, \infty) + (-\infty, t_n]\} \qquad (k = 1, \ldots, n - 1).$$

Let us find the general solution of (5) that satisfies the usual conformality conditions. For this we construct an arbitrary polygon P_w (finite and nondegenerate) with slopes of sides k_i ($i = 1, \ldots, m + n - 2$). After mapping it onto the upper half-plane Im $\zeta > 0$, we get the desired general solution of (5):

$$w = F_w(\zeta) \equiv c_1 e^{i\alpha} \int_{t_1}^{\zeta} \prod_{k=1}^{m+n-2} (\zeta - t_k)^{\beta_k - 1} d\zeta + w_1, \tag{6}$$

where the $\beta_k \pi$ are the interior angles of this polgyon, and c_1 and α are certain real constants.

But

$$\left| \frac{dw}{dt} \right| = |c_1| \prod_{k=1}^{m+n-2} |t - t_k|^{\beta_k - 1} \equiv \frac{dx}{dt} |a_i + ib_i| \,,$$

whence

$$\frac{dx}{dt} = A_i |\Pi_w(t)| = h(t), \qquad t \in [t_i, t_{i+1}] \, (i = n, \ldots, m + n - 2)$$

or

$$h(t) = A(t) |\Pi_w(t)| \,, \qquad t \in [t_n, t_1],$$

where $A(t) = A_i, t \in [t_i, t_{i+1}] \, (i = n, \ldots, m + n - 2)$.

For determining $dz/d\zeta$ we get the boundary-value problem

$$\tan \alpha_i \frac{dx}{dt} - \frac{dy}{dt} = 0, \qquad t \in [t_i, t_{i+1}] \, (i = 1, \ldots, n - 1), \tag{7}$$

$$\frac{dx}{dt} = h(t, t_1, \ldots, t_{n+m-1}), \qquad t \in [t_i, t_{i+1}] \, (i = n, \ldots, m + n - 2). \tag{8}$$

The solution of this problem is found from (1.10):

$$z = F_n(\zeta) \equiv \frac{1}{\pi i} \int_{t_1}^{\zeta} \Pi_z(\zeta) \int_{t_n}^{t_1} \frac{h(t) \, dt \, d\zeta}{\Pi_z(t)(t - \zeta)} + z_1, \tag{9}$$

where

$$\Pi_z(\zeta) = \prod_{k=1}^{n} (\zeta - t_k)^{\gamma_k - 1},$$

the $\gamma_k \pi$ ($k = 2, \ldots, n - 1$) are the interior angles of the polygonal line P_z^1, and $\gamma_1 \pi$ and $\gamma_n \pi$ are the angles between the terminal links of P_z^1 and the OY-axis.

Let us fix the constants t_1 and t_n. Then for determining the constants that remain arbitrary we get the system

$$l_z^i = \int_{t_i}^{t_{i+1}} \left| \frac{dF_z}{dt} \right| dt \qquad (i = 1, \ldots, n - 2), \tag{10}$$

$$l_w^i = \int_{t_i}^{t_{i+1}} \left| \frac{dF_w}{dt} \right| dt \qquad (i = n, \ldots, m + n - 2),$$

where the l_z^i are the lengths of the links of P_z^1, and the l_w^i are the lengths of the links of P_w^2. The system (10) contains $m + n - 3$ equations and just as many unknown constants t_i $(i = 2, \ldots, n + m - 2, i \neq n)$. It is shown in the usual way that this system (10) determines the problem; i.e., if it is solvable, then the contour P_z with equation $z = F_z(t)$ and the contour P_w with equation $w = F_w(t)$ will include the given polygonal lines P_z^1 and P_w^2, respectively. As in Lemma 2 of §2, we establish that the unknown parameters t_i cannot converge. Then the existence and uniqueness of the solution of the auxiliary problem follows by the method of §3 from the existence and uniqueness of a solution for $n = 2$ (in this case a fixed contour is obtained in the w-plane, and the problem reduces to problem (1.1), (1.2) for $\tau = x$, which was considered in §1).

By increasing the number of links of the polygonal lines inscribed in the arcs L_z^1 and L_w^2, we get a sequence of functions

$$z = F_z^{mn}(\zeta) \quad \text{and} \quad w = F_w^{mn}(\zeta).$$

The proof that the family $\{F_z^{mn}\}$ is bounded can be carried out by the following scheme. By increasing the number of links of P_w^2, we arrive at a solution of the problem for the arc L_w^2, and the function $h(t)$ will be bounded for $t \in (t_n, t_1)$, i.e., the properties of the function $z = F_z^{mn}(\zeta)$ will be the same as in problem (1.1), (1.2) (§1.1°). The boundedness of $\{F_z^{mn}\}$ can be proved also directly from the fact that $h(t) = h^*(t)\Pi_n(t)$, where

$$\Pi_n(t) = \prod_{k=n}^{n+m} |t - t_k|^{\beta_k - 1}, \qquad 0 \neq h^*(t) < \infty,$$

and the constants t_i do not converge. Consequently, by means of the arguments in §4 we conclude that the original problem (1), (2) has at least one solution.

The following problems are handled similarly:

$$|dw/dz| = q(\alpha) \quad \text{on } L_z^1, \tag{11}$$

$$\varphi = f_1(s/l), \qquad \psi = f_2(s/l), \qquad s \in [0, l] \text{ on } L_z^2, \tag{12}$$

where s is the arclength parameter for the arc L_z^2 and l is its length, which is to be determined;

$$dw/dz = f(\alpha) \quad \text{on } L_z^1, \tag{13}$$

$$\varphi = f_1(s/l), \qquad \psi = f_2(s/l), \qquad s \in [0, l] \text{ on } L_z^2. \tag{14}$$

2°. Certain planar filtration problems reduce to the following problem. In the plane of the flow z we are given an arc L_z^1, and in the plane of the complex potential w we are given the image L_w^2 of an unknown arc L_z^2 that closes L_z^1. The image L_w^1 of L_z^1 is unknown. The image D_ω of the domain D_z of the flow is known in the velocity hodograph plane of $\omega = \ln(dw/dz)$, where L_ω^1 is the image of L_z^1 and L_ω^2 is the image of L_z^2.

It is required to determine the arc L_z^2 and a function $w = w(z)$ which is analytic in the domain D_z $(L_z = L_z^1 + L_z^2)$ and whose hodograph domain is D_ω, along with the part L_w^1 of the boundary in the potential domain. Let us inscribe in L_z^1 a polygonal line P_z^1 passing through fixed points z_i $(i = 1, \ldots, n)$, and let us inscribe in L_w^2 a polygon P_w^2 passing through fixed points w_i $(i = n, \ldots, m + n - 1)$. We look for functions $w(\zeta)$ and $z(\zeta)$ that map the upper half-plane $\text{Im } \zeta > 0$ conformally onto the respective domains D_w and D_z.

We have

$$dy/dx = k_i, \qquad t \in [t_i, t_{i+1}] \ (i = 1, \ldots, n - 1), \tag{15}$$

$$d\psi/d\varphi = k_i, \qquad t \in [t_i, t_{i+1}] \ (i = n, \ldots, m + n - 2). \tag{16}$$

Let $\omega(\zeta) = \ln q(\zeta) - i\theta(\zeta)$ be a function mapping the domain $\text{Im } \zeta > 0$ conformally onto D_ω. Then

$$\frac{d\psi}{d\varphi} = \frac{\psi_x dx + \psi_y dy}{\varphi_x dx + \varphi_y dy} = \frac{-\sin\theta(t) + k_i\cos\theta(t)}{\cos\theta(t) + k_i\sin\theta(t)}, \qquad t \in [t_i, t_{i+1}] \ (i = 1, \ldots, n - 1). \tag{17}$$

Writing arctan $k_i = \alpha_i$ $(i = n, \ldots, m + n - 2)$ and taking conditions (15) and (16) into account, we get

$$\arg \frac{dw}{dt} = -\theta(t) + \alpha_i, \qquad t \in [t_i, t_{i+1}]\ (i = 1, \ldots, n - 1), \tag{18}$$

$$\arg \frac{dw}{dt} = \alpha_i, \qquad t \in [t_i, t_{i+1}]\ (i = n, \ldots, m + n - 2), \tag{19}$$

from which it is easy to recover the function $w(\zeta)$ and, consequently, also $z(\zeta)$ from the formula

$$z = \int_{t_1}^{\zeta} \frac{dw}{d\zeta} e^{-\omega(\zeta)}\, d\zeta + z_1.$$

The system (10) (see 1°) is obtained for determining the constants; all the subsequent arguments of that subsection remain in force also for this problem.

3°. Let us consider the problem

$$\operatorname{Re} w = \varphi(\alpha) \quad \text{on } L_z^1, \tag{20}$$

$$|\, dw/dz\,| = q(s/l), \qquad s \in [0, l], \tag{21}$$

$$\operatorname{Im} w = c \quad \text{on } L_z^2.$$

Inscribing in the known arc L_z^1 a polygonal line P_z^1 passing through the fixed points z_i $(i = 1, \ldots, n)$ and assuming that the condition (20) on P_z^1 remains in force, we arrive at the following boundary-value problem for the function $w(\zeta)$:

$$\operatorname{Re} w = \varphi(\alpha_i) \quad \text{for } t \in [t_i, t_{i+1}]\ (i = 1, \ldots, n - 1), \tag{22}$$

$$\operatorname{Im} w = c \quad \text{for } t \in [t_n, t_1], \tag{23}$$

the solution of which determines $w(\zeta)$. Then to find $s(t)$ we get a differential equation $\left|\frac{dw}{dt}\right| = q(s)\frac{ds}{dt}$, with separable variables, and from this $s(t)$ is found for $t \in [t_n, t_1]$. Thus, the function $dz/d\zeta$ satisfies the boundary-value problem

$$\arg\left|\frac{dz}{dt}\right| = \alpha_i, \qquad t \in [t_i, t_{i+1}]\ (i = 1, \ldots, n - 1),$$

$$\ln\left|\frac{dz}{dt}\right| = \ln\frac{ds}{dt} = h(t), \qquad t \in [t_n, t_1],$$

whose solution is easy to write; the remaining considerations are analogous to those in 1°.

§9. A brief survey and problems

In the course of this chapter we have considered diverse variants of the problem of constructing a conformal mapping $w = w(z)$ of a domain D_z with partially unknown boundary L_z onto a given domain D_w in the plane of $w = \varphi + i\psi$, which without loss of generality can be taken to be the disk $|\,w\,| < 1$, $w = \rho e^{i\gamma}$. Suppose that the upper semicircle is the image of a given arc $L_z^1 \subset L_z$ with the equation

$$\Phi(x, y) = 0 \quad \text{for } \nu \in [0, \pi], \tag{1}$$

and the lower semicircle (the image of the unknown arc L_z^2) is given in parametric form

$$w = w(\tau), \qquad |\,w(\tau)\,| = 1, \qquad \ln|\,dw/d\tau\,| < \infty, \tag{2}$$

where τ is one of the geometric parameters of the arc L_z^2, for example, $\tau \equiv x = \operatorname{Re} z$.

By finding the functional dependence

$$x = f(\gamma), \qquad \gamma \in [\pi, 2\pi], \tag{3}$$

from the identity $w(x) = e^{i\gamma}$, we arrive at the nonlinear boundary-value problem (1), (3) for an analytic function $z = z(w)$ in the disk $|w| < 1$.

A particular case of the problem is the classical problem of a conformal mapping of one domain D_z onto another D_w, the solvability of which is guaranteed for a large class of domains by the famous Riemann mapping theorem.

Another, simpler, particular case of the original problem is the so-called "inverse boundary-value problem" [135], [32], in which the boundary of D_z is completely unknown. In this case the boundary-value condition (3) is given on the whole circle $|w| = 1$, and the unknown analytic function $z = z(w)$ is recovered by the Schwarz formula. For other values of the parameter τ in (2) the mapping $z = z(w)$ can be constructed just as simply. In the case when the domain D_z turns out to be infinite and τ is the arclength parameter of $L_z = \partial D_z$ (the exterior inverse problem), the uniqueness condition for the function $z = z(w)$ leads to a nonlinear equation in a complex parameter. The solvability of this equation was proved by Gakhov, and Rogozhin constructed examples in which this equation is not uniquely solvable (see [32], §34.7).

A considerable amount of work has been done by specialists in hydromechanics in Kazan on the technical application of inverse boundary-value problems [126].

A problem analogous to the inverse boundary-value problem was first solved by Demchenko [30], who was seeking the boundary L_z from given values on it of a function harmonic in the domain D_z and of its normal derivative, expressed in terms of functions of the arclength parameter of the fixed contour.

In [79]–[84] the author proved the solvability of a mixed boundary-value problem with free boundary, and along with it the solvability of the nonlinear problem (1), (3) and diverse generalizations of it. In particular, in [82] he investigated features of the geometry of the unknown arcs whose occurrence is connected with the unboundedness of $\ln|dw/d\zeta|$ for isolated values of τ (§9).

The idea of the proof that problems like (1), (3) are solvable goes back to the work of Weinstein [134] on jet problems in hydrodynamics (see Chapter IV, §1). According to this idea, we consider a family of polygonal lines $P_n(\gamma_0, \gamma_1, \ldots, \gamma_{n+1})$, whose links are parallel to the links of a fixed polygonal line P_n^0 that is inscribed in the given arc L_z^1. Suppose that the points $t_k = e^{i\gamma_k}$ ($k = 0, \ldots, n+1$) are images of the vertices and endpoints of the polygonal line. For fixed γ_k ($0 = \gamma_0 < \gamma_1 < \cdots < \gamma_{n+1} = \pi$) the solution of the problem (1), (3), with the equation (1) of L_z^1 replaced by the equations of the links of a certain polygonal line P_n, is sought in closed form. A finite system of nonlinear equations analogous to the equations for determining the parameters in the Schwarz-Christoffel formula is obtained for finding the values of the parameters γ_k such that the corresponding polygonal line P_n coincides with the given polygonal line P_n^0. Under the assumption that the polygonal lines P_n are simple, it is not hard to get the following a priori estimates for the solutions of the corresponding system of equations:

(i) $$|\gamma_k - \gamma_m| \geq \varepsilon > 0 \quad \text{for } k \neq m.$$

For the problem (1), (3) and for certain generalizations of it, we can also prove the local uniqueness of solutions of the corresponding system of equations for the parameters γ_k. In other words, we have the following assertion:

(ii) *The Jacobian of the system is nonzero at points where the system is solvable.*

By (i) and (ii), the method of continuity gives us the unique solvability of the corresponding system of equations, and thereby also the unique solvability of the original problem for the polygonal line P_n^0 inscribed in L_z^1. A passage to the limit proves solvability also for the problem when L_z^1 is an arbitrary rectifiable arc.

It should be mentioned that in far from every problem is it possible to establish local uniqueness for solutions of the system for the parameters (property (ii)). In this connection, a method is presented in §3 that enables us to prove the solvability of the system only on the basis of the a priori estimates (i) of its solutions.

We shall indicate one more method for investigating problem (1), (3), a method applicable not only for analytic functions, but also for general quasilinear elliptic systems of equations (§5 in

Chapter VIII). Consider the new unknown function $Z(w) = x + i\Phi(x, y)$, $|w| < 1$, which obviously satisfies the following boundary condition on the circle $|w| = 1$:

$$\text{Im } Z(e^{i\gamma}) = 0, \quad \gamma \in [0, \pi], \qquad \text{Re } Z(e^{i\gamma}) = f(\gamma), \quad \gamma \in [\pi, 2\pi]. \qquad (4)$$

Passing to the new unknown functions $\xi = x$ and $\eta = \Phi(x, y)$ in the Cauchy-Riemann equations for $x(\varphi, \psi)$ and $y(\varphi, \psi)$, we get a system of first-order equations which can be written as follows in the form of a single complex equation:

$$z_{\bar{w}} - \mu_1(z)z_w - \mu_2(z)\bar{z}_{\bar{w}} = 0, \qquad (5)$$

where the coefficients $\mu_i(z)$ can be explicitly expressed in terms of the derivatives Φ'_x and Φ'_y. Under certain assumptions on the $\mu_i(z)$ it will be proved in Chapter VI that the resulting problem (4), (5) is uniquely solvable (see also §5 in Chapter VIII).

We formulate some problems of varying difficulty that remain unsolved.

1. For a fixed polygonal line, investigate the question of the number of independent solutions of the problem (1), (3) that satisfy the conformality conditions (§1.3°). In view of the necessity of the last requirement, the answer to the question posed is obviously not settled by computing the index of the corresponding Hilbert problem (1.8), (1.9).

2. Use the methods of this chapter to investigate the mixed boundary-value problem with free boundary under the assumption that in the boundary condition (2) $(w = w(\tau))$ the geometric parameter τ of the unknown arc L_z^2 takes different values $(\tau = \text{Re } z, \text{Im } z, |z|, \cdots)$ on separate parts of this arc. Certain particular cases of this problem that are encountered in filtration theory are studied by other methods in §5 of Chapter VIII.

3. Study the mixed boundary-value problem with free boundary for a system of arcs in the case when the parameter τ in the boundary condition $w = w(\tau)$ takes one of the following values: $\arg z$, $\tau = |z|$, the arclength parameter s of L_z^2, and so on (the case $\tau = \text{Re } z$ is considered in §6). For $\tau = s$ this problem was partially investigated by Turovskiĭ [127].

4. Consider the various particular cases of the general problem with free boundary when instead of the condition (2) $(w = w(\tau))$ and the known image L_w^1 of L_z^1 we are given relations

$$F_1(z, w, w_z, \cdots) = 0, \quad z \in \left(L_z^1 + L_z^2\right); \qquad F_2(z, w, w_z, \cdots) = 0, \quad z \in L_z^2, \qquad (6)$$

where the F_i are real functions of their arguments. Here the conditions (6) do not, generally speaking, determine the boundary of a domain in the plane of values of the unknown solution $w = w(z)$ or of its derivatives. Some examples of such problems were studied in §8. We present other uninvestigated particular cases of problem (6).

a). The conditions (6) have the form $\text{Re } w = f(\tau)$ on $L_z^1 + L_z^2$ and $\text{Im } w = g(\tau)$ on L_z^2, where $\tau = \text{Re } z$ or $|z|$ or the arclength parameter s of $L_z = L_z^1 + L_z^2$, and so on.

b). The conditions (6) have the form $\text{Re } w = f(x, y)$ on $L_z^1 + L_z^2$ and $\text{Im } w = g(x, y)$ on L_z^2, where $f(x, y)$ and $g(x, y)$ are functions defined on the whole (x, y)-plane.

Investigate, in particular, the solvability of the inverse problem (6) b) when the boundary $L_z = \partial D_z$ is completely unknown, i.e., both conditions (6) are given on the whole unknown boundary L_z. Certain problems (Chapter IX) reduce to problems similar to (4) b).

5. Prove solvability for the mixed boundary-value problem with free boundary in multiply connected domains and on Riemann surfaces.

6. Consider problems in which the right-hand side of the condition (2), in addition to depending on the geometric parameter τ, depends also on some arbitrary parameter t (the time); moreover, even the parameter τ itself may be connected with t, for example, $\tau = dx/dt$.

Investigate the properties of solutions of such problems as the parameter t varies, and, in particular, the univalence of the mapping $z = z(w, t)$. Similar problems arise in the study of time-dependent processes in the mechanics of continuous media (see, for example, §2 in Chapter IX).

CHAPTER IV

FLOWS OF AN INCOMPRESSIBLE FLUID
WITH FREE BOUNDARIES

§1. The continuity method in the problem of flow
with jet separation past obstacles

1°. *Statement of the problem.* We shall consider flow with jet separation past a given polygonal obstacle according to the Kirchhoff scheme, and apply the Weinstein continuity method to the investigation of solvability for this problem.

Let $w(z)$ be the complex flow potential, and $dw/dz = qe^{-i\theta}$ the complex flow velocity. The boundary of the flow domain is the streamline $\text{Im } w = \psi = 0$, the velocity at infinity is $q_\infty = 1$, the separation points A and B are given, and the point P at which the flow branches is to be determined. Let the flow domain and its image in the plane of the complex potential be mapped conformally onto the upper half-plane $\text{Im } \zeta > 0$ in such a way that the points A and B pass into the respective points ± 1 of the real axis, while the unknown branch point of the flow passes into the origin. The figure illustrates the corresponding scheme.

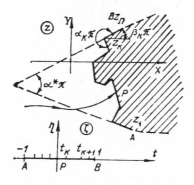

We choose a mapping of the upper half-plane onto the domain in the w-plane in the form $w = K\zeta^2/2$, where K is an unknown constant. For $|t| \geq 1$ (on the curves in the ζ-plane that correspond to jets) we have

$$|dz/dt| = (|dw/dz|)^{-1}|dw/dt| = K|t|.$$

Taking account of the fact that $\arg dz/dt = \beta_k \pi$ is the angle between the kth link of the polygonal line and the OX-axis, we then obtain the following boundary-value problem for the derivative $dz/d\zeta$ of the function $z = z(\zeta)$ mapping the domain $\text{Im } \zeta > 0$ conformally onto the flow domain:

$$\begin{cases} \arg(dz/dt) = \beta_k \pi, & t \in [t_k, t_{k+1}] \ (k = 1, \dots, n-1), \\ \ln|dz/dt| = \ln(K|t|), & |t| \geq 1. \end{cases} \tag{1}$$

Let us look for a solution of this problem in the form

$$\frac{dz}{d\zeta} = Ke^{i\beta} \prod_{k=2}^{n-1} (\zeta - t_k)^{\alpha_k - 1} M(\zeta) \equiv Ke^{i\beta} \Pi(\zeta) M(\zeta),$$

where the constant β is chosen from the condition $\arg[e^{i\beta}\Pi(\zeta)]_{\zeta = t_1 + 0} = \beta_1$. Then for the determination of the function $M(\zeta)$ we get the boundary-value problem

$$\begin{cases} \arg M(t) = 0, & t \in [-1, 1] \equiv L_1, \\ \ln|M(t)| = \ln\left(|t| \prod_{k=2}^{n-1} |t - t_k|^{1-\alpha_k}\right), & |t| \geq 1. \end{cases} \tag{2}$$

The solution of this mixed boundary-value problem that is bounded at the points $t = \pm 1$ has the form

$$\ln M(\zeta) = \frac{\sqrt{1 - \zeta^2}}{\pi i} \int_{|t| \geq 1} \frac{\ln\left(|t| \prod_{k=2}^{n-1} |t - t_k|^{1-\alpha_k}\right) dt}{\sqrt{1 - t^2}(t - \zeta)}. \tag{3}$$

We compute the integrals in (3). Let

$$\ln \frac{dz^k}{d\zeta} = \frac{\sqrt{1 - \zeta^2}}{\pi i} \int_{|t| \geq 1} \frac{\ln|t - t_k| \, dt}{\sqrt{1 - t^2}(t - \zeta)}, \qquad t_k \in (-1, 1).$$

This integral will be interpreted as a solution of the boundary-value problem

$$\arg \frac{dz^k}{dt} = 0, \quad t \in [-1, 1], \qquad \left|\frac{dz^k}{dt}\right| = |t - t_k|, \quad |t| \geq 1,$$

which, in turn, can be regarded as the problem of a nonsymmetric flow past a plate, with the image of the branch point at $t = t_k$. The solution of the latter problem has the form

$$dz^k/d\zeta = \sqrt{(1 - t_k^2)(1 - \zeta^2)} - (\zeta t_k - 1), \qquad dw^k/d\zeta = \zeta - t_k.$$

Obviously, the function $dz^k/d\zeta$ satisfies the boundary conditions written above, and $dz^k/dt \neq 0$ for all $t \in (-\infty, \infty)$ by construction. Thus, since the corresponding flow problem has a unique solution,

$$\frac{\sqrt{1 - \zeta^2}}{\pi i} \int_{|t| \geqslant 1} \frac{\ln|t - t_k|\, dt}{\sqrt{1 - t^2}\,(t - \zeta)} = \ln\left[\sqrt{(1 - t_k^2)(1 - \zeta^2)} - (\zeta t_k - 1)\right],$$

and hence

$$M(\zeta) = \prod_{k=2}^{n-1} \left[\sqrt{(1 - t_k^2)(1 - \zeta^2)} + (-\zeta t_k + 1)\right]^{1-\alpha_k} \left(\sqrt{1 - \zeta^2} + 1\right). \quad (4)$$

Then, integrating $dz/d\zeta = Ke^{i\beta}\Pi(\zeta)M(\zeta)$, we get that

$$z = Ke^{i\beta} \int_{-1}^{\zeta} \prod_k \left[\frac{\sqrt{(1 - t_k^2)(1 - \zeta^2)} + (1 - \zeta t_k)}{\zeta - t_k}\right]^{1-\alpha_k} \left(\sqrt{1 - \zeta^2} + 1\right) d\zeta + z_1.$$

$$(5)$$

It is easy to see that the $dz/d\zeta$ from (5) satisfies the boundary conditions (1). On the basis of (5) it is also easy to find the so-called Brillouin condition, which implies that the curvature at the separation points is finite.

We have

$$\left\{\sqrt{1 - \zeta^2}\,\frac{dM}{d\zeta}\right\}_{\zeta = \pm 1} = \mp \prod_{k=2}^{n-1} (1 \mp t_k)^{1-\alpha_k} \left\{1 + \sum_{j=2}^{n-1} \frac{1 - \alpha_j}{1 \mp t_j}\sqrt{1 - t_j^2}\right\}$$

and

$$\frac{d^2 z}{d\zeta^2} = M(\zeta)\frac{d}{d\zeta}\Pi(\zeta) + \Pi(\zeta)\frac{d}{d\zeta}M(\zeta).$$

Consequently, for $d^2z/d\zeta^2$ to be bounded at the points $\zeta = \pm 1$ we must have

$$1 + \sum_{j=2}^{n-1} (1 - \alpha_j)\sqrt{\frac{1 + t_j}{1 - t_j}} = 0, \qquad 1 + \sum_{j=2}^{n-1} (1 - \alpha_j)\sqrt{\frac{1 - t_j}{1 + t_j}} = 0. \quad (6)$$

The analytic functions represented by (5) form a family depending on $n - 1$ parameters. These parameters $t_2, \ldots, t_{n-1}$, and K must be found from the condition that the polygonal line determined by the equation $z = z(t)$ in (5) for $t \in [-1, 1]$ coincides with the given polygonal obstacle. The corresponding conditions can be written in the form

$$K \int_{t_k}^{t_{k+1}} |\Pi(t)| \, |M(t)| \, dt = l_k \qquad (k = 1, \dots, n-1), \tag{7}$$

where l_k are the lengths of the links of the given polygonal obstacle.

For the solution of (7) we have the a priori estimates

$$K_0^{-1} < K < K_0 < \infty, \qquad t_i - t_{i-1} > \varepsilon > 0 \qquad (i = 1, \dots, n+1),$$

of which the first is obvious, while the rest can be proved in a way completely analogous to Lemma 2* in §2 of Chapter III (these estimates are proved for arbitrary flow schemes in the next section).

2°. *Local uniqueness of solutions.* As usual, we compute the variation δz for fixed ζ due to a variation of t_i $(i = 2, \dots, n-1)$ and of K, assuming that $\delta l_i = 0$ $(i = 1, \dots, n-1)$. Formula (5) gives us that

$$\delta z = \int_{-1}^{\zeta} \frac{dz}{d\zeta} \left\{ \frac{\delta K}{K} + \sum_{k=2}^{n-1} (1 - \alpha_k) \delta \ln \left[\frac{\sqrt{(1 - t_k^2)(1 - \zeta^2)} + 1 - \zeta t_k}{\zeta - t_k} \right] \right\} d\zeta. \tag{8}$$

Elementary computations lead to

$$\frac{d}{dt_k} \left[\ln \left(\sqrt{(1 - t_k^2)(1 - \zeta^2)} + 1 - \zeta t_k \right) - \ln(\zeta - t_k) \right]$$

$$= \frac{\sqrt{1 - \zeta^2} - \sqrt{1 - t_k^2}}{(\zeta - t_k)\sqrt{1 - t_k^2}} + \frac{1}{\zeta - t_k} = \frac{\sqrt{1 - \zeta^2}}{(\zeta - t_k)\sqrt{1 - t_k^2}}.$$

Thus, the formula for δz takes the form

$$\delta z = \int_{-1}^{\zeta} \frac{dz}{d\zeta} \left[\frac{\delta K}{K} - \sqrt{1 - \zeta^2} \sum_k \frac{(\alpha_k - 1)\delta t_k}{(\zeta - t_k)\sqrt{1 - t_k^2}} \right] d\zeta$$

$$\equiv \int_{-1}^{\zeta} \frac{dz}{d\zeta} \left[\frac{\delta K}{K} + i\mu(\zeta) \right] d\zeta.$$

Since the images of the jet separation points do not vary, i.e., $\delta\zeta(\pm 1) = 0$, the equality

$$\delta z + (dz/d\zeta)\delta\zeta = 0,$$

connecting the variations of the direct and inverse functions implies that $\delta z(\pm 1) = 0$. Thus,

$$\delta z = \prod_{k=2}^{n-1} (\zeta - t_k)^{\alpha_k - 1 + \varepsilon_k}(\zeta^2 - 1)R(\zeta), \tag{8*}$$

where the $\varepsilon_k \geq 0$ are integers ($\varepsilon_k > 0$ only when $\delta t_k \neq 0$), and $|R(\zeta)| < \infty$ for $\operatorname{Im} \zeta \geq 0$.

Let us consider the analogues of the Weinstein functions

$$\Lambda(\zeta) = \zeta^2 \delta K - 2K\zeta \frac{\delta z}{z_\zeta}, \tag{9}$$

$$F(\zeta) = \frac{d\Lambda}{d\zeta} + \left(\frac{z_{\zeta\zeta}}{z_\zeta} - \frac{1}{\zeta} \right) \Lambda. \tag{10}$$

Taking the formula (8) for $d\delta z/d\zeta$ into account, we find that

$$\frac{d\Lambda}{d\zeta} = -2K\frac{\delta z}{z_\zeta} + 2K\zeta\delta z \frac{z_{\zeta\zeta}}{z_\zeta^2} - 2iK\zeta\mu(\zeta), \tag{9*}$$

whence

$$F(\zeta) = -\zeta\delta K - 2iK\zeta\mu(\zeta) + \zeta^2 \delta K \frac{z_{\zeta\zeta}}{z_\zeta}.$$

Let $\zeta = t \notin (-1, 1)$. Then

$$\frac{z_{tt}}{z_t} = \left(\ln \frac{dz}{dt} \right)_t' = \frac{d}{dt}\left[\ln(2Kte^{i\theta}) \right] = \frac{1}{t} + i\theta_t,$$

and, consequently,

$$F(t) = it(t\theta_t \delta K - 2\mu K).$$

Since $\operatorname{Im} \mu(t) = 0$ for $|t| \geq 1$ (see (8)), we have

$$\operatorname{Re} F = \operatorname{Re}(d\Lambda/dt + i\theta_t\Lambda) = 0, \qquad |t| \geq 1. \tag{11}$$

On the other hand, δz and $dz/d\zeta$ satisfy for $t \in [-1, 1]$ the same boundary conditions

$$\arg(dz/d\zeta) = \arg \delta z = \beta_i \pi, \qquad t \in [t_1, t_{i+1}]$$

i.e., $\arg(\delta z / z_t) = 0$, and so

$$\operatorname{Im} \Lambda(t) = 0, \qquad t \in [-1, 1]. \tag{12}$$

By the Cauchy-Riemann conditions, for $|t| \geq 1$ we have

$$\partial\lambda/\partial t = \partial p/\partial\eta, \quad \lambda + ip = \Lambda \quad \text{for } \eta = \operatorname{Im} \zeta.$$

Now (11) gives us

$$\text{Re}\{\lambda_t + ip_t + i\theta_t(\lambda + ip)\} = \partial\lambda/\partial t - \theta_t p = \partial p/\partial\eta - \theta_t p = 0.$$

Thus, we arrive at the following boundary-value problem for the harmonic function $p(\zeta) = \text{Im } \Lambda(\zeta)$:

$$p(t) = 0, \quad t \in [-1, 1]; \qquad \partial p/\partial\eta - \theta_t p = 0, \quad |t| \geqslant 1. \tag{13}$$

From (9*) with the equalities $\mu(\pm 1) = 0$, $\delta z = (\zeta^2 - 1)R_1(\zeta)$ (see (8*)), and $z_{\zeta\zeta} = (\zeta^2 - 1)^{-1/2}R_2(\zeta)$ (where $|R_i(\pm 1)| < \infty$) taken into account, we find that $(d\Lambda/d\zeta)(\pm 1) = 0$, and so

$$\frac{\partial\lambda}{\partial t}(\pm 1) = \frac{\partial p}{\partial\eta}(\pm 1) = 0.$$

By construction, $(1/\zeta)dz/d\zeta$ is bounded in a neighborhood of $\zeta = \infty$; consequently,

$$\delta z = \int_{-1}^{\zeta} \zeta\left\{\frac{1}{\zeta}\frac{dz}{d\zeta}\left(\frac{\delta K}{K} + i\mu\right)\right\} d\zeta = c_2\zeta^2 + c_1\zeta + \cdots,$$

and thereby

$$\Lambda(\zeta) = \delta K\zeta^2 - 2K\zeta\frac{\delta z}{z_\zeta} = c_2'\zeta^2 + c_1'\zeta + \cdots.$$

Thus, the harmonic function $p(\zeta) = \text{Im } \Lambda(\zeta)$ satisfies the boundary-value problem (13), is bounded together with its first derivatives at the points $t = \pm 1$, and has order no greater than two at infinity. We prove that the only solution of the boundary-value problem (13) having these properties is $p \equiv 0$. The maximum principle for the harmonic function $p(\zeta)$ implies that it cannot take extremal values at the points $t = \pm 1$ (otherwise it is necessary that $\partial p/\partial\eta \neq 0$). Consequently, the level curves of the harmonic function $p(\zeta)$ with $p = 0$ that start out from the points $t = \pm 1$ must terminate on the boundary. If one of these curves $\Gamma^\pm$, for example, the curve Γ^+ emanating from $t = 1$, terminates at the point $t^* \in (-1, 1)$, then a domain with boundary $\Gamma = \{\Gamma^+ [t^*, 1]\}$ is formed, and $p(\zeta) = 0$ for $\zeta \in \Gamma$; this implies that $p \equiv 0$. Suppose now that $t^* \in (-\infty, -1)$. Then Γ^+ and Γ^- must intersect in the upper half-plane, and again a domain is formed on whose boundary $p = 0$, with the consequence that $p \equiv 0$. If Γ^+ and Γ^- terminate simultaneously at infinity, then the result is an infinite domain with boundary $\Gamma = \{\Gamma^+ + \Gamma^- + [-1, 1]\}$ on which $p(\zeta) = 0$ (in particular, this holds at infinity), and this also implies that $p \equiv 0$. It thus remains to consider the case when at least one of the curves, for example, Γ^+, terminates at a finite point $t^* \in (1, \infty)$, with $\Gamma = \{\Gamma^+ + [1, t^*]\}$ bounding a domain D.

LEMMA 1 (FRIEDRICHS-WEINSTEIN). *Suppose that there exists a harmonic function $f(\zeta)$ in the upper half-plane that does not vanish for $\operatorname{Im}\zeta \geqslant 0$, except possibly for a finite number of points $t_k \in (-1, 1)$, and that satisfies the condition*

$$\partial f/\partial \eta - \theta_t f = 0, \qquad |t| \geqslant 1.$$

Then the harmonic function $p(\zeta) = \operatorname{Im}\Lambda$ satisfying the boundary-value problem

$$p(\zeta) = 0, \quad \zeta \in \Gamma^+, \qquad \partial p/\partial \eta - \theta_t p = 0, \quad t \in [1, t^*], \tag{14}$$

is identically equal to zero in the domain D with boundary $\Gamma = \{\Gamma^+ + [1, t^*]\}$.

For the bounded function $g(\zeta) = p(\zeta)/f(\zeta)$ on $\overline{D}$ we have

$$p^2 \frac{\partial(\ln g)}{\partial n} = 0, \qquad \zeta \in \Gamma,$$

where n is the inward normal to Γ. Indeed, $|\partial \ln g/\partial n| < \infty$ and $p(\zeta) = 0$ for $\zeta \in \Gamma^+$, and this implies the required equality. On the other hand, by the boundary condition for $f(\zeta)$ and (14),

$$\frac{\partial \ln g}{\partial n} = f^{-1}\left(\frac{\partial p}{\partial \eta} - \frac{p}{f}\frac{\partial f}{\partial \eta}\right) = f^{-1}\left(\frac{\partial p}{\partial \eta} - \theta_t p\right) = 0, \qquad t \in [1, t^*].$$

We then form the Weinstein quadratic form

$$I = \int_\Gamma p^2 \frac{\partial(\ln g)}{\partial n}\, ds = 0.$$

Applying Green's formula to it and using the identity $p^2 \nabla^2 \ln g + \nabla p^2 \nabla \ln g = p^2|\nabla g^2|$ for the harmonic functions $f(\zeta) = p(\zeta)/g(\zeta)$ and $p(\zeta)$, we get

$$I = \int_D\int |p^2 \nabla^2 \ln g - \nabla p^2 \nabla \ln g|\, dt d\eta = \int_D\int p^2\left(g_t^2 + g_\eta^2\right) dt d\eta = 0,$$

which implies that $g(\zeta) = \text{const}$ in $\overline{D}$. But $g(1) = p(1)/f(1) = 0$, and so $g(\zeta) = p(\zeta) = 0$ in $\overline{D}$. The lemma is proved.

To prove the existence of a harmonic function $f(\zeta)$ satisfying the conditions of Lemma 1 by the method employed by Weinstein [134], Leray [70], and others it is necessary to impose additional restrictions on the obstacle the fluid flows past. In our case this restriction takes the form

$$\sum_{j=2}^{k} (\alpha_j - 1) < 1 \qquad (k = 2, \ldots, n - 1) \tag{15}$$

and means that the rotation of the tangent along the obstacle (with the jumps at the corner points taken into account) must not exceed π (in the case of symmetric obstacles it is necessary to satisfy condition (15) only for each symmetric part of these obstacles). If condition (15) holds, then there obviously exists an angle α such that $0 < |\theta(t) - \alpha| < \pi$ for $t \in [-1, 1]$, where $\theta(t)$ is the slope angle of the tangent to the polygonal line. But the absolute extremum of the harmonic function $\theta(t) = -\operatorname{Im}\ln((dw/dz)(\zeta))$ cannot be attained on the free boundary, since $q(t) = 1$ for $|t| \geqslant 1$; consequently, $\partial\theta/\partial n = (\partial/\partial t)(\ln q) = 0$. Thus, the inequality $0 \leqslant |\theta(t) - \alpha| < \pi$ holds for all $\zeta = t \in (-\infty, \infty)$ and, hence, everywhere in the upper half-plane. Then the harmonic function

$$f(\zeta) = \operatorname{Im}\left[e^{i\alpha}dw/d\zeta\right] = q(\zeta)\sin[\alpha - \theta]$$

satisfies all the conditions of the Friedrichs-Weinstein lemma. Indeed, $q(\zeta) \neq 0$ for $\zeta \in D^+ \supset D$, where D^+ is any finite domain containing the original domain D with boundary $\Gamma = \{\Gamma^+ + [1, t^*]\}$ and not containing the points t_k $(k = 2,\ldots,n-1)$, while for the chosen α we have that $\sin(\alpha - \theta(\zeta)) \neq 0$ at every point of the upper half-plane. On the other hand, since $q(t) = 1$ for $|t| \geqslant 1$, the functions $\cos(\alpha - \theta(t))$ and $\sin(\alpha - \theta(t))$ are harmonic conjugates for $|t| \geqslant 1$, and the Cauchy-Riemann conditions for them give us that

$$\frac{\partial}{\partial\eta}\sin[\alpha - \theta(t)] = \frac{\partial}{\partial t}\cos[\alpha - \theta(t)] = \theta'_t \sin[\alpha - \theta(t)],$$

i.e.,

$$\frac{\partial \sin[\alpha - \theta(t)]}{\partial\eta} - \theta_t \sin[\alpha - \theta(t)] = \frac{\partial f}{\partial\eta} - \theta_t f = 0.$$

Thus, under the additional condition (15) on the obstacle, there exists a function $f(\zeta)$ satisfying the conditions of Lemma 1. Consequently, $p(\zeta) = \operatorname{Im}\Lambda(\zeta) \equiv 0$ in $\bar{D}$, which implies that $\Lambda(\zeta) \equiv 0$ for $\operatorname{Im}\zeta \geqslant 0$, since $\Lambda(0) = 0$. Taking account of the fact that $\delta z(\pm 1) = 0$, we then get that $\delta K = \Omega(\pm 1) = 0$, and thereby

$$\delta z = \frac{\Omega(\zeta)}{K\zeta}\frac{dz}{d\zeta} \equiv 0.$$

The identity

$$\frac{d\delta z}{d\zeta} = -\sqrt{1 - \zeta^2}\sum_{k=2}^{n-1}\frac{(\alpha_k - 1)\delta t_k}{(\zeta - t_k)\sqrt{1 - t_k^2}} \equiv 0$$

gives us the equalities $\delta t_k = 0$ $(k = 2,\ldots,n-1)$, which conclude the proof of local uniqueness for solutions of (7).

   3°. An existence theorem. The unique solvability of the original flow problem follows by the continuity method from the existence and uniqueness of the solution of the problem of a symmetric flow past a plate, the proven local uniqueness of solutions of (7), and the a priori estimates for K and t_k $(k = 2,\ldots,n-1)$.

THEOREM 1. *The problem stated at the beginning of this section on a flow past a polygonal obstacle according to the Kirchhoff scheme is uniquely solvable when the obstacle satisfies condition* (15).

REMARKS. 1. It is possible not to be given the image of the point P where the flow branches, but to find it from the given angle of inclination of the flow velocity at infinity; this complicates the arguments only in an insignificant way (see §2 of this chapter).

2. If the obstacle flowed past is convex and satisfies (15), then it is possible to find the jet separation points from the Brillouin conditions (6).

We do not dwell on the transition from polygonal boundaries to curvilinear ones, since this is done under more general assumptions in the next section. A uniqueness theorem for curvilinear obstacles will be proved in Chapter VII by other methods.

However, we remark that uniqueness can also be obtained by considering the appropriate Weinstein function [134] under a condition of the type (15) on the rotation of the tangent along the curvilinear obstacle.

§2. An existence theorem in the general hydrodynamics problem with a single free boundary

In this section we study a class of hydrodynamics problems with a single free boundary characterized by the condition that the given and unknown parts of the boundary L_z of the simply connected flow domain D_z have only two common points. This class includes the problems of flow with jet separation past an obstacle according to the schemes of Kirchhoff (see §1) and Thullen and according to the symmetric schemes of Riabouchinsky, Joukowsky and Rozhko, Èfros, and M. A. Lavrent'ev, the problems of fluid flow in channels with partially unknown walls, the problem of fluid flow out of a nozzle, and other problems. A detailed description of these and other such problems in the class under study is given in §3.

$1°$. *Statement of the problem.* Let us consider the general problem of flow of an ideal fluid in a simply connected domain D_z bounded by a single free boundary L_z^2 and a given simple curve L_z^1 with each finite part rectifiable. If L_z^1 is infinite, we assume that in a neighborhood of infinity it consists of semi-infinite line segments and that only one of three values is admitted for the angles at the "vertices" at infinity of D_z: 0 for a confined flow, π for a semiconfined flow, and 2π for a flow that is unconfined in a neighborhood of this vertex.

Suppose that the boundary $L_z = L_z^1 + L_z^2$ of the flow domain D_z consists of streamlines and equipotentials, i.e., on the separate parts of L_z either the real part or the imaginary part of the complex flow potential $w(z) = \varphi(x, y) + i\psi(x, y)$ is constant:

$$\psi = \psi_k = \text{const}(\varphi = \varphi_i = \text{const}) \quad \text{on } L_z = L_z^1 + L_z^2. \tag{1}$$

On the free boundary L_z^2, which is always assumed to be a streamline ($\psi = \text{const}$), we are given the absolute value $q = |\, dw/dz \,|$ of the flow velocity as a function of the arclength parameter s (the abscissa x in 7°) of L_z^2,

$$|\, dw/dz \,| = q(s) \quad \text{on } L_z^2, \tag{2}$$

where $q(s)$ is a continuous function, $|\ln|\,q(s)\,|\,| < N < \infty$, and $q(s)$ is equal to $q(s_0)$ for $s \geqslant |\,s_0\,| > 0$ when L_x^2 is unbounded.

In particular, $q(s) = \text{const}$ in problems of flow with jet separation past obstacles. In the general case the flow domain (which, generally speaking, is multi-sheeted) or the specified part L_z^1 of its boundary L_z contains a finite number of stagnation points $z = a_k$ (branch points or convergence points) of the flow $((dw/dz)(a_k) = 0)$ and points $z = b_i$ at which there are vortices, sources, or sinks $((dw/dz)(b_i) = \infty)$. It is required to determine the free boundary L_z^2 and an analytic function $w(z) = \varphi(x, y) + i\psi(x, y)$ ($w(z)$ is the complex flow potential) in D_z that satisfies conditions (1) and (2) on the boundary $L_z = L_z^1 + L_z^2$. The derivative dw/dz has the representation

$$\frac{dw}{dz} = \frac{\Pi(z - a_k)}{\Pi(z - b_i)}\Pi(z - z_j)^{1-\alpha_j}\Phi(z),$$

where $\Phi(z)$ is a bounded function that is not equal to zero in D_z, the z_j are the corner points of L_z, and $\alpha_j\pi$, $0 < \alpha_j < 2$, are the interior angles at these points.

With the help of the method used previously (which it is natural to call the method of finite-dimensional approximation) we reduce the question of solvability of this complicated nonlinear boundary-value problem (1), (2) to the proof of solvability of certain auxiliary equations in the Euclidean spaces E_n ($n \to \infty$). For this, we draw a polygonal line P_n with angles $\alpha_k\pi$, $0 < \alpha_k < 2$ ($k = 1, \ldots, n$), at the vertices through the junction points z_0 and z_{n+1} of the arcs L_z^1 and L_z^2 and the fixed finite points z_k ($k = 1, \ldots, n$) of L_z^1. If L_z^1 is infinite, then the corresponding polygonal line P_n will have finitely many vertices at infinity, with angles 0, $-\pi$, or -2π.

A polygonal line P_n with m vertices b_j ($j = 1, \ldots, m$) at infinity is said to be simple if each finite part of it is a nondegenerate polygonal line, and at the vertices at infinity with zero angle the distance h_j between the parallel rays is neither zero nor infinity.

Thus, for a simple polygonal line P_n we have the inequalities

$$0 < \varepsilon < \alpha_k \leqslant 2, \quad \varepsilon < |\,z_{k+1} - z_k\,| = l_k < 1/\varepsilon, \quad \varepsilon < h_j < 1/\varepsilon, \tag{3}$$

where the z_k are all the finite vertices of P_n, including its finite endpoints. Let us now consider the original boundary-value problem (1), (2), assuming for convenience that L_z^1 is itself a simple polygonal line ($L_z^1 = P_n$).

2°. Construction of the solution and its properties. We map the upper half-plane Im $T \geqslant 0$ of the parametric variable T conformally onto the domain D_w (which is, generally speaking, not one-sheeted) in the plane of the complex potential $w = \varphi + i\psi$ bounded by the straight lines $\psi = \psi_k = $ const and $\varphi = \varphi_i = $ const. To construct the mapping $w = w(T)$ we must take into account the behavior of its derivative in a neighborhood of the images of all the singular points of the flow (see, for example, [19]). If the finite point $T = T_0$ is the image of the point at infinity in the z-plane of flow, and $(dw/dz)(\infty) \neq 0$, then in a neighborhood of this point

$$dw/dT = (T - T_0)^{-\delta} f(T), \qquad f(T_0) \neq 0, \infty,$$

where $\delta = 1$, 2, or 3, respectively, for confined, semiconfined or unconfined flow in a neighborhood of the point at infinity $z = \infty$. If the point $T = A_k$ is the image of a boundary or interior stagnation point (a branch point or a convergence point) of the flow, then

$$dw/dT = (T - A_k) f(T), \qquad f(A_k) \neq 0, \infty;$$

if $T = B_i$ is the image of a vortex, a source, or a sink located on L_z^1 or in the flow domain D_z, then

$$\frac{dw}{dT} = \frac{f(T)}{T - B_i}, \qquad f(B_i) \neq 0, \infty.$$

And, finally, when there are junction points of the streamlines and the equipotentials, we have in a neighborhood of the images T_j of these points that

$$dw/dT = f(T)(T - T_j)^{\pm 1/2}, \qquad f(T_j) \neq 0, \infty.$$

Thus, in the general case dw/dT can be represented in the form

$$\frac{dw}{dT} = K \frac{\Pi(T - A_k)(T - \overline{A}_k)^{\delta_k}}{\Pi(T - B_i)(T - \overline{B}_i)^{\delta_i}} \Pi(T - T_j)^{\pm 1/2}, \qquad (4)$$

where K is a real constant, the T_j are points on the real axis, $\delta_k = 0$ ($\delta_i = 0$) for Im $A_k = 0$ (Im $B_i = 0$), i.e., when the stagnation point (vortex, source, or sink) lies on the boundary of D_z, and $\delta_k = 1$ ($\delta_i = 1$) otherwise.

Moreover, when the flow is semiconfined in a neighborhood of $z = \infty$, then to the image of this point there corresponds a coalescence of two of the B_i in the denominator of (4), and a coalescence of three when the flow is unconfined. The appearance of the conjugated points $\overline{A}_k$ and $\overline{B}_k$ in (4) is due to the fact that, since the boundaries of D_w are rectilinear, the function $w = w(T)$ can be extended analytically to the lower half-plane by the symmetry principle.

Suppose that the mapping $w = w(T)$ is normalized in such a way that the image of the free boundary L_z^2 lies on $[-1, 1]$, and the images ± 1 of the junction points of L_z^1 and L_z^2 are not points of intersection of equipotentials and streamlines. We fix arbitrarily the images A_k, B_i, and T_j of all the singular points of the flow, and map the upper half-disk D_ζ: $\{|\zeta| < 1, \operatorname{Im} \zeta > 0\}$ of the auxiliary ζ-plane conformally onto the domain $\operatorname{Im} T > 0$ by the Joukowsky function

$$T = -\tfrac{1}{2}(\zeta + 1/\zeta),$$

letting the semicircle L_ζ^2 ($\operatorname{Im} \zeta \geq 0$, $|\zeta| = 1$) correspond to the image of the free boundary L_z^2. Setting $dw/dT = Kf(T)$ in (4), we then get

$$\frac{dw}{d\zeta} = K \frac{1 - \zeta^2}{2\zeta^2} f\left(\frac{1 + \zeta^2}{-2\zeta} \right).$$

Since the free boundary L_z^2 is by assumption a streamline $\psi = \text{const}$, we have $|dw/dz| = |d\varphi/ds|$ for $z \in L_z^2$, where s is the arclength parameter of L_z^2. Comparing the last equality with the boundary condition (2), we find that

$$|dw/dz| = q(s) = \pm d\varphi/ds,$$

whence

$$\pm\varphi = \int_0^s q(s)\, ds \equiv K \int_0^\gamma \sin\gamma\, |f(-\cos\gamma)|\, d\gamma, \qquad \gamma \in [0, \pi], \qquad (5)$$

and, by the condition $q(s) > q_0^{-1} > 0$, the implicit function $s = s(\gamma)$ exists. Therefore, it can be assumed that we know the functional dependence

$$q = q(\gamma) \equiv q[s(\gamma)], \qquad \gamma \in [0, \pi]. \qquad (2^*)$$

The relation (2^*) enables us to construct a boundary-value problem for determining the derivative $dz/d\zeta$ of a conformal mapping of the upper half-disk D_ζ onto the flow domain D_z:

$$\arg \frac{dz}{dt} = \beta_k \pi, \qquad t \in [t_k, t_{k+1}]\ (k = 0,\dots,n),$$

$$\left| \frac{ds}{d\zeta}(e^{i\gamma}) \right| = \frac{k \sin\gamma}{q(\gamma)} |f(-\cos\gamma)|, \qquad \gamma \in [0, \pi], \qquad (6)$$

where the t_k ($k = 1,\dots,n$) are the images of the vertices z_k of the polygonal line L_z^1, $t_0 = -1$ and $t_{n+1} = 1$ are the images of its endpoints z_0 and z_{n+1}, and $\beta_k \pi$ are the angles between its links and the real axis. In order not to complicate the subsequent notation, we assume that the point at infinity always lies on the boundary L_z. Those problems in which this condition fails to hold—for example, problems of flow past finite obstacles with partially unknown boundary—can be assumed symmetric, and we can consider one half or one fourth, correspondingly, of the flow domain. Let us fix the images $\zeta = \tilde{A}_k$, $0 \neq |\tilde{A}_k| < 1$, of the stagnation points $z = a_k$ of the flow, the images $\zeta = \tilde{B}_i \neq \tilde{A}_k$, $0 \neq |\tilde{B}_i| < 1$, of the points $z = b_i$ at which there are vortices, sources, or sinks, and the images $\zeta = \tilde{T}_j$ of the junction points of the streamlines $\psi = \psi_k = \text{const}$ and the equipotentials $\varphi = \varphi_i = \text{const}$.

Some of the constants ψ_k and φ_i, as well as the intensities of all the vortices, sources, and sinks, are determined only after the problem is solved. In concrete hydrodynamic problems we can also investigate the case when all the physical flow parameters are given, while the images of the singular points of the flow are found as a result of solving the problem.

Let us study the properties of the function $(dz/d\zeta)(e^{i\gamma})$, $\gamma \in [0, \pi]$.
For this purpose, we represent $f(T) = K^{-1}dw/dT$ in the form

$$f(T) = \frac{\left(T - T_*\right)^{-\delta_*}(T - 1)^{-\delta_0}(T + 1)^{-\delta_{n+1}}}{\Pi(T - B_j)^{\delta_j}} f_*(T), \qquad (7)$$

where the B_j ($j = 1, \ldots, m$) are the images of the vertices at infinity of L_z, $T_* \neq \pm 1$ is a fixed image of the point at infinity on the free boundary L_z^2, and, moreover, δ_j, δ_0, $\delta_{n+1} = 1, 2, 3$ (respectively, $\delta_* = 1, 2, 3$) if the point z_j ($j = 1, \ldots, m$), z_0, or z_{n+1} of L_z^1 ($z_* \in L_z^2$) lies at infinity and the flow is confined, semiconfined, or unconfined, respectively, in a neighborhood of it, and $\delta_k = 0$ ($\delta_* = 0$) for $|z_k| < \infty$ ($|z_*| < \infty$). Let $T_* = 0$. Then for $T = (1 + \zeta^2)/(-2\zeta)$ formula (7) takes the form

$$f\left(\frac{1 + \zeta^2}{2\zeta}\right) = \tilde{K} \frac{\left(1 + \zeta^2\right)^{-\delta_*}(1 + \zeta)^{-2\delta_0}(1 - \zeta)^{-2\delta_{n+1}}}{\Pi\left[(\zeta - \tau_j)(1 - \zeta\tau_j)\right]^{\delta_j}} f_0(\zeta) \equiv \tilde{K}Q(\zeta)f_0(\zeta),$$

$$(7*)$$

where $\tilde{K} = \Pi\, \tau_j^{\delta_j}$ and the $\tau_j \in [-1, 1]$ are the images of the vertices at infinity of L_z^1.

By construction, $f_0(\zeta)$ is continuously differentiable along with

$$f_*\left((1 + \zeta^2)/(-2\zeta)\right)$$

on the circle $|\zeta| = 1$, and $f_0(e^{i\gamma}) \neq 0, \infty$, as a consequence of the choice of the images $\zeta = \tilde{A}_k$ and $\zeta = \tilde{B}_i$ of the singular points of the flow that appear in these functions. We look for a solution of the boundary-value problem (6) in the form

$$\frac{dz}{d\zeta} = K_0(1 - \zeta^2)Q(\zeta) \prod_{k=1}^{n} \left(\frac{\zeta - t_k}{1 - \zeta t_k}\right)^{\alpha_k - 1} e^{M(\zeta)}, \tag{8}$$

where the $\alpha_k \pi$ are the interior angles of L_z^1, $K_0 = \frac{1}{2}e^{i\beta_n\pi}\prod \tau_j^{\delta_j}K$, and $\beta_n\pi$ is the slope angle of the link $z_n z_{n+1}$. For determining the analytic function $M(\zeta)$ in the upper half-disk D_ζ we arrive at the boundary-value problem

$$\operatorname{Im} M(t) = 0, \qquad t \in [-1, 1],$$

$$\operatorname{Re} M(e^{i\gamma}) = \ln\frac{|f_0(e^{i\gamma})|}{q(\gamma)} = \mu(\gamma), \qquad \gamma \in [0, \pi]. \tag{9}$$

The first of the conditions (9) implies that $M(\zeta)$ can be analytically extended to the lower half-disk by the symmetry principle, and by solving the resulting Schwarz problem we can represent it by the formula

$$M(\zeta) = \frac{1}{2\pi}\int_{-\pi}^{\pi} \frac{\mu_*(\gamma)\left(e^{i\gamma} + \zeta\right)d\gamma}{e^{i\gamma} - \zeta}, \tag{10}$$

where $\mu_*(\gamma) = \mu(\gamma)$, $\gamma \in [0, \pi]$, and $\mu_*(\gamma) = \mu(-\gamma)$, $\gamma \in [-\pi, 0]$, with $\mu_*(\gamma)$ bounded together with $\mu(\gamma)$ for all $\gamma \in [-\pi, \pi]$. Using the equality $\mu_*(\gamma) = \mu_*(-\gamma)$, we can transform (10) to the form

$$M(\zeta) = \frac{1 - \zeta^2}{\pi}\int_0^{\pi}\mu(\gamma)\frac{d\gamma}{1 - 2\zeta\cos\gamma + \zeta^2}. \tag{10*}$$

Formula (10) tells us, in particular, that for the harmonic function $\operatorname{Re} M(\zeta)$

$$|\operatorname{Re} M(\zeta)| \leqslant \max_{\gamma}\left|\ln\frac{f_0(e^{i\gamma})}{q(\gamma)}\right| = N < \infty, \qquad \zeta \in D_\zeta.$$

We remark that the function $q = q(\gamma)$ determined by (2*), (5), and (7) depends continuously on the parameters K and τ_j for $\tau_j \neq \pm 1$. Consequently, according to the representation (10), the analytic function $M(\zeta)$ has the same property for $\zeta \in D_\zeta$.

3°. A priori estimates of the solution. For the time being suppose for definiteness that the junction points $z_0 = 0$ and z_{n+1} of L_z^2 and L_z^1 are finite and that $\delta_j = 1$ ($j = 1, \ldots, m$) and $\delta_* \neq 0$ in (7*), i.e., the flow is confined in a neighborhood of the vertices b_j at infinity of the polygonal line, and the length of the free boundary is infinite. Then the mapping of the upper half-disk D_ζ onto D_z has the form

$$z = K_0 \int_{-1}^{\zeta} \frac{(1 - \zeta^2)(1 + \zeta^2)^{-\delta_*}}{\Pi(\zeta - \tau_j)(1 - \zeta\tau_j)} \prod_{k=1}^{n} \left(\frac{\zeta - t_k}{1 - \zeta t_k} \right)^{\alpha_k - 1} e^{M(\zeta)} \, d\zeta \equiv \Omega(\zeta), \quad (11)$$

where $M(\zeta)$ is defined by (10). Starting from the given geometry of L_z^1, we form equations for determining the unknown constants K_0, t_i, $i = 1, \ldots, n$, and τ_j, $j = 1, \ldots, m$:

$$l_k = |\Omega(t_{k+1}) - \Omega(t_k)| \equiv F_k(K_0, t_i, \tau_j) \qquad (k = 0, \ldots, n),$$

$$h_j = \pi \left| (\zeta - \tau_j) \frac{d\Omega}{d\zeta} \right|_{\zeta = \tau_j} = F_{n+j}(K_0, t_i, \tau_j) \qquad (j = 1, \ldots, m). \quad (12)$$

The quantities h_j ($j = 1, \ldots, m$) denote the given depths of the flow in a neighborhood of the vertices b_j at infinity of L_z^1, and the l_k coincide with the lengths of the links of L_z^1 (Figure 1):

$$l_k = \int_{t_k}^{t_{k+1}} \left| \frac{d\Omega}{dt} \right| dt$$

if the interval $[t_k, t_{k+1}]$ does not contain the image τ_j of a vertex at infinity of L_z, and

$$l_{k_j} = \left| \int_{\Gamma_j} \frac{d\Omega}{d\zeta} d\zeta \right|, \qquad \Gamma_j = \widehat{t_{k_j} t_{k_j+1}} \subset D_\zeta,$$

if $\tau_j \in (t_{k_j}, t_{k_j+1})$. The integral in the last expression can be taken over an arbitrary curve $\Gamma_j \subset D_\zeta$ joining t_{k_j} and t_{k_j+1}, where $\tau_j \notin \Gamma_j = \widehat{t_{k_j} t_{k_j+1}}$. Obviously, the conditions (12) completely determine the given polynomial line L_z^1 with n vertices at the finite points z_k ($k = 1, \ldots, n$) with interior angles $\alpha_k \pi$ ($0 < \alpha_k \leq 2$), with endpoints z_0 ($= 0$) and z_{n+1}, and with m vertices b_j at infinity in a neighborhood of which $h_j \neq 0, \infty$. We shall find a priori estimates for the solution of (12).

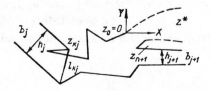

FIGURE 1

Let us first consider the simplest case of a finite polygonal line L_z^1. It suffices to show that $K_0 \neq 0, \infty$, after which the question of the impossibility of the equalities $t_k = t_j$ for $j \neq k$ can be solved completely analogously to Lemma 2 in §2 of Chapter III. Let $\Gamma = (\Gamma_0 + \Gamma_1 + \Gamma_2)$ be a curve joining $\zeta = \pm 1$ in the upper half-disk D_ζ, with $\Gamma_1 = (1, e^{i\varepsilon})$, $\Gamma_2 = (e^{i(\pi-\varepsilon)}, -1)$, $0 < \varepsilon < \pi/2$, arcs of the circle $|\zeta| = 1$, and $\Gamma_0 \subset D_\zeta$ the line segment with endpoints at $e^{i\varepsilon}$ and $e^{i(\pi-\varepsilon)}$. The function $dz/d\zeta = d\Omega/d\zeta$ does not have singular points on Γ and for $K_0 \neq 0$ vanishes only at the endpoints $\zeta = \pm 1$ of this curve; moreover, in a neighborhood of these points $|d\Omega/d\zeta|$ can be represented in the form

$$|d\Omega/d\zeta| = |K_0||1 - \zeta^2||Q(\zeta)||e^{M(\zeta)}| = |K_0||1 - \zeta^2|f_*(\zeta),$$

where $f_*(\zeta) \neq 0, \infty$ for $\zeta \in \Gamma_1, \Gamma_2$. Since the constant K_0 appears linearly in $d\Omega/d\zeta$, we then get that as $K_0 \to 0, \infty$

$$|z_0 - z_{n+1}| = |\Omega(-1) - \Omega(1)| = \left| \int_\Gamma \frac{d\Omega}{d\zeta} d\zeta \right| \to 0, \infty,$$

which contradicts the assumption that L_z^1 is nondegenerate. In the case when L_z^1 is unbounded we first prove that the images τ_j $(j = 1, \ldots, m)$ of the vertices of L_z^1 at infinity cannot converge to the points $\zeta = \pm 1$. Suppose the opposite, $\tau_j \to 1$, and consider a single equation

$$\frac{h_j}{\pi} = \lim_{\zeta \to \tau_j} \left| (\zeta - \tau_j) \frac{d\Omega}{d\zeta} \right|$$

from (12), where the limit on the right-hand side does not depend on the sequence $\zeta \in \overline{D}_\zeta$ of points converging to τ_j. On the circle $|\zeta| = 1$ the function $|(d\Omega/d\zeta)(\zeta - \tau_j)|$ can be represented in the form

$$\left| \frac{d\Omega}{d\zeta}(\zeta - \tau_j) \right| = |K_0||Q(\zeta)||q - \zeta^2||e^{M(\zeta)}|$$

$$= \frac{|K_0||1 - \zeta||\tilde{f}(\zeta)|}{|1 - \zeta\tau_j|\Pi_{i \neq j}|\zeta - \tau_i||1 - \zeta\tau_i|},$$

where the product includes only the τ_i that tend to 1, and $\tilde{f}(\zeta) \neq 0, \infty$ in a neighborhood of $\zeta = 1$, in view of the boundedness $|M(e^{i\gamma})| < N_0(N) < \infty$ proved above. For example, we choose a sequence of values $\gamma_j \to 0$ in such a way that

$$\left| \frac{1 - \zeta}{1 - \zeta\tau_j} \right|^2 = \frac{2(1 - \cos\gamma)}{\tau_j^2 + 1 - 2\tau_j\cos\gamma} = 1,$$

by setting $\cos \gamma_j = (1 + \tau_j)/2$. Then as $\tau_j \to 1$ we have

$$\frac{h_j}{\pi} = \lim_{\zeta_j \to 1} \left| \frac{d\Omega}{d\zeta}(\zeta_j - \tau_j) \right| = |\tilde{f}(1)| \frac{|K_0|}{\Pi_{i \neq j}|1 - \tau_j|^2}, \qquad \zeta_j = e^{i\gamma_j},$$

from which it follows that $|K_0| \neq \infty$. On the other hand, choosing the sequence γ_j from the condition

$$\sin \gamma_j/2 = \frac{1}{4}(1 - \tau_j)^2 \prod_{i \neq j}(1 - \tau_i)^4,$$

we have at the points $\zeta_j = e^{i\gamma_j}$ that

$$R(\zeta) = \frac{(1 - \zeta)^{1/2}}{|1 - \zeta\tau_j|\Pi_{i \neq j}|\zeta - \tau_i||1 - \zeta\tau_i|} \leqslant \frac{2(\sin \gamma_j/2)^{1/2}}{|1 - \tau_j|\Pi_{i \neq j}(1 - \tau_i^2)} = 1.$$

Consequently, by the already proved boundedness of K_0 as $\tau_j \to 1$,

$$\frac{h_j}{\pi} = \lim_{\zeta_j \to 1} \left| \frac{d\Omega}{d\zeta}(\zeta_j - \tau_j) \right| = |K_0||\tilde{f}(1)|, \qquad \lim_{\zeta_j \to 1} \left\{ |1 - \zeta_j|^{1/2} R(\zeta_j) \right\} = 0.$$

The case $\tau_j \to -1$ is handled similarly.

Thus, it has been proved that there is an $\varepsilon > 0$ such that $|\tau_j \pm 1| > \varepsilon > 0$ $(j = 1,\ldots,m)$. In view of the boundedness of the function $(dz/d\zeta)(e^{i\gamma}) = (d\Omega/d\zeta)(e^{i\gamma})$ at $\gamma = 0$ and $\gamma = \pi$ in this case, the inequality $0 < |K_0| < \infty$ can be established completely analogously to the case of a bounded polygonal line L_z^1.

Let us show that $\tau_j \neq t_k$ $(k = 1,\ldots,n; j = 1,\ldots,m)$. Suppose not; namely, suppose that some of the t_k $(k = 1,\ldots,p)$ converge to τ_j, and set

$$t_p + |\tau_j - t_p| \equiv \tilde{t}_p \in (t_p, t_{p+1}),$$

so that as $t_p \to \tau_j$ we also have $\tilde{t}_p \to \tau_j$ $(p \leqslant n)$. We consider the distance

$$\Lambda = \left| \int_{\tilde{t}_p t_{n+1}} \frac{d\Omega}{d\zeta} d\zeta \right| = \left| \int_{\tilde{t}_p t_{n+1}} \left\{ \frac{d\Omega}{d\zeta}(\zeta - \tau_j) \right\} \frac{d\zeta}{\zeta - \tau_j} \right|$$

$(\Lambda \neq 0, \infty)$ between one of the endpoints $z_{n+1} = \Omega(1)$ of L_z^1 and the point $\tilde{z}_p = \Omega(\tilde{t}_p)$, which lies on the side $z_p z_{p+1}$ of L_z^1. The density $(d\Omega/d\zeta)(\zeta - \tau_j)$ of the Cauchy type integral on the right-hand side of the expression for Λ is integrable on the curve $\Gamma = \tilde{t}_p t_{n+1} \in D_\zeta$ with endpoints at $\tilde{t}_p$ and $t_{n+1} = 1$ and is bounded in a neighborhood of the point $\tilde{t}_p \in \Gamma$; moreover, according to (12),

$$\left. \left| \frac{d\Omega}{d\zeta}(\zeta - \tau_j) \right| \right|_{\zeta = \tilde{t}_p} \to \frac{h_j}{\pi} \neq 0 \quad \text{as } \tilde{t}_p \to \tau_j.$$

Consequently, the Cauchy type integral has a logarithmic singularity as $t_p \to \tau_j$, i.e., $\Lambda \to \infty$. This contradiction to the finiteness of Λ proves that the equalities $t_k = \tau_j$ are impossible. Suppose now that the parameters τ_j and τ_{j+1} converge; by the proven inequality $|t_k - \tau_j| > 0$, this can happen only if the interval $[\tau_j, \tau_{j+1}]$ does not contain points t_k. Then, by (12),

$$\frac{h_j}{\pi} = \left| (\zeta - \tau_j) \frac{d\Omega}{d\zeta} \right|_{\zeta = \tau_j} = |f_*(\tau_j)| \frac{1}{|\tau_j - \tau_{j+1}|} \to \infty$$

as $\tau_j \to \tau_{j+1}$, since, by the inequalities $|t_k - \tau_j| \geq 0$ and $|t_k - \tau_{j+1}| > 0$ ($k = 0, \ldots, n + 1$) the function $f_*(\zeta) = (\zeta - \tau_j)(\zeta - \tau_{j+1}) d\Omega / d\zeta$ is neither zero nor infinity in a neighborhood of the points $\zeta = \tau_j$ and τ_{j+1}; in particular, $f_*(\tau_j) \neq 0, \infty$. Thus, it has been proved that $|\tau_j - \tau_i| > \varepsilon > 0$ and $|\tau_j - t_k| > \varepsilon$ for $i \neq j$ and $k = 0, \ldots, n + 1$. The impossibility of two of the parameters converging can now be proved completely analogously to Lemma 2 in §2 of Chapter III.

Let σ_i ($i = 0, \ldots, n + m + 1$) denote the images of all the vertices of the simple polygonal line L_z^1, including those at infinity, i.e., $\sigma_i = t_k$ or τ_i, and $\sigma_0 = t_0 = -1$, $\sigma_{n+m+1} = t_{n+1} = 1$. We state the assertion proved above as a lemma.

LEMMA. *Each solution* $(K_0, \sigma_1, \ldots, \sigma_{n+m})$ *of the system* (12) *corresponding to a simple polygonal line L_z^1 satisfies the inequalities*

$$|\ln|K_0|| < N_0 < \infty, \qquad \sigma_{i+1} - \sigma_i > \varepsilon_0 > 0 \quad (i = 0, \ldots, n + m). \quad (13)$$

In particular, as in the case of the mixed boundary-value problem with free boundary (§2 of Chapter III), this lemma can be used to prove easily the continuous differentiability with respect to t_i ($i = 1, \ldots, n$) of the right-hand sides F_k of (12). And the continuous dependence on the parameters τ_j and K_0 obviously follows from the validity of this property for $M(\zeta)$ (see the end of 2°).

REMARK. If one or both junction points z_0 and z_{n+1} of the polygonal line and the free boundary lie at infinity, then the one or two equations, respectively, corresponding to the lengths $l_0 = |z_0 - z_1|$ and $l_n = |z_{n+1} - z_n|$ of the links vanish from (12). It is easy to see that the assertions of the lemma remain true also in this case.

4°. *An existence theorem in the case of polygonal boundaries.* Let us introduce a one-parameter family of polygonal lines L_z^λ, $\lambda \in [0, 1]$, by transforming continuously the original polygonal line L_z^λ, $\lambda = 1$, into a polygonal line L_z^0 bordered by straight lines parallel to the OX-axis and having finitely many vertices $z_k^* = \infty$ with angles -2π and vertices b_j ($j = 1, \ldots, m$) with angles zero at infinity.

We implement such a transformation in several steps. If between two adjacent vertices b_j and b_{j+1} at infinity there lies a polygonal line with a single finite vertex z_k, or if b_j and b_{j+1} are incident to the single line segment $b_j b_{j+1}$, then the corresponding part of the polygonal line is left unchanged.

If between b_j and b_{j+1} there are two vertices z_k and z_{k+1}, then through the point $z_k^* \in \widetilde{z_k b_j}$ we draw a straight line parallel to the ray $\widetilde{z_{k+1} b_{j+1}}$ and transform the polygonal line $b_j z_k z_{k+1} b_{j+1}$ into the polygonal line $b_j z_k^* b_{j+1}$ (Figure 2).

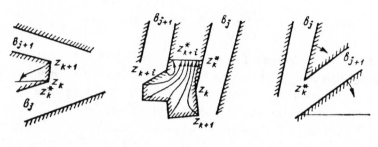

FIGURE 2

The case when there are more than two vertices $z_k, z_{k+1}, \dots, z_{k+i}$, $i \geqslant 3$, between b_j and b_{j+1} is reduced to the case just considered by first transforming the polygonal line $z_k z_{k+1} \cdots z_{k+i}$ into a segment $z_k^* z_{k+i}^*$, $z_k^* \in z_k b_j$ and $z_{k+1}^* \in z_{k+i} b_{k+i}$, by the method used in §3 of Chapter III (see Figure 2). The last step consists in simultaneously transforming all the angles at the remaining vertices z_k^* into the angle 2π and rotating the polygonal line thus obtained so that the slopes of its sides coincide with the direction of the real axis.

In the transformations just described the parameter $\lambda \in [0, 1]$ is introduced in such a way that all the geometric characteristics a_k^λ, l_i^λ, and h_j^λ of the polygonal lines L_z^λ in the corresponding equations (12) are continuous in λ, and, moreover, all the intermediate polygonal lines L_z^λ, $\lambda \in [0, 1]$, are simple (cf. §3 in Chapter III). It is possible to introduce such a λ for each of the elementary transformations considered. Consequently, this is also possible for a finite collection of such transformations. We also introduce the parameter λ as a factor in front of $M(\zeta)$ in (8). Then in the limit we get $\arg e^{\lambda M(\zeta)}\big|_{\lambda=0} = 0$ at $\lambda = 0$, which corresponds to a transformation of the free boundary into a straight line parallel to the OX-axis or into a semi-infinite segment or two semi-infinite segments of this straight line, depending on the size of the angle $-\delta_* \pi$ (formula (7*)) at the point at infinity on the original free boundary. Since the angle $-\delta_* \pi$ is preserved under all our transformations, the domain for

$\lambda = 0$ bounded by the limit polygonal line L_z^0 and by the limit of the free boundary (the polygon P^0) is a strip ($\delta_* = 1$), a half-plane ($\delta_* = 2$), or the plane ($\delta_* = 3$) with finitely many semi-infinite parallel cuts. Let us consider the system (12) corresponding to the polygonal line L_z^0 for $\lambda = 0$, written in the form

$$l_i = \int_{t_i}^{t_{i+1}} \left| \frac{dz^0}{dt} \right| dt, \qquad l_k^* = \left| \int_{t_k^* t_{k+1}} \frac{dz^0}{d\zeta} d\zeta \right|, \qquad (12^*)$$

$$h_j = \left| (\zeta - \tau_j) \frac{dz^0}{d\zeta} \right| \zeta = \tau_j, \qquad \frac{dz^0}{d\zeta} = K_0^0 \frac{\zeta^2 - 1}{(\zeta^2 + i)^{\delta_*}} \frac{\Pi_k(\zeta - t_k^*)}{\Pi_j(\zeta - \tau_j)},$$

where the t_k^* are the images of the remaining vertices L_z^0 with angles 2π, h_j is the distance between two adjacent parallel straight lines in L_z^0, l_i is the distance between fixed points on a single straight line (a cut) in L_z^0, and l_k^* is the distance between the vertices z_k^* and z_{k+1}^* of adjacent cuts or between the vertices of a cut and a fixed point on the parallel straight line in L_z^0 closest to the cut. The parameters K_0^0, τ_j, and t_k^* can be simultaneously determined from the second part of (12*) for arbitrary fixed h_j and l_k^* as parameters in the mapping of the upper half-disk D_ζ onto the polygon P^0 with a fixed correspondence $(z_0^0, z_{n+1}^0, z_*^0) \to (-1, 1, i)$ between three boundary points, where z_0^0 and z_{n+1}^0 are the endpoints of L_z^0, and z_*^0 is a point at infinity on the line obtained from the free boundary when $\lambda = 0$. We substitute the values of K_0^0, τ_j, and t_k^* found by solving the second part of (12*) into $dz^0/d\zeta$, and consider the remaining equations of (12*). The integrand $dz^0/d\zeta$ is nonzero and noninfinite on the intervals $t_i t_{i+1}$, and the unknown constants t_i appear in the expressions for l_i only in the limits of integration. As in the case of the mixed boundary-value problem with free boundary (§2.4 in Chapter III), this implies the unique solvability of the first part of (12*) for the t_i, the images of part of the transformed finite vertices of the original polygonal line L_z^1.

We have thus proved the unique solvability of the sytem (12*) corresponding to the polygon P^0 for arbitrary fixed left-hand sides of these equations with $l_i \neq 0, \infty$, $l_k^* \neq 0, \infty$, and $h_j \neq 0, \infty$.

Let $F_i^\lambda = F_i(\lambda, \alpha^\lambda, \sigma)$ $(i = 0, \ldots, n + m)$ be the right-hand sides in (13), with α_k^λ substituted for α_k and λ introduced in front of $M(\zeta)$, and let $\tilde{l}_k(\lambda)$ $(k = 0, \ldots, n)$ and $\tilde{l}_i(\lambda) = \tilde{h}_{i-n}(\lambda)$ $(i = n + 1, \ldots, n + m)$ be the geometric parameters of L_z^λ, $\lambda \in [0, 1]$, corresponding to a given λ (the left-hand sides of (12)).

Let $\mathfrak{M}$ be the closure of the convex set of elements $\sigma = (K_0, t_k, \tau_j) = (\sigma_0, \ldots, \sigma_{n+m})$ in E_{n+m+1} satisfying (13) with fixed $N_0 < \infty$ and $\varepsilon_0 > 0$. On $\mathfrak{M}$

we define a transformation $s = S(\sigma, \lambda)$ by the relations

$$s_k + 2 = (\sigma_k + 2)\frac{F_k(\lambda, \alpha^\lambda, \sigma)}{\tilde{l}_k(\lambda)} \qquad (k = 0, \ldots, n + m). \tag{14}$$

By construction, the transformation $s = S(\sigma, \lambda)$ is continuous in its arguments (see the end of 3°). In view of the a priori estimates (13), which hold for the simple polygonal lines L_z^λ for all $\lambda \in [0, 1]$, the fixed points $s = \sigma$ of the transformation S on $\mathfrak{M}$ (which coincide by construction with the solutions of the corresponding system (12)) do not belong to the boundary of this set. As was proved above, for $\lambda = 0$ the corresponding system (12*) is uniquely solvable for any left-hand sides l_i and h_j, and this is equivalent to the unique solvability of the equation

$$\sigma - S(\sigma, 0) = \eta \tag{15}$$

for arbitrarily small $\eta \in E_{n+m+1}$.

Thus, all the conditions of the Leray-Schauder fixed point principle (§1 of Chapter II) hold for the transformation $s = S(\sigma, \lambda)$; hence, for each $\lambda \in [0, 1]$ the set $\mathfrak{M}$ contains at least one fixed point $\sigma = S(\sigma, \lambda)$. According to the remark after the lemma, the assertions proved remain true also when one or both of the endpoints z_0 and z_{n+1} of L_z^1 are at infinity.

If the free boundary L_z^2 is finite ($\delta_* = 0$), then one of the unknown parameters K_0, τ_j ($j = 1, \ldots, m$) or t_k ($k = 1, \ldots, n$) must be fixed (it is more convenient to fix K_0), and, respectively, one of the geometric parameters $h_j(j = 1, \ldots, m)$ or l_i ($i = 0, \ldots, n$) of L_z^1 must be regarded as unknown. Here one of the equations disappears from (12), and this obviously does not change the subsequent arguments (examples of such problems are considered in §3).

Suppose now that in a neighborhood of certain vertices b_j at infinity of L_z^1 the flow is semiconfined or unconfined, i.e., at the image τ_j of this vertex the principal term in the expansion of $dz/d\zeta$ has the form

$$\left[(\zeta - \tau_j)(1 - \zeta\tau_j)\right]^{-\delta}, \qquad \delta = 2, 3.$$

In this case all the equations in (12) connected with the presence of such a vertex b_j on L_z^1 are conditions on the lengths l_i of the links of the polygonal line, and, consequently, in obtaining the a priori estimates the corresponding τ_j can be included in the images of the finite vertices of L_z^1.

We remark also that the case of a polygonal line P with a semiconfined or unconfined flow in a neighborhood of a vertex b_j of it can be obtained as a limiting case (as $h \to \infty$) for polygonal lines with a confined flow in a neighborhood of b_j: $P_i \to P$, $i \to \infty$. We use this method in the analogous case of curvilinear boundaries (see the proof of Theorem 2).

Let us formulate our result as a theorem.

THEOREM 1. *Suppose that the flow domain D_z is bounded by a single free boundary L_z^2 and a given simple polygonal line L_z^1 with n vertices z_k ($k = 1,\dots,n$) in the finite plane and m vertices b_j at infinity. Then there is at least one flow in D_z whose complex potential $w(z) = \varphi(x, y) + i\psi(x, y)$ satisfies the boundary conditions* (1) *and* (2). *Moreover, in D_z and on its boundary $L_z = L_z^1 + L_z^2$ it is possible to locate a finite number of singular points* (*stagnation points, vortices, sources, and sinks*) *whose coordinates and physical characteristics* (*the intensities of the vortices, sources, and sinks*) *are determined in the solution process.*

REMARK. If the given function $q(s)$ satisfies the Hölder condition $q(s) \in C^\alpha$, $\alpha > 0$, then $q[s(\gamma)] \in C^\alpha$ by construction. The properties of Cauchy type integrals then give us that $M(\zeta) \in C^\alpha$ in $\overline{D}_\zeta$, and thereby $dz/d\zeta \in C^\alpha$ in $\{\overline{D}_\zeta \backslash O(\tau_j, t_k)\}$, where the $O(\tau_j, t_k)$ are neighborhoods of the images of the finite vertices and those at infinity of the flow domain.

5°. A curvilinear boundary. Suppose now that L_z^1 is a simple curve with m vertices b_j at infinity in a neighborhood of which the flow is confined, $h_j \neq 0, \infty$, and in this neighborhood L_z^1 consists of half-lines emanating from the fixed finite points z_{k_j} and z_{k_j+1}.

On the finite curvilinear part of L_z^1 we choose sufficiently close points z_k^n ($k = 0,\dots,n+1$), including the junction points $z_0^n (= 0)$ and z_{n+1}^n ($=$ a fixed point $\tilde{z}$) of L_z^1 and L_z^2, as well as the junction points z_{k_j} and z_{k_j+1} of the curvilinear and rectilinear parts (in the neighborhood of b_j) of L_z^1. We join the chosen points z_k^n successively by straight lines and consider the resulting simple polygonal line P_n with finite vertices z_k^n ($k = 1,\dots,n$) with interior angles $\alpha_k\pi$, $0 < \alpha_k \leqslant 2$, and with vertices b_j ($j = 1,\dots,m$) at infinity. By adding new vertices located on the curvilinear parts of L_z^1 to the chosen finite vertices, we get in this way a sequence of simple polygonal lines converging to the given simple curve L_z^1 as $n \to \infty$. By Theorem 1, to each polygonal line P_n there corresponds at least one mapping of the form

$$z = K_0^n \int_{-1}^{\zeta} \frac{(1 - \zeta^2)(1 + \zeta^2)^{-\delta_*}}{\prod(\zeta - \tau_j^n)(1 - \zeta\tau_j^n)} \prod_{k=1}^{n} \left(\frac{\zeta - t_k^n}{1 - \zeta t_k^n} \right)^{\alpha_k^n - 1} e^{M^n(\zeta)} \, d\zeta \equiv \Omega^n(\zeta),$$

(11*)

where the τ_j^n ($j = 1,\dots,m$) are the images of the vertices b_j, and the t_k^n are the images of the finite vertices of P_n. The parameters K_0^n, τ_0^n, and t_k^n are solutions of the system of equations

$$l_k^n = |\Omega^n(t_k^n) - \Omega^n(t_{k+1}^n)| \qquad (k = 0,\dots,n),$$

$$h_j^n = \left| (\zeta - \tau_j^n) \frac{d\Omega^n}{d\zeta} \right|_{\zeta = \tau_j^n} \qquad (j = 1,\dots,m). \qquad (16)$$

We remark that the first part of (16) contains, in particular, the equation

$$l_{k_j} = |\,\Omega^n\big(t^n_{k_j+1}\big) - \Omega^n\big(t^n_{k_j}\big)\,| \qquad (j = 1,\dots,m),$$

where $z^n_{k_j+1} = \Omega^n(t^n_{k_j+1})$ and $z^n_{k_j} = \Omega^n(t^n_{k_j})$ are fixed junction points of curvilinear and rectilinear parts of the original curve L^1_z (in a neighborhood of the vertices b_j at infinity), and the distance between these points does not depend on the index n. Remaining equations are the lengths of the links of P_n,

$$l^n_k = \int_{t_k}^{t_{k+1}} \left|\frac{d\Omega^n}{dt}\right| dt,$$

and, in view of the assumed rectifiability of any finite part of the simple curve L^1_z, we have for all n that

$$\sum_{k=0}^{n} l^n_k \leqslant l\big(L^1_z\big) + \sum_{j=1}^{m} l_{k_j}, \tag{17}$$

where $l(L^1_z)$ is the length of all the curvilinear parts of L^1_z joining the points $z^n_{k_j}$ $(j = 1,\dots,m)$ (if L^1_z is finite, then $l(L^1_z)$ is its total length, and $\Sigma l_{k_j} = 0$). Thus, the following inequalities hold independently of n:

$$|\,z^n(t^n_k)\,| \leqslant l\big(L^1_z\big) + \sum_{j=1}^{m} l_{k_j}. \tag{18}$$

On the other hand, according to the a priori estimates (13),

$$|\ln|K^n_0|\,| < N_0, \qquad |\,t^n_{k_j} - \tau^n_i\,| \geqslant \varepsilon_0,$$

$$\tau^n_{j+1} - \tau^n_j \geqslant \varepsilon_0 \qquad (i,\, j = 1,\dots,m), \tag{19}$$

where $N_0 = \sup N^n_0$ and $\varepsilon_0 = \inf \varepsilon^n_0 > 0$.

In proving the a priori estimates (19) for each of the polygonal lines P_n we used only the nondegeneracy of the polygonal lines and the relations

$$l_{k_j} = |\,\Omega^n\big(t^n_{k_j+1}\big) - \Omega^n\big(t^n_{k_j}\big)\,| \neq 0, \infty \qquad (j = 1,\dots,m),$$

$$h_j = \pi \left|\frac{d\Omega^n}{d\zeta}(\zeta - \tau^n_j)\right| \neq 0, \infty,$$

$$|\,z_0 - z^n_{n+1}\,| = |\,\Omega(-1) - \Omega(1)\,| \neq 0, \infty;$$

therefore, it is obvious that

$$N_0 = \sup_n |\ln|K^n_0|\,| < \infty, \qquad \varepsilon_0 = \inf_n \big\{|\,t^n_{k_j} - \tau^n_i\,|, |\,\tau^n_{j+1} - \tau^n_j\,|\big\} > 0.$$

Indeed, in a manner completely analogous to the lemma for Theorem 1 a consideration of the relation

$$\frac{h_j}{\pi} = \lim_{\zeta \to \tau^n_j} \left|(\zeta - \tau^n_j)\frac{d\Omega^n}{d\zeta}\right|$$

gives us the impossibility of the convergence $\tau_j^n \to \pm 1$ ($j = 1, \ldots, m$) independently of n. Further, as $K_0^n \to 0, \infty$,

$$|z_0 - z_{n+1}^n| = \left| \int_\Gamma \frac{d\Omega^n}{d\zeta} d\zeta \right| \to 0, \infty$$

for all n, which contradicts the fact that L_z^1 is simple, and so on.

It is hardly surprising that the dependence of the values of the parameters K_0^n, τ_j^n, and $t_{k_j}^n$ on the index n when the polygonal line is varied does not have any influence on the proof carried out for the inequalities (19) for a fixed polygonal line P_n. We have already considered these parameters as variable quantities, as elements of certain sequences.

Thus, $\varepsilon_0 > 0$ in (19), and, consequently, for all n

$$-1 < \cdots < \tau_{j-1}^n < t_{k_j}^n < \tau_j < t_{k_j+1}^n < \cdots < 1. \tag{20}$$

In the bounded sequences $\{K_0^n\}$, $\{t_{k_j}^n\}$, $\{\tau_j^n\}$, and $\{t_{k_j+1}^n\}$ let us select subsequences that converge as $n \to \infty$ and satisfy together with the original sequences the inequalities (19) and (20).

The existence of such subsequences follows from the relative compactness of the family of cones in Euclidean space that are determined by (19) and (20). In order not to complicate the notation, we assume that the original sequences converge.

Let us consider the family of analytic functions in the upper half-disk given by

$$\Omega_*^n(\zeta) = R(\zeta)\Omega^n(\zeta) \equiv R(\zeta) \int_{-1}^\zeta \frac{d\Omega^n}{d\zeta} d\zeta, \tag{21}$$

where $R(\zeta) = (\zeta^2 + 1)^{\delta_* - 1}$ for $\delta_* = 2$ or 3, and $R(\zeta) = [\ln(\zeta^2 + 1)]$ for $\delta_* = 1$.

According to the properties of the $M(\zeta)$ in (10) studied earlier, the following estimate holds independently of n:

$$|\operatorname{Re} M^n(\zeta)| \leqslant \max_\gamma \left| \ln \frac{h(e^{i\gamma})}{q(\gamma)} \right| = N < \infty, \qquad \zeta \in D_\zeta.$$

Then with the help of (19) we find for $|\zeta| = 1$ that

$$|\Omega_*^n(e^{i\gamma})| \leqslant |R(e^{i\gamma})| |\Omega^n(e^{i\gamma})|$$

$$\leqslant |R(e^{i\gamma})| \left| K_0^n \int_\gamma^\pi \frac{|1 - \zeta^2| |\zeta^2 + 1|^{-\delta_*}}{\prod_{j=1}^m |\zeta - \tau_n^n| |1 - \zeta \tau_n^n|} e^{\operatorname{Re} M^n(\zeta)} d\zeta \right|$$

$$\leqslant R_0 \varepsilon_0^{-2m} e^{N_0 + N} < \infty,$$

where the quantity

$$R_0 = \max_{\gamma \in [0,\pi]} \left\{ |R(e^{i\gamma})| \int_\gamma^\pi \frac{d\gamma}{|\zeta^2 + 1|^{\delta_*}} \right\}$$

is bounded by construction. If $\zeta = t_k$ lies on the interval $[t_k^n, t_{k+1}^n] \ni \tau_j^i$ $(j = 1, \ldots, m)$ corresponding to the side l_k^n of P_n, then

$$|\Omega_*^n(t)| = \left| R(t) \int_{(-1,t)} \frac{d\Omega^n}{d\zeta} d\zeta \right| \leq R_1 \left(|\Omega^n(t_k^n)| + \int_{t_k^n}^t \left| \frac{d\Omega^n}{dt} \right| dt \right)$$

$$\leq R_1 \left[l(L_z^1) + \sum_{j=1}^m l_{k_j} \right], \quad \text{where } 0 < R_1 = \max_{t \in [-1,1]} R(t) < \infty.$$

On the intervals $[t_{k_j}^n, \tau_j^n]$ and $[\tau_j^n, t_{k_j+1}^n]$ which are the images of the parallel half-lines with slope angle $\beta_j \pi$ and with endpoints at the vertices z_{k_j} and z_{k_j+1} of P_n we have, respectively,

$$\arg[\Omega^n(t) - z_{k_i}] = \beta_j \pi, \quad i = j, j+1.$$

Thus, the functions $\Omega_*^n(\zeta)$ satisfy the following Hilbert boundary-value problems with discontinuous coefficients:

$$\text{Re}\left[e^{i\lambda_n(\zeta)} \Omega_*^n(\zeta) \right] = g_n(\zeta), \quad \zeta \in \Gamma, \qquad (22)$$

where Γ is the boundary of the upper half-disk D_ζ, $\lambda_n(\zeta)$ is a piecewise constant function on Γ with $\lambda_n(\zeta) = \left(\frac{1}{2} - \beta_j \right)\pi$, $\zeta = t \in [t_{k_j}, t_{k_j+1}]$, and $\lambda_n(\zeta) = 0$ otherwise, while $g_n(\zeta) = R(t) \text{Re}\{ z_{k_p} e^{-i\beta_j \pi} \}$, $p = j, j+1$, for $\zeta = t \in [t_{k_j}^n, \tau_j^n], [\tau_j^n, t_{k_j+1}^n]$, respectively, and $g_n(\zeta) = \text{Re } \Omega_*^n(\zeta)$ for $\zeta \notin [t_{k_j}^n, t_{k_j+1}^n]$ $(j = 1, \ldots, m)$.

By what was proved above, the function $\Omega_*^n(\zeta)$ is bounded by a constant not depending on n on the boundary Γ for $\zeta \notin [t_{k_j}^n, t_{k_j+1}^n]$; therefore,

$$|g_n(\zeta)| \leq C < \infty, \quad \zeta \in \Gamma,$$

uniformly in n.

In constructing the canonical function $X^n(\zeta)$ of the homogeneous problem (22) we can use the fact that $\lambda_n(\zeta)$ is a piecewise constant function, regarding the constant values of $\lambda_n(\zeta)$ as the slope angles of the sides of a certain polygon. The derivation of the function mapping the upper half-disk onto this polygon can be taken as the canonical function $X^n(\zeta)$:

$$X^n(\zeta) = (1 - \zeta^2) \prod_{j=1}^m (\zeta - t_{k_j})^{\delta_{k_j}} (\zeta - t_{k_j+1})^{\delta_{k_j}+1} X_0(\zeta) \equiv \Pi^n(\zeta) X_0(\zeta),$$

where $|\delta_{k_i}| < 1$, $i = j$, $j + 1$, and $X_0(\zeta) \neq 0, \infty$ in D_ζ. Then the solution of the boundary-value problem (22) can be represented in the form

$$\Omega_*^n(\zeta) = \Pi(\zeta) \sum_{i=0}^{3m+2} R_i^n(\zeta) \int_{\sigma_i \sigma_{i+1}} \frac{g_*^i(t)\, dt}{\Pi(t)(t - \zeta)}, \qquad (23)$$

where $\sigma_{3m+3} = \sigma_0 = -1$, $R_i^n(\zeta)$ and $g_*^i(t)$ are functions uniformly bounded in n, and $\sigma_i \sigma_{i+1}$ is the arc on Γ with endpoints σ_i and σ_{i+1}. We encircle the points σ_k^0 obtained from the convergent sequences σ_k^n in the limit as $n \to \infty$ by the neighborhoods

$$G_k: \left\{ |\zeta - \sigma_k^0| < \varepsilon, \zeta \in \overline{D}_\zeta \right\} \qquad (k = 0, \ldots, 3m + 2).$$

By (19), there exists an $\varepsilon > 0$ such that

$$G_k \cap G_i = \varnothing, \qquad i \neq k,$$

and $|\sigma_k^0 - \sigma_k^n| \leq \varepsilon/2$ for $n \geq n_0(\varepsilon)$. On the curves $\hat{\sigma}_i \hat{\sigma}_{i+1} \subset \sigma_i \sigma_{i+1}$ remaining after removal of the neighborhoods of the singular points σ_k the functions $g_*^i(t)$ have derivatives that are uniformly bounded in n by construction, and $|\sigma_i^n - \sigma_{i+1}^n| \geq \varepsilon > 0$ by (19). Therefore, (23) gives us that

$$|\Omega_*^n(\zeta)| \leq \hat{N} < \infty, \qquad \zeta \in \overline{D}_0, \qquad (24)$$

where $D_0 = D_\zeta \setminus \cup_k G_k$ and the constant $\hat{N} = \hat{N}(\varepsilon)$ does not depend on n.

Then for $\zeta \in \overline{D}_0$ we have

$$|\Omega^n(\zeta)| = \frac{1}{|R(\zeta)|} |\Omega_*^n(\zeta)| \leq \hat{N} \max_{\zeta \in \overline{D}_0} |R(\zeta)|^{-1} \equiv \hat{N}_1(\varepsilon) < \infty,$$

and, consequently, the sequence of mappings $z = \Omega^n(\zeta)$ defined by (11*) is relatively compact in D_0. Let us extract from the sequence $\{\Omega^n(\zeta)\}$ of analytic functions a subsequence $\{\Omega^{n_k}(\zeta)\}$ that converges on compact subsets of D_0.

Since $\varepsilon > 0$ is arbitrary, we see just as in the case of the mixed boundary-value problem with free boundary (§4 in Chapter III) that the limit function $z = \Omega^0(\zeta)$ so constructed satisfies the necessary boundary conditions on the boundary of the upper half-disk D_ζ.

Let us now drop the assumption made above about the flow being confined in a neighborhood of a vertex at infinity of the given curve L_z^1 and thereby of the polygonal lines P^n. Suppose first that L_z^1 contains a vertex b_0 at infinity with the flow semiconfined in a neighborhood of it. We then consider an auxiliary problem, confining the flow by a straight line parallel to the half-lines which give the vertex b_0 at infinity. Accordingly we get two new vertices $\tilde{b}_1$ and $\tilde{b}_2$ with a flow that is confined in a neighborhood of them and has fixed depths $\tilde{h}_i \neq 0, \infty$, $i = 1, 2$. As above, we inscribe in the curvilinear part of the given boundary L_z^1 a polygonal line $\tilde{P}^n$, and as the number of its links is increased we increase also the depths $\tilde{h}_i^n$ in a neighborhood of the vertices $\tilde{b}_i^n$ ($i = 1, 2$) at infinity in such a way that $\tilde{h}_i^n \to \infty$ as $n \to \infty$. Theorem 1 tells us that each of the resulting auxiliary problems for the polygonal lines $\tilde{P}^n$ is solvable, and for each n

$$-1 < \cdots < t_{k_1}^n < \tau_1^n < \tau_2^n < t_{k_2}^n < \cdots < 1, \tag{25}$$

where the $\tilde{\tau}_i^n$ are the images of the vertices $\tilde{b}_i^n$ ($i = 1, 2$), and $\tilde{t}_{k_1}^n$ and $\tilde{t}_{k_2}^n$ are the neighboring images of the finite vertices of $\tilde{P}^n$.

The equations of (12) corresponding to the vertices $\tilde{b}_i^n$ give us that

$$\tilde{h}_i^n = \pi \left| (\zeta - \tau_i^n) \frac{d\tilde{\Omega}^n}{d\zeta} \right|_{\zeta = \tilde{\tau}_i^n}, \qquad i = 1, 2.$$

Since the sequences h_i^n diverge, it no longer follows that the points $\tilde{\tau}_i^n$ do not converge as $n \to \infty$, unlike the case $h_i \neq 0, \infty$ studied earlier.

As above, all the remaining inequalities (20) (respectively, (25)) are preserved also as $n \to \infty$. Consequently, repeating the previous arguments, we arrive at the solvability of the original problem for a curve L_z^1 with a vertex b_0 at infinity in a neighborhood of which the flow is semiconfined. The case when the flow is unconfined in a neighborhood of the vertex b_0 at infinity is handled completely analogously, except that now the flow is confined with the help of a channel that contains the curve L_z^1 and has both walls going to infinity.

Thus, we have proved the following theorem.

THEOREM 2. *In a domain D_z bounded by a simple curve L_z^1 with n vertices b_j at infinity and with free boundary L_z^2 there exists at least one flow whose complex potential satisfies the boundary conditions* (1) *and* (2). *Moreover, on L_z and in D_z it is possible to locate a finite number of singular points of the flow with unknown coordinates and physical characteristics.*

REMARK. Cases when the given magnitude $|\,dw/dz\,| = q(s)$ of the velocity on the free boundary vanishes (stagnation points), is infinite (sources, sinks, vortices), or has a finite discontinuity (flows of Thullen-scheme type) do not differ in any way from the cases analyzed above if the images of such singular points on the plane of the parametric variable are assumed to be given.

However, it should be mentioned that the presence of singular points on the free boundary of the flow domain D_z leads to the failure of the mapping $z = \Omega(\zeta)$ to be conformal at these points. This is really the only difference from the case when the singularities are located in the flow domain or on its given boundaries.

6°. Time-dependent flows. The Wagner-Kármán method ([1], §14) enables us to extend the existence theorems proved to a class of time-dependent motions with free boundary of finite length.

Suppose that in a time-dependent potential flow of an incompressible fluid the free boundaries remain streamlines, and the complex flow potential has the form $W(z, t) = Q(t)w(z)$, where $w(z)$ is an analytic function in the flow domain, and $Q(t) = |dW/dz|_{z=\infty}$.

Then the Cauchy-Lagrange integral on the free boundary takes the form

$$\frac{\partial \Phi}{\partial t} + \frac{1}{2}\left|\frac{dW}{dz}\right|^2 \equiv Q_t\varphi + \frac{Q^2}{2}\left|\frac{dW}{dz}\right|^2 = c(t).$$

Setting $dQ/dt = -kQ^2$ in this equality, i.e., $Q = 1/kt$, we get

$$q^2 = 2k\varphi + c_0 \qquad (q = |dW/dz|),$$

from which, after differentiation with respect to s, we finally find, using the identity $d\varphi/ds = q$, that $q' = k$, or $q = ks + c_1$ ($c_1 = $ const).

Thus, under our assumptions, time-dependent problems on potential flows of a fluid with free boundaries reduce to the steady-state problems studied in this section.

7°. The magnitude of the velocity on the free boundary—a function of one of the coordinates. Let $q(x, x_0)$ be a family of positive continuous functions with $q(x, x_0) = q(x_0, x_0)$ for $x \geqslant x_0$, where x_0 is an unknown parameter, and with $|\ln q(x, x_0)| \leqslant N < \infty$.

We consider the problem stated in 1°, replacing (2) by the following condition:

$$|dw/dz| = q(x, x_0) \quad \text{on } L_z^2. \tag{26}$$

We assume that to the point $z_0 \in L_z^2$ with abscissa $x_0 = x(s_0)$ there corresponds a fixed value $s_0 > 0$ of the arclength parameter, measured from one of the points common to the unknown arc L_z^2 and the given arc L_z^1. If L_z^2 is sought to be finite, then z_0 coincides with the other common point of L_z^2 and L_z^1, and the curve L_z^1 is given, for example, to within a dilation (see the end of 4°).

Let us substitute in the right-hand side of (26) an arbitrary continuous function $x = X(s)$, $s \in [0, s_0]$, satisfying the inequality

$$|X(s)| \leqslant s_0, \qquad 0 \leqslant s \leqslant s_0. \tag{27}$$

The problem so obtained is solvable, by Theorems 1 and 2. Let

$$z = z(X | \zeta), \qquad w = w(X | \zeta), \qquad \zeta \in D_\zeta\{|\zeta| < 1, \operatorname{Im}\zeta > 0\}$$

be a solution of it. Here the notation $F(X \mid \zeta)$ reflects the fact that the values of the operator $F(X)$ depend on the variable ζ. We get that

$$\theta(X \mid s) = \arg(i\zeta \, dz/d\zeta) \quad \text{for } \zeta = \exp\{i\gamma(X \mid s)\},$$

where $\gamma = \gamma(X \mid s)$ is the correspondence determined from (5) between boundary points under the mapping $z = z(X \mid \zeta)$. Let us consider the following transformation:

$$X = \int_0^s \cos \theta(X \mid s)\, ds \equiv F(X \mid s)\, ds, \qquad s \in [0, s_0]. \tag{28}$$

By construction, the continuous transformation F maps the ball $\|X\| \leqslant s_0$ of the Banach space $C[0, s_0]$ into itself, and $F(X \mid s) \in C^1$, i.e., the transformation is compact. Consequently, by Schauder's theorem (§1 in Chapter II), (28) has at least one solution $x = X(s)$, and therefore so does the original problem.

§3. Application of the general existence theorem in problems of flow with jet separation past obstacles

In this section we give a brief physical description, a mathematical formulation, and solution formulas (in the case of polygonal lines) for a number of problems that involve flow with jet separation past obstacles and whose solvability follows from the general existence theroems (Theorems 1 and 2) in the preceding section.

In these problems it is required to find an analytic function $w(z) = \varphi(x, y) + i\psi(x, y)$ (the complex flow potential) and along with it all the hydrodynamic characteristics of the flow in a domain D_z bounded by streamlines $\psi = \psi_k$. Moreover, the boundary $L_z = L_z^1 + L_z^2$ of the flow domain includes an unknown free boundary L_z^2 (the jets) characterized by the constancy of the flow rate on it, $|dw/dz| = q_0 = \text{const}$ ($|dw/dz|$ is a piecewise constant function in the Thullen scheme), and a given simple rectifiable curve L_z^1 which for convenience is assumed to be a simple polygonal line (by Theorem 2 in §2, this is not a restriction). The mathematically distinct flow schemes differ from one another in the geometric structure of the given parts of the boundary L_z, in the behavior of $w = w(z)$ in a neighborhood of the point at infinity of the free boundary L_z^2 (if there is such a point), and, finally, in the number and location of the stagnation points a_i of the flow $((dw/dz)(a_i) = 0)$.

Let us map the half-disk D_ζ: $\{|\zeta| < 1, \text{Im } \zeta > 0\}$ conformally onto a domain D_w in the plane of the complex potential $w = \varphi + i\psi$, which is fixed in concrete problems, in such a way that the image of the free boundary is carried into a semicircle.

We represent the derivatives $dw/d\zeta$ of the mapping function in the form

$$dw/d\zeta = K(1 - \zeta^2)Q(\zeta)f_0(\zeta), \tag{1}$$

where K is a real constant, $f_0(e^{i\gamma}) \neq 0, \infty$, and

$$Q(\zeta) = \frac{(1 - \zeta)^{-2\delta_0}(1 + \zeta)^{-2\delta_{n+1}}(1 + \zeta^2)^{-\delta_*}}{\prod[(\zeta - \tau_j)(1 - \zeta\tau_j)]^{\delta_j}}.$$

Here $\zeta = i$ is the image of the point $z_* \in L_z^2$ at infinity, the τ_j are the images of the vertices b_j ($j = 1, \ldots, m$) at infinity of the given curve L_z^1, and the integers δ are 0 when the boundary $L_z = L_z^1 + L_z^2$ is finite at the corresponding points z_*, b_j, or z_0, z_{n+1} (the endpoints of L_z^1); otherwise, $\delta = 1, 2, 3$, respectively, for a confined, semiconfined, or unconfined flow in a neighborhood of these points. Then, by (2.8) and (2.10), the derivative of the mapping of the upper half-disk D_ζ onto the flow domain D_z has the form

$$\frac{dz}{d\zeta} = K_0(1 - \zeta^2)Q(\zeta) \prod_{k=1}^{n} \left(\frac{\zeta - t_k}{1 - \zeta t_k}\right)^{\alpha_k - 1} e^{M(\zeta)}, \tag{2}$$

where $K_0 = \frac{1}{2}K\prod \tau_j^{\delta_j} e^{i\pi\beta_n}$ ($\beta_n\pi$ is the slope angle of the link $z_n z_{n+1}$), the t_k ($k = 1, \ldots, n$) are the images of the finite vertices z_k of the given polygonal line L_z^1, and the $\alpha_k\pi$ are the interior angles at these vertices.

The function $M(\zeta)$ is a solution of the boundary-value problem

$$\operatorname{Im} M(t) = 0, \qquad t \in [-1, 1],$$

$$\operatorname{Re} M(t) = \ln\left(|f_0(e^{i\gamma})|\left|\frac{dz}{dw}\right|\right) = \mu(e^{i\gamma}), \qquad \gamma \in [0, \pi], \tag{3}$$

where $\mu(e^{i\gamma})$ is a known function.

1°. *The Kirchhoff scheme.* The flow scheme of Kirchhoff was studied in detail in §1 by the continuity method; therefore, we shall not dwell on a description of it. We remark only that Theorems 1 and 2 immediately yield an existence theorem for a Kirchhoff flow near given obstacles without additional restrictions on the value of the rotation of their tangent.

In formula (2), $Q(\zeta) = (\zeta^2 + 1)^{-3}$ and $M(\zeta) = 0$ for the Kirchhoff scheme.

2°. *The Thullen scheme (with two spiral vortices).* In this scheme the free streamlines AD_1 and BD_2 bounding a cavity behind the obstacle are closed by two spiral vortices and then pass into the wake D_1C and D_2C (Figure 1).

At the points D_1 and D_2 the flow rate undergoes a jump from the rate q_0 on the boundary of the cavity to the rate q_∞ of the unperturbed flow. According to Thullen, this discontinuity in the rate can be explained by a loss of pressure arising from turbulence in the zone where the cavity collapses.

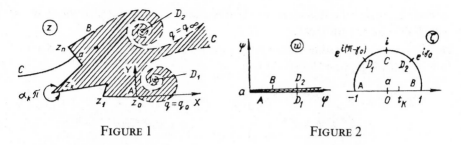

FIGURE 1   FIGURE 2

As in the Kirchhoff scheme, the domain in which the complex flow potential varies is the whole plane with a cut along the positive real axis.

The derivative of the mapping of the upper half-disk D_ζ onto the domain in the plane of the complex potential has the form

$$\frac{dw}{d\zeta} = \frac{K\zeta(1 - \zeta^2)}{(1 + \zeta^2)^3}.$$

In this scheme the function $M(\zeta)$ satisfies the boundary value problem (3), where $\mu(e^{i\gamma}) = \ln q_\infty^{-1}$, $\gamma \in [\gamma_0, \pi - \gamma_0]$, and $\mu(e^{i\gamma}) = \ln q_0^{-1}$, $\gamma \in \{[0, \gamma_0] + [\pi - \gamma_0\pi]\}$, $\zeta = e^{i\gamma}$ and $e^{i(\pi - \gamma_0)}$ are the images of the junction points D_1 and D_2 of the boundary and the wake (see Figures 1 and 2), and $\gamma_0 \in (0, \pi/2)$ is a fixed constant.

Recovering the solution of the last problem by (2.10), we get

$$M(\zeta) = \frac{1}{\pi i}\left(\ln \frac{q_\infty}{q_0}\right) \ln \frac{e^{2i\gamma_0} - \zeta^2}{1 - e^{2i\gamma_0}\zeta^2} - \ln q_\infty.$$

Thus, for the Thullen scheme (2) has the form

$$\frac{dz}{d\zeta} = K_0 \frac{1 - \zeta^2}{(1 + \zeta^2)^3} \prod_{k=1}^n \left(\frac{\zeta - t_k}{1 - \zeta t_k}\right)^{\alpha_k - 1} \left(\frac{e^{2i\gamma_0} - \zeta^2}{1 - e^{2i\gamma_0}\zeta^2}\right)^{(1/\pi i)\ln(q_\infty/q_0)}$$

As in the case of the Kirchhoff scheme, we arrive at the following system of equations for determining the constants K_0 and t_k $(k = 1,\ldots,n)$:

$$l_k = \int_{t_k}^{t_{k+1}} \left|\frac{dz}{dt}\right| dt \qquad (k = 0,\ldots,n);$$

by Theorem 1 in §2, it is solvable. Theorem 2 in §2 also gives the existence of a Thullen flow near curvilinear rectifiable obstacles (see the remark after Theorem 2).

3°. *The symmetric scheme of Éfros.* Jets of the fluid meeting behind the obstacle form a reentrant jet (a physically observable phenomenon), which is regarded as being located on the second sheet of the Riemann surface.

Moreover, inside the flow domain there is a stagnation point a of the flow, at which the velocity of flow is zero.

In view of the assumed symmetry of the flow, it suffices to consider the upper half of the flow domain (Figure 3).

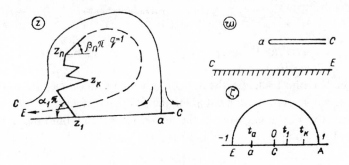

FIGURE 3

Correspondingly, formulas (1) and (2) for this scheme take the form

$$\frac{dw}{dz} = K\frac{(\zeta - t_a)(1 - t_a\zeta)(1 - \zeta)}{\zeta^2(1 + \zeta)},$$

$$\frac{dz}{d\zeta} = K_0\frac{(1 - t_a\zeta)^2(1 - \zeta)}{\zeta^2(1 + \zeta)}\Pi(\zeta),$$

where t_a is the fixed image of the stagnation point of the flow, and

$$\Pi(\zeta) = \prod_{k=1}^{n}\left(\frac{\zeta - t_k}{1 - \zeta t_k}\right)^{\alpha_k - 1}.$$

4°. The Riabouchinsky scheme (the "mirror" scheme). In this scheme the flow has a double symmetry achieved by locating a mirror image of the obstacle at a given distance behind the symmetric obstacle the fluid flows past (Figure 4).

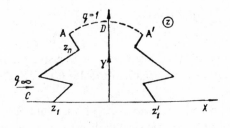

FIGURE 4

The derivatives of the mappings of the upper half-disk D_ζ onto the upper half-plane of $w = \varphi + i\psi$ and onto the flow domain D_z have the form

$$\frac{dw}{d\zeta} = K\frac{1 - \zeta^2}{\zeta^2}, \qquad \frac{dz}{d\zeta} = K_0\frac{1 - \zeta^2}{\zeta^2} \prod_{k=1}^{n} \left(\frac{\zeta^2 - t_k^2}{1 - t_k^2\zeta^2} \right)^{\alpha_k - 1}.$$

5°. The modified Lavrent'ev scheme. There is a flow past a system of two given bodies T and T' according to the "mirror" scheme. Inside the core formed by the boundaries of these bodies and by the streamlines connecting them the fluid finds itself in a rotational motion, and pockets of constant pressure are formed around a system of two other obstacles. Figure 5 represents the upper half of the flow, which is symmetric with respect to the coordinate axes; to obtain a simply connected domain D_z we have made an additional cut.

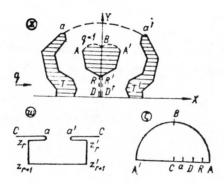

FIGURE 5

The derivatives of the mappings of the upper half-disk onto the domain in the plane of the complex potential $w = \varphi + i\psi$ and onto the flow domain D_z have the form

$$\frac{dw}{d\zeta} = K\frac{(1 - \zeta^2)(\zeta^2 - t_a^2)(1 - \zeta^2 t_a^2)}{\zeta^2} \prod_{i=r}^{r+1} \left[(\zeta^2 - t_i^2)(1 - \zeta^2 t_i^2) \right]^{-1/2},$$

$$\frac{dw}{d\zeta} = K_0\frac{1 - \zeta^2}{\zeta^2} \prod_{k=1}^{n} \left(\frac{\zeta^2 - t_k^2}{1 - \zeta^2 t_k^2} \right)^{\alpha_k - 1} e^{M(\zeta)},$$

where t_a is a fixed parameter, and

$$e^{M(\zeta)} = (1 - \zeta^2 t_a^2) \prod_{i=r}^{r+1} (1 - \zeta^2 t_i^2)^{-1}.$$

The parameter t_a can also be regarded as unknown if the stagnation point a coincides with one of the vertices z_j on T (here $t_a = t_j$).

Let us consider some special cases of the scheme. First of all, we remark that the thickness of T and T' may, in particular, be zero. Moreover, when the length $l = \sum_2^n l_k$ of the obstacle $ARR'A'$ is equal to zero, we obtain the problem of flow past a bubble with a constant rate on its boundary (Figure 6a). In this case it is necessary to set the corresponding t_k $(k = r + 1,\ldots,n)$ equal to 1 in the formulas for $dw/d\zeta$ and $dz/d\zeta$.

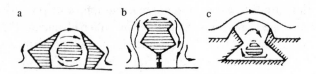

a   b   c

FIGURE 6

If it is assumed in addition that the bodies T and T' are absent (that they degenerate into rectilinear segments of the OX-axis), then we obtain a symmetric flow past a system of two obstacles with pockets of constant pressure (see Figure 6b).

Also of great interest is the particular case of the general scheme pictured in Figure 6c. This case models the vortical flow of a fluid over an uneven bottom surface with a depression.

6°. Flow of a fluid out of a nozzle. For a given geometry of a nozzle (Figure 7) it is required to find the form of the free surface and the physical characteristics of the flow when the rate on the free surface is constant.

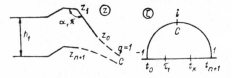

FIGURE 7

In this case (1) and (2) take the form

$$\frac{dw}{d\zeta} = K\frac{1 - \zeta^2}{(1 + \zeta^2)(\zeta - \tau_1)(1 - \zeta\tau_1)},$$

$$\frac{dz}{d\zeta} = K_0\frac{1 - \zeta^2}{(1 + \zeta^2)(\zeta - \tau_1)(1 - \tau_1\zeta)}\prod_{k=1}^{n}\left(\frac{\zeta - t_k}{1 - t_k\zeta}\right)^{\alpha_k - 1}.$$

In particular, if the walls of the nozzle are symmetric, then consideration of half the flow domain leads to the problem of fluid flow out from under a sluice gate onto an impermeable plane.

7°. *Flow in channels with partially unknown walls.* Let us consider the flow of a fluid in a channel (see Figure 8) with flow rate constant on an unknown part of the boundary: $|dw/dz| = q = 1$. Under various particular assumptions this flow scheme can yield several of the schemes already studied, along with some new ones (for example, the so-called Joukowsky-Rozhko scheme).

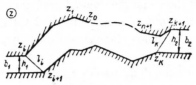

FIGURE 8

For this scheme (1) and (2) have the form

$$\frac{dw}{d\zeta} = K\frac{1 - \zeta^2}{\Pi_{j=1}^2(\zeta - \tau_j)(1 - \zeta\tau_j)},$$

$$\frac{dz}{d\zeta} = K_0\frac{1 - \zeta^2}{\Pi_{j=1}^2(\zeta - \tau_j)(1 - \zeta\tau_j)} \prod_{k=1}^n \left(\frac{\zeta - t_k}{1 - \zeta t_k}\right)^{\alpha_k - 1}.$$

In the case when z_{n+1} and z_0 are finite it is necessary to fix one of the constants $|K_0|$, t_k ($k = 1,\ldots,n$), or τ_j ($j = 1, 2$) (it is more convenient to fix $|K_0|$), dropping one of the equations in (2.12) (for example, the l_i or l_k can be regarded as unknown parameters; see Figure 6). This model scheme yields also the cases of a symmetric flow past symmetric obstacles in a free jet and in a jet flowing out of a symmetric channel if we restrict ourselves to one half of the flow domain.

REMARK. According to Theorems 1 and 2 in §2, in all the schemes considered here we can locate a finite number of singular points of the flow in D_z and on its boundaries, with the arrangement symmetric for a symmetric flow. Moreover, it is possible to specify the rate on the free boundaries as a function of the arclength parameter s, $q = q(s)$, and not to assume, as was done above, that it is constant.

§4. Application of the method of finite-dimensional approximation in problems on potential flows of a fluid in a gravitational field

1°. For approximating the nonlinear boundary-value problems of hydrodynamics studied so far by a sequence of linear problems with additional equations in the parameters, it has been sufficient to approximate the given curvilinear boundaries by polygonal lines. However, this becomes insufficient if the boundary condition on the free boundary is also essentially nonlinear, as is the case, for example, in problems involving flow of a heavy fluid. In this section we present for such problems a method of simultaneous approximation of the curvilinear boundaries by polygonal lines with a linearization of the boundary condition on the free boundary such that the original boundary condition is satisfied at only finitely many points of this boundary. After proving that the auxiliary problems thereby obtained are solvable, we then pass to the limit and show that the original problem is also solvable. It should, however, be noted that not every such method for linearizing the boundary condition on the free boundary leads to solvable auxiliary problems, as it may seem.

2°. Suppose that a heavy fluid flows out from under a polygonal sluice gate $(z_n, \ldots, z_0)$ with given lengths $l_i = |z_{i+1} - z_i|$ $(i = 0, \ldots, n - 1)$ of the sides and with vertex angles $\alpha_i \pi$ $(i = 1, \ldots, n)$ (Figure 1; the force of gravity is directed downwards along the OY-axis) and onto an impermeable plane COB. On the free boundary AB we have the Bernoulli equation

$$q^2 + 2gy = q_\infty^2 + 2gh, \tag{1}$$

where $q = |dw/dz|$ is the magnitude of the flow velocity, g is the acceleration due to gravity, q_∞ and h, respectively, are the magnitude of the velocity and the depth of the flow downstream at infinity, and $w = \varphi + i\psi$ is the complex flow potential. On the impermeable plane we set $\psi = 0$, and on the sluice gate and the free boundary $\psi = Q$ (Q is the amount of outflow). It is required to find a complex flow potential $w = \varphi + i\psi$ in the domain D_z bounded by the given sluice gate $(z_n, \ldots, z_0)$, the impermeable surface COB, and the free boundary AB that satisfies the boundary condition (1) on AB. It will be assumed that we know the dimensionless parameters $a_0 = q_0/q_\infty$ (the ratio of the magnitudes of the velocities at the separation point of the flow and downstream at infinity) and $\mu = h/y_0$ (the ratio of the depths of the flow at the points B of the free boundary ($z = \infty$) and the separation point A ($z = z_0$) of the flow). The so-called Froude number is uniquely determined in terms of the parameters a_0 and μ:

$$\mathrm{Fr} = \frac{q_\infty^2}{gy_0} = \frac{2(1 - \mu)}{1 - a_0^2}.$$

The magnitude q_0 and q_∞ of the velocities, the depths y_0 and h, and the amount of outflow Q are unknown and are to be found along with the complex potential $w = \varphi + i\psi$.

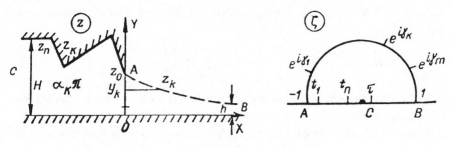

FIGURE 1

It is assumed that the given parameters $\mu = h/y_0$ and $a_0 = q_0/q_\infty$ satisfy the following inequalities:

$$0 < \mu = h/y_0 < 1, \qquad e^{-2\pi} < a_0 = q_0/q_\infty < 1,$$

$$\frac{2(1 - \mu)}{1 - a_0^2} = \frac{q_\infty^2}{qy_0} < \frac{2}{1 + \mu}. \tag{2}$$

The derivatives of the mappings of the upper half-disk D_ζ onto a strip of width Q in the plane of $w = \varphi + i\psi$ and onto the flow domain D_z have the respective forms

$$\frac{dw}{d\zeta} = K\frac{1 + \zeta}{(1 - \zeta)(\zeta - \tau)(1 - \zeta\tau)},$$

$$\frac{dz}{d\zeta} = K_0 e^{M(\zeta)}\frac{1 + \zeta}{(1 - \zeta)(\zeta - \tau)(1 - \tau\zeta)}\prod_{k=1}^{n}\left(\frac{\zeta - t_k}{1 - \zeta t_k}\right)^{\alpha_k - 1}, \tag{3}$$

where

$$K = \frac{(1 - \tau)^2 Q}{\pi}, \qquad K_0 = \frac{K}{q_\infty} = \frac{(1 - \tau)^2 h}{\pi},$$

and $M(\zeta)$ satisfies (2.9) with $\mu(e^{i\gamma}) = -\ln(q_\infty^{-1}q[y(\gamma)])$.
  By (1),

$$[q_\infty^{-1}q(y)]^2 = 1 + 2gq_\infty^{-2}[h - y(\gamma)],$$

where $y = y(\gamma)$ is an unknown function. Taking this into account, we compute $M(\zeta)$ from (2.10*):

$$M(\zeta) = -\frac{1 - \zeta^2}{\pi}\int_0^\pi \ln\left[q_\infty^{-1}q(\gamma)\right]\frac{d\gamma}{1 - 2\zeta\cos\gamma + \zeta^2}. \tag{4}$$

The depth H of the flow in a neighborhood of the point C is determined from (2):

$$H = \pi \left| \frac{dz}{d\zeta} (\zeta - \tau) \right|_{\zeta = \tau} = \frac{\pi K_0}{(1 - \tau)^2} \prod \left| \frac{\tau - t_k}{1 - \tau_k} \right|^{\alpha_k - 1} e^{M(\tau)}.$$

We remark that $H - y_0$ is known, since the geometry of the polygonal line $(z_n, \ldots, z_0)$ is given.

3°. Let us fix the constant K_0 ($> 2 \,|\, H - y_0|$) and, taking into account that $\pi K_0 / (1 - \tau_0)^2 = h$, write the system for determining the unknown parameters t_k ($k = 1, \ldots, n$) and τ in the form

$$l_k = \int_{t_k}^{t_{k+1}} \left| \frac{dz}{dt} \right| dt \qquad (k = 0, \ldots, n - 1),$$

$$l_n = \frac{H}{h} = \prod_{k=1}^{n} \left| \frac{\tau - t_k}{1 - \tau t_k} \right|^{\alpha_k - 1} e^{M(\tau)}, \tag{5}$$

where the quantities l_k ($k = 0, \ldots, n - 1$) are given by the conditions of the problem, and

$$l_n = \mu^{-1} + \frac{(1 + \tau)^2 (H - y_0)}{\pi K_0}$$

is positive and bounded, by the choice of the constant K_0.

We break up the unknown interval $[h, y_0]$ ($\mu = h/y_0 < 1$) by the points

$$y^k = y_0 - k \frac{y_0 - h}{m + 1} \qquad (k = 0, \ldots, m + 1) \tag{6}$$

into the equal intervals $[y^k, y^{k+1}]$ ($y^0 = y_0$ is the height of the output aperture). We denote the images of the points $z^k = x^k + iy^k$ ($k = 0, \ldots, m$) on the free boundary by $\zeta_k = e^{i\gamma_k}$. The equalities (1) and (6) give us that

$$q_{k+1}^2 - q_k^2 = 2gh \frac{1 - \mu}{\mu(m + 1)} \qquad (k = 0, \ldots, m), \tag{7}$$

where $q_k = q(\gamma_k)$ and $q_{m+1} = q_\infty$. With the help of the relations $a_0 = q_0/q_\infty$ and $\mu = h/y_0 < 1$, we express the quantities q_k in terms of the unknown parameter h and the given quantities:

$$q_k^2 = h \frac{2g(1 - \mu)}{\mu} \left(\frac{a_0^2}{1 - a_0^2} + \frac{k}{m + 1} \right) \qquad (k = 0, \ldots, m + 1). \tag{8}$$

For $k = 0, \ldots, m$ let

$$\tilde{q}_k(\gamma) = \exp\left\{ p_\infty^{k+1} + \frac{\cos \gamma - \cos \gamma_{k+1}}{\cos \gamma_k - \cos \gamma_{k+1}} p_{k+1}^k \right\}, \qquad \gamma \in [\gamma_{k+1}, \gamma_k], \tag{9}$$

where $p_i^j = \ln(q_j/q_i)$ $(i, j = 0, \ldots, m + 1)$; the condition $\mu < 1$ and (8) give us that $p_{k+1}^k < 0$, and so

$$\frac{d\tilde{q}_k(\gamma)}{d\gamma} = \frac{\tilde{q}_k(\gamma)p_{k+1}^k \sin \gamma}{\cos \gamma_{k+1} - \cos \gamma_k} < 0.$$

Thus, for $\gamma \in [\gamma_{k+1}, \gamma_k]$ $(k = 0, \ldots, m)$ we have

$$0 < a_0 = \frac{q_0}{q_\infty} \leqslant \tilde{q}_k(\gamma) \leqslant 1, \qquad \left| \frac{d\tilde{q}_k(\gamma)}{d\gamma} \right| \leqslant \sup \frac{-1}{\cos \gamma_k - \cos \gamma_{k+1}}. \tag{10}$$

4°. Let us consider the auxiliary problem of fluid flow in the domain D_z bounded by $\tilde{L}_z = (ACOB + \widetilde{BA})$, with the magnitude $|(dw/dz)(\gamma)|$ of the flow velocity on the image of the free boundary BA given by

$$|(dw/dz)(\gamma)| = q_\infty \tilde{q}(\gamma), \qquad \gamma \in [0, \pi],$$

where $\tilde{q}(\gamma) = \tilde{q}_k(\gamma)$, $\gamma \in [\gamma_{k+1}, \gamma_k]$ $(k = 0, \ldots, m)$, is determined from (9). By construction, $\tilde{q}_k(\gamma_k) = \tilde{q}(\gamma_k) = q_k/q_\infty$; therefore, at the points where the free boundary BA of the main problem and the free boundary $\widetilde{BA}$ of the auxiliary problem intersect the curves $y = y^k$ $(k = 0, \ldots, m + 1)$, the magnitudes of the flow velocity of these problems coincide.

Consequently, the magnitude of the flow velocity of the auxiliary problem satisfies the Bernoulli equation (1) at a finite number of points of the free boundary, including the point at infinity downstream.

Formula (6) gives us that

$$\frac{\Delta y^k}{h} = \frac{y^k - y^{k+1}}{h} = \frac{1 - \mu}{\mu(m + 1)},$$

where $\mu = h/y_0$ is a given quantity. On the other hand, if we determine the quantities

$$l_k = \frac{\Delta y^k}{h} \left(= \frac{1 - \mu}{\mu(m + 1)} \right)$$

from (3), we arrive at the following relations:

$$l_k = \frac{(1 - \tau)^2}{\pi} \int_{\gamma_{k+1}}^{\gamma_k} \frac{(1 + \zeta) \sin \theta(\gamma) \, d\gamma}{|1 - \zeta| |\zeta - \tau|^2 \tilde{q}_k(\gamma)} \qquad (k = 0, \ldots, m - 1), \tag{11}$$

where

$$\theta(\gamma) = \arg \frac{d\tilde{z}}{d\gamma} = \arg \left(i\zeta \frac{d\tilde{z}}{d\zeta} \right),$$

and $d\tilde{z}/d\zeta$ is determined by (3) and (4), in which $q(\gamma) \equiv q_\infty \tilde{q}(\gamma)$.

Thus, we obtain the system (5), (11) (with a total of $m + n + 1$ equations) for determining the unknown parameters t_k $(k = 1,\ldots,n)$, τ, and γ_k $(k = 1,\ldots,m)$ in the auxiliary problem (dz/dt is replaced by $d\tilde{z}/dt$ in (5)).

LEMMA. *The following a priori estimates hold for each solution of* (5), (11) *corresponding to a nondegenerate polygonal line* $P_n \equiv z_0 z_1 \cdots z_n$:

$$\tau - t_k > \varepsilon, \qquad t_{k+1} - t_k > \varepsilon > 0 \qquad (k = 0,\ldots,n),$$

$$1 - \tau > \varepsilon, \qquad \gamma_i - \gamma_{i+1} > \varepsilon > 0 \qquad (i = 1,\ldots,m). \tag{12}$$

The impossibility of the convergence $\tau \to \pm 1$ can be proved completely analogously to the case of the lemma in §2, since, by (10), for $\gamma \in [0, \pi]$ we have

$$|e^{M(e^{i\gamma})}| = 1/\tilde{q}(\gamma) \leqslant q_\infty/q_0 < \infty.$$

Let us now prove that the γ_k cannot converge. We remark first of all that $|z_0 - z^k| < \infty$ for $k = 1,\ldots,m$. Indeed, since $|z_0|$ and the quantities $y^k = \operatorname{Im} z^k$ are bounded, the convergence $|z_0 - z^k| \to \infty$ is possible only if $\operatorname{Re} z^k = x^k \to \infty$. Then the depth of the flow at infinity downstream is equal to $h^* = y_0 - y^k \neq h$, which contradicts the last equation in (5). Suppose first that, beginning with some γ_p, $1 \leqslant p \leqslant m$, we have $\gamma_k \to 0$, $k \geqslant p$. Then

$$|z^p - z^{p-1}| = \frac{(1 - \tau)^2}{\pi}\left|\int_{\gamma_p}^{\gamma_{p-1}} \frac{|1 + \zeta| e^{i\theta(\gamma)} d\gamma}{|1 - \zeta||\zeta - \tau|^2 \tilde{q}_{p-1}(\gamma)}\right| = \left|\int_{\gamma_p}^{\gamma_{p-1}} \frac{f(\gamma) d\gamma}{\sin \gamma/2}\right|,$$

and, since $|f(\gamma)| > 0$ in a neighborhood of $\gamma = 0$, the integral in the last equality diverges as $\gamma_p \to 0$, i.e., $|z^p - z^{p+1}| \to \infty$ as $\gamma_p \to 0$.

Thus, we have proved that $|\gamma_k| > \varepsilon_0 > 0$ $(k = 0,\ldots,m)$. On the other hand, if $\gamma_k - \gamma_{k+1} \to 0$ for some k, then (11) gives us that

$$\frac{1 - \mu}{\mu(m + 1)} \equiv l_k = \frac{(1 - \tau)^2}{\pi}\int_{\gamma_{k+1}}^{\gamma_k} \frac{|1 + \zeta| \sin \theta(\gamma) d\gamma}{(1 - \zeta)(\zeta - \tau)^2 \tilde{q}_k(\gamma)}$$

$$\leqslant \frac{\gamma_k - \gamma_{k+1}}{a_0 \sin(\gamma_{k+1}/2)} \to 0,$$

which contradicts the condition $\mu < 1$. Consequently, there exists an $\varepsilon > 0$ such that $\gamma_{k+1} - \gamma_k > \varepsilon$ $(k = 0,\ldots,m)$. By (10), the last fact guarantees the estimates

$$c_0^{-1} \leqslant |e^{M(\zeta)}| \leqslant c_0 \quad \text{for } \zeta \in D_\zeta\{|\zeta| \leqslant 1, \operatorname{Im} \zeta \geqslant 0\},$$

which, as in the case of the lemma in §2, imply the remaining inequalities in (12). The lemma is proved.

5°. Let us consider the elementary auxiliary problem of a fluid flowing out of a channel of height $y_0 = H$ onto an impermeable surface COB, with the Bernoulli condition required only at the point z_0 where the jet flows out and at the point B at infinity downstream (Figure 2).

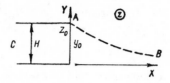

FIGURE 2

In this case (9) obviously takes the form

$$q^0(\gamma) = \exp\left\{\frac{1 - \cos\gamma}{2}\ln\frac{q_0}{q_\infty}\right\}, \tag{13}$$

and, consequently, $M(\zeta) = \frac{1}{2}(\zeta - 1)\ln(q_0/q_\infty)$.

The single equation

$$\frac{h}{H} = e^{-M(\tau)} = \exp\left\{\frac{1 - \tau}{2}\ln\frac{q_0}{q_\infty}\right\}$$

remaining from the system (5), (11) uniquely determines the value of the parameter τ:

$$\tau = 1 - \frac{2\ln\mu}{\ln a_0} \qquad \left(\mu = \frac{h}{H}, a_0 = \frac{q_0}{q_\infty}\right). \tag{14}$$

The last of the inequalities (2), which can be given the form $e^{-2\pi} < a_0 < \mu < 1$ or $|\ln\mu| < |\ln a_0| < 2\pi$, ensure that the value of τ thus found belongs to the necessary interval $(-1, 1)$ and that

$$|\operatorname{Im} M(e^{i\gamma})| = \frac{\sin\gamma}{2}\ln\frac{1}{a_0} < \pi, \qquad \gamma \in [0, \pi].$$

The known τ uniquely determines, in addition, the quantities

$$h = \frac{K_0\pi}{(1-\tau)^2}, \qquad H = \frac{h}{\mu} = \frac{K_0\pi}{\mu(1-\tau)^2},$$

$$q_\infty^2 = \frac{2gh(1-\mu)}{\mu(1-a_0^2)}, \qquad q_0^2 = a_0 q_\infty^2,$$

and, finally, the outflow $Q = q_\infty h$.

6°. Let us take the elementary problem we have investigated as an initial problem in the proof that the original problem is solvable for an arbitrary

polygonal sluice gate and with an arbitrary finite number of points on the free
boundary at which the Bernoulli equation is required to hold.

As usual, we introduce a parameter λ in (5) in such a way that the polygonal
sluice gate $Az_1 \cdots C$ passes into the rectilinear channel wall in the elementary
problem of 5° as λ varies from 1 to 0.

In addition, we introduce the parameter λ also in the boundary condition of
the problem by setting

$$q^\lambda(\gamma) = \lambda \tilde{q}(\gamma) + (1 - \lambda)q^0(\gamma), \qquad \gamma \in [0, \pi],$$

where $\tilde{q}(\gamma)$ and $q^0(\gamma)$ are determined by (9) and (13), respectively.

By construction, the derivative $dz^\lambda/d\zeta$ of the corresponding mapping at
$\lambda = 0$ does not depend on the constants t_k $(k = 1, \ldots, n)$ (the images of the
vertices of the polygonal sluice gate) nor on $e^{i\gamma_k}$ $(k = 1, \ldots, n)$ (the images of
the points $z^k = x^k + iy^k$ $(k = 1, \ldots, m)$ on the free boundary).

Consequently, the equation of the system (5), (12), with the function
$dz^0/d\zeta = dz^\lambda/d\zeta |_{\lambda=0}$ substituted in them, can be used to determine successively all the t_k $(k = 1, \ldots, n)$ and the γ_k $(k = 1, \ldots, m)$ (cf. the case in §2 of
Chapter III). The subsequent proof that the auxiliary problem in 4° is solvable
does not differ at all from the proof of Theorem 1 in §2.

Thus, we have the following theorem.

THEOREM 1. *There exists at least one solution of the auxiliary problem of a
fluid flowing out from under a given polygonal sluice gate (see 4°) such that the
Bernoulli equation* (1) *is satisifed at finitely many points, while on the remaining
part of the free boundary the reduced magnitude of the flow velocity* $\tilde{q} = q_\infty^{-1} dw/dz$
is determined by (9).

7°. According to (10), we have

$$|\ln \tilde{q}(\gamma)| \le \ln \frac{q_\infty}{q_0} = -\ln a_0 < \infty,$$

independently of the number of division points y^k on $[h, y_0]$. Extending, as
usual, the analytic function $M(\zeta)$ to the lower half-disk, we arrive at the
Dirichlet problem for the harmonic function

$$\operatorname{Re} M(\zeta) = \begin{cases} -\ln \tilde{q}(\gamma), & \gamma \in [0, \pi], \\ -\ln \tilde{q}(-\gamma), & \gamma \in [-\pi, 0], \end{cases}$$

which implies that it is bounded in the disk $|\zeta| < 1$.

Consequently, as the number of free boundary points at which Bernoulli's
equation (1) holds is increased without limit, we have that for $\zeta \in D_\zeta$

$$|e^{M(\zeta)}| = e^{\operatorname{Re} M(\zeta)} \le q_\infty/q_0 = 1/a_0 < \infty. \tag{15}$$

In the case of a curvilinear sluice gate we approximate it by polygonal ones and at the same time increase the number of free boundary points z^k at which Bernoulli's equation (1) holds; by (15), we arrive at the problem considered in the general Theorem 2 in §2. We remark also that the case of an impermeable polygonal surface COB only adds new factors in the formula (3) for the derivative $dz/d\zeta$ and does not differ in principle from the example of an impermeable plane which we have studied. Thus, we have proved the following theorem.

THEOREM 2. *There exists at least one flow of a heavy fluid in a domain bounded by given simple rectifiable curves (a sluice gate and an impermeable surface) and by a free boundary Γ that is the limit of Γ_n with (1) satisfied at n points. If $\Gamma \in C^{1+\alpha}$, then the condition (1) is valid everywhere on Γ.*

REMARK. If other parameters are given instead of the ratios $a_0 = q_0/q_\infty$ and $\mu = h/y_0$, then to get a uniquely solvable initial problem it is necessary to fix some additional relation between h and τ, and not simply the parameter K_0, as was done above. The relation $K_0 = f(\tau, h)$ must be chosen in such a way that the system of equations in h and τ

$$\mu = \frac{h}{H} = \exp\left\{ \frac{1-\tau}{2} \ln \frac{q_0(h)}{q_\infty(h)} \right\}, \qquad h = \frac{\pi f(\tau, h)}{(1-\tau)^2}$$

is uniquely solvable on the set $0 < h < \infty$ and $-1 < \tau < 1$.

In choosing such a relation we must, as a rule, impose restrictions on the given parameters of the problem, which, in the final analysis, leads to restrictions like the last of the inequalities (2) on the Froude number.

§5. Detached flows of Riabouchinsky type in a gravitational field with capillary forces taken into account

Joukowsky, McLeod, Beckert, and others (see, for example, [14], [75] and [42]) studied a number of free-boundary problems with surface tension in which the given parts of the boundary of the flow domain are assumed to be rectilinear.

In this section we consider plane irrotational flow of an ideal weightless incompressible fluid past a system of two curvilinear obstacles according to the Riabouchinsky scheme, with free surface acted on by surface tension forces. It is assumed that the flow has a double symmetry, and that at points where branching and coming together take place the obstacles have pointed edges, while at the remaining points they have continuous curvature.

An analogue of such a problem is studied also in the case of a heavy fluid.

1°. Suppose first that the fluid is weightless. Because of the symmetry, it suffices to consider a flow domain bounded by a streamline $CABD$ and an equipotential DC (see Figure 1).

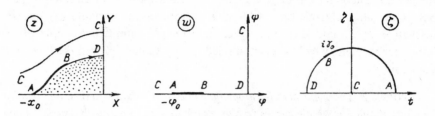

FIGURE 1

Let $\theta = f(s)$, $s \in [0, L]$, be the angle between the tangent to the given curve AB and the OX-axis; s is its arclength parameter, measured from the point $A(-x_0, 0)$, and $f(L) \geqslant 0$. By assumption, $f(s) \in C^1$ and $f(0) = 0$. We map the upper half-disk $|\zeta| < 1$, $\operatorname{Im} \zeta > 0$, conformally onto the quadrant $\varphi < 0$, $\psi > 0$ in the plane of the complex potential $w = \varphi + i\psi$ (the image of one-fourth of the flow domain) in such a way that the semicircle $\zeta = e^{i\gamma}$, $\gamma \in [0, \pi]$, corresponds to the given arc and the free boundary BD (see Figure 1):

$$w(\zeta) = -\frac{\varphi_0}{2}\left(\sqrt{\zeta} + \frac{1}{\sqrt{\zeta}}\right).$$

For determining the analytic function

$$\omega(\zeta) = \theta + i\tau = i \ln \frac{dw}{dz}$$

in the upper half-disk, where θ is the angle between the velocity vector and the OX-axis, $q = |dw/dx|$ is the magnitude of the velocity ($q(\infty) = 1$), and $\tau = \ln q$, we arrive at the following boundary-value problem:

$$\theta = \operatorname{Re}\omega = 0, \qquad \zeta = t, \qquad t \in [-1, +1], \qquad \tau(0) = 0; \tag{1}$$

$$\theta = \operatorname{Re}\omega = f(s(\gamma)), \qquad \zeta = e^{i\gamma}, \qquad \gamma \in [0, \gamma_0]; \tag{2}$$

$$\frac{d\theta}{d\gamma} = \alpha \sin\frac{\gamma}{2}\{e^{\tau(\gamma)} - k\varepsilon^{-\tau(\gamma)}\}, \qquad \zeta = e^{i\gamma}, \qquad \gamma \in [\gamma_0, \pi]. \tag{3}$$

The relation (3) is the Bernoulli equation. Here $\alpha = \varphi_0/4T$, $T > 0$, is the coefficient of surface tension for the fluid, and $k = 2(k_0 - p_0)$ (k_0 is the Bernoulli constant, and p_0 is the internal pressure of the cavity). The dependence $s = s(\gamma)$ is found from the formula

$$s(\gamma) = \frac{\varphi_0}{2}\int_0^\gamma e^{-\tau(\sigma)}\sin\frac{\sigma}{2}\,d\sigma, \tag{4}$$

since

$$\frac{ds}{d\gamma} = \left|\frac{dz}{d\gamma}\right| = \left|e^{i\omega(\zeta)}\frac{dw}{d\zeta}\right| = \frac{\varphi_0}{2}e^{-\tau(\gamma)}\sin\frac{\gamma}{2},$$

and $s(\gamma_0) = L$, wherefore

$$\frac{\varphi_0}{2}\int_0^{\gamma_0} e^{-\tau(\sigma)}\sin\frac{\sigma}{2}\,d\sigma = L. \tag{5}$$

The equality $\operatorname{Re} z(e^{i\gamma}) = 0$ for $\gamma = \pi$ and formula (5) give us that

$$\int_0^{\gamma_0} e^{-\tau(\sigma)}\sin\frac{\sigma}{2}\,d\sigma = \frac{L}{x_0}\int_0^{\pi} e^{-\tau(\sigma)}\cos\theta(\sigma)\sin\frac{\sigma}{2}\,d\sigma. \tag{6}$$

Suppose that the given curve AB satisfies the conditions

$$0 < \delta = \max_s |f(s)| \leq \delta_0 < \frac{\pi}{2}, \qquad \max_s \left|\frac{df}{ds}\right| = \frac{\kappa}{L}\delta < \infty, \tag{7}$$

where $\kappa > 1$ is a fixed number.

Assuming that $L/x_0 < 1$, we define the constants $\eta > 0$ and $\varepsilon > 0$ by the equalities

$$\eta = \frac{1}{2}\ln\left[\frac{x_0}{L}(1 - \sin\beta)\right], \qquad \varepsilon = \min\left\{\frac{L}{x_0}e^{-2\eta}\cos\delta_0,\ 1 - \cos\beta\right\}, \tag{8}$$

where $\beta > 0$ is a sufficiently small number.

In the Banach space $B = \{C \times E_1 \times E_1\}$ let us consider the closed bounded convex set $M = \{M_\tau \times M_\varphi \times M_\gamma\}$, where

$$M_\tau\left\{|\tau(\gamma)| \leq \eta,\ \int_0^\pi \tau(\gamma)\,d\gamma = 0\right\}, \qquad M_\varphi\{Le^{-\eta} \leq \varphi_0 \leq Le^\eta\varepsilon^{-1}\},$$

$$M_\gamma\{\varepsilon_0 \leq \gamma_0 \leq \pi - \varepsilon_0\}, \qquad \varepsilon_0 = 2\arccos(1 - \varepsilon). \tag{9}$$

For arbitrary $(\tau_0, \varphi_0, \gamma_0) \in M$ let

$$\theta_1(\gamma) = \begin{cases} f[\tilde{s}(\gamma)], & \gamma \in [0, \gamma_0], \\ f(L) + \alpha\int_{\gamma_0}^\gamma \sin\frac{\sigma}{2}[e^{\tau_0(\sigma)} - ke^{-\tau_0(\sigma)}]\,d\sigma, & \gamma \in [\gamma_0, \pi], \end{cases} \tag{10}$$

$$\tilde{s}(\gamma) = L\frac{\int_0^\gamma e^{-\tau_0(\sigma)}\sin(\sigma/2)\,d\sigma}{\int_0^{\gamma_0} e^{-\tau_0(\sigma)}\sin(\sigma/2)\,d\sigma}, \qquad k = \frac{f(L) + \alpha\int_{\gamma_0}^\pi e^{\tau_0(\sigma)}\sin(\sigma/2)\,d\sigma}{\alpha\int_{\gamma_0}^\pi e^{-\tau_0(\sigma)}\sin(\sigma/2)\,d\sigma}.$$

$$\tag{11}$$

The function $\theta_1(\gamma)$ is obviously continuous on the whole interval $[0, \pi]$ and continuously differentiable on the intervals $[0, \gamma_0]$ and $[\gamma_0, \pi]$, and $\theta_1(0) = \theta_1(\pi) = 0$. Extending $\theta_1(\gamma)$ antisymmetrically to the whole unit circle ($\theta_1(\gamma) = -\theta_1(2\pi - \gamma)$, $\gamma \in [\pi, 2\pi]$), we recover a function $\tau_1(\gamma)$ conjugate to it from Hilbert's formula:

$$\tau_1(\gamma) = -\frac{1}{2\pi} \int_0^\pi \left[\theta_1(\sigma) \cot \frac{\sigma - \gamma}{2} + \theta_1(\pi - \sigma) \tan \frac{\sigma - \gamma}{2} \right] d\sigma$$

$$\equiv F_\tau(\tau_0, \varphi_0, \gamma_0). \tag{12}$$

By the Fatou-Privalov theorem, $\tau_1(\gamma)$ is a Hölder function with any exponent λ, $0 < \lambda < 1$.

On the set M let us consider the transformation $F(\tau_0, \varphi_0, \gamma_0) = (\tau_1, \varphi_1, \gamma_1)$ whose components F_τ, F_φ, and F_γ are defined, respectively, by (12) and the following relations:

$$\varphi_1 = \frac{2L}{\int_0^{\gamma_0} e^{-\tau_0(\sigma)} \sin(\sigma/2) \, d\sigma} \equiv F_\varphi(\tau_0, \varphi_0, \gamma_0), \tag{13}$$

$$\int_0^{\gamma_1} e^{-\tau_0(\sigma)} \sin \frac{\sigma}{2} \, d\sigma = \frac{L}{x_0} \int_0^\pi e^{-\tau_0(\sigma)} \cos \theta_1(\sigma) \sin \frac{\sigma}{2} \, d\sigma. \tag{14}$$

It is easy to see that the $\gamma_1 = F_\gamma(\tau_0, \varphi_0, \gamma_0)$ in (14) is defined on M for $|\theta_1| \leq \pi/2$. Indeed, the function

$$\Phi(\gamma) = \frac{\int_0^\gamma e^{-\tau_0(\sigma)} \sin(\sigma/2) \, d\sigma}{\int_0^\pi e^{-\tau_0(\sigma)} \cos \theta_1(\sigma) \sin(\sigma/2) \, d\sigma}$$

is continuous and monotonically increasing, and $\Phi(0) = 0$ and $\Phi(\pi) \geq 1$. Consequently, there is a unique value γ_1 such that $\Phi(\gamma_1) = L/x_0 < 1$.

It is not hard to see that the solutions of problem (1)–(6) coincide with the fixed points of the transformation F; the existence of these points will be established with the aid of the Schauder theorem (§1 in Chapter II).

We make the necessary estimates with a view to verifying all the conditions of Schauder's theorem. First, (10) gives us the following inequalities:

$$\left| \frac{d\theta_1}{d\gamma} \right| = \left| \frac{df}{ds} \right| \left| \frac{d\tilde{s}}{d\gamma} \right| \leq \frac{\delta \kappa e^{2\eta}}{2\varepsilon}, \qquad \gamma \in [0, \gamma_0],$$

$$|\theta_1(\gamma)| \leq f(L) + \frac{L e^{2\eta}}{2T\varepsilon},$$

$$\left| \frac{d\theta_1}{d\gamma} \right| \leq \frac{f(L) e^{2\eta}}{2\varepsilon} + \frac{L e^{2\eta}}{4T\varepsilon}, \qquad \gamma \in [\gamma_0, \pi].$$

We assume that

$$L < 2\delta T e^{-2\eta}\varepsilon, \qquad f(L) \le \delta - \frac{Le^{2\eta}}{2T\varepsilon}. \tag{15}$$

Then

$$|\theta_1(\gamma)| \le \delta, \qquad \left|\frac{d\theta_1}{d\gamma}\right| \le \frac{e^{2\eta}}{2\varepsilon}\delta, \qquad \gamma \in [\gamma_0, \pi].$$

Since the estimate

$$|\tau_1(\gamma)| \le \max_{\gamma}|\theta_1| + 2\max_{\gamma}\left|\frac{d\theta_1}{d\gamma}\right|$$

holds for the conjugate function $\tau_1(\gamma)$, it is clear that the inequality $|\tau_1(\gamma)| \le \eta$ holds when

$$\delta \le \min\left\{\delta_0, \frac{\eta}{1 + \kappa\varepsilon^{-1}e^{2\eta}}\right\}. \tag{16}$$

From (13) we next find that $Le^{-\eta} \le \varphi_1 \le Le^{\eta}\varepsilon^{-1}$, and from (14)

$$1 - \frac{L}{x_0}e^{-2\eta} \le \cos\frac{\gamma_1}{2} \le 1 - \frac{L}{x_0}e^{-2\eta}\cos\delta_0,$$

whence

$$\varepsilon_0 \le \gamma_1 \le \pi - \varepsilon_0, \quad \text{i.e.} \quad (\tau_1, \varphi_1, \gamma_1) \in M.$$

Moreover, the transformation F is completely continuous. Thus, we have proved the following theorem.

THEOREM 1. *Under conditions (7), (15), and (16) there exists at least one solution of problem (1)–(3) in the set M.*

2°. Suppose now that the motion of the fluid takes place in a gravitational field perpendicular to the direction of the oncoming flow, with the obstacle AB abutting on a rigid rectilinear wall CA; consequently, as in the preceding problem, we can restrict ourselves to the flow in each quarter of the domain (see Figure 1). According as the direction of the gravitational field coincides with the direction of the OY-axis or with the opposite direction, the boundary condition for the function $\omega(e^{i\gamma})$ with $\gamma \in [\gamma_0, \pi]$ takes the respective forms

$$\frac{d\theta}{d\gamma} = \alpha\sin\frac{\gamma}{2}\left\{e^{\tau(\gamma)} - Ke^{-\tau(\gamma)} \pm g\varphi_0 e^{-\tau(\gamma)}\int_{\gamma_0}^{\gamma}e^{-\tau(\sigma)}\sin\theta(\sigma)\sin\frac{\sigma}{2}d\sigma\right\}$$

$$\equiv \Lambda(\theta, \tau). \tag{17}$$

Proceeding analogously to the case considered above, we define a transformation F on the set M. To find the function $\theta_1(\gamma)$, $\gamma \in [\gamma_0, \pi]$, we obtain the equation

$$\theta_1(\gamma) = f(L) + \int_{\gamma_0}^{\gamma} \Lambda(\theta_1, \tau_0 \,|\, \sigma)\, d\sigma, \tag{18}$$

and the constant K is determined from the condition $\theta_1(\pi) = 0$, i.e.,

$$K = K(\theta_1) = \frac{1}{P(\tau_0)} \left[f(L) + \int_{\gamma_0}^{\pi} \Lambda_0(\theta_1, \tau_0 \,|\, \sigma)\, d\sigma \right],$$

where

$$\Lambda_0(\theta_1, \tau_0 \,|\, \gamma) = \alpha \sin \frac{\gamma}{2} \left\{ e^{\tau_0(\gamma)} \pm g\varphi_0 e^{-\tau_0(\gamma)} \int_{\gamma_0}^{\gamma} e^{-\tau_0(\sigma)} \sin \theta_1(\sigma) \sin \frac{\sigma}{2}\, d\sigma \right\},$$

$$P(\tau_0) = \alpha \int_{\gamma_0}^{\pi} e^{-\tau_0(\sigma)} \sin \frac{\sigma}{2}\, d\sigma.$$

For arbitrary θ_1^i: $|\theta_1^i| \leqslant \delta$, $i = 1, 2$, we have

$$|K(\theta_1^1) - K(\theta_1^2)| \leqslant 2g\varphi_0 e^{\eta} |\theta_1^1 - \theta_1^2|,$$

$$\int_{\gamma_0}^{\gamma} |\Lambda(\theta_1^1, \tau_0 \,|\, \sigma) - \Lambda(\theta_1^2, \tau_0 \,|\, \sigma)|\, d\sigma \leqslant m|\theta_1^1 - \theta_1^2|, \qquad \gamma \in [\gamma_0, \pi],$$

where $m = 3gL^2 e^{4\eta}/2T\varepsilon^2$.

Assuming the relation

$$L^2 < \frac{2}{3} \frac{T\varepsilon^2}{g} e^{-4\eta}, \tag{19}$$

we get that $m < 1$; consequently, (18) is uniquely solvable for $\theta_1(\gamma)$, $\gamma \in [\gamma_0, \pi]$, by the Banach principle.

Moreover, the following estimates hold for $\theta_1(\gamma)$, $\gamma \in [\gamma_0, \pi]$:

$$|\theta_1(\gamma)| \leqslant f(L) + \frac{Le^{2\eta}}{2T\varepsilon} + \frac{gL^2 e^{4\eta}}{T\varepsilon^2}, \qquad \left| \frac{d\theta_1}{d\gamma} \right| \leqslant \frac{f(L)e^{2\eta}}{2\varepsilon} + \frac{Le^{4\eta}}{4T\varepsilon} + \frac{gL^2 e^{4\eta}}{2T\varepsilon^2}.$$

If the given quantities L, T, g, and $f(L)$ are such that

$$\frac{Le^{2\eta}}{2T\varepsilon} \left(1 + \frac{2gLe^{2\eta}}{e} \right) < \delta, \qquad f(L) \leqslant \delta - \left(\frac{Le^{2\eta}}{2T\varepsilon} + \frac{gL^2 e^{4\eta}}{T\varepsilon^2} \right), \tag{20}$$

then by choosing δ according to (16) we get the estimate $|\tau_1(\gamma)| \leqslant \eta$ for $\tau_1(\gamma)$. As before, the components F_φ and F_γ of F are defined by (13) and (14) and are continuous in all arguments. Then the Schauder fixed point principle gives us the solvability of the problem under investigation.

Let us formulate this assertion as a theorem.

THEOREM 2. *If the conditions* (16), (19), *and* (20) *on the given obstacle are satisfied, then there exists at least one solution of the problem* (1), (2), (4)–(6), (17) *in the set M.*

As an example we analyze the case when the equation has the form

$$f(s) = \delta\left(1 - \left|\cos\frac{\pi}{L}s\right|\right).$$

Then $f(L) = 0$, and the second parts of conditions (15) and (20) are automatically satisfied. The curvature df/ds, while remaining bounded, has a discontinuity at the point $s = L/2$, but this does not affect the validity of the assertions in Theorems 1 and 2.

Obviously, $|df/ds| \leqslant \delta\pi/L$, i.e., $\kappa = \pi$.

Suppose for example, that $L/x_0 = \frac{1}{4} < 1$ and $\eta = \ln\sqrt{2}$, while $\delta_0 = \pi/3$. It is easy to see that when $\delta \leqslant \ln\sqrt{2}/(1 + 32\pi)$ all the conditions of Theorems 1 and 2 hold; consequently, solutions of the problems considered exist, and $|\tau(\gamma)| \leqslant \ln\sqrt{2}$. The length of L must be sufficiently small; for example, if the fluid is weightless, then

$$L \leqslant \frac{\ln\sqrt{2}\,T}{16(1 + 32\pi)}.$$

§6. A brief survey and problems

The first global existence theorems for plane steady flows of an ideal incompressible fluid with free boundaries were obtained in 1927–1930 by Weinstein [134], who studied the problem of flow of a weightless fluid out of a curvilinear nozzle. Weinstein's method was illustrated in §1 on a problem involving a jet flow past obstacles. Approximating the curvilinear nozzle by polygonal ones, he reduced the problem to that of determining certain parameters (the images of fixed vertices of the polygonal line under a conformal mapping of the upper half-disk onto the flow domain) from a system of nonlinear equations analogous to the equations for determining the constants in the Schwarz-Christoffel formula for mapping polygons. The method of continuity used by Weinstein to prove that this system of equations can be solved for the parameters is based on a priori estimates of its solutions and on the local uniqueness of these solutions. By analyzing Weinstein's constructions, it is not hard to see that for the validity of the necessary a priori solution estimates it suffices to assume only that the given polygonal boundaries of the flow domain are nondegenerate (this fact holds also in more general problems with free boundaries). But to ensure local uniqueness of the solutions of the system for the parameters it is necessary to impose restrictive conditions on the value $\alpha(L_z^1)$ of the total rotation of the tangent to the given part of the boundary of the flow domain ($\alpha < \pi$). The last restriction has the compensation that the property of local uniqueness of the solutions, together with their global uniqueness for some fixed polygonal line, implies the global uniqueness of solutions also for an arbitrary polygonal line, by the continuity method. Weinstein's proof of solvability for the problem of a weightless fluid flowing out of a curvilinear nozzle is concluded with the aid of the easily established relative compactness of the family of solutions for polygonal lines inscribed in this nozzle.

In 1935 the topological Leray-Schauder principle (a generalization of the continuity method to Banach spaces) enabled Leray [70] to prove significantly more general existence and uniqueness theorems for solutions in the problem of flow with jet separation past curvilinear obstacles according to the Kirchhoff scheme. Leray's results were later improved and extended by Kravtchenko and other authors to jet flows in channels (see the survey in the monograph [19]).

In 1938 Lavrent'ev [63] proposed an original method for proving existence and uniqueness theorems in the problem of a flow with jet separation past symmetric obstacles, based on an idea of a completely different nature and not connected, as the Leray method is, with studying an integrodifferential equation of the problem like the Villat equation. This method uses variational principles developed by Lavrent'ev in the theory of conformal mappings, principles that characterize the behavior of the derivative of a mapping function under small changes in the boundary of the domain. In 1952 Serrin [119], using the methods of Lavrent'ev and Leray, extended the class of obstacles considered by Lavrent'ev and applied these methods to the case of a symmetric flow in a strip, and in 1953 he proved an existence theorem for a symmetric flow according to the Èfros scheme.

In 1963 and 1964 the author in [85]–[90] (see Chapter VII) used a special choice of independent variables to reduce a large class of free-boundary problems in hydrodynamics and subsonic gas dynamics to a mixed boundary-value problem for quasilinear systems of equations, and proved that it is uniquely solvable. Various restrictions have to be imposed on the value $\alpha(L_z^1)$ of the total rotation of the tangent to the given part L_z^1 of the boundary of the flow domain in order to apply the Weinstein continuity method, the topological method of Leray, the variational method of boundary-value problems, or the method proposed by the author to the free-boundary problems described above.

To prove the solvability of the general hydrodynamics problem with a single free boundary we proposed in §§2 and 3 of this chapter the so-called method of finite-dimensional approximation, which involves (as does the continuity method) approximating the given curvilinear boundaries by polygonal lines. This method is based solely on a priori estimates of the solutions of the corresponding systems of equations for the parameters and, unlike the continuity method, does not use local uniqueness of the solutions. Therefore, the existence theorems proved in §§2 and 3 require of the given boundaries only that they be rectifiable and not contain points of self-intersection, with the value $\alpha(L_x^1)$ of the total rotation of the tangent to these boundaries allowed to be arbitrary. If $\alpha(L_z^1) < \pi$, then in most of the problems treated here the Weinstein method can be used to prove local uniqueness of the solutions, and, by this and the established uniqueness of the solutions in the simplest problems, also global uniqueness of the solutions.

While presenting a course of lectures to students at Novosibirsk State University in 1967–68 the author used the method of finite-dimensional approximation to prove the existence of a jet flow according to the Kirchhoff scheme, and extended its application to problems with free boundaries in a gravitational field. In the latter case an auxiliary problem was first analyzed, in which the boundary condition on the free boundary (the Bernoulli condition) was satisfied only at a finite number of points (§4). In 1969, L. G. Guzevskiĭ and G. V. Lavrent'ev [43] (students at Novosibirsk University) used this method to prove an existence theorem for jet flows according to other flow schemes (§3), and the author established an existence theorem for the general hydrodynamics problem with a single free boundary (see [95], as well as §2).

§5 contains results of Kažihov [53], who investigated the solvability of the problem of a jet flow past a system of given curvilinear arcs according to a scheme of Riabouchinsky type with gravitational forces and surface tension forces on the free boundary taken into account. He proved an existence theorem for solutions of this problem with the help of the Schauder fixed point principle under the assumption that the problem parameters (the length of the arcs flowed past, the maximum of the tangent angle and the curvature for these arcs, and the coefficients of surface tension) satisfied a certain system of conditions. It should be pointed out that this is one of the first theorems in which account is taken of the curvilinearity of the given parts of the boundary in the case when capillary forces act on the free boundary.

To the readers interested in the history of the problem and in a bibliography on jet problems in hydrodynamics we recommend consulting the survey articles of Weinstein [134] and Keldysh and

Sedov [55], and the surveys in the monographs of Lavrent'ev [61], Birkhoff and Zarantonello [19], Gurevich [42], and others.

We now state some uninvestigated free-boundary problems in hydrodynamics to which (in our opinion) it is possible to apply the methods of the theory of boundary-value problems for analytic functions and the methods of finite-dimensional approximation (Chapters III and IV).

1°. **Uniqueness theorems.** a. In §1 of Chapter IV the continuity method was used to prove a uniqueness theorem for a Kirchhoff flow. Use this same method to prove a uniqueness theorem for a Thullen flow (§3.2°) and for a flow in channels with partially unknown wall forms (§3.7°).

b. In proving the uniqueness theorem for a Kirchhoff flow by the continuity method (§1 in Chapter IV) an additional condition on the quantity $\alpha(\Gamma)$ arises: the total rotation of the tangent to the given obstacle Γ satisfies $\alpha(\Gamma) < \pi$.

An analogous condition arises also in proving uniqueness theorems by the Lavrent'ev variational method [61]. Consider half of the domain of a symmetric Kirchhoff flow and clear up the question of the value $\alpha(\Gamma) = \alpha_0 \geqslant \pi$ at which branching of solutions in this problem begins (for $\alpha_0 < \pi$ there is a unique solution).

2°. **General existence theorems.** a. Extend Theorems 1 and 2 in §2 to flows with several free boundaries. Examples of such problems are problems of flow past obstacles according to the nonsymmetric schemes of Riabouchinsky (§3.4°), Lavrent'ev (§3.5°), and others.

b. Investigate the question of solvability of the equations for the physical parameters in the general problem of hydrodynamics with a single free boundary (§2). This means that it is necessary to find the images of the singular points of a flow on the plane of the parametric variable from given physical and geometric characteristics of the flow at these points, for example, from given coordinates on the plane of flow and the given intensities of the vortices, sources, and sinks.

c. Apply the methods of finite-dimensional approximation to axisymmetric flows.

d. Consider cases of time-dependent problems that reduce to nonlinear boundary-value problems for analytic functions. In this direction there is work of Kármán, Gilbarg, and others (see, for example, [20] and [19], as well as §2.6°).

3°. **Flows with free boundaries on Riemann surfaces.** The simplest examples of such flows are problems involving flow past obstacles according to schemes of Èfros type (§3.3°). Work has appeared recently in which more general flow schemes on Riemann surfaces are also considered.

On the other hand, there are many articles on the theory of linear boundary-value problems for analytic functions on Riemann surfaces (see the surveys in the monographs [32] and [97]). It is therefore natural to state the following two types of problems, which connect these investigations.

a. Consider nonlinear boundary-value problems on Riemann surfaces.

b. Investigate the solvability of the hydrodynamics free-boundary problems on Riemann surfaces that are analogous to the problems studied for the plane in Chapter IV.

This same circle of problems includes also the question of investigating the univalency of flows such that multivalence is not built into the flow scheme beforehand (the schemes of Kirchhoff, Thullen, Riabouchinsky, and others). Univalence of such flows is usually proved only in the trivial cases when we have the estimate $|\theta| \leqslant \pi/2 - \varepsilon$, $\varepsilon > 0$, for the angle $\theta = \arg(dw/dz)$ of slope of the flow velocity (incidentally, there is also a uniqueness theorem in this case, as will be proved in Chapter VII). There also exist several more subtle results in certain concrete problems ([126], p. 55).

4°. **Flows with free boundaries in multiply connected domains.** The solvability of boundary-value problems with free boundaries in multiply connected domains is closely related to the investigation of these problems on Riemann surfaces (the problems of the preceding subsection).

It is of great interest to extend the methods used in Chapters III and IV for problems in simply connected domains directly to problems in multiply connected domains. In this direction there are a number of references (for example, [19] and [42]) in which only problems that in some sense or another reduce to problems in simply connected domains are investigated.

In Chapter VII we also give examples in which problems in doubly connected domains are investigated without reducing them to problems in simply connected domains. However, so far no sufficiently general methods for investigating such problems have been discovered. In our view, there are no special hindrances to an extension of the methods of finite-dimensional approximation to these problems if it is considered that the theory of linear boundary-value problems for analytic functions in multiply connected domains has been built up to a sufficient degree.

5°. Problems of pasting flows together. An important class of little-investigated hydrodynamic problems consists of those involving fluid flows in which different properties of the medium predominate in different parts of the flow domain D. In these cases an account of the main properties of the medium for each of the subdomains D_i of D leads to different mathematical models of the flow in each of these subdomains. In the final analysis, to this correspond either different boundary conditions on the common boundaries of the domains D_i or different equations for the flow characteristics in these domains. We give some examples of such problems (certain problems of this type are taken up in Chapter IX).

a. The problem of motion of a hydrofoil vessel, in which a system of two geometrically connected bodies T_1 and T_2 moves in fluids of different density. A flow scheme for one possible formulation of such problems is given in Figure 1.

FIGURE 1

In this problem we must take into account the compressibility of the medium in the domain D_1 in which the body T_1 moves when the velocity is large; therefore, the equations for the complex flow potential $w = \varphi + i\psi$ have the form (see Chapter VII)

$$\rho(q)\varphi_x = \psi_y, \qquad -\rho(q)\varphi_y = \psi_x, \qquad q^2 = \varphi_x^2 + \varphi_y^2,$$

while the medium in the domain D_2 in which T_2 moves is incompressible, and $w(z) = \varphi + i\psi$ is an analytic function. Supposing that the pressure on the common boundary Γ of the two media is continuous and assuming a specified shape for the bodies T_1 and T_2 and a specified surface Π joining them (in reality this surface is a system of struts, which is simulated by the condition $\psi = \psi(x)$ of permeability of this surface), find the unknown common boundary Γ of D_1 and D_2 and the complex potential $w = w(z)$ in the domain $D = D_1 + D_2 + \Pi$. Even in the simplest case, when $\rho(q) = \text{const} = \rho_1 = \rho_2$, the problem reduces by means of the Bernoulli condition on Γ to a hitherto unfamiliar nonlinear boundary-value problem in the theory of analytic functions.

b. The problem of pasting together a potential flow of an ideal fluid and a flow of a fluid with constant vorticity. One such problem, posed by Lavrent'ev [61] and investigated by A. B. Shabat [113], Gol'dshtik [40], and Plotnikov [105], is the problem of pasting together a potential flow over an uneven bottom and a fluid flow of constant vorticity in a hole (an analogous scheme, except with a pocket of constant pressure, is pictured in Figure 6 in §3).

In this problem it is interesting to apply the method of finite-dimensional approximation, and to use this method to study the problem whose flow scheme is pictured in Figure 2.

The function $dw/dz = u - iv$, where u and v are the components of the complex flow velocity, is analytic in the flow domain D_1 of an ideal fluid, while in the flow domain D_2 of a fluid with constant vorticity $\omega = \text{const}$ the functions $u(x, y)$ and $v(x, y)$ satisfy the following system of equations:

$$\partial u/\partial x + \partial v/\partial y = 0, \qquad \partial u/\partial y - \partial v/\partial x = \omega, \qquad (x, y) \in D_2. \tag{1}$$

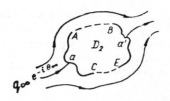

FIGURE 2

It is required to determine the flow on the whole plane in such a way that on the boundaries AB and CE of D_1 and D_2 the velocity components are continuous, assuming that the obstacles CaA and $Ba'E$ are impermeable. This problem can easily be reduced to that of finding the piecewise holomorphic function $\Omega(z) = \tilde{u} - i\tilde{v}$ on the whole plane whose components are determined by the equalities

$$\tilde{u}(x, y) = u(x, y), \quad \text{in } D_1 + D_2,$$

$$\tilde{v}(x, y) = \begin{cases} v(x, y), & (x, y) \in D_1, \\ v(x, y) - \omega x, & (x, y) \in D_2. \end{cases}$$

c. A very interesting problem is that of pasting together a flow of an ideal fluid and a model flow of a viscous fluid described by the boundary layer equations.

(2)   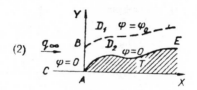

FIGURE 3

Figure 3 gives a flow scheme for a symmetric body T. In the domain D_1 of the ideal fluid flow the function $dw/dz = u - iv$ is analytic, while in the domain D_2, where the viscosity of the fluid is taken into account, the functions $u(x, y)$ and $v(x, y)$ satisfy the system of boundary layer equations

$$u\frac{\partial v}{\partial x'} + v\frac{\partial u}{\partial y'} = \nu\frac{\partial^2 u}{\partial y'^2} - \frac{dp(x)}{dx'}, \qquad \frac{\partial u'}{\partial x'} + \frac{\partial v}{\partial y'} = 0, \qquad (x, y) \in D_2, \qquad (2)$$

where x' and y' are curvilinear coordinates depending on the shape of the body, ν is the coefficient of viscosity, and $p = p(x)$ is that value of the pressure on the unknown streamline $\psi = \psi_0$ associated with the values of $u(x, y)$ and $v(x, y)$ on this line by the Bernoulli condition. It is necessary to find a flow such that its velocity components $u(x, y)$ and $v(x, y)$ are continuous in the whole flow domain $D_1 + D_2$ and the pressure $p = p(x, y)$ satisfies the following boundary conditions on the streamline CAE ($\psi = 0$):

$$u(x, y) = 0 \quad \text{on } CA,$$

$$u = 0, \quad v = v(x) \quad \text{on } AE,$$

where $v = v(x) \equiv 0$ in the absence of swelling or suction on the surface of T.

6°. *Consideration of surface tension forces.* We use the method of finite-dimensional approximation similarly to the case of a heavy fluid (§4). Prove an existence theorem for fluid motion problems with surface tension forces taken into account and without restrictions on the value of the total rotation of the tangent to the known boundaries.

7°. *Schemes of flow past obstacles with minimal wake.* This problem, proposed to the author by Lavrent'ev, led to the necessity of investigating free-boundary flows for which one possible formulation is illustrated in Figure 4.

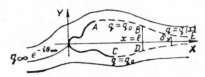

FIGURE 4

The flow scheme in this problem differs from the Kirchhoff scheme in that, beginning with some given value $x = l$ (the necessary size of the cavity), the magnitude $q = q(x) \in C^{1+\alpha}$ of the flow velocity must be determined by minimizing the breadth h of the wake BE, DE, i.e., from the condition

$$h = \inf_{q=q(x)} \left\{ \sup_{x \geqslant l} |\delta y(x)| \right\}.$$

Using the methods of finite-dimensional approximation and the variational methods of Lavrent'ev [61], investigate the solvability of such problems.

Part II
Generalized solutions
of quasilinear systems of equations
in the mechanics of continuous media

CHAPTER V

QUASICONFORMAL MAPPINGS
AND GENERALIZED SOLUTIONS
OF ELLIPTIC SYSTEMS OF EQUATIONS
ON THE PLANE

§1. Generalized derivatives. Sobolev spaces

Let D be a bounded domain in the n-dimensional Euclidean space E^n whose points x are determined by the coordinates $(x_1, \ldots, x_n)$. Denote by $\overset{\circ}{C}{}^l$ the set of compactly supported functions $\varphi(x)$ that are continuous in $\overline{D}$ together with all partial derivatives of order through l and that vanish outside some domain D_φ, $\overline{D}_\varphi \subset D$. Suppose that $f(x)$ has continuous derivatives of order through l inside D. Let us consider a derivative of order l

$$D^l f = \frac{\partial^l f}{\partial x_1^{l_1} \cdots \partial x_n^{l_n}}, \qquad \sum_{i=1}^n l_i = l. \tag{1}$$

Using integration by parts and the fact that $\varphi(x)$ is compactly supported, we get

$$\int_D D^l f(x) \varphi(x) dx = (-1)^l \int_D f(x) D^l[\varphi(x)] dx. \tag{2}$$

The formula (2) can be taken as the definition of the generalized derivative.

Suppose that $f(x)$ and $g(x)$ are summable in any domain D_1, $\overline{D}_1 \subset D$, and satisfy for any $\varphi(x) \in \overset{\circ}{C}{}^l(D)$ the relation

$$\int_D g(x) \varphi(x) dx = (-1)^{|l|} \int_D f(x) D^l \varphi(x) dx. \tag{3}$$

Then $g(x)$ is called the generalized derivative of the form (1) for $f(x)$ in D.

If $f(x) \in C^l(D)$, then the generalized derivative obviously coincides with the ordinary derivative. Generalized derivatives have many properties of ordinary derivatives: $D^1(f_1 + f_2) = D^lf_1 + D^lf_2$; if D^lf_1 exists in D, then it also exists in $D_1 \subset D$ and coincides with the derivative in D; and so on.

By introducing average functions [122] it can be shown that the given definition of generalized derivatives is equivalent to the following one: $g(x)$ is the generalized derivative of the form (1) for $f(x)$ in D if there exists a sequence of functions $f_n(x) \in C^l(D)$ such that $f_n(x)$ and $D^lf_n(x)$ converge in $L_1(D_1)$ to $f(x)$ and $g(x)$, respectively, where D_1 is any subdomain of D: $\overline{D}_1 \subset D$. For example, in the case $l = 2$ the last definition implies that if $f(x, y)$ has second generalized derivatives with respect to x and y that are pth power summable $p > 2$, then it is absolutely continuous in one of the variables for almost all values of the other one, and it has ordinary partial derivatives with respect to x and y everywhere in D. We say that $f(x)$ belongs to the Sobolev space $W_p^l(\overline{D})$ if in $\overline{D}$ it has all generalized partial derivatives D^kf, $|k| \leqslant l$, belonging to the space $L_p(\overline{D})$, $l \geqslant 1$, $p > 1$. This $W_p^l(D)$ is a B-space with the norm

$$\|f\|_{W_p^l} = \sum_{k=0}^{l} \sum_{k_1 + \cdots + k_n = k} \left\| \frac{\partial^k f(x)}{\partial x_1^{k_1} \cdots \partial x_n^{k_n}} \right\|_{L_p(D)}.$$

Naturally, two functions $f(x)$, $g(x) \in W_p^l(\overline{D})$ are regarded as equivalent if all their derivatives of order through l coincide almost everywhere ($l \geqslant 0$). In particular, for $l = 0$ two functions $f(x)$, $g(x) \in L_p(\overline{D})$ are equivalent if $\|f - g\|_{L_p} = 0$. Equivalent functions will not be distinguished in what follows.

A number of so-called imbedding theorems [122] are valid for functions of the classes $W_p^l(\overline{D})$, $p > 1$, $l \geqslant 1$; we give only two.

THEOREM 1 (SOBOLEV). *If all the generalized derivatives of $f(x)$ in $\overline{D}$ with order $l \geqslant 1$ exist and belong to $L_p(\overline{D})$ ($p > 1$) along with $f(x)$, then all the generalized derivatives of $f(x)$ with order $m < l$ also exist and belong to $L_p(\overline{D})$, i.e., $f(x) \in W_p^l(\overline{D})$.*

Examples show that the requirement that all the generalized lth derivatives exist is essential in Theorem 1. Indeed, suppose that $f(x, y) = u(x) + v(y)$, where $u(x)$ and $v(y)$ are continuous in D but not differentiable. Then it is not hard to show that the generalized derivative $\partial^2 f/\partial x \partial y$ exists and is 0 in D, but the generalized first derivatives of $f(x, y)$ do not exist, by construction.

THEOREM 2 (SOBOLEV). *If $p > 1$ and $pl > n$, then every function $f(x) \in W_p^l(\overline{D})$ is equivalent to a function in $C(\overline{D})$, and*

$$\|f\|_{C(\overline{D})} \leqslant M\|f\|_{W_p^l(\overline{D})}, \tag{4}$$

where the constant M depends only on D. Every bounded set $\mathfrak{M}$ in $W_p^l(\overline{D})$ is compact in $C(\overline{D})$.

We can introduce the concept not of a separate generalized derivative of a summable function $f(x)$, but of a generalized linear differential operator of any order, for example,

$$L(f) = \sum_{i,\,k=1}^{n} a_{ik} \frac{\partial^2 f}{\partial x_i \partial x_k} + \sum_{k=1}^{n} b_k \frac{\partial f}{\partial x_k} + cf,$$

where the coefficients are sufficiently smooth functions of the variables $(x_1, \ldots, x_n)$. Such a generalized operator is defined by an equality analogous to (2):

$$\int_D f(x) M(\varphi) \, dx = \int_D L(f) \varphi(x) \, dx,$$

where $M(\varphi)$ is the so-called adjoint differential operator [122], and $\varphi(x) \in \overset{\circ}{C}{}^2(D)$. The existence of the separate derivatives appearing in the operator $L(f)$ is not assumed. The concepts of generalized solutions of differential equations are usually introduced with the aid of such equalities. We dwell separately on the case $n = 2$.

Suppose that $w(x, y) = u(x, y) + iv(x, y)$ is a continuously differentiable function in a domain D, with $w(x, y) = w(z) \in C^1(\overline{D})$, $z = x + iy$. Let

$$w(z, \bar{z}) = w\left(\frac{z + \bar{z}}{2}, \frac{z - \bar{z}}{2i} \right) \equiv w(x, y), \qquad x = \frac{z + \bar{z}}{2}, \qquad y = \frac{z - \bar{z}}{2i}$$

and consider

$$\frac{\partial w}{\partial z} = w_z = \frac{1}{2}\left(w_x + \frac{1}{i} w_y \right) \equiv \frac{1}{2}(u_x + v_y) + \frac{i}{2}(v_x - u_y),$$

$$\frac{\partial w}{\partial \bar{z}} = w_{\bar{z}} = \frac{1}{2}\left(w_x - \frac{1}{i} w_y \right) \equiv \frac{1}{2}(u_x - v_y) + \frac{i}{2}(u_y + v_x).$$

In particular, using these derivatives with respect to complex arguments, we can give the system of Cauchy-Riemann equations

$$u_x = v_y, \qquad u_y = -v_x$$

the following form:

$$\partial w / \partial \bar{z} = 0, \qquad w = u + iv, \qquad z = x + iy.$$

The familiar Green formulas for $w \in C^1(\overline{D})$ and $D \in C^1$ can be written as follows:

$$\iint_D \frac{\partial w}{\partial \bar{z}} \, dx \, dy = \frac{1}{2i} \int_\Gamma w(z) \, dz; \qquad \iint_D \frac{\partial w}{\partial z} \, dx \, dy = -\frac{1}{2i} \int_\Gamma w(z) \, d\bar{z}. \tag{5}$$

The formulas (5) also imply an analogue of the usual Green formula:

$$\iint_D \left(\frac{\partial P}{\partial \bar{z}} - \frac{\partial Q}{\partial z} \right) dx \, dy = \frac{1}{2i} \int_\Gamma P \, dz + Q \, d\bar{z}, \tag{5*}$$

where P, $Q \in C^1(D)$. The last formula shows that the usual definition of a total differential carries over to the case of derivatives with respect to the complex arguments z and $\bar{z}$, these derivatives preserve the usual form of condition for the integral to be independent of the path of integration, and so on.

It is possible to introduce the concept of generalized derivatives with respect to z and $\bar{z}$ similarly to the case of derivatives with respect to a real argument. Let $f(z)$, $g(z) \in L_1(\bar{D})$. If

$$\iint f \frac{\partial \varphi}{\partial \bar{z}} \, dx \, dy + \iint g\varphi \, dx \, dy = 0 \tag{6}$$

for any function $\varphi(z) \in \overset{\circ}{C}{}^1(D)$, then $g(z)$ is called the generalized derivative with respect to $\bar{z}$ of $f(z)$ (respectively, with respect to z).

If $f(z) \in C^1(\bar{D})$, then it is obvious from Green's formula (5) for $w(z) = \{f(z)\varphi(z)\}$, and from (2), that the generalized derivative $g(z)$ coincides with the ordinary derivative $\partial f/\partial z$. Similarly, we introduce the generalized derivatives of higher order and the Sobolev B-spaces $W_p^l(\bar{D})$ with norm

$$\|f\|_{W_p^l(\bar{D})} = \sum_{i,k=0}^{i+k=l} \left\| \frac{\partial^{i+k} f}{\partial z^i \partial \bar{z}^k} \right\|_{L_p(\bar{D})}, \qquad p > 1, l \geqslant 1.$$

Obviously, all the properties of generalized derivatives with respect to real arguments carry over to generalized derivatives with respect to complex arguments, and, in particular, Theorem 2 (Sobolev) can be formulated as follows in this case.

THEOREM 3 (SOBOLEV). *If $f(z) \in W_p^l(\bar{D})$, $p > 2$, $l \geqslant 1$, then $f(z) \in C^{l-1+\alpha}(\bar{D})$, where $\alpha = (p - 2)/p$.*

This theorem will be proved in the next section.

If the boundary of the domain is sufficiently smooth, then $f(z)$ can be extended beyond the boundary with preservation of class (see, for example, [13]), and, consequently, the imbedding theorems, which hold in a domain $D^* \supset D$ in this case, are satisfied, in particular, also in the closed domain $\bar{D}$. Indeed, suppose that the boundary Γ of D belongs to the class C_*^l, $l \geqslant 1$, so that in a neighborhood of each point $z \in \Gamma$ there is a transformation $z = z(\zeta) \in C^l$ with nonzero Jacobian $J(\zeta) = |z_\zeta|^2 - |z_{\bar\zeta}|^2$ such that the boundary points in this neighborhood are images of the points of the line segment $\eta = \operatorname{Im} \zeta = 0$, $\xi = \operatorname{Re} \zeta \geqslant 0$ (or $\xi = 0$, $\eta \geqslant 0$). The fact that the boundary Γ is in the class C_*^l allows us to use the supplementary transformation $z = z(\zeta)$ in a neighborhood of each point $z \in \Gamma$ to reduce the problem of extending functions across a curvilinear boundary to that of extending them across one of the boundaries of a rectangle $D_0 = \{x_0 < x < x_1, 0 < y < y_1\}$. Suppose, therefore, that $f_0(x, y) \in W_p^l(\overline{D}_0)$, where $l \geq 1$. We shall extend $f_0(x, y)$ with preservation of class to the rectangle $D_1 = \{x_0 < x < x_1, -y_1 < y < 0\}$. Let the numbers λ_k be determined from the system of equations

$$\sum_{n=1}^{l+1} \lambda_n \left(-\frac{1}{n}\right)^k = 1, \quad k = 0, \ldots, l,$$

whose determinant is the Vandermonde determinant with the numbers $-1, -1/2, \ldots, -1/(l+1)$. We set $f_1(x, y) = \Sigma \lambda_n f(x, -y/n)$ in D_1. For a smooth function $\psi(x, y)$ with compact support in $D_0 + D_1$ we form the integrals

$$I_j = \iint_{D_j} \psi \frac{\partial^l f_j}{\partial x^s \partial y^{l-s}} dx\, dy, \quad j = 0, 1,$$

and integrate them by parts l times. Then

$$I_j = (-1)^l \iint_{D_j} f_j \frac{\partial^l [\psi]}{\partial x^s \partial y^{l-s}}$$

$$- (-1)^l \int_{x_0}^{x_1} \left[\frac{\partial^{l-1} f_j}{\partial x^s \partial y^{l-s-1}} \psi + \cdots + \frac{\partial^s f_j}{\partial x^s}(-1)^{l-s} \frac{\partial^{l-s-1}\psi}{\partial y^{l-s-1}} \right] dx.$$

The equality

$$\frac{\partial^{l-k} f_1}{\partial x^s \partial y^{l-s-k}}\bigg|_{y=-0} = \frac{\partial^{l-k} f_0}{\partial x^s \partial y^{l-s-k}}\bigg|_{y=+0} \sum_{n=1}^{l+1} \left(-\frac{1}{n}\right)^{l-s-k} \lambda_n = \frac{\partial^{l-k} f_0}{\partial x^s \partial y^{l-s-k}}\bigg|_{y=+0}$$

gives us that

$$I_0 + I_1 = (-1)^l \iint_{D_0 + D_1} f \frac{\partial^l \psi}{\partial x^s \partial y^{l-s}} dx\, dy,$$

and, since (because $f_0 \in W_p^l(\overline{D})$),

$$\iint_{D_0} \left| \frac{\partial^l f_0}{\partial x^s \partial y^{l-s}} \right|^p dx\, dy + \iint_{D_1} \left| \frac{\partial^l f}{\partial x^s \partial y^{l-s}} \right|^p dx\, dy < \infty,$$

it follows that the function $f(x, y) = f_j(x, y)$ for $(x, y) \in D_j, j = 0, 1$, belongs to $W_p^l(\overline{D}_0 + \overline{D}_1)$. Obviously, the existence of an extension of the function and of its generalized derivatives with respect to the real arguments implies the existence of its derivatives along with its generalized derivatives also with respect to the complex arguments.

We formulate the above result as the following theorem [13].

THEOREM 4 (BABICH). *If the boundary* Γ *of a domain* D *belongs to the class* C_*^l, $l \geqslant 1$, *then each function* $f(x, y)$ *of class* W_p^l, $p > 1$, *in the domain* $D(\Gamma)$ *can be extended to a domain* $D_* \supset D$ *with preservation of class.*

REMARK. The usual differentiation formulas hold for composite functions $w(z) = \Omega[f(z)]$ where one (for definiteness let this be $f(z)$) has generalized derivatives, while the other ($\Omega(f)$ in this case) has ordinary derivatives. It is easy to see this if we approximate $f(z)$ by continuously differentiable functions $f^n(z)$.

Indeed, setting

$$w_{\bar{z}}^n = \Omega_f f_{\bar{z}}^n + \Omega_{\bar{f}} \bar{f}_{\bar{z}}^n \tag{7}$$

in the formula (6) defining the generalized derivatives, and passing to the limit in the equality obtained, we arrive at (7) also for $w(z)$. Of course, the arguments are preserved for the derivative $w_z(z)$ and higher-order derivatives. But if the functions $\Omega(f)$ and $f(z)$ are differentiable only in the generalized sense, then $\Omega[f(z)]$ is clearly not always differentiable.

§2. Potential and singular operators with respect to a domain

We formulate as a theorem some properties of the operators (introduced and extensively studied by I. N. Vekua [128])

$$Tf = -\frac{1}{\pi} \iint_D \frac{f(\zeta)\,d\xi\,d\eta}{\zeta - z}, \qquad Sf = -\frac{1}{\pi} \iint_D \frac{f(\zeta)\,d\xi\,d\eta}{(\zeta - z)^2},$$

of which the first is a potential operator, while the second is a singular integral, understood in the sense of the Cauchy principal value.

MAIN THEOREM. *The operators* T *and* S *have the following properties*:

1) $\partial Tf/\partial \bar{z} = f$ *and* $\partial Tf/\partial z = Sf$ *for* $f \in L_p(\overline{D}), p > 1$.

2) *If* $f \in C^{l+\alpha}(\overline{D})$, $0 < \alpha < 1$, *and* $D \in C^{l+1+\alpha}$, *then* $Tf \in C^{l+1+\alpha}(\overline{D})$ *and* $Sf \in C^{l+\alpha}(\overline{D})$.

3) *If* $f \in L_p(\overline{D})$ ($p > 2$) *and* D *is a bounded domain in the plane* E, *then* $Tf \in C^\alpha(E)$, $\alpha = (p - 2)/p$, *and* $Sf \in L_p(E)$.

4) S *is a bounded linear operator in the spaces* $C^{l+\alpha}(\overline{D})$ $(0 < \alpha < 1, l \geqslant 0)$ *and* $L_p(\overline{D})$ ($p > 1$), *and* $\|Sf\|_{L_2} \leqslant \|f\|_{L_2}$, *i.e.*, $\Lambda_2 = \|S\|_{L_2} \leqslant 1$ *in* $L_2(\overline{D})$.

This theorem will be proved stepwise in the course of the whole section; moreover, in passing we shall obtain some important representations for functions having generalized derivatives.

Suppose that $f(\zeta) \in C^1(\overline{D})$, and define $f_1 = f(\zeta)/(\zeta - z)$, where $z \in D$. We apply Green's formula (1.5) to the function $f_1(\zeta)$ in the domain D_ε obtained by deleting the closed disk $|\zeta - z| \le \varepsilon$ from the domain D:

$$\iint_{D_\varepsilon} \frac{f_{\bar\zeta}\, d\xi d\eta}{\zeta - z} = \frac{1}{2i} \int_\Gamma \frac{f\, d\zeta}{\zeta - z} - \frac{1}{2i} \int_{\Gamma_\varepsilon} \frac{f\, d\zeta}{\zeta - z} = \frac{1}{2i} \int_\Gamma \frac{f\, d\zeta}{\zeta - z} - \pi f(z)$$

$$- \frac{1}{2i} \int_{\Gamma_\varepsilon} \frac{f(\zeta) - f(z)}{\zeta - z} d\zeta;$$

as $\varepsilon \to 0$ this gives us

$$f(z) = \frac{1}{2\pi i} \int_\Gamma \frac{f(\zeta)\, d\zeta}{\zeta - z} - \frac{1}{\pi} \iint_D \frac{f_{\bar\zeta}\, d\xi\, d\eta}{\zeta - z} \equiv \Phi(z) + T\left(\frac{\partial f}{\partial \bar\zeta}\right), \qquad (1)$$

which is valid under even more general assumptions (Theorem 6 of the present section). We introduce the notation $T(f|z)$ to reflect the dependence of the values Tf of the operator on the variable z.

THEOREM 1. *Let D be a bounded domain in the plane E. If $f \in L_p(\overline{D})$, $1 \le p \le 2$, then $T(f|z) \in L_s(\overline{D})$, where s is any number satisfying the inequaltiy $1 < s < 2p/(2 - p)$.*

Suppose first that $p < s < 2p/(2 - p)$. Then

$$|Tf| \le \frac{1}{\pi} \iint_D |f(\zeta)|^{p/s} |\zeta - z|^{-2/s + \alpha} |f|^{p(1/p - 1/s)} |\zeta - z|^{-2/q + \alpha}\, d\xi\, d\eta,$$

where $q = p/(p - 1)$ and $\alpha = 1/s - 1/p + 1/2 > 0$. Since $1/s + 1/q + (s - p)/ps = 1$, Hölder's inequality ((1.1) in Chapter II) gives us that

$$|Tf| \le \frac{1}{\pi} \left(\iint_D |f|^p |\zeta - z|^{-2 + s\alpha}\, d\xi\, d\eta \right)^{1/s} \left(\iint_D |\zeta - z|^{-2 + \alpha q}\, d\xi d\eta \right)^{1/q} \|f\|_{L_p}^{(s-p)/s}$$

Since

$$M(\lambda, D) = \sup_{z \in E} \iint_D |z - \zeta|^{-2 + \lambda}\, d\xi\, d\eta < \infty$$

for $\lambda > 0$, we finally find that

$$\iint_D |Tf|^s dx\, dy$$

$$\le \frac{1}{\pi^s} [M(q\alpha, D)]^{s/q} \|f\|_{L_p}^{s-p} \iint_D |f(\zeta)|^p d\xi\, d\eta \iint_D |\zeta - z|^{-2 + s\alpha} dx\, dy$$

$$\le \frac{1}{\pi^s} [M(q\alpha, D)]^{s/q} M(s\alpha, D) \|f\|_{L_p}^s < \infty,$$

as required. As the last estimate shows, the restriction $s > p$ can obviously be removed.

THEOREM 2. *If* $f \in L_1(\overline{D})$, *then for any function* $\varphi \in \overset{\circ}{C}{}^1(D)$

$$\iint\limits_D Tf \frac{\partial \varphi}{\partial \bar{z}} \, dx \, dy + \iint\limits_D f\varphi \, dx \, dy = 0,$$

i.e., the generalized derivative of $T(f \mid z)$ *with respect to* $\bar{z}$ *exists and*

$$\partial T(f \mid z)/\partial \bar{z} = f. \tag{2}$$

Since $\varphi \in \overset{\circ}{C}{}^1(\overline{D})$, according to (1) we have $\varphi(z) = T(\partial \varphi/\partial \bar{\zeta})$, and so

$$\iint\limits_D Tf \frac{\partial \varphi}{\partial \bar{z}} \, dx \, dy = \frac{1}{\pi} \iint\limits_D f(\zeta) \, d\xi \, d\eta \iint\limits_D \frac{\partial \varphi}{\partial \bar{z}} \frac{dx \, dy}{z - \zeta} = -\iint\limits_D f\varphi \, dx \, dy,$$

as required.

THEOREM 3. *If the generalized derivative* $\partial f/\partial \bar{z}$ *is* 0 *in* D, *then* $f(z)$ *is holomorphic in* D.

It suffices to prove that $f(z)$ is holomorphic inside some neighborhood of each fixed point of D, and without loss of generality the point can be taken to be $z_0 = 0$. We take a sufficiently small disk $D_R = \{|z| < R\}$ and consider the function

$$Q(z, \zeta) = 2|\zeta - z|^2 \ln \frac{|R - z\bar{\zeta}|^2}{R \, |z - \zeta|} - \left(R^2 - |z|^2\right)\left(1 - \frac{|\zeta|^2}{R^2}\right), \qquad \zeta, z \in D_R$$

($Q(z, \zeta)$ is the Green function of the biharmonic equation $\Delta\Delta Q = 0$). Simple computations give us that $Q = \partial Q/\partial \bar{z} = \partial Q/\partial z = 0$ for $|z| = R$ and $\zeta \neq z$. Moreover, Q, $Q_{\bar{z}}$, and Q_z are continuous in the closed disk $|z| \leq R$, and

$$\frac{\partial^3 Q}{\partial \bar{z} \partial \zeta \partial \bar{\zeta}} = \frac{1}{\bar{\zeta} - \bar{z}} + \frac{R^2 z - 2R^2\zeta + \bar{z}\zeta^2}{\left(R^2 - \bar{z}\zeta\right)^2} + \frac{z\bar{\zeta}}{R^2 - \bar{z}\zeta}.$$

For a fixed $\zeta \in D_R$ we introduce the function

$$\varphi(z) = \begin{cases} Q(z, \zeta), & |z| \leq R, \\ 0, & |z| > R. \end{cases}$$

Obviously, $\varphi(z) \in \overset{\circ}{C}{}^1(D)$. According to the definition of the generalized derivative $g = \partial f/\partial \bar{z} = 0$ (see (1.6)),

$$\iint\limits_D f \frac{\partial \varphi}{\partial \bar{z}} \, dx \, dy = \iint\limits_{D_R} f(z) \frac{\partial Q(z, \zeta)}{\partial \bar{z}} \, dx \, dy = 0.$$

After differentiating with respect to ζ and $\bar{\zeta}$ under the integral sign (which is allowed because $\partial^3 Q/\partial\bar{z}\partial\zeta\partial\bar{\zeta}$ is bounded), we find that

$$\iint\limits_{D_R} f(z)\,\frac{\partial^3 Q}{\partial\bar{z}\partial\zeta\partial\bar{\zeta}}\,dx\,dy = \iint\limits_{D_R} f(z)\,\frac{dx\,dy}{\bar{\zeta}-\bar{z}} + \Phi(\zeta) + \overline{\Phi}_1(\zeta) = 0,$$

where

$$\Phi(\zeta) = \iint\limits_{D_R} f(z)\,\frac{R^2 z - 2R^2\zeta + \bar{z}\zeta^2}{\left(R^2 - \bar{z}\zeta\right)^2}\,dx\,dy, \qquad \Phi_1(\zeta) = \iint\limits_{D_R} \overline{f(z)}\,\frac{z\bar{\zeta}\,dx\,dy}{R^2 - z\zeta},$$

and $\Phi(\zeta)$ and $\Phi_1(\zeta)$ are holomorphic in D_R, since $|R^2 - \bar{z}\zeta| \neq 0$ for $|z| \leqslant R$ and $|\zeta| < R$. Thus,

$$T\bar{f} = \frac{1}{\pi}\overline{\Phi}(\zeta) + \frac{1}{\pi}\Phi_1(\zeta),$$

which implies that $\bar{f} = \partial T\bar{f}/\partial\bar{\zeta} = \frac{1}{\pi}\overline{\Phi'(\zeta)}$, i.e., $f(\zeta) = \frac{1}{\pi}\Phi'(\zeta)$ is a holomorphic function in D_R, as required.

THEOREM 4. *If the generalized derivative $\partial f/\partial\bar{z}$ exists in $L_1(\overline{D})$, then the function $f(z)$ can be represented in the form*

$$f(z) = \Phi(z) + T(\partial f/\partial\bar{z}), \tag{3}$$

where $\Phi(z)$ is holomorphic in D.

Theorem 4 follows from Theorem 3, since

$$\frac{\partial}{\partial\bar{z}}\left[f - T\left(\frac{\partial f}{\partial\bar{\zeta}}\right)\right] = \frac{\partial f}{\partial\bar{z}} - \frac{\partial f}{\partial\bar{z}} = 0.$$

Theorem 4 gives us immediately that differentiability with respect to $\bar{z}$ in the generalized sense is a local property, i.e., if $g(z) = \partial f/\partial\bar{z}$ is the generalized derivative in D, then it is the generalized derivative also in any subdomain $D_1 \subset D$.

Indeed, the representation (3) implies that

$$f(z) = \Phi(z) - \frac{1}{\pi}\iint\limits_{D-D_1} \frac{g(\zeta)\,d\xi\,d\eta}{\zeta - z} - \frac{1}{\pi}\iint\limits_{D_1} \frac{g(\zeta)\,d\xi\,d\eta}{\zeta - z} \equiv \Phi_1(z) + T_{D_1}(g),$$

and so $\partial f/\partial\bar{z} = g$ in D_1 because $\Phi_1(z)$ is holomorphic in D_1. The converse assertion is proved similarly: *If $f(z)$ has a generalized derivative in a neighborhood of each point of D, then it has a generalized derivative also in the whole domain D.* If $f(z)$ has a generalized derivative $g = \partial f/\partial\bar{z}$, then $\overline{f(z)}$ obviously also has a generalized derivative $\overline{g(z)} = \partial\bar{f}/\partial z$ with respect to z. Consequently, all the properties proved above for derivatives with respect to $\bar{z}$ hold also for derivatives with respect to z. In particular, if the generalized derivative $\partial f/\partial z$ exists and is in $L_1(\overline{D})$, then, for example, (3) takes the form

$$f(z) = \overline{\Phi(z)} + \overline{T}(\partial f/\partial \zeta),$$

where $\Phi(z)$ is a holomorphic function in D, and

$$\overline{T}\varphi = -\frac{1}{\pi} \iint_D \frac{\varphi(\zeta)d\xi\, d\eta}{\overline{\zeta} - \overline{z}}.$$

We study the properties of certain integrals over a domain that depend on parameters. Let

$$I(\alpha, \beta) = \frac{1}{\pi} \iint_D |\zeta - z|^{-\alpha}|\zeta - z_1|^{-\beta}d\xi\, d\eta$$

for $\alpha, \beta < 2$, and

$$I(\alpha, \beta) = \frac{1}{\pi} \iint_D (\zeta - z)^{-\alpha}(\zeta - z_1)^{-\beta}d\xi\, d\eta$$

if α or β is equal to 2.

LEMMA. *Let D be a bounded domain, $z_1 \neq z$, and d the diameter of D. Then*

$$I(\alpha, \beta) \leq \begin{cases} M'_{\alpha,\beta}|z - z_1|^{2-\alpha-\beta}, & \alpha + \beta > 2, \\ M''_{\alpha,\beta}(D) + 8\ln|z - z_1|, & \alpha + \beta = 2, \\ M'''_{\alpha,\beta}d^{2-\alpha-\beta}, & \alpha + \beta < 2, \end{cases} \tag{4}$$

where $M'_{\alpha,\beta}$ and $M'''_{\alpha,\beta}$ do not depend on D, and

$$I(2,0) = \frac{d}{dz}\left(\frac{1}{2\pi i}\int_\Gamma \frac{\overline{\zeta}d\zeta}{\zeta - z}\right) \equiv \frac{d\Phi_\Gamma(z)}{dz}, \tag{5}$$

$$I(2,1) = -\frac{\Phi_\Gamma(z) - \Phi_\Gamma(z_1)}{(z - z_1)^2} + \frac{\overline{z} - \overline{z}_1}{(z - z_1)^2} + \frac{\Phi'_\Gamma(z)}{z - z_1}. \tag{6}$$

We define D_1: $|\zeta - z| < 2|z - z_1| \equiv 2\rho$ and D_0: $|\zeta - z| < 2\rho_0$, with $D \subset D_0$. For $\zeta \in D_1$ the triangle inequality gives us

$$|\zeta - z| \leq |\zeta - z_1| + |z_1 - z| \leq 2|\zeta - z_1|;$$

hence

$$I = \frac{1}{\pi} \iint_{D_0 - D} |\zeta - z|^{-\alpha}|\zeta - z_1|^{-\beta}d\xi\, d\eta \leq 2^{1+\beta}\int_{2\rho}^{2\rho_0} r^{1-\alpha-\beta}\, dr,$$

and this gives us estimate (4) for I_0. To prove the inequality it remains to establish it for

$$I_1 = \frac{1}{\pi} \iint_{D_1} |\zeta - z|^{-\alpha} |\zeta - z_1|^{-\beta} d\xi \, d\eta.$$

In the last integral we set $\zeta - z = \rho w = \rho(u + iv)$, from which we get $\xi - x = \rho u$, $\eta - y = \rho v$, $d\xi d\eta = \rho^2 du \, dv$ and $\zeta - z_1 = \rho(w + e^{i\varphi_0})$, where $\varphi_0 = \arg(z - z_1)$. Then

$$I_1 = \frac{1}{\pi} \rho^{2-\alpha-\beta} \iint_{|w|\leqslant 2} |w|^{-\alpha} |w + e^{i\varphi_0}|^{-\beta} du \, dv \leqslant M(\alpha, \beta) \rho^{2-\alpha-\beta},$$

which finally proves (4).

We set $D_\varepsilon = \{|z - \zeta| < \varepsilon\}$ and use the second of Green's formulas (1.5):

$$I_\varepsilon(2,0) = -\frac{1}{\pi} \iint_{D-D_\varepsilon} \frac{\partial}{\partial \zeta}\left(\frac{1}{\zeta - z}\right) d\xi \, d\eta$$

$$= \frac{1}{2\pi i} \int_\Gamma \frac{d\bar{\zeta}}{d\zeta} \frac{d\zeta}{\zeta - z} - \frac{1}{2\pi i} \int_{\Gamma_\varepsilon} \frac{d\bar{\zeta}}{d\zeta} \frac{d\zeta}{\zeta - z}$$

$$= \frac{d}{dz}\left(\frac{1}{2\pi i} \int_\Gamma \frac{\bar{\zeta} d\zeta}{\zeta - z}\right) + \frac{\varepsilon^2}{2\pi i} \int_{\Gamma_\varepsilon} \frac{1}{\zeta^2} \frac{d\zeta}{\zeta - z}.$$

Since obviously $\int_{\Gamma_\varepsilon} = 0$, we get the required equality (5) in the limit as $\varepsilon = 0$.
Let us represent $I(2, 1)$ in the form

$$I(2,1) = \frac{1}{\pi} \iint_D \frac{d\xi \, d\eta}{(\zeta - z)^2 (\zeta - z_1)}$$

$$= \frac{1}{\pi(z - z_1)} \iint_D \frac{d\xi \, d\eta}{(\zeta - z)^2} - \frac{1}{\pi(z - z_1)^2}\left\{\iint_D \frac{d\xi \, d\eta}{\zeta - z} - \iint_D \frac{d\xi \, d\eta}{\zeta - z_1}\right\}.$$

For the function $f = -\bar{z}$, formula (1) tells us that

$$\frac{1}{\pi} \iint_D \frac{d\xi \, d\eta}{\zeta - z} = -\bar{z} + \frac{1}{2\pi i} \int_\Gamma \frac{\bar{\zeta} d\zeta}{\zeta - z} = -\bar{z} + \Phi_\Gamma(z).$$

Substituting this and the value of $I(2,0)$ into the above representation for $I(2, 1)$, we arrive at (6).

REMARK. It is obvious that if Γ is the circle $|\zeta| = r$, then $\Phi_\Gamma(z) \equiv 0$, and (5) and (6) take an especially simple form.

THEOREM 5. *Let* D *be a bounded domain. If* $f \in L_p(\overline{D})$, $p > 2$, *then the function* $T(f \mid z)$ *satisfies the conditions*

$$|T(f \mid z)| \leqslant M_1 \|f\|_{L_p(\overline{D})},$$

$$|T(f \mid z) - T(f \mid z_1)| \leqslant M_2 \|f\|_{L_p(\overline{D})} |z - z_1|^\alpha, \qquad \alpha = \frac{p-2}{p}, \qquad (7)$$

where z *and* z_1 *are arbitrary points of the plane; the constant* M_1 *depends on* p *and* D, *while* M_2 *depends only on* p.

Applying Hölder's inequality and (4), we get

$$|T(f \mid z)| \leqslant \frac{1}{\pi} \iint_D \frac{|f(\zeta)| \, d\xi \, d\eta}{|\zeta - z|}$$

$$\leqslant \frac{1}{\pi} \|f\|_{L_p(\overline{D})} \left\{ a \iint_D |\zeta - z|^{-q} \, d\xi \, d\eta \right\}^{1/q} \leqslant M_1 \|f\|_{L_p(\overline{D})},$$

since $q = p/(p-1) < 2$ for $p > 2$. Further,

$$|T(f \mid z_1) - T(f \mid z_2)| \leqslant \frac{|z - z_1|}{\pi} \left| \iint_D \frac{f(\zeta) \, d\xi \, d\eta}{(\zeta - z)(\zeta - z_1)} \right|$$

$$\leqslant |z - z_1| \, \|f\|_{L_p(D)} [I(q, q)]^{1/q},$$

and this and (4) lead to the second of the inequalities in (7). The theorem is proved. This theorem implies, in particular, that the operator T is completely continuous in $L_p(\overline{D})$ ($p > 2$), since a bounded set $\{Tf\} \subset C^\alpha(E)$ is obviously relatively compact in $L_p(\overline{D})$. Let $f(z) \in W_p^1(D)$, $p > 2$. Then (3) and Theorem 5 imply that

$$f(z) = \Phi(z) + T(\partial f / \partial \bar{\zeta}) \in C^\alpha(\overline{D}_1), \, D_1 \subset D, \, \alpha = (p-2)/p.$$

Similarly, if $f(z) \in W_p^l(D)$, $p > 2$, then the representation (3) for the generalized derivatives $D^{l-1} f = \Phi(\zeta) + T((\partial/\partial\bar{\zeta})D^{l-1}f)$ gives us that $f(z) \in C^{l-1+\alpha}(D)$, which proves Theorem 3 as stated in the last section (the Sobolev imbedding theorem).

THEOREM 6. *The formula* (1) *holds under the conditions* $D \in C^1, f(z) \in C(\overline{D})$ *and* $\partial f/\partial \bar{z} \in L_p(\overline{D})$, $p > 2$.

Indeed, under these conditions $f(z)$ can be represented by (3), $f(z) = \Phi(z) + T(\partial f/\partial \bar{\zeta})$, and it remains to show that

$$\Phi(z) = \frac{1}{2\pi i} \int \frac{f(\zeta) \, d\zeta}{\zeta - z}.$$

Since $f(z) \in C(\overline{D})$ and $T(\partial f/\partial \bar{\zeta}) \in C^{\alpha}(\overline{D})$, $\alpha = (p-2)/p$, it follows that $\Phi(z) = f(z) - T(\partial f/\partial \bar{\zeta}) \in C(\overline{D})$. Then Cauchy's formula gives us

$$\Phi(z) = \frac{1}{2\pi i} \int_{\Gamma} \frac{[f(\zeta) - g(\zeta)]d\zeta}{\zeta - z} = \frac{1}{2\pi i} \int_{\Gamma} \frac{f(\zeta)d\zeta}{\zeta - z},$$

because

$$\frac{1}{2\pi i} \int_{\Gamma} \frac{g(\zeta)d\zeta}{\zeta - z} = 0$$

as the Cauchy integral of the function $g(z) = T(\partial f/\partial \bar{\zeta})$, which is holomorphic outside D and equal to zero at infinity.

Let us now consider the singular operator

$$Sf = -\frac{1}{\pi} \iint_{D} \frac{f(\zeta)d\xi\, d\eta}{(\zeta - z)^2}$$

$$= -\frac{1}{\pi} \lim_{\varepsilon \to 0} \left[\iint_{D_\varepsilon} \frac{f(\zeta) - f(z)}{(\zeta - z)^2} d\xi\, d\eta + f(z) \iint_{D_\varepsilon} \frac{d\xi\, d\eta}{(\zeta - z)^2} \right].$$

Using (5) for $z \in D$, we obtain

$$Sf = -\frac{1}{\pi} \iint_{D} \frac{f(\zeta) - f(z)}{(\zeta - z)^2} d\xi\, d\eta - f(z) \frac{d\Phi_\Gamma(z)}{dz}.$$

Since the first integral in this formula exists as an improper integral for $f(\zeta) \in C^{\alpha}(\overline{D})$, the singular integral in Sf exists in the principal value sense for $z \in D$. Let us prove that $S(f|z) \in C^{\alpha}(D)$ for $f(z) \in C^{\alpha}(\overline{D})$, $0 < \alpha < 1$.

We have

$$S(f|z) - S(f|z_1)$$

$$= \frac{z_1 - z}{\pi} \left[\iint_{D} \frac{f(\zeta) - f(z)}{(\zeta - z)^2(\zeta - z_1)} d\xi\, d\eta + \iint_{D} \frac{f(\zeta) - f(z_1)}{(\zeta - z)(\zeta - z_1)^2} d\xi\, d\eta \right]$$

$$+ \frac{z_1 - z}{\pi} \left[f(z) \iint_{D} \frac{d\xi\, d\eta}{(\zeta - z)^2(\zeta - z_1)} + f(z_1) \iint_{D} \frac{d\xi\, d\eta}{(\zeta - z)(\zeta - z_1)^2} \right]$$

$$\equiv F_1(z, z_1) + F_2(z, z_1).$$

With the help of the estimates (4) for $z, z_1 \in D$ and $z \neq z_1$ we get

$$|F_1| \leqslant \|f\|_{H^{\alpha}} |z - z_1| \big[|I(2 - \alpha, 1)| + |I(1, 2 - \alpha)| \big] \leqslant M'_{\alpha}(D) \|f\|_{H^{\alpha}} |z - z_1|^{\alpha}.$$

The use of (6) gives us

$$|F_2| = |f(z)(z_1 - z)I(2,1) + f(z_1)(z_1 - z)I(1,2)|$$

$$= |[f(z) - f(z_1)](z_1 - z)I(2,1) + f(z_1)[\Phi'_\Gamma(z) - \Phi'_\Gamma(z_1)]|$$

$$\leqslant |z_1 - z|^\alpha \left(\|f\|_{H^\alpha} \|I(2,1)(z_1 - z)\|_C + \|f\|_C \|\Phi'_\Gamma\|_{H^\alpha} \right) \leqslant M''_\alpha(D) \|f\|_{C^\alpha} |z - z_1|^\alpha.$$

The estimates obtained for $F_i(z, z_1)$ $(i = 1, 2)$ imply that $S(f|z) \in C^\alpha(\bar{D})$ for $f \in C^\alpha(\bar{D})$, i.e., the linear operator S is bounded in C^α, $0 < \alpha < 1$. Moreover,

$$|S(f|z)| \leqslant \tilde{M}'_\alpha(D) \|f\|_{C^\alpha},$$

$$|S(f|z_1) - S(f|z)| \leqslant \tilde{M}''_\alpha(D) \|f\|_{C^\alpha} |z_1 - z|^\alpha, \tag{8}$$

where in the case $D = \{|z| < R\}$ the constant $\tilde{M}'_\alpha(D) = \tilde{M}_\alpha R^\alpha$, and $\tilde{M}_\alpha$ does not depend on R. Indeed, if $D = \{|z| < R\}$, then

$$Sf = -\frac{1}{\pi} \iint_{|\zeta| < 1} \frac{f(\zeta) - f(z)}{(\zeta - z)^2} \, d\xi \, d\eta,$$

and the form of $\tilde{M}'_\alpha(D)$ follows from the last of the inequalities (4).

Let us prove that $\partial Tf/\partial z = Sf$ for $f(z) \in C^\alpha(\bar{D})$. We have

$$\frac{T(f|z_1) - T(f|z)}{z_1 - z} - Sf$$

$$= \frac{z - z_1}{\pi} \iint_D \frac{f(\zeta) - f(z)}{(\zeta - z)^2(\zeta - z_1)} \, d\xi \, d\eta + \frac{f(z)(z - z_1)}{\pi} \iint_D \frac{d\xi \, d\eta}{(\zeta - z)^2(\zeta - z_1)}$$

$$\equiv F'(z, z_1) + F''(z, z_1).$$

Formulas (4) and (6) imply that

$$|F'| \leqslant |z - z_1| \, \|f\|_{H^\alpha} |I(2 - \alpha, 1)| \leqslant M_\alpha(D) \|f\|_{H^\alpha} |z - z_1|^\alpha,$$

$$F'' = f(z) \left[\frac{\bar{z} - \bar{z}_1}{z - z_1} + \Phi'_\Gamma(z) - \frac{\Phi_\Gamma(z) - \Phi_\Gamma(z_1)}{z - z_1} \right].$$

Setting $z - z_1 = |z - z_1| e^{i\theta}$, we arrive at the equality

$$\lim_{|z - z_1| \to 0} \frac{T(f|z) - T(f|z_1)}{z - z_1} = Sf + e^{-2i\theta} f(z),$$

and for $\theta = 0$ and $\theta = \pi/2$ this gives us, respectively,

$$\frac{\partial Tf}{\partial x} = Sf + f \quad \text{and} \quad \frac{1}{i} \frac{\partial Tf}{\partial y} = Sf - f,$$

which is equivalent to the relations $\partial Tf/\partial \bar{z} = f$ and $\partial Tf/\partial z = Sf$.

Thus, for $f(z) \in C^\alpha(\overline{D})$ we have $T(f|z) \in C^{1+\alpha}(\overline{D})$, since

$$\frac{\partial Tf}{\partial \bar{z}} = f \in C^\alpha(\overline{D}) \quad \text{and} \quad \frac{\partial Tf}{\partial z} = S(f|z) \in C^\alpha(\overline{D}).$$

Suppose now that $f(z) \in C^{1+\alpha}(\overline{D})$. Then

$$Sf = -\lim_{\varepsilon \to 0} \frac{1}{\pi} \iint_{D_\varepsilon} \frac{f(\zeta) d\xi\, d\eta}{(\zeta - z)^2} = \lim_{\varepsilon \to 0} \frac{1}{\pi} \iint_{D_\varepsilon} \frac{\partial}{\partial \zeta}\left(\frac{1}{\zeta - z}\right) f(\zeta) d\xi\, d\eta$$

$$= -\frac{1}{\pi} \iint_D \frac{\partial f}{\partial \zeta} \frac{d\xi\, d\eta}{\zeta - z} - \frac{1}{2\pi i} \int_\Gamma \frac{f(\zeta) d\bar{\zeta}}{\zeta - z} + \lim_{\varepsilon \to 0} \frac{1}{2\pi i} \int_{\Gamma_\varepsilon} \frac{f(\zeta) d\bar{\zeta}}{\zeta - z}$$

(here we have used the second of the formulas (1.5)). Taking account of the fact that

$$\lim \int_{\Gamma_\varepsilon} = 0 \quad \text{as } \varepsilon \to 0 \quad \text{and} \quad \frac{d\bar{\zeta}}{d\zeta} = \frac{d\bar{\zeta}}{ds}, \quad \frac{d\zeta}{ds} = \left(\frac{d\bar{\zeta}}{ds}\right)^2,$$

where $\zeta = \zeta(s)$ is the equation of the contour Γ and s is the arclength parameter of this contour, we get, finally, that

$$Sf = T\left(\frac{\partial f}{\partial \zeta}\right) - \frac{1}{2\pi i} \int_\Gamma \frac{f(\zeta)(d\bar{\zeta}/ds)^2\, d\zeta}{\zeta - z}. \tag{9}$$

Differentiation of the last equality with respect to $\bar{z}$ and z gives

$$\frac{\partial Sf}{\partial \bar{z}} = \frac{\partial f}{\partial z}, \quad \frac{\partial Sf}{\partial z} = S\left(\frac{\partial f}{\partial \zeta}\right) - \frac{1}{2\pi i} \int_\Gamma \frac{f(\zeta)(d\bar{\zeta}/ds)^2\, d\zeta}{(\zeta - z)^2},$$

which implies that $S(f|z) \in C^{1+\alpha}(D)$ for $\Gamma \in C^{2+\alpha}$ and $f \in C^{1+\alpha}(D)$. Continuing analogous arguments, we see that if $\Gamma \in C^{l+1+\alpha}$ and $f(z) \in C^{l+1}(\overline{D})$, $0 < \alpha < 1$, then $S(f|z) \in C^{l+1}(\overline{D})$, while $T(f|z) \in C^{l+1+\alpha}(D)$. In particular, this implies that T is a completely continuous operator in $C^{l+1}(\overline{D})$, $0 < \alpha < 1$, $l \geqslant 0$. Thus, all the assertions in the main theorem stated at the beginning of the section have been proved, except for certain properties of the operators T and S in the spaces $L_p(\overline{D})$ ($p > 1$).

We formulate for our particular case one of the main results of Calderón and Zygmund [22], which deals with properties of general singular operators in the spaces L_p ($p > 1$).

THEOREM 7 (CALDERÓN-ZYGMUND). *The integral in Sf exists in the principal value sense for $f \in L_p(\overline{D})$, and $\|Sf\|_{L_p} \leqslant \Lambda_p \|f\|_{L_p}$, where Λ_p is a finite constant depending only on p, i.e., the operator S is linear and bounded in $L_p(\overline{D})$ ($p > 1$).*

Suppose that $f(z) \in \overset{\circ}{C}^{\infty}(E)$.

Following I. N. Vekua [128], let us consider the operators

$$T_*f = \frac{1}{\pi i} \iint_E \frac{f(\zeta) dE_\zeta}{|\zeta - z|}, \qquad S_*f = \frac{\partial T_*f}{d\bar{z}}, \qquad S^*f = \frac{\partial T_*f}{\partial z}. \qquad (10)$$

Since

$$T_*f = \frac{1}{\pi i} \iint_E \frac{f(\zeta + z)}{|\zeta|} dE_\zeta,$$

differentiation with respect to $\bar{z}$ and z under the integral sign (which is possible, since f has compact support) gives us

$$\frac{\partial T_*f}{\partial \bar{z}} = T_*\left(\frac{\partial f}{\partial \bar{\zeta}}\right), \qquad \frac{\partial T_*f}{\partial z} = T_*\left(\frac{\partial f}{\partial \zeta}\right). \qquad (11)$$

By (11),

$$-\pi^2 S_* S^* f = \frac{\partial}{\partial \bar{z}} \iint_E \frac{dE_\zeta}{|\zeta - z|} \iint_E \frac{f_t dE_t}{|t - \zeta|}$$

$$= \frac{\partial}{\partial \bar{z}} \iint_E \frac{dE_\zeta}{|\zeta|} \iint_E \frac{f_t dE_t}{|t - \zeta - z|}$$

$$= \frac{\partial}{\partial \bar{z}} \left\{ \lim_{R \to \infty} \iint_E f_t dE_t \iint_{|\zeta| \leq R} \frac{dE_\zeta}{|\zeta||t - \zeta - z|} \right\},$$

from which, integrating by parts, we find that

$$S_* S^* f = \frac{\partial}{\partial \bar{z}} \iint_E K(z, \xi) f(\zeta) dE_\zeta,$$

where

$$K(z, \zeta) = \lim_{R \to \infty} \frac{1}{\pi^2} \frac{\partial}{\partial \zeta} \iint_{|t| \leq R} \frac{dE_t}{|t||t + (z - \zeta)|}.$$

Performing the change of variables $t = |z - \zeta| \rho e^{i\theta}$, $z \neq \zeta$, we get

$$K(z, \zeta) = \frac{1}{\pi^2} \lim_{R \to \infty} \frac{\partial}{\partial \zeta} \int_0^{R/|z-\zeta|} d\rho \int_0^{2\pi} \frac{d\theta}{|1 - \rho e^{i\theta}|}$$

$$= -\frac{1}{\pi^2} \frac{\partial \ln|z - \zeta|}{\partial \zeta} \lim_{R \to \infty} \int_0^{2\pi} \frac{d\theta}{1 - |z - \zeta| e^{i\theta}/R} = -\frac{1}{\pi(\zeta - z)},$$

because

$$\frac{\partial}{\partial \zeta}(\ln|z - \zeta|) = \frac{1}{2} \frac{\partial}{\partial \zeta} \ln[(z - \zeta)(\bar{z} - \bar{\zeta})] = \frac{1}{2} \frac{1}{\zeta - z}.$$

Consequently,

$$S_*S^*f = \frac{\partial}{\partial\bar{z}}\left(-\frac{1}{\pi}\iint_E \frac{f(\zeta)dE_\zeta}{\zeta - z}\right) = \frac{\partial}{\partial\bar{z}}(Tf) = f. \qquad (12)$$

Application of (1) to the function T_*f gives us

$$T_*f = \frac{1}{2\pi i}\int_{\Gamma_R}\frac{T_*f d\zeta}{\zeta - z} - \frac{1}{\pi}\iint_{|\zeta|\leqslant R}\frac{\partial T_*f}{\partial\bar{\zeta}}\frac{d\zeta\,d\eta}{\zeta - z},$$

which, since $T_*(f\,|\,Re^{i\theta})$ vanishes as $R \to \infty$, gives us that

$$T\left(\frac{\partial}{\partial\bar{\zeta}}T_*f\right) = T_*f. \qquad (13)$$

With the help of (11)–(13) we find that

$$Sf = S(S_*S^*f) = \frac{\partial}{\partial z}T\left(\frac{\partial}{\partial\bar{\zeta}}T_*S^*f\right) = \frac{\partial}{\partial z}(T_*S^*f) = S^*S^*f.$$

In view of the relation between S and S^* it suffices to prove Theorem 7 for the operator S^*:

$$S^*f = -\frac{1}{2\pi i}\iint_E \frac{f(\zeta)dE_\zeta}{(\zeta - z)|\zeta - z|} = -\frac{1}{2\pi i}\iint_E \frac{f(\zeta + z)dE_\zeta}{\zeta|\zeta|}.$$

Setting $\zeta = re^{i\theta}$, we write S^*f in the form

$$S^*(f\,|\,z) = -\frac{1}{2i}\int_0^{2\pi}e^{-i\theta}\,d\theta\frac{1}{\pi}\int_0^\infty\frac{f(z + re^{i\theta})}{r}dr = -\frac{1}{2i}\left(\int_0^\pi + \int_0^{2\pi}\right).$$

Making the substitution $\theta' = \theta - \pi$ in the second integral and returning to the previous notation, we get

$$S^*(f\,|\,z) = -\frac{1}{2i}\int_0^\pi\left(\frac{1}{\pi}\int_0^\infty\frac{f(z + re^{i\theta}) - f(z - re^{i\theta})}{r}dr\right)e^{-i\theta}d\theta,$$

whence

$$\|S^*f\|_{L_p(E)} \leqslant \frac{\pi}{2}\max_\theta\left\|\frac{1}{\pi}\int_0^\infty\frac{f(z + re^{i\theta}) - f(z - re^{i\theta})}{r}dr\right\|_{L_p}$$

$$= \frac{\pi}{2}\max_\theta\left\|\frac{1}{\pi}\int_{-\infty}^\infty\frac{f(z + re^{i\theta})}{r}dr\right\|_{L_p}.$$

Replacement of z by $(\tau + i\sigma)e^{i\theta}$ gives

$$\|S^*f\|_{L_p(E)} \leqslant \frac{\pi}{2}\max_\theta\left\|\frac{1}{\pi}\int_{-\infty}^\infty\frac{f[(\rho + i\sigma)e^{i\theta}]d\rho}{\rho - \tau}\right\|_{L_p(E)} = \frac{\pi}{2}\max_\theta\|U_\theta\|_{L_p},$$

where $f_\theta = f[e^{i\theta}(\rho + i\sigma)]$. But by Theorem 8 (F. Riesz) in §2 of Chapter II,

$$\|Uf_\theta\|_{L_p(E)}^p = \iint_E |Uf_\theta(\tau + i\sigma)|^p \, d\tau \, d\sigma$$

$$\leqslant A_p^p \int_{-\infty}^\infty d\sigma \int_{-\infty}^\infty d\sigma \int_{-\infty}^\infty |f_\theta(\tau + i\sigma)|^p \, d\tau = A_p^p \|f\|_{L_p(E)}^p,$$

which implies that $\|S^*f\|_{L_p(E)} \leqslant (\pi/2)A_p\|f\|_{L_p(E)}$. Since the class $\overset{\circ}{C}{}^\infty(\overline{D})$ is dense in $L_p(\overline{D})$, the last inequality is valid for any function $f(z) \in L_p(\overline{D})$ (Theorem 3 on extension of operators in §1 of Chapter II). Theorem 7 is proved.

Let us now prove that $\partial Tf/\partial z = Sf$ for $f \in L_p(\overline{D})$, $p > 1$. Suppose that $\varphi \in \overset{\circ}{C}_1(\overline{D})$. Then

$$\iint_D Tf \frac{\partial \varphi}{\partial z} \, dx \, dy = -\lim_{\varepsilon \to 0} \iint_D f(\zeta)\left[\frac{1}{\pi} \iint_{D_\varepsilon} \frac{1}{\zeta - z} \frac{\partial \varphi}{\partial z} dx \, dy \right] d\xi \, dy$$

$$= \lim_{\varepsilon \to 0} \left\{ \iint_D f(\zeta)\left[\frac{1}{\pi} \iint_{D_\varepsilon} \frac{\varphi(\zeta)}{(\zeta - z)^2} dx \, dy \right.\right.$$

$$\left.\left. - \frac{1}{2\pi i} \int_{\Gamma_\varepsilon} \frac{\varphi(z) - \varphi(\zeta)}{\zeta - z} d\bar{z} - \frac{\varphi(\zeta)}{2\pi i} \int_{\Gamma_\varepsilon} \frac{d\bar{z}}{\zeta - z} \right] d\xi \, d\eta \right\}$$

$$= -\iint_D \varphi(z)Sf dx \, dy,$$

because

$$\int_{\Gamma_\varepsilon} \frac{d\bar{z}}{\zeta - z} = 0 \quad \text{and} \quad \left| \int_{\Gamma_\varepsilon} \frac{\varphi(z) - \varphi(\zeta)}{\zeta - z} d\bar{z} \right| \leqslant 2\pi\varepsilon\|\varphi\|_{C^1}.$$

To conclude the proof of the main theorem it remains to show that for $f \in L_2(\overline{D})$

$$\|Sf\|_{L_2(\overline{D})} \leqslant \|f\|_{L_2(\overline{D})}.$$

Let $\overline{D}$ be a finite domain, suppose that $f \in \overset{\circ}{C}{}^1(D)$, $f \equiv 0$ outside D, and let $R > 0$ be large enough that $D \subset K_R = \{|z| < R\}$. Then, using Green's formulas (1.5) to integrate twice by parts, we get

$$\|Sf\|_{L_2(\overline{D})}^2 = \iint_D Sf \cdot \overline{sf} \, dx \, dy \leqslant \iint_{K_R} \frac{\partial Tf}{\partial z} \frac{\overline{\partial Tf}}{\partial z} dx \, dy$$

$$= -\iint_{K_R} Tf \frac{\overline{\partial^2 Tf}}{\partial \bar{z} \partial z} dx \, dy + I_R = \iint_{K_R} f(z) \overline{f(z)} \, dx \, dy - \frac{1}{2i} \int_{\Gamma_R} \bar{f} Tf dz + I_R.$$

But, since $f(z)$ is zero outside D,

$$\int_{\Gamma_R} \bar{f} Tf\, dz = 0 \quad \text{and} \quad \iint_{K_R} f(z)\overline{f(z)}\, dx\, dy = \iint_D |f(z)^2|\, dx\, dy.$$

And the integral I_R has the estimate

$$|I_R| = \left| \frac{1}{2i} \int_{\Gamma_R} Tf \cdot \bar{S}f\, d\bar{z} \right| \leq \frac{1}{2} 2\pi R \max|Tf \cdot \bar{S}f| = \frac{M}{R^2}.$$

For $f \in \overset{\circ}{C}{}^1(\overline{D})$ we have in the limit as $R \to \infty$ that

$$\|Sf\|_{L_2(\overline{D})} \leq \|f\|_{L_2(\overline{D})},$$

and, since the set of functions $f(z) \in \overset{\circ}{C}{}^1(\overline{D})$ is dense in $L_2(\overline{D})$, the last inequality holds also for all $f \in L_2(\overline{D})$.

REMARK. Many of the properties of the operators T and S are preserved also for infinite domains [128]. However, in what follows we restrict ourselves to the use of the properties proved above for these operators only in bounded domains.

Let us now consider the operators $\tilde{T}_0$ and $\tilde{T}_1$ acting on functions defined in the disk $K = \{|z| < 1\}$:

$$\tilde{T}_0 = -\frac{1}{\pi} \iint_K \left[\frac{f(t)}{t - z} - \frac{z\overline{f(t)}}{z\bar{t} - 1} \right] dK_t,$$

$$\tilde{T}_1 f = -\frac{1}{\pi} \iint_K \left[\frac{f(t)}{t - z} + \frac{z\overline{f(t)}}{z\bar{t} - 1} \right] dK_t,$$

along with their formal derivatives with respect to z

$$S_i f = -\frac{1}{\pi} \iint_K \left[\frac{f(t)}{(t - z)^2} \pm \frac{\overline{f(t)}}{(z\bar{t} - 1)^2} \right] dK_t$$

for $i = 0, 1$, respectively.

Since $|z\bar{t}| < 1$ for $|z| < 1$, the function

$$\tilde{T}(f|z) = -\frac{1}{\pi} \iint_K \frac{z\overline{f(t)}}{z\bar{t} - 1}\, dK_t$$

is holomorphic in $|z| < 1$, i.e., $\partial \tilde{T}f/\partial \bar{z} = 0$, whence $\partial \tilde{T}_i f/\partial \bar{z} = f$ and

$$\frac{\partial \tilde{T}_i f}{\partial z} = \frac{\partial}{\partial z}[Tf \mp \tilde{T}f] = S_i f, \quad i = 0, 1,$$

since we are allowed to differentiate with respect to z under the integral sign in $\tilde{T}f$. On the other hand, the representation

$$\tilde{T}_i(f|z) = T(f|z) \mp \overline{T(f|1/\bar{z})}, \quad i = 0, 1, \tag{14}$$

implies that the same smoothness properties hold for $\tilde{T}_i f$ and $S_i f$ as for Tf and Sf.

Let us prove that

$$\|S_i f\|_{L_2(\bar{K})} = \|f\|_{L_2(\bar{K})}. \tag{15}$$

Suppose that $f(z) \in \mathring{C}^1(\bar{K})$. Integrating twice by parts, we get

$$\|S_i f\|_{L_2}^2 = \iint\limits_K \frac{\partial \tilde{T}_i f}{\partial z} \frac{\overline{\partial \tilde{T}_i f}}{\partial z} dx\, dy = I - \iint\limits_K \overline{\tilde{T}_i f} \frac{\partial^2 \tilde{T}_i f}{\partial z\, \partial \bar{z}} dx\, dy$$

$$= \iint\limits_K |f(z)|^2 dx\, dy + I.$$

Taking account of the special case of (14) for $z = e^{i\gamma}$, namely

$$\tilde{T}_i(f \mid e^{i\gamma}) = \mp \overline{\tilde{T}_i(f \mid e^{i\gamma})}, \qquad i = 0, 1,$$

and the fact that $\tilde{T}_i f$ is holomorphic near the boundary Γ (since $f \equiv 0$ in a neighborhood of Γ), we have for the integral conjugate to I that

$$\bar{I} = \frac{1}{2i} \int_\Gamma \overline{\tilde{T}_i f} \frac{\partial \tilde{T}_i f}{\partial z} dz = \pm \frac{1}{2i} \int_\Gamma \tilde{T}_i f \frac{\overline{\partial \tilde{T}_i f}}{\partial z} dz = \mp \frac{1}{4i} \int_\Gamma \frac{d}{dz} [\tilde{T}_i f]^2 dz = 0,$$

which proves (15), since the set of all $f(z) \in \mathring{C}^1(\bar{K})$ is dense in $L_2(\bar{K})$.

Thus, we have proved all the assertions of the main theorem also for the operators $\tilde{T}_i$ and S_i studied by I. N. Vekua in [128]. Obviously, all of this is true also for the operators

$$T_i f = \tilde{T}_i(f \mid z) - \tilde{T}_i(f \mid 1), \qquad S_i f = \frac{\partial T_i f}{\partial z} = \frac{\partial \tilde{T}_i f}{\partial z}, \qquad i = 0, 1. \tag{16}$$

For $z = e^{i\gamma}$, $\gamma \in [0, 2\pi]$, formulas (14) and (16) give us

$$\operatorname{Re} T_0(f \mid e^{i\gamma}) = \operatorname{Im} T_1(f \mid e^{i\gamma}) = 0, \qquad T_i(f \mid 1) = 0. \tag{17}$$

REMARK. The assertion of the main theorem about the properties of T and S in the spaces L_p is valid also for the following operators, which are more general than T_0 and T_1 (see [95]):

$$T_n^\Pi \varphi = -\frac{\Pi(z)}{\pi} \iint\limits_K \left[\frac{\varphi(t)}{\Pi(t)(t-z)} + \frac{z^{2n-1} \overline{\varphi(t)}}{\overline{\Pi(t)}\,(1 - z\bar{t})} \right] dK_t, \qquad n \geqslant 2,$$

$$T_n^\Pi \varphi = -\frac{\Pi(z)}{\pi} \iint\limits_K \left[\frac{\varphi(t)}{\Pi(t)(t-z)} + \frac{t^{-2n-1} \overline{\varphi(t)}}{\overline{\Pi(t)}\,(1 - z\bar{t})} \right] dK_t, \qquad n \leqslant -1,$$

where $\Pi(z) = \Pi_1^m (z - z_k)^{\alpha_k}$, $|z_k| \leqslant 1$, $\alpha_k > 0$, and $\Sigma_1^m \alpha_k \leqslant 1$.

The functions $T_n^1(\varphi \,|\, z)$, $\Pi \equiv 1$, satisfy the following boundary conditions for $|z| = 1$:

$$\operatorname{Re}\left[z^{1-n}T_n^1(\varphi \,|\, z)\right] = 0, \quad n \geqslant 2; \qquad \operatorname{Re}\left[\bar{z}^n T_n^1(\varphi \,|\, z)\right] = 0, \quad n \leqslant -1$$

These and more general operators will be studied in §5 of Chapter VI.

§3. Homeomorphisms of the whole plane and representations of generalized solutions of elliptic systems of equations

A one-to-one and bicontinuous mapping $w = w(z) \equiv u(x, y) + iv(x, y)$ in a domain D_z is called a homeomorphic mapping or simply a homeomorphism of D_z. A mapping $w = w(z)$ that is one-to-one and bicontinuous in a neighborhood of each point $z \in D_z$ is called a local homeomorphism.

A differentiable mapping $w = w(z) \in C^1(D_z)$ whose Jacobian $J = u_x v_y - u_y v_x \equiv |w_z|^2 - |w_{\bar{z}}|^2$ is nonzero in D_z is an example of a local homeomorphism.

A continuous mapping $w = w(z)$ is said to be light interior if it can be represented in the form

$$w = \Phi[\zeta(z)],$$

where $\zeta(z)$ is a homeomorphism of a domain D_z and $\Phi(\zeta)$ is an arbitrary analytic function in $D_\zeta \equiv \zeta(D_z)$.

Light interior mappings have all the topological properties of conformal mappings. In particular, these mappings are domain preserving. The definition of a light interior mapping given above differs from that of Stoïlow [125], who first introduced and studied light interior mappings. He calls a continuous mapping $w = w(z)$ interior if

1) it carries any open set $\mathfrak{M} \subset D_z$ to an open set, and
2) no continuum in D_z that is not a point is mapped into a single point w.

Obviously, every homeomorphism is light interior in the sense of the latter definition. On the other hand, any analytic function realizes a light interior mapping, and, thus, the superposition of an analytic function and a homeomorphism is also a light interior mapping.

A fundamental result of Stoïlow is his theorem that any light interior mapping can be obtained by superposition of analytic functions and homemorphisms [125]. Thus, the last property of light interior mappings can actually be used as the definition of such a mapping.

Let us consider a nonlinear system of equations in the functions u and v:

$$F_i(u, v, x, y, u_x, u_y, v_x, v_y) = 0, \qquad i = 1, 2. \tag{2}$$

We call $w(z) = u + iv$ a generalized solution of the system (2) in a domain D_z if $w(z) \in W_p^1(D_z)$, $p \geqslant 2$, and the real and imaginary parts of $w(z)$ satisfy this system almost everywhere in D_z.

DEFINITION. A light interior mapping $w = w(z)$ that is a generalized solution of the system (2) in D_z is called a *quasiconformal mapping satisfying the system* (2), or simply a *quasiconformal mapping*.

Suppose that the system (2) is quasilinear:

$$-v_y + a_{11}u_x + a_{12}u_y + a_0u + b_0v = f,$$

$$v_x + a_{21}u_x + a_{22}u_y + c_0u + d_0v = g, \qquad (3)$$

where a_0, b_0, c_0, d_0 and a_{ij} ($i, j = 1, 2$) are bounded measurable functions of u, v, x, and y. The system (3) is said to be uniformly elliptic in D_z at a solution $w = u(x, y) + iv(x, y)$ if after substitution of $w(z)$ in the coefficients a_{ij} the following inequality holds in D_z:

$$a_{11}a_{22} - \tfrac{1}{4}(a_{12} + a_{21})^2 \geqslant E_0 = \text{const} > 0. \qquad (4)$$

It is more convenient to write the system (3) in the form of a single complex equation

$$w_{\bar{z}} - \mu_1(w, z)w_z - \mu_2(w, z)\overline{w_{\bar{z}}} + Aw + B\bar{w} = F, \qquad (5)$$

where

$$\mu_1 = \frac{2a}{l}, \qquad \mu_2 = -\frac{1}{l}\left(|a|^2 + (1 + \bar{b})(1 - b)\right), \qquad l = |a|^2 - |1 + b|^2,$$

$$2a = a_{11} - a_{22} + i(a_{21} + a_{12}), \qquad 2b = a_{11} + a_{22} - i(a_{12} - a_{21}),$$

and the functions A, B, and F are defined linearly in terms of a_0, b_0, c_0, d_0 and depend on a_{ij} ([128], Chapter III, §17.1).

In complex notation the ellipticity condition (4) takes the form

$$|\mu_1| + |\mu_2| \leqslant \mu_0(E_0) < 1. \qquad (6)$$

For the case when equation (5) is homogeneous in the derivatives, i.e., when $A = B = F \equiv 0$, it will be proved that each generalized solution $w(z)$ of this equation (if (6) is satisfied when it is substituted in the coefficients μ_1 and μ_2) realizes a quasiconformal mapping. We first dwell on a proof of this assertion for the Beltrami equations

$$w_{\bar{z}} - \mu(z)w_z = 0, \qquad (7)$$

$$|\mu(z)| \leqslant \mu_0 < 1, \qquad z \in D. \qquad (8)$$

In accordance with the analytic scheme we have adopted for presenting the theory of quasiconformal mappings we shall construct a mapping that is differentiable and has positive Jacobian $(J(z) = |\zeta_z|^2 = |\zeta_z|^2 - |\zeta_{\bar{z}}|^2 \geqslant \delta > 0)$ in the whole plane E under the assumption that $\mu(z)$ is smooth in E and that $\mu(z) \equiv 0$ outside a finite domain D. Suppose that such a solution $\zeta = \zeta(z)$ has been found. We remark that it has in a neighborhood $K_\varepsilon = \{|z - z_0| < \varepsilon\}$ of each point $z_0 \in E$ an expansion

$$\zeta(z) - \zeta_0 = \zeta_z(z_0)(z - z_0) + \zeta_{\bar{z}}(\bar{z} - \bar{z}_0) + |z - z_0|^{1+\alpha} \cdot A(z),$$

where $A(z) \in C(K_\varepsilon)$ and $0 < \alpha_0 < \alpha$; with the equality $\mu(z_0) = \zeta_{\bar{z}}(z_0)[\zeta_z(z_0)]^{-1}$ taken into account it can be written in the form

$$\zeta(z) - \zeta_0 = [z - z_0 + \mu(z_0)(\bar{z} - \bar{z}_0)] H(z),$$

where $H(z_0) \neq 0$ and $H(z) \in C(K_\varepsilon)$. Moreover, in a neighborhood K_ε of each point z_0 there exists an inverse homeomorphism $z = z(\zeta)$ satisfying the equation

$$z_{\bar{\zeta}} + \mu[z(\zeta)]\bar{z}_{\bar{\zeta}} = 0 \tag{7*}$$

obtained by substituting the expressions $\zeta_z = J(z)\bar{z}_{\bar{\zeta}}$ and $\zeta_{\bar{z}} = -J(z)z_{\bar{\zeta}}$ into (7). Then, considering for an arbitrary solution $w(z) \in C^{1+\alpha}(K_\varepsilon)$ of (7) the composite function $w = w[z(\zeta)] = \Phi(\zeta)$, we find that

$$\frac{\partial\Phi}{\partial\bar{\zeta}} = \frac{\partial w[z(\zeta)]}{\partial\bar{\zeta}} = w_z z_{\bar{\zeta}} + w_{\bar{z}}\bar{z}_{\bar{\zeta}} = w_z(z_{\bar{\zeta}} + \mu\bar{z}_{\bar{\zeta}}) = 0.$$

The last equality proves that $w = \Phi(\zeta)$ is holomorphic on the plane of the homeomorphism $\zeta = \zeta(z)$, and thereby also that the mapping $w = w(z)$ is quasiconformal for an arbitrary differentiable solution $w(z)$ of (7) in a neighborhood of each point z_0. The representation $w = \Phi[\zeta(z)]$ of the solutions of (7), where $\Phi_{\bar{\zeta}} = 0$, in a neighborhood of each zero z_* of $w(z)$ gives us that

$$w(z) = [\zeta(z) - \zeta(z_0)]^n\Phi_0[\zeta(z)] \equiv [z - z_0 + \mu(z_0)(\bar{z} - \bar{z}_0)]^n[H(z)]^n\Phi_0,$$

where $n \geqslant 1$ is an integer, and $\Phi_0[\zeta(z_*)] \neq 0$, $H(z_*) \neq 0$. If the solution $w(z)$ of (7) is nonzero on the boundary of a finite domain D and $w(z) \in C(\bar{D})$, then $w(z)$ obviously has finitely many zeros in $\bar{D}$, and

$$w(z) = w_0(z) \prod_{k=1}^{N} [z - z_k + \mu(z_k)(\bar{z} - \bar{z}_k)],$$

where the zeros z_k are repeated as many times as their multiplicity, $w_0(z) \in C(\bar{D})$, and $w_0(z) \neq 0$ in $\bar{D}$. The latter representation implies the argument principle for $w(z)$, since

$$\frac{1}{2\pi}\{\arg w_0(z)\}_\Gamma = 0 \quad \text{and} \quad \frac{1}{2\pi}\{\arg[z - z_k + \mu(z_k)(\bar{z} - \bar{z}_k)]\} = 1,$$

and therefore $\{\arg w(z)\}_\Gamma/2\pi = N$.

Thus, to prove that the differentiable mappings $w = w(z)$ satisfying the Beltrami equation (7) are quasiconformal and, as a corollary, that the argument principle holds for them it suffices to find a local homeomorphism satisfying this equation in a neighborhood of each point $z_0 \in E$; we pass to the construction of such a mapping. In the case of generalized solutions $w(z) \in W_p^1(\overline{D})$ of (7) we shall establish the quasiconformality of the mappings $w = w(z)$ by approximating them by differentiable mappings.

THEOREM 1. *Suppose that* $\mu(z)$ *is a measurable function in a finite domain* D *in the plane* E, $\mu(z) \equiv 0$ *for* $z \notin D$, *and* $\mu(z)$ *satisfies inequality* (8) *almost everywhere in* D. *Then there exists a unique generalized solution* $w(z)$ *of the Beltrami equation* (7) *such that* $w(0) = 0$ *and* $(w_z - 1) \in L_p(E), p > 2$.

We look for a solution $w(z)$ in the form $w = z + \overset{\circ}{T}f$, where $\overset{\circ}{T}(f|z) = T(f|z) - T(f|0)$, and $f(z) \in L_p(\overline{D})$, $p > 2$, $f(z)$ being an unknown function. Taking the fact that $\partial \overset{\circ}{T}/\partial \bar{z} = f$ and $\partial \overset{\circ}{T}/\partial \bar{z} = Sf$ into account, we get from (7) that

$$f(z) = \mu(z)Sf + \mu(z). \tag{9}$$

Since $\|S\|_{L_p}$ is convex (Theorem 4 in §2 of Chapter II), it is possible to choose $p > 2$ so close to 2 that $k = \|S\|_{L_p}\mu_0 = \Lambda_p\mu_0 < 1$. Then $\|\mu Sf\|_{L_p} \leqslant k\|f\|_{L_p}$, and Banach's principle (Theorem 5 in §1 of Chapter II) can be applied to (9) in $L_p(D)$ with the chosen $p(\mu) > 2$. Consequently, there exists a unique solution $f(z) \in L_p(D)$ of (9), and

$$\|f\|_{L_p} \leqslant \|\mu\|_{L_p} + \|\mu Sf\|_{L_p} \leqslant \mu_0(\text{meas } D)^{1/p} + \mu_0\Lambda_p\|f\|_{L_p},$$

whence

$$\|f\|_{L_p} \leqslant \frac{\mu_0(\text{meas } D)^{1/p}}{1 - \mu_0\Lambda_p} = N. \tag{10}$$

The function $w(z) = z + \overset{\circ}{T}f$ thus constructed satisfies all the requirements of the theorem. Indeed, $w(0) = 0$ by construction, and $w_z - 1 = Sf \in L_p(E)$ by property 3 of S (the Main Theorem in §2).

Generalized solutions of the Beltrami equation that satisfy the conditions $w(0) = 0$ and $w_z - 1 \in L_p(E)$, $p > 2$, will be called *basic*. We prove that a basic solution is unique. Suppose that $w^1(z)$ and $w^2(z)$ are two basic solutions, and consider their difference $F(z) = w^1(z) - w^2(z)$, which obviously satisfies (7) and the conditions $F(0) = 0$ and $F_z = (w_z^1 - w_z^2) \in L_p(E), p > 2$. We form the function $\Omega(z) = F(z) - \overset{\circ}{T}(F_{\bar{z}})$. Since $\Omega_{\bar{z}} = 0$ for $z \in E$, Theorem 3 in §2 tells us that it is holomorphic in E. But $\Omega_z = F_z - S(F_{\bar{z}}) \in L_p(E)$, $p > 2$, which is possible only if $\Omega_z = 0$ at $z = \infty$. Consequently, by Liouville's theorem, $\Omega_z = F_z - S(F_{\bar{z}}) \equiv 0$. Substituting the equality $F_{\bar{z}} = \mu F_z$ from (7) into the last equality, we get the homogeneous equation

$$F_z - S(\mu F_z) = 0,$$

which, since $\|S(\mu F_z)\|_{L_p} \leqslant \Lambda_p \mu_0 \|F_z\|_{L_p}$ and $k = \Lambda_p \mu_0 < 1$, gives us that $F_z \equiv 0$. Then (7) implies that $F_{\bar{z}} \equiv 0$, i.e., F is holomorphic in the whole plane; from this we conclude, as in the case of $\Omega(z)$, that $F(z) = w^1 - w^2 \equiv 0$, and the theorem is proved. Inequality (10) and the estimate for T (inequality (7) in §2) imply that

$$|w(z_1) - w(z_2)| \leqslant M_2 N |z_1 - z_2|^{(p-2)/p} + |z_1 - z_2|. \tag{11}$$

COROLLARY. *If a sequence $\{\mu^n(z)\}$ of functions satisfying the conditions of the theorem converges almost everywhere in D (the constant μ_0 in (8) does not depend on n), then the corresponding sequence $\{w^n - z\}$, where the $w^n(z)$ are the basic solutions, converges in $W_p^1(E)$, $p > 2$.*

Indeed, representing the basic solutions in the form $w^n = z + Tf^n$, we obtain a sequence of functions $f^n(z) \in L_p(\overline{D})$ satisfying the equations

$$f^n = \mu^n Sf^n + \mu^n, \tag{9*}$$

and from this, similarly to the preceding, we find that

$$\|f^n\|_{L_p} \leqslant \frac{\mu_0 (\text{meas } D)^{1/p}}{1 - \mu_0 \Lambda_p} = N. \tag{10*}$$

Next, (9*) gives us that

$$(1 - \mu_0 \Lambda_p) \|f^n - f^m\|_{L_p} \leqslant \|\mu^n - \mu^m\|_{L_p} + \|(\mu^n - \mu^m) Sf^m\|_{L_p}.$$

But

$$\|(\mu^n - \mu^m) Sf^m\|_{L_p} \leqslant \|\mu^n - \mu^m\|_{L_{pp'}} \|Sf^m\|_{L_{pp'}}, \qquad \frac{1}{pp'} + \frac{1}{pq'} = \frac{1}{p}.$$

Therefore, choosing $p' > 1$ close enough to 1 so that $\mu_0 \Lambda_{pp'} < 1$ and taking into account the inequality (10*) in $L_{pp'}$, we get

$$\|f^n - f^m\|_{L_p} \leqslant C(p) \|\mu^n - \mu^m\|_{L_{pq'}},$$

which implies the convergence in $L_p(D)$, $p = p(\mu_0) > 2$.

REMARKS. 1. It is clear that basic solutions $w = z + Tf$ also exist for the general linear equation

$$w_{\bar{z}} - \mu_1(z) w_z - \mu_2(z) \bar{w}_{\bar{z}} = 0 \qquad (|\mu_1| + |\mu_2| \leqslant \mu_0 < 1), \tag{7**}$$

and for the function $f(z)$, as a solution of the equation analogous to (9),

$$f = \mu_1 Sf + \mu_2 \overline{Sf} + \mu_1 + \mu_2,$$

the previous estimate (10) holds. In view of this, the assertions of the corollary also remain in force for the solutions of (7**).

2. For the assertions of the corollary to be true it suffices to assume that $\{\mu^n(z)\}$ converges in measure. Indeed, the above proof remains valid also in this case, since convergence in measure of the functions $\mu^n(z)$, $|\mu^n(z)| < \mu_0$, implies their convergence in $L_p(\overline{D})$ for any $p > 1$, by the Lebesgue dominated convergence theorem ([99], Chapter V, §3).

THEOREM 2. *Suppose that* $\mu(z) \in W_p^2(E)$, $p > 2$, $\mu(z) \equiv 0$ *for* $z \notin D$, *and condition* (8) *holds. Then* $[w(z) - z] \in C^{2+\alpha}(E)$, $\alpha(\mu_0) > 0$, *for the basic solution* $w(z)$, *and* $w = w(z)$ *is a homeomorphism of the whole plane, with Jacobian* $J \geqslant \delta > 0$.

We look for a function $\varphi(z) \in W_p^1(\overline{D})$, $p > 2$, such that the solution of (7) can be represented in the form

$$w_z = e^{T\varphi}; \qquad w_{\bar{z}} = \mu e^{T\varphi}.$$

By the imbedding theorem proved above (Theorem 3 in §1) we have $\mu \in C^{1+\alpha}$, and $\varphi \in C^\alpha(D)$, so that $T\varphi \in C^{1+\alpha}(D)$. Therefore, for the preceding equalities to be possible it suffices that $e^{T\varphi}dz + \mu e^{T\varphi}d\bar{z}$ be a total differential. From this condition we get

$$\varphi = \mu S\varphi + \mu_z. \tag{12}$$

Suppose that the solution $\varphi(z) \in W_p^1(\overline{D})$, $p > 2$, has been constructed. Then the function

$$w(z) = \int_0^z e^{T\varphi}dz + \mu e^{T\varphi}d\bar{z}, \qquad z \in E, \tag{13}$$

where the integral does not depend on the path of integration, is the basic solution of (7). Indeed, $w(0) = 0$ and $w_z = e^{T\varphi}$ is holomorphic outside D and equals 1 at infinity, i.e., $(w_z - 1) \in L_p(E)$, $p > 2$. But (12) gives us $\|\varphi\|_{L_p} \leqslant \|\mu_z\|_{L_p}(1 - \Lambda_p\mu_0)$, whence

$$|T\varphi| < M_1\|\mu_z\|_{L_p}(1 - \mu_0\Lambda_p)^{-1} = M$$

(see (2.7)) and, consequently,

$$J(z) = |w_z|^2 - |w_{\bar{z}}|^2 = (1 - \mu_0^2)e^{2\operatorname{Re}T\varphi} \geqslant (1 - \mu_0^2)e^{-2M},$$

or

$$J(z) \geqslant \delta > 0 \quad \text{on } E, \tag{14}$$

where

$$\delta = (1 - \mu_0^2)\exp\{-M_1\|\mu_{\bar{z}}\|_{L_p}(1 - \Lambda_p\mu_0)^{-1}\}.$$

Thus, the mapping $w = w(z)$ represented by (13) is locally homeomorphic in a neighborhood of each point $z \in E$, and $[w(z) - z] \in C^{1+\alpha}(E)$ for $\varphi \in W_p^1(D)$ by construction.

Let us next consider the function $\Omega(z) = w(z) - w_0$, $w_0 = \text{const}$; obviously, together with $w(z)$ this is a solution of (7), and, since $w(z)$ is single-valued in a neighborhood of $z = \infty$ (there $w(z)$ is holomorphic), $\Omega(z) \neq 0$ on a disk $K_R = \{|z| < R\}$ of sufficiently large radius. Consequently, according to the arguments at the beginning of the section, the argument principle is applicable to the function $\Omega(z)$ in K_R and, since $\Omega(z) = z + O(z^{-1})$ as $z \to \infty$, whence $\{\arg \Omega(z)\}_{|z|=R}/2\pi = 1$, it follows that there is a unique point in K_R such that $\Omega(z_0) = w(z_0) - w_0 = 0$, i.e., the mapping $w = w(z)$ is one-to-one on the whole z-plane. Thus, to prove the theorem it suffices to establish the existence of a solution $\varphi(z) \in W_p^1(\overline{D})$, $p > 2$. If we establish the existence of a solution to (12) of the form $\varphi = Tf, f \in L_p(\overline{D})$, $p > 2$, then the theorem will thereby be proved. We substitute $\varphi = Tf$ in (12) and differentiate both sides of the resulting equality with respect to $\bar{z}$:

$$f = \mu Sf + \mu_{\bar{z}} S(Tf) + \mu_{z\bar{z}}.$$

Let $p_0 = p(\mu_0) > 2$ be such that $\Lambda_{p_0}\mu_0 < 1$. Then, inverting the operator $I - \mu S$ in $L_{p_0}(\overline{D})$, we find that

$$f = (I - \mu S)^{-1}[\mu_{\bar{z}} S(Tf) + \mu_{z\bar{z}}] \equiv Af + F. \qquad (15)$$

Since the operators $(I - \mu S)^{-1}$ and S are linear and bounded in L_{p_0}, while T is completely continuous, the operator A is obviously completely continuous in $L_{p_0}(\overline{D})$. Let us consider the homogeneous equation $f_0 = Af_0$, which is equivalent to the equation

$$f_0 - \mu Sf_0 - \mu_{\bar{z}} S(Tf_0) \equiv \frac{\partial}{\partial \bar{z}}[Tf_0 - \mu S(Tf_0)] = 0.$$

This equation implies that $\Phi(z) = Tf_0 - \mu S(Tf_0)$ is a holomorphic function in the whole plane E, and, since it vanishes at infinity, Liouville's theorem implies that

$$Tf_0 - \mu S(Tf_0) = 0, \qquad z \in E.$$

The last equation, which is homogeneous with respect to Tf_0, has the unique solution $Tf_0 \equiv 0$ because $I - \mu S$ is an invertible operator, and this implies that $f_0 \equiv 0$, too. Thus, the homogeneous equation (15) has only the trivial solution. Therefore, according to Fredholm's theorem (Theorem 10 in §1 of Chapter II), for the operator A the nonhomogeneous equation (15) is uniquely solvable.

Consequently, we have constructed a solution $f(z) \in L_p(\overline{D})$, $p > 2$, of the equation

$$\frac{\partial}{\partial \bar{z}} \left[Tf - \mu S(Tf) - \mu_z \right] = 0,$$

and this implies that the function $\Omega(z) = Tf - \mu S(Tf) - \mu_z$ is holomorphic in the whole plane E; since $\Omega(\infty) = 0$, Liouville's theorem gives us that

$$Tf = \mu S(Tf) + \mu_z,$$

i.e., the function $\varphi(z) = Tf$ satisfies (12). The theorem is proved.

REMARK. It is easy to see that the function $z = z(w)$ inverse to the homeomorphism $w = w(z)$ constructed in Theorem 1 is the basic solution of the Beltrami equation

$$z_{\bar{w}} - \tilde{\mu}(w) z_w = 0, \tag{7***}$$

where $\tilde{\mu}(w) = -\mu[z(w)] \bar{z}_{\bar{w}} / z_w$. By a simple substitution of the expressions $w_z = I(z) \bar{z}_{\bar{w}}$ and $w_{\bar{z}} = -I(z) z_{\bar{w}}$ into (7) we see that the function $z = z(w)$ satisfies (7***). Further,

$$\iint_E |z_w - 1|^p \, dE_w = \iint_E |z_w - 1|^p I(z) \, dE_z = \iint_D + \iint_{E-D}.$$

But the integral over the domain D is obviously bounded, and $w_{\bar{z}} = 0$ for $z \notin D$; therefore, $|w_z|^2 = I(z) \geqslant \delta > 0$. Then

$$\bar{z}_w - 1 = \frac{w_z}{I(z)} - 1 = \frac{w_z}{|w_z|^2} - 1 = \left(\frac{1 - \bar{w}_{\bar{z}}}{\bar{w}_{\bar{z}}} \right) \in L_p(E - D), \qquad p > 2,$$

because $w = w(z)$ is the basic solution and $(w_z - 1) \in L_p(E - D)$, $p > 2$. Thus, $z(0) = w(0) = 0$ and $(z_w - 1) \in L_p(E)$, $p > 2$, i.e., $z = z(w)$ really is the basic solution of the Beltrami equation (7***) and it satisfies, in particular, an inequality of the form (11):

$$|z_1 - z_2| \leqslant M_2 N |w(z_1) - w(z_2)|^{(p-2)/p} + |w(z_1) - w(z_2)|. \tag{16}$$

THEOREM 3. *Suppose that $\mu(z)$ is a measurable function, $\mu(z) \equiv 0$ for $z \notin D$, and $\mu(z)$ satisfies (8). Then the basic solution $w(z) = z + \overset{\circ}{T} f$ of the Beltrami equation (7) is a homeomorphism of the whole plane E.*

Let us consider a set of functions $\mu^n \in C^2(E)$ such that $\mu^n \to \mu$ almost everywhere in D, $\mu^n \equiv 0$ for $z \notin D$, and $|\mu^n| \leqslant \mu_0 < 1$. It is known that such functions can be constructed, for example, by averaging the function $\mu(z)$. Then the basic solutions $w^n = z + \overset{\circ}{T} f^n$ satisfy the Beltrami equations (7) with μ^n in place of μ, and, since (by the corollary to Theorem 1) $f_n \to f$ as $\mu^n \to \mu$, it follows that $w^n \to w(z) = z + \overset{\circ}{T}f$, $f \in L_p(\overline{D})$, $p(\mu_0) > 2$. Passing to the limit in the inequalities (16) for the inverse homeomorphisms $z^n(w)$, we get

$$|z_1 - z_2| \leqslant M_2 N |w(z_1) - w(z_2)|^{(p-2)/p} + |w(z_1) - w(z_2)|,$$

which implies that two different points z_1 and z_2 cannot correspond to a single point w. Consequently, $w(E)$ is a homeomorphism of the whole plane E. Let us now get an estimate for the measure of the image of an open set $\mathfrak{M} \subset D$. For mappings $w^n(z)$ with $\mu^n \in C^2(D)$ we have

$$\text{meas } w^n(\mathfrak{M}) = \int_{\mathfrak{M}} \left(|w_z^n|^2 - |w_{\bar{z}}^n|^2 \right) dx\, dy \leqslant \iint_{\mathfrak{M}} |w_z^n|^2 dx\, dy$$

$$\leqslant 2\|w_z^n\|_{L_p} (\text{meas } \mathfrak{M})^{(p-2)/p},$$

and, since the $\|w_z^n\|_{L_p}$ are bounded independently of n, this inequality holds also in the limit, so that

$$\text{meas } w(\mathfrak{M}) \leqslant 2(1 + \Lambda_p N)(\text{meas } \mathfrak{M})^{(p-2)/p}, \tag{17}$$

where $N = \mu_0 (\text{meas } D)^{1/p} / (1 - \mu_0 \Lambda_p)$ (see (10)). As a direct consequence of (11) and (16) in $\overline{D}$ we have

$$C^{-1/\alpha} |z_1 - z_2|^{1/\alpha} \leqslant |w(z_1) - w(z_2)| \leqslant C |z_1 - z_2|^{\alpha}, \tag{18}$$

where $\alpha = \alpha(\mu_0) > 0$ and $C = C(\mu_0, D)$.

REMARK. Since the basic solutions $w = z + \overset{\circ}{T}f$ of the general linear equation (7**) are also solutions of the Beltrami equation with coefficient

$$\mu(z) = \mu_1(z) + \mu_2(z)\frac{\overline{w_{\bar{z}}}}{w_z} \qquad \left(|\mu| \leqslant \mu_0 < 1\right),$$

these solutions also are homeomorphisms of the whole plane onto itself.

THEOREM 4. *Suppose that $\mu(z) \in W_p^l(E)$, $p > 2$, $l \geqslant 1$, $\mu \equiv 0$ for $z \notin \overline{D}$, and $\mu(z)$ satisfies (8). Then $w(z) = z + \overset{\circ}{T}\varphi$ is a homeomorphism of the whole plane E and is in $W_p^{l+1}(E)$, and its Jacobian $J = |w_z|^2 - |w_{\bar{z}}|^2 \geqslant \delta > 0$ everywhere on E.*

The fact that $w = w_z = z + \overset{\circ}{T}\varphi$ is a homeomorphism follows from the previous theorem. Let us form the equation

$$\varphi - \mu S\varphi = \mu \tag{9'}$$

for $\varphi(z)$ and show that for $\mu(z) \in W_p^1(E)$ (as in the proof of Theorem 2) there is a solution of the form $\varphi = Tf$. If $\mu(z) \in W_p^l(E)$, $l \geqslant 2$, then by repeating the arguments for Theorem 2, we can successively construct solutions of (9) of the form $\varphi = T(T(\ldots T(Tf)\ldots)) \equiv T'f$, where $f \in L_p(\overline{D})$, $p > 2$, which proves the necessary smoothness of $w(z)$:

$$w(z) \in W_p^{l+1}(E), \qquad p > 2.$$

According to the inequality (14) obtained for Theorem 2, if $l \geqslant 2$, then

$$J = |w_z|^2 - |w_{\bar{z}}|^2 \geqslant \delta > 0 \quad \text{on } E.$$

For $l = 1$ we approximate $\mu(z)$ by compactly supported functions $\mu^n(z) \in W_p^2(E)$ in such a way that $\|\mu^n(z) - \mu(z)\|_{W_p^1(\overline{D})} \to 0$ as $n \to \infty$.

Then, as in the corollary to Theorem 1, the convergence of the coefficients in the equation

$$\varphi^n - \mu^n S\varphi^n = \mu_{\bar{z}}^n ST\varphi^n + \mu_{\bar{z}}^n$$

obtained by differentiating both sides of (9) with respect to $\bar{z}$ implies the convergence $\varphi^n \to \varphi$ in $L_p(\overline{D})$, $p > 2$. Consequently, the constructed sequence of homeomorphisms $w^n = z + \overset{\circ}{T}\varphi^n$ with $\varphi^n = Tf^n$ converges in $W_p^2(E)$, $p > 2$, which, by the imbedding theorem, implies that $w^n \to w$ in $C^{1+\alpha}$. By passing to the limit in (14) (in which the δ does not depend on n), we get the required inequality $J \geqslant \delta > 0$ on E also in the case $\mu \in W_p^1(E)$, $p > 2$.

THEOREM 5. *Any generalized solution* $w = w(z)$ *of the Beltrami equation* (7) *realizes a quasiconformal mapping, i.e.,* $w(z)$ *can be represented in the form*

$$w = \Phi[\zeta(z)], \tag{19}$$

where $\zeta = \zeta(z)$ *is a homeomorphism satisfying* (7), *and* $\Phi = \Phi(\zeta)$ *is an analytic function.*

A simple check shows that $\Phi[\zeta(z)]$ satisfies (7), where the basic solution $\zeta = z + \overset{\circ}{T}\varphi$ can be taken as the homeomorphism.

Let us consider the composite function $w = w[z(\zeta)]$, where $z(\zeta)$ is the inverse homeomorphism.

Assuming for the time being that the usual formulas for differentiating composite functions hold for generalized derivatives (both $w(z)$ and $z(\zeta)$ have only generalized derivatives!), we find with the help of (7) for $w(z)$ that

$$\frac{\partial w[z(\zeta)]}{\partial \bar{\zeta}} = \left(\frac{\partial w}{\partial z} \left(\frac{\partial z}{\partial \bar{\zeta}} + \mu \frac{\partial \bar{z}}{\bar{\zeta}} \right) \right).$$

Since $z_{\bar{\zeta}} + \mu \bar{z}_{\bar{\zeta}} = 0$ for $z(\zeta)$, it follows that $w_{\bar{\zeta}}[z(\zeta)] = 0$, and this implies (19).

Let us now prove that it is possible to differentiate the composite function $f = f[w(z)]$ by the usual formulas, where $w(z) \in W_p^1(\overline{D})$, $p > 2$ is a homeomorphism for (7), and $f(w) \in W_p^1$, $p > 2$, is an arbitrary function defined in the domain $w(D)$.

The following assertion holds.

LEMMA. *Suppose that* $w(z) \in W_p^1(K)$, $p > 2$, *and* $w = w(z)$ *is a homeomorphism of the unit disk onto itself that satisfies a Beltrami equation with coefficient* $\mu(z)$, $|\mu(z)| \leqslant \mu_0 < 1$. *If* $f(w) \in W_p^1(\overline{K})$, *then the composite function* $f(z) \equiv f[w(z)]$ *belongs to the set* $W_{p_0}^1(\overline{K})$, $2 < p_0 < p$.

Since $f[w(z)]$ is a bounded function, to prove the lemma it suffices to show that f_z and $f_{\bar{z}}$ are in $L_p(\overline{K})$. We approximate $f(w)$ by smooth functions $f^n(w)$ (for which the usual differentiation formulas are valid) and use Hölder's inequality with $v = p/p_0$ and $s = p/(p - p_0)$ $(1/v + 1/s = 1)$ in the following expression:

$$\iint_K |f_z^n|^{p_0} dK_z = \iint_K |f_w^n w_z + f_{\bar{w}}^n \overline{w}_{\bar{z}}|^{p_0} \left(|z_w|^2 - |z_{\bar{w}}|^2\right) dK_w$$

$$\leqslant \frac{(1 + \mu_0)^{p_0}}{1 - \mu_0^2} \|f^n(w)\|_{W_p^1(\overline{K}_w)} \left\{ \iint_K |w_z|^{(p_0 - 2)p/(p - p_0)} dK_w \right\}^{(p - p_0)/p}$$

$$= M(f^n) \left\{ \iint_K |w_z|^{(p_0 - 2)p/(p - p_0)} \left(|w_z|^2 - |w_{\bar{z}}|^2\right)^2 dK_z \right\}^{(p - p_0)/p}$$

$$\leqslant M\left(1 + \mu_0^2\right)^{(p - p_0)/p} \|w\|_{W_p^1}^{p - p_0},$$

where

$$p_0 = \frac{p^2}{2(p - 1)} < p \quad \text{and} \quad \frac{p_0 - 2}{p - p_0} p + 2 = p, \qquad p_0 > 2.$$

The estimate we have obtained is obviously true also for $f_{\bar{z}}^n$.

In the intermediate computations we used obvious inequalities for the homeomorphisms $w(z)$ and $z = z(w)$:

$$|w_{\bar{z}}| \leqslant \mu_0 |w_z|, \qquad \left(1 - \mu_0^2\right)|w_z|^2 \leqslant |w_z|^2 - |w_{\bar{z}}|^2 \leqslant \left(1 + \mu_0^2\right)|w_z|^2,$$

$$\left(1 + \mu_0^2\right)^{-1}|w_z|^2 \leqslant |z_w|^2 - |z_{\bar{w}}|^2 = \left(|w_z|^2 - |w_{\bar{z}}|^2\right)^{-1} \leqslant \left(1 - \mu_0^2\right)^{-1}|w_z|^{-2}.$$

The lemma is proved.

COROLLARY. *Suppose that* $\varphi(w) \in L_p(\overline{K})$, $p > 2$, *and* $w = w(z)$ *is a homeomorphism satisfying the conditions of the lemma. Then*

$$\varphi[w(z)]\frac{\partial w}{\partial z} = -\left\{ \varphi[w(z)]\frac{z_w}{|z_w|^2 - |z_{\bar{w}}|^2} \right\} \in L_{p_0}(\overline{K}_z), \qquad 2 < p_0 < p.$$

To prove this it suffices to set $f_z = \varphi[w(z)]w_{\bar{z}}$ in the estimates carried out in proving the lemma.

The representation (19) holds also for the solutions of the nonlinear equations

$$w_{\bar{z}} - \mu_1 w_z - \mu_2 \bar{w}_{\bar{z}} = 0, \tag{20}$$

where $\mu_i = \mu_i(z, w, w_z, w_{\bar{z}})$, if after the solution is substituted in the coefficients μ_i the resulting functions $\tilde{\mu}_i(z)$ satisfy the inequality

$$|\tilde{\mu}_1(z)| + |\tilde{\mu}_2(z)| \leq \mu_0 < 1. \tag{21}$$

Indeed, let us write (20) in the form

$$w_{\bar{z}} - \mu(z) w_z = 0,$$

where $\mu(z) = \tilde{\mu}_1 + \tilde{\mu}_2 \bar{w}_{\bar{z}}/w_z$ (for $w_z = 0$ we set $\bar{w}_{\bar{z}}/w_z = 1$); in this expression for $\mu(z)$ we substitute the solution $w(z)$ and get $|\mu(z)| \leq \mu_0 < 1$ by (21). Consequently, each solution of the nonlinear equation (20) for which (21) holds is simultaneously a solution of a certain linear Beltrami equation. Therefore, the representation (19) is valid for such solutions of (20), i.e., they realize quasiconformal mappings. The representation (19) immediately gives many of the properties of analytic functions for quasiconformal mappings: the fact that the zeros in the domain D are isolated, the preservation of domain, the argument principle, the maximum modulus principle, and so on. Moreover, the smoothness assertions of Theorem 4 are true for each solution of the Beltrami equation (7). We construct a representation for solutions of equations that are nonhomogeneous with respect to the derivatives. Let us first consider the simplest such equation

$$w_{\bar{z}} = Aw + B\bar{w}, \tag{22}$$

where $A(z), B(z) \in L_p(\bar{D})$, $p > 2$. The generalized solutions of this equation (according to I. N. Vekua [128], who constructed a systematic theory for them analogous to the theory of analytic functions) are commonly called *generalized analytic functions*. These functions have the representation

$$w(z) = \Phi(z) e^{T(A + B\bar{w}/w)} \tag{23}$$

($\bar{w}/w$ is set equal to 1 for $w = 0$), where $\Phi(z)$ is an analytic function in D. Indeed, suppose that $w(z) \in W_p^1(D)$ $(p > 2)$ is a given solution of (22). Obviously, $A + B\bar{w}/w \equiv \varphi(z) \in L_p(\bar{D})$, $p > 2$, and, therefore, $\Phi(z) = we^{-T\varphi} \in W_p^1(\bar{D})$, $p > 2$. But

$$\Phi_{\bar{z}} = e^{-T\varphi}(w_{\bar{z}} - w\varphi) = e^{-T\varphi}(w_{\bar{z}} - Aw - B\bar{w}),$$

i.e., $\Phi(z)$ is an analytic function, which proves the representation (23). Suppose now that $w = w(z)$ is a generalized solution of the equation

$$w_{\bar{z}} - \mu(z) w_z = Aw + B\bar{w}, \tag{24}$$

$A(z)$, $B(z) \in L_p(\overline{D})$, and $\mu(z)$ satisfies (8) ($|\mu(z)| \leqslant \mu_0 < 1$). Then $w(z)$ can be represented in the form

$$w = \Phi[\zeta(z)]e^{T\varphi}, \tag{25}$$

where $\zeta = \zeta(z)$ is a homeomorphism of the Beltrami equation (equation (24) with $A = B = 0$), Φ is an analytic function, and $\varphi(z) \in L_p(\overline{D})$, $p > 2$, is a solution of the equation

$$\varphi - \mu(z)S\varphi = A + B\frac{\overline{w}}{w}. \tag{26}$$

It is easy to check that the function $F(z) = w(z)e^{-T\varphi}$, where $\varphi(z)$ is a solution of (26), satisfies the Beltrami equation (equation (24) with $A = B = 0$). Consequently, it can be represented as the composition $F(z) = \Phi[\zeta(z)]$ of a homeomorphism and an analytic function, which gives us the representation (26). Since $e^{T\varphi}$ does not have zeros in D, the zeros of the generalized solutions of (24) (in particular, of generalized analytic functions also, in the case $\mu = 0$) are isolated in D, and the argument principle is applicable to the solutions $w(z) = \Phi[\zeta(z)]e^{T\varphi}$. We remark that in (23) and (25) we can replace the operator T by any operator $\tilde{T}$ having the properties

$$(\tilde{T}f)_{\overline{z}} = f \quad \text{and} \quad \|(\tilde{T}f)_z\|_{L_2} \leqslant \|f\|_{L_2}.$$

Consider the quasilinear equation

$$w_{\overline{z}} - \mu_1(w, z)w_z - \mu_2(w, z)\overline{w}_{\overline{z}} = A(w, z)w, \tag{27}$$

where the functions A and μ_i ($i = 1, 2$) are measurable in z for fixed w and are continuous in w for almost all $z \in D$, and for $z \in \overline{D}$ and $w \in [0, \infty]$

$$|\mu_1(w, z)| + |\mu_2(w, z)| \leqslant \mu_0 < 1, \tag{28}$$

while $A(w, z) \in L_p(\overline{D})$, $p > 2$. Under these conditions there is a general representation theorem for the generalized solutions of (27).

THEOREM 6. *Any generalized solution $w(z)$ of (27) in a domain D can be represented in the form*

$$w(z) = \Phi[\zeta(z)]e^{T\varphi}, \tag{29}$$

where $\varphi(z) \in L_p(\overline{D})$, $p > 2$, is some function, $\zeta = \zeta(z)$ is a homeomorphism of D, and $\Phi(\zeta)$ is an analytic function.

The theorem follows from the fact that solutions $w(z)$ of the quasilinear equation (27) satisfy also a linear equation of the form (24):

$$w_{\overline{z}} - \mu(z)w_z = \tilde{A}(z)w,$$

where the solution is substituted in the coefficients $\mu(z) = \mu_1(w, z) + \mu_2(w, z)\bar{w}_{\bar{z}}/w_z$ and $\tilde{A}(z) = A(w, z)$.

REMARK. Obviously, the representation Theorem 6 holds also for solutions $w = w(z)$ of (27) that have finitely many poles and essential singularities in $\bar{D}$ but belong to the class $W_p^1(D_0)$, $p > 2$, in any subdomain D_0 not containing these points. Then, according to (29), the images of the singular points of $w(z)$ are singular points of the analytic function $\Phi(\zeta)$. In particular, this implies that the argument principle is also applicable when $w(z)$ has poles.

It must also be pointed out that the function $\varphi(z)$ and the homeomorphism $\zeta = \zeta(z)$ in the representation (29) are, as a rule, determined by the solution $w = w(z)$ of (27) itself, and cannot be chosen in the same way for all solutions. However, if $\mu_2 \equiv 0$ and the functions $\mu_1(w, z)$ and $A(w, z)$ do not depend on w, i.e., for the solutions of the equation

$$w_{\bar{z}} - \mu(z)w_z = A(z)w, \tag{30}$$

the following remarkable corollary of Theorem 6 holds.

COROLLARY. *The function $\varphi(z)$ and the homeomorphism $\zeta(z)$ in the representation (29) can be chosen the same for all generalized solutions of* (30).

Indeed, we determine $\varphi(z)$ from the equation

$$\varphi - \mu(z)\tilde{S}\varphi = A(z), \tag{31}$$

where $\tilde{S}\varphi = \partial_z\tilde{T}\varphi$ and $\tilde{T}$ is an arbitrary operator of the same type as T in the sense that $\partial_{\bar{z}}\tilde{T}\varphi = \varphi$ and $\|\tilde{S}\varphi\|_{L_2} \leqslant \|\varphi\|_{L_2}$. Then for an arbitrary solution $w(z)$ of (30) the function $\omega(z) = w(z)e^{-\tilde{T}\varphi}$ satisfies the Beltrami equation (equation (30) with $A = 0$), and the representation $w = \Phi[\zeta(z)]$ of a solution of this equation implies the statement of the corollary.

The structure of the basic representation $w(z) = \Phi[\zeta(z)]e^{\tilde{T}\varphi}$ for solutions of (30) enables us, in particular, to reduce boundary-value problems for this equation to boundary-value problems for analytic functions. For example, the Hilbert boundary-value problem

$$\mathrm{Re}\{G(z)w(z)\} = g(z), \qquad |z| = 1, \tag{32}$$

for (30), under the usual assumptions about G and g for boundary-value problems in the theory of analytic functions, reduces to an analogous problem for the analytic function $\Phi(\zeta)$:

$$\mathrm{Re}\{G[z(\zeta)]w[z(\zeta)]\} = ge^{-T_1\varphi}\{z(\zeta)\},$$

where $\mathrm{Im}\, T_1(\varphi\,|\,e^{i\gamma}) = 0$ (see §2), $\varphi(z)$ is a solution of (31), $\zeta = \zeta(z)$ is a fixed homeomorphism satisfying the Beltrami equation (30) with $A = 0$, and $z = z(\zeta)$ is its inverse. A solution $w(z) = \Phi[\zeta(z)]e^{T_1\varphi}$ of the problem (30), (32) can be found from an analytic function $\Phi(\zeta)$ satisfying condition (32).

THEOREM 7 (the symmetry principle). *Suppose that the function* $w(z) \in W_p^1(D), p > 2$, *is continuous in* $\overline{D}$ *and is a generalized solution of* (27).

If the boundary of D contains an arc Γ of the circle $|z| = 1$ and $|w(z)| = 1$ for $z \in \Gamma$, then there exists a domain $D^0 \subset D$ abutting on Γ such that the function $w^*(z)$ defined by $w^*(z) = w(z)$ for $z \in D$, and $w^*(z) = \overline{[w(\overline{z}^{-1})]}^{-1}$ for $z \in D_*^0$, belongs to the class $W_p^1(D + D_*^0) \cap C(D + D_*^0)$ and satisfies an equation of the form (27) *in $D + D_*^0$, where D_*^0 is the mirror image of D^0 with respect to Γ.*

Since $w(z)$ is continuous in D and $|w(z)| = 1$ for $z \in \Gamma$, there exists a domain $D^0 \subset D$ whose boundary contains Γ and $|w(z)| \neq 0$ for $z \in \overline{D^0}$. Then the constructed function $w^*(z)$ is obviously continuous in $D + D_*^0$. If for $\zeta \in D^0$ $(\zeta = 1/\overline{z}, z \in D_*^0)$ we substitute the expressions

$$\frac{\partial \overline{w(\zeta)}}{\partial \overline{\zeta}} = \left[\frac{z}{w^*(z)}\right]^2 \frac{\partial w^*(z)}{\partial z},$$

$$\frac{\partial \overline{w(\zeta)}}{\partial \zeta} = \left[\frac{\overline{z}}{w^*(z)}\right]^2 \frac{\partial w^*(z)}{\partial z}, \qquad w(\zeta) = \overline{[w^*(z)]}^{-1}$$

in (27), we arrive at an analogous equation for the function $w^*(z)$, $z \in D_0^*$, with the coefficients μ_i^* and A^* given by

$$\mu_1^* = \overline{\mu_1(\zeta, w)}\left(\frac{z}{\overline{z}}\right)^2, \qquad \mu_2^* = \overline{\mu_2(\zeta, w)}\left(\frac{w^*}{\overline{w^*}}\right)^2, \qquad A^* = \overline{A(\zeta, w)}\,\overline{z}^{-2}.$$

Thus, $w^*(z)$ satisfies an equation of the form (27) in $D + D_*^0$ and, by (29) (which is valid for the solutions of such equations), belongs to the class $W_p^1(D + D_*^0), p > 2$. The theorem is proved.

COROLLARY. *The assertions of the theorem remain true also in the case when Γ is an arc of the circle $|z - z_0| = r$, and if $A \equiv 0$ in* (27), *then also when $|w(z) - w(z_0)| = \rho, z \in \Gamma$.*

Indeed, this case can be reduced to that considered in the theorem by nonsingular linear transformations of the independent variable z and the unknown function $w(z)$.

The properties of homeomorphisms satisfying a Beltrami equation enable us also to solve the very important problem of reducing differential equations to canonical form. Let us consider the equation

$$w_{\overline{z}} - \mu_1(z)w_z - \mu_2(z)\overline{w}_{\overline{z}} = F(w, z), \qquad (33)$$

where

$$\mu_i(z) \in W_p^1(\overline{D}), \qquad |\mu_1| + |\mu_2| \leq \mu_0 < 1,$$

and $F(w, z) \in L_p(\overline{D})$, $p > 2$, for fixed w (for example, it may be equal to $A_1(z)w + B_1(z)\overline{w}$).

Let $\zeta = \zeta(z)$ be an arbitrary homeomorphism satisfying the Beltrami equation

$$\zeta_{\overline{z}} - \mu(z)\zeta_z = 0, \tag{34}$$

where $\mu(z) \in W_p^1(\overline{D})$, $p > 2$, is some function for which $|\mu(z)| \leqslant \mu_0 < 1$, and let $w = \omega + a\overline{\omega}$, where

$$a(z) \in W_p^1(\overline{D}), \qquad p > 2, |a(z)| \leqslant a_0 < 1. \tag{35}$$

Let us determine $\mu(z)$ and $a(z)$ in such a way that $\omega(\zeta)$ satisfies the equation

$$\omega_{\overline{\zeta}} = A\omega + B\overline{\omega} + \tilde{F}(\omega, \zeta), \tag{36}$$

where $A(\zeta)$, $B(\zeta) \in L_p[\zeta(\overline{D})]$, $p > 2$, and $F = \tilde{\alpha}F + \tilde{\beta}\overline{F}$, $\tilde{\alpha}$, $\tilde{\beta} \in C^\alpha[\zeta(\overline{D})]$. We first use a homeomorphism $\zeta = \zeta(z)$ satisfying (34) to transform (33) by substituting the expressions $w_{\overline{z}} = w_{\overline{\zeta}}\overline{\zeta}_{\overline{z}} + w_\zeta \zeta_z \mu$ and $w_z = w_{\overline{\zeta}}\overline{\zeta}_z \mu + w_\zeta \zeta_z$. into it. Then

$$\alpha_1 w_{\overline{\zeta}} + \beta_1 w_\zeta + \gamma_1 \overline{w}_{\overline{\zeta}} + \delta_1 \overline{w}_\zeta = F.$$

Eliminating the derivative $\overline{w}_\zeta$ from the last equation and its complex conjugate, we find that

$$\alpha w_{\overline{\zeta}} - \beta w_\zeta - \gamma \overline{w}_{\overline{\zeta}} = F_1,$$

where

$$\alpha = |1 - \mu_1 \overline{\mu}|^2 - |\mu_2 \overline{\mu}|^2, \qquad \beta = \left[\overline{\mu}_1 \mu^2 - \left(1 + |\mu_1|^2 - |\mu_2|^2\right)\mu + \mu_1\right]\frac{\zeta_z}{\overline{\zeta}_{\overline{z}}},$$

$$\gamma = \mu_2\left(1 - |\mu|^2\right) \quad \text{and} \quad F_1 = \frac{1}{\overline{\zeta}_{\overline{z}}}\left[F(1 - \overline{\mu}_1 \mu) + \mu_2 \mu \overline{F}\right].$$

Setting $\beta = 0$ in the last equation, we find from the quadratic equation obtained that

$$\mu(z) = 2\mu_1\left(1 + |\mu_1|^2 - |\mu_2|^2 + \sqrt{\Delta}\right)^{-1}, \tag{37}$$

where

$$\Delta = \left(1 + |\mu_1|^2 - |\mu_2|^2\right)^2 - 4|\mu_1|^2$$

$$= \left(1 - |\mu_1|^2 - |\mu_2|^2\right)^2 - 4|\mu_1|^2(\mu_2)^2 \geqslant \sigma(\mu_0) > 0.$$

Since the $\mu(z)$ in the form (37) belongs to $W_p^1(\overline{D})$, $p > 2$, it follows from Theorem 4 that $J = |\zeta_z|^2 - |\zeta_{\overline{z}}|^2 \geqslant \delta > 0$, and this implies that $\zeta_z \neq 0$ in $\overline{D}$.

Thus, (33) takes the form $w_{\bar{\zeta}} - a(\zeta)\bar{w}_{\bar{\zeta}} = \tilde{F}(w, z)$, with

$$a(\zeta) = m^2\mu_2\left(1 - |\mu|^2\right), \qquad F = \frac{m^2}{\bar{\zeta}_{\bar{z}}}\left(F(1 - \bar{\mu}_1\mu) + \mu_2\mu\bar{F}\right), \qquad (38)$$

where $m^2 = (|\,1 - \mu_1\mu|^2 - |\mu_2\mu|^2)^{-1}$, $|a(\zeta)| \leqslant a(\mu_0) < 1$. Replacement of the original function $w(z)$ of the form before (35) with $a(\zeta)$ defined by (38) leads finally to (36), in which

$$A = \frac{a\bar{a}_{\bar{\zeta}}}{1 - |a|^2}, \qquad B = -\frac{a_{\zeta}}{1 - |a|^2}, \qquad \tilde{F} = \frac{\tilde{F}}{1 - |a|^2}. \qquad (39)$$

Thus, we have proved the following theorem.

THEOREM 8. *For $\mu(z) \in W_p^1(\bar{D})$, $p > 2$, equation (33) can be reduced to the canonical form (36), where the coefficients A, B, and F are determined by (38) and (39).*

COROLLARY. *If $\mu(z) \in C^\alpha$, $\alpha > 0$, in a neighborhood of some point $z_0 \in \bar{D}$ where (33) is satisfied, then the equation can be reduced by affine transformations to the previous form, but with coefficients equal to zero at the image of z_0.*

Indeed, we write (33) in the form

$$w_{\bar{z}} - \mathring{\mu}_1 w_z - \mathring{\mu}_2\bar{w}_{\bar{z}} = \Delta\mu_1 w_z + \Delta\mu_2\bar{w}_{\bar{z}} + F \equiv F_0, \qquad (33^*)$$

where $\mathring{\mu}_1 = \mu_1(z_0)$ and $\Delta\mu_i = \mu_i(z) - \mu_i(z_0)$, and reduce it to the canonical form. Since the coefficients of (33) are constant, (38) and (39) imply that $\mu = \text{const}$ and $a = \text{const}$, and the corresponding transformations are affine: $\zeta - \zeta_0 = z - z_0 + \mu(\bar{z} - \bar{z}_0)$ and $w = \omega + \bar{a}\bar{\omega}$.

In (36) we have $A = B = 0$, and the function $\tilde{F}_0$ depends linearly on F_0 and is determined by (38) and (39). Carrying out the transformation of w_z and $\bar{w}_{\bar{z}}$ in the expression (33*) for F_0, we get

$$\omega_{\bar{z}} - \tilde{\mu}_1(\zeta)\omega_z - \tilde{\mu}_2(\zeta)\bar{\omega}_z = \bar{F}(\omega, \zeta), \qquad (40)$$

where

$$\tilde{\mu}_i(\zeta) = \alpha_i\Delta\mu_1 + \beta_i\Delta\mu_2 + \gamma_i\overline{\Delta\mu_1} + \delta_i\overline{\Delta\mu_2}, \qquad i = 1, 2,$$

$$|\tilde{\mu}_1| + |\tilde{\mu}_2| \leqslant \tilde{\mu}_0 = \tilde{\mu}(\mu_0) < 1 \quad \text{and} \quad \tilde{F} = \tilde{\alpha}F + \tilde{\beta}\bar{F},$$

and α, β, γ, and δ are constants. Moreover, by construction $\mu_i(\zeta_0) = 0$, $i = 1, 2$ (ζ_0 is the image of z_0 under the affine transformation $\zeta - \zeta_0 = z - z_0 + \mu(\bar{z} - \bar{z}_0)$, and the smoothness of the coefficients in (33) under such a transformation obviously does not change).

REMARK. Similarly to the preceding, a linear elliptic equation

$$a\frac{\partial^2 u}{\partial x^2} + 2b\frac{\partial^2 u}{\partial x\,\partial y} + c\frac{\partial^2 u}{\partial y^2} + d\frac{\partial u}{\partial x} + e\frac{\partial u}{\partial y} + fu = g \qquad (41)$$

with

$$a, b, c \in W_p^1(\overline{D}), \qquad d, e, f, g \in L_p(\overline{D}), \qquad p > 2,$$

$$\Delta^2 = ac - b^2 \geqslant \Delta_0 > 0$$

can be reduced by means of a homeomorphic solution $\zeta = \zeta(z)$ of the Beltrami equation

$$\zeta_{\bar{z}} - \frac{a\sqrt{\Delta} - i\beta}{a + \sqrt{\Delta} + i\beta}\zeta_z = 0 \qquad (42)$$

to the canonical form

$$\frac{\partial^2 u}{\partial \xi^2} + \frac{\partial^2 u}{\partial \eta^2} + p\frac{\partial u}{\partial \xi} + q\frac{\partial u}{\partial \eta} + ru = h, \qquad \xi + i\eta = \zeta. \qquad (43)$$

It is easy to see this by a simple check.

§4. Construction of homeomorphisms of the unit disk onto itself

An immediate consequence of the results of the preceding section is the possibility (a generalization of the Riemann mapping theorem) of mapping an arbitrary domain D onto the unit disk by a solution of the Beltrami equation

$$w_{\bar{z}} - \mu(z)w_z = 0, \qquad (1)$$

where $\mu(z)$ is a measurable function satisfying almost everywhere

$$|\mu(z)| \leqslant \mu_0 < 1. \qquad (2)$$

Indeed, the function $w = \Phi[\zeta(z)]$ is the desired solution of (1), where $\zeta = z + \mathring{T}f$ is the homeomorphism of the whole plane satisfying (1) that was constructed in §3, and $w = \Phi(\zeta)$ is an analytic function that maps the image $\zeta(D)$ of D conformally onto the disk $|w| < 1$. Under the usual normalization conditions on (1) in the theory of conformal mappings, a uniqueness theorem obviously holds for the homeomorphism of D onto $|w| < 1$. In the case of the general linear equation

$$w_{\bar{z}} - \mu_1(z)w_z - \mu_2(z)\overline{w}_z = 0, \qquad (3)$$

where the functions μ_i are measurable in D and satisfy the inequality

$$|\mu_1(z)| + |\mu_2(z)| \leqslant \mu_0 < 1, \qquad (4)$$

and, all the more so, in the case when equation (3) is quasilinear (the μ_i depend on z and w), the generalized Riemann theorem is no longer a trivial consequence of the result of the previous section, and its proof involves considerable difficulties. We remark that via a preliminary conformal mapping of the disk $|\zeta| < 1$ onto the domain D by the analytic function $z = \Phi(\zeta)$ the construction of a mapping of this domain onto the disk $|w| < 1$ by a solution of (3) with $\mu_i = \mu_i(w, z)$ reduces to the construction of a mapping of the disk $|\zeta| < 1$ onto the disk $|w| < 1$ by a solution of the analogous equation

$$w_{\bar{\zeta}} - \tilde{\mu}_1(w, z)w_\zeta - \tilde{\mu}_2(w, z)\overline{w_\zeta} = 0,$$

where

$$\tilde{\mu}_1(w, z) = \mu_1[w, \Phi(\zeta)]\frac{\overline{\Phi_\zeta}}{\Phi_\zeta}, \qquad \tilde{\mu}_2(w, z) = \mu_2[w, \Phi(\zeta)].$$

Therefore, in proving the generalized Riemann theorem for quasilinear equations of the form (3) we restrict ourselves to the consideration of mappings of the unit disk onto itself.

We first prove two lemmas.

LEMMA 1. *Suppose that the function* $w = \zeta(z) \equiv z + \mathring{T}\varphi$ *is a generalized solution of the linear equation* (3) *in the disk* K ($|z| < 1$). *Then the inequality* $\|f(z)\|_{W^1_p(\overline{K})} \leq C(\mu_0) < \infty, p = p(\mu_0) > 2$, *holds for the function*

$$f(z) = -\frac{1}{2\pi i}\int_{|t|=1}\frac{\ln|t + \mathring{T}(\varphi|t)|(t+z)}{t(t-z)}dt \qquad (5)$$

and, moreover, if $\mu_i^n \to \mu_i$, $i = 1, 2$, *almost everywhere in* $\overline{K}$ *and* $\zeta^n = z + \mathring{T}\varphi^n$ *are the corresponding generalized solutions of* (3) *with the* $\mu_i^n(z)$, *then as* $n \to \infty$ *the sequence* $f^n(z)$ *converges in the spaces* $W^1_p(\overline{K})$ *and* $C^\alpha(\overline{K})$, $p > 2$, $\alpha = (p - 2)/p$.

Since the solution $\zeta = z + \mathring{T}\varphi$ of (3) is also a solution of the Beltrami equation

$$\zeta_{\bar{z}} - \left[\mu_1(z) + \mu_2(z)\frac{\overline{1 + S\varphi}}{1 + S\varphi}\right]\zeta_z = 0,$$

it follows from Theorem 4 in §3 that $\zeta(z) \in W^1_p(\overline{K})$, $p = p(\mu_0) > 2$, and it is a homeomorphism of the whole plane, with

$$C_1^{-1}|z_1 - z_2|^{1/\alpha} \leq |\zeta(z_1) - \zeta(z_2)| \leq C_1|z_1 - z_2|^\alpha, \qquad (6)$$

where z_1, $z_2 \in \overline{K}$, $\alpha = (p - 2)/p > 0$, $p(\mu_0) = p > 2$, and $C_1(\mu_0) < \infty$ (see (18) in §3). Obviously, $g(z) = |\zeta(z)| \in W_p^1(\overline{K})$, $p > 2$; consequently, $g(z) \in C^{(p-2)/p}$, and $g(z)$ has the representation (1) of §2:

$$g(z) = T(\partial_{\overline{z}} g) + \frac{1}{2\pi i} \int_{|t|=1} \frac{g(t)\,dt}{t - z}. \tag{7}$$

Then an analytic function $u(z)$ satisfying the condition $\operatorname{Re} u(t) = g(t) = |\zeta(z)|$, for $z = t = e^{i\gamma}$ can be represented in the form

$$u(z) = \frac{1}{\pi i} \int_{|t|=1} \frac{g(t)\,dt}{t - z} - \frac{1}{2\pi i} \int_{|t|=1} \frac{g(t)\,dt}{t}$$

$$= -2T\left(\frac{\partial g}{\partial \overline{z}}\right) + 2g(z) - \frac{1}{2\pi i} \int_0^{2\pi} g(e^{i\gamma})\,d\gamma,$$

and this implies that $\| u(z) \|_{W_p^1(\overline{K})} \leq C_2(\mu_0) < \infty$.

From the inequality (6) with $z_1 = 0$ and $z_2 = t = e^{i\gamma}$ we have

$$C_1^{-1} \leq |\zeta(t)| = \operatorname{Re}|u(t)| \leq C_1.$$

Consequently, by the maximum modulus theorem, the last estimate holds for the harmonic function $u_1(z) = \operatorname{Re} u(z)$ in the closed disk K, and, therefore, $\|\ln u_1(z)\|_{W_p^1(\overline{K})} \leq C_3(\mu_0) < \infty$. We remark that $\ln|\zeta(t)| \in C^\alpha$ because $\ln|\zeta(t)|$ has been shown to be bounded, and this gives us that $\| f(z) \|_{C^\alpha} \leq C_4$. Since

$$f(z) = -\frac{1}{\pi i} \int_{|t|=1} \frac{g_1(t)\,dt}{t - z} + \frac{1}{2\pi i} \int_{|t|=1} \frac{g_1(t)\,dt}{t},$$

where $g_1(t) = \ln u_1(t) \equiv \ln|t + \hat{T}(\varphi|t)|$, the representation (7) for the function $g_1(z) = \ln u_1(z)$ leads in a similar way to the required inequality $\| f(z) \|_{W_p^1(\overline{K})} \leq C(\mu_0)$. The last assertion of Lemma 1 follows from the obvious convergence of the auxiliary functions $g''(z) = |\zeta''(z)|$, $u''(\zeta)$, and $g_1''(z)$ in $W_p^1(\overline{K})$, $p > 2$, and in C^α, $\alpha = (p - 2)/p$. The lemma is proved.

LEMMA 2. *Let* $w = \zeta(z) \equiv F(z)e^{\omega(z)} \in W_p^1(\overline{K})$, $p > 2$, *be a generalized solution of* (3), *where* $F(z) \in W_p^1(\overline{K})$, $p > 2$, *is homeomorphic in the disk* K $(|z| < 1)$, $F(0) = 0$, *and* $|\zeta(t)| = |F(t)|e^{\operatorname{Re}\omega(t)} = 1$ *for* $t = e^{i\gamma}$, $\gamma \in [0, 2\pi]$. *Then* $\zeta(z) = F(z)e^{\omega(z)}$ *is a homeomorphism of the unit disk onto itself.*

The condition $|\zeta(e^{i\gamma})| = 1$ of the lemma and the maximum modulus principle for the solution $w = \zeta(z) \equiv F(z)e^{\omega(z)}$ of (3) (its validity follows from the representation of solutions in terms of an analytic function of the homeomorphism $\zeta = \Phi[\chi(z)]$) imply that $|\zeta(z)| < 1$ for all $|z| < 1$. Let us consider the function $Z_\lambda = \zeta(z) - \lambda\zeta_0$, where $\lambda \in [0, 1]$, and ζ_0 is an arbitrarily fixed point of the disk K ($|\zeta| < 1$). Obviously, the function $w = Z_\lambda(z) \in W_p^1(\overline{K})$, $p > 2$, together with $\zeta(z) = Fe^\omega$, is a generalized solution of (3), and, since $Z_\lambda(z) \in C^{(p-2)/p}(\overline{K})$ by the imbedding theorem, the argument principle is applicable to it. We compute

$$\kappa(\lambda) = \frac{1}{2\pi}\left[\arg Z_\lambda(e^{i\gamma})\right].$$

The function $Z_\lambda(z)$ is continuous in λ and z, and for all $\lambda \in [0, 1]$ and $|\zeta_0| < 1$. Therefore, for fixed $\lambda \in [0, 1]$ each branch of $\arg Z_\lambda(e^{i\gamma})$ is continuous in γ, and the value of $\kappa(\lambda)$ does not depend on the chosen branch of $\arg Z_\lambda(e^{i\gamma})$. For $\gamma = 0$ we fix the branches of $\arg Z_\lambda(1 + 0) = \theta_\lambda(1 + 0)$ in such a way that $\theta_\lambda(1 + 0)$ is a continuous function of $\lambda \in [0, 1]$. This is possible, because $Z_\lambda(e^{i\gamma}) \neq 0$, $\gamma \in [0, 2\pi]$, and $Z_\lambda(e^{i\gamma})$ is continuous in λ and γ. For the same reason the expression $\theta_\lambda(1 - 0)$ obtained by a continuous motion with respect to γ from the chosen $\theta_\lambda(1 + 0)$ is continuous in λ. Consequently,

$$\kappa(\lambda) = \frac{1}{2\pi}\left[\theta_\lambda(1 - 0) - \theta_\lambda(1 + 0)\right]$$

is also continuous in λ, and, since $\kappa(\lambda)$ is an integer-valued function, this means it is constant, i.e., $\kappa(\lambda) = \kappa(0) = \kappa(1)$. But for $\lambda = 0$

$$\kappa(0) = \frac{1}{2\pi}\left[\arg \zeta(e^{i\gamma})\right] = \frac{1}{2\pi}\left[\arg F(e^{i\gamma})e^{\omega(e^{i\gamma})}\right] = 1,$$

since the fact that $F(z)$ is a homeomorphism implies that for $|z| < 1$ it has a unique zero at $z = 0$, and $e^{\omega(z)}$ does not vanish.

Thus, for $\lambda = 1$ the function $Z_1(z) = \zeta(z) - \zeta_0$ has also a unique zero inside the disk $|z| < 1$ when $|\zeta_0| < 1$, and this, together with the inequality $|\zeta(z)| < 1$ established above for $|z| < 1$, proves the lemma.

We shall construct homeomorphisms of the unit disk onto itself by generalized solutions of the quasilinear equation

$$w_{\bar z} - \mu_1(w, z)w_z - \mu_2(w, z)\bar w_{\bar z} = 0 \tag{8}$$

in the form $w(z) = \zeta(z)e^{\omega(z)}$, which leads to the following relation between $\zeta(z)$ and $\omega(z)$:

$$\zeta_{\bar z} - \mu_1\zeta_z - \mu_2 e^{\bar\omega - \omega}\bar\zeta_{\bar z} + \zeta(z)\left(\omega_{\bar z} - \mu_1\omega_z - \mu_2\frac{\bar\zeta}{\zeta}e^{\bar\omega - \omega}\bar\omega_{\bar\zeta}\right) = 0.$$

Let $\zeta = z + \overset{\circ}{T}\varphi$ and $\omega(z) = f(z) - i\beta + T_0\psi$, where $f(z)$ is defined by (5), the constant $\beta = \operatorname{Im} f(1)$, and the functions φ and ψ are chosen in $L_p(\overline{K})$ ($p > 2$) in such a way that $\zeta(z)$ and $\omega(z)$ satisfy the respective equations

$$\zeta_{\bar z} - \mu_1\zeta_z - \mu_2^\varphi\bar\zeta_{\bar z} = 0, \tag{9}$$

$$\omega_{\bar z} - \mu_1\omega_z - \mu_2^\psi\bar\omega_{\bar z} = 0, \tag{10}$$

where $\mu_1 = \mu_1(\zeta e^\omega, z)$, $\mu_2^\varphi = \mu_2(\zeta e^\omega, z)\exp\{-2i\operatorname{Im}(f(z) - i\beta + T_0\psi)\}$, and

$$\mu_2^\psi = \mu_2^\varphi \overline{(z + \mathring{T}\varphi)}\,(z + \mathring{T}\varphi)^{-1}.$$

From (9) and (10) we get the following system of equations for φ and ψ:

$$\varphi - \mu_1 S\varphi - \mu_2^\varphi \overline{S\varphi} = \mu_1 + \mu_2^\varphi,$$

$$\psi - \mu_1 S_0\psi - \mu_2^\psi \overline{S_0\psi} = \mu_1 \frac{\partial f}{\partial z} + \mu_2^\psi \frac{\partial \bar{f}}{\partial \bar{z}}, \tag{11}$$

where $S\varphi = \partial \mathring{T}\varphi/\partial z \equiv \partial T\varphi/\partial z$ and $S_0\psi = \partial T_0\psi/\partial z$, and, moreover,

$$\operatorname{Re} T_0(\psi \mid e^{i\gamma}) = T_0(\psi \mid 1) = 0$$

by a property of the operator T_0 (§2). Thus, if the solutions $\varphi(z), \psi(z) \in L_p(\overline{K})$, $p > 2$, of (11) have been found, then by construction $\zeta(z)$ and $\omega(z)$ satisfy (9) and (10), and $w(z) = \zeta(z)e^{\omega(z)}$ is thereby a generalized solution of the original equation (8).

THEOREM 1. *Suppose that $\mu_i(w, z)$, $i = 1, 2$, are continuous in w for almost all $z \in K$ and measurable in z for $|w| < \infty$, and that*

$$|\mu_1(w, z)| + |\mu_2(w, z)| \leqslant \mu_0 < 1.$$

Then there exists at least one homeomorphism

$$w(z) = (z + \mathring{T}\varphi)e^{f(z) + T_0\psi - i\beta} \tag{12}$$

satisfying (8) *and mapping the unit disk onto itself with the normalization $w(0) = 0$ and $w(1) = 1$, where the analytic function $f(z)$ is defined by* (5), *$\beta = \operatorname{Im} f(1)$, and φ and ψ are in $L_p(\overline{K})$ ($p > 2$) and form a solution of* (11), *with*

$$\max\left\{\|\varphi\|_{L_p}, \|\psi\|_{L_p}, \left\|\frac{\partial f}{\partial z}\right\|_{L_p}\right\} \leqslant M(\mu_0, p). \tag{13}$$

Conversely, any homeomorphism $w(z)$ of the unit disk onto itself that leaves fixed the points 0 and 1 can be represented in the form (12).

I. Suppose that the number $p > 2$ is chosen from the conditions $\mu_0 \Lambda_p = \mu_0\|S\|_{L_p} < 1$ and $\mu_0 \Lambda_p' = \mu_0\|S_0\|_{L_p} < 1$, that solutions $\varphi, \psi \in L_p(\overline{K})$ of (11) have been found, and along with them a solution of (8) in the form (12). Lemma 1 gives us that $f(z) \in W_p^1(\overline{K}), p > 2$, and by construction $w(0) = 0$, $w(1) = 1$, and $|w(e^{i\gamma})| = 1$, since

$$\operatorname{Re} f(e^{i\gamma}) = -\ln|e^{i\gamma} + \mathring{T}(\varphi \mid e^{i\gamma})|,$$

while $\operatorname{Re} T_0(\psi \mid e^{i\gamma}) = 0$, $\gamma \in [0, 2\pi]$. But, as shown in the proof of Lemma 1, the solution $\zeta(z) = z + \mathring{T}\varphi$ of (9) is a homeomorphism of the whole plane and, consequently, the $w(z)$ in the form (12) maps the unit disk homeomorphically onto itself by Lemma 2. Thus, for the proof of the first part of the theorem it suffices to establish the solvability of (11).

We introduce the Banach space $B = \{L_p \times C^\alpha \times L_p\}$ of vector-valued functions (φ, ρ, ψ) with components in the respective Banach spaces $L_p(\overline{K})$, $C^\alpha(\overline{K})$, and $L_p(\overline{K})$, $p > 2$, $\alpha = (p - 2)/p$, where the norm is defined by

$$\|(\varphi, \rho, \psi)\|_B = \|\varphi\|_{L_p} + \|\rho\|_{C^\alpha} + \|\psi\|_{L_p}.$$

Let us consider the transformation Ω: $(\varphi, \rho, \psi) = \Omega(\Phi, R, \Psi)$ of B into itself given by the equalities

$$\varphi - \mu_1^\Phi S\varphi - \mu_2^\Phi \overline{S\varphi} = \mu_1^\Phi + \mu_2^\Phi, \tag{14_1}$$

$$\rho \equiv f - i\beta = -\frac{1}{2\pi i} \int_{|t|=1} \frac{\ln|t + \mathring{T}(\varphi|t)|}{t} \frac{t+z}{t-z} dt - i\operatorname{Im} f(1), \tag{14_2}$$

$$\psi - \mu_1^\Psi S_0\psi - \mu_2^\Psi \overline{S_0\psi} = \mu_1^\Psi \frac{d\rho}{dz} + \mu_2^\Psi \frac{d\overline{\rho}}{d\overline{z}}, \tag{14_3}$$

where

$$\mu_1^\Phi = \mu_1^\Psi = \tilde{\mu}_1, \qquad \tilde{\mu}_i = \mu_i\big[(z + \mathring{T}\varphi)e^{R+T_0\Psi}, z\big], \quad i = 1, 2,$$

$$\mu_2^\Phi = \tilde{\mu}_2 \exp\{-2i\operatorname{Im}(R + T_0\Psi)\}, \qquad \mu_2^\Psi = \mu_2^\Phi \frac{\overline{z + \mathring{T}\varphi}}{z + \mathring{T}\varphi}.$$

The transformation Ω is determined as follows: To an arbitrary vector $\vec{u} = (\Phi, R, \Psi) \in B$ there corresponds a solution $\varphi = \varphi(\vec{u}\,|\,z)$ of the linear equation (14_1), in terms of which a solution $\rho = \rho(\vec{u}\,|\,z)$ is computed by (14_2), and with them $\psi = \psi(\vec{u}\,|\,z)$ is computed by solving the linear equation (14_3).

By construction, the components φ_0 and ψ_0 of the fixed points $(\varphi_0, \rho_0, \psi_0) = \Omega(\varphi_0, \rho_0, \psi_0)$ of Ω satisfy (11).

II. Let us prove that Ω is continuous. Suppose that $\{\Phi_n\}$, $\{R_n\}$, and $\{\Psi_n\}$ are convergent sequences in the corresponding spaces. Then the $\mu_i^{\Phi_n}$ ($i = 1, 2$) converge almost everywhere in $\overline{K}$, and, consequently, the solutions φ^n of the first equation in (14) converge in L_p (see the corollary to Theorem 1 in §3). Moreover, by Lemma 1, the sequence $f_n(z)$, and with it also $\rho_n(z)$, converges in $C^\alpha(\overline{K})$, $\alpha = (p - 2)/p$. But, by Lemma 1, the convergence of $f_n(z)$ and $\rho_n(z)$ in $W_p^1(\overline{K})$ ($p > 2$) implies the convergence of the right-hand sides of the last equation in (14) in $L_p(\overline{K})$. And this, as in the corollary to Theorem 1 in §3, implies the convergence in $L_p(\overline{K})$ of the solutions ψ_n of the last equation in (14). The continuity of Ω is proved.

III. The compactness of the transformation Ω obviously follows from the relative compactness of the sets $\{\mathring{T}\Phi\}$, $\{R\}$, and $\{\mathring{T}\Psi\}$ in $C(\overline{K})$ when the sets $\{\Phi\}$, $\{R\}$, and $\{\Psi\}$ are bounded in the spaces L_p, C^α, and L_p, respectively.

Indeed, on sequences $(\mathring{T}\Phi)_n$, R_n, and $(T_0\Psi)_n$ converging in $C(\overline{K})$ the coefficients $\mu_i^{\Phi_n}$ ($i = 1, 2$) converge almost everywhere in $\overline{K}$, and this (similarly to the preceding) implies successively the convergence of φ_n in L_p, ρ^n in C^α, and Ψ_n in L_p. Thus, the continuous transformation Ω carries bounded subsets of B into relatively compact sets, i.e., it is completely continuous.

IV. A direct consequence of the first equation in (14) is that

$$\|\varphi\|_{L_p} \leqslant \frac{\|\mu_1^\Phi + \mu_2^\Phi\|}{1 - \mu_0\Lambda_p} \leqslant \frac{\pi\mu_0}{1 - \mu_0\Lambda_p} = M_\varphi(\mu_0).$$

Then the inequality $\| f(z)\|_{C^\alpha} \leqslant C_4(\mu_0)$ established in the proof of Lemma 1 implies that

$$\|\rho(z)\|_{C^\alpha} = \|f(z) + i\operatorname{Im} f(1)\|_{C^\alpha} \leqslant 2C_4 = M_\rho(\mu_0).$$

Further, the last equation in (14) gives us (with the aid of the inequality $\| f\|_{W_p^1(\overline{K})} \leqslant C(\mu_0)$ established in Lemma 1) that

$$\|\psi\|_{L_p} \leqslant \frac{C(\mu_0)\pi\mu_0}{1 - \mu_0\Lambda_p} = M_\psi(\mu_0),$$

which concludes the proof of (13).

V. Thus, the completely continuous transformation maps the closed bounded convex set

$$\mathfrak{M} = \left\{ \|\varphi\|_{L_p} \leqslant M_\varphi, \|\rho\|_{C_\alpha} \leqslant M_\rho, \|\psi\|_{L_p} \leqslant M_\psi \right\} \subset B$$

into itself and, consequently, by Schauder's theorem (Theorem 6 in §1 of Chapter IV), has at least one fixed point. If $w(z)$ is an arbitrary homeomorphism of the unit disk onto itself by a solution of (8) with $w(0) = 0$ and $w(1) = 1$, then (as is easy to see) it can be represented in the form (13) if we solve the system

$$\varphi - \mu_1^\varphi S\varphi - \mu_2^\varphi \overline{S\varphi} = \mu_1^\varphi + \mu_2^\varphi,$$

$$\psi - \mu_1^\psi S\varphi - \mu_2^\psi \overline{S\psi} = \mu_1^\psi \frac{df}{dz} - \mu_2^\psi \frac{\overline{df}}{d\bar{z}}, \qquad (11^*)$$

where

$$\tilde{\mu}_i = \mu_i[w(z), z], \quad i = 1, 2, \qquad \mu_1^\varphi = \mu_1^\psi = \tilde{\mu}_1, \qquad \mu_2^\psi = \tilde{\mu}_2\overline{w}/w,$$

$$\mu_2^\varphi = \tilde{\mu}_2 \exp\{-2i\operatorname{Im}(f - i\beta + T_0\psi)\}.$$

Since this is a simpler variant of (11), it is clearly solvable; hence the last assertion of the theorem is valid, and the theorem is proved.

REMARKS. 1. In view of the symmetry of w and z in the conditions of Theorem 1, its assertions are true also when the functions $\mu_i(w, z)$ are continuous in z for almost all $w \in \overline{K}$ and measurable in w.

2. By what was proved in Theorem 1, the mappings $w = w(z)$ and $z = z(w)$ can be represented in the form (12). Consequently, $w(z)$, $z(w) \in C^\alpha(\overline{K})$, $\alpha(\mu_0) > 0$, and so

$$C^{-1/\alpha}|z_1 - z_2|^{1/\alpha} \leqslant |w(z_1) - w(z_2)| \leqslant C|z_1 - z_2|^\alpha. \tag{15}$$

3. If we set $f = \psi = \beta = 0$ in representation (12) and the equations (14) corresponding to it, then the proof of Theorem 1 tells us that the quasilinear equation (8) has a basic solution of the form

$$w = z + \mathring{T}\varphi, \qquad \varphi \in L_p(\overline{K}), p > 2.$$

Since this solution satisfies at the same time the linear Beltrami equation with

$$\mu(z) = \mu_1[w(z), z] + \mu_2[w(z), z]\frac{\overline{w_{\overline{z}}}}{w_z},$$

it is a homeomorphism of the whole plane and has all the properties of a basic solution that was proved in §1.

We now present a simple proof of a uniqueness theorem for homeomorphisms of the unit disk onto itself under assumptions just as general as those first used in the proof of Bojarski [21].

THEOREM 2. *Under the conditions of the preceding theorem on $\mu_i(w, z)$ and the additional assumption*

$$\left|\mu_i(w^1, z) - \mu_i(w^2, z)\right| \leqslant M_i|w^1 - w^2|, \tag{16}$$

where M_i does not depend on z or w, the homeomorphic solution of (8) with the normalization conditions commonly used in the theory of conformal mappings is unique.

We remark that an additional conformal mapping $\zeta = F(z)$ of the unit disk onto itself (which obviously does not change the properties of the coefficients in (8)) enables us to reduce different normalizations of the mapping to some fixed single normalization. Therefore, in proving the theorem we restrict ourselves to the following normalization:

$$w(e^{i\gamma_k}) = e^{i\gamma_k}, \qquad k = 1, 2, 3, \gamma_k = \frac{2\pi}{3}(k - 1). \tag{17}$$

Suppose that there exist at least two different homeomorphisms $w^1(z)$ and $w^2(z)$ of the disk K onto itself that satisfy (8) and (17). We introduce the notation

$$\omega(z) = w^1(z) - w^2(z); \qquad \mu_i^k = \mu_i(w^k, z), \quad i, k = 1, 2;$$

$$\Delta\mu_i = \mu_i(w^1, z) - \mu_i(w^2, z).$$

By the usual method of taking the difference of the left-hand sides of (8) for $w^1(z)$ and $w^2(z)$, we find the equation

$$\omega_{\bar{z}} - \mu(z)\omega_z = A\omega \tag{18}$$

for $\omega(z) = w_1(z) - w_2(z)$, where

$$\mu(z) = \mu_1^1 + \mu_2^1(\bar{\omega}_{\bar{z}}/\omega_z) \quad \text{and} \quad A(z) = \frac{1}{w^1 - w^2}(w_z^2 \Delta\mu_1 + \bar{w}_z^2 \Delta\mu_2);$$

moreover, the facts that $w_z^2 \in L_p(\bar{K})$, $p > 2$, and $|\Delta\mu_i/\omega| \leqslant M_i$ (see (16)) imply that $A(z) \in L_p(\bar{K}), p > 2$. It is easy to verify that the equality $|w^1(e^{i\gamma})|$ $-|w^2(e^{i\gamma})| = 0$ is equivalent to the following condition:

$$\text{Re}[G(t)\omega(t)] = 0, \qquad t = e^{i\gamma}, \gamma \in [0, 2\pi], \tag{19}$$

where

$$G(t) = \overline{w^1(t)} + \overline{w^2(t)}.$$

With the help of the representation $\omega(z) = \Phi[\zeta(z)]e^{T_1\varphi}$, of the solutions of (18) (where $\varphi \in L_p(\bar{K})$, $p > 2$, is a solution of the equation $\varphi - \mu_1 S\varphi = A$) the homogeneous Hilbert problem (18), (19) can be reduced to an analogous problem for the analytic function $\Phi(\zeta)$ on the ζ-plane (see the corollary to Theorem 6 in §3). The homeomorphism $\zeta = \zeta(z)$ of the unit disk onto itself by a solution of the Beltrami equation (equation (18) with $A \equiv 0$) can be chosen to satisfy the normalization conditions (17). Then (17) and the representation

$$w^1(e^{i\gamma}) - w^2(e^{i\gamma}) = \omega(e^{i\gamma}) = \Phi e^{T_1\varphi}\{e^{i\gamma}\}$$

give us

$$\Phi(e^{i\gamma_k}) = \omega(e^{i\gamma_k}) = 0, \tag{20}$$

where $\gamma_1 = 0$, $\gamma_2 = 2\pi/3$, $\gamma_3 = 4\pi/3$. Thus, to prove the theorem it suffices to show that the boundary-value problem (19), transformed to the plane of the homeomorphism $\zeta = \zeta(z)$, has only the trivial solution in the class of analytic functions under the conditions (20).

Let us compute the index of the boundary-value problem (19) (the possibility of computing it directly was pointed out by S. N. Antoncev). We have

$$G(t) = a - ib = \cos\theta^1(t) + \cos\theta^2(t) - i[\sin\theta^1(t) + \sin\theta^2(t)], \qquad t = e^{i\gamma},$$

whence

$$G(t) = 2\cos\frac{\theta^1 - \theta^2}{2}e^{-i(\theta^1 + \theta^2)/2},$$

where the $\theta^k(t)$ ($k = 1, 2$) are the arguments of the homeomorphisms $w^k(z)$ on the circle $|t| = 1$. We remark first of all that

$$|G(t)|^2 = 4\left|\cos\frac{\theta^1(t) - \theta^2(t)}{2}\right|^2 \geq 1, \qquad t = e^{i\gamma}, \gamma \in [0, 2\pi].$$

Since $w^k(z)$ ($k = 1, 2$) is a homeomorphism, the functions $\theta^k(e^{i\gamma})$ increase as γ increases, i.e.,

$$\theta^k(e^{i\gamma'}) - \theta^k(e^{i\gamma}) > 0 \quad \text{for } \gamma' - \gamma > 0, \tag{21}$$

and the normalization conditions (17) give us

$$\gamma_s \leq \theta^k(e^{i\gamma}) \leq \gamma_{s+1}, \qquad \gamma \in [\gamma_s, \gamma_{s+1}] \tag{22}$$

($\gamma_1 = 0$, $\gamma_2 = 2\pi/3$, $\gamma_3 = 4\pi/3$, $\gamma_4 = 2\pi$). Obviously, (21) and (22) hold also for $\theta(t) = \frac{1}{2}[\theta^1(t) + \theta^2(t)]$. Thus $\theta(e^{i\gamma}) = (\theta^1 + \theta^2)/2$ is monotonically increasing with respect to γ on $[0, 2\pi]$ from $\theta = 0$ to $\theta = 2\pi$; consequently,

$$\kappa = \frac{1}{2\pi}\left[\arg \overline{G(t)}\right]_{|t|=1} = \frac{1}{2\pi}\left[\arg \theta(t)\right]_{|t|=1} = 1.$$

Since the index obviously does not change under continuous homeomorphisms of the disk onto itself, the index $\kappa = \{\arg \overline{G[t(\zeta)]}\}_{|\zeta|=1}/2\pi$ of the Hilbert problem (19), transformed to the plane of the homeomorphism $\zeta = \zeta(z)$, is also equal to 1. Then the general solution of the Hilbert problem (19) on the ζ-plane in the class of analytic functions takes the form (see §4 in Chapter I)

$$\Phi(\zeta) = \zeta(i\beta_0 + c\zeta - \bar{c}\zeta^{-1})e^{\Gamma(\xi)},$$

where $\Phi(\zeta) = \omega e^{-T_1\varphi}\{z(\zeta)\}$. Since the polynomial $\zeta(i\beta_0 + c\zeta - \bar{c}\zeta^{-1})$ has at most two zeros in $\overline{K}$, while (according to (20)) $\Phi(\zeta)$ must vanish at the three distinct points $e^{i\gamma_k}$, $k = 1, 2, 3$, it follows that $\Phi = \omega \equiv w^1 - w^2 = 0$ in $\overline{K}$. The theorem is proved.

We now prove an important theorem on convergence of sequences of homeomorphisms.

THEOREM 3. *Let $w = w^k(z)$ be a sequence of homeomorphisms of the unit disk onto itself that are normalized by $w^k(0) = 0$, $w^k(1) = 1$ and that satisfy the equations*

$$w_{\bar{z}}^k - \mu_1^k(w^k, z)w_z^k - \mu_2^k(w^k, z)\overline{w}_{\bar{z}}^k = 0, \tag{8*}$$

where the $\mu_i^k(w, z)$ are continuous in z for almost all w and satisfy the Lipschitz conditions (16) *and the inequality*

$$\left|\mu_1^k(w, z)\right| + \left|\mu_2^k(\overline{w}, z)\right| \leq \mu_0 < 1.$$

If the sequences $\mu_i^k(w, z)$ converge almost everywhere to $\mu_i^0(w, z)$ for fixed w, then the sequence $w^k(z)$ converges in $W_p^1(\overline{K})$ ($p > 2$) to a homeomorphism $w = w^0(z)$ satisfying the limit equation of (8*).

According to Theorem 1, the homeomorphisms $w^k(z)$ can be represented in the form (12) with φ^k, ψ^k, and df^k/dz uniformly bounded in $L_p(\overline{K})$ ($p > 2$) (see (13)). Let $\{T\varphi^{k_1}\}$, $\{T\psi^{k_1}\}$, and $\{f^{k_1}\}$ be subsequences that converge in $C^\alpha, 0 < \alpha < (p - 2)/p$. Since for the corresponding sequence $\{w^{k_1}\}$ the coefficients of the first of equations (14) converge almost everywhere in $\overline{K}$, the sequence $\{\varphi^{k_1}\}$ converges in $L_p(\overline{K})$ ($p > 2$). Then, by Lemma 1, the sequence $\{f^{k_1}\}$ converges in $W_p^1(\overline{K})$, and, consequently, the last equation in (14) implies also that the sequence $\{\psi^{k_1}\}$ converges in $L_p(K)$.

Moreover, by construction the sequence $\{w^{k_1}(z)\}$ converges in $W_p^1(\overline{K})$, and its limit can be represented in the form (12) and satisfies almost everywhere the limit equation of the form (8*). Consequently, as shown in the proof of Theorem 1, this limit is also a homeomorphism of the unit disk onto itself. But the coefficients of the limit equation of (8*) satisfy a Lipschitz condition with respect to w, so this equation has (by Theorem 2) at most one solution mapping the unit disk homeomorphically onto itself with the normalization $w(0) = 0$, $w(1) = 1$. Therefore, the limits of all the convergent subsequences $w^{k_1}(z)$ coincide, which means that the original sequence $w^k(z)$ converges in $W_p^1(\overline{K})$ ($p > 2$), and the theorem is proved.

REMARKS. 1. If it is not required that the coefficients in (8*) satisfy a Lipschitz condition, then, according to what was proved above, the sequence $w^k(z)$ remains only relatively compact, and the limits of the convergent subsequences will satisfy the limit equation of (8*) as before.

2. If in the proof of Theorem 3 we set $f = \psi = \beta = 0$ in the representations (12) for the homeomorphisms $w^k(z)$, then the assertions of Theorem 3 obviously remain true also for the basic homeomorphisms

$$w^k = z + \hat{T}\varphi^k$$

satisfying (8*).

We study how the smoothness of the homeomorphisms of the unit disk onto itself that we have constructed depends on the smoothness of the coefficients.

THEOREM 4. *Let $w = w(z)$ be a homeomorphism of the unit disk onto itself that satisfies the linear equation (3) with coefficients for which (4) holds. If $\mu_i(z) \in W_p^1(\overline{K})$, $p > 2$, $l \geq 1$, then $w(z) \in W_p^{l+1}(\overline{K})$, and if $\mu_i(z) \in C^{l+\alpha}(\overline{K})$, $\alpha > 0$, $l \geq 0$, then $w(z) \in C^{l+1+\alpha}(\overline{K})$, with $J = |w_z|^2 - |w_{\bar{z}}|^2 \geq \delta_0 > 0$.*

For the Beltrami equation (1) Theorems 4 and 5 in §1 obviously give us that $w(z) \in W_p^{l+1}(\overline{K})$ if $\mu_i(z) \in W_p^l(\overline{K})$. Suppose first that the assertions of the theorem are also valid for the Beltrami equation (1) when $\mu_i(z) \in C^{l+\alpha}(\overline{K})$. Then in the case of the general equation (3) we map the disk $|z| < 1$ onto the disk $|\zeta| < 1$ by a solution of the equation

$$\zeta_z - \mu(z)\zeta_{\bar{z}} = 0, \tag{23}$$

where

$$\mu(z) = 2\mu_1\left(1 + |\mu_1|^2 - |\mu_2|^2 + \sqrt{\Delta}\right)^{-1},$$

$$\Delta = \left(1 - |\mu_1|^2\right)^2 - 2|\mu_1|^2|\mu_2|^2 \geqslant \Delta_0(\mu_0) > 0.$$

Moreover, as shown in the proof of Theorem 7 in §1, the original solution of (3) can be transformed to the form

$$w_{\bar{\zeta}} - a(z)\overline{w}_{\bar{\zeta}} = 0,$$

where $a(z) = \mu_2(1 - |\mu|^2)(|1 - \mu_1\mu|^2 - |\mu_2\mu|^2)^{-1}$ and $|a(z)| \leqslant a_0 < 1$. Turning successively to the Beltrami equation (23) and the Beltrami equation for the inverse function $\zeta = \zeta(w)$

$$\zeta_{\bar{w}} + a[z(w)]\zeta_w = 0, \tag{24}$$

we easily establish the theorem for equation (3). Indeed, if $\mu(z) \in W_p^1(\overline{K})$, then our theorem for (23) gives us that $\zeta(z) \in W_p^2(\overline{K})$. By the imbedding theorem we have $z(w), a(z) \in C^\alpha(\overline{K})$ (because $z(w), a(z) \in W_p^1(\overline{K})$); therefore $a[z(w)] \in C^\alpha(\overline{K})$, and (24) gives us $\zeta(w) \in C^{1+\alpha}(\overline{K})$. Since $\zeta(w) \in C^{1+\alpha}(\overline{K})$ and $\zeta(z) \in W_p^2(\overline{K})$, we get $w(z), z(w) \in C^{1+\alpha}(\overline{K})$, which leads in turn to the relation $a[z(w)] \in W_p^1(\overline{K})$. Finally it follows from (24) that $\zeta(w) \in W_p^2(\overline{K})$. But Theorem 4 in §1 gives us that $J_{(w)} = |\zeta_w|^2 - |\zeta_{\bar{w}}^2| \geqslant \delta > 0$, and, since what has been proved implies that $\zeta(w), w(\zeta) \in C^{1+\alpha}(\overline{K})$, it follows that, for example,

$$(w_{\bar{\zeta}})_{\zeta} = \left(\frac{\bar{\zeta}_{\bar{w}}}{J(w)}\right)_{\zeta} = \left[\frac{\zeta_{\bar{w}\bar{w}}^1 w_{\bar{\zeta}} + \bar{\zeta}_{\bar{w}\bar{w}}\overline{w}_{\bar{\zeta}}}{J(w)} - \frac{\zeta_{\bar{w}}J_w}{J^2(w)}\right] \in L_p(\overline{K}), \qquad p > 2$$

(it can be shown similarly that the remaining second derivatives are in L_p). Thus, $w(\zeta) \in W_p^2(\overline{K})$. We see by a direct check that $w(z) = w[\zeta(z)] \in W_p^2(\overline{K})$ if $w(\zeta), \zeta(w) \in W_p^2(\overline{K})$ (similarly to the way $w(\zeta) \in W_p^2(\overline{K})$ follows from $\zeta(w) \in W_p^2(\overline{K})$). Repeating the arguments, we get that if $\mu_i(z) \in W_p^l(\overline{K})$, then a homeomorphism $w(z)$ satisfying (3) is in $W_p^{l+1}(\overline{K})$. It is clear that all the arguments remain in force also for the case of ordinary derivatives; therefore,

$w(z) \in C^{l+1+\alpha}(\overline{K})$, $\alpha > 0$, $l \geqslant 0$, when $\mu_i(z) \in C^{l+\alpha}(\overline{K})$. The identity

$$J = J(z) \cdot J^{-1}(w)$$

connecting the Jacobians

$$J = |w_z|^2 - |w_{\bar{z}}|^2, \qquad J(z) = |\zeta_z|^2 - |\zeta_{\bar{z}}|^2, \qquad J(w) = |\zeta_w|^2 - |\zeta_{\bar{w}}|^2$$

also implies the inequality

$$J \geqslant \delta_0 = \delta M_0^{-1} > 0,$$

where $0 < \delta < J(z)$, $M_0 = 2\|\zeta(w)\|_{W_p^2} > \max\|J(w)\|$.

Thus, to prove the theorem it suffices to establish that for the Beltrami equation (1) we have

$$w(z) \in C^{l+1+\alpha}(\overline{K}) \quad \text{for } \mu(z) \in C^{l+\alpha}(\overline{K}),$$

which will imply the assertions of the theorem for the full equation (3). Obviously, we can restrict ourselves to a local investigation of the smoothness of some homeomorphic solution $\zeta = \zeta(z)$ of the Beltrami equation (1) in a neighborhood of each particular point $z_0 \in \overline{K}$. Then the basic representation $w = \Phi[\zeta(z)]$ will imply, in particular, that these homeomorphisms of the unit disk onto itself are smooth. Let $|z_0| < 1$. Then, according to the corollary to Theorem 7 in §1, this point is carried by the affine transformation $\zeta = z - z_0 + \mu(z_0)(\bar{z} - \bar{z}_0)$ into the origin, and the coefficient $\mu(\zeta)$ of the new Beltrami equation vanishes at $\zeta = 0$. By an additional conformal mapping of the ellipse that is the image of the disk $|\zeta| < 1$ under the transformation $\zeta = z - z_0 + \mu(z_0)(\bar{z} - \bar{z}_0)$ onto the disk we again arrive at a Beltrami equation in the unit disk, with $\mu(0) = 0$. Thus, without loss of generality we may restrict ourselves to a neighborhood of the origin in investigating the smoothness of homeomorphisms inside the disk, assuming, moreover, that $\mu(0) = 0$. Proceeding similarly also in the case of boundary points, we come to the study of the smoothness of the homeomorphisms in a neighborhood of a particular point, for example, $z = 1$, with $\mu(1) = 0$.

Let us investigate the smoothness of the homeomorphisms in a neighborhood of $z = 1$.

We extend the function $\mu(z)$ in a neighborhood of $z = 1$ outside the unit disk by defining

$$\mu(\theta, r) = \sum_{n=1}^{l+1} \lambda_n \mu\left(\theta, 1 - \frac{r-1}{n}\right)$$

for $r > 1$ ($z = re^{i\theta}$), where the λ_n are determined from the equations

$$\sum_{n=1}^{l+1} \lambda_n \left(-\frac{1}{n}\right)^k = 1 \qquad (k = 0,\dots,l)$$

(see Remark 1 after Theorem 3 in §2).

For $\mu(z) \in C^{l+\alpha}(\overline{K}_\varepsilon)$ we have

$$\frac{\partial^s \mu(\theta, r)}{\partial \theta^k \partial r^{s-k}}\bigg|_{r=1+0} = \frac{\partial^s \mu(\theta, r)}{\partial \theta^k \partial r^{s-k}}\bigg|_{r=1-0} \sum_1^{l+1} \lambda_n \left(-\frac{1}{n}\right)^{s-k} = \frac{\partial^s \mu(\theta, r)}{\partial \theta^k \partial r^{s-k}}\bigg|_{r=1-0};$$

therefore, $\mu(\theta, r) = \mu(z) \in C^{l+\alpha}(\overline{K}_\varepsilon)$ in a full neighborhood K_ε of $z = 1$ for the function so extended, where $K_\varepsilon = \{|z - 1| < \varepsilon\}$, $0 < \varepsilon < 1$. Let us look for a homeomorphic solution of (1)

$$\zeta_{\bar{z}} - \mu(z)\zeta_z = 0$$

in the form $\zeta = ze^{T_0\varphi}$ in the disk $|z - 1| < \varepsilon$, where $\operatorname{Re} T_0(\varphi \,|\, e^{ij}) = 0$ by construction (see §4). For $\varphi(z)$ we get the equation

$$\varphi - \mu S_0\varphi = \mu z^{-1}. \tag{25}$$

Observe that $\zeta(1) = e^{T_0\varphi(1)} = 1$ and $|\zeta(e^{i\gamma})| = e^{\operatorname{Re} T^0\varphi(e^{i\gamma})} = 1$. Let $\mu(z) \in C^\alpha(\overline{K}_\varepsilon)$. Since $\|S_0\varphi\|_{C^\alpha} \leqslant M'_\alpha\|\varphi\|_{C^\alpha}$ and $\|S\varphi\|_{C^\alpha} \leqslant M''_\alpha \varepsilon^\alpha\|\varphi\|_{C^\alpha}$ (see (8) in §3 and the remarks after it, which apply also for S_0), we have

$$\|\mu S_0\varphi\|_{C^\alpha(\overline{K}_\varepsilon)} \leqslant \|\mu\|_C\|S_0\varphi\|_{C^\alpha} + \|\mu\|_{C^\alpha}\|S_0\varphi\|_C \leqslant \varepsilon^\alpha\|\varphi\|_{C^\alpha}(M'_\alpha M + 2MM_\alpha), \tag{26}$$

i.e., $\|\mu S_0\varphi\|_{C^\alpha(\overline{K}_\varepsilon)} \leqslant \varepsilon^\alpha\tilde{M}\|\varphi\|_{C^\alpha(\overline{K}_\varepsilon)}$, where $\tilde{M}$ does not depend on φ or ε, and, therefore, ε can be chosen so small that $\varepsilon^\alpha M < 1$. Consequently, by Banach's theorem (Theorem 5 in §1 of Chapter II), (25) has a unique solution in $C^\alpha(\overline{K}_\varepsilon)$; moreover,

$$\|\varphi\|_{C^\alpha(\overline{K}_\varepsilon)} \leqslant \frac{\|z^{-1}\mu(z)\|_{C^\alpha}}{1 - \varepsilon^\alpha\tilde{M}} \leqslant \frac{2\|\mu(z)\|_{C^\alpha}}{(1 - \varepsilon)^2(1 - \varepsilon^\alpha\tilde{M})} = N < \infty.$$

We have

$$J(z) = |\zeta_z|^2 - |\zeta_{\bar{z}}|^2 \geqslant (1 - |\mu|^2)|\zeta_z|^2 = (1 - |\mu|^2)e^{2\operatorname{Re} T_0\varphi}|1 + zS_0\varphi|^2.$$

But $|zS\varphi| \leqslant M''_\alpha N\varepsilon^\alpha$. Therefore, if ε is chosen to satisfy both the inequalities $M_\alpha\|N\varepsilon^\alpha < 1$ and $\tilde{M}\varepsilon^\alpha < 1$, then (25) is also solvable, and the Jacobian $J(z)$ is positive in the disk $|z - 1| \leqslant \varepsilon$. By construction, $\zeta(z) = ze^{T_0\varphi} \in C^{1+\alpha}(\overline{K}_\varepsilon)$, and it is a homeomorphism satisfying (1) in the neighborhood K_ε. If $\mu(z) \in C^{1+\alpha}(K_\varepsilon)$, then, following the usual scheme, we construct (see the proof of the theorem in §1) a solution of (25) of the form $\varphi = Tf$, where $f \in C^\alpha$; then $\zeta(z) = ze^{T_0\varphi} \in C^{2+\alpha}(\overline{K}_\varepsilon)$, and so on. Thus, for $\mu(z) \in C^{l+\alpha}(\overline{K}_\varepsilon)$, $\alpha > 0$, we have constructed a homeomorphic solution $\zeta(z) = ze^{T_0\varphi} \in C^{l+1+\alpha}(\overline{K}_\varepsilon)$ of the

Beltrami equation in the neighborhood $K_\varepsilon = \{|z - 1| \leq \varepsilon\}$. Let us consider this homeomorphism for $|z| \leq 1$. It satisfies the Beltrami equation with the given $\mu(z)$, and, since $|\zeta(e^{i\gamma})| = 1$, it carries an arc of the circle $|z| = 1$ into an arc of $|\zeta| = |w| = 1$. But any homeomorphism $w = w(z)$ of the unit disk onto itself satisfying (1) can be represented in a neighborhood of $z = 1$ in the form $w = \Phi[\zeta(z)]$ in terms of the homeomorphism $\zeta = \zeta(z)$ constructed in K_ε and a function $\Phi(\zeta)$ that is analytic in the domain $\zeta(\overline{K}_\varepsilon)$ and carries the arc of the circle $|\zeta| = 1$ into an arc of $|\Phi| = 1$. Therefore, if $\mu(z) \in C^{l+\alpha}(\overline{K}_\varepsilon)$, $\alpha > 0$, $l \geq 0$, then $w(z) \in C^{l+1+\alpha}(\overline{K}_\varepsilon)$ together with $\zeta(z)$. The arguments are obviously completely similar for the interior points, except that it is more convenient to look for the auxiliary homeomorphism in the form $\zeta = z + \overset{\circ}{T}\varphi$, and, of course, there is no need to extend $\mu(z)$. The theorem is proved.

REMARKS. 1. The basic representation for solutions of (1) gives us the assertions of Theorem 4 about smoothness in the disk $|z| < 1$ for arbitrary solutions $w(z)$ of the Beltrami equation (1), and even in the closed disk $|z| \leq 1$ if the boundary values of $w(z)$ are sufficiently smooth. For solutions of the full equation (3) the above proof gives us the corresponding smoothness of the solutions only in a neighborhood of each point $|z| < 1$ where the solution is locally homeomorphic. However, the use of (24) for the inverse functions in a neighborhood of branch points of the solutions does not accomplish our purpose.

2. If $\mu_i(w, z) \in C^{l+\alpha}$, $l \geq 0$, $\alpha > 0$, with respect to both variables in the case of the quasilinear equation (8), then $w(z) \in C^{l+1+\alpha}$ inside the disk $|z| < 1$, and even in the closed disk when the boundary values are sufficiently smooth; moreover, according to Remark 1, for $\mu_2 \equiv 0$ the proof that the solutions are smooth extends to all solutions, while for $\mu \not\equiv 0$ it extends only to the solutions that are locally homeomorphic in a given neighborhood. Indeed, let $\mu_i(w, z) \in C^\alpha$. Then, since $w(z) \in \overset{\circ}{W}_p^1(\overline{K})$ $(p > 2)$, we have

$$w(z) \in C_{\alpha_0}(\overline{K}), \qquad \alpha_0 = (p - 2)/p \quad \text{and} \quad \mu_i[w(z), z] \equiv \tilde{\mu}_i(z) \in C^{\alpha\alpha_0}(\overline{K}).$$

Therefore, the validity of Theorem 4 for a solution of the linear equation

$$w_{\bar{z}} - \tilde{\mu}_1(z)w_z - \tilde{\mu}_2(z)\overline{w}_{\bar{z}} = 0$$

implies that $w(z) \in C^{1+\alpha\alpha_0}(\overline{K})$, and so $\mu_i[w(z), z]$, $w(z) \in C^{l+\alpha}(\overline{K})$. Repeating the arguments, we conclude that $w(z) \in C^{l+1+\alpha}(\overline{K})$ for $\mu_i(w, z) \in C^{l+\alpha}(\overline{K})$.

THEOREM 5. *Suppose that the function* $w = w(z)$ *is a solution of* (8) *with coefficients satisfying the conditions of Theorem* 1, *and that it maps the disk* K: $|z| < 1$ *quasiconformally onto a simply connected domain* D *with boundary* $\Gamma \in C^\alpha$. *Then* $w(z) \in C^\beta(\overline{K})$ *and* $z(w) \in C^\beta(\overline{D})$, $0 < \beta = \beta(\alpha, \mu_0) < 1$.

By the Hölder property proved in Theorem 1 for the mapping of the unit disk onto itself, it is obviously sufficient to establish the assertions of the theorem for conformal mappings. A proof of this property of conformal mappings can be found, for example, in [133].

§5. Homeomorphisms of infinite domains

Many problems in the mechanics of continuous media can be reduced to the construction of homeomorphisms of infinite domains onto domains bounded by straight lines parallel to the real axis.

Figure 1 shows shapes typical in aerohydrodynamics problems for domains D_w that are mapped homeomorphically onto corresponding domains D_z by a solution of the equation

$$z_{\bar{w}} - \frac{\rho(q) - 1}{\rho(q) + 1} z_w = 0,$$

where $z \equiv x + iy$, $w = \varphi + i\psi$, $q^2 = \varphi_x^2 + \varphi_y^2$, and $\rho = \rho(q)$ is the density of a gas flow, which satisfies the inequality $\rho^{-1}(d/dq)(q\rho(q)) \equiv 1 - M^2 > 0$ for a subsonic flow. If we assume that the flow characteristics of the gas in the domains D_z are known (in particular, if we assume that a functional dependence $q = q(w)$ is known), then the corresponding equation for the homeomorphisms becomes the Beltrami equation

$$z_{\bar{w}} - \mu(w)z_w = 0, \tag{1}$$

with the coefficient assumed to satisfy the condition

$$|\mu(w)| \leqslant \mu_0 < 1, \qquad w \in \overline{D}_w,$$

along with certain regularity conditions stated below in a neighborhood of $w = \infty$.

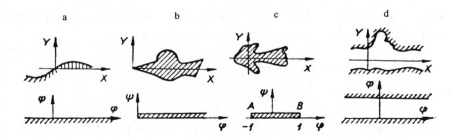

FIGURE 1

The desired homeomorphism $z = z(w)$ that satisfies (1) and maps D_w onto D_z can be represented in the form $z = \Phi[\Omega(w)]$, where $\Omega = \Omega(w)$ is a homeomorphism of D_w onto an analogous domain D_Ω in the Ω-plane, and $z = \Phi(\Omega)$ is an analytic function mapping D_Ω conformally onto D_z. Since the properties of the mapping $z = \Phi(\Omega)$ are known, it suffices to study the homeomorphisms of the domains D_w onto themselves.

The existence of such homeomorphisms can be proved with the aid of Theorem 1 in §2, but for a closer study of their properties we construct such homeomorphisms directly. Let D_w be an arbitrary simply connected domain bounded by straight lines $\psi = \psi_k = $ const parallel to the real axis, and let D_z be an analogous domain in the z-plane. To construct homeomorphisms of D_w onto D_z by a solution of (1) we first map the disk $|\zeta| < 1$ conformally onto the domain D_w by means of the Schwarz-Christoffel formula

$$w = \int_{\zeta_0}^{\zeta} \prod_{k=1}^{n} (\zeta - t_k) \prod_{j=1}^{m} (\zeta - \tau_j)^{-1} (\zeta - \tau_*)^{-2} d\zeta + w_0, \qquad (2)$$

where the t_k are the images of the endpoints w_k of cuts on the plane (for example, of the points A and B in Figure 1c), the τ_j are the images of the vertices of the boundary L_w at infinity, with the coincidence of two or three of the τ_j corresponding to the angles $-\pi$ and -2π at these vertices, and τ_* is the image of the point $w = \infty$ inside D_w.

Some of the kinds of factors in (2) may be missing if the corresponding singular points of D_w or its boundary L_w are missing.

We let $\zeta = \zeta(w)$ be the mapping inverse to (2) and look for a homeomorphism that satisfies (1), maps D_w onto the analogous domain D_z, and has the form

$$z = k\left(w + \mu_* \bar{w}\right) + T_1\left(f \mid \zeta(w)\right) \equiv \Omega(w), \qquad (3)$$

where, if the boundary L_w lies on a single line ($\psi = 0$ for definiteness) we have

$$\mu_* = \mu(\infty), \qquad k = \left(1 + \mu_*\right)^{-1}, \qquad \left(1 + \mu_0\right)^{-1} \leqslant |k| \leqslant \left(1 - \mu_0\right)^{-1} < \infty,$$

and $\mu_* = 0$ and $k = 1$ otherwise. To determine the unknown function $f(\zeta) \in L_p(\overline{K}), p(\mu_0) > 2, K: \{|\zeta| < 1\}$, let us substitute the expressions

$$z_{\bar{w}} = f(\zeta) \frac{d\bar{\zeta}}{d\bar{w}} + k\mu_*, \qquad z_w = S_1(f \mid \zeta) \frac{d\zeta}{dw} + k$$

into (1). Then

$$f(\zeta) - \tilde{\mu}(\zeta) \frac{\overline{w_\zeta}}{w_\zeta} S_1(f \mid \zeta) = k \frac{d\bar{w}}{d\bar{\zeta}} \left[\tilde{\mu}(\zeta) - \mu_*\right] \equiv F(\zeta), \qquad (4)$$

where $\tilde{\mu}(\zeta) = \mu[w(\zeta)]$.

We assume that

$$\|F(\zeta)\|_{L_p(\overline{K})} = \left\|k\frac{d\overline{w}}{d\overline{\zeta}}\left[\tilde{\mu} - \mu_*\right]\right\|_{L_p(\overline{K})} \leqslant N(\mu_0), \tag{5}$$

which ensures that (4) is solvable in the space $L_p(\overline{K})$, $p(\mu_0) > 2$. The inequality (5) is equivalent to the requirement that $[\mu(w) - \mu(\infty)]$ has a zero of sufficient order at $w = \infty$. Taking the behavior of the derivative $dw/d\zeta$ of the conformal mapping (2) into account, we arrive at the following conditions on $\tilde{\mu}(\zeta)$:

$$(\zeta - \tau)^{-\delta + \alpha}\left[\tilde{\mu}(\zeta) - \mu_*\right] < N_\tau(\mu_0), \qquad 0 < \alpha < 1,$$

in a neighborhood of the images τ of the points at infinity on L_w and in D_w, where $\delta = 1$, 2, or 3, respectively, when the angle is 0, $-\pi$, or -2π at a vertex $w(\tau) \in L_w$ at infinity, and $\delta = 2$ when $w(\tau) \in D_w$. Since $|\zeta - \tau|^{-1} \sim e^w$ for $\delta = 1$ and $|\zeta - \tau|^{-1} \sim |w|$, $|w^2|$ respectively, for $\delta = 2$ and $\delta = 3$, the function $[\mu(w) - \mu_*]$ must have an exponential order $e^{-(1-\alpha)}|w|$ of decrease at vertices $w(\tau) \in L_w$ at infinity with zero angle ($\mu_* = \mu(\infty) = 0$ in this case), and a power order for the remaining points $w(\tau) = \infty$.

For the solution $z = \Omega(\zeta)$ of (1) constructed under condition (5) the relations

$$y = \operatorname{Im}\Omega(\psi_k) = \operatorname{Im}w(e^{i\gamma}) = \psi_k$$

hold on the circle $|\zeta| = 1$; these mean that to points of L_w on each line $\psi = \psi_k$ there correspond points of the same line $y = \psi_k$ in the z-plane. Indeed, if the whole boundary L_w lies on the line $\psi = 0$, then

$$\operatorname{Im}z = \operatorname{Im}\left\{k\left(\varphi + \mu_*\varphi\right) + \left(T_1\left(f|e^{i\gamma}\right)\right)\right\} = \varphi\operatorname{Im}k\left(1 + \mu_*\right) = 0;$$

otherwise, $\mu_* = 0$ and $k = 1$, and for $\operatorname{Im}w = \psi_k$ we have

$$\operatorname{Im}z = \operatorname{Im}w + \operatorname{Im}T_1\left(f|e^{i\gamma}\right) = \psi_k.$$

Consequently, when the mapping $z = \Omega(w)$ is a homeomorphism, the image of D_w will be a domain of analogous form in the z-plane. To prove that $z = \Omega(w)$ is a homeomorphism when $\overline{D}_w$ is not the whole plane E we extend $\Omega(w)$ continuously to the domain $E - D_w$ by the formulas

$$z = \Omega_*(w) = \overline{\Omega(\overline{w})}, \qquad \operatorname{Im}(w) \leqslant 0,$$

when $\overline{D}_w$ is the upper half-plane, and

$$z = \Omega_*(w) = \overline{\Omega(\overline{w} + ilh)} + ilh, \qquad |l|h \leqslant \operatorname{Im}w \leqslant (|l| + 1)h$$

in the case of the strip $\overline{D}_w = \{0 \leqslant \operatorname{Im} w \leqslant h\}$, where $l = 0, \pm 1, \pm 2, \ldots$. Then for $w \in E - D_w$

$$\partial_{\overline{w}}\left[\Omega_*(w)\right] = \partial_w \overline{\Omega(w)} = \overline{\partial_{\overline{w}}\Omega}, \qquad \frac{\partial}{\partial w}\Omega_*(w) = \overline{\frac{\partial \Omega}{\partial w}},$$

where $w = \overline{w} + ilh$ (for $\overline{D}_w = \{\operatorname{Im} w \geqslant 0\}$ we set $l = 0$), and so

$$\frac{\partial \Omega_*}{\partial \overline{w}} - \mu(\overline{w} + ilh)\frac{\partial \Omega_*}{\partial w} = \overline{\frac{\partial \Omega}{\partial \overline{w}} - \mu(w)\frac{\partial \Omega}{\partial w}} = 0.$$

Thus, the everywhere-defined function $\tilde{\Omega}(w)$, where $\tilde{\Omega}(w) = \Omega(w)$ for $w \in D_w$ and $\tilde{\Omega}(w) = \Omega_*(w)$ for $w \in E - D_w$, satisfies the following Beltrami equation:

$$\tilde{\Omega}_{\overline{w}} - \tilde{\mu}(w)\tilde{\Omega}_w = 0 \qquad (1^*)$$

($\tilde{\mu}(w)$ is defined similarly to $\tilde{\Omega}(w)$). Next, by the definition of $\Omega_*(w)$, we find from (3) (where we have set $\mu_* = \mu(\infty) = 0$ for definiteness) that

$$\Omega_*(w) - z_0 = \overline{\Omega(\overline{w} + ilh)} + ilh - z_0 + T_1\left[f \mid \varsigma(\overline{w} + ilh)\right]$$

$$= w\left(1 + O\left(\frac{1}{|w|}\right)\right)$$

and, consequently, for sufficiently large $R > 0$ and $z_0 = \text{const}$

$$\operatorname{ind}\{\tilde{\Omega}(w) - z_0\}_{|w|=R} = \operatorname{ind} w\left(1 + O\left(|w|^{-1}\right)\right) = 1.$$

Then the argument principle for the function $\{\tilde{\Omega}(w) - z_0\}$, which is a solution of (1^*), gives us that there is a unique point $w = w_0$ for which $\tilde{\Omega}(w_0) = z_0$. Since z_0 is arbitrary, this last fact means that $z = \tilde{\Omega}(w)$ is a homeomorphic mapping on the whole plane E, and $z = \Omega(w)$ is thereby a homeomorphism for $w \in D_w$.

Let us investigate the smoothness of the homeomorphism $\Omega(w)$ as a function of the smoothness of the coefficient $\mu = \mu(w)$ in (1). Suppose first that $\mu = \mu(w)$ is simply a measurable function satisfying (5) and that $|\mu(w)| \leqslant \mu_0 < 1$, $w \in \overline{D}_w$, while $p = p(\mu_0) > 2$ is a fixed number for which $f(\varsigma) \in L_p(\overline{K})$. Choosing p_1 from the inequalities $1 < (p+2)/4 < p_1 < p/2$, we get

$$\iint_{\overline{D}_w} \left|\Omega_{\overline{w}} - k\mu_*\right|^{p/p_1} dD_w = \iint_{|\varsigma| \leqslant 1} \left|f(\varsigma)\frac{d\varsigma}{dw}\right|^{p/p_1} \left|\frac{dw}{d\varsigma}\right|^2 dK_\varsigma$$

$$\leqslant \left\{\iint_{|\varsigma| \leqslant 1} |f(\varsigma)|^p dK_\varsigma\right\}^{1/p_1}\left\{\iint_{|\varsigma| \leqslant 1}\left|\frac{dw}{d\varsigma}\right|^{-\lambda} dK_\varsigma\right\}^{1/q_1} \leqslant N_0(\mu_0) < \infty,$$

where $1/p_1 + 1/q_1 = 1$ and (by the choice of p_1)

$$\lambda = (p/p_1 - 2)q_1 = (p - 2p_1)/(p_1 - 2) < 2.$$

In a completely analogous way it can be shown that $[\Omega_w - k] \in L_{p_0}(\overline{D}_w)$, where $p_0 = p/p_1 > 2$. Suppose now that $\mu(w) \in C^\alpha(\overline{D}_w)$, i.e., $\mu(w) \in C^\alpha$ for $|w| < \infty$, and $\mu[w(\zeta)] \in C^\alpha$ in a neighborhood of the images $\zeta = \tau$ of the points $w(\tau) = \infty$ that lie in D_w or on its boundary L_w. Then the solution of the Beltrami equation (1) satisfies $z = \Omega(w) \in C^{1+\alpha}$ for all finite w, except possibly for neighborhoods of the endpoints $w_k = w(t_k)$ $(k = 1,\ldots,n)$ of the cuts.

Moreover, $f(\zeta) \in C^\alpha(\overline{K}_\varepsilon)$ everywhere in the disk $|\zeta| < 1$ except for ε-neighborhoods of the images t_k of endpoints of cuts and of the images τ_j of points $w(\tau_j) = \infty$.

But $S_1(f|\zeta) \in L_p(\overline{K})$, $p > 2$, and $S_1(f|\zeta) \in C^\alpha(\overline{K}_\varepsilon)$, i.e., the function $S_1(f|\zeta)$ can only have a singularity of order at most $\lambda = 2/p < 1$ at the points $\zeta = \tau_j$, and, consequently, the limits

$$\lim_{\zeta \to \tau_j}\left[S_1(f|\zeta)\frac{d\zeta}{dw} \right] = 0$$

exist.

Thus, $z = \Omega(w) \in C^{1+\alpha}$ everywhere in D_w with the possible exception of neighborhoods of singular points when $\mu(w) \in C^\alpha(\overline{D}_w)$, and the limits $\Omega_{\overline{w}}(\infty) = k\mu(\infty)$ and $\Omega_w(\infty) = k$ exist.

Let us study the differential properties of the mapping $z = \Omega(w)$ of a full neighborhood $\sigma_w = \sigma_w^1 + \sigma_w^2$ of an endpoint w_0 of a cut (Figure 2).

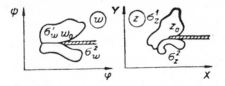

FIGURE 2

Since $\mu(w) \in C^\alpha(\overline{\sigma}_z^i)$, the properties of a solution of the Beltrami equation (1) give us that $\Omega(w) \in C^{1+\alpha}(\overline{\sigma}_w^i)$, from which it follows, in particular, that boundaries of the domains $\sigma_z^i = \Omega(\sigma_w^i)$ are smooth at the point $z = z_0$. By what was proved above, $\Omega(w) \in C^{1+\alpha}(\sigma_w - O_\varepsilon(w_0))$, i.e., the first derivatives of the function $\Omega(w)$, which are continuous in the domains σ_w^i, coincide everywhere on the common boundary of these domains, with the possible exception of the point $w = w_0$. But the fact that $\overline{\Omega}(w)$ is a Hölder function in the full neighborhood $\overline{\sigma}_w = \overline{\sigma}_w^1 + \overline{\sigma}_w^2$ of $w = w_0$ then implies that the derivatives coincide at $w = w_0$. Thus, $\Omega(w) \in C^{1+\alpha}(\overline{\sigma}_w)$. For definiteness we assume that the point $w = 0$ lies on the boundary L_w of D_w, and $\zeta(0) = 1$. Then, according to (3), $\Omega(0) = T_1(f|1) = 0$. We formulate the above results as a theorem.

THEOREM 1. *Suppose that* $|\mu(w)| \leq \mu_0 < 1$ *in* $\bar{D}_w$ *and that condition* (5) *is satisfied. Then there exists a generalized solution* $z = \Omega(w)$ *of* (1) *of the form* (3), *it is a homeomorphism of the domain* D_w *bounded by the straight lines* $\mathrm{Im}\, w = \psi_k$ = const *onto the analogous domain* D_z *in the z-plane,* $\Omega(0) = 0$, *and for* $p = p(\mu_0) > 2$

$$\|\Omega_{\bar{w}} - k\mu(\infty)\|_{L_p(\bar{D}_w)} \leq N_1(\mu_0), \qquad \|\Omega_w - k\|_{L_p(\bar{D}_w)} \leq N_2(\mu_0).$$

If in addition $\mu(w) \in C^\alpha(\bar{D}_w)$, *then* $\Omega(w) \in C^{1+\alpha}$ *for* $|w| < \infty$, *and the derivatives*

$$\Omega_{\bar{w}}(\infty) = k\mu(\infty), \qquad \Omega_w(\infty) = k \equiv [1 + \mu(\infty)]^{-1}$$

exist.

When (1) is a quasilinear equation, i.e., when $\mu = \mu(w, z)$, the existence of a solution $z = \Omega(w)$ of (1) of the form (3) can be proved completely analogously to Theorem 1 in §2, with the difference that instead of a system of equations we obtain only one equation of the form (4) for determining the unknown function $f(\zeta)$.

We mention also that substitution of the solution $z = \Omega(w)$ into the coefficient $\mu(w, z)$ makes equation (1) linear, and, consequently, the properties of $\Omega(w)$ listed in Theorem 1 are also valid in the case of quasilinear equations.

COROLLARY. *Suppose that the slope angle* $\theta = \theta(x)$ *of the tangent to the boundary of the simply connected domain* D_z *is a Hölder function, every x on* L_z *satisfies* $|\theta(x)| \leq \pi/2 - \varepsilon, \varepsilon > 0$, *and*

$$e^{a|x|}|\theta(x)| < C_1, \quad a > 1 \qquad and \qquad |x|^l |\theta(x)| < C_2, \quad l > 2,$$

respectively, in neighborhoods of the "vertices" on L_z *at infinity with zero angle* (*a channel type*) *and with angles* $-\pi$ *or* -2π (*a half-plane or a plane type with a semi-infinite cut*). *Then the homeomorphism* $z = z(w)$ *of the domain* D_w *bounded by the lines* $\mathrm{Im}\, w = \psi_k = $ const *onto* D_z (*which is a similar domain, except bounded by curves*) *that satisfies* (1) *has the properties that* $z = z(w) \in C^{1+\alpha}$ *for* $|w| < \infty$, *and the derivatives* $z_{\bar{w}}(\infty)$ *and* $z_w(\infty)$ *exist and are finite.*

The assertions of the corollary follow from the representation of the homeomorphism $z = z(w)$ in the form $z = \Phi[\Omega(w)]$, where $\Omega(w) = k(w + \mu_* \bar{w}) + T_1[f|\zeta(w)]$ is the homeomorphism of D_w onto itself constructed above, and $z = \Phi(\Omega)$ is a conformal mapping of the domain $\Omega(D_w)$ onto D_z. For the particular cases pictured in Figure 1 the function $\zeta(w)$ in (3) can be chosen in the respective forms: $\zeta(w) = (i - w)/(i + w)$ (see Figure 1a), $\zeta(w) = (i - \sqrt{w})/(i + \sqrt{w})$ (see Figure 1b), $\zeta(w) = w - \sqrt{w^2 - 1}$ (Figure 1c), and, finally, $\zeta(w) = (i - e^w)/(i + e^w)$ for the strip $0 < \mathrm{Im}\, w < \pi$ (Figure 1d), with $\mu_* = \mu(\infty) = 0$ and $k = 1$ in the last case.

REMARK. With the help of specially chosen operators of the same type as T we can construct homeomorphisms also of other domains onto themselves. For example, a homeomorphism of the quadrant $\{\varphi \geqslant 0, \psi \geqslant 0\}$ onto itself by a solution of (1) can be sought in the form

$$z = w + T_2\left(f\Big| \frac{i - w}{i + w} \right) \equiv \Omega(w).$$

According to the properties of T_2 (see [87]),

$$\operatorname{Re} T_2\big(f\,|\, e^{i\gamma} \big) = 0, \quad \gamma \in [0, \pi], \qquad \operatorname{Im} T_2\big(f\,|\, e^{i\gamma} \big) = 0, \quad \gamma \in [\pi, 2\pi];$$

therefore, $\operatorname{Re} \Omega(w) = 0$ when $\operatorname{Re} w = 0$, and $\operatorname{Im} \Omega(w) = 0$ when $\operatorname{Im} w = 0$.

§6. An existence theorem for homeomorphisms of multiply connected domains

Our purpose in this section is to prove the existence of solutions of the quasilinear elliptic system of equations

$$w_{\bar{z}} - \mu_1(w, z)w_z - \mu_2(w, z)\bar{w}_{\bar{z}} = 0 \tag{1}$$

that map a finitely connected domain with order of connectedness $m + 1 \geqslant 1$ homeomorphically onto some canonical domain D_w (see [73] and [104]). In particular, for $m = 0$ the results obtained here will imply the existence theorem proved by another method for homeomorphisms of the unit disk onto itself.

Since equation (1) does not change form under conformal mappings of D_z, we can assume that D_z (as well as D_w) is a multiply connected circular domain with boundary Γ_z (Γ_w) consisting of $m + 1$ disjoint circles Γ_z^i (Γ_w^i), $i = 0, \ldots, m$, of which one, say Γ_z^0 (Γ_w^0), is the unit circle $|z| = 1$ ($|w| = 1$) and contains all the others Γ_z^i (Γ_w^i), $i = 1, \ldots, m$, inside it. It is assumed that the origin $z = 0$ ($w = 0$) lies in D_z (D_w) and the coordinates of the centers and the radii of the circles Γ_z^i, $i = 1, \ldots, m$, are fixed, while the corresponding characteristics of the circles Γ_w^i, $i = 1, \ldots, m$, are not given, of course.

We present without proof a theorem on convergence of analytic functions in domains with variable boundaries [41].

Suppose that $\{D_z^n\}$ is a sequence of arbitrary domains (multiply connected, in general) containing $z = 0$ in the z-plane. The kernel of this sequence with respect to $z = 0$ is defined to be the largest domain D_z containing $z = 0$ such that any closed subset of it belongs to all the D_z^n from some point on. "Largest" is understood in the sense that it contains any other domain having this property. We say that the sequence of domains D_z^n converges to the kernel D_z, written $D_z^n \to D_z$, if any subsequence also has D_z as its kernel.*

THEOREM (CARATHÉODORY). *Suppose that the sequence* $\{D_z^n\}$ *of domains containing* $z = 0$ *converges to its kernel* D_z *with respect to this point, and that the functions* $w = f_n(z)$ *map the domains* D_z^n *univalently onto the domains* D_w^n *in such a way that*

$$f^n(0) = 0, \qquad f_n'(0) > 0.$$

The sequence of functions $f_n(z)$ *converges uniformly on compact subsets of* D_z *to a univalent function if and only if the sequence of domains* D_w^n *has a kernel* D_w *with respect to* $w = 0$ *and converges to its kernel. Then the limit function* $f(z)$ *maps* D_z *univalently onto* D_w.

We remark that in the case of simply connected domains we can take the D_z^n to be a fixed canonical domain D_z, for example, the disk $|z| < 1$ ([41], Chapter V, §5).

THEOREM 2. *Suppose that the functions* $\mu_i(w, z)$, $i = 1, 2$, *are continuous in* w *for almost all* $z \in \overline{D}_z$ *and measurable in* z *for* $0 < |w| < \infty$, *and that*

$$|\mu_1(w, z)| + |\mu_2(w, z)| \leq \mu_0 < 1. \tag{2}$$

Then there exists at least one homeomorphism $w(z) \in W_p^1(\overline{D}_z)$, $p > 2$, *with normalization* $w(0) = 0$ *that satisfies* (1) *and maps the given domain* D_z *onto the canonical domain* D_w.

I. Let us consider the following operators $\overset{\circ}{T}$ and S on the set of functions $\varphi(z) \in L_p(\overline{D}_z)$ ($\varphi(z) \equiv 0$ for $z \notin \overline{D}_z$):

$$\overset{\circ}{T}\varphi = T(\varphi \,|\, z) - T(\varphi \,|\, 0), \qquad S\varphi = \frac{\partial \overset{\circ}{T}\varphi}{\partial z} = \frac{\partial T\varphi}{\partial z},$$

with the fixed numbers $p > 2$ and $p' > 1$ chosen so that

$$\mu_0 \|S\|_{L_{pp'}} = \mu_0 \Lambda_{pp'} < 1. \tag{3}$$

The last part is possible because, by the Calderón-Zygmund theorem (§2), S is linear and bounded in L_p for $p > 1$, and $\|S\|_{L_2} = \Lambda_2 = 1$.

Let $\mathfrak{M}$ be the set of vectors $\vec{\mu}^* = \{\mu_1^*(z), \mu_2^*(z)\}$ such that

$$\vec{\mu}^*(z) \in L_\lambda(\overline{D}_z), \qquad \lambda = \frac{pp'}{p' - 1}, \qquad \vec{\mu}^*(z) \equiv 0, \qquad z \notin \overline{D}_z. \tag{4}$$

Editor's note. For Carathéodory's definition of Kernel, see §122 of his *Conformal representation* (2nd ed., Cambridge Univ. Press, 1952).

Moreover, suppose that almost everywhere in $\overline{D}_z$

$$|\mu_1^*(z)| + |\mu_2^*(z)| \leqslant \mu_0 < 1. \tag{5}$$

To each vector $\vec{\mu}^*(z) \in \mathfrak{M}$ we assign a linear elliptic equation

$$\zeta_{\bar{z}} - \mu_1^*(z)\zeta_z - \mu_2^*(z)\bar{\zeta}_{\bar{z}} = 0, \tag{6}$$

and we construct the basic solutions $\zeta = z + \mathring{T}\varphi$ of these equations.

As we proved in §1 (Theorem 3 accompanying remark), the basic solutions realize homeomorphic mappings of the whole plane onto itself. The functions $\varphi(z)$, as solutions of the equation

$$\varphi - \mu_1^*(z)S\varphi - \mu_2^*(z)\overline{S\varphi} = \mu_1^*(z) + \mu_2^*(z), \tag{7}$$

are continuous operators from L_λ to L_p (the corollary to Theorem 1 in §1):

$$\varphi = (I - \mu_1^*S - \mu_2^*)^{-1}(\mu_1^* + \mu_2^*) \equiv \varphi(\vec{\mu}^* \,|\, z), \tag{8}$$

and

$$\|\varphi(\vec{\mu}^* \,|\, z)\|_{L_p} \leqslant \frac{\mu_0}{1 - \mu_0\Lambda_p}, \qquad \vec{\mu}^* \in \mathfrak{M}. \tag{9}$$

We map the multiply connected image D_ζ of D_z under the homeomorphic mapping $\zeta = z + \mathring{T}\varphi$ conformally onto the canonical domain D_w with the help of the analytic function

$$w = \Phi(\zeta) \equiv \Phi(\vec{\mu}^* \,|\, \zeta), \qquad \Phi(0) = 0, \, \Phi_\zeta'(0) > 0,$$

and on $\mathfrak{M}$ we define the transformation

$$\vec{\mu}(z) = \left\{ \mu_1(\Phi^*, z), \mu_2(\Phi^*, z)\frac{\Phi_\zeta^*}{\overline{\Phi_\zeta^*}} \right\} \equiv \Omega(\vec{\mu}^* \,|\, z), \tag{10}$$

where

$$\Phi^*(\vec{\mu}^* \,|\, z) = \Phi\big(\vec{\mu}^* \,|\, z + \mathring{T}\varphi\big).$$

II. Let us show that to the fixed points $\vec{\mu}(z) = \vec{\mu}^*(z)$ of the transformation $\vec{\mu} = \Omega(\vec{\mu}^*)$ there corresponds the desired generalized solution

$$w(z) = \Phi\big(z + \mathring{T}\varphi\big) \in W_p^1\big(\overline{D}_z\big), \qquad p > p_0 > 2,$$

of (1) which maps the given domain D_z homeomorphically onto the canonical domain D_w with normalization $w(0) = 0$.

The last assertion is obvious, because by construction the function $w = \phi(z + \mathring{T}\varphi)$ is the superposition of the homeomorphic mapping $\zeta = z + \mathring{T}\varphi$ of D_z onto D_ζ and the conformal mapping $w = \Phi(\zeta)$ of D_ζ onto D_w. Further, we observe that for $z \in \overline{D}_z^0 \subset D_z$ the mapping $w = \Phi(z + \mathring{T}\varphi)$ has the generalized derivatives

$$w_{\bar{z}} = \Phi_\zeta \zeta_{\bar{z}} = \varphi \Phi_\zeta, \qquad w_z = \Phi_\zeta \zeta_z = (1 + S\varphi)\Phi_\zeta,$$

and $\Phi_\zeta \neq 0$ for $\zeta \in \zeta(\overline{D}_z^0) \subset D_\zeta$ because $w = \Phi(\zeta)$ is a conformal mapping. If we then substitute the expressions

$$\zeta_{\bar{z}} = w_{\bar{z}}(\Phi_\zeta)^{-1}, \qquad \zeta_z = w_z(\Phi_\zeta)^{-1}$$

into (6) with coefficients

$$\mu_1^* = \mu_1(\Phi, z), \qquad \mu_2^* = \mu_2(\Phi, z)\frac{\Phi_\zeta}{\overline{\Phi}_{\bar{\zeta}}},$$

we see that the homeomorphism $w = \Phi(z + \mathring{T}\varphi)$ is a solution of (1) for $z \in \overline{D}_z^0 \subset D_z$. The fact that $\Phi(z + \mathring{T}\varphi) \in W_p^1(\overline{D}_z)$, $p > p_0 > 2$, follows from the possibility of extending this homeomorphism by the symmetry principle to a solution of the equation analogous to (1) in a domain $D_z^* \supset \overline{D}_z$ (Theorem 7 in §1).

III. We now show that the transformation $\vec{\mu} = \Omega(\vec{\mu}^*)$, which by construction carries the closed convex set $\mathfrak{M} \subset L_\lambda(E)$, $\lambda = pp'/(p' - 1)$, into itself, is completely continuous in $L_\lambda(E)$, and, consequently, by Schauder's theorem, has at least one fixed point in $\mathfrak{M}$. Suppose that $\vec{\mu}_n^* \in \mathfrak{M}$ is a convergent sequence in $L_\lambda(E)$, to which there corresponds the convergent sequence $\varphi_n(z) = \varphi(\vec{\mu}_n^* \mid z)$ in $L_p(\overline{D}_z)$, $p > 1$, since the operator $\varphi(\mu_n^* \mid z)$ defined by (8) is continuous. Moreover, the homeomorphisms $\zeta_n = z + \mathring{T}\varphi_n$ converge in $W_p^1(\overline{D}_z)$, and, consequently, the images $D_\zeta^n = \zeta_n(D_z)$ of D_z converge uniformly to their kernel D_ζ^0. Then, by the Carathéodory theorem for multiply connected domains, the analytic functions $\Phi(\vec{\mu}_n^* \mid \zeta)$ converge uniformly on compact subsets of D_ζ^0, and with them the quotients $(\Phi_n^*)_\zeta/(\overline{\Phi}_n^*)_{\bar{\zeta}}$ in (10) converge almost everywhere in D_ζ^0. Since the functions $\mu_i(w, z)$ are continuous in w, the last fact implies that the expressions $\vec{\mu}_w(z) = \Omega(\vec{\mu}_n^* \mid z)$ converge almost everywhere in D_z and (by virtue of the inequalities $|\vec{\mu}_n(z)| \leq \mu_0 < 1$) also in $L_\lambda(E)$.

To prove that $\vec{\mu} = \Omega(\vec{\mu}^*)$ is a compact transformation it suffices to observe that, by (9), for any sequence $\vec{\mu}_n^* \in \mathfrak{M}$ the corresponding sequence $\varphi_n(z) = \varphi(\vec{\mu}_n^* \mid z)$ is uniformly bounded in $L_p(\overline{D}_z)$ ($p > 2$) and, since $\mathring{T}$ is completely continuous in that space, is carried into a relatively compact sequence of homeomorphisms $\zeta_n = z + \mathring{T}\varphi_n$ in $C^\alpha(\overline{D}_z)$, $0 < \alpha < (p - 2)/p$. The theorem is proved.

THEOREM 3 (a stability criterion). *Suppose that the functions* $w = w^k(z)$, $w^k(0) = w^k(1) - 1 = 0$, *map the given multiply connected circular domain* D_z *quasiconformally onto analogous domains* D_w *and satisfy the equations*

$$w_{\bar{z}}^k - \mu^k(z)w_z^k = 0, \qquad |\mu^k| \leqslant \mu_0 < 1. \tag{11}$$

If $\mu^k(z) = \{\mu^k(z)\}$ *converges to* $\mu(z)$ *in* $L_\lambda(\bar{D}_z)$, $\lambda = p/p'(p' - 1)$, *where* $p > 2$ *and* $p' > 1$ *are chosen from condition* (3), *then* $w^k(z)$ *converges in the norm of* $W_p^1(\bar{D}_z)$, $p > 2$, *to a quasiconformal mapping* $w = w(z)$ *of* D_z *onto* D_w *that is a solution of the limit equation* (11).

This theorem is a direct consequence of the representation of the mappings $w^k(z) \in W_p^1(\bar{D}_z)$ in the form

$$w^k = \Phi^k(z + \mathring{T}\varphi^k) \equiv \Phi^k(\zeta),$$

where $\varphi^k(z) \in L_p(\bar{D}_z)$, $p > 2$, satisfies (7) with $\vec{\mu}^* = \{\mu^k(z), 0\}$, and $w = \Phi^k(\zeta) \in W_{p_0}^1(\bar{D}_\zeta^k)$, $D_0 > z$, is a conformal mapping of the domain D_ζ^k (the image of D_z under the mapping $\zeta = z + \mathring{T}\varphi^k$) onto D_w^k. Indeed, by Corollary 1 in §1, the functions $\varphi^k(z)$ converge in $L_p(\bar{D}_z)$, and with them the images D_ζ^k of D_z converge uniformly to their kernel D_ζ, and so (according to Carathéodory's theorem) the conformal mappings $w = \Phi^k(\zeta)$ also converge uniformly within D_ζ.

§7. Extension of homeomorphisms

We study the problem of extending homeomorphisms of multiply connected domains and their boundary values to homeomorphisms of the plane E onto itself while preserving smoothness.

1°. *Extension of homeomorphisms of multiply connected domains.* Let us first consider an auxiliary problem. Suppose that a quasiconformal mapping $w = w^0(z)$ of the unit disk G_0: $|z| < 1$ onto itself satisfies one of the conditions

   (i) $w = w^0(z) \in C^{n+\alpha}(G_0)$, $n \geqslant 1$, $0 < \alpha < 1$,

   (ii) $w = w^0(z) \in W_p^{n+1}(\bar{G}_0)$, $n \geqslant 0$, $p > 2$,

and that

$$\left|w_{\bar{z}}^0/w_z^0\right| = \left|q^0(z)\right| \leqslant q_0 < 1, \qquad |z| < 1. \tag{1}$$

It is required to construct a homeomorphism $w = F(z)$ of E onto itself with $[F(z) - z] \in C^{n+\alpha}(E)$ (respectively, W_p^{n+1}) that satisfies the conditions

$$\frac{\partial^{k+1}F(z)}{\partial z^k \partial \bar{z}^i} = \frac{\partial^{k+i}w^0(z)}{\partial z^k \partial \bar{z}^i} \quad \text{for } z = e^{i\gamma}, \; k + i = 0,\dots,n, \tag{2}$$

$$F(z) \equiv z \quad \text{for } |r - 1| \geqslant \varepsilon, \; r = |z|. \tag{3}$$

We suppose first that $n \geqslant 1$. In this case, by the imbedding theorem, condition (i) holds simultaneously with (ii). Let

$$\ln w^0(z) = \omega^0(\zeta) = \tau^0 + i\theta^0, \qquad \ln z = \zeta = \xi + i\gamma$$

and observe that on the unit circle, i.e., for $\xi = 0$, we have

$$\tau_\gamma^0(0, \gamma) = \tau^0(0, \gamma) = 0, \qquad \theta_\gamma^0(0, \gamma) \geqslant \delta_0 > 0$$

(the last inequality holds because the boundary correspondence $\theta = \theta^0(0, \gamma)$ is one-to-one). Then inequality (1) for $z = e^{i\gamma}$, namely

$$|q^0(z)| = \left| \frac{\omega_{\bar{\zeta}}^0}{\omega_\zeta^0} \right| = \left| \frac{\tau_\xi^0 - \theta_\gamma^0 + i\left(\theta_\xi^0 + \tau_\gamma^0 \right)}{\tau_\xi^0 + \theta_\gamma^0 + i\left(\theta_\xi^0 - \tau_\gamma^0 \right)} \right| \leqslant q_0 < 1$$

implies that there is a number $\delta_1 > 0$,

$$\delta_1 = \delta_1(q_0, \delta, M), \qquad M = \|\omega(\zeta)\|_{C^{n+\alpha}} \qquad \left(-\tfrac{1}{2} \leqslant \xi = \ln|z| \leqslant 0 \right),$$

such that $\tau_\xi^0 \geqslant \delta_1 > 0$. Consequently, for sufficiently small ξ

$$|\tau^0(\zeta)| \leqslant M|\xi|, \qquad |\tau_\gamma^0(\zeta)| \leqslant M|\xi|^\alpha, \qquad \delta \leqslant \left\{ \theta_\gamma^0(\zeta), \tau_\xi^0(\zeta) \right\} \leqslant M, \qquad (4)$$

where $0 < \delta = \tfrac{1}{2}\min(\delta_0, \delta_1)$ and $\delta < 1$ for definiteness. We extend the function $\omega^0(\zeta)$ given for $\xi = \ln|z| \leqslant 0$ to the domain $\xi > 0$ with preservation of smoothness. Obviously, for sufficiently small ξ the extended function will satisfy (4) as before, with some constants δ and M. Therefore, it will be assumed that the function $\omega^0(\zeta) \in C^{n+\alpha}(\overline{G}_\varepsilon)$ (respectively, W_p^{n+1}) is defined in a fixed annulus G_ε: $-\varepsilon < \xi < \varepsilon_0$ and satisfies (4). Let us consider the function $F(z) = e^{\omega(z)}$ with

$$\omega(\zeta) = \left[1 - l\left(\frac{\xi}{\xi_0} \right) \right] \omega^0(\zeta) + l\left(\frac{\xi}{\xi_0} \right) \omega^1(\zeta), \qquad |\xi| \leqslant |\xi_0|, \qquad (5)$$

where

$$\omega^1(\zeta) = \xi + M\xi_0 + i\theta^0(\zeta),$$

$$l(x) = \int_0^x x^n(1 - x)^n \, dx \left\{ \int_0^1 x^n(1 - x)^n \, dx \right\}^{-1}, \qquad x \in [0, 1],$$

and $0 < |\xi_0|^\alpha \leqslant \delta^2/2M(2\lambda + 1)$, $\lambda = \max l'$. We estimate the first derivatives of this function:

$$\delta \leqslant \tau_\xi = (1 - l)\tau_\xi^0 + l + l'\left(\frac{\xi}{\xi_0} + M - \frac{\tau^0(\xi)}{\xi_0} \right) \leqslant M_0 = 2M(2\lambda + 1),$$

$$|\tau_\gamma| = \left| (1 - l)\tau_\gamma^0 \right| \leqslant M|\xi_0|^\alpha, \qquad \delta \leqslant \theta_\gamma = \theta_\gamma^0 \leqslant M, \qquad |\theta_\xi| = |\theta_\xi^0| \leqslant M.$$

These inequalities enable us to estimate the complex characteristic $\mu(z) = F_{\bar{z}}/F_z$ of the mapping $F(z) = e^{\omega(z)}$:

$$|\mu(z)|^2 = \left| \frac{\tau_\xi - \theta_\gamma + i(\theta_\xi + \tau_\gamma)}{\tau_\xi + \theta_\gamma + i(\theta_\xi - \tau_\gamma)} \right|$$

$$\leqslant 1 - \frac{4\tau_\xi \theta_\gamma - 6M_0 |\xi_0|^\alpha}{(\tau_\xi + \theta_\gamma)^2 + \theta_\xi^2 - 3M_0 |\xi_0|^2} \leqslant 1 - \frac{\delta_0^2}{5M_0^2},$$

i.e.,

$$|\mu(z)| \leqslant \mu_0(\delta, M_0) < 1. \tag{6}$$

The mapping $w = F(z)$ carries the circle $|z| = e^{\xi_0}$ to the circle $|w| = e^{\xi_0(1+M)}$. Consider the function $F(z) = e^{\omega(z)}$:

$$\omega(\zeta) = \left[1 - l\left(\frac{\xi - \xi_0}{\xi_1 - \xi_0} \right) \right] \omega^1(\zeta) + l\left(\frac{\xi - \xi_0}{\xi_1 - \xi_0} \right) \omega^2(\zeta), \quad |\xi_0| \leqslant |\xi| \leqslant |\xi_1|, \tag{7}$$

where $\omega^1(\zeta) = \xi + M\xi_0 + i\theta(\zeta)$, $\omega^2(\zeta) = \xi + i\gamma = \ln z$ and $\xi_1 = \xi_0 + (1 + \lambda)M\xi_0$. Then

$$\tau_\gamma = 0, \qquad \theta_\gamma = (1 - l)\theta_\gamma^0 + l \geqslant \delta > 0, \qquad (|\omega_{\bar{\xi}}|, |\omega_\xi|) \leqslant 2M_0,$$

$$\tau_\xi = 1 - \frac{l'}{\xi_1 - \xi_0} = 1 - \frac{l'}{1 + \lambda} \geqslant \frac{1}{1 + \lambda},$$

and, consequently, the complex characteristic $\mu(z) = F_{\bar{z}}/F_z$ satisfies (6) for $|\xi_0| \leqslant |\xi| \leqslant |\xi_1|$. Thus, $[F(z) - z] = [e^{\omega(\zeta)} - z] \in C^{n+\alpha}(E)$ (respectively, W_p^{n+1}), where $\omega(\zeta)$ is defined by (5) and (7) for $0 \leqslant |\xi| \leqslant |\xi_1|$, and by

$$\omega(\zeta) = \xi + i\gamma = \ln z \quad \text{for } |\xi| \geqslant |\xi_1|. \tag{8}$$

The function $w = F(z)$ maps the plane E quasiconformally onto itself; the complex characteristic $\mu(z) = F_{\bar{z}}/F_z$ of this mapping satisfies (6), and $\mu(z) \equiv 0$ for $|\xi| \geqslant |\xi_1|$.

The mapping $w = F(z)$ thus constructed satisfies (2) on the unit circle, and, in particular,

$$|F(e^{i\gamma})| = \exp\{\tau^0(0, \gamma)\} = 1.$$

In order to satisfy condition (3) outside the given annulus G_ε: $1 - \varepsilon < |z| < 1 + \varepsilon$ simultaneously with (6) it obviously suffices to subject the parameter ξ_0 to the requirement

$$|\xi_0| \leqslant \min\left\{ \left(\frac{\delta^2}{2M_0} \right)^{1/\alpha}, \frac{1}{2M_0} \ln(1 + \varepsilon) \right\},$$

since in that case

$$|\xi_1| = |\xi_0|(1 + M + \lambda M) \leqslant 2M_0|\xi_0| \leqslant \ln(1 + \varepsilon).$$

Thus, it remains to consider the case

$$w^0(z) \in W_p^1(\overline{G}_0), \qquad p > 2.$$

Let $\chi = \chi(z)$ be the quasiconformal mapping of the circular annulus G_z: $\xi_0 < \xi = \ln z < 0$ onto the annulus G_χ: $(1 + M)\xi_0 < \sigma = \ln|\chi| = 0$, where $-1 < M = M(\xi_0) < \infty$, and ξ_0 is a parameter chosen from the condition $0 < |M\xi_0| < \ln(1 + \varepsilon)$. The complex characteristic $q(z) = \chi_{\bar{z}}/\chi_z$ satisfies the conditions

$$q(z) = q^0(z) = w_{\bar{z}}^0/w_z^0, \quad \xi \in \left[\frac{\xi_0}{2}, 0\right] \quad \text{and} \quad q(z) \equiv 0, \xi \in \left[\xi_0, \frac{\xi_0}{2}\right].$$

Let us extend $\ln \chi(\zeta)$ to the annulus by means of (7), in which

$$\theta^0(\zeta) = \arg\chi(\xi_0, \gamma) \quad \text{and} \quad \xi_1 = \xi_0 + (1 + \lambda)|M|\xi_0,$$

while inside the disk $|z| < e^{\xi_1}$ we set $\chi(z) = z$. We extend the homeomorphism $\chi = \chi(z) \in W_p^1(\overline{G}_0)$, $p > 2$, to a homeomorphism of the plane E by setting $\chi = \{\chi(1/\bar{z})\}^{-1}$ for $|z| > 1$.

Then $[\chi(z) - z] \in W_p^1(E)$, $p > 2$, and, by the choice of the parameter ξ_0,

$$q(z) = \chi_{\bar{z}}/\chi_z \equiv 0 \quad \text{for } |r - 1| \geqslant \varepsilon, r = |z|$$

and $|q(z)| \leqslant q_1 < 1$, $z \in E$. We now consider the following homeomorphism of the unit disk onto itself:

$$w = w^0[z(\chi)] \equiv \Phi^0(\chi), \qquad |\chi| < 1.$$

Since $q(z) = \chi_{\bar{z}}/\chi_z = q^0(z) = w_{\bar{z}}^0/w_z^0$ in a neighborhood of the circle $|z| = 1$, the mapping $w = \Phi^0(\chi)$ is conformal in a neighborhood of $|\chi| = 1$. Consequently, the function $w = e^{\omega(\zeta)} = F^0(\chi)$ determined by (5), (7), and (8), where $\zeta = \ln\chi \equiv \sigma + i\varphi$ and $\omega^0 = \ln\Phi^0(0, \varphi)$, is a homeomorphism of the plane E, and, moreover,

$$F^0(e^{i\varphi}) = \Phi^0(e^{i\varphi}) = w^0[\chi(e^{i\gamma})], \qquad [F^0(\chi) - \chi] \in C^n(E)$$

and $n \geqslant 2$ can be fixed arbitrarily.

Then the function

$$w = F^0[\chi(z)] \equiv F(z) \tag{9}$$

obviously realizes a homeomorphic mapping of the plane E, $F(z) - z \in W_p^1(E)$ ($p > 2$), and conditions (2) and (3) are satisfied ($n = 0$).

We proceed to solve the problem of extending homeomorphisms defined in multiply connected domains to homeomorphisms of the whole plane while preserving smoothness.

Let $w = w^0(z) \in C^{n+\alpha}(\overline{G})$ (respectively, W_p^n, $p > 2$), $n \geqslant 1$, $0 < \alpha < 1$, be a quasiconformal mapping of the multiply connected circular domain $G \subset G_0$: $|z| < 1$,

$$G = G_0 \backslash \sum_{k=1}^{m} G_k, \qquad G_k \cap G_i = \varnothing \quad \text{for } k \neq i, \qquad G_k: |z - z_k| < r_k,$$

onto the analogous domain

$$Q = Q_0 \backslash \sum_{k=1}^{m} Q_k, \qquad Q_k: |w - w_k| < \rho_k \quad (w_0 = 0, \rho_0 = 1)$$

with complex characteristic $q^0(z) = w_{\bar{z}}^0 / w_z^0$ satisfying (1). Then the desired homeomorphism $w = F(G, z)$, $(F - z) \in C^{n+\alpha}(E)$ (respectively, W_p^n), $n \geqslant 1$, $0 < \alpha < 1$, of E onto itself with the properties

$$F(G, z) = w^0(z), \quad z \in G \quad \text{and} \quad F(G, z) \equiv z \quad \text{for } |z| > 1 + \delta, \quad (10)$$

can be defined as follows in the rest of the plane:

$$F(G, z) = \begin{cases} \rho_k F\left(\dfrac{z - z_k}{r_k}\right) + w_k, & z \in G_k, k = 1, \ldots, m, \\ F(z), & 1 < |z| < 1 + \varepsilon, \end{cases} \quad (11)$$

where $F(z) = e^{\omega(\zeta)}$ is given by (9) for $w^0(z) \in W_p^1(\overline{G})$, and $\omega(\zeta)$ is defined by (5), (7), and (8) in the remaining cases.

2°. Extension of homeomorphisms from the boundary. Let $\Gamma = \Sigma_0^m \Gamma_k$ be a collection of m disjoint circles Γ_k: $|z - z_k| = r_k$ inside the unit circle Γ_0: $|z| = 1$ ($z_0 = 0$, $r_0 = 1$), and let $w = w(t)$, $t \in \Gamma$, be a complex-valued function that maps each of the circles Γ_k one-to-one onto itself with preservation of orientation. It is assumed that $w(t)$ satisfies one of the following conditions:
   a)

$$\omega(t) \in C^{n+\alpha} \quad (n \geqslant 1, 0 < \alpha < 1), \qquad |\partial w / \partial t| \geqslant \delta > 0.$$

   b) There exists a homeomorphism $w^0(z)$ of a neighborhood of Γ with $|w_z^0| < 1$ such that

$$w(t) = w^0(t), \qquad \omega^0(z) \in W_p^n(\overline{G}), \quad p > 2, n \geqslant 1,$$

where $G(\Gamma)$ is the multiply connected circular domain with boundary Γ.
   Let $\varepsilon > 0$ be small enough that

$$\prod_{k=0}^{m} G_k^\varepsilon = 0, \qquad G_k^\varepsilon: r_k(1 - \varepsilon) < |z - z_k| < r_k(1 + \varepsilon), \qquad k = 0, \ldots, m.$$

Then the desired extension $w = F(\Gamma, z)$, $(F - z) \in C^{n+\alpha}(E)$ (respectively, W_p^n), of the mapping $w = w(t)$, $t \in \Gamma$, to a homeomorphism of the plane E onto itself can be represented in the form

$$
w = F(\Gamma, z) = \begin{cases} r_k F\left(\dfrac{z - z_k}{r_k}\right) + z_k, & z \in G_k^\varepsilon, k = 0,\dots,m, \\ z, & z \in E \setminus \displaystyle\prod_{k=0}^{m} G_k^\varepsilon, \end{cases} \tag{12}
$$

where $F(z) = e^{\omega(z)}$ is defined by (9) for $w^0(z) \in W_p^1(\overline{G})$, and $\omega(z)$ is given by (5), (7), and (8) in the remaining cases, with $\omega^0(z) = \xi + i \arg w(t)$ (respectively, $\omega^0 = \xi + i \arg w_0(z)$; $t = z \in p$, $z = e^\xi$).

An important boundary property of quasiconformal mappings is contained in the following assertion, whose proof requires the use of geometric methods and is therefore outside the scope of this book (see [1], Chapter IV, §§A and B).

THEOREM 1 (AHLFORS). *The boundary values $u = u(x)$ of a quasiconformal mapping $\omega = \omega(\zeta)$, $\omega(\infty) = \infty$, of the upper half-plane $\operatorname{Im} \zeta > 0$ onto itself with complex characteristic $q(\zeta)$, $|q(\zeta)| \leqslant q_0 < 1$, satisfy the condition*

$$
M^{-1} \leqslant \frac{u(x + \xi) - u(x)}{u(x) - u(x - \xi)} \leqslant M(q_0) < \infty, \tag{13}
$$

where ξ is an arbitrary real number. Conversely, every one-to-one mapping $u = u(x)$ of the straight line $y = 0$ onto itself that satisfies (13) can be extended to a quasiconformal mapping of the whole plane E with complex characteristic $q(\zeta)$, $|q(\zeta)| \leqslant q_0(M) < 1$.

We map the upper half-plane $\operatorname{Im} \zeta > 0$ conformally onto the disk $|z - z_0| < r_0$ and denote the respective images of the points $\zeta = \infty$, $x - \xi$, x and $x + \xi$ by t_0, t_1, t_2 and t_3 ($t_k = z_0 + r e^{i\gamma_k}$). Then, since the cross-ratio of four points is invariant under linear fractional transformations, we get

$$
A(t_k) = \frac{t_0 - t_1}{t_0 - t_2} : \frac{t_3 - t_1}{t_3 - t_2} = \frac{1}{2} \qquad (\gamma_0 < \gamma_1 < \gamma_2 < \gamma_3). \tag{14}
$$

On the other hand, by means of a conformal mapping $w = w(\omega)$, $w(\infty) = t_0$, of the upper half-plane $\operatorname{Im} \omega > 0$ onto the disk $|w - z_0| < r_0$, condition (13) on the cross-ratio of the four points $\omega = \infty$, $u(x - \xi)$, $u(x)$ and $u(x + \xi)$ can be transformed to the form

$$
M^{-1} \leqslant A(w_k) = \frac{w(t_0) - w(t_1)}{w(t_0) - w(t_2)} : \frac{w(t_3) - w(t_1)}{w(t_2) - w(t_2)} \leqslant M, \tag{15}
$$

where $w(t) = w\{\omega[\zeta(t)]\}$. Theorem 1 allows us to conclude that the satisfaction of (15) for any quadruple of points t_k connected by relation (14) is necessary and sufficient for the extendibility of the homeomorphism $w = w(t)$ of the circle $|z - z_0| < r_0$ onto itself to a homeomorphism of the whole plane E. Moreover, by (12), this extension can be realized in such a way that outside a fixed neighborhood of the boundary the mapping $w = w(z)$ is the identity. This enables us to use (12) also in the case when $\Gamma = \Sigma_0^m \Gamma_k$ is the boundary of a multiply connected domain. We formulate this as a theorem.

THEOREM 2. *An orientation-preserving homeomorphism $w = w(t)$ of the contour $\Gamma = \Sigma_0^m \Gamma_k$, Γ_k: $|z - z_k| = r_k$, onto itself can be extended to a full homeomorphism $w = F(\Gamma, z)$, $(F - z) \in W_p^1(E)$, $p > 2$, of the whole plane E if and only if (15) holds for each quadruple of points $t_i \in \Gamma_k$ connected by relation (14). Moreover, the mapping $w = F(\Gamma, z)$ can be represented by (12).*

Since any m-connected circular domain can be quasiconformally mapped onto a given similar domain ([17], §9.D), the next assertion is a direct consequence of the previous theorem.

THEOREM 3. *Suppose that $w = w(t)$ satisfies condition (15) for each quadruple of points $t_i \in \Gamma_k$ connected by relation (14), and that it is a one-to-one orientation-preserving mapping of the circles Γ_k: $|z - z_k| = r_k$ onto the circles $w(\Gamma_k) = L_k$: $|w - w_k| < \rho_k$, where the L_k are disjoint and are contained inside the circle $L_0 = w(\Gamma_0)$. Then there exists an extension of the homeomorphism $w = w(t)$ of the contour $\Gamma = \Sigma_0^m \Gamma_k$ onto the contour $L = \Sigma_0^m L_k$ to a full homeomorphism $w = F(\Gamma, z)$ of the whole plane E with complex characteristic $\mu(z)$, $|\mu(z)| \leqslant \mu_0 < 1$, under which $L_k = w(\Gamma_k)$.*

REMARK. There is an example of an increasing function $u(x)$, $x \in (-\infty, \infty)$, that satisfies (13) and is not absolutely continuous. By Theorem 1, this example shows that the boundary correspondence under quasiconformal mappings with uniformly bounded complex characteristic $q(z)$, $|q(z)| \leqslant q_0 < 1$, need not be absolutely continuous.

§8. Mappings of multiply connected domains by solutions of nonlinear L-elliptic systems of equations

The topic of our investigation will be nonlinear systems of partial differential equations

$$F_i(x, y, u, v, u_x, u_y, v_x, v_y) = 0, \qquad i = 1, 2, \tag{1}$$

having the property that each bounded solution is locally homeomorphic in the domain D_z in which the independent variables x and y vary. Such systems were first studied by Lavrent'ev [61], [65], [66], who introduced a constructive definition of ellipticity for them and who extended many important properties of conformal mappings to their solutions.

We use a definition of strong ellipticity (L-ellipticity) different from that introduced by Lavrent'ev, but we prove that the definitions are equivalent. For L-elliptic systems we prove an existence theorem for mappings of multiply connected domains (a generalized Riemann mapping theorem), along with a uniqueness theorem in the case of a simply connected domain.

1°. Definition and properties of L-elliptic systems.

DEFINITION 1. A system (1) is said to be L-*elliptic* in a neighborhood G_0 of some point (x_0, y_0) if it has the following properties:

(i) It is solvable for the derivatives v_x and v_y, and can then be represented in the form

$$v_x = a_{11}u_x + a_{12}u_y, \qquad -v_y = a_{21}u_x + a_{22}u_y; \tag{2}$$

moreover, for $M = \max |\ln(u_x - iu_y)| < \infty$ the inequalities

$$a_{11}a_{22} - \left(\frac{a_{12} + a_{21}}{2}\right)^2 \geqslant \delta(M) > 0, \tag{3}$$

$$\left\| a_{ij}(z, w, \omega) \right\|_{C^1} \leqslant N(M) < \infty \tag{4}$$

hold, where $z = x + iy$, $w = u + iv$, $\omega = \ln(u_x - iu_y)$, and $(z, w, \omega) \in Q_0 = \{G_0 \times w(G_0) \times \omega(G_0)\}$.

(ii) The system (1) is elliptic, i.e., if $\xi^2 + \eta^2 = 1$, then

$$\Lambda(\xi, \eta) = \begin{vmatrix} \dfrac{\partial F_1}{\partial u_x}\xi + \dfrac{\partial F_1}{\partial u_y}\eta, & \dfrac{\partial F_1}{\partial v_x}\xi + \dfrac{\partial F_1}{\partial v_y}\eta \\[2ex] \dfrac{\partial F_2}{\partial u_x}\xi + \dfrac{\partial F_2}{\partial u_y}\eta, & \dfrac{\partial F_2}{\partial v_x}\xi + \dfrac{\partial F_2}{\partial v_y}\eta \end{vmatrix} \neq 0.$$

It is easy to see that for systems of equations of the form (2) condition (ii) is equivalent to the condition of ellipticity for the second-order quasilinear equation in u obtained by eliminating the derivatives v_x and v_y from (2). If in the resulting equation for $u(x, y)$ we pass to the new unknown functions u_x and u_y, then for determining them we also get a quasilinear elliptic system, which can be written as a single complex equation for the function $\omega = \ln(u_x - iu_y)$. If we also write (2) in the form of a complex equation, then we arrive at the following equivalent definition of L-ellipticity.

DEFINITION 2. The system (1) is said to be L-*elliptic* if it can be represented in the form

$$w_{\bar{z}} - q_1(z, w, \omega)w_z - q_2(z, w, \omega)\overline{w_{\bar{z}}} = 0, \qquad |q_1| + |q_2| \leqslant q_0(M) < 1 \tag{5}$$

and the quasilinear equation for $\omega = \ln(w_z + \overline{w}_z)$, obtained by differentiating (5) is elliptic, i.e.,

$$\omega_{\bar{z}} - \mu_1(z, w, \omega)\omega_z - \mu_2(z, w, \omega)\overline{\omega}_{\bar{z}} = F(z, w, \omega),$$

$$|\mu_1| + |\mu_2| \leqslant \mu_0(M) < 1, \tag{6}$$

where μ_i ($i = 1, 2$) and F_j ($j = 1, 2$), $F = F_0 + F_1 w_z + F_2 \overline{w}_{\bar{z}}$, are expressed in terms of the q_i and their derivatives with respect to the arguments, and

$$q_i \in C^1(\overline{Q}_0), \qquad \mu_i, F_j \in C(\overline{Q}_0), \qquad |F_j| \leqslant N_0(M) < \infty. \tag{7}$$

Equation (6) is called the *derived equation* of the nonlinear equation (5). We remark next that the solvability of the elliptic system (1) for the derivatives v_x and v_y implies its solvability for u_x and u_y, and conversely. Indeed, suppose that the elliptic system (1) is represented in the form

$$-u_y = a_{11}^* v_x + a_{12}^* v_y \equiv \tilde{F}_1, \qquad u_x = a_{21}^* v_x + a_{22}^* v_y \equiv \tilde{F}_2, \tag{2*}$$

where $a_{ij}^* = a_{ij}^*(z, w, \sigma)$, $\sigma = \ln(v_x - iv_y)$, satisfies (3) and (4). Substituting the values $F_1 = \tilde{F}_1 + u_y$ and $F_2 = \tilde{F}_2 - u_x$ into the determinant $\Lambda(\xi, \eta)$ in the ellipticity condition (ii), we get

$$\Lambda(\xi, \eta) = \frac{\partial \tilde{F}_1}{\partial v_x}\xi^2 + \left(\frac{\partial \tilde{F}_2}{\partial v_x} + \frac{\partial \tilde{F}_1}{\partial v_y}\right)\xi\eta + \frac{\partial \tilde{F}_2}{\partial v_y}\eta^2,$$

$$\Delta \equiv \left(\frac{\partial \tilde{F}_2}{\partial v_x} + \frac{\partial \tilde{F}_1}{\partial v_y}\right)^2 - 4\frac{\partial \tilde{F}_1}{\partial v_x}\frac{\partial \tilde{F}_2}{\partial v_y} \geqslant 4\frac{D(\tilde{F}_2, \tilde{F}_1)}{D(v_x, v_y)}.$$

The definiteness condition $\Delta < 0$ for the quadratic form $\Lambda(\xi, \eta)$ ensures that the conditions for solvability of the equations (2*) for v_x and v_y are satisfied. The converse assertion can be established in a completely analogous way. This enables us to add one more equivalent form to the definition of L-ellipticity.

DEFINITION 3. The system (1) is *L-elliptic* if it can be represented in the form (5) with $q_i = q_i^*(z, w, \sigma)$, $\sigma = \ln(v_x - iv_y)$, and the function $\sigma = \sigma(z)$ satisfies an elliptic equation of the form (6) with $\mu_i = \mu_i^*(z, w, \sigma)$ and $F = F^*(z, w, \sigma)$ such that the assumptions in (7) are satisfied.

To compute the coefficients of (6) in terms of the $q_i(z, w, \sigma)$ we set

$$w_z = a(z, w, \omega), \qquad w_{\bar{z}} = b(z, w, \omega) = q_1 a + q_2 \bar{a} \tag{8}$$

and with the help of the identity

$$e^\omega = w_z + \overline{w}_z = a(1 + \bar{q}_2) + \overline{aq_1}$$

we express a and b by the formulas

$$a = \frac{(1 + q_2)e^\omega - \bar{q}_1 e^{\overline{\omega}}}{|1 + q_2|^2 - |q_1|^2}, \qquad b = e^{\overline{\omega}} - \bar{a}. \tag{9}$$

The equalities $\partial a/\partial \bar{z} = \partial b/\partial z = w_{z\bar{z}}$ then lead to the following equation for ω:

$$a_\omega \omega_{\bar{z}} + \bar{a}_\omega \omega_z + a_{\bar{\omega}} \overline{\omega_{\bar{z}}} + \left(\bar{a}_{\bar{\omega}} - e^{\bar{\omega}}\right)\bar{\omega}_z + f = 0, \tag{10}$$

where $f = f_0 + f_1 a + f_2 \bar{a}$, $f_0 = a_{\bar{z}} + \bar{a}_z$, $f_1 = \bar{a}_w + a_w q_1 + \bar{a}_{\bar{\omega}} \bar{q}_2$ and $f_2 = \bar{f}_1$. Finally, elimination of the derivative $\bar{\omega}_z$ from this equation and its complex conjugate yields (6) with coefficients computed by the formulas

$$\mu_1 = \Delta_0^{-1} \bar{a}_\omega e^{-\omega}, \qquad \mu_2 = \Delta_0^{-1} a_{\bar{\omega}} e^{-\omega}, \qquad \Delta_0 = 1 - 2\,\mathrm{Re}\big(e^{-\omega} a_\omega\big), \tag{11}$$

REMARK. In determining the homeomorphic solutions of (5) it can be given the following form:

$$w_{\bar{z}} - q(z, w, \omega) w_z = 0, \qquad |q| \leqslant q_0(\mu) < 1, \tag{12}$$

which formally corresponds to the case $q_2 = 0$, and the coefficient

$$q = q_1(z, w, \omega) + q_2(z, w, \omega) \frac{a(z, w, \omega)}{\bar{a}(z, w, \omega)} \tag{13}$$

keeps the properties of the functions $q_i(z, w, \omega)$. We mention an important consequence of (7), namely,

$$\max\left\{ |a_\omega e^{-\omega}|, |a_{\bar{\omega}} e^{-\omega}|, |F_j| \right\} \leqslant N_1(q_0, N_0),$$

which enables us to estimate the right-hand side of (6):

$$|F(z, w, \omega)| \leqslant N_2(q_0, \mu_0, N_0)\big(|w_z| + 1\big). \tag{14}$$

Both properties (i) and (ii) hold for the systems of equations in gas dynamics, where

$$a_{12} = a_{21} = 0, \qquad a_{11}^{-1} = a_{22}^{-1} = \rho(q), \qquad q^2 = u_x^2 + u_y^2.$$

Here the function $\rho(q)$ is bounded and positive for the values $q \leqslant q_{max}$, where $\rho(q_{max}) = 0$, which ensures that (3) and (4) hold. If the gas flow is always subsonic, then the constants q_0 and μ_0 do not depend on the unknown solution.

DEFINITION 4. The system of equations (1) is said to be *uniformly L-elliptic* if the constants δ, N, q_0, μ_0, and N_0 in conditions (3)–(7) do not depend on the unknown solution.

2°. *The generalized Riemann mapping theorem.* In proceeding to the establishment of a generalized Riemann theorem on mappings of multiply connected domains by solutions of uniformly L-elliptic systems (first proved for simply connected domains by Lavrent'ev in [65] and [66]) we remark that it suffices for us to restrict ourselves to the proof of this theorem for mappings of canonical domains. This follows from the next statement.

LEMMA 1. *Properties* (i) *and* (ii) *of nonlinear L-elliptic systems are invariant under conformal transformations*

$$\chi = \chi(z) \in C^2(\overline{D}_z), \qquad \zeta = \zeta(w) \in C^2(D_w).$$

The assertion of the lemma is obvious for χ. Suppose now that a conformal transformation $\zeta = \zeta(w)$ is performed on the unknown function. Then the system (2), regarded as a linear system, is obviously transformed into an analogous system for the new functions $\xi(x, y)$ and $\eta(x, y)$, $\xi + i\eta = \zeta$. But the derivatives u_x and u_y appearing in the coefficients a_{ij} are expressed in terms of ξ_x and ξ_y from the following identity:

$$u_x - iu_y = \frac{dw(\zeta)}{d\zeta}(\xi_x - i\xi_y).$$

Thus property (i) is invariant. On the other hand, by the same identity,

$$\omega = \Omega + \ln \frac{dw}{d\zeta}, \quad \text{where } \Omega = \ln(\xi_x - i\xi_y),$$

and, consequently, the equation for Ω has the same form as the equation (6) for ω. The lemma is proved.

Thus, multiply connected domains D_z (D_w) with smooth finite boundary can be assumed to be circular without loss of generality, i.e.,

$$D_z = D_z^0 \setminus \bigcup_{k=1}^{m} D_z^k, \quad D_z^k: |z - z_k| < r_k, \quad k = 0, 1, 2, \dots, m \ (z_0 = 0, r_0 = 1)$$

(D_w is of similar form). In order to make the problem of mapping multiply connected circular domains D_z and D_w include also the case when the boundaries ∂D_z and ∂D_w do not have sufficient smoothness under the conformal mappings or are not bounded, it suffices to replace the condition (7) of L-ellipticity (Definition 2) by the following relaxed condition:

$$q_i \in C^1(\overline{Q}^\varepsilon), \qquad (\mu_i, F_j) \in C(\overline{Q}^\varepsilon), \qquad |F_j| \le N_0(\varepsilon, M) < \infty, \qquad (7^*)$$

where $Q^\varepsilon = \{D^\varepsilon \times w(D^\varepsilon) \times \omega(D^\varepsilon)\}$, $D^\varepsilon = D_z \setminus \overline{G}_\Gamma^\varepsilon$, and G_Γ^ε is an arbitrary ε-neighborhood of some set $\Gamma \subset \partial D_z$.

DEFINITION 5. A system (1) satisfying conditions (5), (6), and (7*) is said to be *L*-elliptic* or *uniformly L*-elliptic* in D_z in accordance with whether or not the constants q_0, μ_0, and N_0 depend on M.

LEMMA 2. *Suppose that the functions* $w(z)$ *and* $\omega(z)$ *have the properties*

$$w(0) = 0, \qquad \{w(z) - z\} \in W_p^2(E), \qquad \omega(z) \in W_p^1(E), \qquad p > 2, \qquad (15)$$

and form a solution of the system (5), (6) *with coefficients satisfying* (7) *on the plane E, where the* $\mu_i(z, w, \omega)$ *and* $F_j(z, w, \omega)$ *are expressed in terms of*

$q_i(z, w, \omega)$ $(q_i \equiv 0$ for $|z| \geqslant R + 1)$ by the equalities (11). Then for all $z \in E$

$$\omega(z) = \ln(w_z + \overline{w}_z),\tag{16}$$

i.e., the function $w(z)$ is at the same time a solution of the nonlinear equation (5) with $q_i = q_i[z, w, \ln(w_z + \overline{w}_z)]$.

We remark that, by (15), the function $w = w(z)$ is a homeomorphism of the linear equation (5) with $q_i(z) = q_i[z, w(z), \omega(z)] \in W_p^1(E)$. Consequently, by Theorem 4 in §4,

$$J = |w_z|^2 - |w_{\overline{z}}|^2 \geqslant \delta > 0, \qquad |z| < R + 1,$$

and since $J = |\,dw/dz\,|^2 \to 1$ as $R \to \infty$, it follows that $J \geqslant \delta > 0$ also when $|z| \geqslant R + 1$ if R is large enough. But

$$|w_z + \overline{w}_z| \geqslant |w_z|(1 - q_0) \geqslant (1 - q_0)J^{1/2} \geqslant (1 - q_0)\delta^{1/2} > 0,$$

and so $\{\ln(w_z + \overline{w}_z)\} \in W_p^1(E)$. We let

$$w_z = A, \qquad w_{\overline{z}} = -B, \qquad e^{\Omega} = w_z + \overline{w}_z\tag{17}$$

and use (5) to express $A(\omega, \Omega)$ and $B(\omega, \Omega)$ in terms of $q_i(\omega)$ and Ω according to (9), in which e^{ω} and $e^{\overline{\omega}}$ are replaced by e^{Ω} and $e^{\overline{\Omega}}$, respectively. Then, as in the derivation of (10), the identity $A_{\overline{z}} = -B_z = w_{z\overline{z}}$ leads to the relation

$$A_\Omega \Omega_{\overline{z}} + A_\Omega \Omega_z + A_{\overline{\Omega}} \overline{\Omega}_{\overline{z}} + \left(\overline{A}_{\overline{\Omega}} - e^{\Omega}\right)\overline{\Omega}_z$$
$$+ A_\omega \omega_{\overline{z}} + \overline{A}_\omega \omega_z + A_{\overline{\omega}} \overline{\omega}_{\overline{z}} + \overline{A}_{\overline{\omega}} \overline{\omega}_z + g = 0,\tag{18}$$

where $g = g_0 + g_1 w_z + g_2 \overline{w}_{\overline{z}}$, and the g_i are obtained from the f_i in (10) by formally replacing a by A. Let $p^*(\omega)$ denote the values of functions $p(\omega, \Omega)$ for $\Omega = \omega$. With this notation the coefficients of (10) and (18) are connected as follows:

$$a = A^*, \qquad b = B^*, \qquad f = g^*, \qquad a_\omega = (A_\omega + A_\Omega)^*, \qquad a_{\overline{\omega}} = (A_{\overline{\omega}} + A_{\overline{\Omega}})^*.$$

We remark next that the derivatives A_Ω^* and $A_{\overline{\Omega}}^*$ coincide with the derivatives of the function $a^0 = a^0(z, w)$ defined by (9) with respect to ω and $\overline{\omega}$, where the $q_i(z, w, \omega) \equiv q_i^0(z)$ in (9) are computed on the solution $w(z)$, $\omega(z)$ of the system (5), (6).

Taking these facts into account, we subtract (10) and (18) one from the other and as a result get an equation for the difference $\Omega - \omega = \zeta$:

$$a_\omega^0 \zeta_{\overline{z}} + \overline{a}_\omega^0 \zeta_z + a_{\overline{\omega}}^0 \overline{\zeta}_{\overline{z}} + \left(\overline{a}_{\overline{\omega}}^0 - e^{\overline{\omega}}\right)\overline{\zeta}_z + c_0 \zeta = 0,\tag{19}$$

where

$$c_0 = (\Delta\omega)^{-1}\big(\Delta_\omega A + \Delta_\Omega A + \Delta g_0 + w_z \Delta g_1 + \bar{w}_{\bar{z}}\Delta g_2 - \bar{\Omega}_z \Delta e^{\bar{\omega}}\big),$$

$$\Delta\omega = \Omega - \omega, \quad \Delta p(\omega, \Omega) = p - p^*, \quad \Delta_\omega A = \Delta A_\omega \omega_z + \Delta \bar{A}_\omega \omega_{\bar{z}} + \Delta A_{\bar{\omega}}\bar{\omega}_{\bar{z}},$$

and $\Delta_\Omega A$ is defined by formally replacing ω by Ω in $\Delta_\omega A$.

By (15) and the linearity of $A(\omega, \Omega)$ in e^Ω and $e^{\bar{\Omega}}$, for $z \in \bar{K}_n$, $|z| \le R + 1$ we have

$$\|c_0(z)\|_{L_p} \le M\big(q_0, \|\Omega(z)\|_{W_p^1}, \|\omega(z)\|_{W_p^1}\big), \tag{20}$$

and the fact that the coefficients of (5) and (6) are compactly supported when $|z| \ge R + 1$ implies that $\zeta_{\bar{z}} = 0$ for $|z| > R + 1$. Since the coefficients of the derivatives in (19) can be expressed in terms of the $q_i^0(z)$, which do not depend explicitly on ω, this equation reduces to the form

$$\zeta_{\bar{z}} - \mu(z)\zeta_z = c\zeta \qquad \left(\mu(z) = \mu_1^0 + \mu_2^0 \frac{\bar{\zeta}_{\bar{z}}}{\zeta_z}\right), \tag{21}$$

where the $\mu_i^0(z)$ can be computed in terms of the functions $q_i^0(z)$ and $\omega(z)$ by (12) and satisfy the condition (13) for the ellipticity of (19), while $(c\zeta)$ coincides with the F in (11) for $f \equiv c_0\zeta$ and $a = a^0(z, \omega)$:

$$c = -\sigma^{-1}e^{-\omega}\left[c_0\big(1 + \bar{q}_2^0\big) + \bar{c}_0 \frac{\Delta\omega}{\Delta\omega}\big(q_2^0 + 1 - \sigma\big)\right], \qquad \sigma = 1 + \big|q_1^0\big|^2 - \big|q_2^0\big|^2.$$

By construction, $\mu(z) = c(z) \equiv 0$ for $|z| \ge R + 1$, and because of (20) and the inequality $\sigma \ge 1 + q_0^2 > 0$ we have

$$\|c(z)\|_{L_p(E)} \le M < \infty, \qquad p > 2.$$

We represent the generalized analytic function $\zeta = \zeta(z)$ in the form

$$\zeta = \Phi[\chi(z)]e^{T_c},$$

where $\chi = \chi(z)$ is a basic homeomorphic solution of the homogeneous equation (21), and $\Phi(\lambda)$ is an analytic function in the χ-plane. In view of (15) we have for $|z| \ge R + 1$ that $\zeta(z) = \ln(dw/de) - \omega(z) \to 0$ as $z \to \infty$; therefore, $\Phi(\infty) = 0$, which implies that $\Phi(\chi) \equiv 0$, and thereby $\zeta(z) = \ln(w_z + \bar{w}_z) - \omega(z) \equiv 0$. The lemma is proved.

THEOREM 1. *Each uniformly L^*-elliptic system* (1) (*Definition* 5) *admits at least one generalized solution*

$$u + iv = w(z) \in W_p^1(\bar{D}_z) \cap W_p^2(\bar{D}_z \backslash G_\Gamma^\varepsilon), \qquad p > 2,$$

that maps a given multiply connected circular domain D_z quasiconformally onto an analogous domain D_w, with normalization

$$w(0) = w(1) - 1 = 0. \tag{22}$$

For sufficiently small $\nu > 0$ we construct a sequence of domains $G^\nu \subset D_z$,

$$G^\nu = \sum_{k=0}^{m} G_k^\nu, \qquad G_k^\nu : r_k < |z - z_k| < r_k + \nu,$$

$$k = 1, \ldots, m, \qquad G_0^\nu : 1 - \nu < |z| < 1,$$

and we define a family of cutoff functions $g^\nu(z) \in C^{2+\alpha}(D_z)$ in D_z such that

$$g^\nu(z) = 0, z \in G^\nu; \quad g^\nu(z) = 1, z \in D_z \backslash G^{2\nu}; \quad |g^\nu(z)| \leqslant 1, z \in D_z; \qquad (23)$$

moreover, $g^\nu(z) \to 1$ almost everywhere in D_z as $\nu \to 0$. Let

$$q^\nu(z, w, \omega) = g^\nu(z) q(z, w, \omega).$$

We fix a number $\nu > 0$ and, without changing the notation for the present, we consider (5) in the form (12) with the coefficient q^ν, i.e., we assume the conditions

$$q(z, w, \omega) \equiv 0, \qquad z \in G^\nu. \qquad (24)$$

Let the numbers $p > 2$ and $p' > 1$ be such that

$$q_0 S_{L_{pp'}} = \mu_0 \Lambda_{pp'} < 1$$

and let $\mathfrak{M}$ be the set of functions $q^* = \{q_1^*(z)\}$ such that

$$q^* \in L_\lambda(\overline{D}_z), \qquad \lambda = \frac{pp'}{p' - 1}, q^*(z) \equiv 0, z \notin (D_z \backslash \overline{G}^\nu), |q^*(z)| \leqslant q_0 < 1$$

Suppose that $w = w^*(z) \in W_p^1(\overline{D}_z)$, $p > 2$, is a quasiconformal mapping of D_z onto a multiply connected circular domain D_w and that it satisfies the linear equation (12) with coefficient $q^*(z)$ and with the normalization condition indicated in the theorem. We construct a quasiconformal mapping $w = w^0(z) \in W_p^2(\overline{D}_z)$, $p > 2$, $w^0(0) = w^0(1) - 1 = 0$, of D_z onto D_w by a solution of the linear equation (12) with coefficient

$$q^0(z) = q[z, \omega^*(z), \omega^*(z)], \qquad \omega^*(z) = Tf^*,$$

where $f^*(z) \in L_p(\overline{K}_\varepsilon)$ ($p > 2$, $K_\varepsilon : |z| < 1 + \varepsilon$) is an arbitrary function.

Let us extend the mapping $w^0(z)$ according to §7.1° to a homeomorphism $w = w^0(z)$, $w^0 - z \in W_p^2(E)$, of the plane E, and denote by $q^0(z) = w_{\bar{z}}^0 / w_z^0$, $q^0(z) \equiv 0$ for $|z| > 1 + \varepsilon$, the complex characteristic of this mapping.

By (24), $w^0(z)$ is analytic near the boundary of D_z; therefore, the extension can be realized in such a way that $q^0(z) \in C^{2+\alpha}$ (for this it suffices to set $n = 3$ in (7.2)). In the plane E we consider the equation

$$\omega_{\bar{z}} - \mu_1^0(z)\omega_z - \mu_2^0(z)\overline{\omega}_{\bar{z}} + F^0(z), \qquad |\mu_1^0| + |\mu_2^0| \leqslant \mu_0(\varepsilon) < 1, \qquad (6^*)$$

which coincides in D_z with the derived equation (6), in whose coefficients we have substituted the values $w = w^0(z)$ and $\omega = \omega^*(z)$:

$$\mu_i^0(z) = \mu_i[z, w^0(z), \omega^*(z)], \qquad F^0(z) = F[z, w^0(z), \omega^*(z)], \qquad z \in D_z.$$

For $z \notin D_z$, (6) is the derived equation for the mapping $w = w^0(z)$, i.e., its coefficients $\mu_i = \mu_i^0(z)$ and $F = F^0(z)$ are determined by (11), for $q_1 \equiv q^0(z)$, $q_2 \equiv 0$, and $\omega = \omega^*(z)$

$$F^0 = \frac{1 - |q^0|^2}{1 + |q^0|^2} \frac{2}{l} \operatorname{Re} a_{\bar{z}}, \qquad \mu_1^0 = \mu_0^2 \frac{l}{2} = \frac{q^0}{1 + |q^0|^2}, \qquad l = e^{\omega^*}, \quad (25)$$

and, since $q^0(z)$ has compact support,

$$\mu_i^0(z) = F^0(z) \equiv 0, \qquad |z| > 1 + \varepsilon.$$

We remark that, according to (14) and (25),

$$|F^0(z)| \leqslant N_2(|w_2^0| + 1), \quad z \in D_z^{2\nu}, \qquad |F^0(z)| \leqslant N_3(\nu), \quad z \notin D_z^{2\nu}, \quad (26)$$

with constants N_2 and N_3 that do not depend on $\omega^*(z)$, where $D_z^{2\nu} = D_z \backslash G^{2\nu}$.

Let us represent the solution $\omega = \omega(z)$, $\omega(\infty) = 0$, of (6*) in the form

$$\omega = Tf, \quad f \in L_p(E), \quad p > 2; \qquad f(z) \equiv 0, \quad |z| > 1 + \varepsilon.$$

The function $f = f(z)$ satisfies

$$f - \mu_1^0 Sf - \mu_2^0 \overrightarrow{Sf} = F^0, \qquad (27)$$

and the Banach principle is applicable to this equation in $L_p(\overline{K}_\varepsilon)$, where $p = p(q_0, \mu_0, \varepsilon) > 2$ is fixed once and for all so that

$$\max(q_0, \mu_0) \|S\|_{L_p} = k_0 \Lambda_p < 1.$$

The results of §6 on quasiconformal mappings of multiply connected domains give us that

$$\|F^0(z)\|_{W_p^1} < N_4(q_0) < \infty, \qquad z \in D_z \ (p > 2), \qquad (28)$$

and, consequently, (26) implies that

$$\|F^0(z)\|_{L_p(\overline{D}_z^{2\nu})} \leqslant N_5(q_0), \qquad \|F^0(z)\|_{L_p(\overline{K}_\varepsilon)} \leqslant N_6(q_0, \nu). \qquad (26^*)$$

Therefore, the solution of (27) satisfies the estimate

$$\|f(z)\|_{L_p(E)} < \frac{N_6}{1 - k_0 \Lambda_p} = \overline{N}(\nu_0, c_0, \mu_0).$$

For each fixed $q^*(z) \in \mathfrak{M}$ the constructed solution gives rise to a transformation $f = R^0(f^*)$ of the ball $\mathfrak{N} \subset L_p(\overline{K}_\varepsilon)$, $\mathfrak{N}: \|f\|_{L_p} < \overline{N}$, into itself. Consider the following transformation R defined on the set $\mathfrak{N} \times \mathfrak{M}$:

$$(f, q) = R(f^*, q), \qquad R = (R^0, R^1),$$

where $R = q[z, w^0(z)]$.

Suppose that f_n^* and q_n^* converge in L_p and L_λ, respectively. Then, by Theorem 2 in §6, the sequence of mappings $w = w_n^*(z)$ converges in $W_p^1(\overline{D}_z)$, $p > 2$, and with it the functions

$$q_n = q[z_n, w_n^*(z), \omega_n^*(z)], \qquad \omega_n^* = Tf_n^*,$$

also converge in $W_p^1(\overline{D}_z)$. Consequently, μ_{in}^0 and F_n^0 converges in C^α and L_p, respectively, and thereby the solutions f_n of (27) converge in L_p. Thus, R is a continuous transformation. Selecting from the bounded subsets $\{w^*(z)\}$ and $\{\omega^*(z)\}$ of $W_p^1(\overline{D}_z)$, $p > 2$, sequences w_n^* and ω_n^* that converge in C^α, $\alpha = (p-2)/p > 0$, we construct a sequence $q_n(z)$ that converges in C^α, along with a corresponding sequence of homeomorphisms $w = w_n^0(z)$ that converges in $C^{1+\alpha}$. Then μ_{in} and F_n^0 converge in C^α and L_p, respectively, and, by the preceding arguments, the solutions f_n of (27) converge in L_p. Hence, R is completely continuous and maps the closed convex set $\mathfrak{N} \times \mathfrak{M}$ into itself.

Consequently, Schauder's theorem says that R has at least one fixed point $(f, q) = (f^*, q)$ in $\mathfrak{N} \times \mathfrak{M}$. By construction, to this fixed point there corresponds a solution

$$\{w(z), \omega(z)\}, \qquad w(0) = w(1) - 1 = \omega(\infty) = 0,$$

of the system (5), (6), and, by Lemma 2, this means that the mapping $w = w(z) \in W_p^2(\overline{D}_z)$, $p > 2$, satisfies (12) with

$$q = q[z, w, \ln(w_z + \overline{w}_z)]$$

in D_z. Let us investigate the smoothness of the mapping $w = w(z)$. We fix $\nu_0 > 0$ and consider the function

$$\chi(z) = \omega(z)g^{\nu_0}, \qquad z \in D_z,$$

where $g^{\nu_0}(z) \in C^{2+\alpha}$ is the cutoff function defined by (23). From equation (6) for $\omega(z)$ with the coefficients

$$\mu_i^0(z) = \mu_i[z, w(z), \ln(w_z + \overline{w}_z)] \qquad F^0(z) = F[z, w(z), \ln(w_z + \overline{w}_z)],$$

we find that

$$\chi_{\bar z} - \mu_1^0(z)\chi_z - \mu_2^0(z)\overline{\chi}_{\bar z} = F^{\nu_0}(z) \equiv g^{\nu_0}F^0 - \omega\left(g_{\bar z}^{\nu_0} - \mu_1^0 g_z^{\nu_0} - \mu_2^0 \frac{\overline{\omega}}{\omega}\overline{g}_{\bar z}^{\nu_0}\right).$$

Setting $\chi(z) = Tf$, we get the estimate

$$\|\chi(z)\|_{W_p^1(\overline{D}_z)} \leqslant N_7(\nu_0)$$

from the solution of the equation for f, and, since $\chi(z) = \omega(z)$ for $z \in D_z^{z\nu_0} = D_z \backslash G^{2\nu_0}$, it follows that

$$\|\omega(z)\|_{W_p^1(\overline{D}_z^{2\nu_0})} \leqslant N_7(\nu_0), \qquad \|w(z)\|_{W_p^2(D_z^{2\nu_0})} \leqslant N_8(\nu_0). \qquad (29)$$

The last inequality follows from the smoothness of the coefficient of (12):

$$q[z, w(z), \ln(w_z + \overline{w}_z)] = q^0(z) \in W_p^1(\overline{D}_z^{2\nu_0}).$$

We now drop the assumption (24). Let $\{w^{\nu_k}(z)\}$ be a convergent subsequence of the relatively compact (by virtue of (29) and (28)) sequence $\{w^\nu(z)\}$ in $C^{1+\alpha}(\overline{D}_z^{2\nu_0}) \cap C^\alpha(\overline{D}_z)$, $\alpha = (p-2)/p$ (recall that the index ν was omitted for convenience).

Passing directly to the limit in the nonlinear equation (5), we see that the limit function $w = w(z)$ satisfies this equation in $D_z^{2\nu_0}$.

As $\nu \to 0$ we have $q_i^\nu \to q_i$, $|q_1| + |q_2| \leqslant q_0 < 1$, almost everywhere in D_z; therefore, the limit function $w = w(z) \in W_p^1(\overline{D}_z)$, $p > 2$, is a homeomorphism of D_z. To prove the smoothness claimed in the theorem for the limit mapping $w = w(z)$ it remains to investigate its behavior near the boundary.

We map D_z and D_w conformally in such a way that the arc $l_z \subset \Gamma_z$, $l_z \cap \Gamma = \varnothing$ (Γ is the singular set before Definition 5) and its image

$$l_w = w(l_z) \subset \Gamma_w$$

carried into rectilinear segments. As proved in Lemma 1, such mappings do not change the properties of the coefficients of the original system; therefore, we assume (with the previous notation) that the boundary Γ_z contains the rectilinear segment l: $x = 0$, $y \in [-1, 1]$, which is carried into the segment l: $u = 0$, $v \in [-1, 1]$, and that $u(x, y) > 0$ for $x > 0$ in a neighborhood of l_z. Let us fix $\lambda > 0$ in such a way that the half-disk $\sigma_{2\lambda}^+ = \{|z| < 2\lambda, x > 0\}$ lies entirely in D_z (the value of λ is determined only by the geometry of D_z). We suppose that $w(z) \in W_p^2(\overline{\sigma}_{2\lambda}^+)$, $p > 2$ (this can be achieved by averaging the function $q(z, w, \omega)$ over ω), and estimate the norm of $w(z)$ in the neighborhood $\sigma_{2\lambda}^+$ in terms of the problem data. We note that

$$\omega(0, y) = \ln(u_x - iu_y) = \ln u_x,$$

hence, since the maximum principle implies that $u_x(0, y) > 0$ for $u = \operatorname{Re} w(z)$, we have

$$\operatorname{Im} \omega(0, y) = 0, \qquad y \in [-1, 1].$$

Let us extend $\omega(z)$ to the domain $\sigma_{2\lambda}^- = \{|z| < 2\lambda, x < 0\}$ by setting

$$\omega_*(z) = \overline{\omega(-\overline{z})}, \qquad x < 0.$$

The function $\omega = \omega(z)$, which is defined in the full neighborhood $\sigma_{2\lambda} = \sigma_{2\lambda}^+ + \sigma_{2\lambda}^-$, satisfies the equation

$$\omega_{\overline{z}} - \mu_1^0(z)\omega_z - \mu_2^0(z)\overline{\omega}_{\overline{z}} = F^0(z), \qquad |\mu_1^0| + |\mu_2^0| \leqslant \mu_0 < 1, \qquad z \in \sigma_{2\lambda},$$

where

$$\mu_1^0(z) = \overline{\mu_2^0(-\bar{z})}, \qquad \mu_2^0(z) = \overline{\mu_1^0(-\bar{z})}, \qquad F^0(z) = -\overline{F^0(-\bar{z})}, \qquad z \in \sigma_{2\lambda}^-.$$

By (14) and (28), we have in $\bar{D}_z$ (and, in particular, in $\bar{\sigma}_{2\lambda}^+$) that

$$\|F^0(z)\|_{W_p^1(\bar{\sigma}_{2\lambda})} \leq N_2(N_4 + 1) = N_9 < \infty,$$

and it is clear that this inequality is preserved also for $z \in \sigma_{2\lambda}^-$. Consequently, proceeding as in the derivation of (29), we get

$$\|\omega(z)\|_{W_p^1(\bar{\sigma}_\lambda)} \leq N_{10}(\lambda) < \infty, \qquad p > 2.$$

Since the arc l_z is arbitrary, the last inequality is valid in some domain $(G^{\nu_0} \backslash G_\Gamma^\varepsilon)$, $\nu_0 = \nu_0(\lambda) > 0$, which, together with (29), leads to the estimates

$$\|\omega(z)\|_{W_p^1(\bar{D}^\varepsilon)} \leq N_{11} < \infty, \qquad \|w(z)\|_{W_p^2(\bar{D}^\varepsilon)} \leq N_{12} < \infty, \qquad p > 2,$$

and this concludes the proof of the theorem ($D^\varepsilon = D_z \backslash \bar{G}_\Gamma^\varepsilon$).

REMARKS. 1. The last inequalities hold in the closed domain $\bar{D}_z$ in the particular case of uniformly L-elliptic systems (the singular set $\Gamma = \varnothing$).

2. When the system (1) is nonuniformly L-elliptic, we fix arbitrarily a sufficiently large value fo the constant M and extend the coefficients μ_i and q_i of (5) and (7) for any values $\max |\ln(u_x^2 + u_y^2)| > M$ in such a way that (6) and (8) remain valid. If the solution found for the equations satisfies the condition

$$\max|\ln(u_x^2 + u_y^2)| \leq M,$$

then it is by construction also a solution of this same problem for the original system (1).

3°. *A uniqueness theorem.* Let us consider the problem of a mapping of a curvilinear strip D_z in the z-plane onto the rectilinear strip D_w: $0 < u < 1$ in the w-plane, with the following additional conditions assumed:

(a) The system (1) does not depend explicitly on the unknown functions u and v, and, consequently, $w = u + iv$ does not appear explicitly in the coefficients q_i and μ_i of (5) and (7).

(b) There exists a sufficiently large number $R > 0$ such that $q_i = \mu_i = 0$ for $|z| > R + 1$, i.e., the mapping is conformal in a neighborhood of the points at infinity.

(c) The functions

$$q_i, \qquad \frac{\partial q_i}{\partial \omega}, \qquad \frac{\partial q_i}{\partial \bar{\omega}}, \qquad i = 1, 2,$$

satisfy a Lipschitz condition in ω uniformly with respect to z.

Without loss of generality the strip D_z can also be assumed to be the rectilinear strip $0 < x < 1$, since this can be achieved by a conformal mapping with preservation of conditions (a)–(c).

THEOREM 2. *Suppose that the system* (1) *is uniformly L-elliptic and that the additional assumptions* (a)–(c) *hold. Then the mapping* $w = w(z)$ *of the strip* $D_z = \{0 < \operatorname{Re} z < 1\}$ *onto the strip* $D_w = \{0 < \operatorname{Re} w < 1\}$ *by a solution of this system, under the normalization conditions* $\operatorname{Re}\omega(0) = 0$ *and* $\operatorname{Im} w(z) = \pm\infty$ *common in the theory of conformal mappings, is unique when* $\operatorname{Im} z = \pm\infty$.

Let us assume that the mapping problem has at least two solutions $w^1(z)$ and $w^2(z)$, to which there correspond the two distinct functions $\Omega^k = e^{\omega k(z)} = u_x^k - iu_y^k$.

As usual, by subtracting one from the other the equations obtained from (5) by the substitution of unknown functions $\omega^k = \ln \Omega^k$, we get the following differential equation for the difference $\Omega = \Omega^1 - \Omega^2$:

$$\Omega_{\bar{z}} - \mu(z)\Omega_z = A(z)\Omega, \tag{30}$$

where

$$\mu(z) = \mu_1(z, \Omega^1) + \mu_2(z, \Omega^1)\frac{\overline{\Omega^1}}{\Omega^1} \cdot \frac{\Omega_z}{\Omega_{\bar{z}}},$$

$$A(z) = \Omega^{-1}\left(\Omega_z^2 \Delta\mu_1 + \overline{\Omega}_{\bar{z}}^2 \Delta\tilde{\mu}_2 + \Delta A_2\right).$$

Here we have introduced the notation

$$\Delta f = f(z, \Omega^1) - f(z, \Omega^2); \qquad f^k = f(z, \Omega^k);$$

$$\tilde{\mu}_2^k = \mu_2^k \frac{\Omega^k}{\overline{\Omega}^k}; \qquad A_k^0 = F_0^k \Omega^k; \qquad k = 1,2.$$

By assumption (b), $\mu(z) = A(z) \equiv 0$ for $|z| \geqslant R + 1$, and condition (c) tells us that for

$$\Omega^k(z) \in W_p^1\left(\overline{K}_R \cap \overline{D}_z\right), \quad p > 2, \qquad K_R: |z| < R + 1,$$

the function $A(z)$ is pth power summable, $p > 2$, in $\overline{K}_R \cap \overline{D}_z$. Further, for the functions $w^k(z)$ on the boundaries of the strip $D_z = \{0 < x < 1\}$ we have $u^k(x, y) = 0, 1$, whence

$$du^k(x, y) = u_x^k dx + u_y^k dy = u_y^k dy = 0$$

and, consequently,

$$\operatorname{Im}\Omega^k = -u_y^k = 0 \quad \text{for } x = 0, 1.$$

Taking account of the fact that, by the normalization conditions,

$$\operatorname{Re}\omega^k(0) = \ln|\Omega^k(0)| = \operatorname{Re}\Omega^k(0) = 0,$$

we arrive at the following boundary condition for the function $\Omega = \Omega^1 - \Omega^2$:

$$\operatorname{Im}\Omega(z) = 0 \quad \text{for } x = 0, 1; \qquad \operatorname{Re}\Omega(0) = 0. \tag{31}$$

By a homeomorphic mapping $\zeta = \zeta(z)$, $\zeta(0) = 1$, of the strip D_z onto the unit disk $|\zeta| < 1$ by a solution of the homogeneous ($A \equiv 0$) equation (30), the boundary-value problem (30), (31) can be reduced to the homogeneous Schwarz problem

$$\Omega_{\bar\zeta} = A^*(\zeta)\Omega, \qquad \operatorname{Im}\Omega(e^{i\gamma}) = \operatorname{Re}\Omega(1) = 0, \quad \gamma \in [0, 2\pi],$$

where, in view of assumption (b),

$$A\big(\zeta_{\bar z} - \mu\zeta_z\big)^{-1} = A^*(\zeta) \in L_p(\overline{K}), \quad p > 2; \qquad K\colon |\zeta| < 1.$$

The representation

$$\Omega = \Phi(\zeta)\exp\big\{T_1\big(A^* \mid \zeta\big) - T\big(A^* \mid 1\big)\big\}$$

for generalized analytic functions implies that the analytic function $\Phi(\zeta)$ must satisfy the same boundary conditions as $\Omega(\zeta)$. This is possible only if $\Phi(\zeta) \equiv 0$. Consequently, to the distinct mappings $w = w^k(z)$ there can correspond only a single function

$$\omega(z) = \ln\Omega(z) = \ln\big(u_x^k - iu_y^k\big).$$

Substituting it in (7) and mapping D_z and D_w onto the unit disk, we arrive at the problem of mapping by a solution of a linear system of equations, and there is a uniqueness theorem for this problem (Theorem 2 in §4). The theorem is proved.

4°. Systems of equations that are strongly elliptic in the Lavrent'ev sense. Suppose that the mapping $w(z) = u(x, y) + iv(x, y)$ is differentiable in a neighborhood of some point z_0 and that its Jacobian $J(z_0) = |w_z^0|^2 - |w_{\bar z}^0|^2$ is nonzero. Let us consider a unit square in the w-plane with a vertex at the point $w_0 = w(z_0)$ and denote the vertices adjacent to it by w_1 and w_2, with β the angle between the u-axis and the side $\overline{w_0 w_1}$. The preimage of this square under the affine transformation

$$w = w_0 + w_z^0(z - z_0) + w_{\bar z}^0(\bar z - \bar z_0)$$

is a parallelogram with the following geometric characteristics (which completely determine it); α^β is the angle between the side $\overline{z_0 z_1}$ and the x-axis, θ^β is the angle at the vertex z_0, p^β is the length of the side $\overline{z_0 z_1}$, and h^β is the height. By using formulas from differential geometry and elementary trigonometry, the characteristics of the mapping can be expressed in terms of the derivatives u_x, u_y, v_x and v_y by the formulas

$$p^\beta = J^{-1}\big(E\xi^2 - 2F\xi\eta + G\eta^2\big)^{1/2}, \qquad h^\beta = \big(E\xi^{22} - 2\xi F\eta + G\eta^2\big)^{-1/2}, \quad (32)$$

$$\tan\alpha^\beta = -\frac{v_x\eta - u_x\xi}{v_y\eta - u_y\xi}, \qquad \tan\theta^\beta = J\big[(E - G)\xi\eta - F(\eta^2 - \xi^2)\big]^{-1},$$

where we have introduced the notation

$$\xi = \sin \beta, \qquad \eta = \cos \beta, \qquad E = u_x^2 + u_y^2, \qquad G = v_x^2 + v_y^2,$$
$$F = u_x v_x + u_y u_y, \qquad J = u_x v_y - u_y v_x.$$

Conversely, the derivatives u_x, u_y, v_x and v_y can be expressed in terms of the characteristics

$$u_x = l^\beta \left(\frac{\tan \alpha^\beta + \tan \theta^\beta}{p^\beta \tan \theta^\beta} + \frac{\tan \alpha^\beta \tan \beta}{h^\beta} \right),$$

$$u_y = l^\beta \left(\frac{\tan \alpha^\beta \tan \theta^\beta - 1}{p^\beta \tan \theta^\beta} - \frac{\tan \beta}{h^\beta} \right),$$

$$v_x = l^\beta \left(\frac{\tan \alpha^\beta + \tan \theta^\beta}{p^\beta \tan \theta^\beta} \tan \beta - \frac{\tan \alpha^\beta}{h^\beta} \right),$$

$$v_y = l^\beta \left(\frac{\tan \alpha^\beta \tan \theta^\beta - 1}{p^\beta \tan \theta^\beta} \tan \beta + \frac{1}{h^\beta} \right), \qquad (33)$$

where $l^\beta = \cos \alpha^\beta \cos \beta$.

Substituting these expressions in (1), we get the relations

$$G_i(x, y, u, v, p^\beta, h^\beta, \alpha^\beta, \theta^\beta) = 0, \qquad i = 1, 2,$$

commonly called the equations in the characteristics. Suppose in addition that this last system can be solved for h^β and θ^β with any $\beta \in [0, 2\pi]$, i.e.,

$$h^\beta = h^\beta(x, y, u, v, p^\beta, \alpha^\beta), \qquad \theta^\beta = \theta^\beta(x, y, u, v, p^\beta, \alpha^\beta). \qquad (34)$$

In the case $\beta = 0$ the index in the notation for the characteristics will be omitted for convenience. Following Lavrent'ev, we introduce the derived system, which connects the functions $\tau = -\ln p$ and $\alpha = \alpha(u, v)$. We note that in the case when the original system (1) coincides with the Cauchy-Riemann system, i.e., $w_{\bar{z}} = 0$, the derived system is a condition for analyticity of the function

$$\omega(w) = -\ln \frac{dz}{dw}(w) = \tau + i\alpha.$$

In the general case the characteristics $\tau = -\ln p$ and α (for $\beta = 0$) satisfy the system of equations

$$\frac{\partial \tau}{\partial v} = a_1 \frac{\partial \tau}{\partial u} + a_2 \frac{\partial \alpha}{\partial u} + a_3, \qquad \frac{\partial \alpha}{\partial v} = b_1 \frac{\partial \tau}{\partial u} + b_2 \frac{\partial \alpha}{\partial u} + b_3, \qquad (35)$$

whose coefficients are expressed by the formulas

$$a_1 = \frac{\partial h}{\partial p} \cot \theta - \frac{\partial \theta}{\partial p} \frac{h}{\sin^2 \theta}, \qquad a_2 = \frac{1}{p} \left(\frac{\partial \theta}{\partial \alpha} \frac{h}{\sin^2 \theta} - \frac{\partial h}{\partial \alpha} \cot \theta + h \right),$$

$$a_3 = \left(\frac{1}{p} \frac{\partial \theta}{\partial u} + \frac{\partial \theta}{\partial s} \right) \frac{h}{\sin^2 \theta} - \left(\frac{1}{p} \frac{\partial h}{\partial u} + \frac{\partial h}{\partial s} \right) \cot \theta,$$

$$\frac{\partial}{\partial s} = \cos \alpha \frac{\partial}{\partial x} + \sin \alpha \frac{\partial}{\partial y},$$

$$b_1 = -\frac{\partial h}{\partial p}, \qquad b_2 = \frac{1}{p} \left(\frac{\partial h}{\partial \alpha} + h \cot \theta \right), \qquad b_s = \frac{1}{p} \frac{\partial h}{\partial u} + \frac{\partial h}{\partial s}.$$

The derived system can be obtained by substituting the values of u_x, u_y, v_x and v_y from (33) into the identities

$$\frac{\partial}{\partial v} \left(\frac{v_y}{J} \right) = \frac{\partial}{\partial u} \left(\frac{u_y}{J} \right), \qquad \frac{\partial}{\partial v} \left(-\frac{v_x}{J} \right) = \frac{\partial}{\partial u} \left(\frac{u_x}{J} \right),$$

which express the coincidence of the second mixed derivatives of the functions $x(u, v)$ and $y(u, v)$.

DEFINITION 6. The system (1) is said to be *strongly elliptic* (*in Lavrent'ev's sense*) if it has the following properties:

(I) For any $\beta \in [0, 2\pi]$ and any values of the arguments in some domain Q of the space of variables $(x, y, v, u, u_x, u_y, v_x, v_y)$ the system (1) can be written in the form (34) with functions h^β and θ^β that aer single-valued and differentiable in Q.

(II) There exists a constant $\delta_0 > 0$ such that for values of the arguments in Q and for arbitrary $\beta \in [0, 2\pi]$ the inequalities

$$0 < \delta_0 < \theta^\beta < \pi - \delta_0, \qquad \delta_0 < \frac{\partial h\beta}{\partial p\beta} < \delta_0^{-1} \tag{36}$$

hold for the functions that appear in the equations in the characteristics (34).

(III) There is at least one $\beta \in [0, 2\pi]$ for which

$$(p^\beta)^2 \frac{\partial h\beta}{\partial p\beta} \neq p^2 \frac{\partial h}{\partial p}. \tag{37}$$

Condition (III) excludes sytems for which the derived system for (35) is not uniformly elliptic (Lemma 4). The importance of this condition is demonstrated by the following example of a system belonging to the excluded case:

$$u_x v_y - u_y v_x = 1, \qquad v_x - u_y = 0,$$

obtained from the Cauchy-Riemann system for the functions $u = u(x, v)$ and $y = y(x, v)$ by a transformation (one used frequently in hydrodynamics). According to the first equation, the mapping $u + iv = w(z)$ is area-preserving, and, consequently, the solutions of this system a fortiori cannot realize mappings of arbitrary domains onto one another. However, when establishing certain properties of generalized solutions of the system (1), it suffices to assume only the conditions (I) and (II). In particular, this is the case in obtaining the estimate

$$|q(z)| = |w_{\bar{z}}/w_z| \leqslant q_0 < 1, \qquad (38)$$

which ensures that the mapping $w = w(z)$ is topologically similar to a conformal one. It was first proved by Lavrent'ev [65] with a constant q_0 depending on $\max |w_{\bar{z}}|$ and $\min |w_z|$, and by B. V. Šabat [15] with a constant q_0 depending on the maximum modulus of the coefficients a_i and b_i of the derived system (35), which increase like $|w_z|$.

LEMMA 3. *The complex characteristic* $q(z) = w_{\bar{z}}/w_z$ *of each generalized solution* $w = u + iv$ *of a system* (1) *that is strongly elliptic in Lavrent'ev's sense satisfies inequality* (38) *with a constant* $q_0 = q_0(\delta_0)$ *that does not depend on the solution.*

For $\beta = 0$, (32) and (33) give us that

$$E = \frac{1}{p^2 \sin^2 \theta}, \qquad F = -\frac{1}{ph \cot \theta}, \qquad G = \frac{1}{h^2},$$

$$J = \frac{1}{ph}, \qquad u_x = \frac{\sin(\theta + \alpha)}{p \sin \theta}, \qquad u_y = -\frac{\cos(\theta + \alpha)}{p \sin \theta}, \qquad (39)$$

$$v_x = -\frac{\sin \alpha}{h}, \qquad v_y = \frac{\cos \alpha}{h}.$$

Substituting these expressions into the last of equations (32), we get

$$\tan \theta^\beta = 2ph \sin^2 \theta \left[(h^2 - p^2 \sin^2 \theta) \sin 2\beta - ph \sin 2\theta \cos 2\beta \right]^{-1}, \qquad (32^*)$$

which, by the estimate (36) for θ^β, obviously implies that for $\beta = \pi/4$, p and h can be zero or infinite only simultaneously. Thus, the function $h = h(p, \alpha)$ can be represented in the form

$$h = p\frac{\partial h}{\partial p}(\gamma p, \alpha), \qquad 0 \leqslant \gamma \leqslant 1,$$

and inequality (36) for h^β implies the estimates

$$0 < \delta_0 \leqslant \lambda = h/p \leqslant \delta_0^{-1}. \qquad (36^*)$$

With the help of (39) we compute the partial derivative of $w(z)$,

$$2w_{\bar{z}} = u_x - v_y + i(v_x - u_y), \qquad 2w_z = u_x + v_y + i(v_x - v_y),$$

and with them also the ratio $w_{\bar{z}}/w_z$ in terms of the characteristics:

$$2w_{\bar{z}} = -\frac{e^{i\alpha}}{h}(1 - \lambda + i\lambda \cot \theta), \qquad 2w_z = \frac{e^{-i\alpha}}{h}(1 + \lambda + i\lambda \cot \theta), \quad (40)$$

$$q(z) = \frac{w_{\bar{z}}}{w_z} = -e^{2i\alpha}\frac{1 - \lambda + i\lambda \cot \theta}{1 + \lambda + i\lambda \cot \theta}. \qquad (41)$$

The validity of (38) now follows from (41) with the estimates (36*)and (36) for λ and θ taken into account.

REMARK. By (39), the functions

$$E = |\nabla u|^2 = \frac{1}{p^2 \sin^2 \theta} \quad \text{and} \quad G = |\nabla v|^2 = \frac{1}{h^2}$$

can be zero or infinite only simultaneously, since, as we established in the proof of the lemma, p and h have this property. In particular, this implies that the Jacobian

$$J = \frac{D(u, v)}{D(x, y)} = |\nabla u| \cdot |\nabla v| \cdot \sin \theta$$

corresponding to a generalized solution $u + iv = w(z)$ of a strongly elliptic system (1) is zero or infinite only if the function $\omega = \ln(u_x - iu_y)$ is unbounded.

The following important property of strongly elliptic systems of equations was first established by Lavrent'ev [65] (see also [61], Theorem 4.1) and by B. V. Šabat [15], and is proved here in an improved variant (the inequalities (42)).

LEMMA 4. *Suppose that the system* (1) *satisfies condition* (I) *of Definition 6. Then property* (II) *of the equations in the characteristics is necessary and sufficient for the ellipticity of the derived system of* (35), *and its coefficients satisfy the inequalities*

$$\max_{i=1,2} (|a_i|, |b_i|) \leqslant N < \infty, \qquad \Delta = (a_1 - b_2)^2 = 4a_2b_1 \leqslant -N^{-1} \quad (42)$$

with a constant N *not depending on the solution.*

Substituting (39) in (32), we get

$$p^\beta = p^\beta(p, \alpha), \qquad \alpha^\beta = \alpha^\beta(p, \alpha),$$
$$h^\beta = h^\beta = h^\beta(p, \alpha), \qquad \theta^\beta = \theta^\beta(p, \alpha), \qquad (43)$$

where for brevity the dependence of the functions on the arguments x, y, u, v is not indicated. If we solve two of the equations in (43) for p and α and substitute the expressions obtained in the remaining equations, then as a result we get the system (34) in the characteristics for arbitrary $\beta \in [0, 2\pi]$. A condition for this solvability is that the following determinant be nonzero:

$$\frac{D(p^\beta, \alpha^\beta)}{D(p, \alpha)} = \frac{p}{p^\beta} \left[\xi^2 + (a_1 + b_2)\xi\eta + (a_1 b_2 - a_2 b_1)\eta^2 \right] \equiv \frac{p\Lambda_0}{p^\beta}, \quad (44)$$

where a_i and b_i are the coefficients of the derived system of (35), and $\xi = \cos\beta$, $\eta = \sin\beta$. On the other hand, differentiation of the implicit functions in (43) gives

$$\frac{\partial h^\beta}{\partial p^\beta} = \frac{ph}{(p^\beta)^2} \frac{\Lambda_1}{\Lambda_0}, \quad (45)$$

$$\Lambda_1 = -\frac{p}{h} b_1 \xi^2 + \left(b_2 - a_1 - \frac{2b_1}{\tan\theta} \right)\xi\eta = \frac{h}{p}\left(a_2 + \frac{b_2 - a_1}{\tan\theta} - \frac{b_1}{\tan^2\theta} \right)\eta^2.$$

We remark that the discriminants of the quadratic forms $\Lambda_i(\xi, \eta)$ coincide with the discriminant $\Delta = 4a_2 b_1 + (a_1 - b_2)^2$ of the quadratic form corresponding to the derived system of (35).

Suppose that the derived system of (35) is elliptic. Then $\Delta \neq 0$, and, consequentyly, $\Lambda_i \geqslant 0$.

Therefore, if $dh/dp > 0$, then dh^β/dp^β is positive for all $\beta \in [0, 2\pi]$.

Suppose, conversely, that $dh^\beta/dp^\beta > 0$ for all β; then it is clear from (45) that the forms Λ_i have the same sign for all β. Thus, either these forms are definite or their ratio does not depend on β. In the first case this implies that $\Delta \neq 0$, and thereby that the derived system of (35) is elliptic. The second case is the excluded case, and, as is clear from (45), it reduces to systems for which condition (III) does not hold.

To prove (42) we mention first of all that, by (36) and (36*), the ratio

$$\frac{p^\beta}{p} = \left(h^2\xi^2 + ph\sin 2\theta \cdot \xi\eta + p^2\sin^2\theta\eta^2 \right)^{1/2} (p\sin\theta)^{-1}$$

is uniformly bounded below and above. Taking this into account, we find from (45) that

$$N_0^{-1} \leqslant \frac{\Lambda_1(\xi, \eta)}{\Lambda_0(\xi, \eta)} \leqslant N_0(\delta_0) < \infty. \quad (46)$$

Suppose that $p \to 0$ or ∞ and thus, as shown in Lemma 3, that $h \to 0$ or ∞. If the coefficients a_i and b_i $(i = 1,2)$ of the derived system of (35) satisfy the first of the inequalities (42), then, by what was proved above, its discriminant Δ is less than zero, i.e., the last of the inequalities (42) is also satisfied.

Further, we remark that, according to assumption (1), the partial derivatives of the functions defining the equations (34) in the characteristics are uniformly bounded; therefore, we have the boundedness as $p \to 0$ and ∞, respectively, of the expressions pa_i, pb_i and $p^{-1}a_i$, $p^{-1}b_i$, and with them also the coefficients of the corresponding quadratic forms

$$p\Lambda_i(\xi, \eta) \quad \text{and} \quad p^{-1}\Lambda_i(\xi, \eta), \quad i = 0, 1.$$

An elementary analysis shows that if the first of the inequalities (42) does not hold, then the limit as $p \to 0$ or ∞ in the relations (46), which are valid also for the forms $p\Lambda_i$ and $p^{-1}\Lambda_i$, always reduces to the exluded case of systems not satisfying the condition (III). The lemma is proved.

5°. Equivalence of the definitions of L-ellipticity and strong ellipticity in Lavrent'ev's sense.

THEOREM 3. *A nonlinear system* (1) *is uniformly L-elliptic if and only if the conditions* (I)–(III) *forming the definition of strong ellipticity in Lavrent'ev's sense are satisfied.*

Suppose that (I)–(III) hold. Then the last two relations in (39) uniquely determine the characteristics α and h as functions of the variables v_x and v_y. By (36), $p = p(v_x, v_y)$ is determined from the first equation in (34), and then $\theta = \theta(v_x, v_y)$ from the second one. Substituting these values into the first relations in (33) for $\beta = 0$, we get (2*), solved for u_x and u_y. By Lemma 4, the derived system for the functions $-\ln p$ and α is elliptic, and, since the transformation of the variables p and α to the variables h and α is nonsingular (see (36)), the system for the functions $-\ln h$ and α is also elliptic. Therefore, the equation for the function $\sigma = \ln(v_x - iv_y)$ is also elliptic, and its real and imaginary parts are uniquely determined by (39) in terms of $-\ln h$ and α.

Suppose now that the system (1) is L-elliptic. We substitute the values of the derivatives v_x and v_y in terms of the characteristics h and α from (39) into (2*), thereby expressing also u_x and u_y in terms of them. We then get from (32) that

$$h^\beta = h^\beta(h, \alpha), \qquad \theta^\beta = \theta^\beta(h, \alpha), \qquad p^\beta = p^\beta(h, \alpha), \qquad \alpha^\beta = \alpha^\beta(h, \alpha). \tag{43*}$$

Estimating from (2) the ratio

$$\left| \frac{v_x - iv_y}{u_x - iu_y} \right| = \left| \sin(\theta + \alpha)(a_{21} + ia_{11}) - \cos(\theta + \alpha)(a_{22} + ia_{12}) \right| \leqslant N_1,$$

and from (2*) its reciprocal, we find that

$$N_1^{-1} \leqslant \frac{h}{p \sin \theta} = \left| \frac{v_x - iv_y}{u_x - iu_y} \right| \leqslant N_1 < \infty. \tag{47}$$

Then the first inequality $|q(z)| \leqslant q_0(M) < 1$ (condition (5)) for the function $|q(z)| = |q_1| + |q_2|$, computed from (41) leads to (36). But the latter can be satisfied at the same time as (47) only if

$$0 < \delta_1 \leqslant \theta \leqslant \pi - \delta_1. \tag{48}$$

From (32), with (39) substituted into it, the estimates (47), (48), and (36*) give us the first of the inequalities (36) for θ^β (see (32*)) and the boundedness of the following quantities:

$$N_2^{-1} \leqslant \left(\frac{p^\beta}{p}, \frac{h^\beta}{h} \right) \leqslant N_2 < \infty.$$

If we then use (39) to represent $p = p(h, \alpha)$ in the form

$$p = (Jh)^{-1} = \tilde{F}_2 \cos \alpha - \tilde{F}_1 \sin \alpha)^{-1},$$

where the $\tilde{F}_i$ are defined by (2*), we find that

$$\frac{\partial p}{\partial h} = \frac{1}{J^2 h^4} \left[\frac{\partial \tilde{F}_1}{\partial v_x} \sin^2 \alpha - \left(\frac{\partial \tilde{F}_1}{\partial v_y} + \frac{\partial \tilde{F}_2}{\partial v_x} \right) \sin \alpha \cos \alpha + \frac{\partial \tilde{F}_2}{\partial v_y} \cos^2 \alpha \right].$$

By condition (ii) for the system (2*), the discriminant of the quadratic form in the square brackets is negative, and (36*) implies that $\partial p/\partial h > 0$ in a neighborhood of the values $h = 0, \infty$; therefore, $\partial p/\partial h > 0$ for all $h \in [0, \infty]$. Hence, the function $h = h(p, \alpha)$ is uniquely determined from the third equation in (43*) when $\beta = 0$. We remark that the L-ellipticity of system (1) implies the ellipticity of the equation for the function $\sigma = \ln(v_x - iv_y)$, and with it also that of the system for the functions $\varphi = -\ln h$ and α, which are uniquely determined from (39) in terms of v_x and v_y. But since it has been shown that the transformation of the variables $\varphi = -\ln h$ and α to the variables $\tau = -\ln p$ and α is one-to-one, the system for the latter pair of functions is also elliptic.

Lemma 4 then gives us, first, that the first two equations in (43) are solvable for p and α, which enables us to get the system (34) in the characteristics with the help of the remaining equations in (43); and, second, that the second of the inequalities in (36) holds. This concludes the proof of strong ellipticity of system (1) in Lavrent'ev's sense.

REMARK. By (38), the transformation of derived systems from the independent variables u, v to the independent variables x, y obviously does not affect the ellipticity of these systems. However, in such a transition we must also introduce new unknown functions in order to preserve the properties of the free terms of these equations. For example, in the equation for $\sigma = \ln(v_x - iv_y)$

$$\sigma_{\bar{w}} - \tilde{\mu}\sigma_w - \tilde{\mu}_2\overline{w}_{\bar{w}} = \tilde{F}(z, w, \sigma), \qquad |\tilde{\mu}_1| + |\tilde{\mu}_2| \leqslant \tilde{\mu}_0 < 1$$

the function $\tilde{F}$ depends linearly on the coefficients a_3 and b_3 of (35); consequently, $|\tilde{F}| \leqslant N_3 |w_z| + N_4$. It is easy to verify that to preserve this important estimate in passing to the independent variable z it suffices to take $\psi = e^{-\sigma} = (v_x - iv_y)^{-1}$ as a new unknown function.

§9. Problems

We state several unsolved problems in the theory of quasiconformal mappings that arise from applications and are also of independent interest.

$1°$. Prove a uniqueness theorem for a quasiconformal mapping of the disk K ($|z| < 1$) onto a polygon by a solution of the equation

$$w_{\bar{z}} - \mu_1(z)w_z - \mu_2(z)\overline{w}_{\bar{z}} = 0, \tag{1}$$

where the $\mu_i(z)$ are measurable functions, and

$$|\mu_1(z)| + |\mu_2(z)| \leqslant \mu_0 < 1 \tag{2}$$

almost everywhere in K.

Although the problem can be reduced to one studied earlier by a mapping of the polygon onto the disk, the coefficients of the quasilinear equation thus obtained do not satisfy a Lipschitz condition with respect to any of the variables. We indicate one possible way of proving a uniqueness theorem. For the case of a triangle this theorem easily follows from consideration of the boundary-value problem

$$u - k_j v = b_j, \qquad \gamma \in [\gamma_j, \gamma_{j+1}], \qquad j = 1, 2, 3, \tag{3}$$

where the equations (3) are the equations of the sides of a triangle in the plane of $w = u + iv$, and the $t_j = e^{i\gamma_j}$ are the images of its vertices. Indeed, if as usual we form the equation for the difference of two solutions $w = w^1 - w^2$ and perform an additional homeomorphic mapping of the disk $|z| < 1$ onto the disk $|\varsigma| < 1$, then we arrive at the determination of a bounded solution

$$w(\varsigma) = w^1[z(\varsigma)] - w^2[z(\varsigma)]$$

of the homogeneous problem (3) (for $b_j = 0$) in the class of analytic functions. But the last problem obviously cannot have nontrivial solutions, for otherwise the derivative of the function mapping the disk onto the triangle, which satisfies the homogeneous equation (3), would not be uniquely determined (it is also possible to see that $w = w^1 - w^2 = 0$, directly by computing the index of the problem (3)). Thus, if we extend the continuity method to homeomorphic mappings by solutions of (1), then, as for conformal mappings, a uniqueness theorem for any polygon will follow from the uniqueness theorem proved for the triangle (cf. the proof of uniqueness for solutions of the mixed boundary-value problem with free boundary in §5 of Chapter III).

$2°$. Study uniqueness questions for mappings of the disk $|z| < 1$ onto itself by solutions of the quasilinear system of equations

$$w_{\bar{z}} - \mu(z, w, Pw)w_z = 0, \qquad |\mu| \leqslant \mu_0 < 1, \tag{4}$$

where Pw is an operator on the unknown homeomorphism that is defined, for example, in one of the following ways:

a) $Pw = w(z_0)$, where z_0, $|z_0| < 1$, is a fixed point;

b) $Pw = w(\rho_0 e^{i\gamma})$, where ρ_0, $0 \leqslant \rho_0 \leqslant 1$, is a fixed quantity; or

c) $Pw = w[\alpha(z)]$, where $\alpha = \alpha(z)$ is a quasiconformal mapping of the unit disk $|z| < 1$ onto itself.

3°. Prove existence and uniqueness theorems for mappings of the unit disk onto itself by solutions of a degenerate equation (1), i.e., when $|\mu_1(z)| + |\mu_1(z)| = 1$ at isolated points of the circle or on whole curves.

Determine how the order of the degeneracy affects the validity of these theorems, and consider the case of quasilinear equations.

4°. Find conditions on the coefficients $\mu_i(w, z)$ of quasilinear equations of the form (1) that are necessary and sufficient for homeomorphisms satisfying these equations to preserve boundary sets of measure zero.

CHAPTER VI

BOUNDARY-VALUE PROBLEMS

§1. On a class of functions defined on curves

In studying the properties of generalized solutions of differential equations in closed domains it becomes necessary to coordinate the smoothness of the boundary and boundary data with the class of functions in which the generalized solution is sought. The boundary-value problems considered in this chapter are invariant under conformal transformations. Therefore, it will be assumed without loss of generality that $\Gamma = \Sigma_0^m \Gamma_k$, where the Γ_k are disjoint circles contained in the circle Γ_0; D^+ is the multiply connected domain with boundary Γ, and D^- is its complement in the complete plane E. According to Theorem 4 in §1 of Chapter V, a function $f(z)$ in $W_p^1(\overline{D}_k)$ ($p > 2$, D_k is the domain with boundary Γ^k: $|z - z_k| = r_k$) can be extended across Γ_k with preservation of class; therefore, it suffices to study the boundary properties of such functions when the boundary is approached from the domains $D_k \subset D^-$. This circumstance obviously permits us to restrict ourselves to the case when $D_k = K$: $|z| < 1$ and $\Gamma_k = \Gamma$: $|z| = 1$, respectively.

DEFINITION 1. A function $f(t)$, $t = e^{i\gamma}$, defined on an arc Γ of the unit circle is said to *belong to the class* $SW_p^1(\Gamma)$, $p > 2$, if there is a function $F(z)$ of class $W_p^1(\overline{K})$, $p > 2$, that coincides with $f(t)$ on Γ. Such an $F(z)$ will be called an *extension of $f(t)$ inside the disk* $K = \{|z| < 1\}$.

We give some examples of functions $f(t)$ in $SW_p^1(\Gamma)$ and mention some elementary properties of such functions.

1. Suppose that $f(t) \in C^\alpha(\Gamma)$, $\frac{1}{2} < \alpha \le 1$. Then

$$f(t) \in SW_p^1(\Gamma), \qquad 2 < p < 1/(1 - \alpha).$$

Obviously, it can be assumed without loss of generality that $f(t)$ is defined on the whole circle, and that $f(t) \in C^\alpha$ for $|t| = 1$. We define the analytic function

$$\Phi(z) = \frac{1}{2\pi i} \int_{|t|=1} \frac{f(t)(t+z)}{t(t-z)} dt \equiv U(f|z)$$

by the Schwarz formula and use inequality (8) in §2 of Chapter I to estimate its derivative:

$$|d\Phi/dz| \leqslant \frac{M}{\inf_{|t|=1}|z-t|^{1-\alpha}}.$$

This inequality implies that $d\Phi(z)/dz \in L_p(\overline{K})$ for $p(1-\alpha) < 1$, i.e., for $2 < p < 1/(1-\alpha)$. The function

$$F(z) = \operatorname{Re} U(f_1|z) + i \operatorname{Re} U(f_2|z) \tag{1}$$

is the desired extension of $f(t) = f_1(t) + if_2(t)$ inside the disk K.

2. The fact that the set $SW_p^1(\Gamma)$, $p > 2$, is not exhausted by the functions in C^α, $\frac{1}{2} < \alpha \leqslant 1$, is shown by the example of the function $f(t) = (t - t_0)^\alpha \in C^\alpha$, $0 < \alpha \leqslant \frac{1}{2}$. Since the derivative of the analytic function $F(z) = (z - t_0)^\alpha$, which extends $f(t)$ to K, is obviously in $L_p(K)$ for $p(1-\alpha) < 2$, it follows that $f(t) \in SW_p^1$, $2 < p < 2/(1-\alpha)$.

3. In a more general case suppose that $f(t) \in C^\alpha$, $0 < \alpha \leqslant \frac{1}{2}$, $|t| = 1$, is analytic on the open arcs $l_k = \widehat{t_k, t_{k+1}}$, $t_k = e^{i\gamma_k}$. Then $f(t) \in SW_p^1$, $2 < p < 2/(1-\alpha)$, on the circle $l = \Sigma l_k$.

Indeed, we extend $f(t)$ by means of the function $F(z)$ defined by (1) and observe that the derivative $d\Phi/dz$ of the Schwarz integral is analytic in the closed disk with the exclusion of the points t_k, and that, since $f(t)$ is a Hölder function,

$$\Pi|z - t_i|^{1-\alpha}\left|\frac{\Phi(z) - \Phi(t_k)}{z - t_k}\right| \leqslant M < \infty.$$

Thus,

$$\left|\frac{d\Phi}{dz}\right| \leqslant \frac{M_0}{\Pi|z - t_i|^{1-\alpha}}, \qquad |z| < 1,$$

from which, as in the preceding example, it follows that $\Phi(z) \in W_p^1(\overline{K})$, $2 < p < 2/(1-\alpha)$, and in view of this $f(t) \in SW_p^1$ for $t \in l = \Sigma l_k$.

4. From the imbedding theorem $C^{(p-2)/p}(\overline{K}) \supset W_p^1(\overline{K})$, $p > 2$, it follows that every function $f(t)$ in SW_p^1, $p > 2$, satisfies a Hölder condition with exponent $\alpha = (p-2)/p$.

5. If $f(t) \in SW_p^1$, $p > 2$, $|t| = 1$, then its Schwarz integral is a function in $W_p^1(\overline{K})$. Indeed, let $F(z) \in W_p^1(\overline{K})$, $p > 2$, be an extension of $f(t)$ to K. Then, since $F(z) \in C^{(p-2)/p}(\overline{K})$, it can be represented by the formula (see Theorem

6 in §2 of Chapter V)

$$F(z) = T\left(\frac{\partial F}{\partial \bar{z}}\right) + \frac{1}{2\pi i}\int_{|t|=1}\frac{F(t)\,dt}{t-z}.\tag{2}$$

Consequently, the Cauchy type integral on the right-hand side of this formula belongs to $W_p^1(\bar{K})$ ($p > 2$) along with the Schwarz integral.

6. If $f(t)$, $g(t) \in SW_p^1(\Gamma)$, $p > 2$, then

$$[f(t) \pm g(t)] \in SW_p^1(\Gamma), \qquad [f(t)g(t)] \in SW_p^1(\Gamma).$$

If, moreover, $|f(t)| \neq 0$, then $[g(t)/f(t)] \in SW_p^1(\Gamma)$.

These properties are obvious.

7. Let $f(t) \in W_p^1(\Gamma)$, $p > 2$, and suppose that $\tau = \tau(t)$ are the limit values of a homeomorphism $\zeta = \zeta(z) \in W_p^1(\bar{K})$ of the unit disk onto itself. Then $f[t(\tau)] \in W_{p_0}^1(\Gamma_\tau)$, $2 < p_0 < p$, where $\Gamma_\tau = \tau(\Gamma)$ is the image of Γ.

Let us consider a function $F(z) \in W_p^1(\bar{K})$, $p > 2$, that extends $f(t)$ inside the disk. Then, by the lemma in §3 of Chapter V,

$$F[z(\zeta)] \in W_{p_0}^1(\bar{K}_\zeta), \qquad 0 < p_0 < p,$$

and this gives us the required assertion.

8. If $f(t) \in C(l_1 + l_2)$, where l_1 and l_2 are adjoining arcs on the circle and $f(t) \in SW_p^1(l_i)$, $p > 2$, $i = 1, 2$, then $f(t) \in SW_{p_0}^1(l_1 + l_2)$ for any p_0 such that $2 \leqslant p_0 < p$.

Since the properties of $f(t)$ are not changed by subtracting a polynomial from it, we assume that $f(t) = 0$ at the endpoints of the arcs l_i. Further, we remark that among the extensions $F_i(z) \in W_p^1(\bar{K})$ of the functions $f_i(t) \in SW_p^1(l_i)$ there exist extensions $\Phi_i(z)$ whose boundary values are analytic on the complements $Cl_i = (\partial K \setminus l_i)$ of the arcs l_i in the whole circle ∂K. Indeed, let us consider an extension $F_i(z) \in W_p^1(\bar{D})$ to a domain $\bar{D} \supset \bar{K}$ whose boundary $\partial D \in C^2$ includes the arc l_i. And let $\Phi_i(z)$ be a harmonic function in D such that $\Phi_i(z) = F_i(z)$ for $z \in \partial D$. As in the preceding example, it is easy to see that $\Phi_i(z) \in W_p^1(\bar{D})$, and $\Phi_i(z)$ is thus the desired extension for $z \in \bar{K}$. Then an extension $F(z) \in W_{p_0}^1(\bar{K})$, $p_0 > 2$, of the function $f(t)$, $t \in (l_1 + l_2)$, inside the disk K can be constructed according to the formula

$$F(z) = \Omega_1(z) + \Omega_2(z),$$

where $\Omega_i(z)$ ($i = 1, 2$) is a harmonic function in K that equals zero on the corresponding arc l_i and coincides with $\Phi_i(t)$ on the complement $\partial K \setminus l_i$. Since $\Omega_i(t)$ together with $\Phi_i(t)$ belongs to the class $C^{(p-2)/p}$ on the circle and is analytic on the open arcs l_i and $\partial K \setminus l_i$, Property 3 gives us that $\Omega_i(z) \in W_{p_0}^1(K)$.

Let us introduce a norm on $SW_p^1(\Gamma)$, $p > 2$, by setting

$$\| f(t) \|_{SW_p^1(\Gamma)} = \| F(f \mid z) \|_{W_p^1(\bar{K})}, \tag{3}$$

where $F(f \mid z)$ is the Poisson integral with density $f(t)$ (equality (1)).

DEFINITION 2. A sequence $f_n(z)$ is said to *converge in* $SW_p^1(\Gamma)$ if the Poisson integral (1) with density $f_n(t)$ converges in $W_p^1(\bar{K})$, $p > 2$.

We mention some sufficient tests for convergence of functions in $SW_p^1(\Gamma)$.

9. If the sequence $f_n(t)$ converges in $C^\alpha(\Gamma)$, $\alpha > 1/2$, then it converges also in $SW_p^1(\Gamma)$, $2 < p < 1/(1 - \alpha)$.

This fact is an obvious consequence of the representation (1) (see Property 1).

10. If the sequence of functions $f_n(z)$ converges in $W_p^1(\bar{K})$, $p > 2$, then their limit values $f_n(t)$ converge in $SW_p^1(\Gamma)$.

Indeed, we represent $f_n(z)$ in the form (2). Then the Cauchy type integral

$$\Omega_n(z) = \frac{1}{2\pi i} \int_{|t|=1} \frac{f_n(t)\, dt}{t - z} = f_n(z) - T((f_n)_{\bar{z}}) \tag{4}$$

on the right-hand side of this formula converges in $W_p^1(\bar{K})$, since the functions $f_n(z)$ and $T((f_n)_{\bar{z}})$ have this property. But the convergence of the Cauchy type integral $\Omega_n(z)$ implies the convergence of the Schwarz integral with density $f_n(t)$ in $W_p^1(\bar{K})$ and thereby the convergence of the Poisson integral (1).

We shall need an obvious consequence of Property 10.

11. If the sequence $f_n(t)$ converges in $SW_p^1(\Gamma)$, $p > 2$, then the Cauchy type integral $\Omega_n(z)$ with density $f_n(t)$ converges in $W_p^1(D^{\pm})$, $p > 2$, where $D^+ = K$ and $D^- = E \setminus K$.

Suppose now that $f(t) \in SW_p^1(l)$, $p > 2$, $\widehat{t_0 t_1} = l \subset \Gamma$, and let $f_*(t) \in SW_{p_0}^1(\Gamma)$, $2 < p_0 < p$, be the linear extension of $f(t)$ to Γ:

$$f_*(t) = f(t), \quad t \in l;$$

$$f_*(t) = \frac{t_1 - t}{t_1 - t_0} f(t_0) + \frac{t - t_0}{t_1 - t_0} f(t_1), \quad t \in \Gamma \setminus l. \tag{5}$$

With the help of $f_*(t)$ we introduce a norm also on the set of functions $f(t) \in SW_p^1(l)$, $p > 2$,

$$\| f(t) \|_{SW_p^1(l)} = \| f_*(t) \|_{SW_{p_0}^1(\Gamma)}, \quad 2 < p_0 = \frac{p + 2}{2} < p. \tag{6}$$

Definition 2 and Properties 9–11 obviously carry over to sequences of functions $f_n(t) \in SW_p^1(l)$, $l \subset \Gamma$, $p > 2$.

§2. The Hilbert boundary-value problem for quasilinear elliptic systems of equations

Assuming that for all w and almost all $z \in K = \{|z| < 1\}$

$$|\mu_1(w, z)| + |\mu_2(w, z)| \leq \mu_0 < 1, \tag{1}$$

we consider the quasilinear elliptic system of equations

$$w_{\bar{z}} - \mu_1(w, z)w_z - \mu_2(w, z)\overline{w_{\bar{z}}} = A(w, z). \tag{2}$$

The Hilbert boundary-value problem consists in finding a generalized solution $w(z) \in W_p^1(\overline{D})$, $\overline{D} \subset K$, $p > 2$, of (2) whose limit values on $|t| = 1$ satisfy the condition

$$\mathrm{Re}[G(t)w(t)] = g(t), \qquad t = e^{i\gamma}, \gamma \in [0, 2\pi]. \tag{3}$$

In the general case the following condition will be required:

(a) The coefficient $g(t)$ can be represented in the form

$$g = g_*(t)\Pi |t - t_k|^{-\alpha_k^*}, 0 \leq \alpha_k^* < 1, \quad \text{and} \quad G(t) \in C^{\beta_0}[\gamma_k, \gamma_{k+1}], \beta_0 > 0,$$

where $t_k = e^{i\gamma_k}$ are the points of discontinuity of G and g_*, and $|G(t)| \neq 0$ for all $t = e^{i\gamma}$ (in particular, $G, g \in C^{\beta_0}$ for $|t| = 1$).

$1°$. *Properties of solutions of the Hilbert problem for analytic functions.* We recall briefly a method for constructing solutions of the Hilbert problem in the class of analytic functions $w(z)$ (§§4 and 5 of Chapter I).

We compute

$$\lambda_k^+ = \frac{1}{2\pi} \arg \frac{\overline{G(t_k + 0)}}{G(t_k + 0)},$$

after arbitrarily fixing the value of $\arg G(t_k + 0)$ at the point t_k of discontinuity of $G(t)$, and we determine

$$\lambda_{k+1}^- = \frac{1}{2\pi} \arg \frac{\overline{G(t_{k+1} - 0)}}{G(t_{k+1} - 0)}$$

from continuity on the arc $\overparen{t_k t_{k+1}}$.

Let $\alpha_k = \lambda_k^- - \lambda_k^+ - \nu_k \equiv \lambda_k - \nu_k$, where $\nu_k = [\lambda_k]$ (the largest integer not exceeding λ_k) if the unknown solution is bounded at the point t_k, so that $0 \leq \alpha_k < 1$, and $\nu_k = [\lambda_k] + 1$ if the solution is unbounded and integrable, so that $-1 < \alpha_k \leq 0$. We restrict ourselves to the case when $\kappa = \frac{1}{2}\Sigma\nu_k$ is an integer, called the index of the Hilbert problem of the given class (§4 of Chapter I). The class of solutions is fixed by indicating those points t_k of discontinuity of $G(t)$ at which the solution is bounded. The case $\kappa = \kappa_0 + \frac{1}{2}$ (κ_0 an integer) is handled in a completely analogous manner.

With the help of the functions

$$\Pi(z) = \prod_{k_1,\ldots,k_n} (z - t_k)^{\alpha_k} \quad \text{and} \quad \overline{\Pi(1/\bar{z})} = \prod_{k_1,\ldots,k_n} \left(\frac{1}{z} - \bar{t}_k \right)^{\alpha_k},$$

where the t_k, $k = k_1,\ldots,k_n$, are the points of discontinuity of $G(t)$, the homogeneous conjunction problem corresponding to the original homogeneous Hilbert problem can be reduced to a conjunction problem with continuous coefficient. Then the usual formulas for the latter problem (§3 of Chapter I) can be used to determine the canonical solution of the homogeneous conjunction problem, and with it the canonical solution $X(z)$ of the original Hilbert problem of the given class (with a pole or zero of order $|\kappa|$ at the origin):

$$X(z) = e^{\Gamma_0(z)}\Pi(z)z^\kappa, \tag{4}$$

$$\Gamma_0(z) = \frac{1}{4\pi i} \int_{|t|=1} \ln[G_0(t)] \frac{(t + z)\, dt}{(t - z)t}. \tag{5}$$

The coefficient of the auxiliary conjunction problem with continuous boundary condition has the form

$$G_0(t) = -t^{-2\kappa} \frac{\overline{G(t)}}{G(t)} \prod_{k_1,\ldots,k_n} (-tt_k)^{-\alpha_k} \in C^{\beta_0}.$$

In the case $\kappa < 0$ the solution of the nonhomogeneous problem (3) of the given class can be represented by the formula

$$\omega(z) = \frac{X(z)}{2\pi i} \int_{|t|=1} \frac{g(t)(t + z)\, dt}{G(t)X(t)t(t - z)}, \tag{6}$$

and $g(t)$ must satisfy the $2|\kappa| - 1$ real solvability conditions

$$\int_{|t|=1} \frac{g(t)\, dt}{G(t)X(t)t^{k+1}} = 0, \qquad (k = 0, |\kappa| - 1). \tag{7}$$

Taking (7) and the identity

$$\frac{t + z}{t - z} = 1 + 2 \sum_{k=1}^{|\kappa|-1} \left(\frac{z}{t} \right)^k + 2\left(\frac{z}{t} \right)^{|\kappa|} \frac{t}{t - z},$$

into account, we transform $\omega(z)$ to the form

$$\omega(z) = \frac{e^{\Gamma_0(z)}\Pi(z)}{\pi i} \int_{|t|=1} \frac{t^\kappa g(t)\, dt}{G(t)e^{\Gamma_0(t)}\Pi(t)(t - z)}. \tag{8}$$

Observe that the analytic function $\omega(z)$ in the form (8) satisfies the boundary condition (3) only if (7) holds. To single out a unique solution when $\kappa \geq 0$, we require that the unknown analytic function $\omega(z)$ satisfy the conditions

$$\omega(z_k) = 0, \qquad k = 1,\ldots,\kappa, \qquad \operatorname{Im}\omega(1) = 0, \tag{9}$$

$|z_k| < 1$, $z_i \neq z_j$, $i \neq j$, where it is assumed without loss of generality that $z = 1$ is a point of continuity of both $G(t)$ and $g(t)$ and that $\operatorname{Re} G(1) \neq 1$. Then the solution of (3) for $\kappa \geqslant 0$ under the condition $\operatorname{Im}\omega(1) = 0$ has the form

$$\omega(z) = \frac{X(z)(z-1)}{\pi i} \int_{|t|=1} \frac{g(t)\,dt}{G(t)X(t)(t-1)(t-z)} + X(z)P_\kappa(z), \tag{10}$$

where, on the basis of the identity

$$\frac{z-1}{(t-1)(t-z)} \equiv \frac{t+z}{2t(t-z)} - \frac{t+1}{2t(t-1)}$$

the integral is understood as the difference of two integrals in the principal value sense.

Let all the points $t_1,\ldots,t_{k_0}$ of discontinuity of $G(t)$ and $g(t)$ be labeled in increasing order of their arguments, and let $\delta_0 = \inf_k |t_k - t_{k+1}|$, $t_{k_0+1} = t_1$. The points at which the solution $\omega(z)$ is not bounded are denoted by t_k^*, and the order of the singularities at these points by α_k^*, $0 < \alpha_k^* < 1$. If $g(t)$ has a finite discontinuity at a point of continuity of $G(t)$, then the corresponding $\alpha_1^* > 0$ can be taken as small as desired, since the order of the singularity of $\omega(z)$ at this point is logarithmic.

Let β_0, $0 < \beta_0 \leqslant 1$, be the smallest of the Hölder exponents of $G(t)$ and $g_*(t)$ on the intervals of their continuity, and N the largest of the Hölder constants. We formulate as a lemma two properties of solutions of the Hilbert problem proved in §5 of Chapter I during the investigation of the stability of this problem.

LEMMA 1. *Suppose that $G(t)$ and $g(t)$ satisfy assumption* (a). *Then the function*

$$\Omega(z) = \Pi(z - t_k^*)^{\alpha_k^*}\omega(z) = \Pi_*(z)\omega(z), \qquad 0 < \alpha_k^* < 1,$$

where $\omega(z)$ is represented in the form (8) *or* (10) *and the t_k^* are the points where $\omega(z)$ is unbounded, satisfies the inequality*

$$\|\Omega(z)\|_C\beta \leqslant M_\Omega(N, \delta_0), \qquad \beta(\beta_0) > 0. \tag{11}$$

If $t = t^j(\tau) \in C^{\beta_1}$, $0 < \beta_1 \leqslant 1$, is a sequence of homeomorphisms of the circle $|\tau| = 1$ onto itself that converges in the norm of C^{β_1} to the homeomorphism $t = t^0(\tau)$, then the sequence of functions $\Omega^j(z) = \Pi_*^j(z)\omega^j(z)$ converges in the norm of C^{β_2}, $\beta_2 = \beta_2(\beta_0, \beta_1)$, to the function $\Omega^0(z) = \Pi_*^0(z)\omega^0(z)$.

REMARK. If $g(t)$ is bounded and does not have points of discontinuity different from those of $G(t)$, then all the assertions of the lemma are valid for a bounded solution $\omega(z)$ of the form (8) or (10) for the Hilbert problem, i.e., we can set $\Omega(z) \equiv \omega(z)$.

LEMMA 2. *Suppose that the functions $G(t)$ and $g(t)$ in the boundary condition (3) are bounded, have a finite number of points of discontinuity $t_k = e^{i\gamma_k}$, and satisfy on the arcs $l_k = \widehat{t_k t_{k+1}}$ the condition*

$$G(t), g(t) \in SW^1_{p_0}(l_k), \qquad p_0 > 2,$$

and, moreover, that $g(t)$ is continuous on intervals of continuity of $G(t)$ and that (without loss of generality) $|G(t)| = 1$. Then there exists a function $f(t) \in C^1(l_k)$ such that the Poisson integral

$$\Phi(z) = \frac{1}{2\pi i} \int_0^{2\pi} \frac{(1 - r^2)\overline{G(e^{i\gamma})}\left[g(e^{i\gamma}) + if(e^{i\gamma})\right] d\gamma}{1 - 2r\cos(\theta - \gamma) + r^2}, \quad z = re^{i\theta}, \qquad (12)$$

belongs to $W^1_p(\overline{K})$ and satisfies the boundary condition (3).

We let $G(t) = a(t) - ib(t)$ and write the conditions for continuity of $\Phi(t) = (a + ib)(g + if)$ at the points t_k $(k = 1,\ldots,n)$ of discontinuity of $G(t)$ and $g(t)$:

$$(bf)_k^+ - (bf)_k^- = (ag)_k^- - (ag)_k^+, \qquad (af)_k^+ - (af)_k^- = (bg)_k^+ - (bg)_k^-,$$

where we have introduced the notation

$$\varphi(e^{i\gamma_k \pm 0}) = \varphi_k^\pm, \qquad e^{i\gamma_k} = t_k \qquad (k = 1,\ldots,n).$$

The determinant of the system of equations in f_k^+ and f_k^- thus obtained is nonzero. Indeed, otherwise,

$$b_k^+/a_k^+ = b_k^-/a_k^- = \lambda_k,$$

which is equivalent to the absence of a discontinuity of $\arg(a + ib)$ at the corresponding point t_k. Consequently, f_k^+ and f_k^- $(k = 1,\ldots,n)$ are uniquely determined from the last system of equations. Then, for example, we can take the function $f(t)$ we need to be the piecewise linear function

$$f(t) = f_{k+1}^- \frac{t - t_k}{t_{k+1} - t_k} + f_k^+ \frac{t - t_{k+1}}{t_k - t_{k+1}},$$

$$\arg t = \gamma \in [\gamma_k, \gamma_{k+1}] \qquad (k = 1,\ldots,n).$$

By construction,

$$\overline{G(t)}\left[g(t) + if(t)\right] = \Phi(t) \in SW^1_p[\gamma_k, \gamma_{k+1}]$$

and $\Phi(t) \in C^{\alpha}$, $\alpha = (p_0 - 2)/p_0$, on the circle $|t| = 1$; consequently, by Property 8 in §1,

$$\Phi(t) \in SW_p^1, \qquad 2 < p < p_0,$$

for $|t| = 1$, and this means, according to Property 5 in §1, that $\Phi(z) \in W_p^1(\overline{K})$. Since, on the other hand,

$$\mathrm{Re}[G(t)\Phi(t)] = \mathrm{Re}[G\overline{G}(g + if)] = g,$$

it follows that the $\Phi(z)$ defined by (12) satisfies the boundary condition (3). The lemma is proved.

LEMMA 3. *Suppose that*

$$G(t), g_*(t) \in SW_{p_0}^1(l_k), \qquad l_k = \widehat{t_k t_{k+1}}, p_0 > 2.$$

If the $\omega(z)$ in the representations (8) and (10) is bounded, then

$$\|\omega(z)\|_{W_p^1(\overline{K})} \leqslant M_\omega, \qquad 2 < p < p_0, \tag{13}$$

while if $\omega(z)$ is unbounded at the finite number of points t_k^*, $|t_k^*| = 1$, then

$$\left\|\Pi(z - t_k^*)^{\alpha_k^*}\omega(z)\right\|_{W_p^1(\overline{K})} \leqslant M_\Omega'' \qquad (2 < p < p_0, 0 < \alpha_k^* < 1). \tag{14}$$

We note, first of all, that the function

$$G_0(t) = -t^{-2\kappa}\frac{\overline{G(t)}}{G(t)}\Pi(-tt_k)^{-\alpha_k}$$

belongs to the class $SW_p^1(l)$, $2 < p < p_0$, $l = \Sigma l_k$. Indeed, on the arcs $l_k = \widehat{t_k t_{k+1}}$ each of the factors in Π is an analytic function, and $\overline{G}/G \in SW_{p_0}^1(l_k)$. Also, since $G_0(t) \in C^{\beta}$, $\beta > 0$, on $l = \Sigma l_k$ by construction, it follows from Properties 6 and 8 in §1 that $G_0(t) \in SW_p^1(l)$, $2 < p < p_0$. Consequently, the function $\Gamma_0(z)$ in (5) belongs to $W_p^1(\overline{K})$, $2 < p < p_0$.

We represent the solution $\omega(z)$ of the boundary-value problem (3) in the form

$$\omega(z) = z^\kappa \left[X_0(z)T_1\left(X_0^{-1}\Phi_{\overline{\zeta}}\right) - \Phi_0(z) + X_0(z)P_\kappa(z)\right].$$

Here $\Phi_0(z) = \Phi(z) - \mathrm{Im}\,\Phi(1)$, and $\Phi(z)$ is the function constructed in Lemma 2 and satisfying the boundary condition

$$\mathrm{Re}(Gt^{-\kappa}\Phi) = g, \qquad X_0(z) = e^{\Gamma_0(z)}\Pi(z), \qquad P_\kappa = \sum_0^\kappa \left(c_k z^k + \overline{c}_k z^{-k}\right)$$

for $\kappa > 0$, and $P_\kappa = \Sigma_0^{|\kappa|-1} a_k z^k$ for $\kappa < 0$, where in the last case the a_k are chosen from the requirement of boundedness for $\omega(z)$ (the $a_k = 0$ when the solvability conditions are satisfied). It is easy to see that this function $\omega(z)$ is analytic in K and satisfies the boundary condition (3) for $\kappa \geqslant 0$ and arbitrary c_k, while for $\kappa < 0$ it satisfies (3) only if the $2|\kappa| - 1$ solvability conditions hold. Consequently, $\omega(z)$ can also be represented by the respective formulas (8) and (10) for $\kappa < 0$ and $\kappa \geqslant 0$. But what was proved gives us that $\Phi(z)$, $e^{\Gamma_0(z)} \in W_p^1(\overline{K})$, $p > 2$, and from a generalization of the Zygmund-Calderón theorem in §2 of Chapter V to the spaces L_p with a weight (see Theorem 1 in §5) we have

$$X_0 T_1\left(X_0^{-1}\Phi_{\bar{z}}\right) \in W_p^1(\overline{K}), \qquad p > \frac{2}{\max \alpha_k} > 2.$$

On the other hand, by (9) we find that for $\kappa \geqslant 0$

$$-P_\kappa(z_i) = T_1\left(X_0^{-1}\Phi_{\bar{\zeta}} \mid z_i\right) + \Phi_0(z_i)X_0^{-1}(z_i), \quad \|z_i\| < 1, i = 1,\ldots,\kappa, z_0 = 1,$$

i.e., $|P_\kappa(z_i)| < M_p'$, whence $|P_\kappa(z)| \leqslant M_p$ for $|z| \leqslant 1$. An estimate is obtained similarly for $\kappa < 0$.

Thus, (13) is proved in the case when $\omega(z)$ is bounded.

We now establish a simple consequence of (13). Let us replace $g(t)$ in (3) by the function

$$g_k(t) \in SW_p^1 \qquad (l_k, \partial K \backslash l_k), p > 2,$$

$$g_k(t) = \varphi(t)G(t)X_k(t), \qquad t \in l_k = \widehat{t_k t_{k+1}}, g_k(t) = 0, t \in \partial K \backslash l_k,$$

where

$$X_k(t) = X(t)(t - t_k)^{-\alpha_k}(t - t_{k+1})^{-\alpha_{k+1}}, \qquad \varphi(t) \in SW_{p_0}^1(l_k)$$

(k is fixed), and represent the $\omega_k(z) \in C^\beta(\overline{K})$, $\beta > 0$, corresponding to the resulting boundary-value problem in the form (8) or (10), depending on the sign of the index κ. Since $X_k(z) \in W_p^1$, $p > 2$, and since what has been proved implies that $\omega_k(z)$ satisfies (13), we then get that for $2 < p < p_0$

$$\left\| \frac{d}{dz} \int_{l_k} \frac{(z - t_k)^{\alpha_k}(z - t_{k+1})^{\alpha_{k+1}}\varphi(t)\, dt}{(t - t_k)^{\alpha_k}(t - t_{k+1})^{\alpha_{k+1}}(t - z)} \right\|_{L_p} \leqslant M_k, \qquad \varphi(t) \in SW_{p_0}^1(l_k). \quad (15)$$

If $\omega(z)$ is not bounded, multiplying it by $\Pi(z - t_k^*)^{\alpha_k^*}$ leads to the integrals estimated in (15), and, consequently, (14) holds for $\omega(z)$. The lemma is proved.

2°. Solvability of the Hilbert problem. Let us pass to an investigation of problem (2), (3). As in the case of analytic functions, the index of the problem is determined by specifying those points t_k of discontinuity of the coefficient $G(t)$ at which the unknown generalized solution is sought to be unbounded. Because of the natural requirement that $w(z)$ be integrable in a neighborhood of such points, as well as at points where $g(t)$ is unbounded, we require that the corresponding α_k^* satisfy the inequalities

$$0 < \alpha_k^* < \frac{p-2}{p} < 1, \qquad (16)$$

where the quantity $p(\mu_0) > 2$ will be determined later.

Let $W_p^1(K) \cap C_*^\beta(\overline{K})$, $p > 2$, $\beta > 0$, be the set of functions $w(z)$ belonging to $W_p^1(\overline{D})$ in any closed subdomain $\overline{D} \subset K$ and such that

$$\left\{ \Pi(z - t_k^*)^{\alpha_k^*} w(z) \right\} \in C^\beta(\overline{K}),$$

where t_k^* is a point where $w(z)$ is unbounded. If $w(z)$ is bounded, then the set $W_p^1(K) \cap C^\beta(\overline{K})$ is defined similarly.

The set $W_p^*(\overline{K})$ consists of the functions $w(z)$ such that

$$\left\{ \Pi(z - t_k^*)^{\alpha_k^*} w(z) \right\} \in W_p^1(\overline{K}).$$

THEOREM 1. *Suppose that the functions $G(t)$ and $g(t)$ in the boundary condition (3) satisfy assumption* (a), *the coefficients $A(w, z) = A_1(w, z)w + A_2(w, z)\overline{w} + A_0(w, z)$ and $\mu_i(w, z)$ $(i = 1, 2)$ in (2) are continuous in w for almost all z with $|z| < 1$, for some $p_0 > 2$*

$$\| A_i(w, z) \|_{L_{p_0}(\overline{K})} \leqslant M_A, \qquad |\mu_1(w, z)| + |\mu_2(w, z)| \leqslant \mu_0 < 1, \quad (17)$$

and, if $w(z)$ is unbounded, the inequalities (16) also hold.

Then for $\kappa \geqslant 0$ there exists at least one generalized solution $w(z) \in W_p^1(K) \cap C_*^\beta(K)$, $p > 2$, $\beta > 0$, of problem (2), (3) such that the equalities (9) hold at the fixed points z_i of the disk K, $|z_i| < 1$, $z_i \neq z_j$, $i \neq j$.

For $\kappa < 0$ there exists at least one generalized solution $w(z) \in W_p^1(K) \cap C_*^\beta(K)$ only if the solvability conditions hold.

If, in addition,

$$G(t), \qquad g(t) \in SW_{p_0}^1(l_k), \qquad l_k = \widehat{t_k t_{k+1}}, \qquad p_0 > 2,$$

then, if there is a generalized solution $w(z)$ of the problem (2), (3), it belongs to the class $W_p^*(K)$ $(p > 2)$; moreover, if $|w(z)| \leqslant M_w$, $|z| \leqslant 1$, then $w(z) \in W_p^1(\overline{K})$, $p > 2$.

The proof is carried out in several steps.

Construction of a transformation. We look for a solution of problem (2), (3) in the form

$$w(z) = \omega[\zeta(z)] e^{T_1(\varphi|z)} + T_1(\psi|z), \qquad (18)$$

where the potential operator $T_1(f|z)$ has the properties (see §2 of Chapter V)

$$\operatorname{Im} T_1(f|e^{i\gamma}) = T_1(f|1) = 0,$$

$\varphi, \psi \in L_p(\overline{K})$, $p(\mu_0) > 2$, are unknown functions, and $\zeta(z)$ is a homeomorphism of the disk $|z| < 1$ onto itself with normalization $\zeta(0) = 0$, $\zeta(1) = 1$. The analytic function ω in the disk $|\zeta| < 1$ satisfies the boundary condition

$$\text{Re}\{G[t(\tau)]\omega(\tau)\} = g_0(\tau), \qquad \tau = e^{i\gamma}, \gamma \in [0, 2\pi], \qquad (19)$$

where $t = t(\tau)$ are the boundary values of the inverse homeomorphism $z = z(\zeta)$, and $g_0(\tau) = ge^{-T_1\varphi} - T_1\psi e^{-T_1\varphi} \text{Re } G$ for $t = t(\tau)$.

If the homeomorphism $\zeta = \zeta(z)$ and the functions $\varphi(z)$ and $\psi(z)$ satisfy

$$\varphi - \mu_1(w, z)S_1\varphi - \mu_2^\varphi(w, z)\overline{S_1\varphi} = A_3(w, z),$$
$$\psi - \mu_1(w, z)S_1\psi - \mu_2(w, z)\overline{S_1\psi} = A_1T_1\psi + A_2\overline{T_1\psi} + A_0, \qquad (20)$$
$$\zeta_{\bar{z}} - \mu_1(w, z)\zeta_z - \tilde{\mu}_2(w, z)\overline{\zeta_{\bar{z}}} = 0,$$

then it is not hard to see that the function $w = w(z)$ in the form (18) satisfies the original equation (2). Here

$$\mu_2^\varphi = \mu_2 e^{\overline{T_1} - T_1}\frac{\overline{\omega}}{\omega}, \qquad \tilde{\mu}_2 = \mu_2 e^{\overline{T_1} - T}\frac{\overline{\omega_\zeta}}{\omega_\zeta}, \qquad A_3 = A_1 + A_2 e^{\overline{T_1} - T_1}\frac{\overline{\omega}}{\omega}.$$

In analogy to §4 we introduce the set $\mathfrak{M}$ of vectors $\vec{m}^* = \{m_1^*(z), m_2^*(z)\}$ such that

$$\vec{m}^*(z) \in L_\lambda(\overline{K}), \qquad \lambda = \frac{pp'}{p' - 1}, |m_1^*(z)| + |m_2^*(z)| \leqslant \mu_0 < 1, \qquad (21)$$

where $p(\mu_0) > 2$ and $p > 1$ are chosen so that $\mu_0\Lambda_{pp'} < 1$, $\Lambda_q = \|S_1\|_{L_q}$. Further, we construct for each vector $\vec{m}^*(z) \in \mathfrak{M}$ a homeomorphism $\zeta = \zeta^*(z)$, $\zeta(0) = 0$, $\zeta(1) = 1$ of the disk $|z| < 1$ onto itself by a solution of the linear equation

$$\zeta_{\bar{z}} - m_1^*(z)\zeta_z - m_2^*(z)\overline{\zeta_{\bar{z}}} = 0. \qquad (22)$$

Let us take arbitrary functions $\varphi^*(z)$ and $\psi^*(z)$ from some bounded set $\mathfrak{N} \subset (L_p \times L_p)$, $p(\mu_0) > 2$, and substitute them together with the boundary values $t = t^*(\tau)$ of the inverse homeomorphism $z = z^*(\zeta)$ into the boundary condition (19). To the boundary-value problem thus obtained there corresponds a fixed analytic function $\omega = \omega^*(\zeta)$ represented in the form (8) or (10), depending on the sign of the index κ; and to this function there corresponds, according to (18), the known function

$$w^*(z) = \omega^*[\zeta^*(z)]e^{T_1(\varphi^*|z)} + T_1(\psi^* | z).$$

Taking (20) into account, we then construct the following transformation:

$$\varphi - \mu_1(w^*, z)S_1\varphi - \mu_2^\varphi(w^*, z)\overline{S_1\varphi} = A_3^*, \qquad (23_1)$$

$$\psi - \mu_1(w^*, z)S_1\psi - \mu_2(w^*, z)\overline{S_1\psi} = A_1^*T_1\psi + A_2^*\overline{T_1\psi} + A_0^*, \qquad (23_2)$$

$$\vec{m} = \{\mu_1(w^*, z), \tilde{\mu}_2(w^*, z)\}, \qquad (23_3)$$

where $A_i^* = A_i(w^*, z)$, $i = 0, \ldots, 3$. The unknown generalized solution $w(z)$ of the problem (2), (3) represented by (18) obviously corresponds to the fixed points of this transformation on the set $\mathfrak{N} \times \mathfrak{M} \subset B = (L_p \times L_p \times L_\lambda)$.

A priori estimates. The first two components of the transformation

$$(\varphi, \psi, \vec{m}) = V(\varphi^*, \psi^*, \vec{m}^*)$$

for particular φ^*, ψ^*, and $\vec{m}^*$ are solutions of the linear equations (23_1) and (23_2). From (23_1) we find the following estimate for $\varphi(z)$:

$$\|\varphi(z)\|_{L_p(\overline{K})} \leqslant \frac{M_A}{1 - \mu_0\Lambda_p} \equiv M_1, \qquad p(\mu_0) > 2. \qquad (24)$$

With the aim of estimating $\psi(z)$ let us form an equation for $\chi(z) = T_1\psi$,

$$\chi_{\bar{z}} - m_1^*(z)\chi_z - m_2^*\overline{\chi}_{\bar{z}} = A_1^*\chi + A_2^*\overline{\chi} + A_0^*$$

and observe that, by a property of T_1, the function $\chi(z)$ solves the homogeneous Schwarz problem with

$$\operatorname{Im}\chi(e^{i\gamma}) = \chi(1) = 0, \qquad \gamma \in [0, 2\pi].$$

Mapping the disk $|z| < 1$ homeomorphically onto itself by a solution $\zeta = \zeta(z)$ of the Beltrami equation with $\mu(z) = m_1^* + m_2^*\overline{\chi}_{\bar{z}}(\chi_z)^{-1}$, we get

$$\chi_{\bar{\zeta}} = \tilde{A}_1\chi + \tilde{A}_2\overline{\chi} + \tilde{A}_0,$$

where, according to the corollary to the lemma in §3 of Chapter V,

$$\tilde{A}_i(\zeta) = A_i^*(\overline{\zeta}_{\bar{z}} - \mu\overline{\zeta}_z)^{-1} \in L_p(\overline{K}), \qquad p(p_0, \mu_0) > 2.$$

Then from the representation (valid because the homogeneous Schwarz problem is satisfied)

$$\chi(\zeta) = e^{T_1\rho}T_1(e^{-T_1\rho}\tilde{A}_0), \qquad \rho = \tilde{A}_1 + \tilde{A}_2\overline{\chi}\chi^{-1},$$

of the solutions of the last equation we arrive at an estimate for the solutions $\psi(z) = \partial\chi[\zeta(z)]/\partial\bar{z}$ of (23_2):

$$\|\psi(z)\|_{L_p(\overline{K})} \leqslant M_2, \qquad p(\mu_0) > 2, \qquad (25)$$

with a constant M_2 not depending on the solution. The inequalities (24) and (25) determine a closed convex set $\mathfrak{N} \subset (L_p \times L_p)$.

It is now easy to estimate the unknown solution $w(z)$ of problem (2), (3). By (24), (25) and the properties of the homeomorphism $\zeta = \zeta(z)$ of the unit disk onto itself, the functions

$$G[t(\tau)] \quad \text{and} \quad g_0^*(\tau) = g_0(\tau)\Pi\,|\,t(\tau) - t_k^*\,|^{\alpha_k^*}$$

in the boundary condition (19) are uniformly bounded in the norm of C^β, $\beta(\mu_0) > 0$, on the intervals of continuity of $G[t(\tau)]$, and

$$\inf_k |\,\zeta(t_k) - \zeta(t_{k+1})\,| > C_0 \inf_k |\,t_k - t_{k+1}\,|^{1/\alpha} = C_\theta \delta_0^{1/\alpha} > 0, \qquad \alpha = \frac{p-2}{p}.$$

Consequently, inequality (11) of Lemma 1 holds for the function $\omega = \omega(\zeta)$, $|\zeta| \leqslant 1$. If $\kappa \geqslant 0$, then we introduce a polynomial $P_\kappa(\zeta)$ in the representation (10) for $\omega(\zeta) = \omega_0(\zeta) + \zeta^\kappa P_\kappa X_0(\zeta)$, whose coefficients must be chosen from the conditions (9), which take the following form:

$$-\zeta_i^\kappa P_\kappa(\zeta_i)X_0(\zeta_i) = \omega_0(\zeta_i) + T_1(\psi\,|\,z_i)e^{-T_1(\varphi|z_i)},$$

where $\zeta_i = \zeta(z_i)$. Since $\zeta(z) \in C^{(p-2)/p}(\overline{K})$, the properties of the z_i are preserved for $\zeta_i = \zeta(z_i)$, i.e., $|\zeta_i| < 1$ and $\zeta_i \neq \zeta_j$ for $i \neq j$. Then $|P_\kappa(\zeta_i)| < M_p$, which implies that also $|P_\kappa(\zeta)| \leqslant M_p$, $|\zeta| \leqslant 1$, where M_p does not depend on the solution. But, by (24), (25), and (11),

$$T_1\varphi, T_1\psi \in C^{(p-2)/p}(\overline{K}), \qquad \Omega(z) \in C^\beta(\overline{K}), \qquad \beta(\beta_0, \mu_0) > 0,$$

where

$$\Omega(z) = \Pi[\zeta(z) - \zeta_k^*]^{\alpha_k^0}\omega[\zeta(z)].$$

On the other hand, in any subdomain $\overline{D} \subset K$ the function $\omega[\zeta(z)]$ belongs to $W_p^1(\overline{D})$; therefore, the following estimates hold for an unknown generalized solution $w(z)$ of the boundary-value problem (2), (3):

$$\|w(z)\|_{W_p^1(\overline{D})} \leqslant M_3(\overline{D}), \qquad \|w(z)\Pi_*(z)\|_{C^\beta(\overline{K})} \leqslant M_4, \tag{26}$$

where

$$\overline{D} \subset K, \qquad \Pi_*(z) = \Pi(z - t_k^*)^{\alpha_k^*}, \qquad \beta(\beta_0, \mu_0) > 0,$$

and M_4 is an absolute constant.

If in addition $G(t), g_*(t) \in CW_{p_0}^1(l_k), l_k = \widehat{t_k t_{k+1}}$, then, by the invariance of the properties of the classes SW_p^1 under homeomorphisms (Property 7 in §1), we find from estimate (14) of Lemma 3 that

$$\|w(z)\Pi(z - t_k^*)^{\alpha_k^*}\|_{W_p^1(\overline{K})} \leqslant M_5, \qquad p(\mu_0) > 2, \tag{27}$$

and if $w(z)$ is bounded, this holds for $w(z)$ itself (a consequence of estimate (13) in Lemma 3).

Thus, we have established that if there is a generalized solution $w(z)$ of problem (2), (3), then, according to (26) and (27), it has the properties indicated in the theorem.

Complete continuity of the transformation (23). The transformation (23) is continuous on the set $\mathfrak{N} \times \mathfrak{M}$ determined by (21), (24), and (25). Indeed, by Theorem 3 in §2 of Chapter V, to sequences $(\overset{\circ}{\varphi}{}^i, \overset{\circ}{\psi}{}^i)$ and $(\overset{\circ}{m}{}^i_1, \overset{\circ}{m}{}^i_2)$ that converge in $L_p(\overline{K})$ and $L_\lambda(\overline{K})$ there corresponds a sequence of homeomorphisms $\zeta = \zeta_i(z)$ satisfying (22) which converges in $W^1_p(\overline{K})$ ($p(\mu_0) > 2$), and thereby, by Lemma 1, a sequence of analytic functions $\omega = \overset{\circ}{\omega}{}^i(\zeta)$ converging in some $C^\beta_*(\overline{K})$, $\beta > 0$. The last statement follows from the convergence of the coefficients in the boundary condition (19) in C^β_* (i.e., everywhere except in neighborhoods of the points $\zeta^i(t_k)$ where G and g are discontinuous). Together with $\overset{\circ}{\omega}{}^i[\zeta^i(z)] = \overset{\circ}{\omega}{}^i(z)$, the functions $\overset{\circ}{\omega}{}^i(z)$ represented by (18) also converge in $C^\beta_*(\overline{K})$.

Moreover, $\overset{\circ}{\omega}{}^i(z)$ converges also in $W^1_p(\overline{D})$, $\overline{D} \subset K$; therefore, this implies the convergence in $L_\lambda(\overline{K})$ of the components m^i_1 and m^i_2 of the transformation (23) and (by what was proved in the corollary to Theorem 1 in §3 of Chapter V) also the convergence in $L_p(\overline{K})$ of the components φ^i and ψ^i of this transformation.

Compactness of the transformation (23). We observe first of all that to bounded sequences $(\overset{\circ}{\varphi}{}^i, \overset{\circ}{\psi}{}^i)$ and $(\overset{\circ}{m}{}^i_1, \overset{\circ}{m}{}^i_2)$ in L_p and L_λ there correspond compact sequences $T\overset{\circ}{\varphi}{}^i$, $T\overset{\circ}{\psi}{}^i$, and $\zeta = \zeta^i(z)$ in $C^{(p-2)/p}(\overline{K})$. We take convergent subsequences of them, and, in order not to complicate the notation, we assume that the sequences themselves are convergent. But then, by what was proved above, the sequences $\overset{\circ}{\omega}{}^i(z)$ and $\overset{\circ}{w}{}^i(z)$ converge in $C^\beta_*(\overline{K})$, and with them, according to the converge as solutions of (23_1) and (23_2). The convergence of $\overset{\circ}{\omega}{}^i(z)$ in $W^1_p(\overline{D})$, $\overline{D} \subset K$, implies also that of the functions $m^i_1(z)$ and $m^i_2(z)$ in $L_\lambda(\overline{K})$. Thus, the transformation (23) is completely continuous.

By what has been proved, this transformation maps the closed bounded set $(\mathfrak{N} \times \mathfrak{M}) \subset (L_p \times L_\lambda)$ into itself. Therefore, by Schauder's theorem, it has at least one fixed point in this set. This concludes the proof of the theorem.

3°. Uniqueness of solutions of the Hilbert problem.

THEOREM 2 (uniqueness). *Under the conditions of Theorem* 1 *suppose that there exists a generalized solution* $w(z)$ *of problem* (2), (3) *that satisfies condition* (9) *at the particular points* z_i *if* $\kappa \geq 0$. *Then this solution is unique if the coefficients of* (2) *satisfy the conditions*

$$|A(w_1, z) - A(w_2, z)| \leq N(z)|w_1 - w_2|, \qquad N(z) \in L_{p_0}(\overline{K}), p_0 > 2,$$

$$|\mu_i(w_1, z) - \mu_i(w_2, z)| \leq N_i(z)|w_1 - w_2|, \qquad N_i(z) < \infty, \qquad (28)$$

and $w(z) \in W^1_p(\overline{K}), p > 2$.

If $w(z) \in W_p^*(\overline{K})$, $p > 2$, then for the solution to be unique in this class it is necessary to assume in addition that the homogeneous part of (2) is linear in a neighborhood of the points t_k^* where $w(z)$ is unbounded, i.e., $N_i(z) \equiv 0$ for $|z - t_k^*| \leq \varepsilon$, $i = 1, 2$.

In the general case when $w(z) \in W_p^1(K) \cap C_*^\beta(K)$, $p > 2$, $\beta > 0$, the solution in this class is unique if $N_i(z) \equiv 0$ for $1 - \varepsilon \leq |z| \leq 1$, with $\varepsilon > 0$ arbitrarily small.

Suppose not: Suppose that there are at least two solutions $w^1(z)$ and $w^2(z)$ under the conditions of the theorem. Then their difference $\omega(z) = w^1(z) - w^2(z)$ satisfies the homogeneous boundary condition (3) and the equation

$$\omega_{\bar{z}} - \mu(z)\omega_z = B(z)\omega,$$

where

$$\mu(z) = \mu_1^1 + \mu_2^1 \overline{\omega}_{\bar{z}}(\omega_z)^{-1}, \qquad \mu_i^j = \mu_i(w^j, z) \quad (i, j = 1, 2),$$

$$B(z) = \left(w_z^2 \frac{\Delta\mu_1}{\omega} + \overline{\omega}_{\bar{z}}^2 \frac{\Delta\mu_2}{\omega} + \frac{\Delta A}{\omega} \right) \in L_p(\overline{K}),$$

$$\Delta f = f(w^1, z) - f(w^2, z), \qquad p > 2.$$

According to the corollary of Theorem 6 in §3 of Chapter V, the solution of the last equation can be represented in the form $\omega(z) = \Phi[\zeta(z)]e^{T_1\varphi}$, where $\varphi(z) \in L_p(\overline{K})$ is a solution of the equation

$$\varphi - \mu(z)S_1\varphi = B(z),$$

while $\Phi(\zeta)$ is an analytic function on the plane of the homeomorphism $\zeta = \zeta(z)$ of the unit disk onto itself, $\zeta(0) = 0$, $\zeta(1) = 1$. By a property of the operator T_1 we have $\operatorname{Im} T_1(\varphi \,|\, e^{i\gamma}) = 0$; therefore, on the plane of the homeomorphism the analytic function $\Phi(\zeta)$ satisfies the boundary condition

$$\operatorname{Re}(G[t(\zeta)]\Phi(\zeta)) = 0, \qquad \zeta = e^{i\gamma},$$

and, if $\kappa \geq 0$, the additional conditions (9) at the images ζ_k of the points z_k ($k = 1, \ldots, \kappa$) and at the point $\zeta = 1$.

But if $\kappa < 0$, then the homogeneous Hilbert problem for analytic functions has a unique solution that is integrable in $\overline{K}$, namely, $\Phi(\zeta) \equiv 0$, and this gives us the assertion of the theorem, i.e., $w^1 = w^2$.

If $\kappa \geq 0$, then the boundary condition at the point $z = 1$ and the condition (9) imply that

$$w^1(1) = w^2(1) = g(1)/\operatorname{Re} G(1)$$

(by assumption, $z = 1$ is a point of continuity of both $g(t)$ and $G(t)$, and $\operatorname{Re} G(1) \neq 0$).

Thus, the function $\Phi(\zeta) = P_\kappa(\zeta)\zeta^\kappa e^{\Gamma_0(\zeta)}$ must have $\kappa + 1$ zeros in the disk $\tilde{K}$, which is possible only if $\Phi(\zeta) \equiv 0$. The theorem is proved.

If the conditions (9) are not required when $\kappa \geqslant 0$, then under the assumptions of the uniqueness theorem the generalized solution $w(z)$ will depend on the $2\kappa + 1$ arbitrary constants that appear in the function $\omega(\zeta)$ in (18). This circumstance allows us to give Theorems 1 and 2 the following form.

THEOREM 3. *Suppose that the coefficients of* (2) *and of the boundary condition* (3) *satisfy the assumptions of Theorems 1 and 2.*

Then the dimensions of the solution sets of the Hilbert problem for $\kappa \geqslant 0$ (respectively, the numbers of solvability conditions for $\kappa < 0$) in the class of analytic functions and in the class of generalized solutions of (2) *coincide.*

REMARK. The arbitrary constants appearing for $\kappa \geqslant 0$ in the representation (18) of the generalized solution $w(z)$ of problem (2), (3) enable us to solve the Hilbert problem for $\kappa \geqslant 0$ with various additional conditions on the parameters. Suppose here that the corresponding parameter problem can be solved in the class of analytic functions, that there is an a priori bound on these parameters, and that they appear continuously in (18).

Then, adding an extra algebraic equation to the transformation (23) and not changing the arguments carried out in Theorem 1, we deduce the solvability of the parameter problem also in the case of the quasilinear elliptic equation (2). An example of such a problem is presented in the next subsection.

4°. *The mixed boundary-value problem with parameters.* Let the circle $|z| = 1$ be divided by the points $a_k = e^{i\theta_k}$ and $b_k = e^{i\tau_k}$ into $2n$ parts in such a way that

$$\theta_1 < \tau_1 < \theta_2 < \cdots < \theta_k < \tau_k < \cdots < \theta_{n+1} = \theta_1 + 2\pi.$$

We denote by L_1 and L_1^p the collections of the arcs (a_k, b_k) and of p arbitrary arcs in L_1, respectively, and, similarly, by L_2 and L_2^q the collections of the arcs (b_k, a_{k+1}) and of q arcs in L_2.

It is required to find a bounded generalized solution $w(z) \in W_p^1(K)$ $(p > 2)$ of (2) and $n - 1$ real constants u_k and v_k connected by the boundary condition

$$\operatorname{Re}[G(t)w(t)] = g(t) + \delta(t), \qquad t = e^{i\gamma}, \gamma \in [0, 2\pi], \tag{29}$$

where

$$G(t) = \begin{cases} 1, & t \in L_1 \\ -i, & t \in L_2, \end{cases} \qquad \delta(t) = \begin{cases} 0, & t \notin (L_1^p + L_2^q), \\ u_k, & t \in (a_k, b_k) \subset L_1^p, \\ v_k, & t \in (b_k, a_{k+1}) \subset L_2^q, \end{cases} \tag{30}$$

and, moreover, $g(t) \in C^\alpha$, $\alpha > 0$, $t \in L_1$, L_2 and, generally speaking, $g(a_k + 0) \neq g(a_k - 0)$. It should be mentioned that the index of $G(t)$ in our selected class of bounded solutions $w(t)$ is negative.

As before, the solution of the boundary-value problem (2), (29) will be represented in the form (18). Then we get the boundary-value problem P for the analytic function $\omega(\zeta)$ in (18), and, by the results in §4.6° of Chapter I, it is uniquely solvable for arbitrarily fixed $\varphi_*, \psi_* \in L_p(\overline{K})$ and $\zeta_*(z) \in W_p^1(\overline{K})$, $p > 2$.

Since the assertions of Lemmas 1 and 2 of this section hold for the solution $\omega = \omega(\zeta)$ thus obtained, problem (2), (29) has at least one solution that is uniformly bounded in $C^\alpha(\overline{K})$. But if, in addition, the conditions (28) hold and $g(t) \in SW_p^1$, $t \in [a_k, b_k]$, $[b_k, a_{k+1}]$, $k = 1, \ldots, n$, then such a solution is uniformly bounded in $W_p^1(\overline{K})$, $p > 2$, and is unique.

5°. Stability of solutions of the Hilbert problem under small changes in the coefficients of the equation and the boundary condition. We shall analyze in detail the case of bounded solutions, which is important in applications. Let $G^k(t)$ and $g^k(t)$ $(k = 1, 2)$ be bounded piecewise Hölder functions whose points of discontinuity t_j^1 and t_j^2 $(j = 1, \ldots, n)$ coincide, $t_j^1 = t_j^2 \equiv t_j = e^{i\gamma_j}$, and suppose that

$$\|G^1(e^{i\gamma}) - G^2(e^{i\gamma})\|_{C^\beta} < \varepsilon, \qquad \|g^1(e^{i\gamma}) - g^2(e^{i\gamma})\|_{C^\beta} < \varepsilon \qquad (31)$$

for $\gamma \in [\gamma_j, \gamma_{j+1}]$ $(j = 1, \ldots, n)$, $0 < \beta < 1$, and that the $g^k(t)$ do not have points of discontinuity that are not points of discontinuity of $G^k(t)$, $k = 1, 2$. Assume that there exist functions $w^k(z) \in W_p^1(K)$, $p > 2$, satisfying the boundary conditions (3) with $G(t) = G^k(t)$ and $g(t) = g^k(t)$ $(k = 1, 2)$ and the equations (2) with $\mu_i = \mu_i^k(w^k, z)$, $i = 1, 2$, $A = A^k(w^k, z)$, such that (17) holds on the solutions $w^k(z)$. Moreover, the coefficients $\mu_i^k(w^k, z)$ and $A^k(w^k, z)$ satisfy the following inequalities in addition to the Lipschitz conditions (28):

$$|\mu_i^1(w^k, z) - \mu_i^2(w^k, z)| < \varepsilon, \qquad i, k = 1, 2,$$

$$\|A^1(w^k, z) - A^2(w^k, z)\|_{L_p(\overline{K})} < \varepsilon, \qquad p > 2. \qquad (32)$$

As usual, we form an equation for the difference $w^1(z) - w^2(z) \equiv w(z)$:

$$w_{\bar{z}} - \tilde{\mu}(z)w_z = \tilde{A}(z)w + \tilde{F}(z), \qquad (33)$$

where

$$\tilde{\mu}(z) = \mu_1^1[w^1(z), z] + \mu_2[w^1(z), z]\frac{\overline{w_{\bar{z}}}}{w_z},$$

$$\tilde{A}(z) = \frac{1}{w}(\Delta A^1 - w_z^2\Delta\mu_1^2 - \overline{w}_{\bar{z}}^2\Delta\mu_2^2), \qquad \Delta\mu_i^2 = \mu_i^2(w^1) - \mu_i^2(w^2),$$

$$\tilde{F}(z) = A^1(w^2) - A^2(w^2) + w_z^2[\mu_1^2(w^1) - \mu_1^1(w^1)] + \overline{w}_{\bar{z}}^2[\mu_2^2(w^1) - \mu_2^1(w^1)].$$

Under the assumption that the conditions of Theorem 2 hold along with (32), we have

$$\| \tilde{A}(z) \|_{L_p} = \tilde{M}_A < \infty, \qquad \| \tilde{F}(z) \|_L < M_F \varepsilon, \qquad p > 2. \tag{34}$$

On the boundary of the disk $|z| < 1$ the function $w(z)$ satisfies

$$\mathrm{Re}\big[G^1(t) w(t) \big] = g(t), \qquad t = e^{i\gamma}, \tag{35}$$

where

$$\tilde{g}(t) = g^1(t) - g^2(t) - \mathrm{Re}\big[G^1(t) - G^2(t) \big] w^2(t).$$

Mapping the unit disk homeomorphically onto itself by a solution of the equation

$$\zeta_{\bar{z}} - \tilde{\mu}(z) \zeta_z = 0,$$

we arrive at the analogous problem (35) for an equation of the form (33) with $\tilde{\mu}(\zeta) \equiv 0$, and, according to the corollary of the lemma in §3 of Chapter V, the coefficients

$$\tilde{A}(\zeta) = \tilde{A}[z(\zeta)]\big(\bar{\zeta}_{\bar{z}} - \tilde{\mu} \bar{\zeta}_z \big)^{-1}, \qquad \tilde{F}(\zeta) = \tilde{F}[z(\zeta)]\big(\bar{\zeta}_{\bar{z}} - \tilde{\mu}(\zeta_z) \big)^{-1}$$

of the transformed equation belong, as before, to $L_p(\overline{K})$ ($p > 2$). Since

$$\zeta(z) \in C^\alpha(\overline{K}), \qquad \alpha = \frac{p-2}{p} > 0,$$

it obviously follows that a transformation of coordinates does not affect the estimates of the solution $w(z)$ in C^{α_0}, $0 < \alpha_0 < \alpha$, and we can assume without loss of generality that $\tilde{\mu}(z) \equiv 0$ in the original equation (33). Then it is not hard to see that the function

$$w(z) = \Phi(z) e^{T_1 \tilde{A}} + e^{T_1 \tilde{A}} T_1\big(\tilde{F} e^{-T_1 \tilde{A}} \big),$$

where $\Phi(z)$ is holomorphic in the disk $|z| < 1$ and satisfies (33) with $\tilde{\mu} \equiv 0$. But for $|t| = 1$

$$\mathrm{Re}\big[G^1(t) \Phi(t) \big] = \tilde{g} e^{-T_1 \tilde{A}} - T_1\big(\tilde{F} e^{-T_1 \tilde{A}} \big) \mathrm{Re}\, G^1(t) \equiv \tilde{g}(t).$$

Therefore, by (32) and the properties of T_1,

$$\| e^{T_1 \tilde{A}} T_1\big(\tilde{F} e^{-T_1 \tilde{A}} \big) \|_{C^\alpha} \leqslant M_T e^{2 \max |T_1 \tilde{A}|} \| \tilde{F} \|_{L_p} < \tilde{M} \varepsilon,$$

whence

$$\| \tilde{g}(t) \|_{C^\beta [\gamma_j, \gamma_{j+1}]} < \tilde{M}_g \varepsilon, \qquad j = 1, \ldots, n.$$

Thus, for the analytic function $\Phi(z)$ we arrive at the same boundary-value problem as in the study of the stability of problem (3) in the class of analytic functions. Consequently, according to the results in §5 of Chapter I,

$$\|\Phi(z)\|_{C^{\beta_0}(\overline{K})} < M_\Phi \varepsilon^\delta,$$

where $0 < \delta = \delta(\beta, \alpha)$ and $0 < \beta_0 = \beta_0(\beta, \alpha)$. Finally,

$$\|w(z)\|_{C^{\beta_0}(\overline{K})} = \|\Phi e^{T_1 \tilde{A}} + e^{T_1 \tilde{A}} T_1(\tilde{F} e^{-T_1 \tilde{A}})\|_{C^{\beta_0}} < M_0 \varepsilon^\delta, \tag{36}$$

where M_0 depends on the properties of the coefficients of (2) and of the boundary condition (3).

We next remark that if the points of discontinuity coincide as before $(t_j^1 = t_j^2)$, then no essential complications arise in proving the stability of solutions of problem (2), (3) when the $g^k(t)$ are unbounded. It is clear that in studying stability in this case (as in the case of analytic functions—see §5 in Chapter I) we must exclude from consideration neighborhoods of the points where the solution $w(z)$ is not bounded or we must study the stability of the functions $(\Pi_k (z - t_k^*)^{\alpha_k^*} w(z))$.

Let us now prove that a solution of problem (2), (3) is stable in the case when the points of discontinuity of $G^k(t)$ and $g^k(t)$ do not coincide. As in §5 of Chapter I, we construct a homeomorphism $\tau = \tau(t) \in C^{1+\alpha}$, $\alpha > 0$, of the circle $|t| = 1$ onto itself that carries the points t_j^1 of discontinuity of $G^1(t)$ and $g^1(t)$ into the points τ_j^2 ($j = 1, \ldots, n$) of discontinuity of $G^2(\tau)$ and $g^2(\tau)$, with

$$\|\tau(t) - t\|_{C^1} < \varepsilon_1(\varepsilon, \delta_0) \quad \text{if } |t_j^1 - t_j^2| < \varepsilon, \tag{37}$$

where

$$\delta_0 = \inf_{k,j} |t_j^k - t_{j+1}^k|.$$

As usual, it is assumed that

$$\|G^2[\tau(e^{i\gamma})] - G^1(e^{i\gamma})\|_{C^\beta} < \varepsilon, \quad \|g^2[\tau(e^{i\gamma})] - g^1(e^{i\gamma})\|_{C^\beta} < \varepsilon \tag{38}$$

for

$$\gamma \in [\gamma_j^1, \gamma_{j+1}^1], \quad \gamma_j^1 = \arg t_j^1 \quad (j = 1, \ldots, n).$$

We extend the homeomorphism $\tau = \tau(t)$ to a quasiconformal mapping $\sigma = \sigma(z)$ of the disk $|z| < 1$ onto itself with some complex characteristic $\mu(z)$, $|\mu(z)| \leqslant \mu_0(\inf |d\tau/dt|) < 1$ (§7.2° in Chapter V), and we replace z by $\sigma(z)$ in the equation (2) for $w^2(z)$. Then for $\overline{w}^2 \equiv w^2[\sigma(z)]$ we obtain

$$\tilde{w}_{\bar{z}}^2 - \tilde{\mu}_1^2(\tilde{w}^2, z)\tilde{w}_z^2 - \tilde{\mu}_2^2(\tilde{w}^2, z)\overline{\tilde{w}_z^2} = \tilde{A}^2(\tilde{w}^2, z), \tag{39}$$

where

$$\tilde{\mu}_1 = \frac{\varsigma_z}{\overline{\varsigma_{\bar{z}}} a}\left[\overline{\mu}(\mu)^2 - (1 + |\mu_1|^2 - |\mu_2|^2)\mu + \mu_1\right], \quad \tilde{\mu}_2 = \mu_2(1 - |\mu|^2)a^{-1}$$

and

$$\tilde{F} = \left(a \overline{\zeta_{\bar{z}}} \right)^{-1} \left[F(1 - \bar{\mu}_1 \mu) + \mu_2 \mu \bar{F} \right], \qquad a = |1 - \mu_1 \bar{\mu}^2| - |\mu_2 \mu|^2$$

(the index 2 is omitted in the last formulas). It is easy to see that the functions $\tilde{\mu}_i^k$ and A^k ($i, k = 1, 2$) ($\tilde{\mu}_i^1 = \mu_i^1$, $\tilde{A}^1 = A^1$) satisfy (17) and (28). Thus, the general case reduces to the previously studied case where the points of discontinuity coincide ($t_j^1 = t_j^2$, $j = 1, \dots, n$) for the functions

$$\tilde{G}^2(t), \tilde{g}^2(t) \equiv g^2[\tau(t)] \quad \text{and} \quad \tilde{G}^1(t), \tilde{g}^1(t) \equiv g^1(t).$$

We formulate the assertion just proved as a theorem.

THEOREM 4 (stability). *Suppose that the conditions in Theorems 1 and 2 hold along with one of the pairs of inequalities* (31) *or* (38) *for the coefficients* μ_i^k, A^k, G^k, *and* g^k ($i, k = 1, 2$) *of* (2) *and* (3).

Then the difference between the corresponding bounded generalized solutions $w^k(z)$ of problem (2), (3) *satisfies the estimate*

$$\| w^1(z) - w^2(z) \|_{C^{\beta_0}(\overline{K})} \leqslant M_0 \varepsilon^\delta, \qquad \beta_0, \delta > 0, \tag{40}$$

i.e., the bounded generalized solutions of this problem are stable in $C^{\beta_0}(\overline{K})$, $\beta_0 > 0$, under small changes in the coefficients of (2) *and* (3).

6°. The Hilbert problem for nonuniformly elliptic equations. With the help of the Leray-Schauder theorem (Theorem 7 in §1 of Chapter II) the existence theorems we have proved can be generalized to the case of an equation (2) that is elliptic only on the unknown solution. We present the formulation and proof of an existence theorem only for the special case needed in our applications when the solutions of problem (2), (3) are bounded; the task of carrying out the appropriate arguments in the general case is left to the reader. Suppose that the coefficients of (2) are such that

$$|\mu_1(w, z)| + |\mu_2(w, z)| \leqslant \mu_0(\|q_i\|_C) \leqslant 1, \tag{41}$$

where $q_1 + iq_2 = w$, while μ_0 increases with increasing $\|q_i\|_C$, and $\mu_0 = 1$ for $\|q_1\|_C = \overline{N}_q^1$ or $\|q_2\|_C = \overline{N}_q^2$ (in particular, $\overline{N}_q^i$ may be infinite).

Suppose that we know an a priori estimate for the solutions of problem (2), (3) when $|z| \leqslant 1$:

$$|q_i(z)| \leqslant \overline{N}_i(\overline{N}), \qquad i = 1, 2, \tag{42}$$

where the $\overline{N}_i$ increase as the constant $\overline{N}$ increases,

$$\overline{N} = \sup_k \left\{ \|G(e^{i\gamma}), g(e^{i\gamma})\|_{C^{\beta_0}[\gamma_k, \gamma_{k+1}]} \right\}, \qquad \beta_0 > 0,$$

and $t_k = e^{i\gamma_k}$ are the points of discontinuity of the functions $G(t)$ and $g(t)$.

We remark that when the index κ is negative, the estimate (42) is assumed to be valid for $w(z)$ represented in the form (18) and, generally speaking, not satisfying the boundary condition (3).

THEOREM 5. *Suppose that the functions $\mu_i(w, z)$ and $A(w, z)$ in (2) satisfy the conditions of Theorem 2 and (41), while the functions $G(t)$ and $g(t)$ in (3) are bounded, with*

$$G(e^{i\gamma}), g(e^{i\gamma}) \in C^{\beta_0}[\gamma_k, \gamma_{k+1}],$$

where $t_k = e^{i\gamma_k}(k = 1,\ldots,n)$ are the points of discontinuity of $G(t)$ and $g(t)$ ($G(t)$ is also discontinuous at points of discontinuity of $g(t)$).

If the a priori estimates (42) hold with $\overline{N}_i < \overline{N}_q^i$, i.e., if $\mu_0 < 1$ on the unknown solution, then the dimensions of the sets of bounded solutions of the Hilbert problem for $\kappa \geq 0$ (respectively, the numbers of solvability conditions for $\kappa < 0$) in the class of analytic functions and in the class of generalized solutions of (2) coincide.

Let us consider the set V of continuous functions $\tilde{w}(z) = q_1 + iq_2$ satisfying (42) with constants $\overline{N}_i < \overline{N}_q^i$ fixed as in the conditions of the theorem. Introducing a factor $\lambda \in [0, 1]$ in front of $g(t)$ in (3) and substituting an arbitrary function $\tilde{w}^\lambda(z) \in V$ into the coefficients of (2), we get the following one-parameter family of problems:

$$w_{\bar{z}}^\lambda - \mu_1(\overline{w}^\lambda, z)w_z^\lambda - \mu_2(\tilde{w}^\lambda, z)\overline{w}_{\bar{z}}^\lambda = A(\tilde{w}^\lambda, z), \tag{2^λ}$$

$$\text{Re}\left[G(t)w^\lambda(t)\right] = \lambda g(t), \qquad t = e^{i\gamma}, \tag{3^λ}$$

$$w^\lambda(z_k) = w^\lambda(1) = 0, \qquad k = 1,\ldots,\kappa, \kappa \geq 0, \tag{43}$$

where the z_k are fixed points, with $|z_k| < 1$ and $z_i \neq z_j$ if $i \neq j$.

According to Theorem 1, to each fixed $\tilde{w}^\lambda(z) \in V$ there corresponds a generalized solution $w^\lambda(z) \in C^\beta(\overline{K})$, $\beta(\mu_0) > 0$, of the linear equation (2^λ), and it can be represented in the form

$$w^\lambda(z) = \omega^\lambda[\zeta(z)]e^{T_i(\varphi^\lambda|z)} + T_1(\psi^\lambda|z). \tag{18^λ}$$

Then, adding (18^λ) to the equalities (23) corresponding to the linear problem (2^λ), (3^λ), we get a one-parameter family of transformations F^λ of the Banach space $B = \{L_p \times L_p \times L_q \times L_q \times C(\overline{K})\}$ into itself:

$$\sigma_\lambda = F^\lambda \tilde{\sigma}^\lambda, \qquad \sigma^\lambda = (\varphi^\lambda, \psi^\lambda, \vec{m}^\lambda, w^\lambda).$$

By (42), for $\tilde{w} \in V$ we have

$$|\mu_1(\tilde{w}, z)| + |\mu_2(\tilde{w}, z)| \leq \mu_0(\overline{N}_1, \overline{N}_2) = \tilde{\mu}_0 < 1. \tag{44}$$

According to estimates (24) and (25) in Theorem 1, and (42), the transformation F^λ does not have fixed points on the boundary of the set $Q \subset B$ for any λ, where

$$Q: \begin{cases} \|\varphi(z)\|_{L_p(\overline{K})} \leqslant M_1(\mu_0), & \|\psi(z)\|_{L_p(\overline{K})} \leqslant M_2(\tilde{\mu}_0), p(\mu_0) > 2 \\ |\vec{m}(z)| \leqslant \mu_0(\overline{N}_1, \overline{N}_2), & |\operatorname{Re} w^\lambda| \leqslant \overline{N}_1, |\operatorname{Im} w^\lambda| \leqslant \overline{N}_2 \end{cases}.$$

Since F^λ is linear in λ, it is uniformly continuous in λ for fixed $\tilde{\sigma}^\lambda \in Q$. On the other hand, by what was proved in Theorem 1, the transformation (23) is completely continuous in $(L_p \times L_p \times L_q \times L_q)$ for fixed λ and w^λ in the coefficients of problem (2^λ), (3^λ). But the stability of solutions of the Hilbert problem (Theorem 4) implies also that F^λ is continuous in $\tilde{w}^\lambda$ for fixed $\lambda \in [0, 1]$ (stability under a change in the coefficients of the equation), and the fact that to each $\tilde{w}^\lambda \in C(\overline{K})$ there corresponds a $w^\lambda \in C^\beta(\overline{K})$, $\beta > 0$, implies that F^λ is completely continuous on this component.

Further, by the conditions of the theorem when $\lambda = 0$, the solution $w^0(z) \equiv 0$ of the original problem is unique, and F^0 is obviously a homeomorphism in a neighborhood of it. Consequently, all the conditions of the Leray-Schauder theorem hold, and F^λ has at least one fixed point in the set $Q \subset B$ for all $\lambda \in [0, 1]$.

The proof of the theorem is concluded by observing that under the assumptions in the theorem about the coefficients μ_i and A of (2) the solution of problem (2), (3) (with the additional condition (43) when $\kappa \geqslant 0$) is unique.

7°. The Hilbert problem in the case of summable coefficients in the boundary condition. The correspondence due to the representation (18) between analytic functions $\omega(z)$ in the plane of the homeomorphism and generalized solutions of (2) enables us in some cases to solve the Hilbert problem under the same boundary-data assumptions as for analytic functions. However, it is necessary to require that the properties of the coefficients of (2) and (3) be compatible in a certain sense.

The necessity of this compatibility has to do with the fact that it is possible for sets of positive measure to correspond to boundary sets of measure zero under homeomorphisms of the disk onto itself by solutions of the Beltrami equation with an arbitrary measurable coefficient $\mu(z)$ (see, for example, [101]). Consequently, if the singularities of the coefficients $G(t)$ and $g(t)$ in (3) are concentrated on such a set, then in passing to the plane of the homeomorphism these coefficients become, generally speaking, nonmeasurable.

In this connection we assume in the general case of measurable $G(t)$ and $g(t)$ that the conditions of Theorem 1 are supplemented by the following

additional assumption about the coefficients of (2):
(i)

$$\mu_2 = \frac{\partial \mu_1}{\partial w} = \frac{\partial \mu}{\partial \overline{w}} \equiv 0, \qquad \mu_1(z) \in C^\beta, 0 < \rho \leqslant |z| \leqslant 1, \beta > 0.$$

Let us extend the function $\mu_1(z)$ with preservation of smoothness inside the disk $|z| < 1$ in such a way that the constructed function $\mu_1^*(z) \in C^\beta(\overline{K})$ satisfies the condition $|\mu_1^*(z)| \leqslant \mu_0 < 1$, and then map the unit disk homeomorphically onto itself by a solution of the Beltrami equation with $\mu = \mu_1^*(z)$. Then on the ζ-plane of values of this homeomorphism $\zeta = F(z)$ the coefficients $\tilde{\mu}_1$ and $\tilde{\mu}_2$ of the transformed equation (2) turn out to be compactly supported near the circle, and the remaining coefficients of (2) and (3) do not change their properties, since the constructed homeomorphism belongs to the class $C^{1+\beta}(\overline{K})$. Consequently, assumption (i) is equivalent to the following:
(i′)

$$\mu_1(w, z) = \mu_2(w, z) \equiv 0, \qquad 0 < \rho \leqslant |z| \leqslant 1.$$

Suppose that the $G(t)$ and $g(t)$ in (3) satisfy the usual assumptions common in the theory of analytic functions (see Lemma 1 and Theorem 2 in §6 of Chapter II):

(ii) The function $G(t)$, $|G(t)| = 1$, can be represented in the form $G(t) = G_1(t)G_2(t)$, where $G_1(t) \in C^\beta$, $\beta > 0$, and ind $G_1(t) = \kappa$, with

$$|\arg G_2(t)| \leqslant \pi/4p - \delta \qquad (\delta > 0, p > 1).$$

(iii) $g(t) \in L_p, p > 1$.

In analogy to Theorem 1 we define the set $\mathfrak{M}^*$ of vectors $\vec{m}^*(z) = \{m_1^*(z), m_2^*(z)\}$, satisfying (21) and the assumption

$$\vec{m}^*(z) \equiv 0, \qquad 0 < \rho \leqslant |z| \leqslant 1.$$

On this set the hoomeomorphisms $\zeta = \zeta(z)$ satisfying the linear equation

$$\zeta_{\overline{z}} - m_1^*(z)\zeta_z - m_2^*(z)\overline{\zeta}_{\overline{z}} = 0$$

are analytic functions in a neighborhood of the circle. We remark next that if the densities of the Cauchy type integrals in (8) and (10) converge in $L_p, p > 1$, then the analytic functions corresponding to them converge uniformly in any domain $\overline{D} \subset K$. Taking these facts into account, we arrive at the following assertion by literally repeating the arguments in Theorem 1.

THEOREM 6. *Under the conditions of Theorem 1 on the coefficients μ_i and A and the assumptions* (i)–(iii) *in the case* $\kappa \geqslant 0$ *the equation* (2) *has at least one generalized solution that satisfies the Hilbert problem almost everywhere on the circle and the conditions* (9) *at fixed interior points of the disk* $|z| < 1$, *while if*

$\kappa < 0$ *such a solution exists only when the* $2 \, | \, \kappa \, | -1$ *solvability conditions hold. The analytic function* $\omega(\zeta)$ *in the representation* (18) *then belongs to the class* H_p, $p > 1$.

The restriction (i) on $\mu_i(w, z)$ can be discarded if we broaden the class of admissible solutions of problem (2), (3).

DEFINITION. Generalized solutions of (2) will be called *quasisolutions* of the Hilbert problem (3) if on the plane of values of the homeomorphism $\zeta = \zeta(z)$ the analytic function $\omega(\zeta)$ in the representation (18) belongs to H_p, $p > 1$, and satisfies the boundary condition (19) almost everywhere on the circle.

According to this definition, the inverse transition in the boundary condition (3) from the plane of values of the homeomorphism $\zeta = \zeta(z)$ to the plane of the original variable z is not required, which allows us to strengthen the assertions of Theorem 6 as follows without changing the proofs.

THEOREM 7. *Suppose that the conditions of Theorem 1 on the coefficients* μ_i *and* A *of* (2) *are satisfied, while the coefficients* $G(t)$ *and* $g(t)$ *satisfy the respective conditions* (ii) *and* (iii), *and* $g(t)$ *is a measurable function such that* $|g(t)| \leq M < \infty$ *for* $|t| = 1$.

Then for $\kappa \geq 0$ *there exists at least one quasisolution of problem* (2), (3) *that satisfies* (9), *while for* $\kappa < 0$ *such a solution exists only if the* $2 \, | \, \kappa \, | -1$ *solvability conditions are satisfied.*

The fact that the quasisolutions are well-defined is established in the following theorem.

THEOREM 8 (uniqueness). *Suppose that under the conditions of Theorem 7 there exists a quasisolution of problem* (2), (3) *that satisfies for* $\kappa \geq 0$ *the additional conditions* (9) *with fixed* z_k. *Such a quasisolution is unique if the coefficients of* (2) *satisfy the inequalities*

$$|A(w_1, z) - A(w_2, z)| \leq N |w_1 - w_2|, \qquad N = \text{const},$$

$$|\mu_i(w_1, z) - \mu_i(w_2, z)| \leq N_0 |w_1 - w_2|, \qquad N_0 = \text{const}, \, i = 1, 2, \qquad (4^*)$$

and

$$\frac{\partial \mu_1}{\partial w} = \frac{\partial \mu_1}{\partial \overline{w}} = 0, \qquad 0 < \rho \leq |z| \leq 1;$$

$$\mu_2(z, w) \equiv 0 \quad \text{for all } |z| < 1.$$

The method of proving a uniqueness theorem by considering the problem for the difference of solutions is not directly applicable here, since the boundary condition (3) for each of the quasisolutions holds only in the plane of its homeomorphism, and is not defined on the z-plane. Let us then approximate the coefficients $G(t)$ and $g(t)$ in the boundary condition (3) by functions $G^\varepsilon(t)$, $g^\varepsilon(t) \in C^\beta$, $\beta > 0$, and denote by $w^\varepsilon(z)$ the sequence of corresponding generalized solutions of the Hilbert boundary-value problem

$$\operatorname{Re}[G^\varepsilon(t)w^\varepsilon(t)] = g^\varepsilon(t). \tag{46}$$

Suppose that there exist at least two distinct quasisolutions $w = w^i(z)$ of problem (2), (3) and let $\zeta = \zeta^i(z)$ $(i = 1, 2)$ be the corresponding homeomorphisms of the unit disk onto itself. We prove that the sequence $w^\varepsilon(z)$ converges as $\varepsilon \to 0$ to each of these quasisolutions $w^1(z)$ and $w^2(z)$, which implies that they coincide. For fixed $i = 1, 2$ we write the equation for $w^\varepsilon(z)$ in the form

$$w_{\bar z}^\varepsilon - \mu(z)w_z^\varepsilon = A(z)w^\varepsilon - A_0(z)\tilde w^\varepsilon, \tag{2*}$$

where

$$\mu(z) \equiv \mu_1[w^i(z), z], \qquad A(z) = A[w^i(z), z], \qquad \tilde w^\varepsilon = w^i(z) - w^\varepsilon(z),$$

$$-A_0(z) = w_z^\varepsilon \frac{\mu(w^i) - \mu(w^\varepsilon)}{w^i - w^\varepsilon} + \frac{A(w^i) - A(w^\varepsilon)}{w^i - w^\varepsilon} w^\varepsilon.$$

Since $w^i(z)$, $w^\varepsilon(z) \in L_p(\overline K)$, $p > 2$, independently of ε, and since $\mu(w^i, z) - \mu(w^\varepsilon, z) \equiv 0$ in the neighborhood $K_\varepsilon = \{0 < \varepsilon < |z| \leqslant 1\}$, it follows that $A_0(z) \in L_p(\overline K)$, $p > 2$. Using the homeomorphism $\zeta = \zeta^i(z)$ to transform (2) for $w = w^i(z)$ and (2*) and subtracting both sides of the resulting equations one from the other, we obtain

$$\tilde w_{\bar\zeta}^\varepsilon = [A(\zeta) + A_0(\zeta)]\tilde w^\varepsilon, \tag{47}$$

where

$$A(\zeta) = A[z^i(\zeta)] \in L_p, \qquad A_0(\zeta) = A_0[z^i(\zeta)] \in L_p, \qquad p > 2.$$

On the plane of the homeomorphism $\zeta = \zeta^i(z)$ the corresponding quasisolution $w^i = w^i(z)$ satisfies the transformed boundary condition (3) by definition. On the other hand, for $w^\varepsilon(z)$ the corresponding boundary condition (46) with piecewise Hölder coefficients is satisfied in the plane of any homeomorphism in C^β, $\beta > 0$, and, in particular, in the plane of $\zeta = \zeta^i(z)$. Taking the difference of both sides of the boundary conditions for $w^i[z^i(\zeta)]$ and $w^\varepsilon[z^i(\zeta)]$, we get

$$\operatorname{Re}\{G[t^i(\tau)]\tilde w^\varepsilon(\tau)\} = \tilde g^\varepsilon(\tau), \tag{48}$$

where

$$\tilde g^\varepsilon(\tau) = g[t^i(\tau)] - g^\varepsilon[t^i(\tau)] + \operatorname{Re}[(G^\varepsilon - G^i)w^\varepsilon(\tau)].$$

Representing the solution of problem (47), (48) in the form

$$\tilde{w}^{\epsilon}(\zeta) = \Phi(\zeta)e^{T_1(A_0+B_0)},$$

we arrive at a Hilbert problem for the analytic function $\Phi^{\epsilon}(\zeta) \in H_q, q > 1$:

$$\operatorname{Re}\{G[t^i(\tau)]\Phi^{\epsilon}(\tau)\} = \tilde{g}^{\epsilon}e^{-T_1(A_0+B_0)}.$$

By construction, $\|\tilde{g}^{\epsilon}(\tau)\|_{L_q} \to 0$, for some $q > 1$ as $\epsilon \to 0$; therefore, the Cauchy type integrals in the representation (8) or (10) (depending on the sign of the index) for the analytic function $\Phi^{\epsilon}(\zeta)$ converge uniformly in each domain $\overline{D} \subset K$; consequently, $\Phi^{\epsilon}(\zeta) \to 0$ in $\overline{D} \subset K$.

Returning to the z-plane, we obtain a sequence $w^{\epsilon}(z)$ that converges uniformly in $|z| < 1$ to each of the quasisolutions $w^i(z)$. The theorem is proved.

We remark that all the theorems proved in this section are valid, in particular, when $G(t)$ and $g(t)$ are piecewise continuous.

§3. Boundary-value problems of conjunction with a shift in multiply connected domains

1°. Statement of the problem. Let $L = \sum_0^m L_k$ be a collection of $m + 1$ disjoint closed Lyapunov curves contained in the contour L_0. Denote by D^+ the $(m + 1)$-connected domain lying inside L_0 and outside the contours L_k $(k = 1,\ldots,m)$, and by D^- the complement of $D^+ + L$ in the complete plane E. We assume for definiteness that the origin is in D^+.

DEFINITION 1. A homeomorphic mapping $\alpha = \alpha(t)$ of the curve $L = \sum_0^m L_k$ onto itself that preserves orientation on each of the curves L_k is called a *mapping of class SQ* (the trace (Spur) of a quasiconformal mapping) if it extends to a full homeomorphism $\alpha = \alpha(z)$, $\{\alpha(z) - z\} \in W_p^1(E)$, $p > 2$, of the whole plane E with complex characteristic $q(z)$, $|q(z)| \leq q_0 < 1$.

Theorem 2 in §7 of Chapter V gives necessary and sufficient conditions for a mapping α, $\alpha(L) = L$, to be in SQ.

The general boundary-value problem of conjunction with a shift (the generalized Haseman problem) for a quasilinear elliptic system of equations can be formulated as follows.

It is required to find a solution $w = w(z)$ that is bounded at infinity for the equation (defined on the whole plane)

$$w_{\bar{z}} - \mu_1(w, z)w_z - \mu_2(w, z)\bar{w}_{\bar{z}} = A(w, z), \tag{1}$$

and that satisfies on L the boundary condition

$$w^+[\alpha(t)] = G(t)[w^-(t) + \lambda(t)\overline{w^-(t)}] + g(t), \tag{2}$$

whose coefficients have the following properties:

(a) $G(t)$, $g(t)$, $\lambda(t) \in SW_p^1(\widehat{t_k t_{k+1}})$, $p > 2$, $t_k \in L_i$ $(i = 0,\ldots,m;\ k = 0,\ldots,n)$, and $|G(t)| \neq 0$.

(b) $\alpha = \alpha(t)$ is a mapping of class SQ.

(c) $\lambda = \lambda(t) \in SW^1(L)$, and there exists an extension such that $\lambda(z) \in W_p^1(\overline{D^-})$, $\lambda(z) \equiv 0$ if $|z| > R$, and $|\lambda(z)| \leq 1 - \delta_0$, $\delta_0 > 0$, $z \in D^-$.

The coefficients μ_i and $A = A_1(w, z)w + A_2(w, z)\overline{w} + A_0(w, z)$ are assumed to satisfy the conditions

$$|\mu_1(w, z)| + |\mu_2(w, z)| \leq \mu_0 < 1, \tag{3}$$

$$\|A_i(w, z)\|_{L_p(D^\pm)} \leq M \quad \text{for all } w, p > 2, \tag{4}$$

with μ_i and A_i continuous in w for almost all z, and $\mu_i(w, z) = A_i(w, z) \equiv 0$ for $|z| > R$.

By additional conformal mappings of the domains $D^+(L)$ and $D^-(L_k)$ $(k = 0,\ldots,m)$ we can reduce the generalized Haseman boundary-value problem to the same problem for a canonical curve Γ consisting of circles Γ_k: $|z - z_k| = r_k$ $(k = 0,\ldots,m)$, contained in the circle Γ_0: $|z| = 1/R$. Since conformal mappings are smooth, the coefficients of the transformed problem will continue to satisfy conditions (3) and (4) with $R = 1$; therefore, we analyze problem (1), (2) at once for a canonical curve L in what follows, retaining the previous notation.

Note that assumption (c) holds, for example, for functions

$$\lambda(t) = \lambda_1(t) + i\lambda_2(t) \in SW_p^1(L), \qquad p > 2,$$

$$|\lambda_{1,2}(t)| \leq \tfrac{1}{2} - \delta_0, \qquad \delta_0 > 0.$$

Indeed, to construct $\lambda(z)$, $z \in D^-$, it suffices to extend the $\lambda_i(t)$ by harmonic functions $\lambda_i(z)$ $(i = 1, 2)$ into each of the connected components of D^- and, without violating the condition $|\lambda_1| + |\lambda_2| < 1$, to cut off the function $\lambda_1(z) + i\lambda_2(z)$ thus obtained in a smooth fashion so that it has compact support for $|z| > 1$.

Another sufficient condition for (c) is the following:

$$\lambda(t) \in W_p^1(L), \quad p > 2; \qquad |\lambda(t)| < 1 - \delta_0, \quad \delta_0 > 0,$$

since in this case the desired extension $\lambda(z)$, $\lambda(z) \equiv 0$ for $|z| > 1$, can be realized by the formulas

$$\lambda(z) = |z - z_k| \frac{\lambda(t)}{r_k}, \qquad |z - z_k| < r_k,$$

$$\lambda(z) = \frac{(1 - |z|)R}{R - 1}\lambda(t), \qquad \frac{1}{R} < |z| < 1.$$

2°. The boundary-value problem of a jump with a shift. Let us first consider the simplest case of the boundary-value problem (1) for an analytic function $\omega(z)$ in $D^{\pm}$,

$$\omega^+[\alpha(t)] = \omega^-(t) + g(t), \tag{5}$$

with its solution represented in the form

$$u(z) \equiv T(\varphi \mid z) + \Phi(z) + P_n(z) = \begin{cases} \omega^+[\alpha(z)], & z \in D^+, \\ \omega^-(z), & z \in D^-. \end{cases} \tag{6}$$

Here $\alpha = \alpha(z)$ is an extension of the shift $\alpha(t) \in SQ$ to a quasiconformal mapping of D^+ onto itself with complex characteristic $q(z)$, $|q(z)| \leqslant q_0 < 1$, $P_n = \sum_0^n a_k z^k$,

$$T(\varphi \mid z) = -\frac{1}{\pi} \iint_K \frac{\varphi(t) \, dK_t}{t - z}, \qquad \Phi(z) = \frac{1}{2\pi i} \int_K \frac{g(t) \, dt}{t - z}, \qquad K: |z| < 1,$$

and $g(t) \in SW_p^1$, $p > 2$, so that $\Phi(z) \in W_p^1(D^{\pm})$, $p > 2$. By construction, $\omega(z)$ satisfies the boundary condition (5) for any function $\varphi(t) \in L_p(\overline{K})$, $p > 2$, and the assumption that $\omega(z)$ is analytic in $D^{\pm}$ leads to the following integral equation for determining $\varphi(z)$:

$$\varphi(z) - q(z)S(\varphi \mid z) = q(z)\left(\frac{d\Phi^+}{dz} + \frac{dP_n}{dz}\right), \tag{7}$$

where $S\varphi = \partial T\varphi / \partial z$ and $q(z) \equiv 0$ for $z \in D^-$. We find a solution of problem (5) by solving (7) in $L_p(\overline{K})$ ($p(q_0) > 2$) with fixed constants a_k according to (6). We require that $\omega(z)$ have a given expansion $\sum_0^\kappa c_k z^k$ at infinity, i.e.,

$$\lim_{z \to \infty}\left[\omega - (z) - \sum_{k=0}^{\kappa} c_k z^k\right] = 0. \tag{8}$$

For (8) to hold when $\kappa > 0$, it suffices to set $n = \kappa$ and $P_n = \sum_0^\kappa c_k z^k$ ($a_k = c_k$), while for $\kappa < 0$ it must be required that $P_n(z) = c_0$ and that the functions $T(\varphi \mid z)$ and $\Phi^-(z)$ satisfy the relations

$$\iint_K \varphi(t) t^{s-1} \, dK_t - \frac{1}{2i} \int_L g(t) t^{s-1} \, dt = \pi c_s, \qquad s = 1, \ldots, |\kappa|. \tag{9}$$

Let us now investigate the uniqueness of the solution $\omega(z)$. Suppose that $\omega^i(z)$ are two distinct solutions of problem (5), (8). Then the function

$$u(z) = \begin{cases} \omega^1[\alpha(z)] - \omega^2[\alpha(z)], & z \in D^+, \\ \omega^1(z) - \omega^2(z), & z \in D^-, \end{cases}$$

is continuous on the whole plane, vanishes at infinity, and satisfies the Beltrami equation with coefficient $q(z)$, $|q(z)| \leqslant q_0 < 1$ and $q(z) \equiv 0$, $z \in D^-$, which is possible only if $u(z) \equiv 0$, i.e., $\omega^1(z) - \omega^2(z) \equiv 0$ in $D^{\pm}$.

We remark that in the case when $\kappa = 1$ and $c_0 = 1 - c_1 = g(t) \equiv 0$ the functions $u = u_0(z)$ and $\omega = \sigma(z)$ defined by (6) have the form

$$u_0(z) = z + T(\varphi \,|\, z) = \begin{cases} \sigma[\alpha(z)], & z \in D^+, \\ \sigma(z), & z \in D^-, \end{cases} \tag{10}$$

where $\varphi(z)$ satisfies (7) with $\Phi^+(z) \equiv 0$ and $P_n(z) \equiv z$. Consequently, $u_0(z)$ realizes a quasiconformal mapping of the whole plane E onto itself (Theorem 3 in §3 of Chapter V), and $u_0(z) - z \in W_p^1(E)$, $p > 2$. Since the mapping $\alpha = \alpha(z)$ is a homeomorphism, we conclude that the function $\omega = \sigma(z)$ is univalent in the domains D^+ and D^-, and $\sigma^{\pm}(z) - z \in W_p^1(D^{\pm})$, $p > 2$. If in addition it is assumed that $\alpha(t) \in C^{1+\beta}$, $\beta > 0$, then the complex characteristic $q(z)$ of its quasiconformal extension $\alpha = \alpha(z)$ in D^+ (constructed in §7.2° of Chapter V) becomes a Hölder function; hence,

$$\{u_0(z) - z\}, \{\sigma(z) - z\} \in C^{1+\beta}(D^{\pm}), \qquad \beta > 0.$$

We formulate the above assertions as a theorem.

THEOREM 1. *Suppose that $\alpha(t) \in SQ$ and $g(t) \in SW(L)$, $p > 2$. Then for $\kappa \geqslant 0$ the boundary-value problem* (5), (8) *has a unique solution $\omega(z)$ in the class of analytic functions in $D^{\pm}$ such that $\omega(z) - \Sigma_0^\kappa c_k z^k \in W_p^1(\overline{D}^{\pm})$, $p > 2$, while for $\kappa < 0$ the problem has a unique solution only if the solvability conditions* (9) *hold. In the case $\kappa = 1$ and $g(t) \equiv 0$ the solution $\omega = \sigma(z)$ of problem* (5), (8) *can be represented in the form* (10) *and is a univalent function in $D^{\pm}$, and if in addition $\alpha(t) \in C^{1+\beta}$, then $\sigma(z) - z \in C^{1+\beta}(\overline{D}^{\pm})$, $\beta > 0$.*

3°. The connection between the Riemann and Haseman problems.

We prove that the Haseman problem (2) ($\lambda \equiv 0$) can be reduced to the Riemann problem in the class of piecewise analytic functions under the conditions of Theorem 1. Indeed, consider the piecewise analytic functions

$$\Omega^{\pm}(z) = \omega^{\pm}[\sigma^{\pm}(z)],$$

where $\sigma = \sigma(z)$, $\sigma(z) - z \in W_p^1(\overline{D}^{\pm})$, $p > 2$, is the univalent solution of the problem

$$\sigma^+[\alpha(t)] = \sigma^-(t), \qquad t \in L,$$

represented by (10). The boundary condition (2) for $\Omega(z)$ takes the form

$$\Omega^+\{\sigma^+[\alpha(t)]\} = G(t)\Omega^-[\sigma^-(t)] + g(t).$$

Setting $\sigma^+[\alpha(t)] = \sigma^-(t) = \tau$ in this equality, we arrive at the Riemann problem

$$\Omega^+(\tau) = G[\rho(\tau)]\Omega^-(\tau) + g[\rho(\tau)], \qquad \tau \in \Gamma = \sigma^-(L), \qquad (11)$$

where $\rho = \rho(\tau)$ is the mapping inverse to $\sigma = \sigma^-(t)$, i.e., $\sigma^-[\rho(t)] = t$.

However, in the case of an arbitrary shift $\alpha(t) \in SQ$ the curve $\Gamma = \sigma(L)$ on which the boundary condition (11) holds is, generally speaking, not rectifiable (see the remark at the end of §7 in Chapter V). This circumstance prevents us from making direct use of the classical theory of boundary-value problems for analytic functions. But if $\alpha(t) \in C^{1+\beta}(L)$, $\beta > 0$, then the curve $\sigma(L)$ consists of Lyapunov curves, and the results of §3 in Chapter I and §5 in Chapter II are completely applicable to the Riemann problem (11) this obtained.

4°. Stability of solutions of the problem of a jump with a shift. Assume the following conditions hold:

(i) The sequence of multiply connected domains D_k^+ with boundaries Γ_k consisting of disjoint nondegenerate circles Γ_{ki}, $\Gamma_k = \Sigma_0^m \Gamma_{ki}$, converges to its kernel D^+.

(ii) The sequence of complex characteristics

$$q_k^l(z), \quad |q_k^l(z)| \leqslant q_0 < 1 \quad \text{and} \quad q_k^l(z) \equiv 0, \quad z \in D_k^-,$$

of the mappings $\alpha = \alpha_k^l(z)$, $z \in D_k^+$, converges in measure for each fixed k as $l \to \infty$.

(iii) The sequence $g_k^l(t)$ is uniformly bounded in $SW_p^1(\Gamma_k)$,

$$\|g_k^l(t)\|_{SW_p^1(\Gamma_k)} \leqslant M, \qquad (12)$$

and converges in $SW_p^1(\Gamma_k)$ for fixed k as $l \to \infty$ (see Definition 2 in §1).

Then, by Property 11 in §1, the Cauchy type integrals $\Phi_k^l(z)$ with density $g_k^l(t)$ converge in $W_p^1(\overline{K})$ ($p > 2$), which implies that the solutions $\varphi_k^l(\zeta)$ of (7) converge in $L_p(\overline{K})$, and, by (12),

$$\|d\Phi_k^l(z)/dz\|_{L_p(\overline{D}_k^+)} \leqslant M_1(M, q_0). \qquad (13)$$

The next statement follows from (6), since the solutions are obviously stable under a change in the constants c_k in condition (8).

THEOREM 2. *If conditions* (i)–(iii) *hold, then the sequence of solutions* $\omega_k^l(z)$ *of problem* (5), (8) *converges in* $W_p^1(\overline{D}_k^\pm)$ *for each fixed* k *as* $l \to \infty$, *and*

$$\|\omega_k^l(z)\|_{W_p^1(\overline{D}_k^\pm)} \leqslant M_2(M, q_0). \qquad (14)$$

Moreover, as $k, l \to \infty$ *the functions* $\omega_k^l(z)$ *converge in the norm of* $W_p^1(\overline{D}_*^\pm)$, $p > 2$, *where* $\overline{D}_*^\pm \subset D^\pm$, $D^- = E \setminus D^+$, *and* D^+ *is the kernel of the sequence of domains* D_k^*.

5°. The Haseman problem for analytic functions in the case of a nondifferentiable shift. The solution of the Haseman problem (2) (for $\lambda \equiv 0$) in the class of piecewise analytic functions $\omega(z)$ can be obtained, as in the case of the Riemann problem, by successively solving several auxiliary jump problems. Indeed, suppose for the time being that $m = 0$, i.e., the domains $D^{\pm}$ are simply connected and L is $|z| = 1$. Following the usual scheme for solving the Riemann problem with discontinuous coefficients G and g (see §3.8° of Chapter I), we transform condition (2) for $\omega(z)$ to the form

$$\Psi^{+}[\alpha(t)] = G_0(t)\Psi^{-}(t) + g_0(t). \tag{15}$$

Here the following notation is introduced:

$$\Psi^{+}(z) = (\omega\Pi)^{+}, \qquad \Psi^{-}(z) = (\omega\Pi\Pi_0)^{-},$$
$$G_0(t) = G_0\Pi_0^{-}, \qquad g_0(t) = g\Pi^{-},$$
$$\Pi^{\pm}(z) = \prod_{k=1}^{n} \left[\sigma^{\pm}(z) - \sigma^{\pm}(t_k)\right]^{\gamma_k}, \qquad \Pi_0^{-}(z) = \prod_{k=1}^{n} \left[\sigma^{-}(z)\right]^{\gamma_k}, \tag{16}$$

the constants ν_k are determined from $G(t)$ by the formulas in §3.3° of Chapter I, and the univalent function $\sigma = \sigma(z)$ in the domains $D^{\pm}$ is represented by (10). We look for a solution $\omega(z)$ that is unbounded at all points t_k of discontinuity of $G(t)$. Then Re γ_k is obviously positive, and the properties of the functions in the class $SW_p^1(L)$ give us that $G_0(t), g_0(t) \in SW_{p_0}^1(L)$, $p_0(p) > 2$. If we now represent the function $G_0(t)$ as

$$G_0(t) = t^{\kappa} X^{+}[\alpha(t)][X^{-}(t)]^{-1}, \qquad \kappa = \text{ind } G_0(t),$$

in terms of a solution of the jump problem

$$\ln X^{+}[\alpha(t)] = \ln X^{-}(t) + \ln[t^{-\kappa}G_0(t)], \tag{17}$$

then to determine the piecewise analytic function $\Psi(z)$ we come to the analogous problem

$$\left\{\frac{\Psi[\alpha(t)]}{X[\alpha(t)]}\right\}^{+} - \left\{\frac{\Psi(t)}{X(t)}t^{\kappa}\right\}^{-} = g_0(t)\{X^{+}[\alpha(t)]\}^{-1}. \tag{18}$$

By solving the latter problem, with the additional conditions (8) taken into account when $\kappa \geq 0$, we also reconstruct from (16) a solution of the original Haseman problem (2) ($\lambda \equiv 0$) which satisfies (8) if $\kappa \geq 0$. If $\kappa < 0$, then the boundary-value problem (2) has a solution $\omega = \omega(z)$, $\omega(\infty) = c_0$, only when the solvability conditions (9) for problem (18) are satisfied.

Moreover, the solution $\omega = \omega(z)$ is integrable in a neighborhood of the points t_k if

$$0 < \operatorname{Re} \gamma_k < \frac{p-2}{p} < 1, \tag{19}$$

where the quantity $p(q_0) > 2$ is chosen so that $q_0 \|S\|_{L_p} < 1$, which ensures that the equations of the form (7) are solvable (cf. condition (16) in §2). Theorem 1, applied to problems (17) and (18), implies also that the solution obtained for problem (2) is unique.

As is clear from our constructions, the method of reducing problem (2) to jump problems does not depend on the connectivity of D^+, and the case of a multiply connected domain ($m \neq 0$) differs from the case $m = 0$ just considered only in the form of (16).

We next observe that, because the functions $G_0(t)$ and $g_0(t)$ are bounded in the norm of $SW^1_{p_0}(L)$, $p_0 > 2$, $L = \Sigma_0^m L_k$, and because of the properties of the solutions of a jump problem (Theorem 1), the functions $\Psi(z)$, $\sigma(z)$, and $\rho(z)$ ($\sigma[\rho(z)] = z$) are bounded in the norm of $W^1_p(D^\pm)$, $p > 2$. Then (16) gives us the following estimates for the solution of (2):

$$|\omega^\pm(z)| \leqslant M_3 \Pi \, |z - t_k|^{+\gamma_k^*}, \qquad \|\omega^\pm(z)\|_{W^1_p(\bar{D}_\delta^\pm)} \leqslant M_4, \tag{20}$$

where

$$\gamma_k^* = -\frac{p \operatorname{Re} \gamma_k}{p-2} > -1 \qquad \text{(condition (19))},$$

$$\bar{D}_\delta^\pm = D^\pm \backslash \Sigma D_{k_\delta}^\pm, \qquad D_{k_\delta}^\pm = \{z \in D^\pm, |z - t_k| < \delta\}.$$

Thus, the following theorem is proved.

THEOREM 3. *Suppose that the inequalities* (19) *hold along with assumptions* (a) *and* (b) *in* 1°. *Then for* $\kappa \geqslant 0$ *the Haseman problem* (2) ($\lambda \equiv 0$) *has a unique solution in the class of piecewise analytic functions satisfying* (8) *with given* c_k. *For* $\kappa < 0$ *this problem has a unique solution* $\omega = \omega(z)$, $\omega(\infty) = c_0$, *only if solvability conditions of the form* (9) *hold. In both cases the solutions obtained satisfy* (20).

6°. **Stability of solutions of the Haseman problem for analytic functions.** Let us consider the following family of boundary-value problems:

$$\left\{\omega_k^l[\alpha_k^l(t)]\right\}^+ = G_k^l(t)\left[\omega_k^l(t)\right]^+ + g_k^l(t), \qquad t \in \Gamma_k, \tag{21}$$

assuming that conditions (i) and (ii) of 4° are satisfied for the sequences of multiply connected domains D_k^+ and shifts $\alpha = \alpha_k^l(t)$, $t \in \Gamma_k$, and that the

$G_k^l(t)$ and $g_k^l(t)$ are ε-close for each fixed k in the sense of the definition given below (cf. Definition 1 in §5 of Chapter I).

DEFINITION. The functions $f^1(t) \in SW_p^1(\widehat{t_j t_{j+1}})$ and $f^2(\tau) \in SW_p^1(\widehat{\tau_j \tau_{j+1}})$, $t_j, \tau_j \in \Gamma, j = 1, \dots, n$, are said to be ε-close in $SW_p^1(\Gamma)$ if

$$|\tau_j - t_j| \leqslant \varepsilon, \quad \inf_f \{|\tau_j - \tau_{j+1}|, |t_j - t_{j+1}|\} \geqslant \delta_0 > 0, \tag{22}$$

$$\|f^2[\tau_j(t)] - f^1(t)\|_{SW_{p_0}^1(\widehat{t_j t_{j+1}})} \leqslant \varepsilon_1(\varepsilon) \to 0, \tag{23}$$

where $2 < p_0 < p$, and $\tau = \tau_j(t)$ is a differentiable homeomorphism of the circles $\Gamma_{ki} \subset \Gamma_k$ onto itself that carries the points t_j into the points τ_j.

In view of our assumptions, the sequence of functions $\sigma_k^l(t)$ in (16) converges (by Theorem 2) in $SW_p^1(\Gamma_k)$ for each fixed k as $l \to \infty$, and, by Property 8 in §1, the sequence $G_{0k}^l(t)$, $G_0 = G\Pi_0^-$, also converges in $SW_{p_0}^1(\Gamma_k)$, $2 < p_0 < p$ (see (16)). Turning again to Theorem 2, we see that the sequences $\{X_k^l(z)\}$ and $\{\Psi_k^l(z)\}$ of solutions of the respective problems (17) and (18) converge in the norm of $W_{p'}^1(\overline{D}_k^{\pm})$, $p'(q_0, p) > 2$, for fixed k as $l \to \infty$.

We remark that as $k, l \to \infty$ these sequences, and with them also the solutions $\omega_k^l(z)$ of (2), converge in the norm of $W_{p'}^1(\overline{D}_*^{\pm})$, $p' > 2$, in arbitrary domains $D_*^{\pm}$ such that $\overline{D}_*^{\pm} \subset D^{\pm}$, $D^- = E \setminus D^+$. If we consider in addition the estimates (20), then all the foregoing enables us also to deduce the convergence in $C^{\beta_0}(\overline{D}_k^{\pm})$, $\beta_0(q_0) > 2$, of the functions

$$\Omega_k^l(z) = \Pi_k^l(z)\omega_k^l(z), \tag{24}$$

where

$$\Pi_k^l(z) = \Pi |z - t_j^*|^{\gamma_j^*}, \quad t_j^* = t_{jk}^l \in \Gamma_k, \gamma_j^* = \frac{p \operatorname{Re} \gamma_{jk}^l}{p - 2} + \delta, \delta > 0.$$

Summing up the foregoing, we arrive at a theorem.

THEOREM 4. *Suppose that for all k and l the conditions of Theorem 3 hold for the family of problems* (21), *along with conditions* (i) *and* (ii) *of* 4°, *and that the functions $G_k^l(t)$ and $g_k^l(t)$ converge for each fixed k in the sense of Definition 1 as $l = 1/\varepsilon \to \infty$.*

Then as $k, l \to \infty$ the solutions $\omega = \omega_k^l(z)$ of these problems converge in the norm of $W_{p'}^1(D_k^{\pm})$, $p' > 2$, $\overline{D}_*^{\pm} \subset D^{\pm}$, $D^- = E \setminus D^+$, where D^+ is the kernel of the multiply connected domains D_k^+. For each fixed k the functions $\Omega_k^l(z)$ defined by (24) *converge in the norm of $C^{\beta_0}(D_k^{\pm})$, $\beta_0(q_0) > 0$, as $l \to \infty$.*

7°. **The general problem of conjunction with a shift for elliptic systems of equations.**

THEOREM 5. *Suppose that the coefficients of* (1) *and* (2) *satisfy conditions* (3), (4), *and* (a)–(c) *in* 1°. *Moreover, suppose that the inequalities* (19) *hold.*

Then the boundary-value problem (1), (2) *has for* $\kappa \geq 0$ *at least one generalized solution* $w(z)$ *that satisfies* (8). *For* $\kappa < 0$ *this problem has a solution* $w = w(z)$, $w(\infty) = c_0$, *only if the* $|\kappa|$ *solvability conditions of the form* (9) *hold. Moreover, the functions* $w^{\pm}(z)$ *are integrable on* L, *and the estimates* (20) *hold for them.*

If, in addition,

$$|A(w_1, z) - A(w_2, z)| \leq N(z)|w_1 - w_2|, \qquad N(z) \in L_p(\tilde{K}), p > 2,$$

$$|\mu_i(w_1, z) - \mu_i(w_2, z)| \leq N_0|w_1 - w_2|, \qquad N = \text{const}, \qquad (25)$$

then the solution of problem (1), (2) *is unique when the conditions* (8) *hold.*

We suppose first that $\lambda(t) \equiv 0$ in the boundary condition (2), and, as in Theorem 1 in §2, we look for a solution of problem (1), (2) in the form

$$w(z) = \omega[\zeta(z)]e^{T_1(\varphi|z)} + T_1(\psi|z), \qquad (26)$$

$$\zeta(z) = \begin{cases} \zeta^+(z), & z \in D^+, \\ \zeta_k^-(z), & z \in D^-(L_k), k = 0,\ldots,m, \end{cases} \qquad (27)$$

where $\zeta = \zeta^+(z)$ is a homeomorphism of the domain D^+ onto the canonical domain D_ζ^+ with boundary $\Gamma = \Sigma_0^m \Gamma_k$ (see §6 of Chapter V), and the $\zeta = \zeta_k^-(z)$ are homeomorphisms of the domains $D^-(L_k)$ onto the domains $D_\zeta^-(\Gamma_k)$, $D_\zeta^- = E \setminus D_\zeta^+$. The piecewise analytic function $\omega(\zeta)$ in $D_\zeta^{\pm}$ must satisfy the boundary condition

$$\omega^+[\beta(\tau)] = G_*(\tau)\omega^-(\tau) + g_*(\tau), \qquad \tau \in \Gamma. \qquad (28)$$

Here

$$\beta(\tau) = \zeta^+[\alpha(t)], \qquad G_*(\tau) = G(t)\exp\{T_1(\varphi|t) - T_1[\varphi|\alpha(t)]\},$$

$$g_*(\tau) = g(\tau) - T_1[\psi|\alpha(t)] + G(t)T_1(\psi|t)\exp\{-T_1[\varphi|\alpha(t)]\},$$

and the correspondence $t = t(\tau)$ is determined on each contour Γ_k from the equality $\tau = \zeta_k^-(t), t \in L_k$.

It is easy to see that the functions $\beta(\tau)$, $G_*(\tau)$, and $g_*(\tau)$ satisfy conditions (a) and (b) in 1° if

$$\varphi(z), \psi(z) \in L_p(\overline{K}) \quad \text{and} \quad \zeta^{\pm}(z) \in W_p^1(\overline{D}^{\pm}), p > 2.$$

The condition that $w(z)$ in the form (26) is a generalized solution of (1) leads to the system of equations (20) in §2 for determining the functions $\varphi(z)$, $\psi(z)$, and $\zeta(z)$, where the last of these equations holds for each of the homeomorphisms $\zeta^+(z)$ and $\zeta_k^-(z)$ $(k = 0,\ldots,m)$ in its domain D^+ or $D^-(L_k)$. Introducing the set $\mathfrak{M}$ as in §2, we replace the equations for the homeomorphisms by the two equations for their complex characteristics $\{\mu_1^*(z), \mu_2^*(z)\} = \vec{m}^*(z)$

(see (2.23)). As a result we get the transformation

$$(\varphi, \psi, \vec{m}) = V(\varphi^*, \psi^*, \vec{m}^*)$$

determined by (23) for whose components $\varphi(z)$ and $\psi(z)$, as before, (24) and (25) of §2 remain true. Therefore, to prove that the transformation V has fixed points it suffices to prove that it is completely continuous.

In the boundary condition (28) let us substitute sequences of functions φ_k^*, ψ_k^*, and $\vec{m}_k^*$ that converge in the respective sets $\mathfrak{N}$ and $\mathfrak{M}$ of §2, along with sequences of homeomorphisms $\zeta = \zeta_n^{\pm}(\vec{m}_n^* | z)$ that converge in $W_p^1(D^{\pm})$, $p > 2$. Then, by conditions (3) and (4) and Theorem 2, the sequences $\omega_n^*(\zeta)$ converge in the norm of W_p^1 ($p > 2$) in arbitrary domains $\overline{Q}^{\pm} \subset D_{\zeta_0}^{\pm}$ ($D_{\zeta_0}^{\pm}$ is the kernel of the domains $D_{\zeta_n}^{\pm}$), and with them the functions $w_n^*(z)$ in (26) also converge in W_p^1 ($p > 2$) on the preimages $Q_z^{\pm}$ of these domains. As in the case of the Hilbert problem, this ensures that V is continuous. That it is compact follows, as in the case of the Hilbert problem, from the fact that to the sequences φ_k, ψ_k, and $\{z - \zeta_k(z)\}$, bounded in the respective spaces L_p and $W_p^1(D^{\pm})$, there corresponds a sequence of analytic functions $\omega_k(\zeta)$ that is relatively compact in the domains $\overline{Q}^{\pm} \subset D_{\zeta_0}^{\pm}$. Thus, we have proved the existence of at least one fixed point of V, and along with it a solution of problem (1), (2) for $\lambda \equiv 0$. The smoothness of solutions indicated in the theorem follows directly from the estimates (20) for solutions of jump problems and the estimates (24) and (25) for the functions $\varphi(z)$ and $\psi(z)$.

If the additional assumptions (25) hold, then, posing the problem for the difference $w(z) = w_1(z) - w_2(z)$ of two distinct solutions of problem (1), (2), we arrive in the usual way at the homogeneous boundary-value problem (28) ($g_* \equiv 0$) for the piecewise analytic function $\omega(\zeta)$. By Theorem 3, the latter problem has only the trivial solution $\omega(\zeta) \equiv 0$ under condition (8); therefore, also $w = w_1 - w_2 \equiv 0$, which proves the uniqueness asserted in the theorem. Thus, the theorem is completely proved in the case when $\lambda(t) \equiv 0$ in the boundary condition (2). To conclude the proof it remains to see that, by virtue of assumption (c) in 1°, the general boundary-value problem (1), (2) with $\lambda(t) \not\equiv 0$ can be reduced to the Haseman problem already examined with $\lambda(t) \equiv 0$. This is accomplished by introducing the function

$$v(z) = w(z), \quad z \in D^+; \qquad v(z) = w(z) + \lambda(z)\overline{w(z)}, \quad z \in D^-,$$

which obviously satisfies (2) with $\lambda \equiv 0$ and (1) with the coefficients $\tilde{\mu}_i$ and $\tilde{A}_i$. The latter coincide with μ_i and A_i in D^+ and can be expressed explicitly in terms of them in D^-. Since the properties of the coefficients of (1) are obviously invariant under the nonsingular (according to assumption (c)) affine transformation

$$w = \left(1 - |\lambda|^2\right)^{-1}(v - \lambda \bar{v})$$

of the unknown functions, it follows that $\tilde{\mu}_i$ and $\dot{A}_i$ satisfy the conditions (3) and (4) which hold for μ_i and A_i. The theorem is proved.

§4. The Poincaré problem with discontinuous coefficients in the boundary condition for a quasilinear second-order elliptic equation

1°. *The Poincaré problem.* Suppose that in a simply connected domain D it is required to find a generalized solution of the quasilinear second-order elliptic equation

$$\sum_{i,j=1}^{2} a_{ij} \frac{\partial^2 u}{\partial x_i \partial x_j} + \sum_{i=1}^{2} a_i \frac{\partial u}{\partial x_i} + a_0 u = g, \qquad x = (x_1, x_2) \in D, \qquad (1)$$

that satisfies the boundary condition

$$a \frac{\partial u}{\partial x_1} + b \frac{\partial u}{\partial x_2} + cu = d, \qquad a^2 + b^2 = 1, x \in \Gamma, \qquad (2)$$

where a, b, c, and d are given functions on the contour Γ that bounds D.

Under the assumption that $\Gamma \in C^{1+\beta}$, $\frac{1}{2} < \beta < 1$, this problem of Poincaré can be reduced by a conformal mapping to the analogous problem for the disk $|x| < 1$. Therefore, in what follows we assume without loss of generality that $D = \{|x| < 1\}$.

The following conditions are assumed with respect to the problem data:

(i) The coefficients a_{ij}, a_i, and g, which depend on the arguments x and U ($U = \{u, \partial u/\partial x_1, \partial u/\partial x_2\}$), are measurable in x for fixed U and continuous in U for almost all $x \in D$, and they satisfy for all U the inequalities

$$a_{11} a_{22} - \left(a_{12} + a_{21}\right)^2/4 \geqslant \Delta > 0, \qquad a_{11} > 0, |a_{ij}| \leqslant \Delta^{-1},$$

$$\left\{ \|a_i\|_{L_p(\bar{D})}, \|g\|_{L_p(\bar{D})} \right\} \leqslant M, \qquad M = \text{const}, p > 2. \qquad (3)$$

(ii) $\lambda = (a + ib) \in C^\beta$, $0 < \beta < 1$, for $x \in [x_k, x_{k+1}] \subset \Gamma$ $(k = 1, \ldots, n-1)$, with $\lambda(x_k - 0) \neq \lambda(x_k + 0)$, and $c(x)$ and $d(x)$ are representable in the form

$$c = c_0 \prod_1^{m_0} |x - x_k^*|^{-\alpha_k^*}, \qquad d = d_0 \prod_{m_0+1}^{m} |x - x_k^*|^{-\alpha_k^*} \quad (|x_k^*| = 1, 0 < \alpha_k^* < 1),$$

where c_0 and d_0 are bounded piecewise Hölder functions.

An analogous problem was investigated in [128] for the case when equation (1) is linear and the coefficients in the boundary condition (2) are Hölder continuous.

Let x_k^* denote those points of finite or infinite discontinuity of the functions λ, c, and d at which the first derivatives of the unknown solution $u(x)$ are unbounded, and let $W_p^2(D) \cap C_*^{1+\beta}(\overline{D})$ be the space of generalized solutions $u(x) \in W_p^1(D) \cap C^\beta(\overline{D})$ of (1) such that

$$u(x) \in W_p^2(\overline{D}'), \quad \overline{D}' \subset D, \quad \text{and} \quad u(x) \in C_*^{1+\beta}, \quad p > 2, \beta > 0,$$

everywhere except at the points x_k^*.

Introducing the complex-valued function

$$w(z) = \frac{\partial u}{\partial x_1} - i\frac{\partial u}{\partial x_2} \equiv 2\frac{\partial u}{\partial z}, \quad z = x_1 + ix_2,$$

we rewrite problem (1), (2) in the form

$$w_{\overline{z}} + \mu w_z + \overline{\mu}\overline{w}_{\overline{z}} + Aw + \overline{A}\,\overline{w} = -Bu + h \equiv h_*, \quad z \in D, \tag{4}$$

$$\operatorname{Re}[\lambda(t)w(t)] = -c(t)u + d \equiv d_*(t), \quad \lambda = a + ib, t \in \Gamma, \tag{5}$$

where

$$\mu = \frac{a_{11} - a_{22} + i(a_{12} + a_{21})}{2(a_{11} + a_{22})}; \quad A = \frac{a_1 + ia_2}{2(a_{11} + a_{22})};$$

$$B = \frac{a_0}{a_{11} + a_{22}}; \quad h = \frac{g}{a_{11} + a_{22}}.$$

Obviously, the coefficients μ, A, B, and h satisfy condition (i) as before, while the inequalities (3) take the form

$$2|\mu| \leqslant \mu_0(\Delta) < 1, \quad \{\|A\|_{L_p}, \|B\|_{L_p}, \|h\|_{L_p}\} \leqslant M\Delta^{-1}. \tag{6}$$

Suppose that in place of $u(x)$ we substitute an arbitrary real function $v(z) \in W_p^1(\overline{D})$ into the coefficients μ, A, $\overline{A}$, h_*, and d_*. Then we get the Hilbert boundary-value problem (4), (5) for the complex-valued function $w(z)$, and, by Theorem 1 in §2, this problem has for $\kappa \geqslant 0$ at least one generalized solution $w(z) \in W_q^1(D) \cap C_*^\beta(\overline{D})$, $q > 2$, $\beta > 0$, that satisfies the relations

$$w(z_k) = \operatorname{Im}w(1) = 0, \quad k = 0, \ldots, \kappa, \tag{7}$$

at the fixed points z_i ($|z_i| < 1$, $z_j \neq z_i$, $i \neq j$). For $\kappa < 0$ there exists at least one generalized solution $w(z) \in W_p^1(D) \cap C_*^\beta(\overline{D})$ only if the $2|\kappa| - 1$ solvability conditions hold. Moreover, for $w(e^{i\gamma})$ to be integrable it is necessary to require that (16) in §2 holds.

Consequently, for each finite κ the nonlinear operator

$$w = \Omega(v) \tag{8}$$

is defined on the set of functions $v(z) \in W_p^1(\bar{D})$, and, by Lemma 1 in §2,

$$\|\Omega(v)\|_{L_q(\bar{D})} \leqslant M_0 + (M_1\varepsilon_1 + M_2\varepsilon_2) \qquad \|v\|_{W_p^1(\bar{D})}, \qquad q(\mu_0) > 2, \quad (9)$$

$$\|\Omega(v)\|_{W_p^1(\bar{D'})} \leqslant M_3(\|v\|_{W_p^1}), \qquad 2 < p' < \infty, \quad (10)$$

where $\varepsilon_1 = \|B\|_{L_p(\bar{D})}$, $\varepsilon_2 = \|c\|_{C_*^\beta}$, the M_i ($i = 0, 1, 2$) are absolute constants, and D' is any subdomain of D ($\bar{D'} \subset D$). The operator Ω, which acts from $W_p^1(\bar{D})$ to $L_q(\bar{D})$ ($2 < p < q$), is continuous. Indeed, suppose that the sequence $\{v^n(z)\}$ converges in $W_p^1(\bar{D})$. Then, by conditions (i) and (ii), the corresponding sequence $\{\mu^n\}$ converges in measure, the corresponding sequences $\{A^n\}$ and $\{h_*^n\}$ converge in $L_p(\bar{D})$ ($p > 2$), and the corresponding sequence $\{d_*^n\}$ converges in C_*^β, all of which implies that $\Omega(v^n)$ converges in $L_q(\bar{D})$. Next, we remark that the following representation holds for any real function $u(z) \in W_q^1(\bar{D})$, $q > 2$:

$$u(z) = c_0 + \mathrm{Re}\left\{ -\frac{1}{\pi} \iint_D \left[\frac{\overline{w(\zeta)}}{\zeta - z} - \frac{zw(\zeta)}{1 - \bar{\zeta}z} \right] dD_\zeta \right\} \equiv c_0 + Pw, \quad (11)$$

where

$$c_0 = \frac{1}{2\pi} \int_0^{2\pi} u(e^{i\gamma}) \, d\gamma, \qquad w(z) = 2\frac{\partial u}{\partial z}.$$

Indeed, by the properties of T,

$$\partial Pw/\partial\bar{z} = 2w = \partial u/\partial z.$$

On the other hand, for any real function $v(z)$ satisfying the last equation we find that $(u - v)_z = 0$, and this implies that

$$u - v = \overline{\Phi(z)} = \Phi(z) = c_0 \qquad (\Phi_{\bar{z}} = 0).$$

In particular, (11) gives us that

$$\|u(z) - c_0\|_{W_p^1(\bar{D})} = \|Pw\|_{W_p^1(\bar{D})} \leqslant N\|w\|_{L_p(\bar{D})}. \quad (12)$$

Let us fix the constant c_0 arbitrarily. Then (8) and (11) determine a transformation

$$u = c_0 + P\Omega(v) \equiv H(v), \quad (13)$$

and a fixed point of it is obviously a solution of the original problem (1), (2). By construction, the operator H, which acts from $W_p^1(\bar{D})$ to $W_q^1(\bar{D})$, $2 < p < q$, is continuous, and, according to (9) and (12),

$$\|Hv\|_{W_q^1(\bar{D})} \leqslant |c_0| + N\left[M_0 + (M_1\varepsilon_1 + M_2\varepsilon_2)\|v\|_{W_p^1(\bar{D})} \right]. \quad (14)$$

Moreover, it follows from the properties of P and from (10) that $H(v) \in W_p^2(\bar{D'})$ in any interior subdomain $\bar{D'} \subset D$, and this implies that H is a compact transformation in $W_p^1(\overline{D})$. Suppose that the ε_1 and ε_2 in (9) are such that

$$N(M_1\varepsilon_1 + M_2\varepsilon_2) \leqslant \varepsilon_0 < 1. \tag{15}$$

Then the completely continuous transformation H maps the ball $K_R = \{\|u\|_{W_p^1} \leqslant R\}$ into itself for any R such that

$$(c_0 + NM_0)(1 - \varepsilon_0)^{-1} < R,$$

and, consequently, by Schauder's theorem, it has at least one fixed point in the ball $K_R \subset W_p^1$. We formulate this as a theorem.

THEOREM 1. *Suppose that conditions* (i), (ii), *and* (15) *and the inequality* (16) *in* §2 *hold. Then for* $\kappa \geqslant 0$ *the problem* (1), (2) *has at least one generalized solution* $u(x) \in W_p^2(D) \cap C_*^{1+\nu}$, $p > 2$, $\nu > 0$, *satisfying for fixed* z_i ($|z_i| < 1$; $z_i \neq z_j$ *for* $i \neq j$) *the conditions*

$$\frac{\partial u}{\partial z}(z_k) = \operatorname{Im} \frac{\partial u}{\partial z}(1) = 0, \quad k = 1,\ldots,\kappa, \qquad \int_0^{2\pi} u(e^{i\gamma})\, d\gamma = c_0. \tag{16}$$

For $\kappa < 0$ *this problem has a generalized solution satisfying the last of equalities* (16) *only if the* $2|\kappa| - 1$ *solvability conditions are valid.*

REMARK. If the coefficients of (1) and (2) do not depend explicitly on $u(x)$, then the ordinary Hilbert problem is obtained for $w = \partial u/\partial z$, and under the usual assumptions the uniqueness theorem (Theorem 2 in §2) holds for its solutions. An analogous theorem remains valid even when the coefficients in (1) and (2) depend weakly on $u(x)$.

2°. **Maximum principles.** We study some properties of smooth solutions of (1) when $a_0 = g \equiv 0$:

$$L_0(u) = \sum_{i,j} a_{ij} u_{x_i x_j} + \sum_i a_i u_{x_i} = 0, \tag{1*}$$

assuming without loss of generality that $a_{ij} = a_{ji}$.

LEMMA 1. *If the coefficients of the linear uniformly elliptic equation* (1*) *are continuous in the domain* D *and the function* $u(x) \in C^2(D) \cap C(\overline{D})$ *has a minimum* (*maximum*) *at the interior point* $x_0 \in D$, *then* $L_0(u) \geqslant 0$ (*respectively,* $L_0(u) \leqslant 0$) *at this point.*

At x_0 we have

$$u_{x_i} = 0, \qquad \sum_{i,k} u_{x_i x_k} \lambda_i \lambda_k \geqslant 0. \tag{17}$$

Equation (1*) is elliptic, so the quadratic form $\Sigma_{i,k} a_{ik}\lambda_i\lambda_k$ is positive definite and, consequently, can be represented in the form

$$\sum_{i,k} a_{ik}\lambda_i\lambda_k = \sum_i \left(\sum_s b_{is}\lambda_s\right)^2,$$

which is equivalent to the representation $a_{ik} = \Sigma_j b_{ji}b_{jk}$. Then condition (17) gives us that

$$L(u) = \sum_{i,k} a_{ik}u_{x_i x_k} = \sum_j \sum_{i,k} u_{x_i x_k}b_{ji}b_{jk} \geqslant 0, \qquad x = x_0.$$

LEMMA 2. *Under the conditions of Lemma 1 on the coefficients of L_0 suppose that the function $u(x) \in C^2(D) \cap C(\overline{D})$ satisfies the conditions*

$$L_0(u) \geqslant 0, \quad u(x) < u(x_0), x \in D: \qquad |z| < 1 \quad (z = x_1 + ix_2),$$

where $x_0 \in \Gamma = \partial D$ is a particular point. Then

$$\frac{\partial u}{\partial n}(x_0) = \sigma > 0, \tag{18}$$

where the derivative with respect to the outward normal is understood to be the lower limit of the expression $\Delta u/\Delta n$.

Suppose for definiteness that $x_0 = (1,0)$. Let us consider the domains K_0: $|z - \frac{1}{2}| < \frac{1}{2}$, $K_0 \subset D$, and K_1: $|z - 1| < \frac{1}{4}$, and introduce the auxiliary function

$$\omega(x) = e^{-\alpha\rho^2} - e^{-\alpha/4}, \qquad \rho = |z - \tfrac{1}{2}|,$$

which is positive in K_0 and equal to zero on its boundary Γ_0. For sufficiently large $\alpha > 0$ the expression

$$L(\omega) = e^{-\alpha\rho^2}\left[4\alpha^2 \sum_{i,k} a_{ik}\tilde{x}_i\tilde{x}_k - 2\alpha \sum_i (a_{ii} + a_i\tilde{x}_i)\right], \qquad \tilde{x}_i = x_i - \tfrac{1}{2},$$

can be made positive in $K^* = K_0 \cap K_1$, since the form $\Sigma_{i,k} a_{ik}\tilde{x}_i\tilde{x}_k$, is strictly positive here. Denote by Γ_0^* and Γ_1^* the parts of the boundaries of K_0 and K_1 that make up the boundary $\Gamma^* = \Gamma_0^* \cup \Gamma_1^*$ of K^*. By our assumptions, $u(x) < u(x_0)$, $x \in K^*$, and, consequently, for a small fixed $\varepsilon > 0$ we have

$$v(x) = u(x) + \varepsilon\omega(x) < u(x_0), \qquad x \in \Gamma_1^*.$$

Since $L(v) = L(u) + \varepsilon L(\omega) > 0$ for $x \in K^*$, Lemma 1 tells us that v cannot attain an absolute maximum inside K^*. But $v(x) < u(x_0)$ on Γ_1^* and $v(x) < u(x_0)$ on Γ_0^* except for the point x_0, at which $v(x_0) = u(x_0)$. Then at x_0

$$\frac{\partial v}{\partial n} = \frac{\partial u}{\partial n} + \varepsilon\frac{\partial\omega}{\partial n} \geqslant 0,$$

and so

$$\frac{\partial u}{\partial n}(x_0) \geq -\varepsilon \frac{\partial \omega}{\partial n}(x_0) = \varepsilon \alpha e^{-\alpha/4} > 0.$$

REMARK. In some cases the quantity $\sigma = \varepsilon \alpha e^{-\alpha/4}$ can be computed in a fairly efficient manner. Since the constant α depends only on the quantity Δ (the ellipticity constant) in (3), it suffices to indicate a way of choosing the constant ε. If, for example, the inequality

$$u(z) < (1-c)u(x_0), \qquad c = c(|z-x_0|) > 0,$$

holds at each point $z \in D$, then it is obviously sufficient to set $\varepsilon = c(1/4)u(x_0)$.

THEOREM 2 (the Hopf principle). *Each nonconstant solution $u(x) \in C^2(D) \cap C(\overline{D})$ of the uniformly elliptic equation* (1) *with coefficients continuous in D satisfies the inequalities*

$$\min u(s) < u(x) < \max u(s), \qquad x \in D.$$

Suppose that $u(x) \not\equiv \mathrm{const}$ has a maximum at an interior point $x_0 \in D$ and that the disk K_0 lies entirely in D and is such that $x_0 \in \Gamma_0 = \partial K_0$ and $u(x) < u(x_0)$, $x \in K_0$. By Lemma 2, $\partial u/\partial n > 0$ at this point, and this contradicts the equalities $u_{x_i} = 0$, which hold at an interior maximum point.

The next assertion is an immediate corollary of Lemma 2 and Theorem 2.

THEOREM 3 (the Zaremba principle). *Under the conditions of Theorem 2 the inequality* (18) *holds at a boundary point $x_0 \in \partial D$ that is an absolute maximum point.*

With the help of the Hopf and Zaremba principles it is easy to establish uniqueness theorems for solutions of the Dirichlet and Neumann boundary-value problems.

REMARK. If $u(x)$ is a harmonic function, then it is possible to determine a very efficient estimate of the constant in (18). Suppose that $u(z) \geq 0$ is a harmonic function in the upper half-plane D: $\mathrm{Im}\, z > 0$ that has an absolute minimum $u(0) = 0$ at the point $z = 0$. We consider in D the analytic function

$$\omega(z) = u_x - iu_y = \frac{1}{\pi i} \int_{-\infty}^{\infty} \frac{u_x(t)\, dt}{t-z},$$

and, taking into account the fact that $u_x(0) = 0$, we compute its value at $z = 0$:

$$\omega(0) = -iu_y(0) = \frac{1}{\pi i} \int_{-\infty}^{\infty} \frac{u_x(t)\, dt}{t} = -\frac{1}{\pi i} \int_{-\infty}^{\infty} \frac{u(t)}{t^2}\, dt.$$

Consequently,

$$\frac{\partial u}{\partial n}(0) = u_y(0) = \frac{1}{\pi} \int_{-\infty}^{\infty} \frac{u(t)}{t^2}\, dt \equiv \sigma > 0. \tag{19}$$

If $u(x) \geqslant \delta > 0$ on some interval of length λ on the boundary of D, then (19) gives us that $\sigma = \sigma(\delta, \lambda) > 0$.

§5. On certain potential and singular operators in spaces of summable functions

$1°$. We first prove that the formal derivative $S_0^{\Pi}f = \partial T_0^{\Pi}f/\partial z$ of the following operator is bounded in $L_p(\overline{K})$ (where $p > 2$ and $K = \{|z| < 1\}$):

$$T_0^{\Pi}f = -\frac{\Pi(z)}{\pi} \iint_K \left[\frac{f(t)}{\Pi(t)(t - z)} - \frac{\overline{f(t)}\, z}{\overline{\Pi(t)}\,(\bar{z}t - 1)} \right] dK_t, \qquad \frac{\partial T^{\Pi}f}{\partial \bar{z}} = f,$$

where

$$\Pi(z) = \prod_{k=1}^{n} (z - z_k)^{\alpha_k}, \qquad 0 < \alpha_k < 1, |z_k| \leqslant 1.$$

Writing $T^{\Pi}f = \Pi T(f/\Pi)$ and taking the fact that

$$\left(T_0^{\Pi} - T^{\Pi}\right)(f \mid z) = -\Pi(z)\, \overline{T\left(\frac{f}{\Pi} \,\Big|\, \frac{1}{\bar{z}}\right)}$$

into account, we prove that the operators $S^{\Pi}f = \partial T^{\Pi}f/dz$ are bounded in $L_p(\overline{K})$, and this obviously implies that the operators S_0^{Π} are bounded in L_p.

With the help of the identity

$$\frac{\partial \Pi}{\partial z} = \Pi(z) \sum_{k=1}^{n} \frac{\alpha_k}{z - z_k}$$

the operator $S^{\Pi}f$ can be represented in the form

$$S^{\Pi}f = \sum_{k=1}^{n} \alpha_k \tilde{S}_k^{\Pi}f + \tilde{S}^{\Pi}f\left(1 - \sum_{k=1}^{n} \alpha_k\right),$$

where

$$\Pi_k(z) = \Pi(z)(z - z_k)^{-1} \quad \text{and} \quad \tilde{S}_k^{\Pi}f = \Pi_k S\left(\frac{f}{\Pi_k}\right).$$

Thus, to prove that the operator $S^{\Pi}f$ (and with it also $S_0^{\Pi}f$) is bounded it suffices to establish the boundedness in L_p of the operators of the form $\tilde{S}^{\Pi}f = \Pi S(f/\Pi)$, with

$$\Pi(z) = \prod_{k=1}^{m} (z - z_k)^{\alpha_k} \prod_{k=m+1}^{n} (z - z_k)^{-\alpha_k},$$

where $0 < \alpha_k < 1$.

The fact that operators of the form $\tilde{S}^{\Pi}f$ are bounded in L_p can be obtained as a consequence of theorems of Gegelija [37] which generalize some known results of Calderón and Zygmund [22] to a space of functions that are summable with a weight. However, we present a very simple direct proof of this fact, due to S. N. Antoncev.

THEOREM 1. *The operator*

$$\tilde{S}^{\Pi}f = -\frac{\Pi(z)}{\pi}\iint_K \frac{f(t)\,dK_t}{\Pi_{(t)}(t-z)^2} \tag{1}$$

is linear and bounded in $L_p(K)$, *where*

$$\Pi(z) = \prod_{k=1}^m (z-z_k)^{\alpha_k}\prod_{k=m+1}^n (z-z_k)^{-\alpha_k}, \qquad |z_k|\le 1,$$

$$\frac{2}{2-\alpha_k} < p < \infty, \qquad k=1,\dots,m, \tag{2}$$

$$1 < p < \frac{2}{\alpha_k}, \qquad k=m+1,\dots,n.$$

We first set $\Pi(z) = (z-z_0)^{-\alpha}$; then

$$\tilde{S}^{\Pi}f = -\frac{1}{\pi(z-z_0)^\alpha}\iint_K \frac{f(t)(t-z_0)^\alpha}{(t-z)^2}\,dK_t$$

$$= -\frac{1}{\pi}\iint_K \frac{f(t)\,dK_t}{(t-z)^2} - \frac{1}{\pi(z-z_0)^\alpha}\iint_K f(t)\big[(t-z_0)^\alpha - (z-z_0)^\alpha\big]\frac{dK_t}{(t-z)^2}$$

$$\equiv Sf + \tilde{S}f.$$

It is obviously sufficient to estimate $\|\tilde{S}f\|_{L_p}$. The familiar inequality

$$|(t-z_0)^\alpha - (z-z_0)^\alpha| \le |t-z|^\alpha, \qquad 0 < \alpha < 1$$

(see §1 in Chapter I) and the Hölder inequality give us that

$$|\tilde{S}f|^p \le \pi^{-p}|z-z_0|^{-p\alpha}\left[\iint_K |f(t)|\,|t-z|^{-(2-\alpha)}|t-z_0|^{-\alpha/p+\alpha/p}\,dK_t\right]^p$$

$$\le \pi^{-p}|z-z_0|^{-p\alpha}\left[\iint_K |f(t)|^p|t-z|^{-(2-\alpha)}|t-z_0|^\alpha\,dK_t\right]^{p/p}$$

$$\times\left[\iint_K |t-z|^{-(2-\alpha)}|t-z_0|^{-\alpha q/p}\right]^{p/q}.$$

Applying (2.4) of Chapter V to the last integral when $\alpha q/p < 2$, we get

$$|\tilde{S}f|^p \le M_{\alpha,p}|z-z_0|^{-2\alpha}\iint_K |f|^p|t-z|^{-(2-\alpha)}|t-z_0|^\alpha\,dK_t,$$

and so

$$\left[\iint_K |\tilde{S}f|^p\,dK_z\right]^{1/p}$$

$$\le M_{\alpha,p}^{1/p}\left[\iint_K |f|^p|t-z_0|^\alpha\,dK_t\iint_K |z-z_0|^{-2\alpha}|t-z|^{-(2-\alpha)}\,dK_z\right]^{1/p}$$

$$\le \tilde{M}_{\alpha,p}\left[\iint_K |t-z|^{-2-2\alpha-(2-\alpha)}|f(t)|^p|t-z_0|^\alpha\,dK_t\right]^{1/p} = \tilde{M}_{\alpha,p}\|f\|_{L_p}.$$

Here the conditions $\beta < 2$ and $\delta < 2$ for inequality (2.4) in Chapter V to be applicable to the integrals $\iint_K |t - z|^{-\beta} |t - z_0|^{-\delta} dK_t$ hold in our case, because

$$p\alpha, 2\alpha, 2 - \alpha, \frac{\alpha q}{p} = \frac{\alpha}{p - 1} < 2.$$

In the case $\Pi(z) = (z - z_0)^\alpha$ we have

$$\tilde{S}^\Pi f = Sf - \frac{1}{\pi} \iint_K \frac{f(t)\left[(z - z_0)^\alpha - (t - z_0)^\alpha\right]}{(t - z_0)^\alpha (t - z)^2} dK_t \equiv Sf + \tilde{S}f.$$

We next let $\beta = ((2 - \alpha)p - 2)/p^2$ and perform an estimation analogous to the preceding one:

$$|\tilde{S}f|^p \leq \pi^{-p} \left[\iint_K |f(t)| \, |t - z_0|^{-\alpha + \beta - \beta} |t - z|^{-(2-\alpha)} dK_t \right]^p$$

$$\leq \pi^{-p} \left[\iint_K |f(t)|^p |t - z|^{-(2-\alpha)} |t - z_0|^{\beta p} dK_t \right]$$

$$\times \left[\iint_K |t - z|^{-(2-\alpha)} |t - z_0|^{-(\alpha+\beta)q} dK_t \right]^{p/q}$$

$$\leq M_{\alpha,p} \left[\iint_K |f(t)|^p |t - z|^{-(2-\alpha)} |t - z_0|^{\beta p} dK_t \right] |z - z_0|^{-2/q}.$$

Integrating, as previously, under the above integral sign and using (2.4) in Chapter V, we get

$$\|\tilde{S}f\|_{L_p} \leq \tilde{M}_{\alpha,p} \|f\|_{L_p}.$$

It is easy to see that the conditions

$$(\alpha + \beta)q = \alpha + \frac{2}{p} < 2, \qquad 0 < \beta p < 1 - \frac{\alpha}{p} - \frac{1}{p} < 2 \quad \text{and} \quad \frac{2}{q} < 2$$

for applicability of (2.4) in Chapter V follow from (2).

To establish that the operator $\tilde{S}^\Pi f$ is linear and bounded for any $\Pi(z)$ we break up the disk K into a union of domains $K_i, i = 1, \ldots, n + 1$, where for $1 \leq i \leq n$

$$K_i = \{|z - z_i| < \delta, |z| < 1\},$$

$$\text{with } \delta = \frac{1}{4} \inf_{i \neq j} |z_i - z_j|, \text{ and } K_{n+1} = K - \sum_{i=1}^n K_i.$$

It is clear that $K_i \cap K_j = \varnothing$ for $i \neq j$ $(i = 1, \ldots, n)$, and $|z - z_i| \geq \delta$ for $z \in K_{n+1}$. Then

$$\|\tilde{S}^\Pi f\|_{L_p} \leq \frac{1}{\pi} \sum_{i=1}^{n+1} \left[\sum_{j=1}^{n+1} \iint_{K_j} \left| \Pi(z) \iint_{K_i} \frac{f(t) \, dK_t}{\Pi(t)(t - z)^2} \right|^p dK_z \right]^{1/p} \equiv \frac{1}{\pi} \sum_{i=1}^{n+1} \left[\sum_{j=1}^{n+1} J_{ij} \right]^{1/p}.$$

Obviously, $|z - t| \neq 0$ here for $i \neq j$ $(i, j = 1, \ldots, n)$, and, consequently,

$$J_{ji} \leq \iint_{K_j} \left| \Pi(z) \iint_{K_i} \frac{f(t) \, dK_t}{\Pi(t)(t - z)^2} \right|^p dK_z$$

$$\leq \frac{1}{(2\delta)^{2p}} \iint_{K_j} |\Pi(z)|^p \left[\iint_{K_i} \frac{|f(t)|}{|\Pi(t)|} dK_t \right]^p dK_z$$

$$\leq \frac{1}{(2\delta)^{2p}} \iint_{K_j} |\Pi(z)|^p dK_z \left[\iint_{K_i} \frac{dK_t}{|\Pi(t)|^q} \right]^{p/q} \|f\|_{L_p}^p = \tilde{M}_{\alpha,p}^1 \|f\|_{L_p}^p,$$

where $|\Pi(z)|^p$ and $|\Pi(z)|^{-q}$ are summable by virtue of (2), so that $\tilde{M}^1_{\alpha,p} < \infty$. Since $0 < |\Pi(z)| < \infty$ for $i = j = n + 1$, the Calderón-Zygmund theorem (Theorem 7 in §2 of Chapter V) gives us that $J_{n+1,n+1} \leq \tilde{M}^2_{\alpha,p} \| f \|^p_{L_p}$. We have $\Pi(z) = \Pi_*(z)(z - z_i) \pm \alpha_i$, where $0 < |\Pi_*(z)| < \infty$, for $i = j = 1,\ldots, n$; the necessary estimate for J_{ii} follows from the estimates obtained above for the particular cases $\Pi(z) \equiv (z - z_0)^\alpha, (z - z_0)^{-\alpha}$. It remains to estimate the terms $J_{j,n+1}$ and $J_{n+1,i}$. If $\tilde{K}_j = \{|z - z_j| < \delta/2, |z| < 1\}$, then

$$
J_{j,n+1} = \iint_{\tilde{K}_j} \left| \Pi(z) \iint_{K_{n+1}} \frac{f(t)\, dK_t}{\Pi(t)(t - z)^2} \right|^p dK_z = \iint_{K_j - \tilde{K}_j} + \iint_{\tilde{K}_j} \leq \tilde{M}^3_{\alpha,p} \| f \|^p_{L_p},
$$

since $0 < |\Pi(z)| < \infty$ in the first integral, while $|z - t| \neq 0$ in the second one. Similarly, for $J_{n+1,i}$ we have

$$
J_{n+1,i} \leq \iint_{K_{n+1}} \left| \Pi(z) \iint_{K_i - \tilde{K}_i} \frac{f(t)\, dK_t}{\Pi(t)(t - z)^2} \right|^p dK_z + \iint_{K_{n+1}} \left| \Pi(z) \iint_{\tilde{K}_i} \frac{f(t)\, dK_t}{\Pi(t)(t - z)^2} \right|^p dK_z,
$$

whence $J_{n+1,i} \leq \tilde{M}^4_{\alpha,p} \| f \|^p_{L_p}$. The theorem is proved.

$2°$. Let us consider some operators more general than T^Π and S^Π:

$$
T^\Pi_k f = \frac{\Pi(z)}{\pi} \iint_K \left[\frac{f(t)}{\Pi(t)(t - z)} + \frac{z^{2k+1}\overline{f(t)}}{\overline{\Pi(t)}\,(1 - \bar{z}t)} \right] dK_t, \qquad k \geq 0, \tag{3}
$$

$$
T^\Pi_k f = -\frac{\Pi(z)}{\pi} \iint_K \left[\frac{f(t)}{\Pi(t)(t - z)} + \frac{\bar{t}^{-2k-1}\overline{f(t)}}{\overline{\Pi(t)}\,(1 - z\bar{t})} \right] dK_t, \qquad k < 0,
$$

$$
\Pi(z) = \prod_{i=1}^n (z - z_k)^{\alpha_k}, \qquad 0 < \alpha_k < 1, |z_k| \leq 1.
$$

In the case of the operators T^Π_k with $\Pi \equiv 1$ it is easy to see by a direct check that for $k > 0$ and for any function $f \in L_p, p > 2$, the equality

$$
\mathrm{Re}\left[z^{-k} T^1_k f \right] = 0
$$

holds on the circle $|z| = 1$, while for $k < 0$ the equality

$$
\mathrm{Re}\left[z^{|k|} T^1_k f \right] = 0
$$

holds only if the following conditions are satisfied:

$$
a_j(f) = -\frac{1}{\pi} \iint_K \left\{ f(t)t^{j-1} + \overline{f(t)}t^{2|k|-j-1} \right\} dK_t = 0, \qquad j = 1,\ldots,|k|. \tag{4}
$$

We differentiate $T^\Pi_k(f \mid z)$ with respect to z and, after separating out the completely continuous operators in the derivatives, we compute the norm of the remaining singular terms in L_2. In the case $k < 0$ the derivative of $T^\Pi_k(f \mid z)$ does not contain a completely continuous operator. Theorem 1 implies that the singular operators thus obtained are linear and bounded in $L_p, p > 2$. Let us first consider S^Π_k ($k \geq 0$). We introduce the following notation:

$$
A(z) = -\frac{\Pi(z)}{\pi} \iint_K \frac{f(t)}{\Pi(t)(t - z)} dK_t, \qquad B(z) = -\frac{\Pi(z)}{\pi} \iint_K \frac{\overline{f(t)}}{\overline{\Pi(t)}(1 - z\bar{t})} dK_t.
$$

Then $S^\Pi_k f$ can be rewritten as follows:

$$
S^\Pi_k f = \frac{\partial A}{\partial z} + z^{2k+1} \frac{\partial B}{\partial z} + (2k + 1)z^{2k} B(z).
$$

The operator $B(z)$ is completely continuous; therefore, we estimate the norm of the singular operator

$$\mathring{U}_\Pi f = \frac{\partial A}{\partial z} + z^{2k+1}\frac{\partial B}{\partial z}.$$

Suppose first that $f \in \mathring{C}^1(\overline{K})$, i.e., f is continuously differentiable, and $f \equiv 0$ for $1 - \varepsilon \leqslant |z| \leqslant 1$ (ε is dependent on f). Then

$$\left\| \mathring{U}_\Pi f \right\|^2 = \left(\mathring{U}_\Pi f, \overline{\mathring{U}\Pi f} \right)$$

$$= \left(\frac{\partial A}{\partial \zeta}, \frac{\overline{\partial A}}{\partial \zeta} \right) + \left(\zeta^{2n+1}\frac{\partial B}{\partial \zeta}, \frac{\overline{\partial A}}{\partial \zeta} \right) + \left(\zeta^{2n+1}\frac{\overline{\partial B}}{\partial \zeta}, \frac{\partial A}{\partial \zeta} \right)$$

$$+ \left(\zeta^{2n+1}\frac{\partial B}{\partial \zeta}, \bar{\zeta}^{2n+1}\frac{\overline{\partial B}}{\partial \zeta} \right)$$

$$=_1 + E_2 + E_3 + E_4.$$

Applying Green's formula to the integrals over the domain and taking account of the fact that $B(\zeta)$ is analytic in the disk $|\zeta| < 1$, we get

$$E_1 = \left(f, \bar{f} \right) + \frac{1}{2i}\int_\Gamma \overline{A}\frac{\partial A}{\partial \zeta}\,d\zeta + \frac{1}{2i}\int_\Gamma \overline{A}f\,d\zeta$$

$$= \| f \|^2_{L_2} + \frac{1}{2i}\int_\Gamma \overline{A}\frac{\partial A}{\partial \zeta}\,d\zeta = \| f \|^2_{L_2} + J_1,$$

$$\frac{1}{2i}\int_\Gamma \overline{A}f\,d\bar{\zeta} = 0, \qquad \Gamma = \{|\zeta| = 1\},$$

because $f = 0$ close to the boundary, and

$$E_2 = \overline{E}_3 = \frac{1}{2i}\int_\Gamma \overline{A}\zeta^{2n+1}\frac{\partial B}{\partial \zeta}\,d\zeta,$$

$$E_4 = \left(\zeta^{2n+1}\frac{\partial B}{\partial \zeta}, \bar{\zeta}^{2n+1}\frac{\overline{\partial B}}{\partial \zeta} \right) \leqslant \left(\frac{\partial B}{\partial \zeta}, \frac{\overline{\partial B}}{\partial \zeta} \right) = \frac{1}{2i}\int_\Gamma \overline{B}\frac{\partial B}{\partial \zeta}\,d\zeta = J_4.$$

It must be shown that $\operatorname{Re}(J_1 + E_2 + E_3 + E_4) \leqslant 0$, since

$$\operatorname{Im}(J_1 + E_2 + E_3 + J_4) = \operatorname{Im}(J_1 + J_4)$$

because J_1 and J_4 are real. We transform $J_1 + J_4$, taking the following relations

$$A = -\overline{\zeta}B\frac{\Pi}{\overline{\Pi}}, \qquad B = -\bar{\zeta}\frac{\Pi}{\overline{\Pi}}\overline{A}$$

on the boundary $|\zeta| = 1$ into account, as well as the fact that A and B are analytic near the boundary due to f being compactly supported:

$$J_1 + J_4 = \frac{1}{2i}\int_\Gamma \left[\overline{A}\frac{\partial A}{\partial \zeta} + \overline{B}\frac{\partial B}{\partial \zeta} \right]\partial\zeta = \frac{1}{2i}\int_\Gamma |A|^2\left\{ \frac{\partial \ln A}{\partial \zeta} + \frac{\partial \ln B}{\partial \zeta} \right\}d\zeta$$

$$= \frac{1}{2i}\int_\Gamma |A|^2\frac{d}{\partial\zeta}\ln AB\,d\zeta = \frac{1}{2}\int_0^{2\pi}|A|^2\frac{d\arg AB}{d\gamma}\,d\gamma$$

$$= \frac{1}{2}\int_0^{2\pi}|A|^2\frac{d}{d\gamma}\left(-\bar{\zeta}\frac{\Pi}{\overline{\Pi}} \right)d\gamma = \frac{1}{2}\int_0^{2\pi}|A|^2\left\{ 2\frac{d\arg\Pi}{d\gamma} - 1 \right\}d\gamma.$$

We compute $d\arg\Pi/d\gamma$:

$$\frac{d}{d\gamma}\ln\Pi = \sum_{k=1}^s \frac{\alpha_k}{\zeta - \zeta_k}\frac{d\zeta}{d\gamma} = \frac{d}{d\gamma}\ln|\Pi| + i\frac{d\arg\Pi}{d\gamma}.$$

Equating the imaginary parts, we get

$$2 \frac{d \arg \Pi}{d\gamma} = \sum_{k=1}^{s} \alpha_k.$$

Thus, a sufficient condition for $\mathrm{Re}(J_1 + J_4)$ to be nonpositive is that

$$\sum_{k=1}^{s} \alpha_k \leqslant 1. \tag{5}$$

Let us consider E_2 when $\Sigma_1^s \alpha_k = 1$:

$$\bar{E}_2 = \frac{1}{2i} \int_\Gamma A \zeta^{-(2n+3)} \frac{\overline{dB}}{\partial \zeta} \, d\zeta = \frac{1}{\pi} \iint_K \frac{f(t)}{\Pi(t)} \left\{ \frac{1}{2\pi i} \int_\Gamma \frac{\Pi(\zeta) \zeta^{-(2n+3)} \overline{\partial B} / \partial \zeta}{\zeta - t} \, d\zeta \right\} dK_t.$$

We extend $\partial B / \partial \zeta$ outside the disk according to the formula $B^*(\zeta) = \partial B(1/\zeta) / \partial \zeta$. Then $B^*(\zeta)$ is analytic outside the disk and bounded at infinity, and, finally, $B^*(\zeta)\Pi(\zeta)\zeta^{-(2n+3)}$ is a single-value analytic function outside the disk that vanishes at infinity.

Consequently, according to the Cauchy formula, $E_2 = \bar{E}_3 = 0$, which is what was to be proved. In the case $\Sigma_1^s \alpha_k = 1 - \varepsilon$, $\varepsilon > 0$, a straightforward estimate of the integral in E_2 gives us that $E_2 = 0$. We get an analogous estimate for $\| S_k^\Pi f \|_{L_2} = \| \partial T_k^\Pi f / \partial z \|_{L_2}$ when $k < 0$. As usual, suppose first that $f \in \mathring{C}^1$. Then the function $S_k^\Pi(f|z)$ is defined and continuous on the whole plane except for the points z_k, and for any disk K_R of radius $R \geqslant 1$ we have

$$\| S_k^\Pi f \|_{L_2(\bar{K})}^2 \leqslant \| S_k^\Pi f \|_{L_2(\bar{K}_R)}^2.$$

Applying Green's formula in the last integral, we get

$$\| S_k^\Pi f \|_{L_2(\bar{K}_R)}^2 = \| f \|_{L_2(\bar{K}_R)}^2 + \frac{1}{2\pi i} \iint_{\Gamma_R} \overline{T_k^\Pi f} \frac{\partial T_k^\Pi f}{\partial z} \, dz = \| f \|_{L_2(\bar{K})}^2 + J_R.$$

Taking the behavior of $T_k^\Pi(f|z)$ and its derivative for sufficiently large $|z|$ into account, we easily get the estimate

$$|J_R| \leqslant cR^2 \big(\Sigma_1^n \alpha_k - 1\big),$$

where the constant c does not depend on R. Thus, J_R tends to zero if $\Sigma_1^n \alpha_k < 1$, i.e.,

$$\| S_k^\Pi f \|_{L_2} \leqslant \| f \|_{L_2}.$$

Writing out J_R in the case $\Sigma \alpha_k = 1$ and using the Cauchy formula, we have $J_R = 0$. Thus, if (5) holds, then for $f \in \mathring{C}^1$

$$\| S_k^\Pi f \|_{L_2} \leqslant \| f \|_{L_2},$$

and, consequently, $\| S_k^\Pi \|_{L_2} \leqslant 1$, since the family of functions $f \in \mathring{C}^1$ is dense in L_2.

3°. Let us now consider the multicomponent operator T_κ defined in the disk $|z| < 1$ and carrying each vector $f = \{f_1, \ldots, f_n\}$ into a vector of the same dimension:

$$T_\kappa f = -\frac{\Pi(z)}{\pi} \iint_K \left[\frac{\Pi^{-1}(t)f(t)}{t-z} + G(z,t) \frac{\Pi^{-1}(t)f(t)}{1 - \bar{t}z} \right] dK_t,$$

where $G(z, t)$ is a diagonal matrix with elements $G_i(z, t)$,

$$G_i = z^{2\kappa_i + 1}, \, \kappa_i \geqslant 0, \quad \text{and} \quad Gi = \bar{t}^{2|\kappa_i| - 1}, \, \kappa_i < 0,$$

and the κ_i $(i = 1, \ldots, m)$ are fixed real numbers.

The square matrix $\Pi(z)$ is defined as follows:

$$\Pi(z) = \left\| \prod_{k=1}^{n_i} (z - z_k)^{\gamma_{kj}} \right\|,$$

where $0 < \mathrm{Re}\, \gamma_{ij} < 1$ $(i, j \leq m)$ and the n_i are integers $(n_i \leq m)$, and, moreover, $\det \Pi(z) \neq 0$ everywhere in K except for the points $z_k (k = 1,\ldots,n)$, $|z_k| = 1$, where it can have a zero of order less than 1. The operator T_κ was studied by Danilyuk in [28] and [29] for the case $\Pi(z) = J$ (J the identity matrix).

Let $Z^\kappa(z)$ be the diagonal matrix with elements $\{z^{\kappa_i}\}$. Then if all the κ_i $(i = 1,\ldots,m)$ are positive on the boundary $|z| = 1$ of the disk, we have the easily checked equality

$$\mathrm{Re}\left[Z^\kappa(z)\Pi^{-1}(z)T_\kappa f \right] = 0, \qquad |z| = 1. \tag{6}$$

for any vector $f(z) \in L_p$, $p > 2$. But if some of the κ_i are negative (say, κ_i, $i = p,\ldots,m$), then (6) holds only if the following $N = \Sigma_{i=p}^m (2|\kappa_i| - 1)$ conditions that:

$$a_{ji} = -\frac{1}{\pi} \iint_K \left\{ f_i^* t^{j-1} + \overline{f_i^* t^{2|\kappa_i|-j-1}} \right\} dK_i = 0, \tag{7}$$

where f_i^* is the ith component of the vector $f^*(t) = \{\Pi^{-1}(t)\}f(t)$. The conditions (7) arise similarly to the conditions (4) in the expansion of the corresponding components of the vector $\Pi^{-1}(z)Tf$. It follows directly from the construction of T_κ that $\partial T_\kappa f / \partial z \equiv f$ for $f \in L_p(\overline{K})$, $p > 2$.

Applying Theorem 1 to each component of the operator

$$S_\kappa f = \frac{\partial T_\kappa f}{\partial z} = S_\kappa^0 f - \frac{\Pi(z)}{\pi} \iint_K \frac{\partial G}{\partial z} \frac{\Pi^{-1}(t)f(t)}{(1 - \bar{t}z)} dK_t$$

we get that $S_\kappa f$ is linear and bounded in the space of vector-valued functions $f \in L_p(\overline{K})$, $p > 2$. Let us now compute

$$\| S_\kappa^0 f \|_{L_2}^2 = \left(S_\kappa^0 f, \overline{S_\kappa^0 f} \right) = \sum_{i=1}^m \| \rho_i \|_{L_2}^2,$$

where the ρ_i are the components of the vector $S_\kappa^0 f$. Observe that $S_\kappa^0 f = S_\kappa f$ for negative κ_i $(i = 1,\ldots,m)$. We consider this case first. Let $f \in \overset{\circ}{C}{}^1(\overline{K})$ and $K_R = \{|z| < R\}$. Then Green's formula gives us that

$$\left(S_\kappa f, \overline{S_\kappa f} \right)_K \leq \left(S_\kappa f, \overline{S_\kappa f} \right)_{K_R} = (f, \bar{f})_{K_R} + \frac{1}{2i} \int_{\Gamma_R} \overline{T_\kappa f} \frac{\partial T_\kappa f}{\partial z} dz = (f, \bar{f})_{K_R} + J_R,$$

where the expression (g, φ) is understood everywhere to be the inner product of the vectors g and φ, i.e., $(g, \varphi) = \Sigma_1^m g_i\varphi_i$. Consequently, if $\Sigma_{i=1}^k \mathrm{Re}\,\gamma_{ij} < 1$ for $k, j \leq n$, then $I_R \to 0$ as $R \to \infty$.

Thus, if $\kappa_i < 0$, then

$$\| S_\kappa \|_{L_2} \leq 1, \tag{8}$$

since the family of vector-valued functions $f \in \overset{\circ}{C}{}^1$ is dense in L_2. When some of the κ_i are positive, the inequality (8) is established similarly to the case of the operators T_k^Π (see the beginning of the section).

§6. Boundary-value problems for systems of $2m$ ($m \geq 1$) equations

$1°$. Let us consider the problem of finding a bounded vector solution $\omega(z) = \{\omega_1,\ldots,\omega_n\} \in W_p^1$, $p > 2$, of the system of equations

$$\omega_{\bar{z}} + \mu_1(z, \omega)\omega_z + \mu_2(z, \omega)\bar{\omega}_{\bar{z}} = 0 \tag{1}$$

under the boundary condition

$$\mathrm{Re}[(a + ib)\omega(t)] = h(t), \qquad t \in e^{i\gamma}, \gamma \in [0, 2\pi], \tag{2}$$

where μ_1 and $a + ib = \|(a + ib)\|$ are $m \times m$ matrices,

$$h(t) = \{h_1(t),\ldots,h_m(t)\}$$

is a vector, and $\omega_i(a + ib)_{ij}$, $h_i \in SW_{p_0}^1$, $p_0 > 2$, are complex-valued functions; furthermore, $\det(a + ib)$ does not vanish anywhere on Γ, not even at the points of discontinuity t_k $(k = 1,\ldots,p)$.

We also assume the condition

$$\max_i \sum_{j=1}^{m} |\mu_1^{ij}| + \max_i \sum_{j=1}^{m} |\mu_2^{ij}| \leq \mu_0 < 1. \tag{3}$$

The *total index* of the problem (1), (2) is defined to be

$$\kappa = \sum_{i=1}^{m} \kappa_i,$$

where the κ_i are the partial indices of the canonical system of solutions in the class of analytic functions that are bounded at all points of discontinuity. According to [129], the canonical matrix of solutions of the homogeneous problem in the given class can be represented as follows:

$$\Pi(z) = i \left\| \prod_{k=1}^{n_i} (z - t_k)^{\gamma_{ki}} \right\| Z(z),$$

where $Z(z)$ is the diagonal matrix $|z^{\kappa_i}|$, and condition (5.13) holds for the γ_{ki}.

The solution of problem (1), (2) will be sought in the form

$$\omega = T_\kappa f + \Phi(z),$$

where $\Phi(z) = \{\Phi_1, \ldots, \Phi_m\} \in W_p^1(\overline{K})$, $2 < p < p_0$, is an analytic vector which for $\kappa_i \geq 0$ satisfies the nonhomogeneous boundary condition (2) in the class of bounded functions, while (analogously to the case $m = 1$) for $\kappa_i < 0$ it satisfies (2) only if solvability conditions hold (cf. §2).

The operator T_κ has the form

$$T_\kappa f = -\frac{\Pi(z)}{\pi} \iint \left\{ \frac{\Pi(t)^{-1} f(t)}{t - z} + G(z, t) \frac{\left[\overline{\Pi(t)} \right]^{-1} f(t)}{1 - z\bar{t}} \right\} dD_t,$$

where $G(z, t)$ is a diagonal matrix with elements $G_i(z, t) = z^{2\kappa_i + 1}$ for $\kappa_i \geq 0$ and $G_i(z, t) = \bar{t}^{2|\kappa_i| - 1}$, for $\kappa_i < 0$, and $f = \{f_1, \ldots, f_m\}$. For all $\kappa_i \geq 0$ the vector $\omega(z) = T_\kappa f$ satisfies the homogeneous condition (2) ($h \equiv 0$) for any $f \in L_p$, while for $\kappa_i < 0$ ($i = p, \ldots, m$) it satisfies this condition only if the following $N = \Sigma_p^m (2|\kappa_i| - 1)$ conditions hold:

$$c_i^j(f) = -\frac{1}{\pi} \iint_D \left[f_i^* t^{j-1} + \bar{f}_i^* \bar{t}^{2|\kappa_i| - j - 1} \right] dD_t = 0 \qquad (i, j = 1, \ldots, 2|\kappa_i| - 1), \tag{4}$$

where the f_i^* are the components of the vector $f^*(t) = [\Pi(t)]^{-1} f(t)$. According to §5,

$$\frac{\partial}{\partial \bar{z}} (T_\kappa f) = f, \qquad \|S_\kappa^0 f\|_{L_2} \leq \|f\|_{L_2},$$

where

$$\|f\|_{L_2}^2 = (f, f) = \|f_1\|_{L_2}^2 + \cdots + \|f_m\|_{L_2}^2, \qquad \omega(z) = T_\kappa f \in C^\alpha(\overline{D}),$$

for $f \in L_p, p > 2$.

Substituting $\omega(z) = T_\kappa f + \Phi(z)$ in (1), we get

$$f + \mu_1 S_\kappa f + \overline{\mu_2 S_\kappa f} + \mu_1 \frac{d\Phi}{dz} + \mu_2 \frac{d\overline{\Phi}}{d\bar{z}} = 0, \tag{5}$$

and

$$\|\mu_1 S_\kappa^0 f + \mu_2 \overline{S_\kappa^0 f}\|_{L_p} \leq \mu_0 \Lambda_p^\kappa \|f\|_{L_p}, \tag{6}$$

where $\mu_0 \Lambda_p^\kappa < 1$ for p sufficiently closed to 2.

The estimate

$$\|f\|_{L_p} \leq \left\| \frac{d\Phi}{dz} \right\|_{L_p} \left(1 - \mu_0 \Lambda_p^\kappa\right)^{-1} \tag{7}$$

for any solution of (5) follows immediately from (6) and (3) in the case when $\kappa_i \leqslant 0$. A similar estimate holds also in the case when some of the κ_i are positive. By (6), we can rewrite (5) in the equivalent form

$$f = Af = -\left(J + \mu_1 S_\kappa^0 + \mu_2 \overline{S}_\kappa^0\right)^{-1}\left(\mu_1 \frac{d\Phi}{dz} + \mu_2 \frac{d\overline{\Phi}}{d\overline{z}} + \mu_1 T_\kappa^0 f + \mu_2 \overline{T_\kappa^0 f}\right).$$

The a priori estimate (7) and the complete continuity of the operator A gives us

THEOREM 1. *Under the condition* (5.13), *equation* (5) *has at least one solution, and to it there corresponds the solution* $w(z) = T_\kappa f + \Phi(z)$ *of problem* (1), (2) *when* $\kappa_i \geqslant 0$ $(i = 1, \ldots, m)$. *If* $\kappa_i < 0$ $(i = 1, \ldots, m)$, *then this problem is solvable only under the conditions* (4).

Moreover, in the case $\kappa_i \geqslant 0$ $(i = p, \ldots, m)$ *the solution contains as many arbitrary constants as the solution of the nonhomogeneous problem* (2) *for an analytic vector.*

2°. For a quasilinear elliptic system of $2m$ $(m \geqslant 1)$ equations the Riemann problem can be formulated as follows: Find a vector solution $\omega^\pm = (\omega_1^\pm, \ldots, \omega_m^\pm)$ defined on the whole plane $D^+ + \Gamma + D^-$ for the system of equations

$$L^\pm \omega^\pm = \omega_{\overline{z}}^\pm + q_1^\pm(z, \omega^\pm)\omega_z^\pm + q_2^\pm(z, \omega^\pm)\overline{\omega}_{\overline{z}}^\pm = 0 \qquad (8)$$

in the respective domains D^+ and D^-, with the boundary values on Γ satisfying the relation

$$\omega^+(t) = (a + ib)\omega^-(t) + h(t). \qquad (9)$$

Let us consider the m-component vector $u^* = u^*(z)$ defined for $z \in D^+$ by

$$u^*(z) = u^-(1/\overline{z}).$$

Then for the $(2m)$-component vector $W = \{u_1^+, \ldots, u_m^+, u_1^*, \ldots, u_m^*\}$, defined for $z \in D^+$ we arrive at the following Hilbert problem:

$$W_{\overline{z}} + q_1(z, W)W_z + q_2(z, W)\overline{W}_{\overline{z}} = 0,$$
$$\operatorname{Re}\left[(\tilde{a} + i\tilde{b})\right]W(t) = l(t), \qquad t = e^{i\gamma}, \gamma \in [0, 2\pi], \qquad (10)$$

where the $2m \times 2m$ block-diagonal matrices q_i and $\tilde{a} + i\tilde{b}$ have the form

$$q_i(z, W) = \left\|\begin{array}{c} q_i^+(z, u^+), 0 \\ \hline 0, q_i^-[1/z, u^-(1/z)] \end{array}\right\|, \qquad a + ib = \left\|\begin{array}{c} E, -(a - ib) \\ iE, i(a - ib) \end{array}\right\|,$$

and the vector $l(t) = \{\operatorname{Re} h(t), \operatorname{Im} h(t)\}$.

It is not hard to see that the signs of the partial indices of problem (10) for $m = 1$ coincide with the sign of the total index κ, which is equal to $\operatorname{ind}\{\det(\tilde{a} - i\tilde{b})\}$. With the determination of a solution of (10) the vector

$$u(z) = \begin{cases} u^+(z), & z \in D^+ \\ u^*(1/z), & z \in D^- \end{cases}$$

satisfies the original problem (8), (9).

THEOREM 2. *For* $\kappa_i \geqslant 0$ *a solution of problem* (8), (9) *exists and contains as many arbitrary constants as the corresponding analytic vector; for* $\kappa_i < 0$ *it is solvable only if the solvability conditions* (4) *hold.*

§7. Problems

We formulate several unsolved problem which mainly arise in applications and are of independent interest.

1°. Study solvability for boundary-value problems in the case of degenerate linear and quasilinear equations.

Some existence theorems will be proved in §§2 and 3 of Chapter VII for a special type of degenerate equations.

$2°$. In connection with the preceding problem it is of interest to study the solvability of the Hilbert problem for the particular case of degenerate quasilinear equations

$$W_{\bar{z}} - \mu_1(W, z)W_z - \mu_2(W, z)\overline{W_z} = \frac{A(W, z)}{z^s},$$

where $|\mu_1| + |\mu_2| \leqslant \mu_0 < 1$, and the function $A(W, z)$ is summable in Z to the power $p > 2$, or

$$|A(W, z)| \leqslant A_0(z)|W|^\beta \quad \text{and} \quad A_0(z) \in L_p(\overline{K}), p > 2, \beta > 0.$$

The problem has been investigated by Mihaĭlov [77] for the case when $\mu_1 = \mu_2 = 0$ and $A(W, z)z^{-s} \equiv a(z)Wz^{-1} + b(z)\overline{W}z^{-1}$.

$3°$. Consider the boundary-value problems

$$\text{Re}[G(t, \lambda)W_\lambda(t)] = g(t, \lambda), \qquad t = e^{i\gamma}, \tag{1}$$

for functions $W_\lambda(z)$ analytic in the disk $|z| < 1$ under the assumption that $G(t, \lambda), g(t, \lambda) \in C^\alpha$ for $\lambda \in [0, 1]$ and that the index $\kappa(\lambda) = \text{ind } \overline{G(t, \lambda)}$ of the problem changes value when λ passes through some $\lambda_0 \in [0, 1]$. Suppose that $G(t, \lambda_0 \pm \varepsilon) \neq 0$ for every t, $|t| = 1$, when $\varepsilon > 0$ is sufficiently small, and that the problem has bounded solutions for all λ, $0 \neq |\lambda - \lambda_0| < \varepsilon$. Study the question of stability for solutions of the problem (1) as a function of the change in the index $\kappa(\lambda)$; in particular, determine conditions under which there exist solutions $W_{\lambda_1}(z)$ and $W_{\lambda_2}(z)$ that are close in C^α, where $\lambda_1 < \lambda_0 < \lambda_2$, and $\lambda_2 - \lambda_1$ is arbitrarily small. In a way similar to §2, extend the stability results obtained for analytic functions in problem (1) to quasilinear equations and discontinuous functions $G(t, \lambda)$ and $g(t, \lambda)$.

$4°$. Study the solvability of boundary-value problems for quasilinear systems of equations with derivatives in the boundary condition:

$$\text{Re}\left[a(t)\frac{\partial w}{\partial t} + b(t)w\right] = g(t), \qquad t = e^{i\gamma},$$

where

$$a(t), b(t)uq^*(t) = g(t)\Pi(t - t_k)^{\alpha_k}, \qquad 0 \leqslant \alpha_k < 1,$$

are complex-valued piecewise Hölder functions. In the case when the functions $a(t)$, $b(t)$, and $g(t)$ are continuous the most complete results have been obtained by Daniljuk [28], [29].

$5°$. Investigate the conjunction problem with a shift $\alpha(t) \in SQ$ for quasilinear systems when the coefficients of the boundary condition are Hölder functions and do not belong to SW_p^1, $p > 2$. This problem involves essential difficulties even for analytic functions, since, after reduction of it to a Riemann problem on another contour, the contour turns out not to be a rectifiable curve in the general case. This fact prevents us from applying the well-developed apparatus of Cauchy type integrals directly to this problem.

$6°$. Study the conjunction problem with shifts

$$\vec{\Phi}^+[\vec{a}(t)] = G(t)\vec{\Phi}^-(t) + \vec{g}(t), \qquad t = e^{i\gamma}, \gamma \in [0, 2\pi],$$

for a system of analytic functions $\vec{\Phi}(z) = \{\Phi_1(z), \ldots, \Phi_m(z)\}$, where

$$\vec{\Phi}(\vec{a}) = \{\Phi_1[a_1(t)], \ldots, \Phi_m[a_m(t)]\}, \qquad a_i(t) \in SQ.$$

For example, the usual conjunction problem for a system of equations

$$\vec{w}_{\bar{z}} - \mu(\vec{w}, z)\vec{w}_z = 0$$

with a diagonal matrix $\mu(\vec{w}, z)$ can be reduced to such a problem by quasiconformal transformations of the unit disk $|z| < 1$ onto itself.

$7°$. Study boundary-value problems with continuous and discontinuous coefficients in the boundary conditions for quasilinear systems of equations on Riemann surfaces.

CHAPTER VII

BOUNDARY-VALUE PROBLEMS IN HYDRODYNAMICS
AND SUBSONIC GAS DYNAMICS

§1. The equations of gas dynamics and their transformation
to different variables

1°. Let us first consider a steady irrotational (potential) flow of a nonviscous compressible fluid in the plane of $z = x + iy$. Let ρ be its density, u and v the projections of the velocity vector on the OX- and OY-axis, $q = (u^2 + v^2)^{1/2}$ the magnitude of the velocity, and p the pressure. Then the equation for the absence of vortices and the equation of continuity are written in the respective forms

$$\frac{\partial v}{\partial x} - \frac{\partial u}{\partial y} = 0, \qquad \frac{\partial(\rho u)}{\partial x} + \frac{\partial(\rho v)}{\partial y} = 0. \tag{1}$$

The fluid is assumed to be barotropic, i.e. for the whole flow there is a uniquely determined relation between the pressure and the density

$$p = p(\rho), \tag{2}$$

where $p(\rho)$ is a smooth increasing function. For example, in the case when the fluid moves adiabatically we have $p = p_0 \rho^\gamma$, where $\gamma > 1$; for air $\gamma \approx 1.4$. The Bernoulli integral, which for a suitable choice of measurement units can be given the form

$$\frac{q^2}{2} + \int_1^\rho \frac{1}{\rho} \frac{dp}{d\rho} d\rho = 0, \tag{3}$$

where $dp/d\rho$ is found from (2), determines a relation

$$\rho = \rho(q). \tag{4}$$

In particular, for an adiabatic motion

$$\rho = \left(1 - \frac{\gamma - 1}{2} q^2\right)^{1/(\gamma - 1)}.$$

We introduce a potential $\varphi(x, y)$ and a stream function $\psi(x, y)$ by setting $\varphi_x = u$, $\varphi_y = v$ and $\psi_x = -\rho v$, $\psi_y = \rho u$. Then the following equations are obtained for the velocity potential and the stream function:

$$\rho(q)\varphi_x = \psi_y, \qquad \rho(q)\varphi_y = -\psi_x, \tag{5}$$

where $q^2 = \varphi_x^2 + \varphi_y^2$. This system is the basic one in our subsequent investigations. If

$$1 - M^2 = \frac{1}{\rho}\frac{d}{dq}(q\rho) > 0$$

in the flow domain, i.e., if the Mach number M is less than 1, then the flow is said to be subsonic, while if $M > 1$, it is supersonic, if $M < 1$ in part of the flow domain and $M > 1$ in the other part of it, then the flow is called *transonic* or *mixed*. It is easy to see that the system (5) is elliptic, hyperbolic, and mixed, respectively, in these cases. A value q_{cr} of the speed q at which the Mach number is equal to 1 is called a critical speed. We shall consider uniformly subsonic flows, for which $q \leq q_{cr} - \varepsilon$, $\varepsilon > 0$, in the whole flow domain, i.e.,

$$1 - M^2 = \frac{1}{\rho}\frac{d}{dq}(q\rho) \geq a > 0. \tag{6}$$

Since condition (6) can be verified only after finding a solution, in theoretical investigations and in numerical calculations of subsonic flows we introduce the fictitious density

$$\rho_\varepsilon(q) = \begin{cases} \rho(q) & \text{for } q \leq q_{cr} - \varepsilon, \\ \tilde{\rho}(q) & \text{for } q > q_{cr} - \varepsilon, q \leq q_{cr}, \\ 1 & \text{for } q \geq q_{cr}, \end{cases}$$

where $\tilde{\rho}(q) > 0$ extends $\rho(q)$ with preservation of smoothness in such a way that $\rho_\varepsilon(q)$ satisfies (6) for all $q \geq 0$, and $\rho_\varepsilon(q_{cr}) = \tilde{\rho}(q_{cr}) = 1$. Flows with the density $\rho_\varepsilon(q)$ are commonly called constantly subsonic flows. After constructing a constantly subsonic flow with $q_m = \max q < \infty$, it is possible to choose the flow parameters (the magnitude q_∞ of the velocity at infinity will play the role of one such parameter in what follows) in such a way that in the flow domain the inequality $q_m \leq q_{cr} - \varepsilon$ holds and thereby the fictitious density $\rho_\varepsilon(q)$ coincides with the flow density $\rho(q)$. Without changing the notation we assume that $\rho(q)$ satisfies (6) for $q \geq 0$ and that $\rho(q) \in C^{l+\alpha}$, $\alpha > 0$, $l \geq 1$, when we consider constantly subsonic flows. To the case $\rho \equiv 1$, $M \equiv 0$ corresponds a flow of an ideal fluid, described by the Cauchy-Riemann system

$$\varphi_x = \psi_y, \qquad \varphi_y = -\psi_x.$$

Let us perform certain transformations of the equations of gas dynamics (5). If $\theta = \arctan(\varphi_y/\varphi_x)$ is the angle between the flow velocity and the OX-axis, then $\varphi_x = u = q\cos\theta$ and $\varphi_y = v = q\sin\theta$. With the aid of (5) we determine the derivatives of the inverse functions $x = x(\varphi, \psi)$ and $y = y(\varphi, \psi)$:

$$\frac{\partial x}{\partial \varphi} = \frac{u}{q^2} = \frac{\cos\theta}{q}, \qquad \frac{\partial x}{\partial \psi} = -\frac{v}{\rho q^2} = -\frac{\sin\theta}{\rho q},$$

$$\frac{\partial y}{\partial \varphi} = \frac{v}{q^2} = \frac{\sin\theta}{q}, \qquad \frac{\partial y}{\partial \psi} = \frac{u}{\rho q^2} = \frac{\cos\theta}{\rho q}. \qquad (7)$$

Then the equality of the mixed second derivatives of the functions $x(\varphi, \psi)$ and $y(\varphi, \psi)$ gives us the equations

$$\left(\frac{\cos\theta}{q}\right)_\psi = -\left(\frac{\sin\theta}{\rho q}\right)_\varphi, \qquad \left(\frac{\sin\theta}{q}\right)_\psi = \left(\frac{\cos\theta}{\rho q}\right)_\varphi, \qquad (7^*)$$

which lead to the equations for the inverse functions

$$\varphi_\theta = \frac{q}{\rho}\psi_q, \qquad \varphi_q = -\frac{1 - M^2}{\rho q}\psi_\theta. \qquad (8)$$

We introduce a fictitious speed and a fictitious density by the formulas

$$q^* = \int_1^q \frac{\sqrt{1 - M^2(q)}}{q}\,dq, \qquad \rho^*(q^*) = \frac{\rho(q)}{\sqrt{1 - M^2(q)}} \qquad (9)$$

($q^*(q)$ has a logarithmic singularity for $q = 0$ and $q = \infty$) and write (8) and the corresponding system for the inverse functions in the form

$$\rho^*\varphi_0 = \psi_{q^*}, \qquad -\rho^*\varphi_{q^*} = \psi_\theta, \qquad (10)$$

$$\rho^*\frac{\partial q^*}{\partial \varphi} = \frac{\partial \theta}{\partial \varphi}, \qquad \rho^*\frac{\partial \theta}{\partial \psi} = -\frac{\partial q^*}{\partial \varphi}. \qquad (11)$$

The system (11) can be written in complex form for the generalized Joukowsky function $\omega = q^* - i\theta$:

$$\omega_{\bar{w}} - \mu^w(\omega)\omega_w = 0, \qquad (12)$$

where $w = \varphi + i\psi$ and $\mu^w(\omega) = (\rho^* - 1)/(\rho^* + 1)$. Note that $\omega = \ln q - i\theta \equiv \ln(dw/dz)$ in the case of an incompressible fluid, i.e., when $M \equiv 0$ and $\rho \equiv 1$. From (5) we find an equation for the inverse functions

$$z_{\bar{w}} - \frac{\rho - 1}{\rho + 1}z_w = 0. \qquad (13)$$

Substituting the expressions

$$\omega_{\overline{w}} = \omega_{\overline{z}} \cdot \overline{z}_{\overline{w}} + \omega_z \frac{\rho - 1}{\rho + 1} z_2, \qquad \omega_w = \omega_{\overline{z}} \frac{\rho - 1}{\rho + 1} \overline{z}_{\overline{w}} + \omega_z z_w$$

in (12), we get

$$\omega_{\overline{z}} - \mu^z(\omega)\omega_z = 0, \qquad (14)$$

where

$$\mu^z(\omega) = e^{2i\theta} \frac{1 - \sqrt{1 - M^2}}{1 + \sqrt{1 - M^2}}, \qquad \theta = \arctan \frac{\varphi_v}{\varphi_x}.$$

Let us consider $\theta(x, \psi) \equiv \theta[x(\varphi, \psi), \psi]$ and $q^*(x, \psi) \equiv q^*[x(\varphi, \psi), \psi]$, and denote by θ_x and θ_y the derivatives of $\theta(x, \psi)$ and by $\partial\theta/\partial\varphi$ and $\partial\theta/\partial\psi$ the derivatives of $\alpha[x(\varphi, \psi), \psi]$ (and similarly for q^*). Then (7) gives us that

$$\frac{\partial\theta}{\partial\varphi} = \theta_x \frac{\partial x}{\partial\varphi} = \theta_x \frac{u}{q^2}, \qquad \frac{\partial q^*}{\partial\varphi} = q_x^* \frac{u}{q^2},$$

$$\frac{\partial\theta}{\partial\psi} = \theta_y + \theta_x \frac{\partial x}{\partial\psi} = \theta_y - \theta_x \frac{v}{\rho q^2}, \qquad \frac{\partial q^*}{\partial\psi} = q_\psi^* - q_\varphi^* \frac{v}{\rho q^2}.$$

We substitute the last formulas into (11) and write the resulting system in the complex form

$$\omega_{\overline{\tau}} - \tilde{\mu}(\omega)\omega_\tau = 0, \qquad (15)$$

where

$$\tau = x + i\psi \quad \text{and} \quad \tilde{\mu}(\omega) = \frac{\rho q^2 - u\sqrt{1 - M^2} + iv}{\rho q^2 + u\sqrt{1 - M^2} - iv}.$$

Since the flow is constantly subsonic (condition (6)), equations (12) and (14) are uniformly elliptic, i.e.,

$$|\mu^z(\omega)|, |\mu^w(\omega)| \leqslant \hat{\mu}_0(a) < 1. \qquad (16)$$

But the last equation (15) is elliptic only on solutions $\omega(z)$ for which $|q^*| < N < \infty$ and $|\theta| < \pi/2 - \delta$ (i.e., when $q \neq 0, \infty$ and $|\theta| \neq \pi/2$) in the flow domain, and this implies that

$$|\tilde{\mu}(\omega)| \leqslant \mu_0(N, \delta) < 1. \qquad (17)$$

In the case of an incompressible fluid, we have $\rho \equiv 1$, $M \equiv 0$ and $\omega = \ln q - i\theta$; therefore,

$$\tilde{\mu}(\omega) = \frac{q^2 - qe^{-i\theta}}{q^2 + qe^{-i\theta}} = \frac{qe^{i\theta} - 1}{qe^{i\theta} + 1} = \frac{e^{\overline{\omega}} - 1}{e^{\overline{\omega}} + 1}.$$

It is easy to see that the transformation $z = z(\tau)$, $\tau = x + i\psi$, is quasiconformal under the conditions (17). Indeed, according to (15) and (17), the mapping $\omega = \omega(\tau)$ is quasiconformal, while the mapping $\omega = \omega(z)$ is quasiconformal because it is a solution of (14); therefore, $z = z[\omega(\tau)]$ is also quasiconformal.

Let us transform (11) by passing to the variable $\zeta = s + i\psi$, where s is the arclength parameter of the point $w = \varphi$ on the streamline $\psi = 0$. Taking the fact that $d\varphi/ds = q(s, 0)$ into account, we get that

$$\omega_{\bar{\zeta}} - \frac{q(s,0)\rho^* - 1}{q(s,0)\rho^* + 1}\omega_{\zeta} = 0. \tag{18}$$

Along with ω we consider the function

$$\Omega = e^{\omega} \equiv eq^{*-i\theta}, \tag{19}$$

which obviously satisfies the same equations (12), (14), and (15) as ω, where the coefficients of these equations can be given the form $\mu^w(\ln \Omega)$, $\mu^z(\ln \Omega)$ and $\bar{\mu}(\ln \Omega)$. Representing $q^* = \int_1^q q^{-1}\sqrt{1 - M^2}\, dq$ in the form $q^* \equiv \ln q + \sigma(q)$, we arrive at the following representation for $\Omega = e^{\omega}$:

$$\Omega = qe^{-i\theta + \sigma(q)} \equiv (u - iv)e^{\sigma(q)}, \tag{20}$$

where

$$\int_1^q q^{-1}\left(\sqrt{1 - M^2} - 1\right) dq \equiv \sigma(q) \in C^{l-1+\alpha}, \qquad \alpha > 0, l \geqslant 1,$$

for $0 \leqslant q \leqslant \infty$. Indeed, by the assumption that $\rho(q) \in C^{1+\alpha}$ and $M(0) = M(\infty) = 0$ $(1 - M^2 = \rho^{-1}d(q\rho)/dq)$, the mean value theorem can be applied to the integrand in $\sigma(q)$:

$$\sqrt{1 - M^2} - 1 = -\frac{M^2(q)}{2}\left[1 - M^2(sq)\right]^{-1/2} \equiv g(q)q^{2\alpha}, \qquad 0 \leqslant s \leqslant 1,$$

and this implies that $\sigma(q) \in C^{\alpha} \in$ for finite $q \geqslant 0$. Similarly, for $q \to \infty$ the representation

$$\sqrt{1 - M^2} - 1 \equiv g(1/Q)Q^{-2\alpha}$$

implies that $\sigma(q) \equiv \sigma(1/Q) \in C^{\alpha}$ in a neighborhood of $Q = 0$. Obviously, $\sigma(q) \in C^{l-1+\alpha}$ for all $q \in [0, \infty]$ if $\rho(q) \in C^{l+\alpha}, l > 1$. Note also that $\sigma(q) \equiv 0$ in the case of an incompressible fluid.

2°. In the case of rotational flows of a compressible fluid the components of the complex velocity $W = u - iv = qe^{-i\theta}$ satisfy the system

$$\frac{\partial v}{\partial x} - \frac{\partial u}{\partial y} = \rho A(\psi), \qquad \frac{\partial(\rho u)}{\partial x} + \frac{\partial(\rho v)}{\partial y} = 0, \tag{21}$$

which becomes the following equation in complex form:

$$W_{\bar{z}} - \mu_1(W)W_z - \mu_2(W)\overline{W}_{\bar{z}} = \frac{i}{2}A(\psi), \tag{22}$$

where $A(\psi)$ is the value of the vorticity, $\mu_1 = \bar{\mu}_2 = e^{2i\theta}M^2/2(2 - M^2)$, and, by (6),

$$|\mu_1| + |\mu_2| \leqslant \frac{M^2}{2 - M^2} \leqslant \mu_0 < 1.$$

By substituting the expressions

$$u = q(q^*)\cos\theta, \qquad v = q(q^*)\sin\theta, \qquad dq/dq^* = q(1 - M^2)^{-1/2}$$

in (21), we arrive at a system for q^* and θ with the complex form

$$\omega_{\bar{z}} - \mu^z(\omega)\omega_z = iA\left(\frac{\rho}{1 + \sqrt{1 - M^2}}\right)W^{-1} \equiv A_1. \tag{23}$$

In the plane of $\tau = x + i\psi$ we have

$$\omega_{\bar{\tau}} - \tilde{\mu}(\omega)\omega_\tau = A_2(\omega, \psi), \tag{24}$$

where

$$A_2(\omega, \psi) = A_\tau \left[1 + \rho W + \mu^z(1 - \mu^z\rho\overline{W}^2)\right].$$

We note also that the relation $(\rho q)^2 = \psi_x^2 = \psi_y^2$, obtained as a consequence of (5), uniquely determines the implicit function $q = q(\psi_x, \psi_y)$ by virtue of (6). This enables us to get the following quasilinear equation for the stream function by eliminating $\varphi(x, y)$ from (5):

$$\left(\rho^2 c^2 - \psi_y^2\right)\psi_{xx} + 2\psi_x\psi_y\psi_{xy} + \left(\rho^2 c^2 - \psi_x^2\right)\psi_{yy} = 0, \tag{25}$$

where $c^2 = -\rho q(d\rho/dq)^{-1}$.

§2. General properties of solutions of boundary-value problems in hydrodynamics and gas dynamics. A priori estimates

1°. *Statement of the problems.* Most of the problems discussed below are formulated as problems on constantly subsonic flows of a gas such that $\rho = \rho(q) \in C^{l+\alpha}$, $\alpha > 0$, $l \geqslant 2$, relating the density to the speed satisfies condition (1.6) and is otherwise arbitrary. In particular, for $\rho(q) \equiv 1$ the case of flows of an incompressible fluid is also included.

As a rule the boundaries of the flow domain are assumed to be streamlines, along with $\psi = \text{const}$.

If some part of the boundary of the flow domain is given, and $\tilde{\theta} = \tilde{\theta}(x)$ is the tangent angle along it, then the relation $\psi = \text{const}$ and the equations (1.1) imply that $\varphi_x \, dx - \varphi_y \, dy = 0$, or $\theta(x) = \arctan(\varphi_y/\varphi_x) = \tilde{\theta}(x) + k\pi$ (the integer k is equal to 0 or ± 1), i.e., to within a constant term the angle $\theta = \theta(x)$ between the flow velocity and the OX-axis is known. On the unknown parts of the boundary of the flow domain we assume that in addition to the condition $\psi = \text{const}$ we are given some condition relating the boundary values of the real and imaginary parts of the generalized Joukowsky function $\omega = q^*(q) - i\theta$,

$$q^* = \int_1^q q^{-1}\sqrt{1 - M^2} \, dq, \qquad q^2 = \varphi_x^2 + \varphi_y^2$$

being the magnitude of the flow velocity. The form of such a boundary condition for $\omega(z)$ is determined by the original physical assumptions. For example, in jet problems the boundary condition can have the form

$$\partial\theta/\partial\varphi = aq + bq^{-1}, \qquad (a, b = \text{const}), \, q^2 + 2gy = \text{const}, \text{ and } q = \text{const}$$

depending upon whether or not the forces of surface tension or the forces of gravity are taken into account.

Thus, suppose that on the streamlines $\psi = \text{const}$ bounding the flow domain D_z we are given one of the relations

$$\text{Re}[G(x)\omega(x)] = g(x, y) \tag{1}$$

or

$$\text{Re}[G(x)\Omega(x)] = g(x, y), \tag{2}$$

where

$$\Omega = e^\omega \equiv q e^{-i\theta + \sigma(q)}, \qquad \sigma(q) = \int_1^q q^{-1}\left(\sqrt{1 - M^2} - 1\right) dq$$

(see (1.19) and (1.20)), while $G(x)$ and $g(x, y)$ are single-valued functions on a finite number of curves L_z^1 making up the boundary L (for example, on the upper and the lower walls of a channel), and $|G(x)| \equiv 1$. It is required to find the partially or completely unknown boundary of the flow domain D_z formed from streamlines $\psi = \text{const}$ and a generalized Joukowsky function $\omega = q^* - i\theta$ that satisfies the equation

$$\omega_{\bar{z}} - e^{2i\theta}\frac{1 - \sqrt{1 - M^2}}{1 + \sqrt{1 - M^2}}\omega_z = 0 \tag{3}$$

in D_z and one of the boundary conditions (1) or (2) on its boundary L_z. A large class of problems, including the majority of classical hydrodynamics and aerodynamics problems, is described by the boundary conditions (1) or (2). For example, if $G(x) = i$ on L and $g(x, y) \equiv -\theta(x)$, then the boundary conditions (1) describe problems on subsonic gas flows in given domains, while if $G(x) \equiv 1$ and $g(x, y) = q^*(x)$ on L_z, then they describe gas flow problems in domains with completely unknown boundaries to be determined from a specified velocity distribution on them. If $G(x) = i$ and $g(x, y) \equiv -\theta(x)$ on part of the boundary L_z while $G(x) \equiv 1$ and $g(x, y) = q^*(x)$ or $g(x, y) = q^*(y)$ on the other part, then the boundary conditions (1) describe classes of mixed problems in hydrodynamics and aerodynamics in flow domains with partially known boundaries. The class of mixed problems includes, in particular, problems of cavitational flow past an obstacle according to various schemes (Kirchhoff, Raibouchinsky, Èfros, and others), problems on motions of a fluid in a gravitional field, and other important problems in hydrodynamics and aerodynamics. In many cases it is more convenient to describe these problems and a number of other problems by the boundary condition (2).

The conditions of hydrodynamics and aerodynamics problems in the plane of the complex potential $w = \varphi + i\psi$ and the plane of the variable $\tau = x + i\psi$ determine the images D_w and D_τ of the flow domains D_z, whose main forms are illustrated in the figure ((a) the upper half-plane; (b) the exterior of a semi-infinite cut; (c) the exterior of a finite cut; (d) a strip; (e) a strip with a semi-infinite cut, which is the upper half-plane with a semi-infinite cut when the depth is infinite, i.e., $h = \infty$, $\psi_0 = \infty$).

We assume first that $g(x, y)$ in the boundary condition (1) does not depend on y, i.e., $g(x, y) = g(x)$, which is the case in most boundary-value problems in hydrodynamics and aerodynamics. In this case it is completely natural to pass to the independent variables x and ψ, in the plane of which the image D_τ $(\tau = x + i\psi)$ of the flow domain D_z is known and has boundary L_τ consisting of lines $\psi = $ const. Unlike in the plane of the complex potential $w = \varphi + i\psi$, where the image D_w of D_z can also be assumed known (see the figure), the coefficients of the boundary condition (1) on the plane of $\tau = x + i\psi$ are given point functions on the boundary L_τ of D_τ. In the general case when $g(x, y)$ depends on y, we shall also use the plane $\tau = x + i\psi$, representing y in the form

$$y = \int_{x_1}^{x} \tan \theta(x) \, dx + y_1 \equiv y[\theta \mid x]. \qquad (4)$$

Thus, we come to the Hilbert boundary-value problem (1) for a generalized Joukowsky function $\omega = \omega(\tau)$ satisfying (1.15):

$$\omega_{\bar{\tau}} - \tilde{\mu}(\omega)\omega_\tau = 0, \qquad \tau \in D_\tau, \qquad (5)$$

on the plane of $\tau = x + i\psi$, where

$$\tilde{\mu}(\omega) = \frac{\rho q^2 - u\sqrt{1 - M^2} + iv}{\rho q^2 + u\sqrt{1 - M^2} - iv}.$$

However, as follows from the properties of the coefficient $\tilde{\mu}(\omega)$ (see §1), equation (5) is uniformly elliptic only if the following inequalities hold in $\overline{D}_\tau$ ($\overline{D}_z$):

$$|\operatorname{Re}\omega| \leqslant N < \infty, \qquad |\operatorname{Im}\omega| \leqslant \pi/2 - \delta, \quad \delta > 0. \tag{6}$$

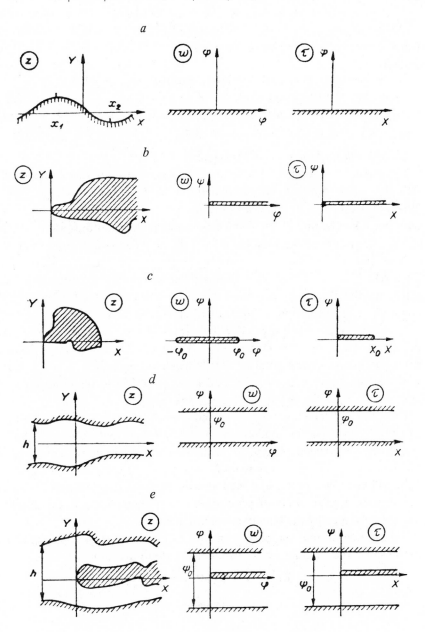

We suppose for the time being that $g(x, y) = g(x)$ and, if nothing special is mentioned, we assume the following conditions on the coefficients of the boundary condition (1):

a) $G(x), g(x) \in C_*^\alpha$, $\alpha > 0$, for all finite x, and $g(x)$ does not have points of discontinuity that are not such points for $G(x)$.

b) $G(x) \equiv e^{i\beta_0}$, $\beta_0 = 0, \pi/2$, and $G[x(\xi)]$, $g[x(\xi)] \in C^\alpha$ in a neighborhood of vertices $b_j \in L_z$ at infinity, where $\xi = |x|^{-k}$ for the angles $-\pi$, -2π and $\xi = e^{-|x|}$ for zero angles, and the quantity $k > 0$ depends on the original problem.

In the case when $g(x, y)$ depends on y explicitly we assume that

c) $g(x, y) = g_0(x) + \varepsilon \tilde{g}(x, y)$, where $g_0(x)$ and $\tilde{g}(x, y)$ (for each fixed y) satisfy conditions (a) and (b), and $\tilde{g}(x, y) \equiv 0$ for $|x| \geqslant X$ (X sufficiently large), and $\varepsilon > 0$ is an arbitrary parameter.

2°. Solvability of problems in the class $\mathfrak{M}(N, \delta)$.

DEFINITION 1. We say that functions $G(x)$ and $g(y)$ (and the corresponding boundary-value problems of gas dynamics) are *functions* (*problems*) *of class* $\mathfrak{M}(N, \delta)$ if they satisfy conditions (a)–(c), and the a priori estimates (6) hold with fixed $N < \infty$ and $\delta > 0$ in the closed domain $\bar{D}_z$ (and thus also in $\bar{D}_\tau$ and $\bar{D}_w$) if there is a bounded solution $\omega(z)$ of the boundary-value problem (1), (3).

Note that if the index of problem (1) is negative, then condition (6) is assumed to hold in the class of bounded solutions for a function $\omega(z)$ satisfying the boundary condition (1) only when the $2|\kappa| - 1$ solvability conditions (representations of the form (2.7) in Chapter VI) hold.

If the boundary condition (1) allows (6) to be violated at finitely many points z_k of L_z, then the functions $G(x)$ and $g(x, y)$ are "corrected" in an ε-neighborhood of these points z_k (($1/\varepsilon$)-neighborhoods, if $z_k = \infty$) in such a way that the new functions $G_\varepsilon(x)$ and $g_\varepsilon(x, y)$ belong to the class $\mathfrak{M}(N_\varepsilon, \delta_\varepsilon)$ and $\{G_\varepsilon(x), g_\varepsilon(x, y)\} \to \{G(x), g(x, y)\}$ as $\varepsilon \to 0$.

DEFINITION 2. The functions $G(x)$ and $g(x, y)$ *belong to the class* $\mathfrak{M}^*$ if they can be approximated by functions $G_\varepsilon(x)$ and $g_\varepsilon(x, y)$ that coincide with $G(x)$ and $g(x, y)$ everywhere except in ε-neighborhoods of finitely many points of the boundary L_z of the flow domain D_z, and $G_\varepsilon(x), g_\varepsilon(x, y) \in \mathfrak{M}(N_\varepsilon, \delta_\varepsilon)$.

As will be proved in this and the following sections, the classes $\mathfrak{M}(N, \delta)$ and $\mathfrak{M}^*$ contain a large circle of problems of hydrodynamics and gas dynamics that are described by the boundary condition (1).

We use a conformal mapping of the disk $|\zeta| < 1$ onto the known domain D_τ to transform the boundary-value problem (1), (5) to the form

$$\text{Re}\big[G_0(e^{i\gamma}) \omega(e^{i\gamma}) \big] = g_0(e^{i\gamma}), \tag{7}$$

$$\omega_{\bar{\zeta}} - \mu(\omega, \zeta)\omega_\zeta = 0, \tag{8}$$

where

$$G_0(e^{i\gamma}) = G[x(\gamma)], \qquad g_0(e^{i\gamma}) = g(x(\gamma), y[\theta \mid x(\gamma)])$$

and

$$\mu(\omega, \zeta) = \tilde{\mu}(\omega) \exp\{-2i \arg(d\tau/d\zeta)\}.$$

If (6) holds, then

$$|\mu(\omega, \zeta)| = |\tilde{\mu}(\omega)| \leqslant \mu_0(N, \delta) < 1,$$

where the constant $\mu_0(N, \delta) \to 1$ if $N \to \infty$ or $\delta \to 0$, and $\mu(\omega, \zeta)$ satisfies a Lipschitz condition with respect to ω, because of the assumption that $\rho(q) \in C^{2+\alpha}$, $\alpha > 0$.

Consequently, Theorem 5 in §2 of Chapter VI is directly applicable to the boundary-value problem (7), (8) when $g(x, y)$ does not depend explicitly on y. But if $g(x, y)$ depends on y and satisfies condition (c), then this theorem gives us directly only an existence theorem, since in this case the function $g_0(e^{i\gamma})$ is not known beforehand and depends on the unknown solution. Thus, for gas dynamics problems of the class $\mathfrak{M}(N, \delta)$ the following important theorem holds.

THEOREM 1. *Suppose that the domain D_τ is known and the boundary-value problem (1), (5) belongs to $\mathfrak{M}(N, \delta)$. Then for $\partial g(x, y)/\partial y \equiv 0$ the boundary-value problem (7), (8) has as many bounded solutions as in the case of analytic functions. For $\partial g(x, y)/\partial y \not\equiv 0$ it has at least $2\kappa + 1$ bounded solutions in the case of a nonnegative index $\kappa \geqslant 0$, while for a negative index it has at least one bounded solution only if the $2|\kappa| - 1$ solvability conditions hold.*

The fact, observed in Theorem 1 that the number of solutions for boundary-value problems of the class $\mathfrak{M}(N, \delta)$ is independent of the properties of the fluid is a quantitative expression of the similarity principle for flows of an ideal and compressible fluid when the mode of flow is constantly subsonic. On the other hand, the complex velocity and the complex potential of a subsonic flow, like the solution of a quasilinear elliptic Beltrami equation, can be represented as the superposition of an analytic function and a homeomorphism of the flow domain onto itself. This property, which is obviously shared by all solutions of the equations of subsonic gas dynamics, reflects the fact that flows of an ideal and compressible fluid are topologically similar (the fact that streamlines in the flow domain do not intersect, the disjointness of equipotentials, preservation of domain under mappings by solutions of the equations of gas dynamics, and so on). We remark that Theorem 1 bears a conditional character until effectively verifiable conditions are given which ensure the a priori estimates (6) and thereby the membership of the boundary-value problems (1), (5) in the class $\mathfrak{M}(N, \delta)$.

THEOREM 2. *Suppose that the domain D_τ is known, the boundary-value problem* (1), (5) *belongs to the class* $\mathfrak{M}^*$, *and* $z_k \in L_z$ $(k = 1,\ldots,n)$ *are singular points near which the estimates* (6) *are violated. Then for* $\kappa > 0$ *the boundary-value problem* (7), (8) *has at least* $2\kappa + 1$ *solutions that are bounded in* $\overline{D}_z$ *everywhere except possibly at the points* z_k, *while for* $\kappa < 0$ *there exists at least one such solution only if the* $2|\kappa| - 1$ *solvability conditions hold.*

According to Definition 2, there exist functions $G_\varepsilon(x)$, $g_\varepsilon(x, y) \in \mathfrak{M}(N_\varepsilon, \delta_\varepsilon)$ that converge to the original coefficients $G(x)$ and $g(x, y)$ of the boundary condition (1). Let $\varepsilon = n^{-1}$ and consider the sequence $\omega^n(\tau)$, $\tau = x + i\psi$, of solutions of the boundary-value problems (1), (5) with the functions $G_n(x)$ and $g_n(x, y)$ in the boundary condition (1). By (6), the mapping from the plane of $\tau = x + i\psi$ onto the plane of $w = \varphi + i\psi$ is quasiconformal; therefore, there exists a sequence $\omega^n[\tau(w)] \equiv \omega^n(w)$ of solutions of (1.12):

$$\omega_{\overline{w}}^n - \mu^w(\omega^n)\omega_w^n = 0, \tag{9}$$

where

$$\mu^w(\omega^n) = \frac{\rho^*(q^n) - 1}{\rho^*(q^n) + 1}.$$

We map the unit disk $|\zeta| < 1$ conformally onto D_w and set $\omega^n(\zeta) \equiv \omega^n[w(\zeta)]$, with

$$\omega_{\overline{\zeta}}^n - \hat{\mu}^n(\omega^n, \zeta)\omega_\zeta^n = 0, \tag{9*}$$

where

$$\hat{\mu}^n(\omega^n, \zeta) = \frac{\rho^* - 1}{\rho^* + 1} \exp\left\{-2i \arg \frac{dw^n}{d\zeta}\right\} \quad \text{and} \quad |\hat{\mu}^n| \leqslant \mu_0 < 1$$

for $|\zeta| < 1$ and arbitrary ω^n. Let $\Omega^n(\zeta) = 1/(\omega^n - i\pi)$; by (6) ($|\operatorname{Im} \omega^n| \leqslant \pi/2$), these functions satisfy the inequality $|\Omega^n(\zeta)| \leqslant 2/\pi$ for $|\zeta| \leqslant 1$. The functions $\Omega^n(\zeta)$ are solutions of the equations

$$\Omega_{\overline{\zeta}}^n - \hat{\mu}^n\left(\frac{1}{\Omega^n} + i\pi, \zeta\right)\Omega_\zeta^n = 0$$

and, therefore, can be represented in the form

$$\Omega^n(\zeta) = \Phi^n[\chi^n(\zeta)],$$

where $\chi = \chi^n(\zeta)$ is a homeomorphism of the unit disk onto itself with the normalization $\chi^n(0) = 0$, $\chi^n(1) = 1$, and $\Phi^n(\chi)$ is an analytic function in the disk $|\chi| < 1$. In succession we choose simultaneously convergent subsequences $\{\arg(dw^{n_k}/d\zeta)\}$ and $\{\Phi^{n_k}(\chi)\}$ from the family $\{\arg(dw^n/d\zeta)\}$ or harmonic functions and the family $\{\Phi^n(\chi)\}$ of analytic functions (which are uniformly bounded in the respective domains $|\zeta| < 1$ and $|\chi| < 1$), and we consider the corresponding subsequence of homeomorphisms $\chi = \chi^{n_k}(\zeta)$ of the unit disk onto itself that satisfy the equations

$$\chi^{n_k}_{\bar{\zeta}} - \hat{\mu}^{n_k}\left(\frac{1}{\Phi^{n_k}(\chi)} + i\pi, \zeta\right)\chi^{n_k}_{\zeta} = 0.$$

The uniform convergence of the sequences $\arg(dw^{n_k}/d\zeta)$ and Φ^{n_k} to the functions $f^0(\zeta)$ and $\Phi^0(\chi)$ implies also the uniform convergence $\hat{\mu}^{n_k} \to \hat{\mu}(1/\Phi^0(\chi) + i\pi, \zeta)$ for $|\chi| < 1$ and thereby, according to the remark after Theorem 3 in §4 of Chapter V, also the convergence of the homeomorphisms $\chi^{n_k}(\zeta)$ to a homeomorphism $\chi^0(\zeta)$ satisfying the limit equation. It has thus been proved that the sequence $\Omega^{n_k}(\zeta) = 1/(\omega^{n_k}(\zeta) - i\pi)$ converges to a function $\Omega^0(\zeta)$ in $|\zeta| < 1$. But the sequence $\omega^{n_k}(\zeta)$ is uniformly bounded in $|\zeta| < 1$, except possibly in ε-neighborhoods of the images of the singular points z_k ($k = 1, \ldots, n$); therefore, outside these ε-neighborhoods

$$\omega^{n_k}(\zeta) = \frac{1}{\Omega^{n_k}(\zeta)} + i\pi \to \frac{1}{\Omega^0(\zeta)} + i\pi \equiv \omega^0(\zeta).$$

Limit values exist for $\omega^0(\zeta)$ almost everywhere on the circle (by Fatou's theorem in §2 of Chapter II for the analytic functon $\Phi^0(\chi)$). Therefore, since each function $\omega^{n_k}(\zeta)$ satisfies the boundary condition (1) in the z-plane outside ε-neighborhoods of the points z_k, it follows that this holds also for the limit function $\omega^0(\zeta)$. This proves the theorem.

THEOREM 3. *Suppose that the domain D_τ is known that the piecewise continuous functions $G(x)$ and $g(x, y)$ in the boundary conditions (1) and (2) can be approximated by functions $G_\varepsilon(x)$ and $g_\varepsilon(x, y)$ of the class $\mathfrak{M}(N_\varepsilon, \delta_\varepsilon)$. Then for $\kappa \geqslant 0$ there exist at least $2\kappa + 1$ quasisolutions $\omega(\tau)$ and $\Omega(\tau) = e^{\omega(\tau)}$ in each of the respective boundary-value problems (1), (3) and (2), (3), whereas for $\kappa < 0$ these problems each have at least one quasisolution only if the $2|\kappa| - 1$ solvability conditions hold.*

For definiteness we take only the problem (2), (3), since the problem (1), (3) is handled in a completely analogous fashion. By mapping the disk $|\zeta| < 1$ conformally onto the domain D_τ, we arrive at the sequence of boundary-value problems

$$\text{Re}\left[G_\varepsilon(e^{i\gamma})\Omega(e^{i\gamma})\right] = g_\varepsilon(e^{i\gamma}) \tag{2*}$$

for functions $\Omega(\zeta) = \Omega[\tau(\zeta)]$ satisfying an equation of the form (8):

$$\Omega_{\bar{\zeta}} - \mu(\ln \Omega, \zeta)\Omega_\zeta = 0, \tag{8*}$$

where

$$\mu(\ln \Omega, \zeta) = \hat{\mu}(\ln \Omega) \exp\{-2i \arg(d\tau/d\zeta)\},$$

and $|\mu| \leqslant \mu_0(N_\varepsilon, \delta_\varepsilon) < 1$. Since the equations (8*) are uniformly elliptic on the unknown solutions for each fixed $\varepsilon > 0$, Theorem 5 in §2 of Chapter VI gives us that the boundary-value problem (2*), (8*) has at least as many solutions as in the class of analytic functions. We set $\varepsilon = 1/n$ and (as for Theorem 2) consider the sequence of functions

$$F^n(\zeta) = \frac{1}{\Omega^n + 1}, \qquad \Omega^n \equiv \exp \omega^n = \Omega^n\{\tau[w(\zeta)]\},$$

where the Ω^n satisfy an equation of the form (9*). As in the proof of Theorem 2, it is easy to see that this sequence is relatively compact, and we choose a subsequence $F^{n_k}(\zeta)$ of it that converges uniformly inside the disk $|\zeta| < 1$. By what was proved in Theorem 7 of §2 in Chapter VI, the limit $\Omega^0(\zeta)$ of the corresponding subsequence Ω^{n_k} is a quasisolution of the boundary-value problem (2), (3), transformed to the disk. The theorem is proved. We remark that if the limit functions $G(x)$ and $g(x, y)$ are piecewise Hölder, then the quasisolution is an ordinary solution of class $\mathfrak{M}^*$.

$3°$. *A priori estimates of the flow velocity.* For the boundary-value problems (1), (5) of class $\mathfrak{M}(N, \delta)$ it turns out that an a priori estimate on the slope angle of the flow velocity (the first of the inequalities in (6)). Such an a prior estimate of the flow velocity in terms of its slope angle was first obtained for jet flows of an ideal fluid around curvilinear obstacles in the fundamental investigations of Leray [70], and was then generalized to the case of flows in channels by Kravtchenko [58] and others. Unfortunately, however, not only the estimates of Leray and Kravtchenko but also methods for obtaining similar estimates fail to be applicable in problems as general as those we are studying, not to mention problems in gas dynamics. The method presented below for proving a priori estimates on the flow velocity differs essentially from Leray's method and is based on general properties of solutions of quasilinear elliptic equations and, in particular, on the properties of quasiconformal mappings (Chapter V).

THEOREM 4. *For the conformal mapping $z = F(\zeta)$ ($|\zeta^\nu(dF/d\zeta)(\zeta)|_{\zeta=0} = |c|$, where c is a fixed constant and $\nu = 0$ or 2) of the unit disk $|\zeta| < 1$ onto the simply connected domain of the constantly subsonic flow of a gas, the slope angle $\theta(t) = \theta[x(t)]$ of the flow velocity is a measurable function on the circle, and*

$$|\theta(e^{i\gamma})| \leqslant \pi/2 - \delta, \qquad \delta > 0, \gamma \in [0, 2\pi]. \tag{10}$$

Then the inequality

$$\| z(e^{i\gamma}) \|_{C^\beta} \leqslant C_0(\varepsilon, \delta), \qquad \beta(\delta) > 0, \tag{11}$$

holds everywhere on the circle $|\zeta| = 1$ *except in* ε-*neighborhoods of the fixed images* t_j *of the vertices* $b_j \in L_z$ $(j = 1, \ldots, m)$ *at infinity, while the inequality*

$$\| [x(e^{i\gamma})]^{-k} \|_{C^\beta} \leqslant C_1(\varepsilon, \delta), \qquad 0 < k(\delta) < \infty, \tag{12}$$

holds in ε-*neighborhoods of the points* t_j *corresponding to the vertices* $b_j \in L_z$ *with angle* $-\pi$ *(a semiconfined flow) or* -2π *(an unconfined flow).*

Let us first consider the case of a flow of an ideal fluid. We represent the derivative of the conformal mapping $w = w(\zeta)$ of $|\zeta| < 1$ onto the domain D_w in the complex potential plane of the flow by the Schwarz-Christoffel formula

$$\frac{dw}{d\zeta} = K_0 \prod_{j=1}^{m} (\zeta - t_j)^{-1} \prod_{i=1}^{n} (\zeta - \sigma_i) \zeta^{-\nu}, \tag{13}$$

where the t_j are the images of the vertices $b_j \in L_z$ at infinity (to the angles $-\pi$ and -2π correspond the cases where two and three of the t_j respectively coincide), the σ_j are the images of endpoints of cuts (the t_j and the σ_j are assumed to be known), K_0 is a constant, $\nu = 2$ if the point at infinity lies in D_z, and $\nu = 0$ otherwise. If we now use the Schwarz formula to recover the complex flow velocity

$$\frac{dw}{dz}(\zeta) = q(\zeta) e^{-i\theta(\zeta)},$$

$$\frac{dw}{dz} = q_0 \exp\left\{ \frac{i}{2\pi} \int_0^{2\pi} \theta(e^{i\gamma}) \frac{e^{i\gamma} + \zeta}{e^{i\gamma} - \zeta} d\gamma \right\} \equiv q_0 e^{U(\theta|\zeta)}, \tag{14}$$

then we arrive at the following representation for the derivative of the mapping $z = F(\zeta)$:

$$\frac{dz}{d\zeta} = c e^{-U(\theta|\zeta)} \prod_{j=1}^{m} (\zeta - t_j)^{-1} \prod_{i=1}^{n} (\zeta - \sigma_i) \zeta^{-\nu}, \tag{15}$$

where $q_0 = |dw/dz|_{\zeta=0}$ and $|c| = |\zeta^\nu dz/d\zeta|_{\zeta=0}$. An inequality of Zygmund (Theorem 7 in §2 of Chapter II) for the analytic function $(dw/dz)(\zeta) \equiv q(\zeta) e^{-i\theta(\zeta)}$ gives us that

$$\| q^{\pm 1}(e^{i\gamma}) \|_{L_p} = q_0^{\pm 1} \| e^{\pm U(\theta|\gamma)} \|_{L_p} \leqslant q_0^{\pm 1} \left[\frac{2\pi}{\cos(\pi/2 - \delta)} \right]^{1/p}, \tag{16}$$

where $1 < p < \pi/(\pi - 2\delta)$. With the help of (15) and (16) we find that

$$|z(e^{i\gamma}) - z(e^{i\gamma'})| = \left| \int_{t'}^{t} \frac{dz}{dt} dt \right| \leqslant \frac{2^n c}{\varepsilon^m} \int_{\gamma'}^{\gamma} |e^{-U(\theta|\gamma)}| \, d\gamma \leqslant C_0(\varepsilon, \delta) |\gamma - \gamma'|^{(p-1)/p}$$

everywhere on the circle $|\zeta| = 1$ except in ε-neighborhoods of the points t_j ($j = 1, \ldots, m$), which proves (11). Suppose now that $t = e^{i\gamma}$ is an arbitrary point of the ε-neighborhood of the image t_j of a vertex $b_j \in L_z$ at infinity with angle $-\pi$ or -2π, and let $t' = e^{i\gamma'}$, $\gamma - \gamma' = \gamma_j - \gamma_0$. Using (16) and the converse Hölder inequality ((14) in Chapter II), and assuming for definiteness that $\operatorname{Re} z(\gamma) \equiv x(\gamma) \geqslant x(\gamma') > 0$, we get

$$x(\gamma) \geqslant x(\gamma) - x(\gamma') = \int_{\gamma'}^{\gamma} \frac{dx}{d\varphi} \frac{d\varphi}{d\gamma} d\gamma = c \int_{\gamma'}^{\gamma} e^{-U(\theta|\gamma)} \cos \theta(\gamma) \frac{d\varphi}{d\gamma} d\gamma$$

$$\geqslant c' \int_{\gamma'}^{\gamma} e^{-U(\theta|\gamma)} \frac{d\gamma}{(\gamma_j - \gamma)^2} \geqslant c' \left\{ \int_{\gamma'}^{\gamma} e^{pU(\theta|\gamma)} d\gamma \right\}^{-1/p} \left\{ \int_{\gamma'}^{\gamma} \frac{d\gamma}{(\gamma_j - \gamma)^{\beta_1 + 1}} \right\}^{(p_1 + 1)/p}$$

$$= \tilde{c}(\varepsilon, \delta) |\gamma_j - \gamma|^{(1-p)/p}, \qquad \beta_1 = \frac{p-1}{p+1}.$$

Then for arbitrary $\gamma, \gamma' \in [\gamma_j - \varepsilon, \gamma_j + \varepsilon]$ we have

$$|x^{-k}(\gamma) - x^{-k}(\gamma')| = \left| \int_{\gamma'}^{\gamma} \frac{d}{d\gamma} \left(\frac{1}{x^k} \right) d\gamma \right| = k \left| \int_{\gamma'}^{\gamma} \frac{1}{x^{k+1}} \frac{dx}{d\varphi} \frac{d\varphi}{d\gamma} d\gamma \right|$$

$$\leqslant \frac{N}{\tilde{c}} \int_{\gamma'}^{\gamma} e^{-U(\theta|\gamma)} |\gamma_j - \gamma|^{\beta_2} d\gamma,$$

where $\beta_2 = -2 + (k+1)(p-1)/p$ for a semiconfined flow in a neighborhood of $z(e^{i\gamma_j}) = \infty$, and $\beta_2 = -3 + (k+1)(p-1)/p$ for an unconfined flow. Setting $k(\delta) = (2p+1)/(p-1)$, where $1 < p < \pi/(\pi - 2\delta)$, we get that $\max_{\gamma} |\gamma_j - \gamma|^{\beta_2} = N_1 < \infty$, and, consequently,

$$|x^{-k}(\gamma) - x^{-k}(\gamma')| \leqslant \frac{NN_1}{\tilde{c}} \int_{\gamma'}^{\gamma} e^{-U(\theta|\gamma)} d\gamma$$

$$\leqslant N_2 \left\{ \int_{\gamma'}^{\gamma} e^{-pU(\theta|\gamma)} d\gamma \right\}^{1/p} |\gamma - \gamma'|^{(p-1)/p}. \qquad (17)$$

The assertions of the theorem are proved for the case of an ideal fluid. Suppose now that $z = F(\zeta)$ is a conformal mapping of the domain D_z of a constantly subsonic gas flow and $q = q(w)$ is the known speed of flow. Mapping D_w homeomorphically onto an analogous domain by a solution of the equation

$$W_{\bar{w}} - \frac{\rho[q(w)] - 1}{\rho[q(w)] + 1} W_w = 0, \qquad (18)$$

we find from (1.13) an analytic function $z = \Phi[W(w)]$ on the plane of the homeomorphism W. The inverse function $W = \Phi^{-1}(z)$ can be regarded as the complex potential of an ideal fluid flow in the z-plane, with the slope angle

$\tilde{\theta}(z)$ of the velocity obviously the same as the slope angle $\theta(x)$ of the velocity of the original gas flow. The proof of the theorem can be completed by applying (11) and (12) to the analytic function $z = F(\zeta) = \Phi[w(\zeta)]$, where $W = W(\zeta)$ is a conformal mapping of the disk $|\zeta| < 1$ onto the domain in the W-plane.

REMARKS. 1. The estimate (11) holds also locally for a conformal mapping $z = F_0(\zeta)$ of $|\zeta| < 1$ onto a neighborhood $D_0 \subset D_z$ of an arbitrary finite arc $l_0 \subset L_z$ $(\partial D_0 = L_0 \in H^1)$ on which the slope angle $\theta(x) = -\mathrm{Im}\,\omega$ of the flow velocity satisfies the inequality in (6).

Indeed, in this case the equation of l_0 can be represented in the form

$$y = y_0(x), \qquad |dy_0/dx| = |\tan\theta(x)| \leqslant |\tan(\pi/2 - \delta)| < \infty,$$

i.e., $l_0 \in H^1$. Therefore, estimate (11) with a constant $C_0 = C_0(\delta, D_0)$ follows immediately from Theorem 5 in §4 of Chapter V.

2. When the normalization of the mapping $z = F(\zeta)$ is different from that chosen in the theorem, the constant c can obviously be estimated from above and from below on the basis of the Lindelöf principle (Theorem 2 in §3 of Chapter II) if there exist fixed simply connected domains D^+ and D^- such that $D^- \subset D \subset D^+$.

THEOREM 5. *Suppose that the a priori estimate in* (6) *holds for the slope angle* $\theta = -\mathrm{Im}\,\omega$ *of the velocity in the case of bounded solutions* $\omega = \omega(z)$ *of the boundary-value problem* (1), (3), *and that the boundary* L_z *of* D_z *is nondegenerate in the sense that for adjacent streamlines* $\psi = \psi_k$ *belonging to this boundary with equations* $y = y_k(x)$, $x \in I_k$, *the inequality*

$$0 < h < |y_{k+1}(x) - y_k(x)| < h^{-1}, \qquad x \in I_{k+1} \cap I_k,$$

holds with a fixed constant $h > 0$. *Then* $\omega = \omega(z)$ *satisfies the a priori estimate*

$$|\mathrm{Re}\,\omega(z)| \leqslant N(\delta, h), \qquad z \in \overline{D}_z, \tag{19}$$

and for $G \equiv i$ *the constant* N *depends in addition on the magnitude* q_0 *of the velocity at a fixed point* $z_0 \in L_z$.

By the maximum principle, it suffices to prove (19) for bounded solutions of (3) only on the boundary of D_z. We fix $\varepsilon > 0$ and choose quantities x_k, $-X \leqslant x_k < x_{k+1} \leqslant X < \infty$, and a $\sigma > 0$ such that each branch point of the flow and each point of discontinuity of $G(x)$ falls inside one of the intervals (x_k, x_{k+1}), and such that the oscillation of $\arg G(x)$ on its intervals of continuity in $[x_k - \sigma, x_k + \sigma]$ does not exceed ε.

Let $l_k^i \subset L_z$ be the segment of the streamline $\psi = \psi_i$ lying in the strip

$$x_k - \sigma < x < x_{k+1} + \sigma,$$

and $D_k^i \subset D_z$, $l_k^i \subset \partial D_k^i = L_k^i \in H^1$, a neighborhood of this arc of diameter $d = \min(\sigma, h)$.

We map the disk $|\zeta| < 1$ conformally onto D_k^i by the function $z = F_k^i(\zeta)$, carrying the arc l_k^i to the upper semicircle, and we consider the boundary-value problem

$$\mathrm{Re}\left[G_0(e^{i\gamma})\omega(e^{i\gamma})\right] = g_0(e^{i\gamma}), \qquad \gamma \in [0, \pi],$$

$$\mathrm{Im}\,\omega(e^{i\gamma}) = \theta_0(e^{i\gamma}), \qquad \gamma \in [\pi, 2\pi], \qquad (1^*)$$

where $G_0 = G[x(e^{i\gamma})]$, $g_0 = g[x(e^{i\gamma})]$, $\theta_0 = \theta[z(e^{i\gamma})]$, and the value of $\theta(z)$ on the arc L_k^i/l_k^i is found from the solution of the original boundary-value problem (1), (3). Here (3) is transformed to the equation

$$\omega_{\bar\zeta} - \tilde\mu(\zeta)\omega_\zeta = 0, \qquad \tilde\mu = \mu^z(\omega)\exp\left(-2i\,\arg\frac{dF_k^i}{d\zeta}\right). \qquad (3^*)$$

By construction, the function $\omega = \omega[F_k^i(\zeta)]$, where $\omega = \omega(z)$ is a solution of the boundary-value problem (1), (3), solves the boundary-value problem (1^*), (3^*). The case of a uniformly subsonic flow can be reduced to that of an incompressible fluid by a quasiconformal mapping (which obviously does not change the properties of the coefficients of the boundary condition (1^*)) of the unit disk onto itself by a solution of the linear equation (3^*).

Therefore, it can be assumed without loss of generality that $\omega = \omega(\zeta)$ is an analytic function. Since

$$z = F_k^i(\zeta) \in C^\beta(\overline{K}), \qquad \beta > 0, K: |\zeta| < 1$$

(Remark 1 after Theorem 4), it follows that G_0, $g_0 \in C_*^{\beta_0}$, $\beta_0 = \beta_0(\beta, \alpha) > 0$. Suppose also that $\theta_0 \in C^{\beta_1}$, $\beta_1 > 0$. We remark that the index of problem (1) in the class of bounded functions is negative when $G_0 \not\equiv i$, because the oscillation of $\arg G_0$ is small, and the coefficients of the boundary condition (1^*) have at least two points of discontinuity.

Asuming the solvability conditions for problem (1), we represent (for $G_0 \not\equiv i$) a bounded solution of it in the form

$$\omega = \frac{\Pi(\zeta)}{\pi i}\left(\int_0^\pi \frac{ig_0(t)t\,d\gamma}{\Pi(t)G_0(t)(t-\zeta)} + \int_\pi^{2\pi} \frac{\theta_0(t)t\,d\gamma}{\Pi(t)(t-\zeta)}\right) \equiv \Pi(\omega_0 + \omega_1),$$

where

$$\Pi(\zeta) = \Pi_1^3(\zeta - \zeta_k)^{\alpha_k}e^{\Gamma_0(\zeta)}, \qquad 0 \leqslant \alpha_k < 1, \zeta_{1,2} = \pm 1, \zeta_3 = e^{i\gamma_3}\,(0 < \gamma_3 < \pi)$$

is a point of discontinuity of $G_0(t)$ $(t = e^{i\gamma})$, and $\|\Gamma_0(\zeta)\|_{C^\beta} \leqslant M_0\|G_0(e^{i\gamma})\|_{C_*^\beta}$ (see Chapter I, §4). Let us consider the domain

$$K_0: \{|\zeta| < 1, |\zeta - i| < 1 - \varepsilon_0\}, \qquad \varepsilon_0 = \min(|t_k + 1|, |t_k - 1|),$$

where the t_k are the preimages of the points $z_k = x_k + iy_k \subset l_k^i$. Since for $\zeta \in \overline{K}_0$ the quantities

$$\|\Pi(\zeta)\|_{C^{\beta_2}}, \qquad \|\Pi(\zeta)\omega_0(\zeta)\|_{C^{\beta_2}}, \qquad \left\|\frac{d\omega_1}{d\zeta}\right\|_C, \qquad \beta_2 > 0,$$

are bounded by constants depending only on h, δ, and $\|G(x)\|_{C_*^\alpha}$, $\|g(x)\|_{C_*^\alpha}$, it follows that (19) is proved on the part $\tilde{l}_k^i$ of the arc l_k^i lying in the strip $x_k < x < x_{k+1}$. In the case $G_0 \equiv i$ the same estimate as in the case $G_0 \not\equiv i$ are obviously preserved for the solution of the Dirichlet problem (1*) in the domain $\overline{K}_0$, but the constant in (19) now also depends linearly on the quantity $g_k^* = \operatorname{Re} \omega(t_k)$. By condition (b) on $G(x)$, on the arcs $l_j \subset L_z$ ($|\operatorname{Re} z(l_j)| \geqslant X - \sigma$) lying in a neighborhood of the vertices b_j at infinity we are given either the magnitude of the flow velocity (in which case (19) holds automatically) or the quantity $\theta(x) \in C^\beta$, and then, as on the finite arcs $\tilde{l}_k^i$, (19) holds on the arc $\tilde{l}_j$ ($|\operatorname{Re} z(\tilde{l}_j)| \geqslant X$) when $G \equiv i$ with a constant depending on $q_0^* = \operatorname{Re} \omega(t_0)$, where t_0 is the preimage of the point $z_0 \in L_z$, $|\operatorname{Re} z_0| = X$.

The estimates of the form (19) obtained on the finite number of pairwise adjoining arcs $\tilde{l}_k^i$ ($x_k < \operatorname{Re} z(\tilde{l}_k^i) < x_{k+1}$) and $\tilde{l}_j$ obviously imply (19) on the whole boundary L_z, and thereby in D_z.

REMARK. If the boundary of D_z does not contain vertices at infinity with zero angle, then (19) can be obtained immediately by applying Theorem 4 to a mapping of the disk $|z| < 1$ directly onto the whole domain D_z.

COROLLARY. *Under the conditions of Theorem 5 suppose that the piecewise Hölder functions $G(x)$ and $g(x, y)$ belong to the space $C^{l+\alpha}$, $l \geqslant 0$, on their intervals of continuity, and that $\rho(q) \in C^{l+1+\alpha}$. Then the generalized Joukowsky function $\omega[\tau(\zeta)] \equiv q^*(\zeta) - i\theta(\zeta)$ belongs to the space C^{α_0}, $\alpha_0(\alpha, \delta) > 0$, in the closed disk $|\zeta| \leqslant 1$, and $\omega[\tau(\zeta)] \in C^{l+\alpha_0}(\overline{K}_\varepsilon)$ for $l \geqslant 1$, where K_ε is the disk $|\zeta| \leqslant 1$ with deleted ε-neighborhoods of the images of the points of discontinuity of $G(x)$ and $g(x, y)$ and the images of the vertices $b_j \in L_z$ at infinity.*

Indeed, since the boundary-value problem (1), (5) belongs to the class $\mathfrak{M}(N, \delta)$ under the conditions of Theorem 5, it can be transformed by a conformal mapping of $|\zeta| < 1$ onto D_τ to the boundary-value problems (7), (8) with $G_0(e^{i\gamma})$, $g_0(e^{i\gamma}) \in C^{l+\alpha}$ and $|\mu(\omega, \zeta)| \leqslant \mu_0(N, \delta) < 1$. If we then map the disk $|\zeta| < 1$ homeomorphically onto itself by a solution of the equation

$$\chi_{\bar{\zeta}} - \mu[\omega(\zeta), \zeta]\chi_\zeta = 0$$

and use the properties of homeomorphisms (§§3 and 4 of Chapter V) and the properties of solutions of the resulting boundary-value problem for the analytic function $\omega(\zeta) = \omega[\zeta(\chi)]$, then we arrive at the assertion of the corollary. We remark, further, that since under the conditions of Theorem 5 the problem (1), (5) belongs to the class $\mathfrak{M}(N, \delta)$, Theorem 1 gives us that when D_τ is known this problem can be solved in the same way as for analytic functions.

THEOREM 6. *Suppose that the distances* $h_k = r(\Gamma_k, \Gamma_{k+1})$ *between the adjacent streamlines* Γ_k: $\psi = \psi_k$ *on the boundary* $L_z = \Sigma \Gamma_k$ *of* D_z *are uniformly bounded below* ($h_k \geqslant h > 0$) *and that on some arc* $l \subset L_z$ *the boundary condition* (1) *has the form*

$$\operatorname{Re} \omega(z) = q^*(x), \qquad \| q^*(x) \|_{C^{1+\alpha}} = M < \infty.$$

If the slope angle $\theta(z)$ *of the velocity on* $L_z \backslash l$ *satisfies inequality* (6), *then this inequality holds also on the arc* l *with some constant* δ_0,

$$\delta_0 = \delta_0(\delta, M, h, \lambda), \qquad \lambda = \min\{1, \operatorname{meas}(L_z \backslash l)\}.$$

Let us consider a neighborhood of the points where the arc l joins the remaining part of L_z, denoting its parts by $l_0 \subset l$ and $L_0 \subset L_z$, respectively. Since the smoothness of $q^*(x)$ implies that the quantities

$$|z_w| = \frac{1}{2q}\left|1 + \frac{1}{\rho}\right|, \qquad |z_{\bar{w}}| = \frac{1}{2q}\left|1 - \frac{1}{\rho}\right|,$$

and along with them also $|dx/d\varphi|$, $|dy/d\varphi|$, are uniformly bounded on l, it follows that $l \in H^1$. Then $q^*[x(\varphi)]$, $dx/d\varphi$, $dy/d\varphi \in C^\alpha$, i.e., $l \in C^{1+\alpha}$. But the equation of L_0 can be represented in the form

$$y = y_0(x), \qquad \left|\frac{dy_0}{dx}\right| \leqslant |\tan(\pi/2 - \delta)| < \infty,$$

and, consequently, $L_0 \subset H^1$; moreover, the arcs l_0 and L_0 obviously join at an angle $\theta_0 \neq 0, 2\pi$. Mapping the disk $|\zeta| < 1$ onto a neighborhood of the arc $L_0 + l$, we get (as in the preceding theorem) the inequality (19) on the arc $L_0 + l_0$, and along with it also an estimate of the Hölder constant of $\omega(\zeta)$ on the preimage of this arc. Thus, we can find a sufficiently small arc $l_0 \subset l$ on which (6) holds for $\theta = -\operatorname{Im} \omega$ with the constant $\delta/2$. We substitute the known solution $\omega = \omega(z)$ in (3) and map D_z quasiconformally onto itself by a solution of the Beltrami equation

$$Z_{\bar{z}} - \mu^z[\omega(z)] Z_z = 0, \qquad |\mu^z| \leqslant \mu^0 < 1,$$

leaving the endpoints of l fixed. The function $\omega(Z) \equiv \omega[z(Z)]$ is analytic on the Z-plane, and $\omega(Z) \in C^{1+\alpha}(\overline{D}_0)$ by virtue of the smoothness of l and the differential properties of the homeomorphisms (§§1–3 of Chapter V), where D_0 is a neighborhood of the arc $(l \backslash \tilde{l}_0)$, $\tilde{l}_0 = Z(l_0)$, that is at a fixed distance from the endpoints of l.

If the harmonic function θ attains an absolute extremum at the point $Z_0 \in (l \setminus \tilde{l}_0)$, then

$$|\partial\theta/\partial n(Z_0)| \geqslant \sigma(\delta, \lambda) > 0$$

(Theorem 3 in §4 of Chapter VI and the remark after it). On the other hand, the Cauchy-Riemann equations at this point gives us that

$$\left|\frac{\partial\theta}{\partial N}(Z_0)\right| = \left|\frac{\partial q^*}{\partial s}\right| = \left|\cos\theta(Z_0)\frac{dq^*}{dx}\right|\frac{ds}{dS} = 0,$$

which yields the assertion of the theorem, since the boundary dilation ds/dS is uniformly bounded because of the proven smoothness of the homeomorphism $Z = Z(z)$, and $dq^*/dx \in C^\alpha$ by an assumption of the theorem.

4°. The parameter problem.

THEOREM 7. *Suppose that the coefficients $G(x)$ and $\lambda g(x, y)$ of the boundary-value problem (1) belong to the class $\mathfrak{M}(N, \delta)$ for $\lambda \in [0, 1]$, and that in a neighborhood of each vertex $b_j \in L_z$ with zero angle the flow depth $h_j \neq 0, \infty$ is given. Then for all $\lambda \in [0, 1]$ this problem for (5) has at least as many bounded solutions as it has in the class of analytic functions for the given domain.*

If there are no vertices b_j with zero angle on L_z and D_τ is known, then the assertions of the theorem follow immediately from Theorem 1. In the general case the matter is complicated by the fact that the values $\psi(x, y) = \psi_k = \text{const}$ of the stream function on L_z cannot be assumed to be fixed, since otherwise we would come to an overdetermined problem for the given h_j. Thus, when there are vertices b_j with zero angle on L_z, the domain D_τ is, generally speaking, unknown. According to the a priori estimates (6), $q(z) \neq 0, \infty$ for $z \in \bar{D}_z$, i.e., the dilation under the quasiconformal mapping from the z-plane to the w-plane is different from zero and infinity. In particular, this implies that to the flow depth h_j ($\neq 0, \infty$) at a vertex $b_j \in L_z$ with zero angle there corresponds the quantity (outflow) $Q_j = (\psi_{j+1} - \psi_j) \neq 0, \infty$ on the w-plane, where the constants ψ_j and ψ_{j+1} are the values of the stream function on L_z in a neighborhood of the vertex b_j. Consequently, if under the conditions of the theorem there is a solution of problem (1), (5), then corresponding to it is a nondegenerate domain D_τ, i.e., $|\psi_{j+1} - \psi_j| \geqslant \varepsilon_0 > 0$, $|\psi_j| \leqslant 1/\varepsilon_0$ for a fixed $\varepsilon_0 > 0$. We map the disk $|\zeta| < 1$ conformally onto D_τ (the derivative of such a mapping can be represented by (13)) and consider the resulting boundary-value problem (7), (8). Note that, since $dy/dx = \tan\theta(x)$ on streamlines,

$$y = \int_0^\gamma \tan\theta(\gamma)\frac{d\tau}{d\gamma}d\gamma + y_0 \equiv y(\theta \,|\, \gamma).$$

Unlike the case studied in Theorem 1, we must adjoin to the boundary conditions (7) the following equations in the images t_j of the vertices $b_j \in L_z$ at infinity and the images R_i of the points σ_i where the flow branches:

$$h_j = \lim_{\varepsilon \to 0} |y(\theta \,|\, \gamma_j + \varepsilon) - y(\theta \,|\, \gamma_j + \varepsilon)| \equiv H_j(t_k, \sigma_l),$$

$$x_i = \mathrm{Re}\left\{ \int_0^{\beta_i} \frac{d\tau}{d\gamma} d\gamma \right\} + x_0 = X(t_k, \sigma_l), \tag{20}$$

where the x_i are the fixed abscissas of the points R_i, $\gamma_j = \arg t_j$ and $\beta_i = \arg \sigma_i$. If the parameter λ introduced in front of the right-hand side of the boundary condition (1) vanishes, then the homogeneous boundary-value problem (7), (8) obviously can only have the trivial solution $\omega = 0$. Corresponding to this solution is a flow in the domain D_z bounded by lines parallel to the real axis, and, since the flow is uniform ($\omega \equiv 0$ and $q \equiv 1$), we have $|\psi_{j+1} - \psi_j| = h_j$. Thus, for $\lambda = 0$ the domain D_τ is fixed and the system (20) is uniquely solvable for the parameters of its mapping onto the disk $|\zeta| < 1$. Consequently, the initial problem has a unique solution. By what was proved above, D_τ is nondegenerate; therefore, all the images ζ_k of the points of discontinuity of $G(x)$ and $g(x, y)$ and the parameters t_j and σ_i, which are isolated on L_z, are also isolated on the circle $|\zeta| = 1$. We fix arbitrarily two adjacent parameters on the circle and let $t_* = e^{i\gamma_*}$ be any point between them. Then $t_j, \sigma_i, \zeta_k \neq t_*$ and, consequently, assuming that the solution $\omega(\zeta)$ is known, we can introduce the following continuous transformation of the Euclidean space into itself with the help of (20):

$$(t_j - t_*) = \frac{1}{h_j}(T_j - t_*)H_j(T_k, S_l), \qquad (\sigma_i - t_*) = \frac{1}{x_i}(S_i - t_*)X_i(T_k, S_l). \tag{20*}$$

The fixed points $t_j = T_j$ and $\sigma_i = S_i$ of this transformation on the set

$$|\sigma_{i+1} - \sigma_i| > \varepsilon_0, \qquad |t_{j+1} - t_j| > \varepsilon_0, \qquad |\sigma_i - t_j| > \varepsilon_0$$

obviously coincide with the solutions of (20). Thus, adjoining the relations (20*) to the equations that determine the transformation F_λ (Theorem 5 in §2 of Chapter VI) corresponding to the boundary-value problem (7), (8) does not change the properties of this transformation necessary for the Leray-Schauder theorem to be applicable. Consequently, for any $\lambda \in [0, 1]$ this problem with the supplementary conditions (20) on the parameters has (by Theorem 5 in §2 of Chapter VI) at least as many solutions as the problem (7) with fixed parameters has in the class of analytic functions. The theorem is proved.

REMARKS. 1. By the preceding theorem, for the assertions of Theorem 7 it suffices to assume only the a priori estimate (10) on the slope angle of the velocity and the isolation of the images of the points of discontinuity of $G(x)$ and $g(x, y)$ on the circle.

2. If the coefficients $G(x)$ and $\lambda g(x, y)$ of the boundary-value problem (1) belong to $\mathfrak{M}^*$, i.e., can be approximated by functions in the class $\mathfrak{M}(N_\varepsilon, \delta_\varepsilon)$, then the assertions of Theorem 7 remain in force (cf. Theorem 2), but now the unknown solution of the boundary-value problem (1), (3) is bounded everywhere except at the singular points of $G(x)$ and $g(x, y)$.

3. The assertions of Theorem 7 also remain true if the abscissas x_k of some of the points of discontinuity of $G(x)$ and $g(x, y)$ are not fixed, but the preimages t_k of these points on the circle $|\zeta| = 1$ under the conformal mapping of the disk $|\zeta| < 1$ onto the unknown domain D_z are given. In this case, nonsingular equations of the same form as for the branch points of the flow are appended to the system (20); for this reason the added equations do not change the properties of (20).

LEMMA 1 (comparison principle). *Suppose that the domains D_z and $\tilde{D}_z$ of two uniformly subsonic gas flows are bounded by the streamlines $\psi = 0$ (Γ_0 and $\tilde{\Gamma}_0$) and $\psi = 1$ (Γ_1 and $\tilde{\Gamma}_1$), with Γ_i, $\tilde{\Gamma}_i \in C^{2+\alpha}$, $\alpha > 0$, and, moreover, $\tilde{\Gamma}_0 \subset D_z$ and $\Gamma_1 \subset \tilde{D}_z$, $\tilde{\Gamma}_0 \cap \Gamma_1 = 0$. Then at the common points of the curves Γ_0, $\tilde{\Gamma}_0$ and the curves Γ_1, $\tilde{\Gamma}_1$ the flow speeds $q(z)$ and $\tilde{q}(z)$ corresponding to D_z and $\tilde{D}_z$ are related by the inequalities*

$$q(z_0) \leqslant \tilde{q}(z_0), \qquad z_0 \in \Gamma_0 \cap \tilde{\Gamma}_0, \tag{21}$$

$$q(z_1) \geqslant \tilde{q}(z_1), \qquad z_1 \in \Gamma_1 \cap \tilde{\Gamma}_1, \tag{22}$$

in which the equality sign is attained only if the flows coincide completely.

The stream functions $\psi(z)$ and $\tilde{\psi}(z)$ in D_z and $\tilde{D}_z$ satisfy the quasilinear equation (25) of §1.

Let us form the equation for the difference $\Psi(z) = \psi(z) - \tilde{\psi}(z)$, which is defined in the domain $D = D_z \cap \tilde{D}_z$ with boundary $\Gamma = \Gamma_0 + \tilde{\Gamma}_1$:

$$A\Psi_{xx} + 2B\Psi_{xy} + C\Psi_{yy} + E_1\Psi_x + E_2\Psi_y = 0, \tag{23}$$

where A, B, and C are the coefficients of (1.25) for $\psi(z)$,

$$E_i = A_i\tilde{\psi}_{xx} + 2B_i\tilde{\psi}_{xy} + C_i\tilde{\psi}_{yy},$$

$$A_2 = \frac{A - \tilde{A}}{\psi_y - \tilde{\psi}_y}, \qquad A_1 = \frac{A - \tilde{A}}{\psi_x - \tilde{\psi}_x}, \qquad \tilde{A} = A(\tilde{\psi}_x, \tilde{\psi}_y),$$

and B_i and C_i are defined similarly. By the assumed smoothness of the boundaries $L_z = \partial D_z$ and $\tilde{L}_z = \partial \tilde{D}_z$, the coefficients of (23) are obviously Hölder continuous in D. Taking account of the fact that the mappings of D_z and $\tilde{D}_z$ onto the strip $0 < \psi < 1$ are quasiconformal, we get on the boundary of D that $\Psi = \psi - 1 \leqslant 0$ on $\tilde{\Gamma}_1$ and $\Psi = -\psi \leqslant 0$ on Γ_0. Consequently, according to the maximum principle for the solutions of (23) (Theorem 2 in §4 of Chapter VI), we have $\Psi(z) < 0$ if $z \in D$. Then Zaremba's theorem (Theorem 3 in §4 of Chapter VI) gives us that at points $z_0 \in \Gamma_0 \cap \tilde{\Gamma}_0$ and $z_1 \in \Gamma_1 \cap \tilde{\Gamma}_1$ where $\Psi(z) = 0$ we have

$$\frac{\partial \Psi}{\partial n} = \frac{\partial \psi}{\partial n} - \frac{\partial \tilde{\psi}}{\partial n} > 0.$$

From (1.5) we find at the point z_0 (assuming without loss of generality that the tangent to Γ_0 and $\tilde{\Gamma}_0$ at this point is directed along the OX-axis) that

$$-\frac{\partial \psi}{\partial n} = \psi_y = q\rho(q) < \tilde{q}\rho(\tilde{q}) = -\frac{\partial \tilde{\psi}}{\partial n}, \qquad z = z_0,$$

and, since

$$\frac{d}{dq}(q\rho) = \rho(1 - M^2) > 0$$

for a uniformly subsonic flow, it follows that the inequality obtained above is equivalent to (21). The proof of (22) is similar.

REMARK. It is obviously sufficient to keep the assumptions in the lemma about the high degree of smoothness of the curves Γ_i and $\tilde{\Gamma}_i$ only in a neighborhood of the points z_0 and z_1 at which the flow speeds are compared.

Suppose that a domain D in the complex plane is bounded by disjoint curves Γ_1 and Γ_0 having the following property: For each $z \in \Gamma_0$ there is a line segment $I(z)$ with endpoint at z and length not less than a fixed number $d > 0$ that lies outside D. Let h be the distance between the curves Γ_0 and Γ_1, and let the points $z_j \in \Gamma_j$ be such that $|z_1 - z_0| = h$.

LEMMA 2 (local comparison principle). *Let $\psi(z)$ be the stream function of a uniformly subsonic flow in D having a normal derivative at the point $z_1 \in \Gamma_1$ ($\psi = 0$, Γ_0: $\psi = 1$, Γ_1). There exists a constant $\sigma(d) > 0$ such that*

$$(\partial \psi / \partial n)(z_1) > \sigma(d)/h.$$

Without loss of generality we assume that $z_0 = 0$. Let Q be the doubly connected domain bounded by the segment $I(z_0)$ and the curve Γ, the envelope of the family of circles of unit radius with centers on $I(z_0)$.

We consider a solution $v(z)$ of the quasilinear equation

$$A\left(\frac{\Delta v}{h}\right)v_{xx} + 2B\left(\frac{\Delta v}{h}\right)v_{xy} + C\left(\frac{\Delta v}{h}\right)v_{yy} = 0$$

that is continuous in $\overline{Q}$ and satisfies on the boundary ∂Q the conditions

$$v(z) = 0, \quad z \in I(z_0), \qquad v(z) = 1, \quad z \in \Gamma,$$

where A, B, and C are the coefficients of (25) in §1 for the stream function $\psi(z)$. Observing that $v(z)$ satisfies the system

$$\rho\left(\frac{|\nabla u|}{h}\right)u_x = v_y, \qquad \rho\left(\frac{|\nabla u|}{h}\right)u_y = -v_x,$$

we map Q by a solution of the linear Beltrami equation

$$\zeta_{\bar{z}} - \frac{\rho - 1}{\rho + 1}\zeta_z = 0, \qquad \rho = \rho\left(\frac{|\nabla u|}{h}\right),$$

onto the annulus K_h: $0 < r(d) < |\zeta| < 1$. Then the harmonic function $v[z(\zeta)]$ in K_h is determined by

$$v = 1 - \ln|\zeta|/\ln r,$$

which, since the image of a domain under a quasiconformal mapping is nondegenerate, gives us

$$v(z) \leq c_1(\lambda) < 1,$$

where λ is the distance from z to Γ. The last fact, by the remark after Lemma 2 in §4 of Chapter VI, implies the inequality

$$(\partial v/\partial n)(z) \geq \sigma(d) > 0, \qquad z \in \Gamma.$$

The function $v_h(z) = v(z/h)$ satisfies (25) in §1 and is defined in the domain Q_h which is the image of Q under the transformation $z^1 = hz$. By construction, the function $\psi - v_h$ is nonnegative on the boundary of $D \cap Q_h$ and vanishes at z_1. From this, as in the proof of Lemma 1, we have

$$\frac{\partial \psi}{\partial n}(z_1) \geq \frac{\partial v_k}{\partial n}(z_1) = \frac{1}{h}\frac{\partial v}{\partial n}\left(\frac{z_1}{h}\right) \geq \frac{\sigma(d)}{h}.$$

The comparison principles enable us in many problems to establish that the boundary of the flow domain is nondegenerate and thereby to estimate the constant h in the estimate (19) of the flow velocity. One important problem in which this can be done is the mixed boundary-value problem with free boundary in which the boundary condition (1) takes the form

$$\text{Im}\,\omega(z) = -\theta(x), \quad z \in L_z^1, \qquad \text{Re}\,\omega(z) = q^*(x), \quad z \in L_z^1. \qquad (24)$$

We make the following assumptions:

(i) *The specified curve L_z^1 is nondegenerate in the sense of the definition given in Theorem 5 and has two points of contact with the unknown boundary L_z^2.*

(ii) *The functions $\theta(x)$ and $q^*(x)$ satisfy requirements* (a) *and* (b) *on the coefficients of the boundary condition* (1), *and, in addition,*

$$q^*(x) \in C^{1+\alpha}, \quad \alpha > 0, \qquad |\theta(x)| \leq \pi/2 - \delta, \quad \delta > 0.$$

(iii) *The value of the outflow $Q_k = \psi_{k+1} - \psi_k$ between adjacent streamlines in the boundary of the flow domain is given if at least one of them belongs to L_z^2, i.e., its equation is unknown.*

THEOREM 8. *Under conditions* (i)–(iii) *suppose that there exists a bounded solution $\omega = \omega(z)$ of the mixed boundary-value problem with free boundary* (3), (24), *for which solution $|\theta(z)| \leqslant \pi/2 - \delta$, $z \in \bar{D}_z$ ($\delta > 0$). Then the boundary $L_z = L_z^1 + L_z^2$ of D_z is nondegenerate.*

Let us first consider the case when the image of D_z is the strip D_w: $0 < \operatorname{Im} w = \psi < Q$. We estimate upper and lower bounds for the value of the outflow Q, assuming that neither of the streamlines Γ_0: $\psi(z) = 0$ nor Γ_1: $\psi(z) = Q$ belongs entirely to the free boundary L_z^2, for otherwise Q is assumed to be given, according to condition (iii).

Let $\Gamma_i^1 \subset \Gamma_i$, $i = 0, 1$, denote the given disjoint curves ($\Gamma_0^1 + \Gamma_1^1 = L_z^1$) that form one of the vertices $b_j \in L_z$ at infinity (for definiteness, the one at which $x = -\infty$), and let $z_i^1 \in \Gamma_i^1$ ($\operatorname{Re} z_0^1 \leqslant \operatorname{Re} z_1^1$) be the points where L_z^1 is in contact with L_z^2. Fix $\varepsilon > 0$ so that the disk K_ε: $|z - z_0^1| < \varepsilon$ does not contain points of Γ_1^1. We estimate the magnitude and the slope angle of the flow velocity locally on the part of the boundary of D_z in K_ε, as in the proofs of Theorems 5 and 6. If we then move along Γ_0^1 from z_0^1 upstream, we thus get an estimate for the magnitude of the velocity everywhere on Γ_0^1, including a neighborhood of the point at infinity, and thereby also an estimate for the value of the outflow Q, since the flow depth at infinity is given.

Returning to the general problem, we remark that a completely analogous situation also holds in this case. Indeed, if the known boundary is not located on a single streamline, then its parts must necessarily form a vertex at infinity (by assumption, there are only two points of contact of the unknown boundary and the given boundary!). The values of the outflows between all the given streamlines on the boundary L_z can then be estimated according to the scheme indicated above. But if a streamline belongs entirely to the free boundary, then the outflow between it and an adjacent streamline is given. Thus, we have proved that the image L_w of L_z is nondegenerate under the assumptions of the theorem.

A uniform lower estimate for the distance $h_k = r(\Gamma_k, \Gamma_{k+1})$ between adjacent streamlines Γ_k: $\psi = \psi_k$ can be obtained by applying Lemma 2 directly. Indeed, if the distance $h_k = |z_k - z_{k+1}|$ is realized at points $z_i \in \Gamma_i \cap L_z^1$ ($i = k, k + 1$), then it is given, and $h_k \geqslant h > 0$, since the given boundary L_z^1 is nondegenerate. But if $z_i \in \Gamma_0 \cap L_z^2$ for at least one of the points, then the magnitude $q_i = q(x_i)$, $x_i = \operatorname{Re} z_i$, of the velocity is known at this point, and then, by Lemma 2,

$$h_k > \sigma(1) \Big/ \left| \frac{\partial \psi}{\partial n}(z_i) \right| = \frac{\sigma(1)}{q_i \rho(q_i)} \geqslant h(q_0^*) > 0,$$

where $q_0^* = \max_{z \in L_z^2} q^*(z)$. To prove that the boundary $L_z = \Sigma \Gamma_k$, $\Gamma_k \colon \psi = \psi_k$ is nondegenerate it now suffices to get a uniform upper estimate for the quantities

$$H_{k,i} = \max_{x \in I_k \cap I_i} |y_k(x) - y_i(x)|,$$

where $y = y_k(x)$, $x \in I_k$, is the equation of the streamline Γ_k. Let us begin again with the case where $D_w \colon 0 < \operatorname{Im} w = \psi < Q$ ($\neq 0, \infty$). We first investigate the smoothness of L_z. By mapping fixed neighborhoods of the points of contact of the given and the unknown arcs L_z^1 and L_z^2 conformally onto the disk $|\zeta| < 1$, as in the proof of Theorem 5 we find that $L_z \in C^{1+\alpha}$ in these neighborhoods. We next consider the following boundary-value problem for (9):

$$\operatorname{Re} \omega(w) = q^*[x(\varphi)], \quad w \in L_w^2, \qquad \operatorname{Im} \omega(w) = -\theta[x(\varphi)], \quad w \in L_w^1.$$

Even if one of the endpoints of L_z^2 lies in the finite part of the plane, its image on the boundary of D_w is assumed to be fixed.

By the condition of the theorem, $\theta(x)$ is bounded, and $q^*[x(\varphi)] \in H^1(L_w^2)$ because $|dx/d\varphi|$ and $|dy/d\varphi|$ are bounded; therefore, the solutions of the boundary-value problem (24), (9) are Hölder continuous in any domain not containing points of L_w^1.

Then $x(\varphi) \in C^{1+\beta}(\tilde{L}_w^2)$ ($\tilde{L}_w^2 \subset L_w^2$, $\tilde{L}_w^2 \cap L_w^1 = 0$), and thus $\omega(\varphi) \in C^{1+\alpha}(\tilde{L}_w^2)$. Consequently, the image $\tilde{L}_z^2$ of $\tilde{L}_w^2$ under the mapping $z = z(w)$ is a smooth curve of class $C^{2+\alpha}$, and the whole boundary L_z belongs to the class $C^{1+\alpha}$.

Suppose for definiteness that $\Gamma_1 \colon \psi(z) = Q$ contains all or part of the free boundary L_z^2. We take a copy of the curve Γ_0 and, moving it along the OY-axis, bring it into contact with the curve from outside the domain D_z. Let $\tilde{\Gamma}_1$ be the new position of Γ_0, and $\tilde{D}_z$ the domain with boundary $\tilde{L}_z = \Gamma_0 + \tilde{\Gamma}_1$. If the contact takes place at the point $z_1 \in \Gamma_1 \cap L_z^1$, then the distance $H = r(\Gamma_0, \tilde{\Gamma}_1)$ is given, and, consequently, the quantity $H_{0,1}$ introduced earlier is bounded above by the fixed constant H. In the case where $z_1 \in \Gamma_1 \subset L_z^2$, we consider auxiliary gas flows in D_z and $\tilde{D}_z$ with complex potential $w^0 = \lambda w$, $\lambda = H/Q$, that map the flow domains onto the strip $0 < \psi^0 < H$. Here the magnitude $\tilde{q}_0$ of the flow velocity in $\tilde{D}_z$ is obviously determined only by the smoothness of the boundaries and does not depend on H.

The lemma gives us that

$$q^0(z_1) \leqslant \tilde{q}^0(z_1), \qquad z_1 \in \Gamma_1 \cap \tilde{\Gamma}_1.$$

Returning to the actual flow, we get from the estimate of the corresponding quasiconformal mapping of the strip $0 < \psi^0 < H$ onto the strip $0 < \psi < Q$ that

$$\tilde{q}^0(z_1) \geqslant K \frac{H}{Q} \min_{L_z^2} |q(x)|,$$

where the constant K does not depend on H. Thus, we have obtained an estimate of H, and thereby also of $H_{0,1}$. In the general case the quantities $H_{k,k+1}$ for adjacent streamlines Γ_k: $\psi = Q_k$ can be estimated in a completely analogous way by applying the comparison principle to the domains D_k, $\partial D_k = \Gamma_k \cup \Gamma_{k+1}$, and to the smoothly curved half-strips $\tilde{D}_k$. The domains $\tilde{D}_k$ are constructed on the basis of one of the streamlines $\psi = Q_k$ or $\psi = Q_{k+1}$ that has a branch point, similarly to the construction of $\tilde{D}_z$. The theorem is proved.

5°. Univalence. We now investigate the univalence of the mapping $z = z(w)$. Since $z = z(w)$ can be represented in the form $z = \Phi[W(w)]$, where $W(w)$ is a homeomorphism of D_w onto itself and $\Phi(W)$ is an analytic function, it suffices for us to restrict ourselves to the case of an ideal fluid. Suppose first that D_w is a half-plane or strip and the original problem belongs to the class $\mathfrak{M}(N, \delta)$. Choosing α so that

$$\left| \alpha - \arg \frac{dz}{dw} \right| < \frac{\pi}{2}, \qquad w \in \partial D_w = L_w$$

(this is possible because of (6)), we get

$$\mathrm{Re}\left[e^{i\alpha} \frac{dz}{dw}(w) \right] \geqslant 0, \qquad w \in L_w.$$

By assumption, D_w is convex; therefore, Theorem 5 in §3 of Chapter II gives us that $z = z(w)$ is univalent. In the case of an arbitrary domain D_w bounded by straight lines $\psi = \mathrm{const}$ we make additional cuts along these lines and come to a mapping of the strips and half-planes D_w^k $(k = 1, \ldots, n)$ so obtained onto the corresponding subdomains D_z^k of the flow domain.

By what has been proved, the mapping $z = z(w)$ is univalent in each of the domains $\overline{D}_w^k$ $(k = 1, \ldots, n)$ and, generally speaking, is not univalent in $\overline{D}_w$. Thus, we have proved the following theorem.

THEOREM 9. *In the boundary-value problems* (1), (3) *and* (2), (3) *of class* $\mathfrak{M}(N, \delta)$ *in hydrodynamics and gas dynamics the mapping* $z = z(w)$ *is univalent in the domain* $\overline{D}_w$ *if this domain is a strip or half-plane, and in the general case it* *is univalent in each of the strips and half-planes $\overline{D}_w^k$ into which D_w is partitioned by the lines $\psi = \psi_k$ (the constants ψ_k are the values of the function $\psi = \psi(x, y)$ on L_z).*

REMARKS. 1. Since $x(w)$ is a strictly monotone function of $\operatorname{Re} w$ in boundary-value problems of class $\mathfrak{M}(N, \delta)$, the inequality

$$\tau(w_1) - \tau(w_2) \geqslant \min\{| x(w_1) - x(w_2) | , | \operatorname{Im} w_1 - \operatorname{Im} w_2 |\}$$

implies the assertion of Theorem 9 for the mapping $\tau = \tau(w)$ of D_w onto D_τ and, consequently, also for the mapping $z = z(\tau)$ of D_τ onto D_z.

2. If the coefficients $G(s)$ and $g(s)$ in the boundary conditions (1) and (2) are given as functions of the arclength parameter of the boundary L_z of the flow domain, then it is more convenient to study these problems in the plane of $\zeta = s + i\psi$ instead of in the plane of $\tau = x + i\psi$. However, it should be pointed out that, depsite the uniform ellipticity of (1.18), it is not possible to get the needed a priori estimates (not depending on $\theta(z)$) for solutions in the case of a compressible fluid if only the single condition $| q^* |< N < \infty$ is satisfied. We do not go into details but restrict ourselves to the remarks made, also taking account of the fact that such problems have already been studied for an incompressible fluid in §§2 and 3 of Chapter IV.

§3. Existence and uniqueness theorems in boundary-value problems in hydrodynamics and subsonic gas dynamics

A flow is assumed to be a potential flow in what follows if nothing special is stipulated. The simplest problems to which the results of the preceding chapter can be applied are the following.

1°. *Problems on constantly subsonic gas flows in given domains bounded by rigid walls.*

1. *A subsonic gas flow in a channel or over an uneven bottom (a channel of infinite depth).* The quantities q_∞, θ_∞, and h_∞ are assumed to be given. Problems on a symmetric subsonic flow past a contour in a channel or with an unbounded flow obviouly can also be reduced to these problems if we consider half the flow domain (Figure 1).

FIGURE 1

2. *Subsonic gas flow without separation past a given curvilinear profile.* We assume that the coordinates of the points R_1 and R_2 (where the flow divides and comes together) are given, along with the magnitude q_∞ of the velocity at infinity. The slope angle θ_∞ of the velocity at infinity and the circulation $\Gamma = \int_{L_z} d\varphi$ are determined after the problem is solved (Figure 2).

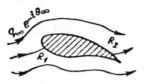

FIGURE 2

3. *The general problem of a constantly subsonic gas flow in a given domain D_z with finitely many vertices b_j ($j = 1,\dots,m$) at infinity and branch points R_i ($i = 1,\dots,n$) on L_z, where $n = m - 2$, $m - 1$, or m when the simply connected single-sheeted domain D_w in the plane of $w = \varphi + i\psi$ has the shape of a rectilinear strip with cuts, a half-plane with cuts, or a plane with cuts, respectively.* The geometry of the boundary L_z of the flow domain is assumed to be completely known, along with the coordinates of all the points R_i ($i = 1,\dots,n$) where the flow branches and the magnitude q_∞ of the flow velocity at one of the vertices b_j at infinity (Figure 3):

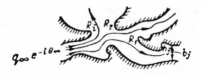

FIGURE 3

It is assumed in all the above problems that the slope angle $\theta = \theta(x)$ of the flow velocity on L_z is a piecewise Hölder function, and that $e^{|x|}[\theta(x) - \theta(\infty)]$ $\to 0$ as $|x| \to \infty$ in a neighborhood of all the vertices b_j. Mapping the disk $|\zeta| < 1$ conformally onto the given domain D_z and taking into account the fact that all the boundaries of the flow domain are streamlines, we arrive at the following boundary-value problem:

$$\operatorname{Re}\left[ie^{i\theta_0(\gamma)}\Omega(e^{i\gamma})\right] = 0, \qquad \gamma \in [0, 2\pi],$$
$$\operatorname{Re}\left[e^{i\theta_\infty}\Omega(\zeta_0)\right] = \exp(q_\infty^*), \tag{1}$$

where $\Omega(\zeta) = \exp\{\omega(\zeta)\}$, $\theta_0(\gamma) = \theta[x(\gamma)]$, and ζ_0 is the fixed image of the point at infinity at which the magnitude q_∞ of the flow velocity (and thus also q_∞^*) is given. The generalized complex flow velocity $\Omega = \Omega(\zeta) \equiv (u + iv)e^{\sigma(q)}$ (the formula (1.20)) satisfies a transformed equation of the form (1.14)

$$\Omega_{\bar\zeta} - \hat\mu(\ln\Omega, \zeta)\Omega_\zeta = 0, \tag{2}$$

where

$$\hat\mu(\ln\Omega, \zeta) = \mu^z(\ln\Omega)\exp\left\{-2i\arg\frac{dz}{d\zeta}\right\}, \qquad |\hat\mu(\ln\Omega, \zeta)| \leqslant \mu_0 < 1.$$

We must look for a solution of the boundary-value problem (1), (2) that is bounded at the points of discontinuity of $\theta(\gamma)$ corresponding to branch points and corner points of L_z with angle less than π, and bounded at the remaining points of discontinuity of $\theta(\gamma)$. If the fluid is ideal, then $\Omega = e^\omega \equiv dw/dz$ (the complex flow velocity), and the boundary-value problem (1) is uniquely solvable as the problem of finding the derivative of a conformal mapping of the given domain D_z onto a domain D_w bounded by straight lines parallel to the real axis. Consequently, since the boundary-value problem (1) is uniquely solvable for analytic functions, Theorem 1 in §2 of Chapter VI tells us that it has at least one solution for the quasilinear equation (2). Moreover, if there are no corner points on the boundary of D_z with angles greater than π, then the unknown solution is bounded and, hence (by Theorem 2 in §2 of Chapter VI), unique.

THEOREM 1. *Suppose that a constantly subsonic gas flow takes place in a simply connected domain D_z bounded by given rigid walls on which there are finitely many vertices b_j ($j = 1,\ldots,m$) at infinity and given branch points R_i ($i = 1,\ldots,n$) of the flow, where $n = m - 2, m - 1$, or m, depending on the geometry of D_z (see problem 3). If the slope angle $\theta(x)$ of the velocity is a piecewise Hölder function and $(e^{|x|}|\theta(x) - \theta(\infty)|) \to 0$ as $|x| \to \infty$, then problems $1°.1$–$1°.3$ each have at least one solution, and these solutions are unique when there are no corner points with angles greater than π on L_z.*

REMARK. Theorem 1 can be interpreted as an existence theorem for a quasiconformal mapping of D_z onto a domain D_w bounded by lines parallel to the real axis by a solution of the nonlinear (and not quasilinear!) system of equations of gas dynamics (1.1). In this sense Theorem 1, applied to problem $1°.1$, is a generalization of a known theorem of Lavrent'ev [61] on the existence of mappings by solutions of nonlinear systems of equations. Theorem 1 applied to Problem $1°.2$, generalizes significantly results of Bers [16] for an analogous problem. It should also be mentioned that the solvability of the mapping problem for L-elliptic systems of equations (of which a particular case is the system of equations of gas dynamics) was proved by another method in §8 of Chapter V.

The problems discussed next are considerably more difficult than the problem studied above about gas flow in known domains.

2°. Problems of constantly subsonic gas flows with partially unknown (free) boundaries of the flow domain.

1. *The problem of a semiconfined subsonic gas flow in a domain with a partially unknown boundary determined from a given distribution of the flow speed* $q = q(x)$ *on it.* We are given θ_∞ if the boundary of the flow domain is known in a neighborhood of $z = \infty$, and we are given q_∞ otherwise (Figure 4).

FIGURE 4

2. *A subsonic gas flow with jet separation past a given curvilinear arc according to the Kirchhoff scheme.* The coordinates of the points A, B, and R are given along with the quantity q_∞ ($= 1$), and θ_∞ is sought (Figure 5).

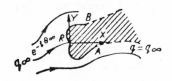

FIGURE 5

3. *The mixed problem of determining a wing profile from a chord diagram when the speed distribution* $q = q(x)$ *of a subsonic flow is specified on the unknown part of this profile.* It is assumed that the coordinates of R_1 and the abscissa x_2 of R_2 are known, while q_∞ and θ_∞ are sought (Figure 6).

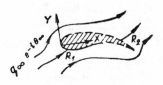

FIGURE 6

4. *The general problem of a subsonic gas flow in an arbitrary domain D_z with finitely many branch points $R_i \in L_z$ $(i = 1, \ldots, n)$ of the flow and vertices b_j $(j = 1, \ldots, m)$ at infinity with angles 0, $-\pi$, and -2π, (m and n are related as in problem $1°.3$), where the speed distribution $q = q(x)$ of the flow is given on the unknown part of L_z (see Figure 3).* It is assumed that the free boundary has two points in common with the known part L_z^1 of L_z, and the outflow between adjacent streamlines $\Gamma_k \subset L_z$ is given if $\Gamma_k \cap L_z^1 = \varnothing$ for one of them.

5. *A cavitational subsonic gas flow past a system of nonsymmetric curvilinear obstacles according to a Riabouchinsky scheme.* It is assumed that the arcs AR_1B and CR_2D are joined by a partition R_1R_2 whose length is unknown, as is the quantity l. Different constant pressures p_0 and p_1 are maintained in the cavities $R_1R_2DAR_1$ and $R_1BCR_2R_1$, and, by Bernoulli's law, constant speeds $q = q_0$ and $q = 1$ correspond to these pressures. The quantities l, q_0, q_∞, and θ_∞ are sought along with the solution of the problem. This problem is a special particular case of problem $2°.3$ in which two parts of the wing profile flowed past are specified (Figure 7).

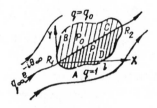

FIGURE 7

6. *Subsonic gas flow in a domain on whose boundary finitely many nonclosed arcs are given.* This problem is formulated in the same way as problem 4, the only difference being that here the boundary L_z of the flow domain D_z consists of finite families of arcs L_1^k and L_2^k $(k = 1, \ldots, n)$ that alternate on L_z in a specified order.

On the arcs L_1^k we are given the slope angle of the tangent to the OX-axis:

$$\theta = \theta(\xi), \qquad l_k^0 x + l_k = \xi \in [\xi_k, \xi_{k+1}],$$

where the ξ_k are given and the l_k^0 and l_k are unknown parameters, of which two, l_i^0 and l_i, are fixed arbitrarily. Thus, the arcs L_1^k are given to within a dilation and a relative translation along the real axis.

The magnitude of the velocity on the free boundary $L_2 = \Sigma_1^n L_2^k$ is given in such a way that

$$q^* = q_0^*(x) + q_k, \qquad z \in L_2^k \, (k = 1, \ldots, n),$$

where the function $q_0^*(x)$ is defined on the whole boundary $L_z = \Sigma_1^n (L_1^k + L_2^k)$ of the flow domain ($q_0^*(x) \equiv 0$ in cavitation problems), and the q_k are parameters to be determined, except for one which can be fixed arbitrarily. It is assumed that the arcs L_2^k are arranged in a finite part of the plane, and only one of them can be infinite, in which case the corresponding parameter q_k is fixed, $q_k = 0$.

Obviously, the preceding problem is a particular case of that just formulated. We shall dwell later on the investigation of problems $2°.5$ and $2°.6$. Let us first consider problems $2°.1$–$2°.4$, in each of which the generalized Joukowsky function $\omega(z) = q^*(z) - i\theta(z)$ satisfies the boundary conditions

$$\operatorname{Im}\omega(x) = -\theta(x) \quad \text{on } L_z^1, \qquad \operatorname{Re}\omega(x) = q^*(x) \quad \text{on } L_z^2, \qquad (3)$$

where L_z^1 is the given part of the boundary L_z of D_z, and L_z^2 is the free boundary. The functions $\theta(x)$ and $q^*(x)$ in the boundary condition (3) are assumed to satisfy the following requirements (cf. §2):

(a) $\theta(x) \in C^\alpha$ and $|\theta(x)| \leqslant \pi/2 - \delta_0$, $\delta_0 > 0$, on L_z^1, while $q^*(x) \in C^{1+\alpha}$, $\alpha > 0$, on L_z^2.

(b) In a neighborhood of the vertices $b_j \in L_z$ at infinity

$$G(x) \equiv e^{i\beta_0}, \qquad \beta_0 = 0, \pi/2, G[x(\xi)], g[x(\xi)] \in C^\alpha,$$

where $\xi = |x|^{-k}$ for the angles $-\pi$ and -2π and $\xi = e^{-|x|}$ for zero angles, the quantity $k > 0$ depending on the original problem. According to condition (b), each branch point R_i of the flow is a cusp point with the angle 2π, i.e., at such points the obstacles flowed past have a "beak" directed toward the flow domain.

THEOREM 2. *Under conditions* (a) *and* (b) *on the given quantities* $\theta(x)$ *and* $q^*(x)$, *if there exists a generalized Joukowsky function* $\omega(z) = q^*(z) - i\theta(z)$ *that is bounded in* D_z, *then the slope angle of the flow velocity satisfies the inequality*

$$|\theta(z)| \leqslant \pi/2 - \delta, \qquad z \in D_z \ (\delta > 0). \qquad (4)$$

By Theorem 8, L_z is nondegenerate; therefore, (4) follows immediately from Theorem 6.

REMARK. The boundedness of dq/dx (condition (a)) is an essential assumption about the given magnitude $q(x)$ of the velocity: it ensures that L_z^2 does not have stagnation points ($\theta = \pi/2$). Elementary examples show that the violation of this condition can lead to the appearance of stagnation points of the flow. Indeed, let us consider the problem of an irrotational flow of an ideal fluid past the unit disk, with $e^{i(\pi+\alpha)}$ and $e^{i\alpha}$ the points at which the flow divides and comes together, respectively, and with $q_\infty = 1$ the speed at infinity.

The complex potential of such a flow has the form $w(z) = e^{i\alpha}z + e^{-i\alpha}\bar{z}^{-1}$, and

$$q(x) = |dw/dz| = 2|\sin \alpha + \arccos x|,$$

which implies that $dq/dx|_{x=1} = \infty$.

THEOREM 3. *Under conditions* (a) *and* (b) *on the given quantities* $\theta = \theta(x)$ *and* $q = q(x)$ *problems* $2°.1$–$2°.3$ *are uniquely solvable, and problem* $2°.4$ *has at least one solution. If in problem* $2°.4$ *the magnitude* $q = q(x)$ *of the velocity is given in a neighborhood of all the vertices* $b_j \in L_z$ *at infinity with zero angle, then the solution of this problem is also unique.*

Since the coefficients of the boundary condition (3) in problems $2°.1$–$2°.4$ have only two points of discontinuity (the junction points if L_z^1 and L_z^2), the images of these points under a conformal mapping of the disk $|\xi| < 1$ onto D_z can be located at fixed points, for example, at $\zeta = \pm 1$. Then the a priori estimate (4) in Theorem 4 implies (by Theorem 5 in §2) that the function $q^*(z)$ is bounded in $\bar{D}_z$, and thereby that problems $2°.1$–$2°.4$ belong to the class $\mathfrak{M}(N, \delta)$. Consequently, these problems have at least one solution each, by Theorems 1 and 7 in §2. Since the domain D_τ in the plane of $\tau = x + i\psi$ in problems $2°.1$–$2°.3$ is fixed, Theorem 1 in §2 gives us that the solution of each is unique. If in problem $2°.4$ the magnitude $q = q(z)$ of the velocity is given in a neighborhood of all the vertices $b_j \in L_z$ at infinity with zero angle, then the outflow $Q_j = \psi_{j+1} - \psi_j$ can be computed explicitly in terms of the given magnitude $q_\infty^j = q(\infty)$ of the flow velocity and the flow depth h_j. Thus, in this case D_τ is known also for problem $2°.4$, which ensures (by Theorem 1 in §2) that its solution is unique. This proves the theorem.

Let us now consider the case where the magnitude q of the velocity in problems $2°.1$–$2°.4$ depends on x and y. We require in addition to conditions (a) and (b) that

(c)

$$q^*(x, y) = q_0^*(x) + \varepsilon g(x, y), \qquad \varepsilon > 0,$$

where $q_0^*(x)$ satisfies (a) and (b), $g(x, y) \in C^{1+\alpha}$, $\alpha > 0$, with respect to x and y, $g(x, y) \equiv 0$ when $N_0 > x$, and $g(x, y)$ satisfies conditions (a) and (b) for fixed y. Suppose that for all $\varepsilon > 0$ there exists a bounded generalized Joukowsky function $\omega(z) = q^*(z) - i\theta(z)$ satisfying the boundary condition (3) with $q^*(x, y)$. As in the proof of Theorem 2, the continuous differentiability of $q(x, y)$ on L_z^2 implies the smoothness of L_z^2 and, consequently, the validity of the a priori estimate (4) for sufficiently small $\varepsilon > 0$. Indeed, supposing the contrary, we arrive at once at a contradiction to the estimate on $\theta(z)$ proved in Theorem 2, which corresponds to the case $\varepsilon = 0$. According to Theorem 5 in §2, inequality (4) for $\theta(z)$ implies that problems $2°.1$–$2°.4$ belong to the class

$\mathfrak{M}(N, \delta)$ and thereby, by Theorems 1 and 7 in §2, that these problems are solvable. We have thus proved the following theorem.

THEOREM 4. *Suppose that the magnitude of the flow velocity on the free boundary in problems* $2°.1$–$2°.4$ *depends on the two variables* x *and* y *and that the functions* $q(x, y)$ *and* $\theta(x)$ *satisfy conditions* (a)–(c). *Then each of the problems* 1–4 *has at least one solution for sufficiently small* $\varepsilon > 0$.

We pass to an investigation of problems $2°.5$ and $2°.6$ in which the generalized Joukowsky function $\omega(z) = q^*(z) - i\theta(z)$ is a solution of the mixed problem with parameters

$$\operatorname{Re}\omega(x) = q_k^*(x) + q_k, \quad z \in L_2^k, \quad \operatorname{Im}\omega(x) = -\theta\big(l_k^0 x - l_k\big), \quad z \in L_1^k, \quad (5)$$

where the q_k, l_k^0 and l_k are unknown quantities.

As in problems $2°.1$–$2°.4$, we assume that the given functions $\theta_k(x)$ and $q^*(x)$ $(= q_k^*(x)$ for $z \in L_2^k)$ satisfy conditions (a) and (b) on the respective arcs L_1^k and on $L_z = \Sigma_1^n(L_1^k + L_2^k)$. According to Theorem 2, if there exists a bounded function $\omega = \omega(z)$, then these conditions ensure the a priori estimate (4) in $\overline{D}_z$. We map the disk $|\zeta| < 1$ conformally onto the unknown domain D_z and determine the parameters l_k^0 and l_k by specifying the preimages t_k $(k = 1, \ldots, 2n)$ on the circle of the junction points of the given and unknown arcs L_1^k and L_2^k. By Theorem 5 of §2, the unknown solution $\omega(\zeta)$ of the boundary-value problem (5) for the uniformly elliptic equation (14) in §1, transformed to the ζ-plane, is then uniformly bounded in $C(\overline{K})$, which together with (4) implies that problems $2°.5$ and $2°.6$ belong to the class $\mathfrak{M}(N, \delta)$. We remark that the proof that $\omega(\zeta)$ is bounded is essentially based on uniform estimates in $C^\alpha(\overline{K})$ for solutions of a mixed boundary-value problem of the form (5) with Hölder coefficients for uniformly elliptic systems of equations (§2.4° in Chapter VI). Finally, using an additional conformal mapping of D_τ onto the disk $|\zeta| < 1$ to transform the boundary-value problem (5) for equation (15) in §1 in the plane of $\tau = x + i\psi$, we get, according to Theorem 7 in §2 (see also Remark 3 after this theorem), that there is at least one solution for each of the problems $2°.5$ and $2°.6$, which belong to the class $\mathfrak{M}(N, \delta)$. The following theorem has thus been proved.

THEOREM 5. *Under conditions* (a) *and* (b) *there is at least one solution to each of the problems* $2°.5$ *and* $2°.6$ *on subsonic cavitational flow according to a scheme of Riabouchinsky type.*

Let us now study problems $2°.1$–$2°.6$ when condition (a) or (b) is violated. Suppose that the slope angle $\theta = \theta(x)$ of the velocity on the given parts L_z^1 of

L_z and the magnitude $q = q(x)$ of the flow velocity on the free boundary satisfy condition (b), but that (a) is replaced by the condition

(a*) $\theta(x) \in C^\alpha$ and $|\theta(x)| \leqslant \pi/2$ on L_z^1,

and, moreover, that there are only finitely many points of discontinuity of $\theta(x)$ and points where $\theta(x) = \pm\pi/2$, while $q^*(x) \in C^{1+\alpha}$ everywhere on L_z^2 except for finitely many points at which either the functions only fail to be differentiable, or $q^*(x_0 + 0) \neq q^*(x_0 - 0)$ (the cavitational scheme of Thullen) or $q^*(x_0) = \infty$.

We show that the boundary-value problems 2°.1–2°.6 belong to $\mathfrak{M}^*$ in this case, i.e., their solutions can be approximated by solutions in the classes $\mathfrak{M}(N_e, \delta_e)$. Indeed, suppose that $|\theta(x_0)| = \pi/2$ or $\theta(x_0 + 0) \neq \theta(x_0 - 0)$ at a point that is not a point where the flow divides or comes together. We introduce a function

$$\theta_n(x) = \begin{cases} \theta(x), & x \notin \left[x_0 - \dfrac{1}{n}, x_0 + \dfrac{1}{n}\right], \\[2mm] \tilde{\theta}(x), & x \in \left[x_0 - \dfrac{1}{n}, x_0 + \dfrac{1}{n}\right], \end{cases}$$

with $|\theta_n(x)| \leqslant \pi/2 - \delta_n$, $\delta_n > 0$, and $\theta_n(x) \in C^\alpha$ in a neighborhood of $x = x_0$ (Figure 8). If $|\theta(x_0)| = \pi/2$ at a branch point of the flow, then $\theta(x)$ is replaced by such a function $\theta_n(x)$ with $\theta_n(x_0) = 0$ and $\theta_n(x) \in C^\alpha$. We proceed similarly also when $q = q(x)$ fails to be smooth at isolated points, approximating it by a smooth function. According to Theorems 5 and 6 in §2, the approximating functions $\theta_n(x)$ and $q_n^*(x)$ belong to the class $\mathfrak{M}(N, \delta)$; therefore, the original functions $\theta(x)$ and $q^*(x)$ belong to $\mathfrak{M}^*$, and, consequently, problems 2°.1–2°.6 are solvable by Theorems 2 and 7 in §2 (see also Remark 2 after the latter theorem).

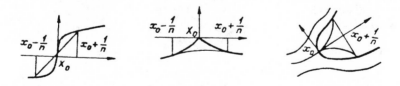

FIGURE 8

If the given functions $\theta(x)$ and $q^*(x)$ are piecewise continuous (but not piecewise Hölder!) and if

$$|\theta(x)| \leqslant \pi/2 \quad \text{on } L_z^1, \tag{6}$$

then such functions can obviously be approximated by functions of the class $\mathfrak{M}(N_\varepsilon, \delta_\varepsilon)$; however, in this case we can assert (according to Theorem 3 in §2) only the existence of quasisolutions of problems $2°.1-2°.6$ corresponding to the given functions $\theta(x)$ and $q^*(x)$. We thus arrive at the following theorem.

THEOREM 6. *If the given functions $\theta(x)$ and $q^*(x)$ are piecewise continuous and satisfy condition* (b) *and inequality* (6), *then problems $2°.1-2°.6$ have at least one quasisolution in this case. These quasisolutions are ordinary solutions, bounded everywhere in $\overline{D}_z$ except possibly at the singular points of $\theta(x)$ and $q^*(x)$, if these functions satisfy conditions* (a) *and* (b).

REMARK. If the magnitude of the flow velocity in problems $2°.1-2°.6$ depends on the two variables x and y and the dependence on y is weak (condition (c)), then all the assertions of Theorem 6 obviously remain in force.

As an example of the application of Theorem 6, we mention an existence theorem in the problem of a cavitational flow past a curvilinear obstacle according to the Thullen scheme (§3 in Chapter IV), which differs from the Kirchhoff scheme (problem $2°.2$) only by the presence of a finite jump in the magnitude of the velocity on the jets. A larger field of application of Theorem 6 is given in what follows.

$3°$. Problems on the motion of a fluid in a gravitational field.

1. *The problems of the flow of a heavy fluid over an inclined impermeable surface, out from under a curvilinear sluice gate onto an inclined impermeable surface, and out of an inclined curvilinear nozzle* (*Figure* 9). The outflow ψ_0 and the Bernoulli constant C are assumed to be given, and the known boundaries have horizontal asymptotes.

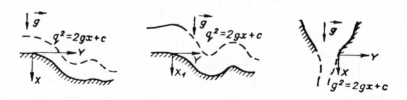

FIGURE 9

2. *A flow past obstacles according to schemes of Kirchhoff type and according to the symmetric Riabouchinsky scheme in a vertical gravity field* (*Figure* 10). The Bernoulli constant C is given.

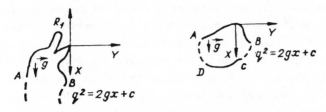

FIGURE 10

3. *Problems of a fluid flow over an uneven bottom and out from under a sluice gate onto a curvilinear impermeable surface when there are horizontal parts on the rigid walls* (*Figure* 11). In this case, unlike 2°.1, the OY-axis is directed parallel to the force of gravity.

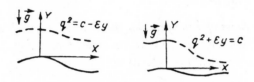

FIGURE 11

4. *The problem of periodic surface waves of a fluid on an uneven periodic bottom, with a quarter wave period considered* (*Figure* 12). The outflow ψ_0 and the constant c are given, and $\varepsilon = 2g/q_\infty^2$.

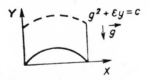

FIGURE 12

The following theorem is an immediate consequence of Theorem 6 and the remark after it.

THEOREM 7. *If conditions* (a*), (b), *and* (c) *hold for the slope angle* $\theta = \theta(x)$ *of the velocity on the given parts of the boundary of the flow domain and the* $\varepsilon > 0$ *in problems* 3°.3 *and* 3°.4 *is sufficiently small, then each of the problems* 3°.1–3°.4 *has at least one solution that is bounded everywhere in* $\overline{D}_z$ *with the possible exception of the points at infinity and the points of discontinuity of the given function* $\theta(x)$.

4°. Problems with completely unknown boundaries of the flow domain. This group of problems includes all the problems 1–5 in 2° when the boundary L_z of the flow domain is completely unknown and is determined from the given flow speed $q = q(x) \in C^{1+\alpha}$ on it. The smoothness of $q(x)$ and Theorem 2 give us the a priori estimate (4) on the slope angle of the flow velocity for the problems of this group, and, consequently, they belong to the class $\mathfrak{M}(N, \delta)$. On the other hand, D_τ is known in all the problems of this group, and on the plane of $\tau = x + i\psi$ they obviously reduce to the Dirichlet problem, so all these problems are uniquely solvable, by Theorem 1 in §2.

THEOREM 8. *Suppose in problems* 2°.1–2°.6 *that the boundary of the flow domain is completely unknown and is determined from a given distribution* $q = q(x)$ *of the flow speed on it. If* $q = q(x)$ *satisfies conditions* (a) *and* (b), *then all these problems are uniquely solvable.*

5°. Rotational flows of a compressible fluid. In solving problems on rotational flows of a compressible fluid, as in the case of potential flows, we use the equations for the complex flow velocity $w = u + iv$

$$w_{\bar{\zeta}} - \mu_1 \frac{\bar{z}_{\bar{\zeta}}}{z_\zeta} w_\zeta - \mu_2 \overline{w_{\bar{\zeta}}} = \bar{z}_{\bar{\zeta}} A(\psi) \equiv B(w \,|\, \zeta) \tag{7}$$

and for the generalized Joukowsky function $\omega = q^* - i\theta$

$$\omega_{\bar{\zeta}} - \mu(\omega, \zeta)\omega_\zeta = \bar{\tau}_{\bar{\zeta}} f(\omega) A(\psi) \equiv \varepsilon B_1(\omega \,|\, \zeta). \tag{8}$$

Here

$$f(\omega) = iw^{-1}\left(\frac{\rho}{1 + \sqrt{1 - M^2}} \right),$$

the coefficients $\mu_i(w)$ and $\tilde{\mu}(\omega) = (\tau_\zeta / \bar{\tau}_{\bar{\zeta}})\mu(\omega, \zeta)$ were defined in the derivation of (1.22) and (1.24), and $z = z(\zeta)$ and $\tau = x + i\psi = \tau(\zeta)$ are conformal mappings of the disk $|\zeta| < 1$ onto the respective domains D_z and D_τ. It is assumed that the vorticity $A(\psi)$ satisfies the condition

(i)

$$A(\psi) = \delta(|z|)A_0(\psi), \qquad A_0(\psi) \in C^{1+\alpha}(-\infty, \infty), \alpha > 0,$$

where $\delta(|z|) = 0$ for $|z| > R$ and $\delta(|z|) = \varepsilon = $ const for $|z| \leqslant R$, R a sufficiently large number, and $\varepsilon = 1$ in the problems in 1° (D_z is given), while ε is the small parameter in the problems in 2° (D_z is unknown).

We remark that the quantity ψ in (8) is an independent variable, while in (7) it must be expressed in terms of the complex flow velocity w from the relation

$$2\frac{\partial \psi}{\partial \bar{\zeta}} = i\rho \frac{d\bar{z}}{d\bar{\zeta}}\bar{w},$$

which is equivalent to the equalities $\psi_y = \rho u$ and $\psi_x = -\rho v$. For solutions of the last equation we have the representation (11) of §4 in Chapter VI:

$$\psi(w\,|\,\zeta) = \tfrac{1}{2}\operatorname{Re} T_1(i\rho \bar{z}_{\bar{\zeta}}\,|\,w\zeta) \tag{9}$$

everywhere except in neighborhoods of the images ζ_k of the points of D_z at infinity, where $w \in C^{\alpha}(\bar{K})$, $\alpha > 0$, and the singularities at the points ζ_k are isolated. Suppose, for example, that

$$\frac{dz}{d\zeta} = F_0(\zeta)\bar{\zeta}^2, \qquad |\ln F_0(\zeta)| \leqslant M, \qquad |\zeta| \leqslant 1,$$

which corresponds to a mapping of the disk $|\zeta| < 1$ onto the exterior of a finite smooth profile. Then the integrals in (9) are understood in the sense of the identity

$$T_1\left(\frac{f}{t^2}\,\bigg|\,\zeta\right) = T_1\left(\frac{f(t) - f(0)}{t^2}\,\bigg|\,\zeta\right) + \frac{f(0)}{\zeta} + \overline{f(0)}\,\zeta, \qquad f(\zeta) \in C^{\alpha}(\bar{K}).$$

By assumption (i),

$$|B(w\,|\,\zeta)| \leqslant M\left(\|A\|, \max_{|z| \leqslant R}\left|\frac{dz}{d\zeta}\right|\right) < \infty,$$

from which, taking the properties of the operator T_1 into account, we conclude that $B(w\,|\,\zeta)$ is a completely continuous operator from $C^{\alpha}(\bar{K})$ to $L_p(K)$ for any $p > 1$. Consequently, by Theorem 2 in §2 of Chapter VI, the unique solvability of the Hilbert problem (1) in the class of analytic functions implies its solvability also for (7), as in the case of potential flows ($B \equiv 0$). We have thereby proved that the problems in 1° are solvable for rotational flows of a compressible fluid.

Let us now consider the problems with free boundaries 2°.1–2°.6. In front of the right-hand side of (3) and in front of q_k^* and θ in (5) we introduce a parameter $\lambda \in [0, 1]$ and form the transformation

$$\omega = F_\lambda(\omega), \qquad \omega \in \mathfrak{M} \equiv \{|\operatorname{Re}\omega| < N, |\operatorname{Im}\omega| < \pi/2 - \delta\}$$

correponding to problems (3), (8) and (5), (8) in the case of a potential flow ($B_1 \equiv 0$). As proved above, problems 2°.1–2°.6 belong to the class $\mathfrak{M}(N, \delta)$ under assumptions (a)–(c), and, consequently, by §2.6° of Chapter VI, the transformations F_λ satisfy all the conditions of the Leray-Schauder theorem.

The transformation corresponding to the case $B_1 \not\equiv 0$ can be represented in the form

$$\omega = F_\lambda(\omega) + \varepsilon Q(\omega), \tag{10}$$

where $\omega^0 = Q(\omega)$ is an operator realizing a solution of the linear equation of form (8)

$$\omega^0_{\zeta\zeta} - \mu(\omega, \zeta)\omega^0_{\zeta} = B_1(\omega \mid \zeta)$$

under the homogeneous boundary condition (3) when $q_q^* = \theta = 0$ in (5). Taking the fact that the function $B_1[\omega(t) \mid \zeta]$ is bounded in modulus for $\omega(t) \in \mathfrak{M}$ into account, we conclude that the operator $Q(\omega)$, $\omega(t) \in \mathfrak{M}$, is completely continuous in $C^\alpha(\overline{K})$. Consequently, all the conditions of the theorem on small perturbations (Theorem 8 in §1 of Chapter II) hold for the transformation (10), so this transformation has at least one fixed point $\omega \in \mathfrak{M}$ for sufficiently small $\varepsilon > 0$.

THEOREM 9. *Under assumption* (i) *and conditions* (a) *and* (b), *the problems* $1°.1-1°.3$ *on rotational flows of a compressible fluid in specified domains and the free-boundary problems* $2°.1-2°.6$ *each have at least one solution, where in the last problems the number* $\varepsilon > 0$ *in the determination of the vorticity* $A(\psi)$ *is sufficiently small.*

Let us consider the problem of the change in the characteristics of a potential flow of a liquid or gas as it passes through a fixed domain in which a vortex motion is maintained (with the help of jet devices, condensers, or other devices) [5]. It is assumed that the vorticity A varies in jumps on the boundary Γ of the vortex domain and takes a zero value outside this domain (this can be realized, for example, by vortex-quelling devices: gratings, etc.). Inside the vortex domain the vorticity A is constant along each fixed streamline, i.e., $A = A(\psi)$. The boundary of the flow domain, which in the general case is not a streamline, is assumed to be fixed, and it is natural to suppose that the momentum vector is continuous in passing across it, which implies that the velocity vector is continuous in passing across it, which implies that the velocity vector is continuous for the motion of a fluid with constant speed.

When there are suction or pressuring devices on the boundary of the vortex zone, the momentum vector can have a given discontinuity which is defined, generally speaking, for the distinct points t and $\alpha(t)$ of the boundary upon approach from the domain of the rotational flow and from the domain of the potential flow, respectively.

The assumptions about the behavior of the momentum vector imply the following: When it is continuous, the streamlines are continuous smooth curves on the whole plane of flow; but when there is a discontinuity and a shift, the streamlines have a finite discontinuity $\{t - \alpha(t)\}$ together with their tangent direction.

The maintenance of a fixed vortex zone with given vorticity $A = A(\psi)$ in a potential flow is the basis for the operation of heat-exchange and mass-exchange devices.

In a free flow an interface is, as a rule, a common streamline and must be determined from the coincidence on Γ of the normal components of the velocities and of the pressures, or from the equality of the velocity vectors in the case of an incompressible fluid.

The complex flow velocity $w(z) = u - iv$, defined on the whole plane E, satisfies (22) in §1 with the right-hand side identically zero in the domain D^- of the potential motion. We assume that the vorticity $A(\psi)$ satisfies assumption (i) with $\varepsilon = 1$ in the domain $D^+ = E \setminus D^-$ bounded by Γ. The condition on the momentum vector in passing across the boundary Γ of the vortex zone has the form

$$\rho^+(q)w^+[\alpha(t)] = \rho^-(q)w^-(t) + h(t), \qquad t \in \Gamma, \tag{11}$$

where $h(t) \equiv 0$ when there are no suction or pressuring devices. In the case when a specified discontinuity in the velocity components is maintained on the interface between two media, the condition (11) is transformed to the form

$$w^+(t) = w^-(t) + h(t), \qquad t \in \Gamma. \tag{12}$$

Thus, the original problem of conjunction of subsonic rotational flows and subsonic potential flows has been reduced to the conjunction problem with a shift for a uniformly elliptic system of equations, which was studied in §3 of Chatper VI.

§4. Some problems in gas dynamics with free boundaries in doubly connected domains

In this section we consider examples of hydrodynamics problems in doubly connected domains to which the results of §§1–3 are directly applicable.

I. *The problem of the flow of a subsonic gas jet past a given profile when a certain distance from the profile the jet breaks up into several jets going off to infinity.* The distribution $q = q(x)$ of the magnitude of the velocity is given on the free jets; in particular, $q(x) = $ const. The coordinates of the point R_1, the abscissas of the points R_i, and the values S_i of the outflow in the jets are also assumed to be given. We get a problem on the motion of a given profile under a free surface when there is only one free boundary (Figure 1).

FIGURE 1

II. *The problem of the motion of an unknown profile over an uneven bottom in a flow of infinite depth.* As before, the coordinates of R_1, the abscissa of R_2, the distribution of the magnitude of the velocity on the profile, and the outflow S_1 in the jet are given (Figure 2).

FIGURE 2

Similarly to §§1–3, we introduce the planes of the variables $w = \varphi + i\psi$ and $\tau = x + i\psi$. Let us map the domain D_τ, which is known in each of the problems I and II, conformally onto the annulus $K = \{\rho < |\zeta| < R\}$ in such a way that the image of the unknown boundary of the flow domain is the circle $|\zeta| = R$. Then for the function

$$\omega = \omega[\tau(\zeta)] = q^* - i\theta$$

we get the following boundary-value problem (Figure 3):

$$\omega_{\bar\zeta} - \tilde\mu(\zeta, \omega)\omega_\zeta = 0, \qquad \rho < |\zeta| < R, \tag{1}$$

$$\mathrm{Re}[G(\gamma)\omega(\gamma)] = g(\gamma), \tag{2}$$

where

$$G = \begin{cases} 1, & |\zeta| = R, \\ -i, & |\zeta| = \rho, \end{cases} \qquad g = \begin{cases} g^*[x(\gamma)], & |\zeta| = R, \\ \theta[x(\gamma)], & |\zeta| = \rho. \end{cases}$$

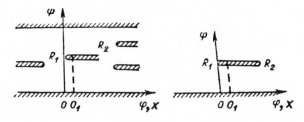

FIGURE 3

The function is defined by (1.15), and if the a priori estimates (2.6) hold for the slope angle and magnitude of the velocity, then

$$|\tilde{\mu}(\zeta, \omega)| \leqslant \mu_0(N, \delta) < 1. \tag{3}$$

In the remainder of this section we assume that the slope angle $\theta(x)$ of the velocity on the given parts of L_z satisfies a Hölder condition with exponent $\alpha > 1/2$, while the magnitude of the flow velocity on the unknown parts of L_z satisfies the condition $q(x) \in C^{1+\beta}$, $\beta > 0$. In studying the boundary-value problem (1), (2) we shall need properties of certain integral operators in doubly connected domains.

In the domain $K = \{\rho < |z| < R\}$ let us consider the integral operator

$$T_\varphi = -\frac{1}{\pi} \iint_K \left[\frac{\varphi(t)}{t - z} + \frac{z\,\overline{\varphi(t)}}{\bar{t}z - \rho^2} \right] dK_t, \tag{4}$$

According to Theorems 5 and 7 in §2 of Chapter V, for any function $\varphi \in L_p(\bar{K}), p > 2$,

$$T(z) \equiv T(\varphi \,|\, z) \in C^{(p-2)/p}(E),$$

and the generalized derivatives

$$\partial T_\varphi/\partial z = S\varphi, \qquad \partial T_\varphi/\partial\bar{z} = \varphi,$$

exist and are pth power summable, $p > 2$, in $\bar{K}$. By construction,

$$\operatorname{Im} T(\varphi \,|\, \rho e^{i\gamma}) = -\frac{1}{\pi} \iint_K 2\operatorname{Re}\left\{ \frac{\varphi(t)}{t - \rho e^{i\gamma}} \right\} dK_t = 0, \qquad \gamma \in [0, 2\pi]. \tag{5}$$

We determine a function $\Phi(\varphi \,|\, z)$ that is analytic in K, continuous in $\bar{K}$, and satisfies the boundary condition

$$\operatorname{Re} \Phi(\varphi \,|\, \gamma) = -\operatorname{Re} T(\varphi \,|\, R e^{i\gamma}) = u(\gamma), \qquad |z| = R,$$

$$\operatorname{Im} \Phi(\varphi \,|\, \gamma) = -\operatorname{Im} T(\varphi \,|\, \rho e^{i\gamma}) = v(\gamma) \equiv 0, \qquad |z| = \rho, \gamma \in [0, 2\pi]. \tag{6}$$

Let us cut the annulus along the positive part of the real axis and map the resulting simply connected domain conformally by the transformation $z = \rho e^{2\pi i \zeta / \omega_1}(R\rho^{-1}e^{\pi i \omega_2/\omega_1})$ onto a rectangle with sides $\omega_2/2$ and ω_1 in the ζ-plane. If we then use Villat's formula (see [117]), we get

$$\Phi(\varphi \mid z(\zeta)) = \frac{\omega_1}{2\pi^2 i}\left\{ \int_0^{2\pi} Ru(\gamma)g_R(\zeta, t)\,d\gamma - i\rho\int_0^{2\pi} v(\gamma)g_\rho(\zeta, t)\,d\gamma \right\}, \quad (7)$$

where

$$g(\zeta, t) = \frac{\sigma(t - \zeta + \omega_1/2)}{\sigma(\omega_1/2)\sigma(t - z)}\exp\left\{\frac{\eta}{2}(t - \zeta)\right\}, \qquad \eta = \text{const},$$

and

$$g_\tau(\zeta, t) \equiv g[\zeta(z), t(\tau e^{i\gamma})],$$

while $\sigma = \sigma(\zeta)$ is the Weierstrass function with periods ω_1 and ω_2, and

$$\max|\Phi| \leqslant c(\|u\|_{C^\alpha}, \|v\|_{C^\alpha}). \qquad (8)$$

By (6), the function $\Phi(\varphi \mid z)$ admits the following continuous extension to the annulus $\rho^2/R^2 < |z| < R$:

$$\Phi^*(z) = \begin{cases} \Phi(z), & \rho < |z| < R, \\ \overline{\Phi\left(\dfrac{\rho^2}{\bar{z}}\right)}, & \dfrac{\rho^2}{R^2} < |z| < R. \end{cases}$$

Since the limit values of the single-valued harmonic function $u = \operatorname{Re}\Phi^*(z)$ are the limit values of a function of class W_p^1, property 5 in §1 of Chapter VI gives us that $u = u(z) \in W_p^1$, and so $\Phi(\varphi \mid z) \in W_p^1(\overline{K})$, $p > 2$. We introduce the operator

$$\Lambda\varphi = T\varphi + \Phi(\varphi \mid z). \qquad (9)$$

By construction,

$$\Lambda\varphi \equiv \Lambda(\varphi \mid z) \in W_p^1(\overline{K}), \qquad p > 2,$$

for

$$\varphi(z) \in L_p(\overline{K}), \qquad \partial\Lambda\varphi/\partial\bar{z} = \partial T_\varphi/\partial\bar{z} \equiv \varphi(z),$$

and $\Lambda(\varphi \mid z)$ satisfies the homogeneous boundary condition (6) on the boundary of the domain; consequently, $\Xi\varphi = \partial\Lambda\varphi/\partial z$ is a bounded linear operator on the space of functions $\varphi(z) \in L_p(\overline{K})$, $p > 2$. Let us compute its norm in L_2.

Suppose that $\varphi \in C^1$. Then, applying Green's formula in the doubly connected domain K, we get

$$\|\Xi\varphi\|_{L_2}^2 = \|\varphi\|_{L_2}^2 + \frac{1}{2i} \int_\Gamma \overline{\Lambda\varphi} \, \frac{\partial \Lambda_\varphi}{\partial z} dz = \|\varphi\|_{L_2}^2 + I_\Gamma. \tag{10}$$

Since $\varphi(z)$ has compact support, the function $\Lambda\varphi = \Lambda(\varphi \,|\, z)$ is holomorphic near the boundary and is continuous along with its first derivatives. Therefore, $\partial\Lambda\varphi/dz = d\Lambda\varphi/dz$, and the integral I_Γ can be represented in the form

$$I_\Gamma = \frac{1}{2i} \int_\Gamma \overline{\Lambda\varphi} \frac{d\overline{\Lambda\varphi}}{\partial z} dz = \frac{1}{2i} \int_\Gamma \overline{\Lambda\varphi} \, d\left(\overline{\Lambda}_\varphi\right).$$

Since either $\operatorname{Re}\Lambda\varphi = d(\operatorname{Re}\Lambda\varphi) = 0, \operatorname{Im}\Lambda\varphi = d(\operatorname{Im}\Lambda\varphi) = 0$ on Γ, it follows that $I_\Gamma = 0$. Consequently, $\|\Xi\|_{L_2} = 1$. The solution of the boundary-value problem (1), (2) will be sought in the form

$$\omega(z) = \Lambda\varphi + \Phi(\zeta), \tag{11}$$

where $\varphi(z) \in L_p(\overline{K})$, $p > 2$, is an unknown density, and $\Phi(\zeta)$ is an analytic function satisfying the boundary condition (2).

THEOREM 1. *The gas dynamics problems* I *and* II *are solvable if they belong to the class* $\mathfrak{M}(N, \delta)$.

Let us introduce a parameter $\lambda \in [0, 1]$ on the right-hand side of the boundary conditions (2) and represent the solutions of the corresponding problems in the form (11). We come to the thoroughly studied integral equation

$$\varphi - \hat{\mu}(\zeta, \Lambda\varphi + \lambda\Phi)\Xi\varphi = \lambda\frac{d\Phi}{d\zeta}\hat{\mu}(\zeta, \Lambda\varphi + \lambda\Phi) \tag{12}$$

for determining the unknown density $\varphi = \varphi(z)$, where Λ is a completely continuous operator on L_p, $p > 2$, $\|\Xi\|_{L_2} = 1$, and for $d\Phi/d\varphi$ we have the estimate

$$\left\|\frac{d\Phi}{d\zeta}\right\|_{L_p(\overline{K})} \leqslant M(\|q\|_{C^\alpha}), \quad 2 < p < \frac{1}{1 - \alpha}, \frac{1}{2} < \alpha < 1.$$

The proof that (12) is solvable is completely analogous to that for Lemma 2 in §4 of Chapter V, and the assertion of the theorem follows.

In the domain $K_\zeta = \{\rho < |\zeta| < R\}$ let us consider the equation

$$\omega_{\overline{\zeta}} - \mu(\zeta)\omega_\zeta = 0, \quad |\mu(\zeta)| \leqslant \mu_0 < 1.$$

Analogously to the case of a simply connected domain, any solution of this equation in K_ζ has the representation $\omega = \Phi[z(\zeta)]$, where $\Phi = \Phi(z)$ is analytic in K_z, and $z = z(\zeta)$ is a homeomorphism of K_ζ onto the annulus

$$K_z = \{\tilde{\rho} < |z| < R\} \qquad \left(0 < \tilde{\rho}\left(\frac{R}{\rho}, \mu_0\right) < R\right)$$

satisfying the equation

$$z_{\bar{\zeta}} - \mu(\zeta)z_\zeta = 0.$$

Indeed, we extend the coefficient $\mu(\zeta)$ arbitrarily inside the unit disk in such a way that $|\tilde{\mu}(\zeta)| \leqslant \mu_0 < 1$, $|\zeta| < R$, and we map the disk $|\zeta| < R$ homeomorphically onto itself by the solution of the equation

$$\chi_{\bar{\zeta}} - \tilde{\mu}(\zeta)\chi_\zeta = 0$$

with the normalization $\chi(0) = 0$, $\chi(R) = R$. Under this mapping the circle $|\zeta| = \rho$ is mapped homeomorphically to some curve $L_\rho \in C^\alpha$, $\alpha = \alpha(\mu_0)$. We arrive at the representation $\omega = \Phi[z(\zeta)]$ by a conformal mapping of the resulting domain D_χ onto the annulus $K_z = \{\tilde{\rho} < |z| < R\}$. The same smoothness assertions as for a simply connected domain are valid for the homeomorphism $z(\zeta)$ because of their local nature. We remark that, as in the case of a simply connected domain, the estimate (2.4)

$$\{|\theta(z)| \leqslant \pi/2 - \delta, \delta > 0\}$$

on the slope angle implies a uniform estimate

$$\{|q^*(z)| \leqslant N(\delta)\}$$

on the magnitude of the velocity. Indeed, by making an additional cut $(0, R_1)$ in each of the domains D_w, we come to a simply connected domain, from which, as in a theorem in §2, we get that $\theta(\gamma)$, $q^*(\gamma) \in C^\beta$, $\beta > 0$. Then inequality (8) for the solution of the corresponding boundary-value problem (6) immediately gives us the needed estimate on the magnitude of the flow velocity. The restrictions we have imposed on the given quantities $\theta(x)$ and $q^*(x)$ also ensure (with the remark about the smoothness of the homeomorphisms of doubly connected domaisn taken into account) that an a priori estimate holds for the slope angle of the velocity, by Theorem 2 in §3. This finally proves that problems I and II belong to the class $\mathfrak{M}(N, \delta)$ and thereby, by Theorem 1, that they are solvable. We formulate this as a theorem.

THEOREM 2. *The gas dynamics problems in doubly connected domains stated at the beginning of the section are solvable under the conditions* (a)–(c) *of* §3 *on the given quantities $\theta(x)$ and $q^*(x)$.*

§5. Variational formulas in hydrodynamics problems with free boundaries

1°. We give formulas obtained by Lavrent'ev's variational method ([69], Paragraph 65) for a function which maps a domain close to the upper half-plane conformally onto the upper half-plane (Figure 1).

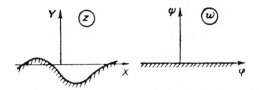

FIGURE 1

Suppose that the boundary L_z of the domain in the z-plane is sufficiently close to the OX-axis. Namely, the function $\gamma(x) = \arctan dy/dx$ satisfies the following conditions: $\gamma(x) \in C^{1+\alpha}$ for all $x \in (-\infty, \infty)$ except in arbitrarily small neighborhoods δ_i of finitely many isolated points x_i, at which $\gamma(x) \in C^\alpha$; moreover, $xy|_{x=\pm\infty} = 0$, $\gamma(\pm\infty) = 0$, and $|\gamma(x)| < \varepsilon$ for all x, while $|\gamma'(x)| < M\varepsilon$ outside the neighborhoods of δ_i of the points x_i.

Then, according to the Lavrent'ev formulas

$$w = z + \frac{1}{\pi} \int_{-\infty}^{\infty} \frac{y(t)}{z-t} dt + O(\varepsilon^2), \tag{1}$$

$$\frac{dw}{dz} = 1 + \frac{1}{\pi} \int_{-\infty}^{\infty} \frac{y'(t)}{z-t} dt + O(\varepsilon^2), \tag{2}$$

$$x = \varphi - \frac{1}{\pi} \int_{-\infty}^{\infty} \frac{y(t)}{\varphi-t} dt + O(\varepsilon^2), \tag{3}$$

$$\left| \frac{dw}{dz} \right| = 1 + \frac{1}{\pi} \int_{-\infty}^{\infty} \frac{y'(t)}{t-\varphi} dt + O(\varepsilon^2). \tag{4}$$

These formulas were derived under the assumption that the curvature of L_z is bounded at all points, but they remain true also when the curvature of L_z is unbounded at isolated points (see, for example, [69], Paragraph 34). The validity of (1)–(4) can also be checked directly. It is clear from these formulas that the function $\ln|dw/dz| \equiv \ln Q(x)$ satisfies the inequality $|\ln Q(x)| < M_1\varepsilon$, and $\ln Q(x) \in C^{1+\alpha}$.

$2°$. We obtain formulas that are inverses of (1)–(4) in a certain sense. Suppose that the quantity $\ln Q(x) = \ln|dw/dz|$ is given on $[-1, 1]$, with $\ln Q(x) \in C^{1+\alpha}$,

$$\left| \frac{d^k}{dx^k} \ln Q(x) \right| < \varepsilon, \qquad k = 0, 1, 2,$$

and $y(x) = 0$ for $x \notin [-1, 1]$ (Figure 2). It is required to construct the mapping function $w = w(z)$ of D_z onto the upper half-plane. Suppose for the time being that the function $\gamma(x)$ satisfies the conditions under which (1)–(4) were obtained: The curve L_z with equation $y = y(x)$ is sufficiently close to the OX-axis.

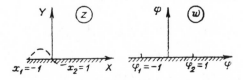

FIGURE 2

Taking into account (3), we then arrive at the boundary-value problem

$$\operatorname{Re}\ln\frac{dw}{dz}=\ln Q[\varphi-\sigma(\varphi)]\quad\text{for }\varphi\in[\varphi_1,\varphi_2],$$

$$\operatorname{Im}\ln\frac{dw}{dz}=0\quad\text{for }\varphi\in[\varphi_1,\varphi_2],\tag{5}$$

where the φ_i are the images of the points ± 1, while

$$\sigma(\varphi)=\frac{1}{\pi}\int_{-\infty}^{\infty}\frac{y(t)\,dt}{\varphi-t}\quad\text{and}\quad|\sigma(\varphi)|<M_2\varepsilon,$$

and we can always assume that $\varphi_i=\pm 1$ $(i=1,2)$, i.e., $\sigma(\varphi_i)=0$. The solution of the last problem can be written in the form

$$\ln\frac{dw}{dz}=\frac{R(w)}{\pi i}\int_{-1}^{+1}\frac{\ln Q[t-\sigma(t)]}{R(t)(t-w)}\,dt,$$

where $R(w)=\sqrt{1-w^2}$. But

$$\left|\ln[Q(t-\sigma)/Q(t)]\right|=\left|\frac{d}{dt}\ln Q(t^*)\right|\cdot|\sigma(t)\leqslant M_2\varepsilon^2\quad\text{for }t\in[-1,1].$$

Consequently, to within second-order infinitesimals we have

$$\ln\frac{dw}{dz}=\frac{R(w)}{\pi i}\int_{-1}^{+1}\frac{\ln Q(t)\,dt}{R(t)(t-w)},\tag{6}$$

and this implies that, to the same accuracy,

$$\frac{dw}{dz}=1+\frac{R(w)}{\pi i}\int_{-1}^{+1}\frac{\ln Q(t)\,dt}{R(t)(t-w)}.\tag{6*}$$

Similarly, if $|dw/dz|=Q(x)$ is known on the whole OX-axis, and $|d^k\ln Q(x)/dx^k|<\varepsilon$, $k=0,1,2$, and if $\ln Q(\pm\infty)=0$ and $Q(\pm\infty)=y(\pm\infty)=0$, then the solution of the corresponding Dirichlet problem gives us that

$$\ln\frac{dw}{dz}=\frac{1}{\pi i}\int_{-\infty}^{+\infty}\frac{\ln Q(t)\,dt}{t-w}.\tag{7}$$

Formulas (6) and (7) determine curves L_z with equation $y=y(x)$ that are sufficiently close to the OX-axis, while on $[-1,1]$ we have formula (3) which we used to get the simplified formula (6). Indeed, (6*) gives us

$$\frac{dz}{dw}=1-\frac{R(w)}{\pi i}\int_{-1}^{+1}\frac{\ln Q(t)\,dt}{R(t)(t-w)},\tag{6**}$$

whence $dx/d\varphi=1-\ln Q(\varphi)$ for $\varphi\in[-1,1]$, which leads us to the necessary equality $x=\varphi-\sigma(\varphi)$, where $|\sigma(\varphi)|<M\varepsilon$. Now (6) implies that

$$\gamma(\varphi)=-\operatorname{Im}\ln\frac{dw}{dz}=\frac{R(\varphi)}{\pi}\int_{-1}^{+1}\frac{\ln Q(t)-\ln Q(\varphi)}{R(t)(t-\varphi)}\,dt,\qquad\varphi\in[-1,1],$$

since $\int_{-1}^{1}(1/R(t)(t-\varphi))\,dt=0$ on this interval. Then

$$|\gamma(\varphi)|\leqslant\max_{t\in[-1,1]}\left|\frac{d}{dt}[\ln Q(t)]\right|\cdot\frac{1}{\pi}\int_{-1}^{1}\left|\frac{R(\varphi)}{R(t)}\right|\,dt\leqslant\hat{M}\varepsilon;$$

moreover, $\gamma(\varphi)=0$, $|\varphi|\geqslant 1$, and $\gamma(\varphi)\in C^\alpha$. For $\varphi\in[-1+\delta,1-\delta]$ we get

$$\left|\frac{d\gamma}{d\varphi}\right|=\left|\frac{R'(\varphi)}{\pi}\int_{-1}^{1}\frac{\ln Q(t)\,dt}{R(t)(t-\varphi)}+\frac{R(\varphi)}{\pi}\int_{-1}^{+1}\frac{\ln Q(t)\,dt}{R(t)(t-\varphi)^2}\right|\leqslant M(\delta)\varepsilon,$$

where $M(\delta)$ can be computed explicitly. On the other hand, for $\varphi \in [-1, 1]$

$$\left| \frac{d\varphi}{dx} \right| = \left| \frac{1}{1 - \delta(\varphi)} \right| \leqslant M_1 < \infty,$$

and, hence,

$$| \gamma(x) | = | \gamma[x(\varphi)] | < \hat{M}\varepsilon \quad \text{for } \hat{x} \in [-1, 1],$$

$$\left| \frac{d\gamma}{dx} \right| = \left| \frac{d\gamma}{d\varphi} \right| \cdot \left| \frac{d\varphi}{dx} \right| < M(\delta) M_1 \varepsilon \quad \text{for } \varphi \in [-1 + \delta, 1 - \delta].$$

Thus, the curve $y = y(x)$ corresponding to (6) satisfies the conditions under which (1)–(4) were derived. Corresponding estimates can be carried out also for (7) in a completely analogous way, and the result is that

$$| \gamma(x) | < \hat{M}\varepsilon, \qquad | d\gamma/dx | < \hat{M}\varepsilon$$

for all $x \in (-\infty, \infty)$.

The problem considered in this subsection is also of independent interest as a hydrodynamics problem. Indeed, suppose that we are also seeking the flow of an incompressible fluid near a thin symmetric profile L_z whose form is unknown and must be determined from the velocity magnitude $| dw/dz | = Q(x)$, $x \in [-1, 1]$, which is given on it. Considering half the flow domain, we arrive at the problem formulated at the beginning of this subsection.

3°. Let us now proceed to the case of arbitrary close contours. Suppose that D_z is the domain bounded by a finite or infinite contour L_z, and $w = w(z)$ is a function mapping this domain onto the upper half-plane. In gas dynamics problems this corresponds to the case when the domain in the complex potential plane is the upper half-plane. As we shall see, cases in which D_w is not the upper half-plane can be reduced to this case. Let

$$\omega(z) = \ln \left| \frac{dw}{dz} \right| + i \arg \frac{dw}{dz} = \ln q - i\theta,$$

with $| \theta(z) | < \pi/2 - \varepsilon_0 \, (\varepsilon_0 > 0)$ on L_z. Suppose that the contour L_z is changed by varying

$$\operatorname{Re} \omega(x) = \ln q(x) \in C^{2+\alpha}(L_z^2)$$

(in particular, $| dq/dx | < \infty$) on some part L_z^2 of L_z with $\operatorname{Im} \omega(x) = - \theta(x)$ fixed on $L_z^1 = L_z - L_z^2$. Thus, the part L_z^1 of L_z does not change, while L_z^2 changes because $q(x)$ is varied. Suppose that the varied part L_z^2 of L_z passes into the segment $[-1, 1]$ of the real axis. We denote the contour obtained after the variation by $\tilde{L}_z^2$, and the corresponding analytic function mapping D_z conformally onto the upper half-plane by $w = \tilde{w}(z)$; $\ln(d\tilde{w}/dz) = \tilde{\omega} = \ln \tilde{q} - i\tilde{\theta}$. Suppose that the points $\hat{z}$ and $\check{z}$ correspond to the point $w = \varphi$ on the real axis in the w-plane under the mapping $w = \tilde{w}(\hat{z})$ and $w = \tilde{w}(\check{z})$. The curve $\tilde{L}_z$ passes into some curve $\tilde{L}_w$ under the mapping of D_z onto the upper half-plane (Figure 3). We let z be a point of L_z^1 such that $\operatorname{Re} z = \operatorname{Re} \hat{z}$, $\hat{z} \in \tilde{L}_z^2$, and we compute the variation $\delta y(x)$ as a function of the variation $\delta q = \tilde{q}(x) - q(x)$.

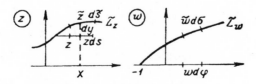

FIGURE 3

To do this let us consider the variation $\delta[\ln(dw/dz)(z)]$ as L_z^2 is varied. We find that

$$\Delta\left[\ln\left|\frac{dw}{dz}(\tilde{z})\right|\right] = \ln\left|\frac{dw}{dz}(\tilde{z})\right| - \ln\left|\frac{d\tilde{w}}{dz}(\tilde{z})\right|,$$

where $\operatorname{Re} z = \operatorname{Re} \tilde{z} = x$. We have

$$\Delta\left[\ln\left|\frac{dw}{dz}(\tilde{z})\right|\right] = \ln\left|\frac{d\sigma}{d\tilde{s}}\right| - \ln\left|\frac{d\varphi}{d\tilde{s}}\right| = \ln\left|\frac{dw}{dz}(\tilde{z})\right| - \ln\tilde{q}(x)$$

on $\tilde{L}_z^2$, but

$$\frac{dw}{dz}(\tilde{z}) - \frac{dw}{dz}(z) = \frac{d^2w}{dz^2}(z)i\delta y + O(|\delta y|^2).$$

Then

$$\left|\frac{dw}{dz}(\tilde{z})\right| = \left|\frac{dw}{dz}(z) + \frac{d^2w}{dz^2}(z)i\delta(y) + O(|\delta y|^2)\right| = q(x)\,|\,1 + l(x)\delta y\,|,$$

where $l(x) = i\omega_x e^{-i\theta}\cos\theta$. Thus,

$$-\ln Q(x) \equiv \delta\left[\delta\left|\frac{dw}{dz}(\tilde{z})\right|\right] = \ln\left|\frac{d\sigma}{d\varphi}(x)\right|$$

$$= \ln\left[\frac{q(x)}{\tilde{q}(x)}[1 + l(x)\delta y]\right] = -\ln\frac{\tilde{q}(x)}{q(x)} + \ln|\,1 + l(x)\delta y\,|,$$

and

$$\ln\frac{\tilde{q}(x)}{q(x)} = \ln\left(1 + \frac{\delta q}{q}\right) = \frac{\delta q}{q}$$

to within small quantities of second order in δy. To the same accuracy in δy we have

$$\ln|\,1 + l(x)\delta y\,| = \tfrac{1}{2}\ln|\,1 + (l_1 + il_2)\delta y\,|^2$$

$$= \tfrac{1}{2}\ln\left[1 + 2l_1\delta y + (l_1^2 + l_2^2)(\delta y)^2\right] = l_1\delta y,$$

where

$$l_1(x) = \operatorname{Re} l(x) = \frac{\sin\theta(x)\cos\theta(x)}{q}q_x + \cos^2\theta\cdot\theta_x.$$

Consequently, to within small quantities of second order in δq and δy,

$$\ln Q(x) + l_1(x)\delta y = \delta q/q.$$

As is clear from Figure 4, with the previous accuracy we have

$$|\,\tilde{\psi}\,| = |\operatorname{Im}\tilde{w}| = |\sin(\pi/2 - \theta)\delta w| = |\cos\theta\delta w| = q(x)\,|\cos\theta\delta y\,|,$$

i.e., $\delta y = \tilde{\psi}/q(x)\cos\theta(x)$, and so

$$\ln Q(x) + \left(\sin\theta\frac{q_x}{q^2} + \cos\theta\frac{\theta_x}{q}\right)\tilde{\psi} = \frac{\delta q}{q},$$

where $\tilde{\psi} = \operatorname{Im}\tilde{w}$. It follows from (2) that

$$\ln Q[x(\varphi)] = \ln\left[1 - \frac{1}{\pi}\int_{-1}^{+1}\frac{(d\tilde{\psi}/dt)\,dt}{t - \varphi}\right] \approx \frac{1}{\pi}\int_{-1}^{+1}\frac{(d\tilde{\psi}/dt)\,dt}{t - \varphi}.$$

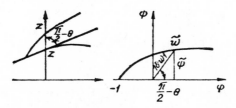

FIGURE 4

Then we get the following singular integrodifferential equation for determining $\tilde{\psi}$:

$$\left(\sin\theta\,\frac{q_x}{q^2} + \cos\theta\,\frac{\theta_x}{q}\right)\tilde{\psi} + \frac{1}{\pi}\int_{-1}^{1}\frac{d\tilde{\psi}/dt}{t-\varphi}\,dt = \frac{\delta q}{q},$$

which, since

$$\frac{1}{\pi}\int_{-1}^{1}\frac{(d\tilde{\psi}/dt)\,dt}{t-\varphi} = \frac{1}{\pi}\int_{-1}^{1}\frac{\tilde{\psi}\,dt}{(t-\varphi)^2}, \qquad \frac{d^k\tilde{\psi}}{d\varphi^k}(\pm1) = 0, \quad k = 0,1,$$

can be rewritten in the form

$$\left(\sin\theta\,\frac{q_x}{q^2} + \cos\theta\,\frac{\theta_x}{q}\right)\tilde{\psi} + \frac{1}{\pi}\int_{-1}^{+1}\frac{\tilde{\psi}\,dt}{(t-\varphi)^2} = \frac{\delta q}{q}.$$

According to the preceding subsection, the mapping of the domain close to the upper half-plane is determined for given $Q(x) = |\,dw/dz\,|$ by the formula

$$\frac{d\tilde{w}}{dw}(w) = 1 - \frac{R(w)}{\pi i}\int_{-1}^{1}\frac{\ln Q[x(t)]\,dt}{R(t)(t-w)},$$

where $R(w) = (1 - w^2)^{1/2}$ and $x = x(\varphi)$ is the correspondence between the points of the arcs $\tilde{L}_z^2$ and $\tilde{L}_w^2$ under the mapping $w = w(z)$ (see (6**)). For $\varphi \in [-1, 1]$

$$\frac{d\tilde{\psi}}{d\varphi} = -\frac{R(\varphi)}{\pi}\int_{-1}^{+1}\frac{\ln Q[x(t)]\,dt}{R(t)(t-\varphi)};$$

therefore,

$$\tilde{\psi} = \int_{-1}^{+1}K_1(\varphi, t)\ln Q[x(t)]\,dt, \tag{8}$$

where

$$K_1(\varphi, t) = -\frac{1}{\pi R(t)}\int_{-1}^{\varphi}\frac{R(\varphi)\,d\varphi}{t-\varphi}.$$

The last integral is easy to compute:

$$K_1(\varphi, t) = \frac{2}{\pi}\sqrt{\frac{1-\varphi^2}{1-t^2}} - \frac{t}{\pi\sqrt{1-t^2}}[\arcsin\varphi + \pi]$$

$$+ \frac{1}{2}\ln\left|\frac{1-\varphi t - \sqrt{(1-t^2)(1-\varphi^2)}}{1-\varphi t + \sqrt{(1-t^2)(1-\varphi^2)}}\right|.$$

Thus, we arrive at the integral equation

$$\ln Q[x(\varphi)] + \int_{-1}^{+1} K(\varphi, t) \ln Q[x(t)] \, dt = \frac{\delta q[x(\varphi)]}{q}, \tag{9}$$

where

$$K(\varphi, t) = K_1(\varphi, t) \left[\sin \theta(x) \frac{q_x}{q^2} + \cos \theta(x) \frac{\theta_x}{q} \right].$$

The relation between the variation

$$\delta\left[\ln\left| \frac{dw}{dz}(\tilde{z}) \right| \right] = -\ln Q(x)$$

and δy is found from (8):

$$\tilde{\psi} = q(x) \cos \theta \, \delta y = -\int_{-1}^{+1} K_1(\varphi, t) \delta\left[\ln\left| \frac{dw}{dz}[\tilde{z}(t)] \right| \right] dt,$$

and so

$$\delta y = \int_{-1}^{+1} K_2(\varphi, t) \delta\left(\ln\left| \frac{dw}{dz}[\tilde{z}(t)] \right| \right) dt,$$

where

$$K_2(\varphi, t) = \frac{1}{q[x(\varphi)] \cos \theta[x(\varphi)]} K_1(\varphi, t).$$

REMARKS. 1. Up to this point we have assumed that the image of D_z in the w-plane is the upper half-plane. Suppose now that the domain D_w, which is bounded by lines $\psi = \text{const}$, is not the upper half-plane (Figure 5). Similarly to the preceding case, we get

$$\ln Q(x) + \left(\sin \theta \frac{q_x}{q^2} + \cos \theta \frac{\theta_x}{q} \right)(\tilde{\psi} - C_1) = \frac{\delta q}{q}.$$

Mapping D_w conformally onto the upper half-plane, we get an equation analogous to (9):

$$\ln Q\{x[\varphi(\xi)]\} + \int_{-1}^{+1} K[\varphi(\xi), \varphi(\xi')] \ln Q\{x[\varphi(\xi')]\} \, d\xi' = \frac{\delta q\{x[\varphi(\xi)]\}}{q}.$$

In the case where L_z is not completely specified formula (9) has the similar form, though the integrals in it are taken along an infinite interval.

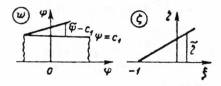

FIGURE 5

2. As follows from the considerations in §3, there is a uniqueness theorem for hydrodynamics problems with a single free boundary if $q(x) \in C^{1+\alpha}$ on L_z^2, $\theta(x) \in C^\alpha$ on L_z^1, and $|\theta(x)| \leqslant \pi/2 - \varepsilon_0$, $\varepsilon_0 > 0$. In particular, this means that the homogeneous equation (9), with the corresponding values of $q(t)$ and $\theta(t)$ substituted in its kernel, has only the trivial solution. Therefore, for sufficiently small δq we can use the nonhomogeneous equation (9) to compute an approximate solution of a problem close to the original one, then the resulting values $q^1(t)$ and $\theta^1(t)$ can again

be substituted in the kernel $K(\varphi, t)$ of (9), and so on. Since the properties of $\theta^k(x)$ and $q^k(x)$ are obviously preserved in such iterations, this method enables us to construct an approximate solution for arbitrary $\theta(x) \in C^\alpha$ on L_z^1 ($|\theta| \leqslant \pi/2 - \varepsilon_0, \varepsilon_0 > 0$) and $q(x) \in C^{1+\alpha}$ on L_z^2.

§6. Inverse boundary-value problems for elliptic systems of equations

1°. In this section we prove the existence and uniqueness of a subsonic flow in inverse boundary-value problems in gas dynamics when the boundary data are given as functions of the arclength parameter s of the unknown boundary L_z^2 of D_z. At the end of the section we pose a general intrinsic inverse boundary-value problem for nonlinear elliptic systems, and prove the existence of a unique univalent solution of it. Recall that the complex potential $w = \varphi + i\psi$ of a subsonic gas flow satisfies the nonlinear elliptic system

$$\rho(q)\varphi_x = \psi_y, \qquad \rho(q)\varphi_y = -\psi_x, \tag{1}$$

where $q = \sqrt{\varphi_x^2 + \varphi_y^2}$ is the flow speed. The condition that the flow is subsonic can be written in the form

$$\frac{1}{\rho} \frac{d(q\rho)}{dq} \geqslant \beta > 0. \tag{2}$$

This condition is also a condition for the system (1) to be elliptic. As in the previous sections, we assume that the flow is constantly subsonic, i.e., that (2) holds for any q, with $0 < \rho_0 \leqslant \rho(q) \leqslant R_0 < \infty$ and $\rho(q) \in C^\alpha$. We shall study the following gas dynamics problems with free boundaries:

1. *The problem of an unconfined irrotational gas flow past an unknown profile L_z when on it we are given the speed distribution*

$$q = f(s), \qquad \theta_\infty = 0, \tag{3}$$

where $s \in [0, s_1]$ is the arclength parameter of L_z, s_1 is its length, θ_∞ is the slope angle of the velocity at infinity, $|\ln f(s)| < M$, and the function $df(s)/ds$ satisfies a Hölder condition. Assuming that $\varphi(0) = \varphi(s_0) = 0$ at the points $s = 0$ and $s = s_0$ where the flow branches and comes together, we find that

$$\varphi = \int_0^s f(s)\,ds, \quad s \in [0, s_0], \qquad \varphi = \int_s^{s_1} f(s)\,ds, \quad s \in [s_0, s_1]. \tag{4}$$

The condition that the circulation is zero can be written in the form

$$\int_0^{s_0} f(s)\,ds = \int_{s_0}^{s_1} f(s)\,ds = \varphi_0.$$

Corresponding to the flow domain D_z in the complex potential plane there is a domain D_w, namely the plane with a cut along the segment $[0, \varphi_0]$ of the real axis.

2. *The problem of an unconfined gas flow with jet separation past an unknown arc L_z when condition (3) is satisfied on it for $s \in [0, s_1]$, where the points $s = 0$ and $s = s_1$ are points of jet separation and $f(0) = f(s_1) = 1$ (in other respects $f(s)$ satisfies the conditions of the preceding problem).* We set $q = 1$ on the unknown jets L_1 and L_2. Let $s_0 \in (0, s_1)$ be a branch point of the flow and define the function $\varphi(s)$ by

$$\varphi = \int_0^{s_0} f(s)\,ds, \quad s \in [0, s_0], \qquad \varphi = \int_{s_0}^{s} f(s)\,ds, \quad s \in [s_0, s_1]. \tag{5}$$

Corresponding to the flow domain in the complex potential plane there is a domain D_w, namely the plane with a cut along the semi-axis $\varphi \geqslant 0$. Moreover, the points $\varphi_1 = \int_0^{s_0} f(s)\,ds$ and $\varphi_2 = \int_{s_0}^{s} f(s)\,ds$ are the images of the points $s = 0$ and $s = s_1$ of jet separation.

3. *The problem of a fluid flow in a channel with unknown shapes of the walls l_i ($i = 1, 2$) when on them we are given the conditions*

$$q = f_i(s), \qquad s \in (-\infty, \infty), \qquad \theta_\infty = 0. \tag{6}$$

It is assumed that $f_1(\pm\infty) = f_2(\pm\infty)$, $|\ln f_i(s)| < \infty$, and the function $df_i(s)/ds$ satisfies a Hölder condition on each finite interval. Moreover, on each of the walls l_i ($i = 1, 2$) we take some point as the origin $s = 0$. Then

$$\varphi = \int_0^s f_1(s)\, ds, \qquad s \in (-\infty, \infty). \tag{7}$$

Suppose for definiteness that $\psi = \pm\pi/2$ on the upper and lower walls, respectively. Then corresponding to the flow domain is the strip D_w, $-\pi/2 < \psi < \pi/2$, in the complex potential plane.

4. *The problem of a jet flow past an unknown obstacle L_z in a channel with unknown shapes of the walls l_1 and l_2 when the respective conditions* (3) *and* (6) *are satisfied, on L_z and l_i ($i = 1, 2$) *and the speed is constant* ($q = 1$) *on the free jets L_1 and L_2.* The image of the flow domain in the complex potential plane is known to be the domain D_w, namely the strip $-\pi/2 < \psi < \pi/2$ with a cut along the semi-axis $\varphi \geqslant 0$.

2°. According to §1, we introduce the fictitious speed

$$q^* = \int^q \frac{\sqrt{1 - M^2}}{q}\, dq$$

and the fictitious density

$$\rho^* = \rho/\sqrt{1 - M^2}.$$

Then the function $\omega[z(w)] \equiv \omega(w)$ satisfies the following complex equation (1.12):

$$\omega_{\bar{w}} - \mu^w(\omega)\omega_w = 0, \tag{8}$$

where $\mu^w(\omega) = (\rho^* - 1)/(\rho^* + 1)$, and $|\mu^w(\omega)| \leqslant \mu_0 < 1$ for any ω, since the flow is constantly subsonic. In each of the problems 1–4 we map D_w conformally onto the disk $|\zeta| < 1$ by an analytic function $\zeta = \zeta(w)$ and let ζ_∞ be the image of the point at infinity under this mapping. This establishes a correspondence

$$\varphi = \varphi(\gamma), \qquad \gamma \in [0, 2\pi]$$

between the boundary points of the domains $|\zeta| < 1$ and D_w. Comparing the resulting function $\varphi = \varphi(\gamma)$ with the function $\varphi = \varphi(s)$ given by (4), (5), and (7), we find a functional dependence $s = s(\gamma)$. Substitution of $s(\gamma)$ into the boundary conditions (3) and (6) gives the speed $q[s(\gamma)]$ on the boundary of $|\zeta| < 1$, whence

$$\mathrm{Re}\, \omega = q^*[s(\gamma)] = q^*(\gamma), \qquad \gamma \in [0, 2\pi],$$
$$\mathrm{Im}\, \omega(\zeta_\infty) = -\theta_\infty = 0, \tag{9}$$

moreover, by the assumptions made about the speed $q(\zeta)$ in problems 1–4, the function $q^*(\zeta)$ is bounded and its derivative satisfies a Hölder condition. The change of variables $\zeta = \zeta(w)$ reduces (1) to the form

$$\omega_{\bar{\zeta}} - \hat{\mu}\left(\omega, \bar{\zeta}\right)\omega_\zeta = 0, \tag{10}$$

where $\hat{\mu}(\omega, \zeta) = \mu^w(\omega)\exp\{-2i\arg(dw/d\zeta)\}$. Thus, the original gas dynamics problems have been reduced to the Dirichlet problem for equation (10) with the boundary condition (9). Since $|\hat{\mu}(\omega, \zeta)| \leqslant \mu_0 < 1$ for any ω, the unique solvability of problem (9), (10) follows from Theorem 3 in §4 of Chapter V. Knowing the function $\omega(w) = \omega[\zeta(w)]$ and, consequently, also the functions $q(w)$ and $q(\omega)$, $\omega \in \overline{D}_u$, we find that

$$z(w) = \int e^{i\theta} q^{-1}\left(d\varphi + \frac{i}{\rho}\, d\psi\right). \tag{11}$$

This formula can be rewritten also in the form

$$z(\zeta) = \int \frac{e^{i\theta(\xi)}}{q(\xi)} \left[\left(\varphi_\xi + i\rho^{-1}\psi_\xi \right) d\xi + \left(\varphi_\eta + i\rho^{-1}\psi_\eta \right) \right] d\eta, \tag{12}$$

where $\xi + i\eta = \zeta$.

The equation of the boundary L_z of D_z can be written as

$$z = \int_0^\gamma \frac{e^{i\theta(\gamma)}}{q(\gamma)} \frac{d\varphi}{d\gamma} d\gamma + K \equiv F(\gamma). \tag{13}$$

The contour L_z is obviously closed if $F(0) = F(2\pi)$ in (13). Thus we have proved that there is a unique gas flow in all the gas dynamics problems listed in 1°. However, since the assignment of the speed distribution $q = q(s)$ on the boundary of D_z is arbitrary, these flows may not satisfy physical condiitons: The unknown contour L_z flowed past may turn out to be nonclosed or multisheeted, and the jets or walls of the channels may intersect at finite points. Therefore, we must find conditions on $q = q(s)$ for the original problems to be physically realizable. However, as in the case of incompressible fluid flows, these conditions are not expressed explicitly in terms of the boundary value values $q = q(s)$, and they must be imposed on the unknown function $z = z(\zeta)$ mapping the canonical domain $|\zeta| < 1$ onto the flow domain D_z. But since for incompressible fluid flows this mapping is quasiconformal in the ζ-plane in which we solve the Dirichlet problem (9), (10) and satisfies a quasilinear elliptic system, it is difficult to write out explicitly conditions for the physical realizability of the flow even for the function $z(\zeta)$. We thus proceed in the following way. The quasiconformal mapping $z = z(w)$ defined by (11) can be represented in the form $z = z[\chi(w)]$, where $\chi = \chi(w)$ is a homeomorphism of D_w onto the disk $|\chi| < 1$, and $z = z(\chi)$ is an analytic function. The homeomorphism $\chi(w)$ establishes a correspondence $\varphi = \varphi(\alpha)$, $\alpha \in [0, 2\pi]$, $\alpha = \arg \chi$, between the boundary points of the domains $|\chi| < 1$ and D_w. Comparing the functions $\varphi = \varphi(\alpha)$ and $\varphi = \varphi(s)$ at the boundary points of the domain, we find a functional dependence $s = s(\alpha)$, $\alpha \in [0, 2\pi]$. Since the function $z = z(\chi)$ is analytic, we then get that $|dz/d\chi| = ds/d\alpha$ for $\alpha \in [0, 2\pi]$, and so

$$z = e^{i\beta_0} \int_0^\chi \exp \left\{ \frac{1}{2\pi} \int_0^{2\pi} \ln \frac{ds}{d\alpha} \cdot \frac{e^{i\alpha} + \chi}{e^{i\alpha} - \chi} d\alpha \right\} d\chi + K. \tag{14}$$

Formula (14) coincides with the corresponding formula for the solutions of the problems 1–4 for an incompresible fluid (see, for example, [126], pp. 21–22 and 50, 51). Consequently, the conditions for physical realizability of the flow will have the same form as in the case of an incompressible fluid.

3°. Let us now consider the unique solvability of the inverse boundary-value problem for elliptic systems of equations. Suppose that we have a complex equation of the form

$$w_{\bar{z}} - \mu_1(w, z) w_z - \mu_2(w, z) \bar{w}_{\bar{z}} = 0, \tag{15}$$

where μ_1 and μ_2 are measurable functions of their arguments, and

$$|\mu_1| + |\mu_2| \le \mu_0 < 1. \tag{16}$$

We consider an inverse boundary-value problem for this equation: Find a contour $L_z = L_z^1 + L_z^2$ that passes through a fixed point $M(x_1, 0)$ and bounds a finite domain D_z, along with a function $w(z)$ that satisfies (15) in this domain and the following boundary conditions on the arcs L_z^i $(i = 1, 2)$:

$$\varphi = f_1^i(x), \qquad \psi = f_2^i(x), \qquad x \in [x_1, x_2] \; (i = 1, 2). \tag{17}$$

It is assumed that the functions df_k^i/dx $(i, k = 1, 2)$ satisfy a Hölder condition, and that $f_j^1(x_k) = f_j^2(x_k)$ and

$$|f_1^i(\bar{x}_1) - f_1^j(\bar{x}_2)|^2 + |f_2^i(\bar{x}_1) - f_2^j(\bar{x}_2)|^2 \ne 0 \qquad (i, j, k = 1, 2)$$

for any $\bar{x}_1$ and $\bar{x}_2$ in $[x_1, x_2]$ except when $\bar{x}_1 = \bar{x}_2 = x_k$ ($k = 1, 2$). Under these conditions the plane of $w = \varphi + i\psi$ contains a simple contour $L_w = L_w^1 + L_w^2$ with the equations (17) that bound some domain D_w. The boundary condition (17) gives us that

$$x = f(\sigma), \qquad (18)$$

where σ is the arclength parameter of L_w (here x increases monotonically on one of the arcs L_w^1 or L_w^2 and decreases monotonically on the other as the arclength σ increases).

Thus, we arrive at the Dirichlet problem with the boundary condition (18) and with the additional condition

$$y(w_1) = y[f_1^1(x_1), f_2^1(x_1)] = 0$$

for the equation

$$-z_{\bar{w}} + \mu_2(w, z)z_w + \mu_1(w, z)\bar{z}_{\bar{w}} = 0. \qquad (19)$$

The unique solvability of the boundary-value problem (18), (19) follows from Theorem 3 in §4 of Chapter V, by (16). The mapping $z = z(w)$ is quasiconformal because the system corresponding to the complex equation (19) is elliptic; consequently, the inverse function $w(z)$ satisfies (15) and (by construction) the boundary conditions (17). The mapping $z = z(w)$ can be represented in the form $z = z[\zeta(w)]$, where $\zeta = \zeta(w)$ is a homeomorphism of D_w onto the disk $|\zeta| < 1$ that carries the arcs L_w^1 and L_w^2 into the upper and lower halves of $|\zeta| = 1$, respectively, and $z = z(\zeta)$ is an analytic function. The homeomorphism $\zeta = \zeta(w)$ establishes a correspondence $\sigma = \sigma(\gamma)$, $\gamma \in [0, 2\pi]$ ($\gamma = \arg\gamma$), between the boundary points on the contours L_w and $|\zeta| = 1$. Then condition (18) gives us a correspondence $x = f[\sigma(\gamma)] = g(\gamma)$, $\gamma \in [0, 2\pi]$, between the abscissas of the points of L_z and the arclength on the circle $|\zeta| = 1$. The function $z = z(\zeta)$ can be represented in the form

$$z = \frac{1}{2\pi} \int_0^{2\pi} g(\gamma) \frac{e^{i\gamma} + \zeta}{e^{i\gamma} - \zeta} d\gamma + iy_0. \qquad (20)$$

It follows from the properties of the function $x = f(\sigma)$ that $g(\gamma) = f[\sigma(\gamma)]$ increases monotonically on one of the intervals $[0, \pi]$ or $[\pi, 2\pi]$, and decreases monotonically on the other. According to a theorem of Kaplan (Theorem 6 in §3 of Chapter II), this property of $g(\gamma)$ ensures that the function $z(\zeta)$ represented by (20) is univalent.

We have thus proved that the inverse boundary-value problem formulated above has a unique univalent solution.

4°. Let us now consider the inverse boundary-value problem with the same boundary conditions (17) for the nonlinear system of equations

$$\Phi_i(x, y, \varphi, \psi, \varphi_x, \varphi_y, \psi_x, \psi_y) = 0 \qquad (i = 1, 2). \qquad (21)$$

At points where the Jacobian $J = x_\varphi y_\psi - x_\psi y_\varphi$ is nonzero we pass to the inverse system of equations

$$\Phi_i(x, y, \varphi, \psi, J^{-1}y_\psi, J^{-1}x_\psi, J^{-1}y_\varphi, J^{-1}x_\varphi) = 0 \qquad (i = 1, 2). \qquad (22)$$

In the domain D_w we get the Dirichlet boundary-value problem for the system (22) with the boundary condition (18) and the additional condition

$$y(w_1) = y[f_1^1(x_1), f_2^1(x_1)] = 0.$$

The inverse boundary-value problem is obviously uniquely solvable for the nonlinear equation (21) under the conditions (17) if:

(a) the mapping $w = w(z)$ by a solution of the original problem is quasiconformal, and

(b) the boundary-value problem (22), (18) is uniquely solvable.

Condition (a) holds for systems that are elliptic in Lavrent'ev's sense. As was shown in §8 of Chapter V (where the definition of ellipticity in Lavrent'ev's sense was introduced), such systems can be written in the form

$$w_{\bar{z}} - \mu(z, w, w_z, w_{\bar{z}})w_z = 0, \qquad (23)$$

where $|\mu| \leqslant \mu_0 < 1$ for all values of the arguments. This implies that each solution of a system elliptic in Lavrent'ev's sense realizes a quasiconformal mapping.

Condition (b) is also satisfied by a large class of elliptic systems. For example, consider a system (21) in which y does not appear explicitly, and suppose that the inverse system (22) can be solved for the derivatives y_φ and y_ψ. Then we rewrite (22) in the form

$$y_\varphi = F_1(x, \varphi, \psi, x_\varphi, x_\psi), \qquad -y_\psi = F_2(x, \varphi, \psi, x_\varphi, x_\psi), \tag{24}$$

where the functions F_i are continuously differentiable with respect to their arguments. Differentiating the first equation in (24) with respect to ψ and the second with respect to φ, and adding the resulting relations, we obtain

$$A x_{\varphi\varphi} + 2B x_{\varphi\psi} + C x_{\psi\psi} = F, \tag{25}$$

where

$$A = \frac{\partial F_2}{\partial x_\varphi}, \qquad B = \frac{1}{2}\left(\frac{\partial F_1}{\partial x_\varphi} + \frac{\partial F_2}{\partial x_\psi} \right), \qquad C = \frac{\partial F_1}{\partial x_\psi}$$

and

$$-F = \frac{\partial F_1}{\partial \psi} + \frac{\partial F_2}{\partial \varphi}.$$

Thus, the boundary-value problem (22), (18) has been reduced to the Dirichlet problem for the quasilinear elliptic equation (25) with the boundary condition (18), and the latter is uniquely solvable (by familiar theorems about elliptic partial differential equations; see, for example, [59]) under very minimal assumptions on the smoothness of the functions A, B, C, and F. In particular, the system inverse to the system of equations of gas dynamics can be written in the form (24). Indeed, this system has the form

$$x_\varphi = \rho(q)y_\psi, \qquad -y_\varphi = \rho(q)x_\psi, \tag{26}$$

where $q^2 = (x_\psi^2 + y_\psi^2)/y^2$, and $J = x_\varphi y_\psi - x_\psi y_\varphi$ is the Jacobian of the mapping $z = z(w)$. Multiplying the first equation in (26) by y_ψ and the second by x_φ, and adding the resulting equations, we find that

$$J = \rho\left(x_\varphi^2 + y_\psi^2 \right) \quad \text{or} \quad J^{-1} = q\rho(q).$$

On the other hand, $J = x_\varphi y_\psi - x_\psi y_\varphi = x_\varphi^2/\rho + \rho x_\psi^2$. Since $d(\rho q)/dq \geqslant \beta > 0$ in the subsonic domain, the implicit function

$$q = F(J) = F\left(x_\varphi^2 \rho^{-1} + x_\psi^2 \rho \right) \equiv \tilde{F}(x_\varphi, x_\psi)$$

in the equation $q\rho(q) = J^{-1}$ exists. Consequently, (26) can be written in the form

$$y_\varphi = -\rho\left[\tilde{F}(x_\varphi, x_\psi) \right] x_\psi, \qquad y_\psi = \left(\rho\left[\tilde{F}(x_\varphi, x_\psi) \right] \right)^{-1} x_\varphi.$$

Conditions (a) and (b) thus hold simultaneously for the system of equations of gas dynamics in the subsonic domain; consequently, the inverse boundary-value problem for this system is uniquely solvable when $\rho(q) \in C^{1+\alpha}$. If the inverse boundary-value problem is solvable for the nonlinear system (21) and condition (a) holds for the solution, then we can show that this solution is univalent, completely analogously to the case of the inverse boundary-value problem for the quasilinear elliptic system (15).

§7. Problems

In this section we restrict ourselves to brief formulations for the case of subsonic gas flows, since many of the problems discussed below were described in detail for ideal fluid flows in §6 of Chapter IV.

$1°$. Lavrent'ev [64] first investigated the spatial problem of ideal fluid flows with free boundaries in a rigorous formulation. Apply his investigations to problems involving spatial subsonic gas flows past bodies. Extend the methods of the theory of boundary-value problems to flows that are asymptotically close to planar flows.

2°. Consider boundary-value problems with free boundaries in magnetohydrodynamics.

3°. Consider the problems formulated in §6 of Chapter IV about conjunction of flows in different hydrodynamical models also in the case of subsonic gas flows.

4°. In §3 of this chapter it was proved that the problems of cavitational flow past curvilinear obstacles according to the schemes of Kirchhoff, Riabouchinsky, and Thullen are solvable. Consider also the problem of subsonic cavitational flows past bodies according to the Lavrent'ev scheme (§3 of Chapter IV).

5°. Extend the methods of finite-dimensional approximation (Chapters III and IV) to subsonic gas flows.

6°. The absence of stagnation points of the flow on the boundary of the flow domain was used in an essential way in the uniqueness theorems proved in §§2 and 3. Considering boundary-value problems for the function $\Omega(z) = e^{\omega(z)}$, where $\omega(z) = q^*(z) - i\theta(z)$ is the generalized Joukowsky function, prove uniqueness theorems also when stagnation points are present.

7°. In this chapter we have studied only problems for uniformly subsonic gas flows. Generalize the existence and uniqueness theorems proved in §§2 and 3 to the case of transonic gas flows with free boundaries. Many articles have dealt with this question; surveys can be found, for example, in [16] and [19]. However, as a rule, only approximate models of the flows are considered in these articles. Such problems still remain to be investigated for the exact equations of gas dynamics in a direct formulation that does not use the hodograph.

8°. Extend the methods of §§2 and 3 to problems of subsonic gas dynamics with free boundaries in multiply connected domains (examples of such problems were considered in §4).

CHAPTER VIII

PROBLEMS IN FILTRATION THEORY
FOR A FLUID WITH FREE BOUNDARIES

§1. On certain transformations of the equations of motion
of a fluid in porous media

The motion of a fluid in a porous medium is characterized by the filtration velocity vector $\vec{v}$, whose magnitude v coincides with the specific outflow of fluid per unit time across a porous area element normal to $\vec{v}$ when the pressure is p and the density is ρ.

The filtration velocity $\vec{v}$ (the specific outflow vector) is related to the mean velocity $\vec{V}$ of particle motion by the formula

$$\vec{v} = m\vec{V} = \left\{ m\frac{dx}{dt}, m\frac{dy}{dt}, m\frac{dz}{dt} \right\},$$

where $m = s_1/s$ is the porosity of the soil, which is equal to the ratio of the area s_1 of the pores on the area element s to the area of s.

The equation of continuity of the flow (the equation of mass conservation) in the case of an incompressible fluid has the form

$$\operatorname{div} \vec{v} = 0. \tag{1}$$

Taking into account the resistance of the porous medium to the motion of the fluid in it, we arrive at the following relation in the general case:

$$\frac{\Phi(v)}{v}\vec{v} = -\operatorname{grad}(p + \rho g\eta), \tag{2}$$

where η is the height of a point over a fixed level (usually, $\eta \equiv z$, or, in the planar case, $\eta = y$), $\Phi(v) > 0$, and $v^{-1}\Phi'(v) > 0$ for all $v \geqslant 0$. The law of resistance of the porous medium can be experimentally confirmed to be linear with high accuracy, and for this law the relations (2) take the form

$$\vec{v} = -\frac{k_0}{\mu}\operatorname{grad}(p + \rho g\eta), \tag{2*}$$

where μ is the viscosity of the fluid, and k_0 is the permeability coefficient (having the dimension of area), which characterizes the filtration properties of the porous medium. In this form the relation (2*) is called Darcy's law. The permeability coefficient in the case of a nonhomogeneous soil depends on the spatial coordinates, and for a compressible medium also on the pressure, i.e., $k_0 = k_0(x, y, z, p)$. If the soil is anisotropic, then $k_0 = \{k_{ij}\}$ is a symmetric tensor.

REMARK. In the case of gas filtration, when $\rho = \rho(p) \neq$ const, it is convenient to introduce the filtration velocity by the formula $\vec{v} = m\rho\vec{V}$. Then the equation of continuity (1) keeps the previous form, but Darcy's law (2*) is written in the form

$$\vec{v} = \frac{k(p)}{\mu} \operatorname{grad}(P + gh),$$

where $k = k_0\rho^2$ and $P = \int_{p_0}^p dp/\rho(p)$, which corresponds to the filtration of an incompressible fluid in a compressible soil. However, if we neglect the effect of gravity (a natural assumption in filtration problems for a gas), then Darcy's law can be given the form

$$\vec{v} = -\frac{k_0}{\mu} \operatorname{grad} P, \qquad P = \int_{p_0}^p \rho(p)\, dp.$$

Thus, taking the compressibility of the fluid into account does not affect the form of (1) and (2*). This fact holds also in the case of a nonlinear filtration law (2). Therefore, without loss of generality the fluid being filtered is always assumed to be imcompressible in what follows.

1°. *Darcy's law.* Let us first consider the case of steady planar filtration of a fluid. Letting u and v be the components of the velocity vector, we introduce the flow potential $\varphi(x, y)$ and the stream function $\psi(x, y)$ by the formulas

$$\varphi = -k\left[\frac{p(x, y)}{\rho g} + y\right] + \text{const}, \qquad \frac{\partial\psi}{\partial y} = u, \quad \frac{\partial\psi}{\partial x} = -v.$$

The equations (1) and (2*) take the form $\varphi_x = \psi_y$ and $\varphi_y = -\psi_x$. We compute the Jacobian J of the mapping with $\varphi = \varphi(x, \psi)$ and $y = y(x, \psi)$:

$$J = \frac{D(\varphi, y)}{D(x, \psi)} = \frac{D(\varphi, y)}{D(x, y)}\left[\frac{D(x, \psi)}{D(x, y)}\right]^{-1} = \varphi_x\psi_y^{-1} = 1.$$

On the other hand,

$$\frac{\partial\varphi}{\partial y}[x, \psi(x, y)] = \frac{\partial\varphi(x, \psi)}{\partial\psi} \cdot \frac{\partial\psi(x, y)}{\partial y},$$

$$\frac{\partial\psi}{\partial x}[\varphi(x, y), y)] = \frac{\partial\psi}{\partial\varphi} \cdot \frac{\partial\varphi(x, y)}{\partial x},$$

and

$$\frac{\partial \psi(\varphi, y)}{\partial \varphi} = \frac{D(\psi, y)}{D(\varphi, y)} \cdot \frac{D(x, \psi)}{D(x, \psi)} = -J^{-1} \frac{\partial y(x, \psi)}{\partial x}.$$

Substituting the above formulas in the equation $\varphi_y = -\psi_x$, we come to the system of equation for $\varphi(x, \psi)$ and $y(x, \psi)$:

$$J \equiv \varphi_x y_\psi - \varphi_\psi y_x = 1, \qquad \varphi_\psi = y_x, \tag{3}$$

or, in another form,

$$\varphi_x = \frac{1 + y_x^2}{y_\psi}, \qquad \varphi_\psi = y_x. \tag{3*}$$

After eliminating $\varphi(x, \psi)$ from (3*), we find that

$$y_{xx} - 2\frac{y_x}{y_\psi} y_{x\psi} + \frac{1 + y_x^2}{y_\psi^2} y_{\psi\psi} = 0. \tag{4}$$

By introduction of the new unknown functions $r(x, \psi) = y_\psi$ and $t(x, \psi) = -y_x$ and their inverse functions, (4) can be transformed to the respective systems

$$r_x + t_\psi = 0, \qquad t_x + 2\frac{t}{r}t_\psi - \frac{1 + t^2}{r^2}r_\psi = 0, \tag{5}$$

$$\psi_t + x_r = 0, \qquad \psi_r + 2\frac{t}{r}\psi_t - \frac{1 + t^2}{r^2}x_t = 0. \tag{6}$$

Passing to the inverse functions in (3) and eliminating $\psi(\varphi, y)$ from the resulting system

$$x_y = \psi_\varphi, \qquad x_\varphi \psi_y - x_y \psi_\varphi = 1, \tag{7}$$

we arrive at an equation analogous to (4):

$$\frac{1 + x_y^2}{x_\psi^2} x_{\varphi\varphi} - 2\frac{x_y}{x_\varphi} x_{\varphi y} + x_{yy} = 0. \tag{8}$$

Taking account of the fact that

$$\frac{\partial \psi(\varphi, y)}{\partial y} = \frac{\partial y(x, \psi)}{\partial \psi} = r, \qquad \frac{\partial x(\varphi, y)}{\partial y} = -\frac{\partial y(x, \psi)}{\partial x} = t,$$

we pass, as before, from (8) to the system

$$r_y - t_\varphi = 0, \qquad \frac{1 + t^2}{r^2}r_\varphi - 2\frac{t}{r}r_y + t_y = 0, \tag{9}$$

$$\varphi_t - y_r = 0, \qquad \varphi_r + 2\frac{t}{r}y_r + \frac{1 + t^2}{r^2}y_t = 0. \tag{10}$$

The systems (3) and (7) are elliptic; equation (4) and the systems (5) and (6) are elliptic if $0 < y_\psi^2 = r^2 < \infty$; and equation (8) and the systems (9) and (10) are elliptic if $0 < x_\varphi^2 = r^2 < \infty$. Suppose that these last conditions hold. Then the mappings $x = x(t, r)$, $\psi = \psi(t, r)$ and $\varphi = \varphi(t, r)$, $y = y(t, r)$ are quasi-conformal because the corresponding systems (5), (6) and (9), (10) are elliptic.

We remark that the variables r and t of the generalized hodograph are connected with the variables u and v of the velocity hodograph by the formulas

$$r = \frac{\partial y(x, \psi)}{\partial \psi} = \frac{1}{u}, \qquad t = -\frac{\partial y(x, \psi)}{\partial x} = -\frac{v}{u}.$$

These formulas enable us to get directly from the systems

$$x_u = -y_v, \, x_v = y_u, \qquad \varphi_u = -\psi_v, \, \varphi_v = \psi_u$$

the following equations, which are satisfied by the mappings $z = z(t, r)$ and $w = w(t, r)$:

$$-x_r = \frac{1 + t^2}{r} y_t + t y_r, \qquad x_t = t y_t + r y_r; \tag{11}$$

$$\psi_r = \frac{1 + \psi^2}{r} \psi_t + t \psi_r, \qquad -\psi_t = t \varphi_t + r \varphi_r. \tag{12}$$

Consequently, the mappings $z = z(t, r)$ and $w = w(t, r)$ are quasiconformal when the ellipticity condition $0 \neq |r| < \infty$ holds for the systems (11) and (12).

Thus, in a domain where the condition $0 \neq |r| < \infty$ holds we have Scheme 1 pictured below, which makes it clear, in particular, that the mapping $x = x(\varphi, y)$, $\psi = \psi(\varphi, y)$ is also quasiconformal under the condition $0 \neq |r| < \infty$. If the last condition is not satisfied, then, as the form of the system (7) shows, this mapping (in view of the fact that the Jacobian is constant) may fail to be quasiconformal; in particular, the generalized Riemann mapping theorem on the possibility of a quasiconformal mapping of any domain onto the unit disk may not be valid for it. The system (7) is an example of an elliptic system that is not elliptic in Lavrent'ev's sense.

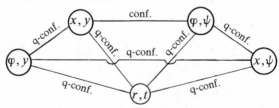

SCHEME 1

We remark that the transformations made in this section are legitimate under the condition that $0 < |r| < \infty$.

2°. *A nonlinear law of filtration.* In many practically important filtration problems Darcy's law agrees badly with experiment, so that we must consider more complicated models for describing the motion of a fluid in porous media. According to Hristianovič [46], the nonlinear relations (2) connect the velocity magnitude $q(x, y) = (u^2 + v^2)^{1/2}$ of the motion and the piezometric head $h = p(x, y)/\rho g + y$:

$$\frac{\partial h}{\partial x} = -\frac{\Phi(q)}{q}u, \qquad \frac{\partial h}{\partial y} = -\frac{\Phi(q)}{q}v. \tag{13}$$

The equation of continuity, as usual, has the form

$$\frac{\partial u}{\partial x} + \frac{\partial v}{\partial y} = 0$$

and shows that there exists a stream function ψ such that

$$\frac{\partial \psi}{\partial x} = -v, \qquad \frac{\partial \psi}{\partial y} = u. \tag{14}$$

We introduce the potential $\varphi = -kh + c$, where k is a fixed constant. Then (13) gives us that

$$u = \frac{q}{k\Phi(q)}\varphi_x, \qquad v = \frac{q}{k\Phi(q)}\varphi_y.$$

Substituting this in (14), we get

$$\varphi_x = \rho(q)\psi_y, \qquad \varphi_y = -\rho(q)\psi_x, \tag{15}$$

where

$$\rho(q) = \frac{k\Phi(q)}{q}, \qquad q^2 = \psi_x^2 + \psi_y^2.$$

This is one of the forms of the filtration equations in the case of a nonlinear filtration law.

The system (15) is elliptic under the condition $\rho(q\rho)'_q \geq \beta > 0$ or

$$\frac{k^2\Phi}{q}\Phi'_q \geq \beta > 0.$$

Performing the transformations of 1° on system (15), we get in a completely analogous way that

$$\varphi_x y_\psi - y_x \varphi_\psi = \rho(q), \qquad \varphi_\psi = \rho(q)y_x, \tag{16}$$

or

$$\varphi_x = q^2 \rho y_\psi, \qquad \varphi_\psi = \rho y_x, \tag{17}$$

where

$$q^2 = \frac{1 + y_x^2}{y_\psi^2}, \quad \varphi = \varphi(x, \psi) \quad \text{and} \quad y = y(x, \psi).$$

Eliminating $\varphi(x, \psi)$ from (17) leads to

$$\left(q^2 \rho y_\psi\right)_\psi - \left(\rho y_x\right)_x = 0.$$

But

$$\rho_x = \rho_q \frac{(q^2)_x}{2q}, \qquad \rho_\psi = \rho_q \frac{(q^2)_\psi}{2q},$$

$$\left(q^2\right)_x = \left(\frac{1 + y_x^2}{y_\psi^2}\right)_x, \qquad \left(q^2\right)_\psi = \left(\frac{1 + y_x^2}{y_\psi^2}\right)_\psi.$$

Taking these relations into account, we get

$$\left(\rho + \frac{\rho'}{q} \frac{y_x^2}{y_\psi^2}\right) y_{xx} - 2\frac{y_x}{y_\psi}(\rho + q\rho')y_{x\psi} + q^2(\rho + q\rho')y_{\psi\psi} = 0. \qquad (18)$$

The ellipticity condition for the last equation has the form

$$\Delta = \frac{y_x^2}{y_\psi^2}(\rho + q\rho')^2 - q^2\left(\rho + \frac{\rho'}{q} \frac{y_x^2}{y_\psi^2}\right)(\rho + q\rho') < 0,$$

where Δ is the discriminant of the corresponding quadratic form. Transformation of the expression for Δ gives

$$-\Delta = \rho(q\rho')_q/y_\psi^2 \geqslant \beta/y_\psi^2 \geqslant 0,$$

since $\rho(q\rho')'_q \geqslant \beta > 0$ by virtue of the ellipticity of (15). Consequently, (18) is uniformly elliptic in the domain where $0 < y_\psi^2 < \infty$. We write (18) in the form

$$Ay_{xx} - 2By_{x\psi} + Cy_{\psi\psi} = 0. \qquad (18^*)$$

Setting $y_\psi = r$ and $y_x = -t$, we arrive at the system

$$r_x + t_\psi = 0, \qquad At_x + 2Br_x - Cr_\psi = 0, \qquad (19)$$

which for the inverse functions $x = x(t, r)$ and $\psi = \psi(t, r)$ gives us

$$\psi_t + x_r = 0, \qquad A(r, t)\psi_r + 2B(r, t)x_r - C(r, t)x_t = 0. \qquad (20)$$

The systems (19) and (20) are elliptic in the domain where $0 \neq |r| < \infty$. Let us rewrite the nonlinear filtration equations (15) in a somewhat different form. Squaring both sides of (15) and adding these equalities termwise, we get $Q^2 = q^2\rho^2$, where $Q^2 = \varphi_x^2 + \varphi_y^2$. Since (15) is uniformly elliptic, it follows that $\rho \neq 0$ and $(\rho q)_q \neq 0$. Suppose for definiteness that $\rho > 0$; then $Q = q\rho$, and since $(\rho q)_q \neq 0$, it follows that the implicit function $q = q(Q)$ exists. We call Q the distorted filtration velocity. We have that $\rho = \rho(q) = \rho[q(Q)] \equiv 1/\tilde{\rho}(Q)$. Then (15) can be written in the form

$$\tilde{\rho}(Q)\varphi_x = \psi_y, \qquad \tilde{\rho}(Q)\varphi_y = -\psi_x, \qquad (21)$$

where $Q^2 = \varphi_x^2 + \varphi_y^2$. Let us transform it to the new functions x and ψ:

$$x_\varphi \psi_y - \psi_\varphi x_y = \tilde{\rho}(Q), \qquad \psi_\varphi = \tilde{\rho}(Q)x_y, \qquad (22)$$

$$\psi_\varphi = \tilde{\rho}(Q)x_y, \qquad \psi_y = Q^2\tilde{\rho}(Q)x_\varphi, \qquad (23)$$

where $Q^2 = (1 + x_y^2)/x_\varphi^2$. Elimination of ψ from (23) gives us $[\tilde{\rho}(Q)x_y]_y - [Q^2\tilde{\rho}(Q)x_\varphi]_\varphi = 0$. Comparing systems (23) and (17) and taking into account the symmetry of these equations with respect to the dependent and independent variables, we find that

$$\left(\tilde{\rho} + \frac{\tilde{\rho}'x_y^2}{Qx_\varphi^2}\right)x_{yy} - 2\frac{x_y}{x_\varphi}(\tilde{\rho} + Q\tilde{\rho}')x_{y\varphi} + Q^2(\tilde{\rho} + Q\tilde{\rho}')x_{\varphi\varphi} = 0. \qquad (24)$$

The condition for this equation to be elliptic has the form

$$-\tilde{\Delta} = \tilde{\rho}(Q\tilde{\rho}')_Q/x_\varphi^2 \geqslant \beta_2 > 0.$$

But $\tilde{\rho}(Q) \equiv 1/\rho(q)$, $Q = q\rho$, and then

$$-\tilde{\Delta} = \left\{x_\varphi^2\rho(q)[q\rho(q)]_q'\right\}^{-1}.$$

The ellipticity of (15) implies that $\rho(\rho q')q > 0$; therefore, (24) is elliptic in the domain where $0 \neq x_\varphi^2 < \infty$. We write this equation in the form

$$\tilde{A}(x_\varphi, x_y)x_{yy} - 2\tilde{B}(x_\varphi, x_y)x_{\varphi y} + \tilde{C}(x_\varphi, x_y)x_{\varphi\varphi} = 0. \qquad (24^*)$$

Taking $\partial x(\varphi, y)/\partial \varphi = r$ and $\partial x(\varphi, y)/\partial y = t$, into account, we get

$$\tilde{A}(t, r)t_y - 2\tilde{B}(t, r)r_y + \tilde{C}(t, r)r_\varphi = 0, \qquad r_y = t_\varphi, \qquad (25)$$

$$\tilde{A}(t, r)\varphi_r + 2\tilde{B}(t, r)y_r + \tilde{C}(t, r)y_t = 0, \qquad \varphi_t = y_r. \qquad (26)$$

Systems (25) and (26) are also elliptic in the domain where $0 \neq |r| < \infty$. System (21) can be rewritten in the form

$$L(\varphi, y) = \frac{\tilde{\rho}'}{Qr^2(\tilde{\rho}Q)_Q'}\varphi_r, \qquad \varphi_t = y_r, \qquad (21^*)$$

where $L(\varphi, y) \equiv \varphi_r + 2ty_r/r + (1 + t^2)y_t/r^2$ is the operator corresponding to the linear system. Indeed, the second equation in (26) can be written in the form $(\tilde{\rho} + \tilde{\rho}'Q)L(\varphi, y) = \tilde{\rho}'y/Qr^2$, which gives us the necessary relation.

Scheme 1 holds for the transformations of this subsection, and, consequently, they are valid under the same condition $0 < |r| < \infty$ as in 1°.

§2. Solvability of a certain class of problems in the linear
theory of filtration with free boundaries

1°. *Boundary conditions.* Figure 1 shows a sketch of the domain of filtration of a fluid in an earthen dam with vertical walls on an impermeable base.

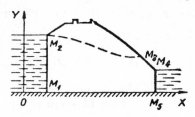

FIGURE 1

We encounter four forms of boundaries of the region of motion when we consider problems involving filtration of a fluid under a hydraulic structure or in the body of an earthen dam, or when there is water seepage from canals into the soil, and so on. The impermeable boundaries (underground contours of a hydraulic structure, impermeable soil) are streamlines $\psi = $ const (the base M_1M_5 in Figure 1). On the boundaries of water reservoirs with the soil (the walls M_1M_2 and M_4M_5 of the dam) the hydrostatic law of Bernoulli gives us for the fluid that

$$h(x, y) + q^2/2g = \text{const},$$

and it is possible to set $h(x, y) = $ const, which is equivalent to $\varphi(x, y) = $ const, due to the smallness of the filtration velocities. The free boundary (the curve M_2M_3 in Figure 1) is the contour separating wet and dry soil. Since the pores communicate with one another and with the atmospheric air, there is constant pressure $p(x, y) = $ const on the free surface (the depression curve), which leads to the condition $\varphi + ky = $ const (k is the permeability coefficient). If there is no influx of moisture to the free boundary and no evaporation from its surface, then it is a streamline $\psi = $ const. Otherwise, the flux across any part of this boundary is assumed to be proportional to the horizontal projection of the arc of this curve, i.e., we set $\psi - \psi_0 = \varepsilon(x - x_0)$, $\varepsilon > 0$, where $\varepsilon > 0$ for infiltration (precipitation or percolation of water) and $\varepsilon < 0$ for evaporation from the free surface, and $|\varepsilon|$ is the amount of water per unit time seeping into (respectively, evaporating from) a unit length of the horizontal projection of the depression curve. Along a seepage interval, where the water from the body of the dam trickles down in streams along its surface, the pressure inside the filtration domain is usually taken to be equal to atmospheric pressure, i.e., $\varphi = -ky + $ const (the part M_3M_4 in Figure 1).

If we consider a problem with an inclined seepage interval, it is natural to assume that infiltration (or evaporation) takes place at the same time on the free surface and on the seepage interval. Therefore, the condition $\psi = \varepsilon x +$ const can be used on the seepage interval in place of the condition $\varphi = -ky +$ const usually adopted, i.e., the two problems can actually be regarded on an equal footing.

Indeed, taking the pressure on the seepage interval to be the same as atmospheric pressure, we assume that the water surface coincides with the boundary of the interval, i.e., the seepage interval is actually a free surface. But it is impossible to require both conditions $\varphi = -ky +$ const and $\psi = \varepsilon x +$ const at once on the known boundary of the seepage interval; therefore, it is natural to solve the two different problems corresponding to each of these conditions.

Thus, the following conditions hold on the boundary of the filtration domain:

$$\psi = 0, \quad y = 0 \quad \text{on } M_1 M_5; \tag{1}$$

$$\varphi = \varphi_i, \quad x = x_i \quad \text{on } M_1 M_2 \, (i = 1), \quad M_4 M_5 \, (i = 2); \tag{2}$$

$$\varphi = -ky + \varphi_0, \quad \psi = \varepsilon x + \psi_0 \quad \text{on } M_2 M_3; \tag{3}$$

$$\varphi = -ky + \varphi_0, \quad x = ay + x_0 \quad \text{on } M_3 M_4 \tag{4}$$

or

$$\psi = \varepsilon x + \psi_0, \quad x + ay + x_0 \quad \text{on } M_3 M_4. \tag{4*}$$

Moreover it cannot be assumed that all the constants φ_i, x_i ($i = 0, 1, 2$), and ψ_0 simultaneously are arbitrary. Some of them must be determined in the solution process.

The question of which constants can be chosen arbitrarily depends on the problem being solved, and will therefore not be discussed at present. We must use conditions (1)–(4) and (4*) to determine the form of the depression curve $M_2 M_3$ and to find in the domain bounded by this curve and the polygonal line $M_2 M_3 M_4 M_5$ a complex filtration potential $w = \varphi + i\psi$ which satisfies the system of Cauchy-Riemann equations or one of the systems (1.17) or (1.23).

The intervals over which the variables x and y vary can be given in the boundary conditions (3) and (4). If this is done for both variables, then the coordinates of M_3 are completely specified, i.e., x_0 is fixed. Then the y-coordinates of the points M_2', M_3', and M_4' will be known in the plane of the variables (φ, y), and, consequently, the constants φ_1 and φ_2 will also be fixed. If one of the intervals is not given, it is possible to choose either the constant x_0 or one of the constants φ_i ($i = 1, 2$) arbitrarily. Similar considerations apply to the boundary-value problem (1)–(4*).

2°. *The problem on the generalized hodograph plane.* Let us first dwell on the boundary-value problem (1)–(4) for the analytic function $w = w(z)$. As is clear from the boundary conditions (1)–(4), the images of the boundaries of the filtration domain are known in the plane of the variables (φ, y). We assume that the constants φ_0, φ_1 and φ_2 are fixed, and we construct the contour in the (φ, y)-plane. The position of the point M_3' on the line $M_2' M_4'$ is obviously not fixed by conditions (1)–(4). It is natural to take advantage of the presence of the closed contour in the (φ, y)-plane (Figure 2). Therefore, we express x and ψ as functions of φ and y. Such a transformation was implemented in §1, where the system (1.7) was obtained for $x = x(\varphi, y)$ and $\psi = \psi(\varphi, y)$; this system can be rewritten in the form

$$\psi_\varphi = x_y, \qquad \psi_y = \left(1 + x_y^2\right)/x_\varphi, \tag{5}$$

and $|dw/dz| = (1 + x_y^2)/x_\varphi^2$. The boundary conditions (1)–(4) give us that on the sides of the polygon in the (φ, y)-plane the following conditions hold for the functions $x = x(\varphi, y)$ and $\psi = \psi(\varphi, y)$:

$$\begin{aligned}
\psi &= 0 \quad \text{on } M_1' M_5'; \\
x &= x_i \quad \text{on } M_1' M_2', \, M_4' M_5'; \\
\psi &= \varepsilon x + \psi_0 \quad \text{on } M_2' M_3'; \\
x &= ay + x_0 \quad \text{on } M_3' M_4'.
\end{aligned} \tag{6}$$

Thus, we get a Hilbert boundary-value problem with discontinuous coefficients for system (5). Since conditions (1)–(4) are linear, this boundary-value problem can be considerably simplified.

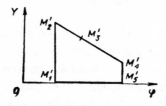

FIGURE 2

To do this, we eliminate ψ from (5) and write $x_\varphi = r$ and $x_y = t$. Then, according to §1, we obtain

$$t_y - 2\frac{t}{r}r_y + \frac{1 + t^2}{r^2}r_\varphi = 0, \qquad r_y = t_\varphi \tag{7}$$

or, for the inverse functions,

$$\varphi_r + 2\frac{t}{r}y_r + \frac{1 + t^2}{r^2}y_t = 0, \qquad \varphi_t = y_r. \tag{8}$$

Let us find the images of the sides of the polygon $M_1' M_2' M_4' M_5' M_1'$. Suppose that the functions $\psi = \psi(\varphi, y)$ and $x = x(\varphi, y)$ are continuously differentiable up to the boundary. Then $d\psi/d\varphi = \psi_\varphi + \psi_y dy/d\varphi$ on $M_1' M_5'$, but $d\psi/d\varphi = dy/d\varphi = 0$ along $M_1' M_5'$, which implies that $\psi_\varphi = x_y = t = 0$. On $M_1' M_2'$ and $M_5' M_4'$ we have $d\psi/dy = x_y + x_\varphi d\varphi/dy$, but $d\varphi/dy = 0$, whence $x_y = t = 0$. Along $M_2' M_3'$ we have $d\psi/dy = \varepsilon dx/dy$, and, since

$$\frac{d\psi}{dy} = \psi_y + \psi_\varphi \frac{d\varphi}{dy} = \frac{1 + t^2}{r} - kt, \qquad \frac{dx}{d\varphi} = x_y + x_\varphi \frac{d\varphi}{dy} = t - kr,$$

it follows that

$$\frac{1 + t^2}{r} - kt = \varepsilon t - k\varepsilon r \quad \text{or} \quad t(\varepsilon + k) - k\varepsilon r - \frac{1 + t^2}{r} = 0,$$

and $dx/dy = a$ on $M_3' M_4'$. On the other hand,

$$\frac{dx}{dy} = x_y + x_\varphi \frac{d\varphi}{dy} = t - kr,$$

and so $t - kr = a$. Thus, we arrive at the following equalities:

$$
\begin{aligned}
t &= 0 \quad \text{on } M_2' M_1' M_5' M_4'; \\
t - kr &= a \quad \text{on } M_3' M_4'; \\
t(\varepsilon + k) - k\varepsilon r - (1 + t^2)/r &= 0 \quad \text{on } M_2' M_3'.
\end{aligned}
\tag{9}
$$

By the equality $|dw/dz|^2 = (1 + t^2)/r^2$, it is necessary to look for a solution of problem (5), (6) such that $x_\varphi^2 = r^2 \neq 0$ inside the polygon in the (φ, y)-plane, i.e., to find out whether there is a domain bounded by curves with equations (9) in the (t, r)-plane which does not contain points of the straight line $r = 0$ inside it.

3°. *Selection and properties of the domain in the generalized hodograph plane.* The curve

$$rt(\varepsilon + k) - k\varepsilon r^2 - (1 + t^2) = 0$$

is a hyperbola. We find the point of intersection of this hyperbola with the straight line $t = kr + a$:

$$r_3 = \frac{1 + a^2}{a(\varepsilon - k)}, \qquad t_3 = \frac{k + a^3\varepsilon}{a(\varepsilon - k)}.$$

Then the point of intersection of the straight line $t = kr + a$ with the axis $t = 0$ is $r_4 = -a/k$, $t_4 = 0$. For the points (r_3, t_3) and (r_4, t_4) to be images of the points M_3' and M_4', they must lie on one side of the axis Ot, since otherwise points of the straight line $r = 0$ fall inside the domain bounded by the curves (9). Consequently, the inequality $(1 + a^2)/k(k - \varepsilon) > 0$ must hold, and, since

$k > 0$, the inequality $k > \varepsilon$ must also hold. Otherwise, i.e., when $k < \varepsilon$, the (t, r)-plane does not contain a domain bounded by the curves (9) and satisfying the necessary conditions. Thus, suppose that $k > \varepsilon$. Let us find the points of intersection of the hyperbola with the Or-axis. For $t = 0$ we have $r^2 = -1/k\varepsilon$; consequently, if $\varepsilon > 0$, then there are no intersection points, and for $\varepsilon < 0$ there are two intersection points with the straight line $t = 0$: $(-\sqrt{-1/k\varepsilon}, 0)$ and $(\sqrt{-1/k\varepsilon}, 0)$. The slopes Q of the asymptotes satisfy the equation

$$1 - (\varepsilon + k)Q + Q^2 k\varepsilon = 0,$$

where $Q_1 = 1/\varepsilon$ and $Q_2 = 1/k$; and its center is at the origin. Thus, there are different forms of domains in the (t, r)-plane, depending on the signs of ε and a. For $\varepsilon > 0$ (i.e., for infiltration) and $a < 0$ the inequalities $r_3 > 0$, $t_3 > 0$ and $r_4 > 0$ hold. We give a sketch of one of the domains bounded by the curves (9), with $\tau \equiv t$ and $R \equiv r$.

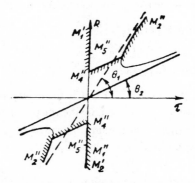

FIGURE 3

In Figure 3 this is the shaded domain in the right half-plane. It is clear that the circuits on the boundary of the shaded domain and on the polygon in the (φ, y)-plane agree, and

$$\theta_1 = \arctan \frac{1}{\varepsilon}, \qquad \theta_2 = \arctan \frac{1}{k}.$$

The inequalities $r_3 < 0$, $t_3 < 0$, and $r_4 < 0$ hold when $\varepsilon > 0$ and $a > 0$, and a similar domain is obtained in the left half-plane (see Figure 3). We remark that the case $a > 0$ is seldom encountered, since the seepage interval must have the form pictured in Figure 4, which is impossible in practice.

FIGURE 4

If $\varepsilon < 0$ (i.e., when there is evaporation from the free surface) and $a < 0$, then $r_3 > 0$ and $r_4 > 0$. Suppose, moreover, that $k + a^2\varepsilon > 0$; then $t_3 > 0$. The corresponding domain is pictured in Figure 5 as the shaded domain lying in the right half-plane ($\tau \equiv t$ and $R \equiv r$).

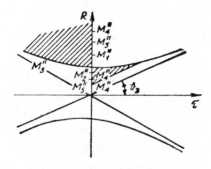

FIGURE 5

If $\varepsilon < 0$, $a > 0$, and $k + a^2\varepsilon > 0$, then $r_3 > 0$, $r_4 > 0$, and $t_3 < 0$.

The shaded domain lying in the left half-plane in Figure 5 corresponds to this case: if $\varepsilon < 0$ and $a > 0$, then $r_3 < 0$ and $r_4 < 0$, while t_3 can have any sign. The domains corresponding to the cases $t_3 > 0$ or $t_3 < 0$ are contiguous to the lower branch of the hyperbola (below the Ot-axis in Figure 5) and have the same form as the corresponding domains in this figure ($\tau \equiv t$ and $R \equiv r$).

Let us consider particular cases. If $a = 0$, which corresponds to a vertical seepage interval, then the line $t = kr + a$ coincides with one of the asymptotes of the hyperbola, and the domains pictured in Figure 3 take the respective forms pictured in Figure 6. Note that the equation of the hyperbola can be written in the form

$$(kr - t)(\varepsilon r - t) + 1 = 0.$$

From this it follows that the case $k = \varepsilon$ does not correspond to any curve, i.e., this case is impossible. Figures 5 and 6 also change in an obvious way. For

$\varepsilon = 0$ the second asymptote of the hyperbola coincides with the axis $t = 0$ (Figures 1–6) and the obvious changes occur without changing in essence the form of the domains.

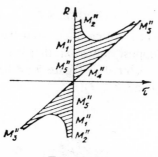

FIGURE 6

4°. Construction of the quasiconformal mapping. Thus, in the cases analyzed in 3° the (t, r)-plane, which will be called the generalized hodograph plane, contains a domain D_r bounded by a contour L_r given by (9). This contour, as is clear from the constructions of 3°, has corner points M_2'' and M_3'' in the finite part of the generalized hodograph plane or at infinity, along with a corner point M_4'' that for $a = 0$ can coincide with the origin, which is a singularity in the problem being solved. Since the functions $\varphi = \varphi(t, r)$ and $y = y(t, r)$ satisfy the linear elliptic system (8), the mapping from the (t, r)-plane onto the (φ, y)-plane is quasiconformal. Consequently, by the generalized Riemann mapping theorem (§4 in Chapter V), there is a solution $\varphi = M(t, r), y = N(t, r)$ that maps the domain D_r in the (t, r)-plane quasiconformally onto the domain D_φ of the (φ, y)-plane bounded by the polygon $L_\varphi = M_1' M_2' M_3' M_4' M_5' M_1'$, with the points M_2'' and M_4'' corresponding to the points M_2' and M_4' (the mapping is not yet completely fixed, since it is still possible to specify a correspondence, for example, between the point M_3'' and an arbitrary point of the interval $M_2' M_4'$).

Considering the inverse mapping $r = f_1(\varphi, y), t = f_2(\varphi, y)$, we find that

$$x_\varphi = f_1(\varphi, y), \qquad x_y = f_2(\varphi, y),$$

which can be used to reconstruct the function $x = x(\varphi, y)$ by a curvilinear integral:

$$x = \int_{(\varphi_2, 0)}^{(\varphi, y)} f_1(\varphi, y)\, d\varphi + f_2(\varphi, y)\, dy + x_2 \equiv F_1(\varphi, y). \qquad (10)$$

From system (5) we obtain

$$\psi_\varphi = f_2(\varphi, y), \qquad \psi_y = \frac{1 + [f_2(\varphi, y)]^2}{f_1(\varphi, y)},$$

and so

$$\psi = \int_{(\varphi_2, 0)}^{(\varphi, y)} f_2(\varphi, y)\, d\varphi + \frac{1 + [f_2(\varphi, y)]^2}{f_1(\varphi, y)}\, dy \equiv F_2(\varphi, y). \qquad (11)$$

By construction, the functions $x = F_1(\varphi, y)$ and $\psi = F_2(\varphi, y)$ satisfy the boundary conditions (6) for certain values of the constants x_0, x_1, x_2, and ψ_0. The integrals in (10) and (11) do not depend on the path of integration, since the satisfaction of (7) for the functions $t = f_2(\varphi, y)$ and $r = f_1(\varphi, y)$ (which is true by construction) is a condition ensuring this independence. Consequently, corresponding to the closed contour L_φ in the (φ, y)-plane is the closed contour L_ψ with equations (10) and (11) in the (x, ψ)-plane for $(\varphi, y) \in L_\varphi$. Since $x_\varphi = r \neq 0$ in D_φ by construction (except for the point M_4', where $x_\varphi = 0$ when $a = 0$), the implicit function $\varphi = G_1(x, y)$ exists in (10). Substitution of it in (11) gives

$$\psi = F_2[G_1(x, y), y] \equiv G_2(x, y).$$

The function $w = G(x, y) = G_1(x, y) + iG_2(x, y)$ is analytic by construction, and it satisfies the boundary conditions (1)–(4). Since

$$0 \neq \left| \frac{dw}{dz} \right|^2 = \frac{1 + x_y^2}{x_\varphi^2} < \infty,$$

except possibly at the points corresponding to M_2'', M_3'' and M_4'', we can find the inverse function

$$z = g(w) = g_1(\varphi, \psi) + ig_2(\varphi, \psi).$$

The contours L_φ and L_ψ are closed, so the formulas $x = g_1(\varphi, \psi)$ and $y = g_2(\varphi, \psi)$ give a closed contour L_z in the z-plane when φ and ψ run through the values along the contours L_φ and L_ψ. This establishes a correspondence $\varphi = \varphi(\psi)$ given by the mapping $x = F_1(\varphi, y)$, $\psi = F_2(\varphi, y)$. As has been mentioned, one constant is arbitrary in the construction of the quasiconformal mapping $\varphi = M(t, r)$, $y = N(t, r)$. Therefore, it is possible, for example, to fix the position of the point M_3' on the contour L_φ by specifying the y-coordinate of $M_3(x, y)$. Then the points M_2'', M_3'', and M_4'' must be carried into the points M_2', M_3', and M_4' by the mapping of D_r onto D_φ.

Thus, of the constants x_i, φ_i $(i = 0, 1, 2)$, ψ_0 and y_3 ($y = y_3$ is the ordinate of M_3) in the boundary conditions (1)–(4) it is possible to fix the constants y_3 and

φ_i $(i = 0, 1, 2)$ and to determine the remaining ones in the solution process. And such a choice of constants fixes at once the ordinates y_2 and y_4 of the points M_2 and M_4. It should be pointed out that, according to (10), the constant x_2 can be fixed arbitrarily. Then the constant x_0 is also found, since the line $x = ay + x_0$ passes through M_4, and, moreover, x_3 is also determined, i.e., the position of M_3 is completely fixed. Hence, the constants x_1 and ψ_0 remain undetermined, and they are found after the solution of the problem. It is also possible to fix any four of the above constants and to obtain equations from which the remaining ones must be determined.

We now clear up the nature of the corner points of the contours L_z and L_w bounding the filtration domain and the range of the complex potential, respectively. If the boundary of the generalized hodograph domain does not contain points of the axis $r = 0$, then we have $0 \neq |dw/dz| < \infty$ up to the boundary. This means that the sizes of the angles of the contours L_z and L_w coincide at points corresponding to the corner points of the boundary of the generalized hodograph. If at least one boundary point of the generalized hodograph falls on the axis $r = 0$, then L_z and L_w have different angles at the corresponding point.

We remark that the boundary L_z of the filtration domain is one-sheeted. This could fail to be the case only if there were a "loop" in the free boundary, as pictured in Figure 7. However, the passage from the (φ, y)-plane to the (x, ψ)-plane is quasiconformal, and the matching between the points of the segments $M_2' M_3'$ and $\tilde{M}_2' \tilde{M}_3'$ corresponding to the free boundary $M_2 M_3$ can only be one-to-one (Figure 8). Consequently, to each y there can correspond only one x, and thus there cannot be a "loop" in the (x, y)-plane.

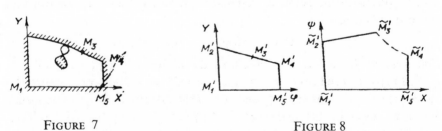

FIGURE 7   FIGURE 8

We consider the filtration problem in the case where $M_1 M_2$ and $M_4 M_5$ are inclined and have the respective equations

$$x = b_1 y + x_3 \qquad (M_1 M_2),$$

and

$$x = -b_2 y + x_4 \qquad (M_4 M_5),$$

where $b_1 > 0$ and $b_2 > 0$, and x_3 and x_4 are unknown constants. Corresponding to the sides of the quadrangle L_φ are the curves in the hodograph plane of (t, r) with equations

$$t = b_1 \quad \text{on } M_1' M_2',$$
$$t = -b_2 \quad \text{on } M_4' M_5',$$
$$t = 0 \quad \text{on } M_1' M_5',$$
$$t - kr = a \quad \text{on } M_3' M_4',$$
$$t(\varepsilon + k) - k\varepsilon r - \frac{1 + t^2}{r} = 0 \quad \text{on } M_2' M_3'.$$

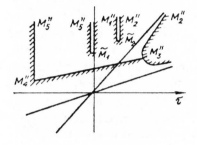

FIGURE 9

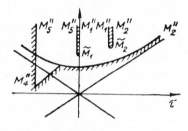

FIGURE 10

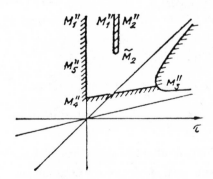

FIGURE 11

Figures 9 and 10 provide sketches of forms of domains in the plane of the variables $t \equiv \tau$, r in dependence on the signs of ε and a. The shaded domains correspond to the domains lying in the left half-planes in Figures 3 and 4. To prove that problems of this type are solvable we must show that it is possible to construct a quasiconformal mapping of D_r onto D_φ with a correspondence between the points M_i'' and the points M_i' ($i = 1, \ldots, 5$) by varying the position of the points $\tilde{M}_1$ and $\tilde{M}_2$ at the ends of the cuts. Let us dwell on the particular case of this problem when only the wall $M_1 M_2$ is inclined, while the wall $M_4 M_5$ is vertical. Then the coordinates of M_3' are not fixed on the boundary of the domain in the (φ, y)-plane: the domain in the (t, r)-plane has the form pictured in Figure 11. The position of the point M_5'' on the line $t = 0$ is unknown. Mapping D_z quasiconformally onto D_φ by solutions of (8) with the points M_i'' corresponding to the points M_i' ($i = 1, 2, 4$), we find the desired solution of the problem.

5°. *Other boundary conditions.* Let us now consider the boundary-value problem (1)–(3), (4*). In this problem the image of the filtration domain in the (x, ψ)-plane is known. The functions $x(t, r)$ and $\psi(t, r)$, where $t = -y_x$ and $r = y_\psi$, satisfy system (6) in §1. This system differs from the system (10) corresponding to problem (1)–(4) only in the notation for the independent variables and the functions. The boundary conditions for problem (1)–(3), (4*) in the (t, r)-plane differ also only in notation from the boundary conditions corresponding to problem (1)–(4). Therefore, we do not repeat all the computations carried out in the preceding subsections for problem (1)–(4), but only mention that all the results obtained for it remain in force for the problem (1)–(3), (4*).

We remark that the methods presented in this section can be used to prove solvability for linear boundary-value problems more general than filtration problems, for example, for the problem of determining an analytic function $w(z) = \varphi + i\psi$ from the following boundary conditions on the boundary $L_z = \Sigma_1^n L_z^k$ of this domain:

$$a_k w + b_k \overline{w} + c_k z + d_k \overline{z} + l_k \quad \text{for } z \in L_z^k.$$

The solvability of this problem by the method of the usual hodograph will be proved in §4.

§3. On solvability of filtration problems with free boundaries
in the case of a nonlinear filtration law

1°. *Reduction to a boundary-value problem.* For the system of equations

$$\psi_x = -\tilde{\rho}(Q)\varphi_y, \qquad \psi_y = \tilde{\rho}(Q)\psi_x, \qquad Q^2 = \varphi_x^2 + \varphi_y^2 \tag{1}$$

let us consider the boundary-value problem with boundary conditions (1)–(4) in §2. Just as in the linear case, conditions (1)–(4) determine a polygon in the

(φ, y)-plane, as pictured in Figure 2 of §2. As proved in §1, the functions $x = x(\varphi, y)$ and $\psi = \psi(y, \varphi)$ satisfy the system

$$\psi_\varphi = -\tilde{\rho}(Q)x_y, \qquad \psi_y = Q^2\tilde{\rho}(Q)x_\varphi, \qquad Q^2 = \frac{1 + x_y^2}{x_\varphi^2}. \qquad (2)$$

Setting $x_\varphi = r$ and $x_y = t$, we get for the functions $\varphi = \varphi(t, r)$ and $y = y(t, r)$ the system

$$\tilde{A}(t, r)\varphi_r + 2\tilde{B}(t, r)y_r + \tilde{C}(t, r)\varphi_t = 0, \qquad \varphi_t = y_r, \qquad (3)$$

where

$$\tilde{A} = \tilde{\rho} + \frac{\tilde{\rho}'t^2}{Qr^2}, \qquad \tilde{B} = -\frac{t}{r}(\tilde{\rho} + Q\tilde{\rho}'), \qquad \tilde{C} = Q^2(\tilde{\rho} + Q\tilde{\rho}').$$

Let us obtain the equations for the curves in the (t, r)-plane that are the images of the sides of the polygon in the (φ, y)-plane. The boundary conditions (1)–(4) in §2 give us that $d\psi/d\varphi = \psi_\varphi + \psi_y dy/d\varphi$ on $M_1'M_5'$, but $d\psi/d\varphi = dy/d\varphi = 0$ along $M_1'M_5'$, which implies that

$$\psi_\varphi = x_y = t = 0.$$

On $M_1'M_2'$ and $M_5'M_4'$ we have $dx/dy = x_y + x_\varphi d\varphi/dy = 0$, but $d\varphi/dy = 0$, and so $x_y = t = 0$. Along $M_2'M_3'$ we have $d\psi/dy = \varepsilon dx/dy$, and since

$$\frac{d\psi}{dy} = \psi_y + \psi_\varphi \frac{d\varphi}{dy} = \tilde{\rho}(Q)\left[\frac{1 + t^2}{r} - kt\right]$$

and

$$\frac{dx}{dy} = x_y + x_\varphi \frac{d\varphi}{dy} = t - kr,$$

we obtain

$$\tilde{\rho}\left(\frac{1 + t^2}{r} - kt\right) = \varepsilon t - k\varepsilon r.$$

On $M_3'M_4'$ the equality $dx/dy = a$ holds, and, on the other hand, $dx/dy = x_y + x_\varphi d\varphi/dy = t - kr$, which gives $t = kr + a$. Thus, we arrive at the equalities

1. $\qquad\qquad\qquad t = 0 \quad$ on $M_2'M_1'M_5'M_4'$;

2. $\qquad\qquad\qquad t - kr = a \quad$ on $M_3'M_4'$; $\qquad\qquad\qquad (4)$

3. $\qquad t(\varepsilon + k\tilde{\rho}) - k\varepsilon r - \frac{1 + t^2}{r}\tilde{\rho} = 0 \quad$ on $M_2'M_3'$,

where $\tilde{\rho} = \tilde{\rho}(Q)$ and $Q^2 = (1 + t^2)/r^2$. Suppose that the (t, r)-plane contains a domain D_r that satisfies the following conditions:

(a) The contour L_r formed by the contours with equations (4) has finitely many corner points, some of which may lie at infinity, and it contains only finitely many points of the line $r = 0$.

(b) D_r does not contain points of the line $r = 0$ in its interior.

Then the proof that the boundary-value problem (1)–(4) is solvable is completely analogous to the case of the linear filtration law. Therefore, it remains to investigate the existence of a domain satisfying (a) and (b) in the (t, r)-plane of the generalized hodograph.

$2°$. **Properties of the boundary of the domain in the generalized hodograph plane.** Let us investigate the form of the curve (4_3):

$$t(\varepsilon + k\tilde{\rho}) - k\varepsilon r - \frac{1 + t^2}{r}\tilde{\rho} = 0.$$

This curve is obviously symmetric with respect to the origin. Since $\tilde{\rho} \neq 0$ by an assumption, it does not intersect the axis $r = 0$. We find its asymptotes. Let l be the slope of the asymptotes. To determine l we get the equation

$$\varepsilon + k\tilde{\rho}\left(\frac{1}{l}\right) - k\varepsilon l - \frac{1}{l}\tilde{\rho}\left(\frac{1}{l}\right) = 0,$$

from which it follows that

$$l_1 = \frac{1}{k}, \qquad l_2 = \frac{1}{\varepsilon}\tilde{\rho}\left(\frac{1}{|l_2|}\right). \tag{5}$$

It is clear from (5) that l_2 has the same sign as ε, since $\tilde{\rho} > 0$ by assumption. Let $1/|l_2| = Q$. Equation (5) can be rewritten:

$$Q\tilde{\rho}(Q) = |\varepsilon|, \tag{5*}$$

and $l_2 = Q \operatorname{sgn} \varepsilon$. The quantity Q is the distorted speed at the point corresponding to unbounded t and r. Then

$$\frac{d}{dQ}[Q\tilde{\rho}(Q)] = \frac{d}{dQ}\left[Q\frac{1}{\rho(q)}\right] = \frac{d}{dQ}\left[q\rho(q)\frac{1}{\rho(q)}\right] = \frac{dq}{dQ},$$

or

$$\frac{d}{dQ}[Q\tilde{\rho}(Q)] = \left(\frac{d}{dq}[q\rho(q)]\right)^{-1};$$

but $0 < d[q\rho(q)]/dq < \infty$. Thus, $d[Q\tilde{\rho}(Q)]/dQ > 0$, and, consequently, the implicit function in (5) exists. Hence, (5) is uniquely solvable for l_2 for any ε. For definiteness we require that the solution $l_2 \neq 1/k$ when $k \neq \varepsilon$. Then the curve (4_3) has two distinct asymptotes passing through the origin when $\varepsilon \neq k$. To find the ordinates of M_2'' we set $t = 0$ in (4_3), which gives

$$r^2 = -\frac{1}{k\varepsilon}\tilde{\rho}\left(\frac{1}{|r|}\right). \tag{6}$$

Writing $1/|r| = Q_2$, we get $Q_2^2\tilde{\rho}(Q_2) = -k\varepsilon$ (Q_2 is the distorted speed at M_2). For $k\varepsilon > 0$ equation (6) does not have any solutions, while for $k\varepsilon < 0$

$$Q\sqrt{\tilde{\rho}(Q_2)} = \sqrt{-k\varepsilon}. \tag{7}$$

Then

$$\frac{d}{dQ_2}\left[Q_2\sqrt{\tilde{\rho}(Q_2)}\right] = \frac{1}{2}Q_2^{-1/2}\sqrt{Q_2\tilde{\rho}(Q_2)} + \frac{Q_2^{1/2}(d/dQ_2)[Q_2\tilde{\rho}(Q_2)]}{2\sqrt{Q_2\tilde{\rho}(Q_2)}},$$

since $Q_2 > 0$, $Q_2\tilde{\rho}(Q_2) > 0$, and, by the preceding,

$$\frac{d}{dQ_2}[Q_2\tilde{\rho}(Q_2)] = \frac{1}{(d/dq_2)[q_2\rho(q_2)]} > 0.$$

Consequently, $(d/dQ_2)[Q_2\sqrt{\tilde{\rho}(Q_2)}] > 0$, and (7) has a unique solution. Thus, (6) has two solutions:

$$r_2' = \frac{1}{Q_2} > 0 \quad \text{and} \quad r_2'' = -\frac{1}{Q_2} < 0.$$

Let us find the points where the curve (4_3) intersects the line (4_2), $t = kr + a$. For $a = 0$ the line (4_2) coincides with one of the asymptotes of the curve (4_3). Setting $t = kr$ in (4_3), we get

$$kr(\varepsilon + k\tilde{\rho}) - k\varepsilon r - \frac{1 + k^2r^2}{r}\tilde{\rho} = 0,$$

from which, after simplifying, we arrive at the equation $-(1/r)\tilde{\rho} = 0$, which does not have a solution. Thus, for $a = 0$ the line $t = kr$ does not intersect the curve (4_3). Suppose that $a \neq 0$. Then, setting $t = kr + a$ in (4_3), we get

$$r = \frac{\tilde{\rho}(1 + a^2)\{r^{-2} + (ar^{-1} + k)^2\}^{1/2}}{a(\varepsilon - k\tilde{\rho})}, \tag{8}$$

or $\tilde{\rho} = \varepsilon ar/(1 + a^2 + akr)$. Let $1/r^2 + (a/r + k)^2 = z^2$; then

$$\frac{1}{r} = \frac{-ak \pm \sqrt{z^2(1 + a^2) - k^2}}{1 + a^2}.$$

Substituting the value of r into (8), we get

$$\frac{a\varepsilon}{\pm\sqrt{1 + a^2 - (k/z)^2}} = z\tilde{\rho}(z),$$

where

$$z = \sqrt{\frac{1}{r^2} + \left(\frac{a}{r} + k\right)^2} > 0, \rho > 0,$$

and, therefore, the sign in front of the root must be chosen to coincide with the sign of $a\varepsilon$. Hence, z satisfies

$$\frac{|a\varepsilon|}{\sqrt{1 + a^2 - (k/z)^2}} = z\tilde{\rho}(z). \tag{9}$$

Since $d[z\tilde{\rho}(z)]/dz > 0$, the curve $z\tilde{\rho}(z) = u$ is defined for $z \geqslant 0$, and $z(u)$ is monotonically increasing.

Let us consider the curve

$$u = \frac{|a\varepsilon|}{\sqrt{1 + a^2 - (k/z)^2}} \equiv f(z),$$

defined for $z \geqslant k^2/(1 + a^2)$. We have that

$$\frac{du}{dz} = \frac{-|a\varepsilon|k^2}{z^3\left(1 + a^2 - k^2/z^2\right)^{3/2}} < 0,$$

i.e., the function $u(z)$ decreases monotonically to the value

$$u_0 = |a\varepsilon|(1 + a^2)^{-1/2}$$

(see the sketch), and, consequently, the curves have only a single point of intersection. Then (9) is uniquely solvable for z. Therefore, (8) also has the unique solution

$$r = \frac{1 + a^2}{-ak + \left[z^2(1 + q^2) - k^2\right]^{1/2} \operatorname{sgn} a}.$$

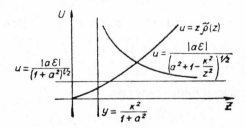

By what has been proved, the line $t = kr + a$, which is parallel to one of the asymptotes of the curve (4_3), necessarily intersects this curve for any $a \neq 0$, and only in a single point. Moreover, the curve (4_3) is symmetric with respect to the origin and does not intersect the axis $r = 0$. Consequently, it splits into two continuous branches lying in the upper and lower half-planes, respectively. Therefore, the (t, r)-plane contains domains bounded by curves described by (4), and their form is completely analogous to the domains pictured in Figures 3–5 in §2. Thus, from the point of view of our proposed method for solving filtration problems there is no fundamental difference between the linear case and the nonlinear case.

$3°$. *Another form for the boundary conditions.* For the system of equations

$$\varphi_x = -\rho(q)\psi_y, \qquad \varphi_y = \rho(q)\psi_x, \qquad q^2 = \psi_x^2 + \psi_y^2 \qquad (10)$$

let us now consider the boundary-value problem (1)–(3), (4*) in §2. The (x, ψ)-plane contains a polygon determined by the given boundary conditions. The functions $\psi(t, r)$ and $x(t, r)$, where $r = y_\psi$ and $t = -y_x$, satisfy (1.22) which differs from the system (1.23) corresponding to problem (1)–(4) only in the notation for the independent variables and the functions and by the fact that $1/\rho(q)$ plays the role of $\bar{\rho}(Q)$ in the coefficients of (1.22). The boundary conditions for problem (1)–(3), (4*) in the (t, r)-plane differ only in notation from the corresponding boundary conditions for the problem (1)–(4). Therefore, all the results obtained for problem (1)–(4) remain in force for problem (1)–(3), (4*).

§4. Reduction of problems of filtration type in earthen dams to the Dirichlet problem

$1°$. *The case of Darcy's law.* In the preceding sections we used the generalized hodograph method to prove the solvability of problems with free boundaries. It would also be possible not to introduce the generalized hodograph, but to use the ordinary hodograph, determining the image D_ω of the filtration domain in the hodograph plane of $\omega = u - iv$ from physical considerations. Then the filtration problems could again be reduced to the problem of mapping the domain D_φ in the (φ, y)- (or (x, ψ)-) plane quasiconformally onto D_ω in the hodograph plane. The corresponding system of equations is not hard to construct. However, this mapping is required to establish a correspondence between four or more specified boundary points of the domains D_ω and D_φ, which can be achieved only by varying D_ω or D_φ. In this sense the generalized hodograph is more convenient for proving that the corresponding filtration problems are solvable, since for the mapping of D_φ onto the domain in the generalized hodograph plane it is required in most problems to establish a correspondence for not more than three specified boundary points. In certain cases, when the domain D_ω in the ordinary hodograph plane is completely fixed as a result of particular considerations, it makes sense to reduce the filtration problem not to the problem of mapping it quasiconformally onto a modified domain D_φ, but to a Dirichlet problem in the domain D_ω; this is what will be done below.

Let us consider the following class of boundary-value problems with free boundaries: Find an analytic function $w(z) = \varphi + i\psi$ in a domain D_z from the boundary conditions $a_k w + b_k \bar{w} + c_k z + d_k \bar{z} + l_k = 0$ for $z \in L_z^k$ on the boundary $L_z = \sum_1^n L_z^k$ of this domain, where a_k, b_k, c_k, and d_k $(k = 1,\ldots,n)$ are fixed complex constants, and the complex constants l_k are to be determined. Condition (1) can be rewritten in the form

$$a_k^i \varphi + b_k^i \psi + c_k^i x + d_k^i y + l_k^i = 0 \ (i = 0, 1) \quad \text{on } L_z^k, \tag{2}$$

where the real constants a_k^i, b_k^i, c_k^i, d_k^i are fixed, and the l_k^i are to be found.

Suppose that the ranks of the matrices

$$\begin{pmatrix} a_k^1 & b_k^1 \\ a_k^2 & b_k^2 \end{pmatrix}, \quad \begin{pmatrix} c_k^1 & d_k^1 \\ c_k^2 & d_k^2 \end{pmatrix}, \quad \begin{pmatrix} a_k^i & b_k^i \\ c_k^i & d_k^i \end{pmatrix} \quad (k = 1,\ldots,n; i = 1,2)$$

are nonzero. Obviously, if two of the constants a_k^i, b_k^i, c_k^i, d_k^i are equal to zero, then one of the conditions (2) determines the equation of a line in one of the planes

$$x + iy, \quad \varphi + iy, \quad \varphi + i\psi, \quad x + i\psi, \quad \varphi + ix, \quad y + i\psi.$$

In particular, in the problems formulated in §2 on the theory of filtration in earthen dams we know the equations of the rectilinear parts of the boundaries in the planes $x + iy$, $\varphi + i\psi$, $\varphi + iy$ and $x + i\psi$. Let us introduce the hodograph plane of $\omega = u - iv \equiv dw/dz$, where $u = \partial\varphi/\partial x = \partial\psi/\partial y$ and $v = \partial\varphi/\partial y = \partial\psi/\partial x$. The boundary conditions (2) will be transformed, under the assumption that the unknown analytic function $w = \varphi + i\psi$ is continuous along with its derivative dw/dz up to the boundary of L_z except at a finite number of corner points of L_z.

We differentiate the boundary conditions (2), taking account of the fact that $d\varphi = u\,dx + v\,dy$ and $d\psi = -v\,dx + u\,dy$, which gives us

$$\left(a_k^i u - b_k^i v + c_k^i\right)dx + \left(b_k^i u + a_k^i v + d_k^i\right)dy = 0 \quad (i = 1, 2). \tag{3}$$

Elimination of dx and dy from (3) gives

$$A_k(u^2 + v^2) + 2D_k u + 2E_k v + F_k = 0 \quad (k = 1,\ldots,n), \tag{4}$$

where

$$A_k = a_k^1 b_k^2 - \partial_k^2 b_k^1, \qquad 2D_k = a_k^1 d_k^2 + c_k^1 b_k^2 - a_k^2 d_k^1 - c_k^2 b_k^1,$$

$$2E_k = a_k^2 c_k^1 - b_k^1 d_k^2 + b_k^2 d_k^1 - a_k^1 c_k^2, \qquad F_k = c_k^1 d_k^2 - c_k^2 d_k^1.$$

Assume that in the plane of $\omega = u - iv$ we have chosen a fixed domain D_ω bounded by arcs of the circles with equations (4) and corresponding to the domain D_z. Observe that some of the circles may degenerate into straight lines when $A_k = 0$. We map the disk $|\zeta| < 1$ conformally onto this domain by the function $\omega = F(\zeta)$ and introduce the analytic function $dz/d\omega = -y_v + iy_u$. We find the limit values of $\arg(dz/d\omega)$ on the boundary $|\zeta| = 1$ of the disk from the conditions (3):

$$\arg \frac{dz}{dw} = \arctan \frac{-l_k u_\gamma + m_k v_\gamma}{m_k u_\gamma + l_k v_\gamma} = g_k(\gamma), \qquad \gamma \in [\gamma_k, \gamma_{k+1}], \qquad (5)$$

where

$$l_k = a_k^1 u - b_k^1 v + c_k^1, \quad m_k = b_k^1 u + a_k^1 v + d_k^1 \qquad (k = 1, \ldots, n).$$

Here the $e^{i\gamma_k}$ are points of the circle corresponding to the junction points of adjacent circular arcs of the boundary of D_ω, and

$$\gamma_{n+1} = \gamma_1 + 2\pi, \qquad u_\gamma = du/d\gamma, \qquad v_\gamma = dv/d\gamma.$$

Thus, we come to the problem of determining the analytic function

$$-i \ln \frac{dz}{d\omega} = \arg \frac{dz}{q\omega} - i \ln \left| \frac{dz}{d\omega} \right|$$

from the known real part on the boundary of the disk. Consequently,

$$\frac{dz}{d\omega} = k \exp \left\{ \frac{i}{2\pi} \int_{\gamma_k}^{\gamma_{k+1}} g_k(\gamma) \frac{e^{i\gamma} + \zeta}{e^{i\gamma} - \zeta} d\gamma \right\} \equiv kG(\zeta). \qquad (6)$$

But $d\omega/d\zeta = dF/d\zeta$ is a known function, and so

$$z = k \int_0^\zeta Q(\zeta) \frac{dF^{\cdot}}{d\zeta} d\zeta + c \equiv \Omega(\zeta) \qquad (7)$$

is a function mapping the disk $|\zeta| < 1$ conformally onto D_z. The unknown analytic function $w(z) = \varphi + i\psi$ is determined by

$$w = \int_0^z F[\Omega^{-1}(z)] \, dz + c_1, \qquad (8)$$

where $\zeta = \Omega^{-1}(z)$ is the function inverse to $z = \Omega(\zeta)$. We mention that the solution of the boundary-value problem (2) in our formulation is not unique, due to the nonuniqueness of the choice of the domain D_ω bounded by circular arcs with the equations (4). However, if the domain D_ω in the hodograph plane is chosen from certain considerations, then D_z and D_w are determined to within a translation and dilations. In particular, in the theory of filtration in earthen dams it is frequently possible to choose D_ω from physical considerations. In the case where the seepage intervals and the equipotential curves $\varphi = $ const are not parallel to the OY-axis the corresponding problem (5) can be written in the form

$$
\begin{cases}
\arg \dfrac{dz}{d\omega} = \arctan \dfrac{(\varepsilon - k)[v(\gamma) + k]}{4v(\gamma) + 3k + \varepsilon} = f(\gamma), & \gamma \in [0, \pi], \\[4mm]
\arg \dfrac{dz}{dw} = 0, & \gamma \in [\pi, 2\pi],
\end{cases}
\tag{9}
$$

where k is the permeability coefficient, and ε is the coefficient of infiltration or evaporation. Here the conformal mapping $\omega = F(\zeta)$ of the disk onto D_ω is constructed in such a way that the image of the depression curve passes into the upper half of the circle $|\zeta| = 1$. Then

$$
z = k \int_0^\zeta \frac{dF}{d\zeta} \exp\left\{ \frac{i}{2\pi} \int_0^\pi f(\gamma) \frac{e^{i\gamma} + \zeta}{e^{i\gamma} - \zeta} d\gamma \right\} d\zeta + c,
$$

$$
w = k \int_0^\zeta F \frac{dF}{d\zeta} \exp\left\{ \frac{1}{2\pi} \int_0^\pi f(\gamma) \frac{e^{i\gamma} + \zeta}{e^{i\gamma} - \zeta} d\gamma \right\} d\zeta + c_1.
$$

By suitably choosing the real constant k, which affects the dimensions of D_z, it is possible, for example, to choose an arbitrary value for the length of the impermeable base of the dam. In the case where the seepage interval or the equipotential curve is parallel to the imaginary axis the condition (9) can be written on the corresponding arc of the circle $|\zeta| = 1$ as

$$
\arg \frac{dz}{d\omega} = \pm \frac{\pi}{2}, \qquad \gamma \in [\gamma_i, \gamma_{i+1}],
$$

and consequently the form of the solution does not change in an essential way.

We remark that the class of problems considered contains the inverse boundary-value problem

$$
\varphi = a_k x + b_k y + f \quad \text{on } L_k, \qquad \psi = c_k x + d_k y + l_k \quad \text{on } L_k. \tag{10}
$$

2°. *Nonlinear law of filtration.* Let us now consider the problem stated in 1° for the nonlinear system of equations

$$
\rho(q)\psi_y = \varphi_x, \qquad \rho(q)\psi_x = -\varphi_y, \qquad \left(q^2 = \psi_x^2 + \psi_y^2\right). \tag{11}
$$

Introducing, as earlier, the fictitious speed $Q = (\varphi_x^2 + \varphi_y^2)^{1/2}$, we rewrite this system in the form

$$
\tilde{\rho}(Q)\varphi_x = \psi_y, \qquad \tilde{\rho}(Q)\varphi_y = -\psi_x, \qquad Q^2 = \varphi_x^2 + \varphi_y^2. \tag{12}
$$

The boundary conditions (2) will be transformed, under the assumption that $\varphi(x, y)$ and $\psi(x, y)$ are continuous along with their first-order partial derivatives up to the boundary L_z except possibly at a finite number of corner points of L_z.

We differentiate the boundary conditions (2), taking into account that

$$d\varphi = u\,dx + v\,dy, \qquad d\psi = -\tilde{\rho}(Q)(v\,dx + u\,dy),$$

where $u = \varphi_x$ and $v = \varphi_y$. Then

$$\left(a_k^i u + \tilde{\rho} b_k^i v + c_k^i\right)dx + \left(\tilde{\rho} b_k^i u + a_k^i v + d_k^i\right)dy = 0 \qquad (i = 1, 2). \quad (13)$$

Eliminating dx and dy from (13), we get

$$A_k(u^2 + v^2) + 2D_k u + 2E_k v + F_k = 0 \qquad (k = 1, \ldots, n), \quad (14)$$

where

$$A_k(u, v) = \left(a_k^1 b_k^2 - a_k^2 b_k^1\right)\tilde{\rho},$$

$$2D_k(u, v) = a_k^1 d_k^2 - d_k^2 d_k^1 + \tilde{\rho}\left(c_k^1 b_k^2 - c_k^2 b_k^1\right),$$

$$2E_k(u, v) = a_k^2 c_k^1 - a_k^1 c_k^2 + \tilde{\rho}\left(b_k^2 d_k^1 - b_k^1 d_k^2\right), \qquad F_k = c_k^1 d_k^2 - c_k^2 d_k^1.$$

Suppose that in the plane of $\omega = u - iv$ a fixed domain D_ω is chosen, bounded by curves L_ω^k ($k = 1, \ldots, n$) with the equations (14). Note that when $\tilde{\rho} = 1$ the curves L_ω^k become circular arcs. In filtration problems this happens when Darcy's law holds.

We introduce the function $p(u, v) = y_v(u, v)/y_u(u, v)$. Let $(dv/du)_k$ be the tangent slope at the points of the arc L_ω^k. Then, taking

$$x_v = y_u, \qquad x_u = \frac{2uv}{v^2 - c^2}y_u + \frac{c^2 - u^2}{v^2 - c^2}y_v \quad (15)$$

into account, where $c^2 = -(d\tilde{\rho}/dq)^{-1}Q\tilde{\rho}(Q)$, we find from (13) the value of $p(u, v)$ on the arcs L_ω^k ($k = 1, \ldots, n$):

$$p(u, v) = \frac{M_k(c^2 - v^2) - 2uv + (c^2 - v^2)(dv/du)_k}{c^2 - u^2 - M_k(c^2 - v^2)(dv/du)_k}, \quad (16)$$

where

$$M_k = \frac{\tilde{\rho} b_k^1 u + a_k^1 v + d_k^1}{a_k^1 u - \tilde{\rho} b_k^1 + c_k^1}.$$

Eliminating x_u and x_v from (15), we can write it in the form

$$\frac{\partial}{\partial v}(\ln y_u) = \frac{p_u - bpp_v - pa_v - b_v p^2}{1 + ap + bp^2} \equiv m_2(u, v),$$

$$-\frac{\partial}{\partial u}(\ln y_u) = \frac{(a + bp)p_u + bp_v + a_v + b_v p}{1 + ap + bp^2} \equiv m_2(u, v),$$

(17)

where

$$a = \frac{2uv}{c^2 - v^2}, \qquad b = \frac{c^2 - u^2}{c^2 - v^2}, \qquad m_i(u, v) = m_i[u, v, p, p_u, p_v].$$

Differentiating the first equation in (17) with respect to u and the second with respect to v and adding the resulting equalities, we obtain a second-order elliptic equation for $p(u, v)$:

$$\frac{\partial^2 p}{\partial u^2} + \frac{2uv}{c^2 - v^2} \frac{\partial^2 p}{\partial u \partial v} + \frac{c^2 - u^2}{c^2 - v^2} \cdot \frac{\partial^2 p}{\partial v^2} + F(u, v, p, p_u, p_v) = 0, \quad (18)$$

where the function F is computed explicitly from (17). Consequently, the original boundary-value problem (12), (2) has been reduced to the Dirichlet problem for the quasilinear elliptic equation (18) with the boundary condition (16), and a proof of the unique solvability of the latter problem can be found, for example, in the monograph [128] or [59]. It is also not hard to prove the solvability of problem (16), (18) by a direct application of the methods of §4 in Chapter V. Suppose that the function $p(u, v) = y_v/y_u$ has been found in D_ω. Then from (17) we determine $\ln y_u$ with the help of a curvilinear integral which obviously does not depend on the path of integration:

$$\ln y_u = \int_{(u_0, v_0)}^{(u, v)} m_1(u, v) \, du + m_2(u, v) \, dv + \ln k \equiv \ln[kl(u, v)],$$

where $k = \text{const}$. Thus

$$y_u = kl(u, v), \qquad y_v = kp(u, v)l(u, v),$$

and so

$$y(u, v) = k \int_{(u_0, v_0)}^{(u, v)} l(u, v) \, du + p(u, v)l(u, v) \, dv + y_0,$$

$$x(u, v) = k \int_{(u_0, v_0)}^{(u, v)} \left(\frac{2uvl}{v^2 - c^2} + \frac{c^2 - u^2}{v^2 - c^2} pl \right) du + l \, dv + x_0.$$

If the original domain in the plane of $\omega = u - iv$ was chosen to be finite, then the quasiconformality of the mapping from the ω-plane to the z-plane, which follows from the ellipticity and homogeneity of system (15), gives us that D_z is a finite one-sheeted domain in the z-plane, determined to within a dilation and a translation.

The unknown functions $\varphi(u, v)$ and $\psi(u, v)$ are found by the formulas

$$\varphi = \int_{(u_0, v_0)}^{(u, v)} (ux_u + vy_u) \, du + (ux_0 + vy_v) \, dv + \varphi_0,$$

$$\psi = \int_{(u_0, v_0)}^{(u, v)} \tilde{\rho}(uy_u - vx_u) \, du + \tilde{\rho}(uy_v - vx_v) \, dv + \psi_0.$$

Thus, for the cases when the domain D_ω in the hodograph plane can be chosen, we have proved, in particular, that the problem of filtration of a fluid in an earthen dam is solvable for a nonlinear filtration law, which problem is a special case of problem (1), (2). We remark that by suitably choosing the constant k we can assign, for example, an arbitrary value to the length of the impermeable base of the dam.

§5. Planar and axisymmetric problems involving filtration of a fluid in nonhomogeneous anisotropic media

1°. Problems in given domains. It follows from (1) and (2) that in the case of filtration of a fluid in nonhomogeneous anisotropic media the piezometric head $h(x, y) = p/\rho g + y$ satisfies the equation

$$(\nabla, (k, \nabla h)) = 0,$$

where $k = \{k_{ij}(x, y, h)\}$ is the symmetric filtration tensor, with

$$\delta = k_{11}k_{22} - k_{12}^2 \geqslant \delta_0 > 0 \quad \text{and} \quad |k_{ij}| \leqslant M < \infty.$$

Similar formulas hold also in the axisymmetric case $(r \equiv x)$, except that $k = xk(x, y, h)$, and, therefore, the equation for the head is uniformly elliptic only in a domain not containing points with $x = 0$, which we shall assume in what follows. Introducing the stream function $\psi = \psi(x, y)$ and the generalized potential $\varphi = -k_0 h$ ($k_0 = $ const), we arrive at the respective equations for the function $w(z) = \varphi + i\psi$ and its inverse $z = x + iy = z(w)$:

$$w_{\bar{z}} + \mu_2(w, z)w_z + \mu_1(w, z)\bar{w}_{\bar{z}} = 0, \tag{1}$$

$$z_{\bar{w}} - \mu_1(w, z)z_w - \mu_2(w, z)\bar{z}_{\bar{w}} = 0, \tag{2}$$

where

$$\mu_1 = \frac{k_{22}^0 - k_{11}^0 - 2ik_{12}^0}{1 + k_{11}^0 + k_{22}^0 + k_{12}^0(k_{22}^0 - k_{11}^0)},$$

$$\mu_2 = \frac{1 - k_{22}^0 k_{11}^0 - 2k_{12}^0}{1 + k_{11}^0 + k_{22}^0 + k_{12}^0(k_{22}^0 - k_{11}^0)},$$

and $k^0 = k/k_0 = \{k_{ij}/k_0\}$ is the reduced filtration tensor. It should be mentioned that (1) is the complex expression for the following elliptic system of equations for the potential $\varphi = \varphi(x, y)$ and the stream function $\psi = \psi(x, y)$:

$$-\psi_y + k_{11}^0\varphi_x + k_{12}^0\varphi_y = 0, \qquad \psi_x + k_{21}^0\varphi_x + k_{22}^0\varphi_y = 0.$$

Because of the assumptions made above about the filtration tensor, equations (1) and (2) are uniformly elliptic, i.e.,

$$|\mu_1(w, z)| + |\mu_2(w, z)| \leqslant \mu_0(\delta, M) < 1. \tag{3}$$

As shown in §2, the generalized complex filtration potential $w = \varphi + i\psi$ and its inverse function $z = z(w)$ satisfy the following conditions on the boundaries of the filtration domain D_z. On the specified (unknown) impermeable boundaries

$$f(x, y) = 0, \qquad (\varphi = -k_0 h(x)), \qquad \psi = \text{const}, \qquad (4)$$

where $f(x, y) = 0$ (the equation of the boundary), and $h = h(x)$ is the given head diagram* on this boundary if the impermeable boundary is sought. On the boundaries of reservoirs (the boundaries of the filtration domain with the fluid medium) we have

$$f(x, y) = 0, \qquad \varphi = \text{const}. \qquad (5)$$

On seepage intervals we assume, as a rule, the condition (4) (this is justified in §2), but in problems with a given domain D_z we sometimes replace (4) by the condition

$$f(x, y) = 0, \qquad \varphi + k_0 y = \text{const} \quad (k_0 = \text{const}). \qquad (6)$$

On the free boundary (the interface between the wet and dry soil or the interface between fluids with different physical properties), which is a streamline ($\psi = \text{const}$), it is usually assumed that the pressure coincides with the atmospheric pressure or is less than it by a constant quantity, and, consequently, $\varphi + k_0 y = \text{const}$ on the free boundary. However, if the free boundary is an interface of two fluids or if the soil is compressible, and also if the filtration velocity is large, so that it is impossible to ignore the velocity head, then we assume the conditions

$$\psi = \text{const}, \qquad \varphi + k_0 y = \text{const} + S(\varphi)x, \qquad (7)$$

where $S(\varphi) \in C^\alpha$, $\alpha > 0$, $S(\varphi_1) = S(\varphi_2) = 0$, and φ_1 and φ_2 are the value of the potential at the endpoints of the free boundary. We remark that the right-hand side of (7) is

$$S(\varphi)x + \text{const} = -\frac{p(x, y)}{\rho g} k_0,$$

where $p(x, y)$ is the pressure. In particular, for $S(\varphi) \equiv 0$ we arrive at the relation used in the previous sections.

Let us first consider two variants of the general problem of filtration of a fluid in an arbitrary simply connected domain D_z when there is no free boundary. The given boundary L_z is assumed to be a piecewise smooth curve

---

* *Editor's note.* In this content the term 'head diagram' denotes a diagram of forces of the pressure exerted by the fluid.

having finitely many vertices b_i $(i = 1,\ldots,m)$ at infinity, in a neighborhood of which the filtration flow is assumed to be bounded. For example, the filtration domain (which is infinite, in general) may contain finitely many impermeable levels (Figure 1).

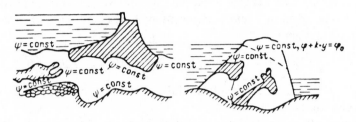

FIGURE 1

PROBLEM $1°.1$. Suppose that on the boundary L_z of the given filtration domain D_z we know the location of the impermeable parts of the boundary, the seepage intervals, and the reservoir boundaries. Let z_k $(k = 1,\ldots,n)$ denote the junction points of these parts, if different boundary conditions for the complex potential $w = \varphi + i\psi$ hold on them. By a conformal mapping of the disk $|\zeta| < 1$ onto D_z we arrive at the following mixed boundary-value problem:

$$\operatorname{Re}\left[w(e^{i\gamma})\right] = \varphi_k(\gamma), \qquad \gamma \in [\gamma_k, \gamma_{k+1}],$$
$$\operatorname{Im}\left[w(e^{i\gamma})\right] = \psi_{k-1}(\gamma), \qquad \gamma \in [\gamma_{k-1}, \gamma_k]. \tag{8}$$

where the $t_k = e^{i\gamma_k}$ are fixed images of the junction points z_k $(k = 1,\ldots,n)$ of the streamlines and the reservoir boundary (then $\varphi_k(\gamma) = \varphi_k = \text{const}$) or the seepage intervals (then $\varphi_k(\gamma) = -k_0 y(\gamma) + \varphi_k$, $\varphi_k = \text{const}$). The complex generalized filtration potential $w = \varphi + i\psi$ satisfies (1), transformed to the ζ-plane:

$$w_{\bar\zeta} + \tilde\mu_2(w, \zeta)w_\zeta + \tilde\mu_1(w, \zeta)\overline{w_\zeta} = 0, \tag{1*}$$

where

$$\tilde\mu_2(w, \zeta) = \mu_2[w, z(\zeta)]\exp\left\{-2i\arg\frac{dz}{d\zeta}\right\}, \qquad \tilde\mu_1(w, \zeta) = \mu_1[w, z(\zeta)].$$

If there are more than three points t_k $(k = 1,\ldots,n)$ in the mixed problem (8) where the right-hand side of the boundary condition is discontinuous, then $n - 3$ of the constants ψ_k and φ_k in it can be regarded as arbitrary, while otherwise they are all fixed. We assume that the filtration tensor $k(x, y, h) = \{k_{ij}(x, y, h)\}$ has the property

(a) $k(x, y, h)$ is continuous in h for all x and y, and is measurable in x and y.

Since the coefficients $\tilde{u}_i(w, \zeta)$ of (1*) are continuous in w and measurable in ζ, what was proved in §2.4° of Chapter VI tells us that problem (1), (8), and with it the original filtration problem, has at least one solution.

This solution is also unique if the filtration tensor satisfies the condition

(b) $k(x, y, h) = \{k_{ij}\}$ is continuous in h for almost all x and y, and satisfies a Lipschitz condition in x and y uniformly with respect to h, and, if D_z is an infinite domain, then

$$k_{11}(x, y, h) = k_{22}(x, y, h) = k_0, \qquad k_{12}(x, y, h) \equiv 0$$

when $|z| \geqslant R > 0$, i.e., outside a fixed finite domain the medium is assumed to be homogeneous and isotropic.

PROBLEM 1°.2. Suppose either that the boundary L_z of a given filtration domain D_z lacks not only a free boundary part but also seepage intervals, or that, if they are present, they can be assumed to be streamlines.

In this problem the domain D_w in the plane of the complex potential $w = \varphi + i\psi$ is bounded by equipotentials $\varphi = \varphi_k$ and streamlines $\psi = \psi_i$. Thus, Problem 1°.2 reduces to the problem of constructing a homeomorphism $z = z(w)$ that satisfies (2) and maps D_w onto a domain D_z bounded by straight lines parallel to the coordinate axes. According to the results of §§4 and 5 of Chapter V, if condition (b) on the filtration tensor (and thereby also on the coefficients of (2)) holds, then this mapping problem has as many solutions as in the class of conformal mappings (homogeneous isotropic soil), while if (a) holds, then it has no fewer than for conformal mappings. Thus, we arrive at the following theorem.

THEOREM 1. *If the symmetric filtration tensor* $k(x, y, h) = \{k_{ij}\}$ *satisfies condition* (a), *then Problems* 1°.1 *and* 1°.2 *on filtration of a fluid in a nonhomogeneous anisotropic medium in the planar and the axisymmetric cases have no fewer solutions than for the planar case of a homogeneous isotropic medium, while if* (b) *holds, then they have just as many solutions.*

2°. *Problems with free boundaries.* Let us now pass to the study of filtration problems with free boundaries under the assumption that seepage intervals are either absent in general or that those present are streamlines. Thus, in problems with free boundaries the domain D_w in the complex potential plane is, according to the boundary conditions (4), (5), and (7), bounded by equipotentials and streamlines, i.e., its boundary L_w is determined by these conditions to within arbitrary parameters (this is not so if the boundary condition (6) holds on the seepage intervals). We assume that D_w is fixed and let $w = w(\zeta)$ be a conformal mapping of the disk $|\zeta| < 1$ onto D_w. This transformation takes (2) to the form

$$z_{\bar{\zeta}} - \hat{\mu}_1(\zeta, z)z_{\zeta} - \hat{\mu}_2(\zeta, z)\bar{z}_{\bar{\zeta}} = 0, \tag{9}$$

where

$$\hat{\mu}_1 = \mu_1[w(\zeta, \zeta)] \exp\left\{-2i \arg \frac{dw}{d\zeta}\right\}, \qquad \hat{\mu}_2 = \mu_2[w(\zeta), \zeta].$$

The unknown mapping $z = z(\zeta)$ of $|\zeta| < 1$ onto D_z satisfies (9) and the boundary conditions

$$f[x(\gamma), y(\gamma)] = 0, \qquad \gamma \in [\gamma_k, \gamma_{k+1}], \tag{10}$$

if the corresponding part of the boundary L_z is given (this amounts to the impermeable parts, the reservoir boundaries, and the seepage intervals), and

$$\operatorname{Re}[G(\gamma)z] = C(\gamma), \qquad \gamma \in [\gamma_{k-1}, \gamma_k], \tag{11}$$

if this part is unknown, where the $t_k = e^{i\gamma_k}$ are fixed images of the junction points of the known and unknown parts of L_z. The last form of the conditions (11) includes, besides the relation (7) on the depression curve, for which

$$G(\gamma) = S[\varphi(\gamma)] + ik_0, \qquad c(\gamma) = \varphi(\gamma) - \varphi_k,$$

also the conditions on the unknown parts of the underground contour of the hydraulic structure which are determined by the head diagram.* Then

$$G(\gamma) \equiv 1, \qquad c(\gamma) = h^{-1}\left[-\frac{\varphi(\gamma)}{k_0}\right],$$

where $\varphi = \varphi(\gamma)$ is the correspondence between the boundary points under the conformal mapping $w = w(\zeta)$. We remark that if the given parts of the boundary L_z are polygonal lines and the function $z = z(\zeta)$ is analytic (the case of a homogeneous isotropic soil), then the results in Chapter II imply that the problem is solvable. In the case when the lengths of the links of the polygonal lines in L_z are unknown and are determined from the specified head at their vertices (or from the outflow on these parts when $h = $ const), Theorem 1 of §2 in Chapter VI implies the solvability of the problem also for equation (9). Therefore, it remains for us to consider only the case when the lengths of the links of the polygonal lines are fixed or the given parts of the boundary L_z are curvilinear. We assume that the given parts of L_z are piecewise smooth curves

---

* *Editor's note.* See the footnote in subsection 1° above.

that can be broken up into parts in such a way that on each part one of the inequalities

$$\left|\frac{\partial f(x, y)}{\partial x}\right| \geq \varepsilon_0 > 0, \qquad \left|\frac{\partial f(x, y)}{\partial y}\right| \geq \varepsilon_2 > 0 \qquad (12)$$

holds and the partial derivatives f_x and f_y are bounded. Corresponding to this, let us make the following substitutions of the unknown function $z = z(\zeta)$, setting

$$Z = \xi + i\eta \equiv f(x, y) + x_k + iy \qquad (13)$$

in the domain contiguous to the given part of L_z where the first of the inequalities (12) holds, while

$$Z = \xi + i\eta \equiv f(x, y) + x_k + ix \qquad (14)$$

where the second holds, and $-Z - x_k = z$ or iz near the unknown boundaries. In the adjacent domains we have $-Z = f_0(x, y) + if(x, y)$ if $DZ/Dz \geq \alpha > 0$. Thus, different substitutions of the unknown function are implemented in the domains $D_z^i \subset D_z$ and are adjusted so that $Z = Z(z)$ is continuous in D_z. For definiteness let us take the case of the transformation defined by (13). Since the Jacobian of this transformation has the form

$$\frac{DZ}{Dz} = |Z_z|^2 - |Z_{\bar z}|^2 = \frac{1}{4}\left\{|f_x' - if_y' + 1|^2 - |f_x' + if_y' - 1|^2\right\} = f_x',$$

the transformation $Z = f(x, y) + x_k + iy$ is one-to-one by the first of the conditions (12). Differentiation of the equality $Z = Z[z(\zeta)] \equiv Z(\zeta)$ gives us that

$$Z_\zeta = Z_z \bar z_\zeta + Z_z z_\zeta, \qquad Z_{\bar\zeta} = Z_{\bar z}\bar z_{\bar\zeta} + Z_z z_{\bar\zeta},$$

which, after determining z_ζ and $z_{\bar\zeta}$ and substituting them in (9), leads to the following equation for the unknown function $Z = Z(\zeta)$:

$$Z_{\bar\zeta} = q_1(\zeta, Z)Z_\zeta - q_2(\zeta, Z)\bar Z_{\bar\zeta} = 0, \qquad (15)$$

where

$$q_1 = \left[(Z_z)^2 - (Z_{\bar z})^2 - 2iZ_z Z_{\bar z}\operatorname{Im}\hat\mu_2\right]Q(\zeta),$$

$$q_2 = \left\{\hat\mu_2(|Z_z|^2 + |Z_{\bar z}|^2 - \hat\mu_2 Z_z Z_{\bar z}) + \bar Z_{\bar z}Z_{\bar z}\left(|\hat\mu_1| - \frac{Z_z\bar Z_z}{\bar Z_{\bar z}Z_{\bar z}}\right)\right\}Q(\zeta),$$

$$\frac{1}{Q(\zeta)} = |Z_z - \mu_2 Z_{\bar z}|^2 - |\hat\mu_1 Z_{\bar z}|^2.$$

Moreover, it is easy to see, by the assumption that $f_x' \neq 0, \infty$ and $f_y' \neq \infty$, that

$$|q_1| + |q_2| \leqslant q_0(|f_x'|, |f_y'|, \mu_0) < 1.$$

Equation (9) is also transformed similarly in the case of the substitution $Z = f(x, y) + x_k + ix$. As a result of this substitution of the unknown function, the boundary condition (10) corresponding to the given parts of L_z is transformed to the form

$$\operatorname{Re} Z(\gamma) = x_k, \qquad \gamma \in [\gamma_k, \gamma_{k+1}], \tag{10*}$$

and condition (11) remains unchanged. Thus, we come to the following boundary-value problem for determining a function $Z = Z(\zeta)$ satisfying (15):

$$\operatorname{Re}[G_0(\gamma)]Z(\gamma)] = g(\gamma), \qquad \gamma \in [0, 2\pi], \tag{16}$$

where

$$G_0(\gamma) = (1 - x_k)S[\zeta(\gamma)] + ik_0, \qquad g(\gamma) = \varphi(\gamma) - \varphi_k$$

on the image of the depression curve (in the case of the substitution $Z = z + z_k$),

$$G_0(\gamma) = 1, \qquad g(\gamma) = h^{-1}\left[-\frac{\varphi(\gamma)}{k_0}\right]$$

on the images of the unknown parts of L_z determined by the head diagram $h = h(x)$ given on them, and, finally, $G_0(\gamma) = 1$ and $g(\gamma) = x_k = \text{const}$ on the images of the given parts of L_z. We choose the parameters x_k introduced in (10) in such a way that the coefficient $g(\gamma)$ in (16) is left with finite discontinuities only at points of discontinuity of $G_0(\gamma)$ and at images of points at infinity on L_z, where (as in Problems $1°.1$ and $1°.2$) the filtration flow is assumed to be finite. If a given part l_z of L_z meets on two sides with unknown curves determined by the head diagram, then for the function $g(\gamma)$ to be continuous at the images of the endpoints of l_z we require the necessary condition that the head diagrams be consistent on the parts abutting on l_z. The function $G_0(\gamma)$ has discontinuities only at the images of junction points of the depression curve and known parts of L_z or parts determined from the given head diagram $h = h(x)$. But $G_0(\gamma)$ does not have discontinuities at images of junction points of the unknown parts of the boundary on which $h = h(x)$ is specified and the depression curves. Therefore, the boundary-value problem (16) has a unique almost bounded solution ($Z(\zeta)$ has a logarithmic singularity at the images of the points at infinity on L_z) in the class of analytic functions when L_z contains only one depression curve part, while it has a unique almost bounded solution in the case of m ($m \geqslant 2$) such parts only if $m - 1$ solvability conditions hold. Then under the corresponding conditions on the coefficients

$q_i(\zeta, Z)$ of (15) (i.e., in the final analysis, on the filtration tensor), Theorem 1 in §2 of Chapter VI tells us that the boundary-value problem (15), (16) has no fewer almost bounded solutions than has problem (16) in the class of analytic functions. We thus arrive at the following theorem.

THEOREM 2. *Suppose that the boundary L_z of the filtration domain D_z consists of finitely many specified piecewise smooth curves, unknown impermeable parts on which the head diagram $h = h(x)$ is given, and depression curve parts, and that the filtration tensor $k(x, y, h)$ in D_z is continuous in x and y and measurable in h. If L_z contains no more than one depression curve part, then the corresponding filtration problem has at least one solution, while if there are m $(m > 1)$ such parts, then it has at least one solution only if $m - 1$ solvability conditions hold.*

REMARK. The absence in Theorem 2 of any assertions about the uniqueness of the solutions is easily explained by the fact that, when several different substitutions of the unknown function are carried out simultaneously, the coefficients of the equation (15) obtained for determining the function $Z = Z(\zeta)$ cannot satisfy a Lipschitz condition in Z. However, as will be shown below, under certain conditions the nonlinear boundary condition (10) can be transformed into a linear condition by a single substitution of the unknown function. In this case there is a uniqueness theorem when the given curves and the filtration tensor are sufficiently smooth.

EXAMPLES. We shall use some simple examples to clarify the method used in proving Theorem 2 for linearizing the boundary conditions in filtration problems with free boundaries.

$2°.1$. *The planar problem of filtration of a fluid from a reservoir in the case where the free boundary ends at a drain (Figure 2).*

In the respective domain D_1 and D_2 we set

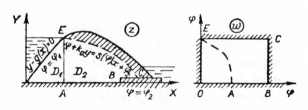

FIGURE 2

$$Z = \xi + i\eta \equiv g - y(x) + ix, \qquad Z = y_E - y + ix.$$

For determining the function $Z = Z(w)$ in D_w we arrive at the boundary-value problem

$$\operatorname{Re} Z = 0 \quad \text{on } OE, \qquad \operatorname{Im} Z = 0 \quad \text{on } OABC,$$
$$S(\varphi)\eta - k_0(y_E - \xi) = \varphi - \varphi_2 \quad \text{on } CE.$$

According to the definition of $S(\varphi)$ we have $S(\varphi_1) = S(\varphi_2) = 0$; therefore, C is not a point of discontinuity of the coefficients of the boundary condition, and only two such points remain: O and E. Consequently, by Theorem 2, Problem $2°.1$ has at least one solution.

$2°.2.$ *The problem of filtration of water in nonhomogeneous anisotropic soil from an axisymmetric well* (*Figure* 3). By the same substitutions in D_1 and D_2 as in the preceding problem, this problem can be reduced to a mixed boundary-value problem with two points of discontinuity.

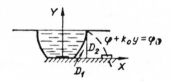

FIGURE 3

$2°.3.$ *The problem of constructing part of the underground contour of a hydraulic structure from a given head diagram* (*Figure* 4).The problem reduces to a mixed boundary-value problem with two singular points.

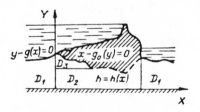

FIGURE 4

We illustrate methods for investigating uniqueness questions in two examples of fairly general problems in the theory of filtration with free boundaries.

$2°.4.$ *The problem of filtration of ground water from an inclined water-bearing layer* (*Figure* 5).

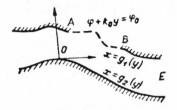

FIGURE 5

The flow domain D_2 in this problem is bounded by the streamlines $CABE$ ($\psi = 0$) and COE ($\psi = \psi_0$). We join the points A and B and assume accordingly that $x - g_1(y) = 0$ is the equation of the whole curve $CABE$. Suppose that $g_i(y) \in C^2$, and that $|g_1(y) - g_2(y)| \geqslant \varepsilon > 0$ for all y. On the z-plane let us define the transformation

$$Z = \frac{x - g_1(y)}{g_2(y) - g_1(y)} + iy \equiv f(x, y) + iy.$$

Then we arrive at the following mixed boundary-value problem for determining the function $Z(w) = Z[z(w)]$:

$$\operatorname{Re} Z = 1 \quad \text{on } COE; \quad \operatorname{Re} Z = 0 \quad \text{on } CA, BE; \quad \operatorname{Im} Z = \frac{\varphi - y}{k_0} \quad \text{on } AB.$$

Mapping the strip in the w-plane onto the disk $|\zeta| < 1$, we now come to a mixed boundary-value problem for (15), with the coefficients satisfying a Lipschitz condition in Z because of the assumptions made about the filtration tensor and the equations of the given boundaries. Consequently, problem $2°.4$ is uniquely solvable, by Theorem 3 of §4 in Chapter V.

$2°.5.$ *The problem of filtration of a fluid in an earthen dam on a permeable base of finite depth (Figure 6).*

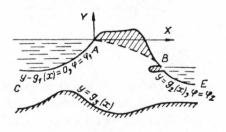

FIGURE 6

It is assumed that the given functions $g_1(x)$, $x \in (-\infty, 0]$, $g_2(x)$, $x \in [x_0, \infty)$, and $g_3(x)$, $x \in (-\infty, \infty)$ in the equations of the curvilinear boundaries of the reservoirs CA and BE and the curvilinear impermeable surface CE ($\psi = 0$ on it) satisfy the conditions $g_i(x) \in C^2$, and, moreover,

$$\frac{dg_i(x)}{dx} \to 0, \quad |x| \to \infty, \quad \frac{dg_1(0)}{dx} = 0.$$

At its end B the free boundary comes to a horizontal drain ($dg_2(x)/dx = 0$, $x \in [x_0, x_*]$) whose length is to be found (in particular, $|x_0 - x_*| = 0$). On the free boundary AB we have $\psi = \psi_0$ and $\varphi + k_0 y = \varphi_0 + S(\varphi)x$, where $S(\varphi_1) = S(\varphi_2) = 0$, and φ_1 and φ_2 are the respective values of the potential on CA and BD. Suppose that the problem has two solutions, to which correspond the mappings $z = z^1(w)$ and $z = z^2(w)$ of the fixed domain D_w onto the filtration domain D_z that are solutions of (2). The difference $z(w) = z^1(w) - z^2(w)$ between these solutions satisfies the equation

$$z_{\overline{w}} - \mu_1(w, z^1)z_w - \mu_2(w, z^1)\overline{z}_{\overline{w}} = F(w)z, \tag{17}$$

and the boundary condition

$$a(\gamma)x - y = 0 \quad \text{on } L_w, \tag{18}$$

where

$$F(w) = z_w^2 \frac{\mu_1(w, z^1) - \mu_1(w, z^2)}{z^1 - z^2} + \overline{z}_w^2 \frac{\mu_2(w, z^1) - \mu_2(w, z^2)}{z^1 - z^2},$$

$$a(\gamma) = \frac{g_i[x^1(\gamma)] - g_i[x^2(\gamma)]}{x^1 - x^2}, \quad i = 1, 2, 3,$$

on the fixed images of CA, BE, and CE, and $a(\gamma) = S(\varphi)/k_0$ on the image of AB, γ being the arclength parameter of the boundary L_w of the rectangle $D_w = \{\varphi_1 < \varphi < \varphi_2, \ 0 < \psi < \psi_0\}$. But, by the assumptions made, $a(\gamma)$ is continuous and satisfies a Hölder condition, while the index of the boundary-value problem (18) is zero, since $\text{ind}(a + i) = 0$, and $z^1 = z^2 = z = 0$ at A. Thus, Theorem 3 of §4 in Chapter V is also applicable in this case, and it tells us that the original problem $2°.5$ is uniquely solvable. We formulate this as a theorem.

THEOREM 3. *The filtration problems $2°.4$ and $2°.5$ are uniquely solvable if the filtration tensor $k(x, y, h)$ satisfies condition (b) and if the equations of the given parts of the boundary of the filtration domain are twice continuously differentiable.*

§6. Problems involving filtration of a nonhomogeneous fluid
with free boundaries for a given time-dependent distribution
of temperature and component concentration

In this section we study problems of filtration of multi-component fluids bordering along unknown curves on a homogeneous fluid at rest (with moisture laden bottom layer and, in the case of unforced motion, with atmospheric air), in nonhomogeneous anisotropic soils. This class of problems includes problems on forcing out immiscible (for example, water and oil) or miscible (oil and a solvent) fluids, filtration of an aerated fluid (oil and a gas), salt diffusion in filtration under hydraulic structures, and so on. Moreover, we also take into account the temperature change of a fluid as time changes (for example, the cooling of a petroleum layer when a cold fluid is pumped into it).

Such processes are described by a system of quasilinear equations consisting of second-order parabolic equations for the temperature and the concentrations of the components of the nonhomogeneous fluid, and of an elliptic system of first-order equations for the generalized complex filtration potential. A large class of steady-state and time-dependent filtration problems for a nonhomogeneous fluid in domains with a given or partially unknown boundary has been studied in joint work of S. N. Antontsev and the author (see [7]–[11]).

In order not to invoke the theory of parabolic equations, we assume that the temperature $u_0(x, y, t)$ and the concentrations $u_i = u_i(x, y, t)$ $(i = 1,\ldots,s)$ of the components of the nonhomogeneous fluid are known. In this case the original problems reduce to the construction of quasiconformal mappings of a given domain onto a domain with a partially unknown boundary, where the characteristics of this mapping depend on the time as a parameter. Existence and uniqueness theorems are proved for such mappings, and their differential properties are studied.

1°. Differential equations. Let p, $\rho = \rho(\vec{u})$, $\nu = \nu(\vec{u})$, and $\vec{v} = (v^1, v^2)$ denote the pressure, density, viscosity, and filtration velocity of a mixture $\vec{u} = (u_0, u_1,\ldots,u_s)$. These variables are connected by the following generalized Darcy's law and the equation of continuity:

$$-\vec{v} = k\nu^{-1}\,\text{grad}\,\varphi, \qquad \text{div}(\rho\vec{v}) = 0,$$

where $k = \{k_{ij}(z, t, p, \vec{u})\}$ is the symmetric filtration tensor, $z = x + iy$, and $\varphi = p + \rho gy + q(\vec{u}, z)$ is the filtration potential of the mixture, which depends (via q) on the physical and chemical properties of the fluid and the soil ($q \equiv 0$ in salt diffusion problems).

The system (1) is equivalent to the following complex equation for the generalized complex potential $w = \varphi + i\psi$, $\psi_x + i\psi_y = \rho(v^2 - iv^1)$:

$$w_{\bar{z}} + \mu_1(w, z, t)\bar{w}_{\bar{z}} + \mu_2(w, z, t)w_z = 0, \qquad (2)$$

whose coefficients can be expressed explicitly in terms of the components of the symmetric tensor $\rho \nu^{-1} k = \{\rho \nu^{-1} k^{ij}\}$ (see §5).

As in the previous section, we assume that

$$\rho \nu^{-1}\left(k_{11} k_{22} = k_{12}^2\right) \geqslant \delta_0 > 0, \qquad \rho \nu^{-1} |k_{ij}| \leqslant M < \infty;$$

then

$$|\mu_1| + |\mu_2| \leqslant \mu_0(\delta_0, M) < 1, \tag{3}$$

i.e., equation (2) is uniformly elliptic.

In the axisymmetric case $(x - x_0 \equiv r)$ the filtration tensor can be represented in the form $k = (x - x_0)\tilde{k}$, and, therefore, (2) is uniformly elliptic only in a domain not containing points of the line $x = x_0$, a condition we shall assume.

2°. *Boundary conditions.* Suppose that filtration of a nonhomogeneous fluid takes place in a domain D_z bounded by unknown interfaces Γ_i $(i = 1, 2)$ with a fluid at rest and by the straight lines $x = 0$ and $x = l(t)$, to which and from which, respectively, a homogeneous mixture flows parallel to the OX-axis. In forcing out oil by water these lines are the drill hole for exerting pressure and the production well, and in problems of diffusion removal of soluble substances from hydraulic structures these are the interfaces with the reservoir.

The continuity of the complex potential on the boundaries (interfaces) with the homogeneous flow implies that the boundaries are equipotential, i.e.,

$$\varphi(z, t) = \varphi_{0,l}(t) \quad \text{for } x = 0, l(t). \tag{4}$$

We remark that for $\rho = \text{const}$ and $q \equiv 0$ the last conditions are the usual consequence of the Bernoulli law on the side of uniform flow.

On the interface Γ_1 with the fluid at rest we have $\rho = \rho_0 = \text{const}$ (this is the bottom moisture in forcing-out problems), and, assuming continuous pressure, we get from the hydrostatic Bernoulli law that $p + \rho_0 gy = \text{const}$, while $p = p_{\text{atm}}$ on the depression curve Γ_2 (the boundary of the wet and nonwet soil).

Substituting these values of p into the formula for the potential φ on Γ_i (streamlines), we arrive at the boundary conditions

$$\varphi = p + \rho(\vec{u})gy + q(\vec{u}, z, t) \equiv \varphi_i(x, y, t), \qquad \psi = \psi_i(t), z \in \Gamma_i. \tag{5}$$

Let $Cd_{n+\beta_0}^{m+\beta}$ denote the set of functions whose mth derivatives with respect to x and y and whose nth derivatives with respect to t satisfy Hölder conditions with respective exponents β and β_0; along with the previously adopted notation $\|f\|_{C^\alpha(\Omega)}$ for the norm of a function we use also the expression $C^\alpha(f, \Omega)$.

Assume that the following conditions hold for the given functions:

(i) The $\mu_i(z, w, t)$ are twice continuously differentiable with respect to all arguments.

(ii) $\varphi_i(x, y, t) \in C_3^3, \psi_i(t), \varphi_{0,l}(t) \subset C_3[O, T]$, and

$$\frac{\partial \varphi_i}{dx} = 0, \qquad \frac{\partial \varphi_1}{\partial y} \frac{\partial \varphi_2}{\partial y} \geqslant \delta_0 > 0.$$

Without loss of generality the quantities $\varphi_{0,l}(t)$ and $\varphi_{1,2}(t)$ can be assumed to be given, since by nonsingular affine transformations

$$Z = \alpha_1(t)z + \beta_1(t)\bar{z} + \gamma_1(t), \qquad W = \alpha_2(t)w + \beta_2(t)\bar{w} + \gamma_2(t),$$

which do not change the form of (2), the boundary-value conditions (4) and (5) for Z and W can be transformed to analogous conditions with fixed $\tilde{\varphi}_{0,l}$ and $\tilde{\psi}_{1,2}$.

3°. *Quasiconformal mappings dependent on parameters.* Let $z = z(w, t)$ be the function inverse to $w = w(z, t)$, which is defined in the domain

$$D_w = \{0 < \varphi < \varphi_l, 0 < \psi < \psi_2\} \qquad (\varphi_0 = \psi_1 = 0)$$

and satisfies there the elliptic equation

$$\bar{z}_w - \mu_1(w, z, t)z_w - \mu_2(w, z, t)\bar{z}_{\bar{w}} = 0. \tag{6}$$

On the boundary L_w the unknown function $z(w, t)$ is subject to the boundary conditions

$$\text{Im } z \big|_{\psi=0,\psi_2} = y_{1,2}(\varphi, t) \, \text{Re } z \big|_{\varphi=0,\varphi_l} = 0, l(t), \tag{7}$$

where the $y_i(\varphi, t)$ are the functions inverse to $\varphi_i(y, t)$. Thus, the original filtration problems have been transformed into the problem of finding a solution of (6) which maps a given rectangle D_w quasiconformally onto the unknown domain D_z and satisfies the boundary condition (7) on the boundary of D_w.

The above mapping problem is obviously also equivalent to the filtration problem for a homogeneous fluid in a nonhomogeneous anisotropic soil with filtration tensor

$$\kappa = \rho(u)\vec{v}^{-1}(\vec{u})k.$$

Depending on the physical formulation of the original problem, we can consider different ways of assigning the function parameters $l(t)$ and the ordinates $y_i^0(t)$, $y_i^l(t)$ of the points where the unknown boundaries Γ_i ($i = 1, 2$) intersect the straight lines $x = 0$ and $x = l(t)$.

For definiteness let us consider the problem of a rational choice of the distance $l(t)$ between the vertical parts of the boundary of the filtration domain D_z; corresponding to this choice are the given pressures for $x = 0$ and $x = l(t)$, and along with them also the ordinates $y_i^0(t)$, $y_i^l(t) \in C_3[O, T]$ of the endpoints of Γ_i, which are connected by the inequalities

$$0 < y_1^0 < y_1' < y_2', \qquad 0 < y_2^0 < y_2'. \tag{8}$$

If the distance between these parts is fixed, then the position of the initial point y_i^0 (for $x = 0$) of one of the free boundaries, say Γ_1, must be taken as unknown. The problems of finding $l(t)$ or $y_i^0(t)$ do not differ mathematically.

THEOREM 1. *For each* $t \in [0, T]$ *there exists a unique solution* $z(w, t)$ *of the boundary-value problem* (6), (7) *that satisfies the inequalities*

$$W_p^1\{z(w, t), \overline{D}_w\} \leq M, \qquad |\ln l(t)| \leq M, \tag{9}$$

where the constant M *depends on* μ_0 *and* α_0, $0 < \alpha_0 = \alpha_0(p, \alpha) < \alpha \leq 1$, *and on* $C^2\{y_i(\varphi, t)\}$, *and* $p(\mu_0) > 2$.

The solvability of the boundary-value problem (6), (7) and the first of the estimates in (9) follow from the validity of these assertions for the more general mixed boundary-value problem with parameters (see §2.4° in Chapter VI). To prove the second estimate in (9) we remark that for the analytic function

$$z(\zeta, t) \equiv z[w(\zeta, t), t], \qquad w = \zeta^{-1}(\zeta, t),$$

in the plane of the homeomorphism $\zeta = \zeta(w, t)$, $\operatorname{Im} \zeta > 0$, the functional l can be written in the explicit form

$$l = \left\{ \int_{\xi_1}^{\xi_2} y_1[\varphi(\xi), t] \Pi(\xi) \, d\xi + \int_{\xi_3}^{\xi_4} y_2[\varphi(\xi), t] \Pi(\xi) \, d\xi \right\} \left\{ \int_{\xi_2}^{\xi_3} \Pi(\xi) \, d\xi \right\}^{-1},$$

where $\Pi(\xi) = \Pi_1^4 |\xi - \xi_k|^{-1/2}$, and the ξ_k are the images of the corner points of the boundary of D_w. Then the inequality $|\ln l(t)| < M$ follows from the fact that the functions $y_i(\varphi, t)$ are positive and from the inequalities

$$C^{-1/\alpha} |w_k - w_{k-1}|^{1/\alpha} \leq |\xi_k - \xi_{k-1}| \leq C |w_k - w_{k-1}|^{\alpha} \qquad (k = 1, \dots, 4),$$

which are valid by known properties of the mappings $\zeta = \zeta(\overline{w}, t)$. The theorem is proved.

THEOREM 2. *The solution* $z = z(w, t)$ *of the boundary-value problem* (6), (7) *is univalent in* D_w *for each fixed* $t \in [0, T]$, *and the inverse mapping* $w = w(z, t)$ *satisfies the inequalities*

$$W_{p'}^1[w(z, t), D_z] \leq M_2, \qquad C^{\beta}[w(z, t), D_z] \leq M_2, \tag{10}$$

where $2 < p' < p$ *and* $\beta = (p' - 2)/p'$.

Let $\mu_i^\varepsilon(w, z, t)$ be functions twice continuously differentiable with respect to all arguments, coinciding with μ_i in the domain $D_w^{2\varepsilon} = \{2\varepsilon < \varphi < \varphi_l - 2\varepsilon, 2\varepsilon < \psi < \psi_2 - 2\varepsilon\}$, and equal to zero in $D_w \backslash D_w^\varepsilon$, and let $z = z^\varepsilon(w, t)$ be the solutions of the corresponding equation (6) with μ_i^ε. Let $D_{\delta w}$ ($\subset D_w^\varepsilon$) be a domain whose boundary $\Gamma_{\delta w} \subset C^{2+\alpha}$ coincides with $\Gamma_w = \partial D_w$ everywhere except in δ-neighborhoods of the corner points of Γ_w ($0 < \delta < \varepsilon/2$).

The index ε will be omitted in what follows, when this does not lead to confusion. We represent the solution of the boundary-value problem (6), (7) by the formula

$$z = \Psi[\zeta(w, t), t], \qquad \Psi_{\bar{\zeta}} = 0,$$

where $\zeta = \zeta(w, t)$ is a homeomorphism of the rectangle onto $K = \{|\zeta| < 1\}$ that is conformal in an ε-neighborhood of ∂D_w because μ_i^ε has compact support here. Thus, to prove that the mappings $z = z^\varepsilon(w, t)$ are univalent it suffices to establish that the analytic functions $\Psi(\zeta, t)$ in K are univalent.

By the boundary conditions (7) and condition (ii),

$$\left.\frac{\partial y_1}{\partial \varphi} \frac{\partial y_2}{\partial \varphi}\right|_{\psi=0,\psi_2} > 0, \qquad \frac{\partial y}{\partial \varphi} = -\frac{\partial x}{\partial \varphi} = 0 \quad \text{for } \varphi = 0, \varphi_l,$$

i.e., the quantity

$$\operatorname{Im} \frac{dz}{d\zeta} = \frac{dy}{d\varphi} \left| \frac{dw}{d\zeta} \right|$$

preserves its sign on the circle $|\zeta| = 1$. This ensures that $z = \Psi(\zeta, t)$ is univalent in K and, since $\Psi(\zeta, t)$ can be continued analytically across the lines $\varphi = 0, \varphi_l$ (Re $\Psi = 0, l(t)$ on them) and the inequality Im $dz/d\zeta > 0$ holds on the images of the lines $\psi = 0, \psi_2$, the function $z = \Psi(\zeta, t)$ is also univalent on the closed domain $\overline{K} \cap \Sigma_i Q_i$, where $Q_i = \{|\zeta - \zeta_i| < \delta\}$, the ζ_i ($i = 1, \ldots, 4$) are the images of the corner points of D_w, and $\delta > 0$ is arbitrary. Further, according to (9) and the properties of $y_1(\varphi, t)$ and $y_2(\varphi, t)$, we have for $\psi = 0$, ψ_2 that

$$\min \left| \frac{dy_i}{d\varphi} \right| |\varphi_1 - \varphi_2| \leqslant |z(\varphi_1) - z(\varphi_2)| \leqslant C |\varphi_1 - \varphi_2|^\alpha, \qquad \varphi_{1,2} \in [0, \varphi_l].$$

Consequently, the image of the boundary of the rectangle D_w under the mapping $z = z(w, t)$ is a simple closed Jordan curve for all $t \in [0, T]$. Theorem 5 in §4 of Chapter V then gives us for the analytic function $\zeta = \zeta(z, t)$ inverse to $z = \Psi(\zeta, t) \in C^\alpha(\overline{K})$ the estimate

$$|\zeta(z_1, t) - \zeta(z_2, t)| \leqslant M_0 |z_1 - z_2|^\alpha, \qquad z_1, z_2 \in D_z,$$

with the help of which we find that, for all $z \in D_z$ and α_0, $0 < \alpha_0 < \alpha$,

$$\left| \frac{d\zeta}{dz} \right| \leq M_1 \inf_{z_\Gamma \in D_z} |z - z_\Gamma|^{\alpha_0 - 1} \leq M_2 \inf |\zeta(z) - \zeta(z_\Gamma)|^{(\alpha_0 - 1)/\alpha}$$

$$\leq M_3 \inf |w(z) - w(z_\Gamma)|^{(\alpha_0 - 1)/\alpha}.$$

Using the usual methods for estimating derivatives of quasiconformal mappings (see §3 of Chapter V), we now get

$$\iint_{D_z} \left| \frac{dw}{dz} \right|^{p_0} dD_z \leq \frac{(1 + \mu_0^2)^{1/q}}{(1 - \mu_0^2)^{1 - p_0}} \left\{ \iint_K \left| \frac{dw}{d\zeta} \right|^{(p_0 - 2)q + 2} dK_\zeta \right\}$$

$$\cdot \left\{ \iint_{D_w} \left| \frac{d\zeta}{dz} \right|^{(p_0 - 2)/(q - 1)} dD_w \right\}^{(q-1)/q}.$$

If in this inequality we choose

$$q = \frac{p - 2}{p_0 - 2}, \qquad 2 < p_0 < \frac{2(p - 2)(1 - \alpha_0) + \alpha^2 p}{(p - 2)(1 - \alpha_0) + \alpha^2},$$

then the integrand in the last integral has an integrable singularity on ∂D_w,

$$\left| \frac{d\zeta}{dz} \right|^{(p_0 - 2)/(q - 1) \cdot q} \leq M_4 \inf |w - w_\Gamma|^{-\lambda},$$

$$\lambda = \frac{(p - 2)(1 - \alpha_0)(p - 2)}{(p - p_0)\alpha^2} < 1,$$

which ensures inequalities (10) with constants not depending on ε, and Theorem 2 is proved.

THEOREM 3. *Suppose that* $\mu_i(w, z, t) \in C^{m+\alpha}$ ($m \geq 2, 0 < \alpha < 1$), *that* $y_i(\varphi, t)$, $\varphi_l(t)$, $\psi_2(t) \in C^{m+\alpha+2}$ *with respect to all their arguments, and that the norms of these functions in the indicated spaces are bounded by a constant* $N > 0$. *Then the following estimate holds for the solution of the boundary-value problem* (6), (7):

$$C_{m+\beta}^{m+1+\alpha}[z(w, t), Q_{\delta w}] \leq M_6(N), \tag{11}$$

where $Q_{\delta w} = \overline{D}_{\delta w} \times [0, T]$.

The assertions about the smoothness of the mapping $z = z(w, t)$ with respect to w follow immediately from the properties of quasiconformal mappings (Theorem 4 in §5 of Chapter V). We next differentiate (6) and both sides of (7) with respect to t and consider the boundary-value problem obtained for the function $v = z_t(w, t)$, which satisfies the equation

$$v_{\bar{w}} - \mu_1 v_w - \mu_2 \bar{v}_{\bar{w}} = Av + F,$$

where

$$A = z_w\left(\frac{\partial \mu_1}{\partial z} + \frac{\partial \mu_1}{\partial \bar{z}}\frac{\bar{v}}{v}\right) + \bar{z}_{\bar{w}}\left(\frac{\partial \mu_2}{\partial z} + \frac{\partial \mu_2}{\partial \bar{z}}\frac{\bar{v}}{v}\right), \qquad F = z_w\left(\frac{\partial \mu_1}{\partial t} + \bar{z}_{\bar{w}}\frac{\partial \mu_2}{\partial t}\right).$$

Since

$$\frac{\partial y_i}{\partial t}(\varphi, t) \in C_{m+1+\alpha}^{m+1+\alpha}, \quad A \text{ and } F \in L_p(\bar{D}_w), \qquad p > 2,$$

a solution $v(w, t)$ of this problem that is unbounded at the corner points of D_w satisfies a Hölder condition in $\bar{D}_{\delta w}$, as shown in §2 of Chapter VI for the general Hilbert problem. Finally, (11) can be proved by repeated differentiation of (6) and (7) with respect to t.

§7. Problems

We formulate some mathematical problems in the theory of filtration of a fluid with free boundaries.

1°. Steady-state problems.

1. As a rule, filtration problems are multi-parameter problems, and, therefore, the question of a unique choice of parameters is one of the most important and difficult in this theory. In the practical solution of such problems one frequently resorts to the hodograph method, the application of which usually allows neither the geometric characteristics of the filtration domain nor the physical parameters of the filtration flow to be specified in advance. When filtration problems are solved in this way, the statement of the question of unique solutions is naturally devoid of meaning in general. However, the solution of direct problems takes on all the more significance when the geometrical and physical characteristics of the filtration flow are specified beforehand, and the parameters of the problem are determined as a result of its solution. A fairly large class of such problems was described in Chapter II, where we studied the mixed boundary-value problem with free boundary for analytic functions, which is a filtration problem in a homogeneous isotropic soil under particular assumptions about the initial data (§5.1° of Chapter III). Theorem 2 of this chapter establishes solvability for an even larger class of filtration problems with free boundaries in nonhomogeneous anisotropic media (the planar and axisymmetric cases). The question of determining conditions ensuring the uniqueness of the solutions constructed naturally arises here. In §5 two possible methods for investigating uniqueness were analyzed in two examples (Theorem 3).

2. The question of investigating the univalence of solutions of filtration problems is a necessary complement to the preceding problem. As a rule, it is very simple to establish that the filtration domain is one-sheeted in the cases when the hodograph is used for solving problems (§§1–4). A number of interesting articles deal with the investigation of univalence for special classes of filtration problems (see the survey in the monograph [106]). Considering auxiliary boundary-value problems for the unknown functions.

$$\theta = \arctan\left(\frac{\partial y(\varphi, \psi)}{\partial \varphi}\bigg/\frac{\partial x(\varphi, \psi)}{\partial \varphi}\right) \quad \text{and} \quad v = \frac{\partial y(\varphi, \psi)}{\partial \varphi}$$

in the case of the problems studied in §5 and assuming that the parameters of the problem are unknown, determine the class of solutions to which one-sheeted filtration domains correspond (cf. §2.3° in Chapter X).

3. Extend Theorem 2 in §5 to the filtration problems with free boundaries in which the relations (5.6) hold on the seepage interval, and, moreover, precipitation can fall on the free boundary and evaporation can take place from its surface (cf. §§1–4).

4. In §§1–4 we studied filtration problems for a nonlinear law connecting the velocity with the gradient of the head, and in §3 filtration problems with the compressibility and anisotropy of the medium taken into account, which corresponds to a dependence of the filtration tensor on the value of the head and on the space variables. It is interesting to consider the general case when the filtration tensor is a given function of the head $h(x, y)$, the filtration speed $q(x, y)$, and the spatial variables x and y: $k = \{k_{ij}(x, y, h, q)\}$.

5. *Conjunction problems.* In these problems it is required to determine the interface between filtering fluids with different physical properties (for example, water and oil, fluids with different salinity, etc.).

We gave various formulations of a hydrodynamics problem of a similar type in §5 of Chapter III.

6. The question of the asymptotic behavior of the depression curve as it goes to infinity is important in many problems involving filtration of a fluid. However, if the filtration flow is finite at infinity and the impermeable surface goes out to a horizontal asymptote, which is a natural assumption in these problems, then such a problem becomes ill-posed. This has to do with the fact that as we go to infinity in such a filtration scheme (an example of which is pictured in Figure 1) the piezometric head $h = h(x, y)$ (and thereby also the potential $\varphi = -k_0 h(x, y) + \text{const}$) tends to a constant value, and, consequently, the influence of the velocity head begins to take effect. In the fluid filtration scheme suggested in Figure 1 the fact that the boundary condition for the piezometric head $h = h(x, y)$ is replaced by Bernoulli's equation for the filtration speed $q = q(x, y)$ beginning with some value $x = x_0$ ensures that the effect of the velocity head at infinity is taken into account. We remark that under various assumptions both conditions on the free boundary are a consequence of the single Bernoulli equation. Investigate the solvability of the problem stated under the assumptions of Theorem 2 in §5.

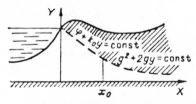

FIGURE 1

7. As mentioned more than once in our investigation of filtration problems in §§1–5, it is a simple consequence of the homogeneity and uniform ellipticity of the equations describing them that filtration flows with compressibility and nonhomogeneity of the medium taken into account (nonlinear equations) are topologically similar to flows in a homogeneous isotropic medium (linear equations). Moreover, the passage from nonlinear problems to linear problems can in many cases be realized constructively: by constructing a quasiconformal mapping of one domain onto another by a solution of a certain linear elliptic system of equations.

Consider the following example. Suppose that the components of the filtration tensor $k = k(x, y, h)$ depend only on the value of the head and do not depend explicitly on the space variables. Then the mapping $z = z(w)$ of the domain D_w, which we assume to be known, onto the unknown domain D_z satisfies the linear equation (5.2). If we construct a homeomorphic mapping of D_w onto itself by a solution of this linear equation, then the unknown function $z(\zeta) = z[w(\zeta)]$ becomes analytic, independently of the boundary conditions determining it. Consequently, as a result of constructing such a mapping, the original filtration problem is reduced to an analogous problem in a homogeneous isotropic medium. Therefore, numerical methods for constructing quasiconformal mappings of domains by solutions of linear (and, all the more so, nonlinear) systems of elliptic equations take on great significance in connection with the solution of filtration problems with free boundaries.

2°. *Time-dependent problems.* The equations for time-dependent filtration of a fluid in a porous medium have the form

$$-\operatorname{div}(\overrightarrow{\rho v}) = \frac{\partial(m\rho)}{\partial t} + \frac{ab}{(1+ap)^2}\frac{\partial p}{\partial t} \equiv \frac{\partial A(p)}{\partial t}, \tag{1}$$

$$\frac{\Phi(v)}{v}\vec{v} = -\operatorname{grad}(p + \rho gh), \qquad \rho = \rho(p), \tag{2}$$

where $\vec{v}$ is the volumetric filtration velocity, p is the pressure, ρ is the density, m is the porosity of the soil, a and b are constants ($a = b = 0$ in the absence of gas adsorption), h is the height of a point above a fixed level ($h \equiv y$ in the planar case), and $\Phi(v) > 0$ and $v^{-1}\Phi'(v) > 0$ for $v \geqslant 0$.

In the case of a linear filtration law (Darcy's law) relation (2) takes the form

$$\vec{v} = -\frac{k}{\mu}\operatorname{grad}(p + \rho gh), \qquad k = k(x, y, z, p), \rho = \rho(p), \tag{3}$$

where μ is the viscosity of the fluid, and k is the permeability coefficient.

Suppose $\Phi(v)$ is such that (2) is solvable for the components of the vector $\vec{v}$ (write out these conditions!), i.e., system (2) is equivalent to the following:

$$\vec{v} = \vec{B}\left(\frac{\partial p}{\partial x}, \frac{\partial p}{\partial y}, \frac{\partial p}{\partial z}, p\right).$$

Then the system (1), (2) reduces to a single equation for the pressure:

$$-\operatorname{div}(\rho\vec{B}) = \partial A(p)/\partial t. \tag{4}$$

The initial and boundary conditions for (4) can be written in the form

$$p\Big|_{t=0} = p_0(x, y, z); \qquad \left| l\frac{\partial p}{\partial n} + l_0 p \right|_\Gamma = l_1, \tag{5}$$

where Γ is the boundary of the filtration domain and n is its normal.

If the filtration takes place when the soil is not completely saturated with the fluid, then the equation of continuity (1) takes the form

$$-\operatorname{div}(\rho\vec{v}) = \partial(m\rho s)/\partial t, \tag{1*}$$

where s is the saturation of the interstices by the fluid, $a = b = 0$, $p =$ constant, and $k = k(s)$, $m(s)$, and $p(s)$ (the capillary pressure) are functions determined by experiment. Completely analogously to the preceding we arrive at the single equation for s

$$-\operatorname{div}\vec{B} = \partial(ms)/\partial t \tag{4*}$$

also in the case of incomplete saturation, and we get initial and boundary conditions for $s(x, y, z, t)$ of the same form as conditions (5) for p. Equation (4*) describes, in particular, the filtration process of a fluid during watering.

8. Consider the main boundary-value problems for equation (4) ((4*)), as well as problems with a free (unknown) boundary. Here one can first use known results for the steady case to study the case of weakly unsteady filtration or filtration for a weakly compressible fluid, when the right-hand side of (4) is small. Study also the case of closeness to Darcy's law. In the general case use, for example, Rothe's method, replacing the time derivatives by finite differences.

9. In the case $h \equiv 0$ prove the existence of self-similar solutions of the main boundary-value problems for the one-dimensional equation (4) ((4*))

$$\frac{\partial A(p)}{\partial t} = -\frac{\partial}{\partial x}\left[\rho(p)B\left(\frac{\partial p}{\partial x}, p\right)\right],$$

i.e., find the solutions of the form $p = p(x/t^\alpha)$ and $p = p(x - At)$.

10. Study the solvability of the problem of patching together solutions of system (1), (2) and system (1*), (2) along an unknown surface in the general and the self-similar cases, i.e., consider the problem with an unknown saturation front moving with a finite velocity. Here the patching surface is determined by the conjunction conditions which connect the solutions of systems (1), (2) and (1*), (2) and their gradients on the patching boundary. For example, when Darcy's law holds in the one-dimensional case, the interface $x = l(t)$ can be determined by the conditions

$$p_1(t, l) = p_2(t, l), \qquad \frac{k}{\mu_1} \frac{\partial p_1(t, l)}{\partial x} = \frac{k}{\mu_2} \frac{\partial p_2(t, l)}{\partial x} = -\sigma \frac{\partial l}{\partial t},$$

where δ is the saturation deficiency (a problem of Verigin type), or by the conditions

$$p_1(t, l) = p_2(t, l) = p_0, \qquad \alpha \frac{\partial p_1(t, l)}{\partial x} - \frac{\partial p_2(t, l)}{\partial x} = \beta \frac{\partial l}{\partial t},$$

where α and β are constants (a problem of Stefan type).

11. Study the problem of patching together solutions of system (1), (2) (respectively, (1*), (2)) in the steady-state and time-dependent cases. In the one-dimensional case the boundary conditions on the unknown boundary $x = l(t)$ have, for example, the form

$$p(t, l) = p_0(t), \qquad \frac{\partial p(t, l)}{\partial x} = f(t, l),$$

where $p_0(t)$ and $f(t, l)$ are known functions (a problem of Florin type).

In the hydraulic approximation Boussinesq's equation for the free boundary $h = h(x, y, t)$ (the boundary of the wet and dry soil) has the form

$$\frac{\partial(sh)}{\partial t} = \operatorname{div}[k(h - z_0)\operatorname{grad} h] - \frac{k_1}{M_0(h - h_1)} + f, \tag{5}$$

where $z_0 = z_0(x, y)$ is the equation of a weakly permeable surface, while $M_0 = M_0(x, y)$ and k are its thickness and permeability coefficient, respectively, $f = f(x, y, t)$ is the amount of infiltration per unit area per unit time, and s is the water output or the saturation deficiency. If $z_0 = \mathrm{const}$ (the surface is impermeable), $f \equiv 0$ (there is no infiltration), and $s = \mathrm{const}$, then (5) becomes the usual Boussinesq equation

$$\frac{\partial h}{\partial t} = a^2\left[\frac{\partial^2(h^2)}{\partial x^2} + \frac{\partial^2(h^2)}{\partial y^2}\right]. \tag{5*}$$

A motion described by (5) is said to be unpressurized. We remark that the filtration potential φ is connected with the function h by the formula

$$\varphi = -k(h + p/\rho g)$$

(in the given case $p = p_{\mathrm{atm}} = \mathrm{const}$), and $\varphi = \varphi(x, y, z, t)$ is a harmonic function of the space variable in the filtration domain.

Suppose that an infinitely long plane layer is exposed by a perfect well through which the outflow is $Q(t)$.

As the fluid is pumped out through the well its level changes according to some law $y_A = h_0(t)$, and in correspondence to this $\varphi_0(t) = -k(h_0 + p_a/\rho g)$. Then the filtration domain can by convention be broken up into two parts: the domain $OABD$ of unpressurized filtration with free boundary AB, and the domain BDC of pressurized motion.

12. Study the problem of pressurized-unpressurized motion in the following variant:

(a) First find the form $y = h(x, t)$ of the free boundary AB, which depends on the function parameters $h_0(t)$ and $x_0(t)$, from the solution of the one-dimensional (in x) Boussinesq equation (5), and find the coordinate $x_0(t)$, for example, from the condition that the curve ABC has a tangent at B.

(b) Then construct a conformal mapping $w = w(z, t)$ of the domain D_z thus obtained onto D_w (Chapter VII).

13. Consider the Florin problem (cf. problems 6 and 7) for equation (5).

In studying the filtration of miscible fluids when they interact physically and chemically with a porous medium, we consider as simultaneous the filtration equations (1), (2) and the equations of motion (Fick's law) and mass conservation of the substance contained in the liquid and solid phases:

$$u_i = cv_i - D_i \frac{\partial c}{\partial x_i}, \qquad \operatorname{div} \vec{u} + \frac{\partial N}{\partial t} = -\frac{\partial(mc)}{\partial t}, \tag{6}$$

where $\vec{u} = (u_1, u_2, u_3)$ is the mass velocity vector of the substance associated with the fluid, c and N are the mass concentrations of the substance in the liquid (per unit volume of solution) and the solid (per unit volume of porous medium) phases, and

$$\vec{D}(c, \partial c/\partial x_i) = (D_1, D_2, D_3)$$

is the convective diffusion vector, which takes into account the molecular diffusion and the hydrodynamic dispersion. The kinetic equation

$$\partial N/\partial t = f(c, N) \tag{7}$$

must be put together with the equations (1), (2) and (6), along with the equations of state of the fluid and of the porous medium

$$\rho = \rho_0[1 + \alpha_0 + (p - p_0)], \qquad m = m_0 + \alpha_1(p - p_0). \tag{8}$$

The coefficient of convective diffusion is ordinarily assumed to depend linearly on the filtration velocity, i.e.,

$$D_i = D_M + \lambda_i v, \tag{9}$$

where D_M is the coefficient of molecular diffusion in the porous medium, and λ_i is the dispersion parameter. In the processes of dissolution (salination) and crystallization the rate of mass exchange $\partial N/\partial t$ is determined for surface and volume dissolution by the respective expressions

$$\partial N/\partial t = -\gamma(c_* - c), \qquad \partial N/\partial t = -\gamma(c_* - c)N^{1/2}, \tag{10}$$

where γ is a constant depending on the physical and mechanical properties of the medium and the liquid.

In the one-dimensional case equations (4), (6), and (7) take the form

$$\frac{\partial A(p)}{\partial t} = \frac{\partial}{\partial x}\left[\rho(p)B\left(c, p, \frac{\partial p}{\partial x}\right)\right], \tag{11}$$

$$\frac{\partial}{\partial x}\left(D\frac{\partial c}{\partial x} - cB\right) - \frac{\partial N}{\partial t} = \frac{\partial(mc)}{\partial t}, \tag{12}$$

$$\frac{\partial N}{\partial t} = f(c, N), \tag{13}$$

where $B = B(c, p, \partial p/\partial x)$ is the function inverse to $\Phi = \Phi(c, p, v)$ in (2). In the case of a homogeneous medium and a linear filtration law we have

$$-\frac{\partial p}{\partial x} = \Phi(c, p, v) \equiv \frac{\mu}{k}(c)v, \qquad B = -\frac{k}{\mu}\frac{\partial p}{\partial x}. \tag{14}$$

It is usually assumed that the initial distributions of pressure p and mass concentrations c and N are given, i.e.,

$$p(0, x) = p_0(x), \qquad c(0, x) = c_0(x), \qquad N(0, x) = N_0(x), \qquad x \in [x_1, x_2],$$

and also that one each of the following conditions is given for the functions p and c when $x = x_i$:

$$p(t, x_i) = p_i(t), \qquad c(t, x_i) = c_i(t), \tag{15}$$

$$\frac{\partial p(t, x_i)}{\partial x} = q_i(t), \qquad D\frac{\partial c}{\partial x} - c\frac{k}{\mu}\frac{\partial p}{\partial x}\bigg|_{x=x_i} = -\mu_i(t). \tag{16}$$

The conditions (15) mean that given values of the pressure and concentration of the liquid phase are maintained at the given end x_i. For example, $x = x_i$ may be the boundaries of the filtration domain with reservoirs filled with a homogeneous fluid. The conditions (16) arise in the case when the filtration speed or the concentration flow of the liquid phase is given on the boundary.

Suppose that the linear filtration law (14) holds, and that

$$f(c, N) = f_0(c)e^{-N}, \qquad D = D(c, p). \tag{17}$$

Then system (11)–(13) admits self-similar solutions of the form

$$p = p(\xi), \qquad c = c(\xi), \qquad N = \ln[(t+1)n(\xi)], \qquad \xi = \frac{x}{\sqrt{1+t}},$$

where $p(\xi)$, $c(\xi)$, and $n(\xi)$ satisfy

$$\frac{1}{2}\xi\frac{dA(p)}{d\xi} = -\frac{d}{d\xi}\left[\rho\frac{k}{\mu}\frac{dp}{d\xi}\right], \tag{11*}$$

$$\frac{d}{d\xi}\left(D\frac{dc}{d\xi} + c\frac{k}{\mu}\frac{dp}{d\xi}\right) + \frac{\xi}{2n}\frac{dn}{d\xi} - 1 = -\frac{\xi}{2}\frac{d(mc)}{d\xi}, \tag{12*}$$

$$1 - \frac{\xi}{2}\frac{dn}{d\xi} = \frac{f_0(c)}{n(\xi)}. \tag{13*}$$

14. Prove that the boundary-value problems formulated above are solvable for system (11)–(13).

15. Prove the solvability of the corresponding boundary-value problems for system (11*)–(13*) and for the analogous system for finding the self-similar solutions of (11)–(13) of traveling wave type, i.e., the solutions of the form

$$c = c(x - Vt), \quad p = p(x - Vt), \quad N = N(x - Vt).$$

16. Study the solvability of system (11)–(13) in the steady case, and investigate the solvability of the problem of Florin type (cf. problem 6).

17. Study the problem of determining the boundary $x = x_0$ where two flows, each described by its own system of equations (11)–(13), are patched together, assuming conditions analogous to those in the problems of Stefan or Verigin type (cf. problem 5) at the patching point, or assuming that the unknown functions and their first derivatives are continuous at x_0.

CHAPTER IX

SOME PLANAR PROBLEMS
WITH AN UNKNOWN BOUNDARY
IN ELASTICITY THEORY

As is well known (see [2] or [97]), in two-dimensional steady-state problems in the linear theory of elasticity the components σ_x, σ_y, and τ_{xy} of the symmetric stress tensor P can be expressed according to the Kolosov-Muskhelishvili formulas in terms of two analytic functions $\Phi(z)$ and $\Psi(z)$:

$$\sigma_x + \sigma_y = 4 \operatorname{Re} \Phi(z), \qquad \tfrac{1}{2}(\sigma_y - \sigma_x) + i\tau_{xy} = \bar{z}\Phi_z' + \Psi(z).$$

In order to determine $\Phi(z)$ and $\Psi(z)$ (perhaps to within arbitrary complex constants) in a simply connected domain $D(\Gamma)$ with a given boundary Γ, it is necessary to specify two relations connecting the boundary values of these functions on Γ. If the boundary Γ is unknown, i.e., along with $\Phi(z)$ and $\Psi(z)$ a conformal mapping $\zeta = \Omega(z)$ of $D(\Gamma)$ onto a canonical domain (for example, onto the unit disk) is also sought, then for the problem to be determined it is now obviously necessary to specify three relations between the boundary values of the functions Φ, Ψ, and Ω on Γ.

Thus, planar static problems in the linear theory of elasticity with an unknown boundary consist in the determination of the contour Γ and the components σ_x, σ_y, and τ_{xy} of the symmetric stress tensor P in $D(\Gamma)$ (as a rule, $D(\Gamma) = D^-$, the domain outside Γ) from boundary conditions of the form

$$G_i\left(\xi, \sigma_x, \sigma_y, \tau_{xy}, \frac{\partial \sigma_x}{\partial x}, \ldots, \frac{\partial \tau_{xy}}{\partial y} \right) = 0, \qquad i = 1, 2, 3, \tag{1}$$

where ξ is a set of geometric parameters associated with the unknown contour Γ. The following parameters can appear in ξ: $x = \operatorname{Re} z$, $y = \operatorname{Im} z$, $\beta = \arg z$, $r = |z|$, the arclength parameter s of Γ, the slope angle $\theta = \arg(dz/ds)$ of the tangent to Γ, the curvature $\kappa = d\theta/ds$, and so on.

Problems in elastico-plasticity provide an important example of problems in elasticity theory with unknown boundary. In these problems the unknown contour Γ is the boundary separating the plastic zones from the elastic domain, and in passing across it the stress tensor P remains continuous or has a specified jump. One of the boundary conditions (1) on Γ in this case is the generalized fluidity condition

$$(\sigma_y - \sigma_x)^2 + 4\tau_{xy}^2 = 4k^2(\varphi), \tag{2}$$

where $\varphi = \frac{1}{4}(\sigma_x + \sigma_y)$ and $k^2(\varphi) \geqslant k_0^2 > 0$. A problem in elastico-plasticity can be regarded as solved if the corresponding problem in the theory of elasticity with an unknown boundary is solved. Indeed, the stress tensor P in the domain of plasticity can then easily be constructed from the solution of the Cauchy problem, which exists at least in a neighborhood of Γ. A number of elastico-plasticity problems have been studied by Galin [35], Cherepanov [24], Annin [2], and others.

§1. Determination of the unknown boundary from a stress tensor given on it

Suppose that the values of the components σ_x, σ_y, and τ_{xy} of the stress tensor P are determined on the unknown boundary Γ from the relations (1) as functions of some set ξ of geometric parameters of Γ. Then, taking the Kolosov-Muskhelishvili formulas [97] into account, we can represent the boundary conditions in the form

$$\text{Re}\, \Phi(z) \equiv \frac{1}{4}(\sigma_x + \sigma_y) = \varphi(\xi), \tag{3}$$

$$f(z) = \bar{z}\Phi_z' + \Psi(z) = \sigma(\xi) + i\tau(\xi), \tag{4}$$

where $\sigma \equiv \frac{1}{2}(\sigma_y - \sigma_x)$, $\tau = \tau_{xy}$, $\Phi(z)$ and $\Psi(z)$ are unknown analytic functions in $D(\Gamma) = D^-$, and $\varphi(\xi)$, $\sigma(\xi)$, $\tau(\xi) \in C^{2+\beta}$ are given functions on Γ. In this case the formation of the boundary conditions (3) and (4) in terms of functions of the parameter ξ can be realized, for example, by the preliminary solution of a one-dimensional plasticity problem with the unknown variable ξ.

In this section we consider several different formulations of the problems, depending on the properties of the functions appearing in the boundary conditions (3) and (4). From the mathematical point of view the most typical formulation for elastico-plasticity problems, as problems on patching together elasticity and plasticity solutions, is that suggested in the next section. However, problems A and B of the present section, which we present first, are considerably closer to the elastico-plasticity problems already studied by other authors. The assignment in problems A and B of the components of the stress tensor as functions of a single geometric parameter of the unknown contour Γ

leads to the appearance of constant solutions for the problem, which make it difficult to prove that A and B are solvable in a precise formulation even when the parameter is small.

In the next problem of this section (problem C) we obtain an existence theorem without assuming that any of the given quantities are small. The existence of solutions of problem D is proved by reducing it to a boundary-value problem with free boundary on a Riemann surface which is solvable for small values of a certain parameter.

A. Suppose that $\xi \in [\xi_0, \xi_1]$ in (3) and (4) coincides with one of the geometric parameters of the unknown finite contour Γ; for definiteness let $\xi = x$. The strip $x_0 \leq x \leq x_1$, with each side tangent to Γ at one or several points, is assumed to be unknown, and x_0 and x_1 are sought along with the solution of the problem. Suppose that one of the functions $\sigma = \sigma(x)$ or $\tau = \tau(x)$ has an inverse function, for example,

$$x = h(\tau), \qquad \tau \in [a, b], \tag{5}$$

where $h(\tau) \in C^{2+\beta}$, $\beta > 0$, and a and b are unknown constants for which $M_0 \leq a < b \leq M_1$.

We look for an unknown contour Γ close to a fixed boundary Γ_0, representing the corresponding functional dependence $x = h(\tau)$ in the form

$$x = x_0(t) + \delta h_0(\tau(t)), \qquad t = e^{i\gamma}, \tag{5*}$$

where $x_0(t) = \operatorname{Re} \omega_0(t) \in C^{1+\beta}$, and $z = \omega_0(\zeta)$ is a conformal mapping of the unit disk $|\zeta| < 1$ onto the domain $D_0^- = D_0(\Gamma_0)$ outside Γ_0. Then a conformal mapping $z = \omega(\zeta)$ of the unit disk onto the domain $D^- = D(\Gamma)$ outside the unknown boundary Γ and its derivative $d\omega/d\zeta$ can be computed by the formulas

$$\omega(\zeta) = \frac{1}{2\pi i} \int_{|t|=1} \{x_0(t) + \delta h_0[\tau(t)]\} \frac{(t+\zeta)\,dt}{t(t-\zeta)} + R(\zeta) \equiv H(\tau|\zeta), \tag{6}$$

$$\omega'(\zeta) = \frac{1}{\pi i} \int_{|t|=1} \frac{d}{dt}[x_0(t) + \delta h_0(\tau)] \frac{dt}{t-\zeta} + R'(\zeta) \equiv H'(\tau|\zeta), \tag{7}$$

where $R(\zeta) = c_1/\zeta - \bar{c}_1\zeta + ic_0$, the constants $c_1 = c_1' + ic_1''$, and c_0 are fixed arbitrarily for the time being, $\tau = \tau(t) \in C^{1+\beta}$ $(0 < \beta < 1)$ is an unknown function, and the notation $\tau|\zeta$ reflects the fact that the operators H and H' acting on τ are functions of ζ, $|\zeta| \leq 1$. Differentiating both sides of (4) with respect to $\bar{z}$ in $D^- = D(\Gamma)$, we find that

$$\partial f/\partial \bar{z} = \Phi_z', \qquad z \in D^-,$$

which, after transformation to the unit disk, gives us

$$\frac{\partial f}{\partial \bar{\zeta}} = \frac{\overline{\omega'}}{\omega'} \Phi'_\zeta \equiv \Omega_\delta(\tau \mid \zeta), \qquad |\zeta| \leqslant 1, \tag{8}$$

where $\overline{\omega'}/\omega' = \exp\{-2i \arg H'(\tau \mid \zeta)\}$. The analytic function $\Phi(\zeta)$ can be determined by the Schwarz formula from its real part, which is given by (3), and so

$$\Phi'(\zeta) = \frac{1}{\pi i} \int_{|t|=1} \frac{d}{dt} \varphi[x_0(t) + \delta h_0(\tau)] \frac{dt}{t - \zeta}.$$

We represent a solution of (8) satisfying the condition

$$\operatorname{Re} f(t) = \sigma[x_0(t) + \delta h_0(\tau)]$$

on the circle $|t| = 1$ in the form

$$f(\zeta) = T_0(\Omega_\delta \mid \zeta) + Q_\delta(\tau \mid \zeta) + i\tau_0, \tag{9}$$

where

$$Q_\delta(\tau \mid \zeta) = \frac{1}{2\pi i} \int_{|t|=1} \sigma[x_0(t) + \delta h_0(\tau)] \frac{(t + \zeta)\,dt}{t(t - \zeta)}, \qquad \tau_0 = \text{const},$$

$$T_0(\Omega_\delta \mid \zeta) = -\frac{1}{\pi} \iint_{|z|<1} \left\{ \frac{\Omega_\delta}{z - \zeta} - \frac{\overline{\Omega}_\delta \zeta}{\bar{z}\zeta - 1} \right\} dx\,dy,$$

and $\operatorname{Re} T_0(\Omega_\delta \mid e^{i\gamma}) = 0$ by construction (see Chapter V, §4). Then, taking the boundary condition (4) into account, by which $\operatorname{Im} f(e^{i\gamma}) = \tau$, we get the following equation for determining the unknown dependence $\tau = \tau(t)$, $t = e^{i\gamma}$:

$$\tau(t) = -iT_0(\Omega_\delta \mid t) - iQ_\delta(\tau \mid t) + \tau_0 \equiv \Lambda_\delta(\tau \mid t). \tag{10}$$

If $\tau(t) \in C^{1+\beta}$, $\beta > 0$, then $\sigma(x_0 + \delta h_0)$, $\varphi(x_0 + \delta h_0) \in C^{1+\beta}$, and, consequently, in the disk $K = \{|\zeta| < 1\}$ we have

$$\Omega_\delta(\tau \mid \zeta) \in C^\beta(\bar{K}), \qquad Q_\delta(\tau \mid \zeta) \in C^{1+\beta}(\bar{K}),$$

which, by a property of T_0, gives us that $T_0(\Omega \mid \zeta) \in C^{1+\beta}(\bar{K})$.

Thus, $\Lambda_\delta(\tau \mid t)$ maps $C^{1+\beta}$ into itself and, by the assumed smoothness of the given functions $\varphi(x)$, $\sigma(x)$, $h_0(\tau) \in C^{1+\beta}$, $\beta > 0$, it satisfies in each fixed ball S_M ($\|\tau(t)\|_{C^{1+\beta}} \leqslant M$) the Lipschitz condition

$$\|\Lambda_\delta(\tau^1 \mid t) - \Lambda_\delta(\tau^2 \mid t)\|_{C^{1+\beta}} \leqslant \delta N(M, \delta) \|\tau^1 - \tau^2\|_{C^{1+\beta}}, \tag{11}$$

where the constant $N = N(M, \delta)$ is bounded for finite M and δ. Since for $\delta = 0$ the right-hand side $\Lambda_0(\tau \mid t)$ of (10) does not depend on τ, it determines a known function $\tau = \Lambda_0(t) \in C^{1+\beta}$, $\beta > 0$, corresponding to the elastic state of a plane with a hole Γ_0 of give shape. For arbitrarily fixed values $\tau_0 = \text{const}$

and $\varepsilon > 0$ we consider the ball $S_M \subset C^{1+\beta}$:

$$\|\tau(t)\|_{C^{1+\beta}} \leqslant M \equiv \|\Lambda_0(t)\|_{C^{1+\beta}} + \varepsilon. \tag{12}$$

Let $\delta_0 > 0$ be small enough that

$$\|\Lambda_{\delta_0}(\tau \mid t) - \Lambda_0(t)\|_{C^{1+\beta}} < \varepsilon.$$

Then for all $\delta \in [0, \delta_0]$ the operator $\Lambda_\delta(\tau \mid t)$ obviously maps the chosen ball $S_M \subset C^{1+\beta}$ into itself. On the other hand, for values of the parameter δ satisfying the condition

$$0 < \delta \sup_{\delta \in [0,\delta_0]} N(M, \delta) < 1$$

Λ_δ is, by (11), a contraction in S_M; consequently, the Banach principle gives us that (10) has a unique solution $\tau = \tau(t)$ in $S_M \subset C^{1+\beta}$, $\beta > 0$. By (6), this $\tau(t) \in C^{1+\beta}$ determines the mapping function $z = \omega(\zeta)$ and its inverse $\zeta = \omega^{-1}(z)$, and with them the contour Γ. Then

$$\frac{1}{4}(\sigma_x + \sigma_y) = \frac{1}{2\pi} \operatorname{Re}\left\{\frac{1}{i} \int_{|t|=1} \frac{\varphi[\operatorname{Re} \omega(t)][t + \omega^{-1}(z)]\, dt}{t[t - \omega^{-1}(z)]}\right\} + \varphi_0,$$

$$f = T_0\big(\Omega_\delta \mid \omega^{-1}(z)\big) + Q_\delta\big(\tau \mid \omega^{-1}(z)\big) + i\tau_0,$$

from which the components of the stress tensor can be determined. When ξ coincides with other geometric parameters of Γ, the difference from the case $\xi = x$ amounts only to changing the formulas (6) and (7) that determine the mapping function $z = \omega(\zeta)$ from the known correspondence

$$\xi = \xi_0(t) + \delta h[\tau(t)].$$

The arbitrary constant c_0 in (6) determines only a translation of the unknown contour Γ along the OX-axis and can be chosen together with the constant τ_0 in (10) from the following normalization condition for the mapping $z = \omega(\zeta)$:

$$\omega(t_0) = x_0 + iy_0 \equiv z_0,$$

where $t_0 = e^{i\gamma_0}$ and z_0 are fixed constants.

Thus, in the case $\xi = x$ the unknown solution, i.e., the shape of the contour Γ and the stress tensor P, depends essentially on the single arbitrary complex constant c_1 (see (6) and (7)), which can be determined, for example, from the conditions $\sigma_x = \sigma_y$ and $\tau_{xy} = 0$ as $z \to \infty$.

B. Let us now consider the case where the inverse function in (4) exists for one of the functional dependences $k = k(\xi)$ or $\alpha = \alpha(\xi)$, where

$$f = \sigma(\xi) + i\tau(\xi) \equiv k(\xi)e^{i\alpha(\xi)}. \tag{4}$$

Suppose for definiteness that $\xi = x$ and, similarly to the preceding problem,

$$x = x_0(t) + \delta h_0[k(t)], \qquad t = e^{i\gamma}, \tag{13}$$

where $x_0(t) = \mathrm{Re}\,\omega_0(t) \in C^{1+\beta}$, $\beta > 0$, and $z = \omega_0(\zeta)$ is a conformal mapping of the unit disk $|\zeta| < 1$ onto the domain $D^- = D_0(\Gamma_0)$ outside a given contour Γ_0. Let

$$\Omega_\delta(k \mid \zeta) = \frac{1}{\pi i} e^{-2\arg H'(k \mid \zeta)} \int_{|t|=1} \frac{d\varphi(x_0 + \delta h_0)}{dt} \frac{dt}{t - \zeta}. \tag{14}$$

Here $H'(k \mid \zeta)$ is defined by (7), with $k(t)$ playing the role of $\tau(t)$. We form the corresponding equation (8) and look for its solution in the form

$$f(\zeta) = R_\delta(k \mid \zeta)\big[T_1(\Omega_\delta R_\delta^{-1} \mid \zeta) + c\big], \tag{15}$$

where

$$T_1\sigma = -\frac{1}{\pi} \iint_{|z| \leq 1} \left\{ \frac{\sigma(z)}{z - \zeta} + \frac{\overline{\sigma(z)}\,\zeta}{\bar{z}\zeta - 1} \right\} dx\, dy \qquad \left(\mathrm{Im}\, T_1(\sigma/e^{i\gamma})\right) = 0,$$

c is an arbitrarily fixed real constant, and

$$R_\delta(k \mid \zeta) = \exp\left\{ \frac{1}{2\pi i} \int_{|t|=1} \frac{i\alpha[x_0(t) + \delta h_0(k)](t + \zeta)\, dt}{t(t - \zeta)} \right\}.$$

If $[c + T_1(\Omega_\delta R_\delta^{-1} \mid e^{i\gamma})] > 0$, then, by construction,

$$\arg f(t) = \alpha[x_0(t) + \delta h_0(k)],$$

and for determining the unknown function $k = k(t) \in C^{1+\beta}$, $t = e^{i\gamma}$, we come to the equation

$$k(t) = \big| R_\delta(k \mid t)\big[T_1(\Omega_\delta R_\delta^{-1} \mid t)\big] \big| \equiv \Lambda_\delta(k \mid t). \tag{16}$$

As in the preceding problem, the operator Λ_δ maps $C^{1+\beta}$ into itself. For $\delta = 0$ the right-hand side of (16) does not depend on k, and it determines the known function $k = \Lambda_0(t) \in C^{1+\beta}$, $\beta > 0$. We assume that $\Lambda_0(t) \geq k_0 > 0$ ($T_1(\Omega_0\Lambda_0 \mid 1) > 0$) and consider in $C^{1+\beta}$ the ball S_M

$$\|k(t)\|_{C^{1+\beta}} \leq M = \|\Lambda_0(t)\|_{C^{1+\beta}} + \varepsilon, \tag{17}$$

where $\varepsilon > 0$ is a fixed number. Since

$$\Lambda_\delta(k \mid t) \geq k_0 - \varepsilon_0(M \mid \delta) > 0$$

on S_M for sufficiently small $\delta > 0$, and since the operator $\Lambda_\delta(k \mid t)$ (as in problem A) is a contraction of S_M into itself, the equation (16) has a unique solution in S_M by the Banach principle.

The problem of the arbitrary constants C_0, C_1, and C is solved completely analogously to the preceding problem.

Consider now the case where instead of (13) we have

$$x = x_0(t) + \delta h_0[\alpha(t)], \qquad t = e^{i\gamma}. \tag{13*}$$

We look for a solution $f = f(\zeta)$ ($\neq 0$ for $|\zeta| \leqslant 1$) of the corresponding equation (8) with the coefficient $\Omega_\delta = \Omega_\delta(\alpha \,|\, \zeta)$.

Let us introduce a new unknown function $F(\zeta) = 1/f(\zeta)$ satisfying the equation

$$\partial F/\partial \bar{\zeta} = -[\Omega_\delta(\alpha \,|\, \zeta)F] F. \tag{18}$$

Considering the solution of the last equation as a generalized analytic function, we represent it in the form

$$F(\zeta) = \exp\{-T_1(\Omega_\delta F \,|\, \zeta) - Q_\delta(\alpha \,|\, \zeta)\} \equiv \Lambda_1(F, \alpha), \tag{19}$$

where

$$Q_\delta(\alpha \,|\, \zeta) = \frac{1}{2\pi i} \int_{|t|=1} k[x_0(t) + \delta h_0(\alpha)] \frac{(t + \zeta)\, dt}{t(t - \zeta)} + i\alpha_0, \; \alpha_0 = \text{const.}$$

By construction,

$$|F(t)| = \left| \frac{1}{f(t)} \right| = k^{-1}(x_0 + \delta h_0).$$

Equating the argument of the right-hand side of (19) to $-\alpha(t)$, we get

$$\alpha(t) = T_1(\Omega_\delta F \,|\, t) + Q_\delta(\alpha \,|\, t) \equiv \Lambda_2(F, \alpha). \tag{20}$$

If for $\delta = 0$ the corresponding function $f = f_0(\zeta)$ is nonzero for all ζ, $|\zeta| \leqslant 1$, then this holds also for the unknown function $f(\zeta)$ when $\delta > 0$ is sufficiently small. Then, applying the Banach principle as before to the transformation

$$\{F, \alpha\} = \{\Lambda_1(F, \alpha), \Lambda_2(F, \alpha)\},$$

we get the existence of a unique solution of the system (19), (20).

C. Unlike the preceding problems, the problem stated below is solvable without assuming that any of the given quantities are small. Suppose that on the unknown boundary Γ the components σ_x and σ_y of the stress tensor are given functions of the parameter $x = \text{Re}\, z$, while the component $\tau = \tau_{xy}$ depends on the slope angle θ of Γ.

Then the boundary conditions (3) and (4) can be written in the form

$$\tfrac{1}{4}(\sigma_x + \sigma_y) = \text{Re}\,\Phi(z) = \varphi(x), \qquad f = \bar{z}\Phi_z' + \Psi(z) = \sigma(x) + i\tau(\cos\theta), \tag{21}$$

where $\varphi(\xi)$, $\sigma(\xi)$, $\tau(\xi) \in C^{1+\beta}$, $\beta > 0$. Suppose that the function $\tau = \tau(\cos\theta)$ has an inverse:

$$\cos\theta = h(\tau), \qquad \tau \in [a, b]. \tag{22}$$

If the correspondence $\tau = \tau(t) \in C^{1+\beta}$, $\beta > 0$, is known, then (22) gives us also $\cos \theta(t) = h[\tau(t)]$.

To determine the dependence $\theta(t)$ from the known $\cos \theta(t)$ we must impose some conditions on the unknown contour Γ.

It is required that Γ contain two cusps with interior angles 2π, that their images be located at the points $t = \pm 1$ on the circle, and that $\theta = \theta(e^{i\gamma})$ increase as $\gamma > 0$ increases in a neighborhood of $\gamma = 0$. These assumptions enable us to find $\theta = \theta(t)$ by the formula

$$\theta = \theta[\tau(t)] = \arg\left\{h[\tau(t)] + i\sqrt{1 - h^2[\tau(t)]}\right\}.$$

The condition $|\cos \theta| \geq \mu_0 > 0$ on Γ gives us that $|\theta(t)| \leq \pi/2m$ ($m > 1$). Moreover, if $\tau(t) \in C^\beta$, $\beta > 0$, then $\theta[\tau(t)] \in C^\beta$ on the intervals $[0, \pi]$ and $[\pi, 2\pi]$ of variation of $\gamma = \arg t$, and

$$\theta[\tau(\pm 1 + 0)] - \theta[\tau(\pm 1 - 0)] = \pi.$$

Then an arbitrary analytic function $z = \omega(\zeta)$ conformally mapping $|\zeta| < 1$ onto D^- has the form

$$\omega'(\zeta) = \frac{c_0(\zeta^2 - 1)}{\zeta^2} \exp\left\{\frac{i}{2\pi} \int_0^\pi \frac{\theta^*(\tau \mid e^{i\gamma})(e^{i\gamma} + \zeta)\, d\gamma}{e^{i\gamma} - \zeta}\right\} \equiv H'(\tau \mid \zeta),$$

$$(22')*$$

from which the dependence $x(t)$ can be determined:

$$x = \operatorname{Re} \int_0^t H'(\tau \mid t)\, dt + x_0 \equiv x(\tau \mid t), \tag{23}$$

where c_0 and x_0 are arbitrarily fixed real constants, and $\theta^*(\tau \mid e^{i\gamma}) = \theta(\tau \mid e^{i\gamma}) + \gamma - \arg(e^{2\gamma} \pm 1)$. The unknown function $\tau = \tau(t)$ satisfies equation (10) of problem A:

$$\tau(t) = -iT_0(\Omega \mid t) - iQ(\tau \mid t) + \tau_0 \equiv \Lambda(\tau \mid t). \tag{24}$$

Since, by a condition of the problem,

$$|\theta[\tau(t)]| \leq \pi/2m < \pi/2,$$

Zygmund's theorem (Theorem 7 in §2 of Chapter II) gives us from (22') that

$$\|H'(\tau \mid t)\|_{L_p} \leq M(m) < \infty, \qquad 1 < p < m.$$

Then

$$|x(\tau \mid t_1) - x(\tau \mid t_2)| = \left|\operatorname{Re} \int_{t_1}^{t_2} H'(\tau \mid t)\, dt\right| \leq M \mid \gamma_2 - \gamma_1 \mid^{(p-1)/p}$$

---

* *Editor's note.* To distinguish between the two different formulas assigned the same number (22) in the original Russian text, the second is labeled (22') in this translation.

and, consequently, for $0 < \beta = (p - 1)/p$,

$$\|x(\tau \mid t)\|_{C^\beta} \leqslant M_1, \qquad \|Q(\tau \mid \zeta)\|_{C^\beta} \leqslant N_0 \|x(\tau \mid t)\|_{C^\beta} \leqslant M_2.$$

On the other hand, for $|\zeta| < 1$

$$|\Omega(\tau \mid \zeta)|^{p_0} = \frac{1}{\pi} \left| \int_{|t|=1} \frac{\varphi_x'(dx/dt)\,dt}{t - \zeta} \right|^{p_0}$$

$$\leqslant N_1^{p_0} \left\{ \int_0^{2\pi} \frac{|dx/dt|^p \, d\gamma}{|t - \zeta|^{(1-\varepsilon)p}} \right\}^{p_0/p} \left\{ \int_0^{2\pi} \frac{d\gamma}{|t - \gamma|^{\varepsilon q}} \right\}^{p_0/q}.$$

Choosing ε and p_0 such that

$$0 < \varepsilon < \frac{p-1}{p} = \frac{1}{q}, \qquad 2 < p_0 < \frac{2}{1-\varepsilon},$$

we get

$$\left\{ \int_0^{2\pi} \frac{d\gamma}{|t - \zeta|^{\varepsilon q}} \right\}^{p_0/q} \leqslant N_2(\varepsilon),$$

and so

$$\iint_{|\zeta|<1} |\Omega(\tau \mid \zeta)|^{p_0} \, dK_\zeta \leqslant (N_1 N_2)^{p_0} \left\| \frac{dx}{dt} \right\|_{L_p}^{p_0} \iint_{|\zeta|<1} \frac{dK_\zeta}{|t_* - \zeta|^{(1-\varepsilon)p_0}} \leqslant N_3(\varepsilon).$$

Thus, we arrive at the following uniform estimate for the operator $\Lambda(\tau \mid t)$:

$$\|\Lambda(\tau \mid t)\|_{C^{\beta_0}} \leqslant M_\Omega < \infty, \qquad 0 < \beta_0 = \min\left(\frac{p-1}{p}, \frac{p_0 - 2}{p_0} \right),$$

from which we get, in particular, that $\Lambda(\tau \mid t)$ is completely continuous in $L_p(0, 2\pi)$, $p > 1$. An application of Schauder's theorem concludes the proof that (24) is solvable in L_p, $p > 1$, and $\tau(t) \in C^{\beta_0}$, $\beta_0 < 0$, by a property of $\Lambda(\tau \mid t)$. The solution we have constructed depends on the arbitrary constants x_0, τ_0, and c_0.

As in problem A, the constants x_0 and τ_0 can be chosen from the normalization conditions for the mapping $z = \omega(\zeta)$. In the same way we can prove solvability of elastico-plasticity problems with boundary conditions depending on two different geometric parameters of Γ, to which different differential properties of Γ correspond. For example, one of the parameters $y = \operatorname{Im} z$, $r = |z|$, or $\beta = \arg z$ can be taken along with the parameter $\theta = \arg(dz/ds)$ instead of $x = \operatorname{Re} z$, the two parameters $\kappa = d\theta/ds$ and $\theta = \arg(dz/ds)$ can be considered at the same time, and so on.

D. In the case of static problems in the theory of elasticity it is natural to associate impact loads not so much with the magnitudes of the stress tensor components as with their gradients. We consider an example of such a model problem in which the boundary conditions on a known contour Γ contain the derivatives of the stress tensor components.

Suppose that in the domain D^- outside the unknown finite contour Γ we are looking for the symmetric stress tensor P with components σ_x, σ_y, τ_{xy} satisfying on Γ the boundary conditions

$$\frac{\partial \varphi}{\partial x} = u(\xi), \qquad \frac{\partial \varphi}{\partial y} = v(\xi), \qquad |f|^2 \equiv \frac{1}{4}(\sigma_y - \sigma_x)^2 + \tau_{xy}^2 = k^2(\varphi), \qquad (25)$$

where $u(\xi)$, $v(\xi)$, $k(\xi) \in C^{1+\beta}$, $\varphi = \frac{1}{4}(\sigma_x + \sigma_y)$, $k(\varphi) \geqslant k_0 > 0$, and $\xi \in [\xi_1, \xi_2]$ is one of the geometric parameters, taken to be $|z| = r \in [r_1, r_2]$ for definiteness.

We assume that $0 < r_1 < r_2 < \infty$ and that the origin is outside $D^- = D(\Gamma)$. For $\xi = r$ the contour is broken up into two arcs Γ^1 and Γ^2, on each of which the value of r changes monotonically. Corresponding to this, $u(r)$ and $v(r)$ are given two-valued functions of the argument r, i.e.,

$$w^j \equiv u - iv = u^j(r) - iv^j(r) \quad \text{on } \Gamma^j \ (j = 1, 2), r \in [r_1, r_2].$$

Suppose that $|dw^j/dr| \neq 0$ and $w^1(r_i) = w^2(r_i)$, $i = 1, 2$, with $w^j(r) \in C^{1+\beta}$ by assumption. We explore what additional requirements must be imposed on $w(r)$. Since $\varphi(z) = \operatorname{Re} \Phi(z) = \frac{1}{4}(\sigma_x + \sigma_y)$, by the Kolosov-Muskhelishvili formulas, where $\Phi(z)$ is an analytic function in D^-, it follows that $w(r)$ is the limit value of the analytic function $\partial \Phi/\partial z = \partial \varphi/\partial x - i\partial \varphi/\partial y$.

Let $z = \omega(\zeta)$ be an analytic function mapping a finite domain, say the disk $|\zeta| < 1$, onto D^-. Then $d\Phi/dz = \Phi'_\zeta/\omega'_\zeta$, and, as before, Φ'_ζ has a zero of order at least two at $\zeta = 0$, and w'_ζ has a pole of order two at this point. Thus, $d\Phi/dz$ must have a zero of order four at $z = \infty$. Consequently, corresponding to the boundary values $d\Phi/dz|_\Gamma = w(r)$ there must be a contour Γ_w in the w-plane that describes four loops in the positive direction around the origin as Γ is traversed in the negative direction, i.e., $\operatorname{ind} w(r)|\Gamma = -4$. Thereby Γ must break up into four adjoining arcs with one and the same contour Γ_w in the w-plane corresponding to each of them, i.e., the values $w = w(r)$ coincide on these arcs. Thus, it is assumed that to the given function $w = w(r)$ there corresponds on the plane a simple closed contour Γ_w traversed four times in the positive direction for one negative circuit, and the four-sheeted finite domain D_w contains the origin in its interior.

We map the unit disk $|\zeta| < 1$ onto the four-sheeted domain D_w with branch point $w = 0$ by the analytic function

$$w = [\zeta\Omega(\zeta)]^4 \equiv (d\Phi/dz)(\zeta), \qquad (26)$$

where $\Omega(\zeta) \in C^{1+\beta}$ and $\Omega(\zeta) \neq 0$ for $|\zeta| \leqslant 1$. The identity $u(r) - iv(r) \equiv [t\Omega(t)]^4$ gives us a dependence $r = \exp\{R(t)\}$, with the help of which we can reconstruct the analytic function $z = \omega(\zeta)$ mapping $|\zeta| < 1$ onto D^-:

$$\omega(\zeta) = \frac{1}{\zeta} \exp\left\{ \frac{1}{2\pi i} \int_{|t|=1} \frac{h(t)(t+\zeta)\, dt}{t(t-\zeta)} + ic_0 \right\}, \tag{27}$$

where c_0 is an arbitrarily fixed real constant.

Thus, we have found the unknown contour Γ and the functions

$$\frac{d\Phi}{dz} = [\zeta\Phi(\zeta)]^4, \qquad \varphi(\zeta) = \mathrm{Re}\left\{ \int \frac{d\Phi}{dz}(\zeta)\omega'(\zeta)\, d\zeta \right\} \equiv \frac{1}{4}(\sigma_x + \sigma_y).$$

On the other hand, the equality

$$f \equiv \frac{\sigma_y - \sigma_x}{2} + i\tau_{xy} \equiv \bar{z}\Phi'_z + \Psi(z)$$

gives us that $\partial f/\partial \bar{z} = \Phi'_z$, and so

$$\partial f/\partial \bar{\zeta} = A(\zeta), \tag{28}$$

where $A(\zeta) = \{\omega'(\zeta)[\zeta\Omega(\zeta)]^4\} \in C^{1+\beta}$ is a known function. From the last of the relations (25) we find that on $|\zeta| = 1$

$$|f(t)| = k[\varphi(t)] \equiv g_0(t), \qquad t = e^{i\gamma}, \tag{29}$$

where $g_0(t) \in C^{1+\beta}$, $\beta > 0$, and $g_0(t) \geqslant k_0 > 0$.

Thus, we have come to the boundary-value problem (28), (29) for determining the function $f(\zeta)$ and with it the components of the unknown stress tensor P.

In order to prove that the latter problem is solvable, we impose the following condition on the coefficients in (25):

$$\{u^2(r) + v^2(r)\}^{1/2} \leqslant k_0/2r_2 - \varepsilon, \qquad r \in [r_1, r_2], \tag{30}$$

where $k_0 = \min k(\varphi) > 0$ and $\varepsilon > 0$ is an arbitrarily small fixed number. The inequality (30) ensures that the solution $f(\zeta)$ does not have zeros on $|\zeta| = 1$. Indeed, on $|t| = 1$

$$0 < k_0 \leqslant |\bar{\omega}(t)\Phi'_z(t) + \Psi(t)| = |f| \leqslant r_2(u^2 + v^2)^{1/2} + |\Psi(t)|,$$

and so, by (30) and the maximum modulus principle for analytic functions,

$$k_0/2 + \varepsilon r_2 \leqslant |\Psi(\zeta)|, \qquad |\omega(\zeta)\Phi'_z(\zeta)| \leqslant k_0/2 - \varepsilon.$$

Consequently, for all ζ, $|\zeta| < 1$, we have

$$|f(\zeta)| \geqslant |\Psi(\zeta)| - |\omega\Phi'_z| \geqslant \varepsilon(r_2 + 1). \tag{31}$$

The function $F = 1/f$, by virtue of the equation

$$\partial F / \partial \bar{\zeta} = - (A(\zeta)F)F,$$

can be regarded as a generalized analytic function and, consequently, can be represented in the form

$$F(\zeta) = \exp\{ -T_0(AF|\zeta) - Q(\zeta)\} \equiv \Lambda(F|\zeta),$$ (32)

where

$$Q(\zeta) = \frac{1}{2\pi i} \int_{|t|=1} \frac{\ln g_0(t)(t + \zeta) \, dt}{t(t - \zeta)} + i\alpha_0, \quad \alpha_0 = \text{const},$$

and, by construction, $|F(t)| = 1/|f| = [g_0(t)]^{-1}$. If $F \in L_p(\overline{K})$ ($p > 2$), then $\Lambda(F|\zeta) \in C^\beta$, $\beta = (p - 2)/p$; therefore, the continuous operator Λ is compact in L_p.

We include $\Lambda(F|\zeta)$ in a one-parameter family $\{\Lambda_\delta(F|\zeta)\}$ of transformations uniformly continuous in $\delta \in [0, 1]$ by introducing a parameter δ in front of the function $A(\zeta)$ in (32). For sufficiently small $\delta > 0$ the equation $F = \Lambda_\delta(F)$ is then uniquely solvable by the Banach principle (cf. the solvability of (19)).

On the other hand, by (31), the a priori estimates

$$|\Lambda_\delta(F|\zeta)| < \exp\{ M_1 \varepsilon^{-1} + \max|Q|\} = M$$

hold for the solutions of (32) with a constant M not depending on F or δ.

Thus, all the conditions of the Leray-Schauder theorem hold for the transformation $F = \Lambda_\delta(F)$, and, consequently, the equation (32) corresponding to the value $\delta = 1$ has at least one solution. The properties of $\Lambda(F|\zeta)$ imply that each solution of (32) in the class $L_p(K)$ ($p > 2$) is in the space $C^{1+\beta}$.

§2. Determination of the unknown boundary from a stress vector given on it

In the investigation of many practically important elastico-plasticity problems it is natural to specify the stress vector $\vec{P}(p_x, p_y)$ on the unknown elastico-plastic boundary Γ, in addition to the plasticity condition of the components. Such problems include, for example, the problem of punching out a hole of specified shape in the plane. In this problem the stress vector $\vec{P}$ on the boundary L of the hole is normal to it, and it is natural to assume that its direction does not change when a load is distributed in the plastic domain. If the law of variation of $|\vec{P}| = (p_x^2 + p_y^2)^{1/2}$ as we go away from the hole L is found theoretically (by solving an elementary plasticity problem) or experimentally, then the stress vector $\vec{P}$ is thereby determined on the elastico-plastic boundary as a function of the distance to L. Some of the problems considered in this section can be reduced to the problems studied in §1 with boundary conditions depending, as a rule, on several geometric parameters of Γ. However, the special form of the boundary conditions enables us to apply other methods in investigating these problems, allowing us to prove their solvability under less stringent restrictions on the boundary data. Suppose that the following conditions hold on the boundary Γ of the elastic and the plastic domains:

$$p_n = p_1(\xi), \qquad p_s = p_2(\xi), \qquad (\sigma_y - \sigma_x)^2 + 4\tau_{xy}^2 = 4k^2(\varphi),$$

where $\varphi = \frac{1}{4}(\sigma_x + \sigma_y)$, p_n and p_s are the projections of the stress vector $\vec{P}$ on the normal and the tangent to Γ, respectively, σ_x, σ_y, and τ_{xy} are the components of the symmetric stress tensor P, and ξ is a set of geometric parameters of Γ which may contain $x = \mathrm{Re}\, z$, $y = \mathrm{Im}\, z$, $\beta = \arg z$, $r = |z|$, and the arclength parameter s of Γ, i.e., ξ is, generally speaking, a vector with components (x, y, β, r, s). It is assumed that $P_i \in C^{2+\beta}$ ($\beta > 0$, $i = 1, 2$) for finite values of the parameters in ξ, while $k(\varphi) \in C^{2+\beta}$ in a sufficiently small neighborhood of the fixed value φ_0, and

$$p_1^2(\xi) + p_2^2(\xi) \leqslant k_0^2 = \min k^2(\varphi) > 0. \tag{2}$$

We remark that the second of the equalities in (1) is the von Mises plasticity condition, when $k(\varphi) = k = \mathrm{const}$. If it is noted that p_x and p_y can be expressed in terms of σ_x, σ_y, and τ_{xy} by the formulas

$$p_x = -\sigma_x \sin\theta + \tau_{xy}\cos\theta, \qquad p_y = \tau_{xy}\sin\theta + \sigma_y\cos\theta, \qquad \theta = \arg\frac{dz}{ds}$$

and, on the other hand, that

$$p_n = -p_x\sin\theta + p_y\cos\theta, \qquad p_s = p_x\cos\theta + p_y\sin\theta,$$

then this gives

$$p_n = \sigma_x\sin^2\theta + \sigma_y\cos^2\theta - \tau_{xy}\sin 2\theta \equiv p_1(\xi),$$

$$p_s = \frac{\sigma_y - \sigma_x}{2}\sin 2\theta + \tau_{xy}\cos 2\theta \equiv p_2(\xi). \tag{3}$$

Putting the fluidity condition (the last of the conditions (1)) together with (3) and solving the resulting system for σ_x, σ_y, and τ_{xy}, we find that

$$\varphi \equiv \frac{1}{4}(\sigma_x + \sigma_y) = \frac{p_n(\xi)}{2} \pm \frac{k(\varphi)}{4}\cos\alpha(\xi),$$

$$f \equiv \frac{\sigma_y - \sigma_x}{2} + i\tau_{xy} = k(\varphi)e^{i[\alpha(\xi) - 2\theta]}, \tag{4}$$

where $\alpha = \arcsin(p_s/k)$ ($|p_s| \leqslant k$ by condition (2)), and, by construction, the equalities (7) are equivalent to the original boundary conditions (1). As (4) shows, if the components p_n and p_s of the stress vector are given as functions of $\theta = \arg(dz/ds)$, then the components σ_x, σ_y, and τ_{xy} of the stress tensor also depend only on $\theta \equiv \xi$. As in problems A and B of the preceding section, we assume that the function $\varphi(\xi) \equiv \frac{1}{4}(\sigma_x + \sigma_y)$ is sufficiently close to a constant on Γ, and, consequently, the $k(\varphi)$ in the fluidity condition is also close to a constant k_0. Let

$$k(\varphi) = k_0[1 + \delta_0 h(\varphi)], \qquad h(\varphi) \in C^{2+\beta}, \delta_0 > 0.$$

Suppose that the given functions $p_n = p_1(\xi)$ and $p_s = p_2(\xi)$ are such that the function $\varphi(\xi)$ found from the first formula in (4) can be represented in the form

$$\varphi \equiv \tfrac{1}{4}(\sigma_x + \sigma_y) = \varphi_0 + \delta_0 k_0 g(\xi),$$

where $g(\xi) \in C^{2+\beta}$.

Let us consider a function $z = \omega(\zeta)$ mapping the domain $|\zeta| < 1$ conformally onto the domain D^- outside Γ. Obviously, on the circle $|t| = 1$

$$e^{-2i\theta(t)} = \frac{\overline{\omega'(t)}}{t^2 \omega'}, \qquad t = e^{i\gamma};$$

therefore, with the help of the Kolosov-Muskhelishvili formulas we can give (4) the form

$$\varphi \equiv \operatorname{Re} \Phi(t) = \varphi_0 + \delta_0 k_0 g(\xi),$$

$$f \equiv \frac{\overline{\omega}' \Phi_t'}{\omega'} + \Psi = k_0[1 + \delta_0 h(\varphi)] \frac{\overline{\omega}'}{t^2 \omega'} e^{i\alpha(\xi)}, \tag{5}$$

where $\Phi(t)$ and $\Psi(t)$ are the limit values of functions analytic in $|\zeta| < 1$, with $\Phi(\zeta)$ bounded at infinity, and $\Psi(\infty) = 0$. If the functional dependence $z = \omega(t) \in C^{1+\beta}$ is known, then all the geometric parameters in ξ become known functions of $t = e^{i\gamma}$, $x = \operatorname{Re} \omega(t)$, $y = \operatorname{Im} \omega(t)$, $r = |\omega(t)|$, $\beta = \arg \omega(t)$ and $s = \int_0^\gamma |\omega'(e^{i\gamma})| \, d\gamma + s_0$. Here $\Phi(\zeta)$ and $\Phi'(\zeta)$ are found from the formulas

$$\Phi(\zeta) = -\frac{\delta_0 k_0}{2\pi i} \int_{|t|=1} \frac{g[\xi(t)](t+\zeta)}{t(t-\zeta)} \, dt + \varphi_0 + i\varphi_1,$$

$$\Phi'(\zeta) = -\frac{\delta_0 k_0}{\pi i} \int_{|t|=1} \frac{(d/dt)g[\xi(t)]}{t-\zeta} \, dt \equiv \delta_0 k_0 H(\omega|\zeta). \tag{6}$$

We set

$$2F(\omega|\zeta) = \frac{1}{2\pi i} \int_{|t|=1} \frac{i\alpha[\xi(t)](t+\zeta)}{t(t-\zeta)} \, dt$$

and rewrite the second of the conditions (5) in the form

$$\frac{t^2}{k_0}\Psi\omega'e^{-F} - e^{\bar{F}}\bar{\omega}' = \delta_0\rho(\omega\,|\,t), \qquad (7)$$

where

$$\rho(\omega\,|\,t) = \bar{\omega}H(\omega\,|\,t)e^{-F(t)} + h[\operatorname{Re}\Phi(t)]\bar{\omega}'e^{\overline{F(t)}}.$$

Let us consider the piecewise holomorphic function

$$\Omega(\zeta) = \begin{cases} \overline{\omega'(1/\bar{\zeta}\,)}\,\exp\{\overline{F(\omega\,|\,1/\bar{\zeta}\,)}\}, & |\zeta| < 1, \\[2mm] \dfrac{\zeta^2}{k_0}\Psi(\zeta)\omega'e^{-F}, & |\zeta| > 1, \end{cases}$$

which has a first-order pole at infinity and obviously satisfies the condition

$$\Omega^+ - \Omega^- = \delta_0\rho(\omega\,|\,t)$$

for $|\zeta| = 1$. From the solution of the last jump problem we have

$$\Omega(\zeta) = \frac{\delta_0}{2\pi i}\int_{|t|=1}\frac{\rho(\omega\,|\,t)}{t-\zeta}\,dt + c_1\zeta + c_0,$$

where $c_0 \neq 0$ and c_1 are arbitrary complex constants. If the mapping function $z = \omega(\zeta)$ is to be univalent, then we must require the coefficient of $1/\zeta$ in the expansion of the function

$$\omega'(\zeta) = e^{-F(\zeta)}\,\overline{\Omega(1/\bar{\zeta}\,)}$$

in a neighborhood of infinity to be zero, which is equivalent to the condition

$$\lim_{\zeta\to 0}\frac{1}{\zeta}[\Omega(\zeta) - \Omega(0)] = 0.$$

But

$$\Omega(\zeta) = c + \zeta\left(c_1 + \frac{\delta_0}{2\pi i}\int_{|t|=1}\frac{\rho\,dt}{t^2} + \frac{\delta_0\zeta^2}{2\pi i}\int_{|t|=1}\frac{\rho\,dt}{t^2(t-\zeta)}\right),$$

where

$$c = c_0 + \frac{\delta_0}{2\pi i}\int_{|t|=1}\frac{\rho\,dt}{t}$$

is an arbitrary constant. Setting

$$c_1 + \frac{\delta}{2\pi i}\int_{|t|=1}\frac{\rho\,dt}{t^2} = 0,$$

we get, finally,

$$\Omega(\zeta) = \frac{\delta_0 \zeta^2}{2\pi i} \int_{|t|=1} \frac{\rho(\omega\,|\,t)\,dt}{t^2(t-\zeta)} + c.$$

The formula for the limit value $\Omega^+(t) = e^t \overline{w'(t)}$ gives us

$$\omega'(t) = \frac{\delta_0}{2} e^{-F} \overline{\rho} - \frac{\delta_0 e^{-F} \overline{t^2}}{2\pi i} \int_{|t|=1} \frac{\overline{\rho}\,dt_0}{\overline{t}_0^2(\overline{t}_0 - t)} + \overline{c}e^{-F} = \delta_0 \Lambda_0(\omega\,|\,t) + \overline{c}e^{-F}.$$

Taking the equality $\omega'(t) = (1/it)\,d\omega/d\gamma$ into account, we transform this equation to the form

$$\omega(e^{i\gamma}) = -i\delta_0 \int_0^\gamma e^{i\gamma} \Lambda_0(\omega\,|\,e^{i\gamma})\,d\gamma + Q(\omega\,|\,e^{i\gamma}) \equiv \delta_0 \Lambda(\omega\,|\,t) + Q(\omega\,|\,t),$$

$$(8)$$

where

$$Q(\omega\,|\,e^{i\gamma}) = ci \int_0^\gamma e^{i\gamma} e^{-F(\omega|e^{i\gamma})}\,d\gamma + 1$$

and for definiteness we set $\omega(1) = 1$. Let $M > 2\pi\,|\,c\,| + 1$ be a fixed number. We take $\|\,\omega(t)\,\|_{C^{1+\beta}} \leqslant M$. Then, obviously,

$$\|\,H(\omega\,|\,t)\,\|_{C^\beta} \leqslant M_H(M), \|\,F(\omega\,|\,t)\,\|_{C^{1+\beta}} \leqslant M_F(M),$$

whence $\|\,\rho(\omega\,|\,t)\,\|_{C^\beta} \leqslant M_\rho(M)$, and thereby also

$$\|\,\Lambda(\omega\,|\,t)\,\|_{C^{1+\beta}} \leqslant M_\Lambda(M), \|\,Q(\omega\,|\,t)\,\|_{C^{2+\beta}} \leqslant 2\pi\,|\,c\,|\,e^{M_F(M)} + 1.$$

On the other hand, taking the smoothness of the given functions $\varphi(\xi)$, $h(\varphi)$, and $\alpha(\xi)$ into account, we find as in problem A of §1 that

$$\|\,\Lambda(\omega_1\,|\,t) - \Lambda(\omega_2\,|\,t)\,\|_{C^{1+\beta}} \leqslant M_{1,2}(M)\|\,\omega_1 - \omega_2\,\|_{C^{1+\beta}}.$$

Suppose that $\|\,Q(\omega\,|\,t)\,\|_{C^{1+\beta}} < M$, i.e.,

$$M_0 = 2\pi\,|\,c\,|\,e^{M_F(M)} + 1 < M. \tag{9}$$

This inequality is automatically satisfied when the arbitrary constant c is small. It holds also for an arbitrary choice of c if we require in addition that the given function $\alpha(\xi)$ be close to a fixed constant α_0 on Γ, i.e., if we set $\alpha(\xi) = \alpha_0 + \delta_1 h_1(\xi)$, where $\delta_1 > 0$ is a small parameter. In this case $\exp(-F) = \exp(-i\alpha_0 + \delta_1 F_1)$; therefore, $\|\,e^{-F}\,\|_{C^{1+\beta}} \leqslant e^{\delta_1 M_F(M)}$. And since $2\pi\,|\,c\,| + 1 < M$, we also have $M_0 = 2\pi\,|\,c\,|\,e^{\delta_1 M_F(M)} + 1 < M$ for sufficiently small $\delta_1 > 0$. Next, choose δ_0 such that

$$M \equiv \delta_0 \max\left\{ \frac{M_\Lambda(M)}{M - M_0}, M_{1,2}(M) \right\} < 1. \tag{10}$$

Then for $\omega, \omega_i \in \overline{K}_M = \{\|\omega\|_{C^{1+\beta}} \leqslant M\}$ we get

$$\|\delta_0 \Lambda(\omega \mid t) + Q(\omega \mid t)\|_{C^{1+\beta}} \leqslant M,$$

$$\delta_0 \|\Lambda(\omega_1 \mid t) - \Lambda(\omega_2 \mid t)\|_{C^{1+\beta}} \leqslant M \|\omega_1 - \omega_2\|_{C^{1+\beta}}. \qquad (11)$$

Let us substitute an arbitrary function $\hat{\omega}(t) \in \overline{K}_M$ in the operator $Q(\omega \mid t)$ and consider the following equation in $\omega(t)$:

$$\omega(t) = \delta_0 \Lambda(\omega \mid t) + Q(\hat{\omega} \mid t). \qquad (12)$$

According to (10) and (11), the continuous operator $I - \delta\Lambda$ is invertible in the ball $\overline{K}_M$, and, consequently, (12) is equivalent to the relation

$$\omega(t) = (I - \delta_0 \Lambda)^{-1} Q(\hat{\omega} \mid t) \equiv \Lambda_1(\hat{\omega} \mid t). \qquad (13)$$

The continuous transformation $\Lambda_1(\omega \mid t)$ is compact in $\overline{K}_M$. Indeed, let $\hat{\omega}_i \in \overline{K}_M$ be an arbitrary bounded sequence. Then the sequence $Q_i = Q(\hat{\omega}_i \mid t)$ is relatively compact in $C^{1+\beta}$, and on a convergent subsequence $Q_{i_n}(t)$ the corresponding solutions ω_{i_n} of (12) also converge. Next, the first of the inequalities (11) implies that the completely continuous operator $\Lambda_1(\omega \mid t)$ maps the ball $\overline{K}_M$ into itself; therefore, (13) has at least one solution by Schauder's theorem.

If we impose additional conditions on the small parameter c for arbitrary $\alpha(\xi)$ or on the $\delta_1 > 0$ in the expansion $\alpha(\xi) = \alpha_0 + \delta_1 h_1(\xi)$, for arbitrary c, then we can make the whole operator $\{\delta_0 \Lambda(\omega \mid t) + Q(\omega \mid t)\}$ a contraction. In this case the iteration method for equation (8) converges in $\overline{K}_M$. Thus, the existence of a solution $\omega(t)$ of (8) has been proved. Then $\omega(t)$ can be used to reconstruct the functions $\Phi(\zeta)$, $\rho(\omega \mid t)$, $F(\omega \mid t)$, and with them also

$$\Omega(\zeta), \qquad \omega'(\zeta) = e^{-F(\zeta)} \overline{\Omega(1/\bar{\zeta})}, \qquad \Psi(\zeta) = k_0 e^{-F(\zeta)} / \zeta^2 \omega'(\zeta).$$

By construction, the function

$$\frac{\sigma_y - \sigma_x}{2} + i\tau_{xy} \equiv \frac{\overline{\omega}'_\zeta \Phi'_\zeta}{\omega'_\zeta} + \Psi(\zeta),$$

vanishes at infinity, i.e., $\tau_{xy}(\infty) = 0$ and $\sigma_x(\infty) = \sigma_y(\infty)$. The arbitrary complex constant c appears in the solution. This constant can be manipulated, for example, in determining the solution in the plastic domain D^+.

§3. The semi-inverse problem with unknown boundary

In this section we investigate elastico-plasticity problems by means of the semi-inverse method well known for problems in the mechanics of a continuous medium with free boundaries, when the boundary values are given on the unknown contour as functions of the boundary arclength of a canonical domain. As the canonical domain we choose the exterior of the unit disk and assume that the components of the stress tensor P are given on the unknown elastico-plastic boundary Γ as functions of the arclength of the unit circle, i.e.,

$$\frac{1}{4}(\sigma_x + \sigma_y) \equiv \operatorname{Re} \Phi(t) = \frac{k_0}{\delta_0} g(t) + \varphi_0, \qquad t = e^{i\gamma},$$

$$\frac{\sigma_y - \sigma_x}{2} + i\tau_{xy} \equiv \frac{\overline{\omega}'}{\omega'} \Phi' + \Psi = k(t) e^{i\alpha(t)}, \tag{1}$$

where $z = \omega(\zeta)$ is an unknown analytic function mapping $|\zeta| > 1$ conformally onto the domain D^-, $g(t) \in C^{4+\beta}$, $k(t)$, $\alpha(t) \in C^{3+\beta}$, $\beta > 0$, $|k(t)| \le k_0$, and δ_0 is a small parameter. Thus, we must determine three analytic functions ω_0, Ψ and Φ in the domain $|\zeta| > 1$ that satisfy (1), with $\omega(\infty) = \infty$, $|\Phi(\infty)| \le \infty$, $|k(t)| \le k_0$, and $\Psi(\infty) = 0$. The function $\Phi(\zeta)$ can be determined directly from the first condition in (1) by the Schwarz formula:

$$\Phi(\zeta) = -\frac{k_0}{2\pi i \delta_0} \int_{|t|=1} \frac{g(t)(t + \zeta)}{t(t - \zeta)} dt + \varphi_0 + i\varphi_1,$$

whence

$$\Phi'(\zeta) = -\frac{k_0}{\delta_0 \pi i} \int_{|t|=1} \frac{g'_t \, dt}{t - \zeta} = +\frac{k_0}{2\delta_0 \pi \zeta} \int_{|t|=1} \frac{g'_\gamma(t)(t + \zeta)}{t(t - \zeta)} dt, \tag{2}$$

since $\int_0^{2\pi} g'_\gamma(e^{i\gamma}) \, d\gamma = 0$. We assume that the function $d^2g/d\gamma^2$ vanishes at two points $e^{i\gamma_0}$ and $e^{i\gamma_1}$ of the circle, $\gamma_0 < \gamma_1$, and has different signs on the intervals (γ_0, γ_1) and $(\gamma_1, 2\pi + \gamma_0)$, say for definiteness that

$$d^2g/d\gamma^2 > 0, \quad \gamma \in (\gamma_0, \gamma_1); \qquad d^2g/d\gamma^2 < 0, \quad \gamma \in (\gamma_1, 2\pi + \gamma_0). \tag{3}$$

In the domain $|\zeta| > 1$ and $|\zeta| < 1$ let us consider the respective analytic functions

$$w(\zeta) = \frac{\delta_0}{k_0} i\zeta \Phi'(\zeta), \qquad w_*(\zeta) = \overline{w\left(\frac{1}{\overline{\zeta}}\right)},$$

where

$$\operatorname{Re} w(e^{i\gamma}) = \operatorname{Re} w_*(e^{i\gamma}) = g'_\gamma(e^{i\gamma}) \quad \text{and} \quad w(\infty) = w_*(0) = 0.$$

Representing $w_*(\zeta)$ by a Schwarz integral with density $dg/d\gamma$ satisfying (3), we conclude from Kaplan's theorem (Theorem 6 in §3 of Chapter II) that $w_*(\zeta)$ is univalent in $|\zeta| < 1$. Consequently, this function does not vanish in $|\zeta| \le 1$ except at $\zeta = 0$, where $w_*(0) = 0$ by construction, and the function $\Phi'(\zeta) = k_0 w(\zeta)/\delta_0 i\zeta$ thus does not have zeros in any domain with $1 < |\zeta| < \infty$.

Let us consider in the ζ-plane the piecewise holomorphic function

$$\Omega(\zeta) = \begin{cases} \overline{\omega(1/\bar{\zeta})}, & |\zeta| < 1, \\ -\Psi(\zeta)\omega'/\Phi'(\zeta), & |\zeta| > 1, \end{cases}$$

which, according to the second relation in (1), satisfies on $|\zeta| = 1$ the condition

$$\Omega^+(t) - \Omega^-(t) = \frac{\omega'(t)}{\Phi'(t)}k(t)e^{i\alpha(t)} \equiv \delta_0 F(t)\omega'(t), \tag{4}$$

where $F(t) \in C^{3+\beta}$. Since $\Phi'(\zeta)$ has a zero of second order at $\zeta = \infty$, and $\Psi(\zeta)$ has a zero of first order, $\Omega(\zeta)$ has a pole of first order there, and at the same time $\omega(1/\bar{\zeta})$ also has a pole of first order at $\zeta = 0$. Consequently, from the solution of the jump problem (4) we have

$$\Omega(\zeta) = \frac{\delta_0}{2\pi i}\int_{|t|=1}\frac{F(t)\omega'(t)\,dt}{t-\zeta} + c_{-1}\zeta^{-1} + c_0 + c_1\zeta,$$

where $c_{-1} \neq 0$, c_0, and c_1 are arbitrarily fixed complex constants. We extend $F(t)$ inside the unit disk by means of the Poisson integral

$$F(\zeta) = \frac{1}{2\pi}\int_0^{2\pi}\frac{(1-r^2)F(e^{i\gamma_0})\,d\gamma_0}{1-2r\cos(\gamma_0-\gamma)+r^2}, \qquad \zeta = re^{i\gamma}.$$

After obvious transformations we get

$$\frac{1}{2\pi i}\int_{|t|=1}\frac{F(t)\omega'(t)\,dt}{t-\zeta} = \frac{F(\zeta)}{2\pi i}\int_{|t|=1}\frac{\omega'(t)\,dt}{t-\zeta} + \int_{|t|=1}H(t\,|\,\zeta)\omega(t)\,dt,$$

where

$$H(t,\zeta) = \frac{1}{2\pi i}\left[\frac{F(t)-F(\zeta)}{t-\zeta} - F'(t)\right]\frac{1}{t-\zeta} \equiv \frac{1}{2\pi i}F''[t+s(t-\zeta)],$$

$$|s| < 1.$$

But the equality $\overline{\omega(1/\bar{\zeta})} = \Omega(\zeta)$ gives us $\overline{w'(\infty)} = c_{-1}$, and, consequently,

$$\frac{1}{2\pi i}\int_{|t|=1}\frac{\omega'(t)\,dt}{t-\zeta} = \overline{\omega'(\infty)} = \bar{c}_{-1}.$$

Substituting this into the representation for $\Omega(\zeta)$ and passing from the domain $|\zeta| < 1$ to the limit values on $|\zeta| = 1$ in the resulting equality, we get

$$\overline{\omega(t)} = \delta_0\int_{|t|=1}H_0(t_0,t)\omega(t_0)\,dt_0 + Q(t), \tag{5}$$

where $Q(t) = \delta_0\bar{c}_{-1}F(t) + c_{-1}t + c_0 + c_1 t \in C^{3+\beta}$ and, by construction, $H(t_0,t) \in C^{1+\beta}$ with respect to both arguments. Obviously, the Banach fixed-point principle is applicable to the linear equation (5) in $C^{1+\beta}$ $(\beta > 0)$ for sufficiently small $\delta_0 > 0$, and, consequently, for fixed constants c_{-1}, c_0, and c_1 there exists a unique solution $\omega(t) \in C^{1+\beta}$ of (5). Substituting this solution $\omega(t)$ of (5) into the formula for $\Omega(\zeta)$, we find that $\omega(\zeta) = \overline{\Omega(1/\bar{\zeta})}$ and, along with it,

$$\Psi(\zeta) = \Omega(\zeta)\Phi'(\zeta)\left[\frac{d}{d\zeta}\overline{\Omega\left(\frac{1}{\bar{\zeta}}\right)}^{-1}\right].$$

By construction, $f(\zeta) = (\bar{\omega}/\omega)\Phi' + \Psi$ vanishes at infinity, i.e., $\sigma_x(\infty) = \sigma_y(\infty)$ and $\tau_{xy}(\infty) = 0$. The solution thus obtained depends on the three arbitrary complex constants $c_{-1} \neq 0$, c_0, and c_1; and c_0 is a translation constant and can be fixed, in particular, from the condition $\omega(1) = z_0$, where z_0 is a given point on Γ. The constants $c_{-1} = \omega'(\infty)$ and c_1 have an essential influence on the solution and can be exploited, for example, in solving the problem in the plastic domain. As an illustration let us consider the particular case of a semi-inverse problem when $f(t) = k(t)e^{i\alpha(t)}$ is the limit value of an analytic function $f(\zeta)$ in $|\zeta| > 1$ that has at infinity the expansion $f(\zeta) = \sum_{-\infty}^{\infty} a_k \zeta^k$. Then, according to the second of the relations (1), for $|t| = 1$ the function

$$\Omega(\zeta) = \begin{cases} \overline{\omega(1/\bar{\zeta})}, & |\zeta| \leqslant 1, \\ [f(\zeta) - \Psi(\zeta)]\omega'/\Phi', & |\zeta| > 1 \end{cases}$$

satisfies the condition $\Omega^+(t) - \Omega^-(t) = 0$, i.e., it is holomorphic in any finite part of the plane not containing the origin. Taking the expansion of $f(\zeta)$ into account, by the generalized Liouville theorem we have

$$\Omega(\zeta) = \sum_{k=-1}^{n+2} c_k \zeta^k, \qquad z = \omega(\zeta) = \Omega\left(\frac{1}{\zeta}\right).$$

We should point out the essential difference between the small parameters in the semi-inverse problem in §3 and the problems in §§1 and 2, in which we assumed the smallness of the magnitude of the ratio

$$\left|\frac{\varphi - \varphi_0}{f}\right| < \delta_0, \qquad \varphi \equiv \frac{1}{4}(\sigma_x + \sigma_y), \qquad f = \frac{\sigma_x - \sigma_y}{2} + i\tau_{xy},$$

and the magnitude of its inverse ratio $|f/(\varphi - \varphi_0)| < \delta_0$, respectively.

§4. Problems

We do not attempt the hopeless task of listing all the unsolved problems in elastico-plasticity (it would be easier to list the solved ones), but only formulate some of them which are close to those studied in this chapter with regard to construction and possible methods of investigation. The problems stated below are not of uniform difficulty: Some are simple exercises on the material in this chapter, while others represent fairly complicated research problems.

1. Because of the existence of a constant solution in problems A and B of §1, we constructed only a first-approximation theory for them. It is a very important and difficult problem to prove that when $X_0(t) \equiv 0$ the problems A and B have, along with the constant solution of (1.13), also solutions $\tau(t) \neq$ const for some $\delta_0 > 0$. If the operator $\Lambda(\tau | t)$ in (1.13) were completely continuous, then the investigation of the existence of nonconstant solutions could be reduced to an investigation of the spectrum of its Fréchet differential (see [57]). The operator $\Lambda(\tau | t)$ is differentiable by virtue of the assumptions about the initial data of the problem.

2. Extend the results of this chapter to domains with boundary Γ passing through the point at infinity. This generalization can be realized without modifying the methods used in this chapter, and requires only a supplementary investigation of the unknown functions at the image on the circle $|\varphi| = 1$ of the point $z = \infty$ on Γ.

3. Obtain the boundary conditions in the problems studied in §§1 and 2 by solving elementary problems in the theory of plasticity. Conversely, after solving the problem in elasticity theory with an unknown elastico-plastic boundary Γ, extend the solution found into the domain of plasticity (there are similar investigations in [2] and [124]).

4. Study mixed elastico-plasticity problems in which part of the boundary Γ, which in general is not entirely the elastico-plastic boundary, is known and only two of the three relations (1) and (2) at the beginning of the chapter are given on it, while the other part of Γ is found from all the relations (1) and (2) given on it.

5. Generalize the results of this chapter by considering spaces of functions having generalized derivatives.

6. Investigate elastico-plasticity problems in spaces of infinitely differentiable functions; this should enable us in some problems to get rid of the assumptions about smallness of some of the given quantities. Moreover, the well-developed apparatus in the analytical theory of nonlinear integral equations can be applied to the investigation.

7. Construct variational formulas describing the change in the elastico-plastic boundary Γ under a small change in the boundary data, in particular, under a change in the small parameter δ_0. Since the solutions of elastico-plasticity problems can, as a rule, be written out in explicit form when $\delta_0 = 0$, it would be interesting, in particular, to find variational formulas close to this trivial solution.

8. Determine conditions on the boundary data under which the solution found in the domain D^- has no zones of plasticity. Is it possible to ensure that there are no zones of plasticity by a suitable choice of the arbitrary constants in the solution?

9. In the elastico-plasticity problems studied in this chapter a solution was constructed in such a way that the function $z = \omega(\zeta)$ mapping the canonical domain D_ζ conformally onto the domain D^- outside the unknown boundary Γ is univalent in D_ζ. It would be interesting to investigate the question of univalence of this function in the closed domain, i.e., to find when the unknown contour Γ is one-sheeted. In some problems it is possible to get a one-sheeted contour by a special choice of the arbitrary constants appearing in the solution. In other problems the question of univalence can be solved by applying the theorem of Kaplan cited at the beginning of §3. However, in most elastico-plasticity problems it is difficult to predict in advance a method for investigating the univalence question for the unknown contour Γ.

10. Prove the solvability of the elastico-plasticity problem in which we are given

$$\frac{\partial \varphi}{\partial x} = \text{Re } \Phi_z' = \frac{k_0}{\delta_0} h(\xi), \qquad \bar{z}\Phi_z' + \Psi = k(\xi)e^{i\alpha(\xi)} = f(\xi) \tag{1}$$

on the unknown contour Γ, where ξ is one of the geometric parameters of Γ, say $\text{Re } z = x \in [x_1, x_2]$ for definiteness. As always, let $z = \omega(\zeta)$ be an unknown analytic function mapping the domain $|\zeta| > 1$ onto D^-:

$$\Phi_\omega'(\zeta) = \frac{k_0}{2\pi\delta_0 i} \int_{|t|=1} \frac{h[\text{Re } \omega(t)](t + \zeta)\, dt}{t(t - \zeta)}. \tag{2}$$

Suppose that $h(x)$ is such that Φ'_ω does not vanish anywhere in $|\zeta| \geqslant 1$ except at the point $\xi = \infty$, where it has a zero of second order. We introduce the piecewise holomorphic function

$$\Omega(\zeta) = \begin{cases} \overline{\omega(1/\bar{\zeta}\,)}, & |\zeta| \leqslant 1, \\ -\Psi(\zeta)[\Phi'_\omega(\zeta)]^{-1}, & |\zeta| > 1, \end{cases}$$

which for $|t| = 1$ satisfies the condition

$$\Omega^+(t) - \Omega^-(t) = \frac{f[\operatorname{Re}\omega(t)]}{\Phi'_\omega(t)} \equiv \delta_0 H(\omega \,|\, t),$$

from which, similarly to §3, we come to the equation

$$\overline{\omega(t)} = \frac{\delta_0}{2} H(\omega \,|\, t) + \frac{\delta_0}{2\pi i} \int_{|t|=1} \frac{H(\omega \,|\, t_0)\, dt_0}{t_0 - t} + Q(t) \equiv \Lambda(\omega \,|\, t), \tag{3}$$

where $Q(t) = c_{-1}\bar{t} + c_0 + c_1 t$, and we can prove the solvability of this by the Banach principle for sufficiently small $\delta_0 > 0$. Thus, the solvability of problem (1) will be completely proved when conditions on $h(x)$ are found which ensure the existence of a second-order zero of $\Phi'_\omega(\zeta)$ at $\zeta = \infty$ which is its only zero in $|\zeta| \geqslant 1$. If, similarly to the problem in §3, the derivative

$$\frac{dh(\gamma)}{d\gamma} \equiv \frac{d}{d\gamma} h[x(\gamma)] = \frac{dh}{dx} \cdot \frac{dx}{d\gamma}$$

had only two zeros on the circle $|\zeta| = 1$, with preservation of different signs in the corresponding intervals, and if $\int_0^{2\pi} h[x(\gamma)]\, d\gamma = 0$, then $\Phi'_\omega(\zeta)$ would satisfy the conditions necessary in the problem. Therefore, if $dh/dx \neq 0$, $x \in [x_1, x_2]$, then for the required properties of $\Phi'_\omega(\zeta)$ to hold it suffices that $dx/d\gamma$ satisfy the conditions imposed on $dh(\gamma)/d\gamma$, and, consequently, the unknown contour Γ would then be one-sheeted by Kaplan's theorem. Thus, in this problem it is necessary to find conditions under which the operator $\Lambda\omega(t)$ given by (3) guarantees that the contour Γ, is one-sheeted. In this sense the problem is close to the preceding one.

BIBLIOGRAPHY

1. Lars V. Ahlfors, *Lectures in quasiconformal mappings*, Van Nostrand, Princeton, N.J., 1966.

2. B. D. Annin, *Two-dimensional problems of elastico-plasticity*, Izdat. Novosibirsk. Gos. Univ., Novosibirsk, 1968. (Russian)

3. S. N. Antontsev, *Subsonic gas flows in multiply connected regions*, Fluid Dynamics Trans. Vol. 5, Part II (Ninth Sympos. Adv. Problems and Methods in Fluid Mech., Kazimierz, 1969), PWN, Warsaw, 1971, pp. 7–18.

4. S. N. Antontsev and N. A. Kucher, *The Poincaré problem with discontinuous coefficients of the boundary condition for a quasilinear elliptic equation*, Dinamika Sploshnoĭ Sredy Vyp. 5 (1970), 118–124. (Russian)

5. S. N. Antontsev and V. D. Lelyukh, *Some conjunction problems of rotational and potential subsonic flows*, Dinamika Sploshnoĭ Sredy Vyp. 1 (1969), 134–153. (Russian)

6. S. N. Antontsev and V. N. Monakhov, *On the solvability of a class of conjunction problems with shift*, Dokl. Akad. Nauk SSSR **205** (1972), 263–266; English transl. in Soviet Math. Dokl. **13** (1972).

7. _____, *On some problems in the filtration of two-phased incompressible fluids*, Dinamika Sploshnoĭ Sredy Vyp. 2 (1969), 156–167. (Russian)

8. _____, *On a general quasilinear model of the filtration of immiscible fluids*, Dinamika Sploshnoĭ Sredy Vyp. 3 (1969), 5–17. (Russian)

9. _____, *Some nonstationary problems in the filtration of nonhomogeneous fluids with free (unknown) boundaries*, Dinamika Sploshnoĭ Sredy Vyp. 3 (1969), 18–32. (Russian)

10. _____, *On some nonstationary problems with unknown boundaries*, Some Problems of Math. and Mech. (M. A. Lavrent'ev Seventieth Birthday Vol.), "Nauka", Leningrad, 1970, pp. 75–87; English transl. in Amer. Math. Soc. Transl. (2) **104** (1976).

11. _____, *The Riemann-Hilbert boundary value problem with discontinuous boundary conditions for quasilinear elliptic systems of equations*, Dokl. Akad. Nauk SSSR **175** (1967), 511–513; English transl. in Soviet Math. Dokl. **8** (1967).

12. _____, *Boundary value problems with discontinuous boundary conditions for quasilinear systems of $2m$ ($m \geqslant 1$) first order equations*, Izv. Sibirsk. Otdel. Akad. Nauk SSSR **1967**, no. 8, 65–73. (Russian)

13. V. M. Babich, *On the extension of functions*, Uspekhi Mat. Nauk **8** (1953), no. 2(54), 111–113. (Russian)

14. Herbert Beckert, *Existenzbeweis für permanente Kapillarwellen einer schweren Flüssigkeit entlang eines Karals*, Arch. Rational Mech. Anal. **13** (1963), 15–45.

15. Paul W. Berg, *The existence of subsonic Helmholtz flows of a compressible fluid*, Comm. Pure Appl. Math. **15** (1962), 289–347.

16. Lipman Bers, *Mathematical aspects of subsonic and transonic gas dynamics*, Wiley, New York, and Chapman & Hall, London, 1958.

17. _____, *Quasiconformal mappings and Teichmuller's theorem*, Analytic Functions, Princeton Univ. Press, Princeton, N.J., 1960, pp. 89–119.

18. A. V. Bitsadze, *On the problem of equations of mixed type*, Trudy Mat. Inst. Steklov. **41** (1953); German transl., VEB Deutscher Verlag Wiss., Berlin, 1957.

19. Garrett Birkhoff, *Hydrodynamics. A study in logic, fact and similitude*, Princeton Univ. Press, Princeton, N.J., 1950.

20. Garrett Birkhoff and E. H. Zarantonello, *Jets, wakes, and cavities*, Academic Press, 1957.

21. B. V. Boyarskiĭ [Bogdan Bojarski], *Generalized solutions of a system of differential equations of first order and of elliptic type with discontinuous coefficients*, Mat. Sb. **43(85)** (1957), 451–503. (Russian)

22. A. P. Calderón and A. Zygmund, *On the existence of certain singular integrals*, Acta Math. **88** (1952), 85–139.

23. I. A. Charnyĭ, *Underground hydrodynamics and gas dynamics*, Gostoplekhizdat, Moscow, 1963. (Russian)

24. G. P. Cherepanov, *On a method of solving a problem of elastico-plasticity*, Prikl. Mat. Mekh. **27** (1963), 428–435; English transl. in J. Appl. Math. Mech. **27** (1963).

25. Royal Eugene Collins, *Flow of fluids through porous materials*, Reinhold, New York, 1961.

26. R. Courant, *Dirichlet's principle, conformal mapping, and minimal surfaces*, Interscience, 1950.

27. I. I. Danilyuk, *On Hilbert's problem with measurable coefficients*, Sibirsk. Mat. Zh. **1** (1960), 171–197. (Russian)

28. _____, *A problem with oblique derivative*, Sibirsk. Mat. Zh. **3** (1962), 17–55. (Russian)

29. _____, *A generalized Cauchy formula for axisymmetric fields*, Sibirsk. Mat. Zh. **4** (1963), 48–85. (Russian)

30. Basile Demtchenko, *Problèmes mixtes harmoniques en hydrodynamique des fluides parfaits*, Gauthier-Villars, Paris, 1933.

31. Nelson Dunford and Jacob T. Schwartz, *Linear operators*. Vol. I, Interscience, 1958.

32. F. D. Gakhov, *Boundary-value problems*, Fizmatgiz, Moscow, 1958; English transl. of 2nd (1963) ed., Pergamon Press, Oxford, and Addison-Wesley, Reading, Mass., 1966.

33. _____, *Riemann's boundary value problem for a system of n pairs of functions*, Uspekhi Mat. Nauk **7** (1952), no. 4(50), 3–54. (Russian)

34. F. D. Gakhov and I. M. Mel'nik, *Singular contour points in the inverse boundary value problem of the theory of analytic functions*, Ukrain. Mat. Zh. **11** (1959), 25–37. (Russian)

35. L. A. Galin, *Plane elastico-plastic problem. Plastic zones in the vicinity of circular apertures in plates and girders*, Prikl. Mat. Mekh. **10** (1946), 367–386. (Russian; English summary)

36. P. R. Garabedian, H. Lewy and M. Schiffer, *Axially symmetric cavitational flow*, Ann. of Math. (2) **56** (1952), 560–602.

37. T. G. Gegeliya, *Boundedness of singular operators*, Soobshch. Akad. Nauk Gruzin. SSR **20** (1958), 517–523. (Russian)

38. Robert Gerber, *Sur les solutions exactes des équations du mouvement avec surface libre d'un liquide pesant*, J. Math. Pures Appl. (9) **34** (1955), 185–299.

39. David Gilbarg, *Unsteady flows with free boundaries*, Z. Angew. Math. Phys. **3** (1952), 34–42.

40. M. A. Gol'dshtik, *A mathematical model of separated flows in an incompressible liquid*, Dokl. Akad. Nauk SSSR **147** (1962), 1310–1313; English transl. in Soviet Phys. Dokl. **7** (1962/63).

41. G. M. Goluzin, *Geometric theory of functions of a complex variable*, GITTL, Moscow, 1952; English transl. of 2nd (1966) ed., Transl. Math. Mono., vol. 26, Amer. Math. Soc., Providence, R.I., 1969.

42. M. I. Gurevich, *Theory of jets in ideal fluids*, Fizmatgiz, Moscow, 1961; English transl., Academic Press, 1965.

43. L. G. Guzevskiĭ and G. V. Lavrent'ev, *Application of the method of finite-dimensional approximation in problems of jet flows*, Dinamika Sploshnoĭ Sredy Vyp. 1 (1969), 75–91. (Russian)

44. M. I. Kaĭkin, *Existence theorems for a class of inverse mixed boundary-value problems of the theory of analytic functions*, Trudy Kazan. Aviats. Inst. Vyp. 64 (1961), 3–24. (Russian)

45. G. H. Hardy, J. E. Littlewood and G. Pólya, *Inequalities*, Cambridge Univ. Press, 1934.

46. S. A. Kristianovich, *Motion of ground water which does not conform to Darcy's law*, Prikl. Mat. Mekh. 4 (1940), no. 1, 33–52. (Russian; English summary)

47. R. Huron, *Contribution à l'étude de l'unicité des solutions du problème de représentation conforme de Helmholtz*, Ann. Fac. Sci. Univ. Toulouse (4) 15 (1951), 5–78.

48. N. V. Kusnutdinova, *On the behavior of solutions of the Stefan problem as time increases without bound*, Dinamika Sploshnoĭ Sredy Vyp. 2 (1969), 168–177. (Russian)

49. B. V. Kvedelidze, *Linear discontinuous boundary value problems in the theory of functions, singular integral equations and some of their applications*, Akad. Nauk Gruzin. SSR Trudy Tbiliss. Mat. Inst. Razmadze 23 (1956), 3–158. (Russian)

50. N. E. Zhukovskiĭ [Joukowsky], *A modification of Kirchhoff's method of determining a two-dimensional motion of a fluid, given a constant velocity along an unknown streamline*, Mat. Sb. 15 (1890/91), 121–276; reprinted as Trudy Tsentral. Aèro-gidrodinam. Inst. Vyp. 41 (1930); also reprinted in at least three collections of his *Works*. (Russian)

51. Wilfred Kaplan, *Close-to-convex schlicht functions*, Michigan Math. J. 1 (1952), 169–185.

52. A. V. Kazhikhov, *Some self-similar problems of time-dependent filtration, and their numerical solution*, Dinamika Sploshnoĭ Sredy Vyp. 3 (1969), 33–44. (Russian)

53. _____, *On the existence of a separation flow of Ryabushinskiĭ type in a gravitational field with capillary forces taken into account*, Dinamika Sploshnoĭ Sredy Vyp. 1 (1969), 92–99. (Russian)

54. M. Keldysh and F. Frankl, *Die äussere Neumann'sche Aufgabe für nichtlineare elliptische Differentialgleichungen mit Anwendung auf die Theorie des Flügels im kompressiblen Gas*, Izv. Akad. Nauk SSSR (7) Otd. Mat. Estestv. Nauk 1934, 561–601. (Russian; German summary)

55. M. V. Keldysh and L. I. Sedov, *Applications of the theory of functions of a complex variable to hydrodynamics and aerodynamics (review of some works of the Moscow school)*, Appl. Theory of Functions in Continuum Mech. (Proc. Internat. Sympos., Tbilisi, 1963), Part II: Fluid and Gas Mech., Math. Methods, "Nauka", Moscow, 1965, pp. 13–42; English transl., ibid., pp. 43–64.

56. N. E. Kochin, I. A. Kibel' and N. V. Roze, *Theoretical hydromechanics*. I, 5th ed., GITTL, Moscow, 1955; English transl., Wiley, 1964.

57. M. A. Krasnosel'skiĭ, *Topological methods in the theory of nonlinear integral equations*, GITTL, Moscow, 1956; English transl., Macmillan, 1964.

58. Julien Kravtchenko, *Sur le problème de représentation conforme de Helmholtz; théorie des sillages et des proues*, J. Math. Pures Appl. (9) 20 (1941), 35–234, 235–303.

59. O. A. Ladyzhenskaya and N. N. Ural'tseva, *Linear and quasilinear elliptic equations*, "Nauka", Moscow, 1964; English transl., Academic Press, 1968.

60. O. A. Ladyzhenskaya, V. A. Solonnikov and N. N. Ural'tseva, *Linear and quasilinear equations of parabolic type*, "Nauka", Moscow, 1967; English transl., Transl. Math. Mono., vol. 23, Amer. Math. Soc., Providence, R.I., 1968.

61. M. A. Lavrent'ev, *Variational methods for boundary value problems for systems of elliptic equations*, Izdat. Akad. Nauk SSSR, Moscow, 1962; English transl., Noordhoff, 1963.

62. _____, *Boundary value problems in the theory of univalent functions*, Mat. Sb. 1(43) (1936), 815–844; English transl. in Amer. Math. Soc. Transl. (2) 32 (1963).

63. _____, *Sur certaines propriétés des fonctions univalentes et leurs applications à la théorie des sillages*, Mat. Sb. 4 (1938), 391–458. (Russian; French summary)

64. _____, *Certain boundary-value problems for systems of elliptic type*, Sibirsk. Mat. Zh. 3 (1962), 715–728. (Russian)

65. _____, *The general problem of the theory of quasiconformal mappings of plane regions*, Mat. Sb. 21(63) (1947), 285–320; English transl. in Amer. Math. Soc. Transl. (1) 2 (1962).

66. _____, *A fundamental theorem of the theory of quasiconformal mappings of two-dimensional regions*, Izv. Akad. Nauk SSSR Ser. Mat. **12** (1948), 513–554; English transl. in Amer. Math. Soc. Transl. (1) **2** (1962).

67. M. A. Lavrent'ev and A. V. Bitsadze, *On the problem of equations of mixed type*, Dokl. Akad. Nauk SSSR **70** (1950), 373–376. (Russian)

68. M. A. Lavrent'ev and B. V. Shabat, *Geometrical properties of solutions of nonlinear systems of partial differential equations*, Dokl. Akad. Nauk SSSR **112** (1957), 810–811. (Russian)

69. _____, *Methods of the theory of functions of a complex variable*, 2nd ed., GITTL, Moscow, 1958; German transl. of 3rd (1965) ed., VEB Deutscher Verlag Wiss., Berlin, 1967.

70. Jean Leray, *Les problèmes de représentation conforme de Helmholtz; théories des sillages et des proues*, Comment. Math. Helv. **8** (1935/36), 149–185, 250–263.

71. Jean Leray and Jules Schauder, *Topologie et équations fonctionnelles*, Ann. Sci. École Norm. Sup. (3) **51** (1934), 45–78.

72. Jean Leray and Alexandre Weinstein, *Sur un problème de représentation conforme posé par la théorie de Helmholtz*, C. R. Acad. Sci. Paris **198** (1934), 430–432.

73. Li Chung [Li Zhong], *On the existence of homeomorphic solutions of a system of quasilinear partial differential equations of elliptic type*, Acta Math. Sinica **13** (1963), 454–461; English transl. in Chinese Math. Acta **4** (1963).

74. M. J. Lighthill, *A new method of two-dimensional aerodynamic design*, Ministry of Supply [London], Aeronaut. Res. Council, Reports and Memoranda, No. 2112 (1945); reprinted in *Aeronaut. Res. Council Tech. Rep. for the Year* 1945, Vol. I (65), HMSO, London, 1955, pp. 105–157.

75. E. B. McLeod, Jr., *The explicit solution of a free boundary problem involving surface tension*, J. Rational Mech. Anal. **4** (1955), 557–567.

76. I. M. Mel'nik, *An exceptional case in the Riemann boundary value problem*, Akad. Nauk Gruzin. SSR Trudy Tbiliss. Mat. Inst. Razmadze **24** (1957), 149–162. (Russian)

77. L. G. Mikhaĭlov, *A new class of singular integral equations and its application to differential equations with singular coefficients*, Akad. Nauk Todzhik. SSR, Dushanbe, 1963; English transl., Noordhoff, 1970.

78. N. N. Moiseev and A. M. Ter-Krikorov, *Study of the motion of a heavy fluid at speeds close to critical*, Moskov. Fiz.-Tekhn. Inst. Trudy Vyp. 3 (1959), 25–59. (Russian)

79. V. N. Monakhov, *A boundary value problem in function theory*, Izv. Vyssh. Uchebn. Zaved. Matematika **1960**, no. 1(14), 154–165; erratum, ibid., no. 4(17), 218. (Russian)

80. _____, *On a boundary value problem in function theory*, Trudy Kazan. Aviats. Inst. Vyp. 61 (1960), 13–21. (Russian)

81. _____, *On the inverse mixed boundary value problem*, Studies in Contemporary Problems of the Theory of Functions of a Complex Variable, Fizmatgiz, Moscow, 1961, pp. 375–380. (Russian)

82. _____, *On special cases in the inverse mixed boundary value problem*, Trudy Kazan. Aviats. Inst. Vyp. 64 (1961), 25–45. (Russian)

83. _____, *Inverse mixed boundary value problem for several unknown arcs*, Dokl. Akad. Nauk SSSR **141** (1961), 800–802; English transl. in Soviet Math. Dokl. **2** (1961).

84. _____, *Solvability questions of inverse mixed boundary value problems for analytic functions*, Functional Anal. and Theory of Functions, No. 1, Izdat. Kazan. Univ., Kazan, 1963, pp. 56–71. (Russian)

85. _____, *On uniqueness theorems in hydrodynamics problems with free boundaries*, Trudy Sem. Obratn. Kraev. Zadacham **1** (1964), 81–87. (Russian)

86. _____, *On theorems on existence of solutions in hydrodynamic problems with free boundaries*, Trudy Sem. Obratn. Kraev. Zadacham **2** (1964), 142–152. (Russian)

87. _____, *Unique solvability of two-dimensional problems of gas dynamics with free boundaries*, Dokl. Akad. Nauk SSSR **164** (1965), 982–984; English transl. in Soviet Math. Dokl. **6** (1965).

88. _____, *On boundary value problems with free boundaries for elliptic systems of equations*, Trudy Sem. Obratn. Kraev. Zadacham **1** (1964), 88–92. (Russian)

89. _____, *Some properties of solutions of nonlinear systems of equations, which are elliptic in the sense of Lavrent'ev*, Dinamika Sploshnoĭ Sredy Vyp. 15 (1974), 89–103. (Russian)

90. _____, *On a class of boundary value problems with free boundaries for elliptic systems*, Trudy Sem. Obratn. Kraev. Zadacham **1** (1964), 93–95. (Russian)

91. _____, *Solvability of filtration problems with free boundaries*, Dokl. Akad. Nauk SSSR **156** (1964), 1320–1322; English transl. in Soviet Phys. Dokl. **9** (1964/65).

92. _____, *On some plane problems of elasticity theory with unknown boundaries*, Dinamika Sploshnoĭ Sredy Vyp. 1 (1969), 242–257. (Russian)

93. _____, *Boundary value problems for subsonic gas dynamics with free boundaries*, Mechanics of Continuous Media (Materials Internat. Conf., 1966), Bolgar. Akad. Nauk, Sofia, 1968, pp. 47–61. (Russian; English summary)

94. _____, *Solvability and the principle of topological similarity for free boundary problems in gas dynamics and filtration theory*, Fluid Dynamics Trans., Vol. 4 (Proc. Eighth Sympos., Tarda, 1967), PWN, Warsaw, 1969, pp. 91–104.

95. _____, *Boundary value problems with free boundaries for elliptic systems of equations*. Parts I, II, Novosibirsk. Gos. Univ., Novosibirsk, 1968, 1969. (Russian)

96. V. N. Monakhov and S. N. Antontsev, *On some problems with discontinuous boundaries in the theory of motion of a fluid or gas with free boundaries*, Mechanics of Continuous Media (Materials Internat. Conf., 1966), Bolgar. Akad. Nauk, Sofia, 1968, pp. 63–73. (Russian; English summary)

97. N. I. Muskhelishvili, *Singular integral equations*, 2nd ed., Fizmatgiz, Moscow, 1962; English transl. of 1st ed., Noordhoff, 1953; reprinted, 1972.

98. Mitio Nagumo, *A theory of degree of mapping based on infinitesimal analysis*, Amer. J. Math. **73** (1951), 485–496.

99. I. P. Natanson, *Theory of functions of a real variable*, 2nd rev. ed., GITTL, Moscow, 1957; English transl., Vols. 1, 2, Ungar, New York, 1955, 1961.

100. Nguen Din Či [Nguyen Dinh Thi], *On a problem with free boundary for a parabolic equation*, Vestnik Moskov. Univ. Ser. I Mat. Mekh. **1966**, no. 2, 40–54. (Russian)

101. Kiyoshi Noshiro, *Cluster sets*, Springer-Verlag, 1960.

102. M. T. Nuzhin and N. B. Il'inskiĭ, *A method of constructing the underground contour of hydraulic structures*, Kazan. Gos. Univ., Kazan, 1963. (Russian)

103. O. A. Oleĭnik, *Mathematical problems of boundary layer theory*, Uspekhi Mat. Nauk **23** (1968), no. 3 (141), 3–65; English transl. in Russian Math. Surveys **23** (1968).

104. Seymour V. Parter, *On mappings of multiply connected domains by solutions of partial differential equations*, Comm. Pure Appl. Math. **13** (1960), 167–182.

105. P. I. Plotnikov, *Some problems on the flow of a turbulent fluid*, Dinamika Sploshnoĭ Sredy Vyp. 1 (1969), 124–133. (Russian)

106. P. Ja. Polubarinova-Kochina, *Theory of ground water movement*, GITTL, Moscow, 1952; English transl., Princeton Univ. Press, Princeton, N. J., 1962.

107. _____ (editor), *The development of research on filtration theory in the USSR* (1917–1967), "Nauka", Moscow, 1969. (Russian)

108. G. Pólya and G. Szegö, *Isoperimetric inequalities in mathematical physics*, Princeton Univ. Press, Princeton, N. J., 1951.

109. I. I. Privalov, *Boundary properties of analytic functions*, 2nd ed., GITTL, Moscow, 1950; German transl., VEB Deutscher Verlag Wiss., Berlin, 1956.

110. G. N. Pykhteev, *Solution of the inverse problem for a plane cavitational flow past a curvilinear arc*, Prikl. Mat. Mekh. **20** (1956), 373–381. (Russian)

111. D. Riabouchinsky [Ryabushinskiĭ], *Sur la détermination d'une surface d'après les données qu'elle porte*, C. R. Acad. Sci. Paris **189** (1929), 629–632.

112. V. S. Rogozhin, *On the uniqueness of the solution of the exterior inverse boundary value problem*, Uchen. Zap. Kazan. Gos. Univ. **117** (1957), no. 2, 38–41. (Russian)

113. A. B. Shabat, *On a scheme for the plane motion of a fluid when there is a trench on the bottom*, Ž. Prikl. Mekh. Tekhn. Fiz. **1962**, no. 4, 68–80. (Russian)

114. _____, *Two problems in splicing solutions of Dirichlet's problem*, Dokl. Akad. Nauk SSSR **150** (1963), 1242–1245; English transl. in Soviet Phys. Dokl. **8** (1963/64).

115. B. V. Shabat, *Mappings effected by solutions of nonlinear systems of partial differential equations*, Studies in Contemporary Problems of the Theory of Functions of a Complex Variable, Fizmatgiz, Moscow, 1960, pp. 451–461. (Russian)

116. Z. Schapiro [Z. Va. Shapiro], *Sur l'existence des représentations quasi-conformes*, C. R. (Dokl.) Acad. Sci. URSS **30** (1941), 690–692.

117. L. I. Sedov, *Two-dimensional problems in hydrodynamics and aerodynamics*, GITTL, Moscow, 1950; English transl., Interscience, 1965.

118. Ya. I. Sekerzh-Zen'kovich, *Sur la théorie des sillages*, Trudy Tsentral. Aèro-gidrodinam. Inst. Vyp. 299 (1937). (Russian; French summary)

119. James Serrin, *On plane and axially symmetric free boundary problems*, J. Rational Mech. Anal. **2** (1953), 563–575.

120. Max Shiffman, *On the existence of subsonic flows of a compressible fluid*, J. Rational Mech. Anal. **1** (1952), 605–652.

121. I. B. Simonenko, *The Riemann boundary value problem for n pairs of functions with measurable coefficients and its application to the study of singular integrals in L_p spaces with weights*, Izv. Akad. Nauk SSSR Ser. Mat. **28** (1964), 277–306. (Russian)

122. S. L. Sobolev, *Applications of functional analysis in mathematical physics*, Izdat. Leningrad. Gos. Univ., Leningrad, 1950; English transl., Transl. Math. Mono., vol. 7, Amer. Math. Soc., Providence, R. I., 1963.

123. V. V. Sokolovskiĭ, *On nonlinear filtration of ground water*, Prikl. Mat. Mekh. **13** (1949), 525–536.

124. _____, *Some shapes of uniformly strong masses*, Inzh. Zh. Mekh. Tverd. Tela **1968**, no. 2, 44–51; English transl. in Mech. Solids **3** (1968) (1972).

125. S. Stoïlow, *Leçons sur les principes topologiques de la théorie des fonctions analytiques*, 2nd ed., Gauthier-Villars, Paris, 1956.

126. G. G. Tumashev and M. T. Nuzhin, *Inverse boundary value problems and their applications*, 2nd ed., Izdat. Kazan. Univ., Kazan, 1965. (Russian)

127. O. M. Turovskiĭ, *A case of the inverse mixed boundary value problem*, Dokl. Akad. Nauk SSSR **168** (1966), 292–295; English transl. in Soviet Math. Dokl. **7** (1966).

128. I. N. Vekua, *Generalized analytic functions*, Fizmatgiz, Moscow, 1959; English transl., Pergamon Press, Oxford, and Addison-Wesley, Reading, Mass., 1962.

129. N. P. Vekua, *Systems of singular integral equations and some boundary value problems*, GITTL, Moscow, 1950; English transl., Noordhoff, 1967.

130. V. S. Vinogradov, *On a boundary value problem for linear elliptic systems of first order in the plane*, Dokl. Akad. Nauk SSSR **118** (1958), 1059–1062. (Russian)

131. _____, *On some boundary value problems for quasilinear elliptic systems of first order on the plane*, Dokl. Akad. Nauk SSSR **121** (1958), 579–581. (Russian)

132. L. I. Volkovyskiĭ, *Quasiconformal mappings*, Izdat. L'vov. Gos. Univ., L'vov, 1954. (Russian)

133. Stefan Warschawski, *Über das Randverhalten der Ableitung der Abbildungsfunktion bei konformer Abbildung*, Math. Z. **35** (1932), 322–456.

134. Alexander Weinstein, *Non-linear problems in the theory of fluid motion with free boundaries*, Proc. Sympos. Appl. Math., Vol. 1, Amer. Math. Soc., Providence, R. I., 1949, pp. 1–18.

135. L. C. Woods, *Compressible subsonic flow in two-dimensional channels with mixed boundary conditions*, Quart J. Mech. Appl. Math. **7** (1954), 263–282.

136. A. Zygmund, *Trigonometric series*, 2nd rev. ed., Vols. I, II, Cambridge Univ. Press, 1959.

ABCDEFGHIJ–CM–89876543